Medium/Heavy Duty Truck Technician Certification Test Preparation Manual

Medium/Heavy Duty Truck Technician Certification Test Preparation Manual

Don Knowles

Africa • Australia • Canada • Denmark • Japan • Mexico • New Zealand • Philippines
Puerto Rico • Singapore • Spain • United Kingdom • United States

Delmar Staff

Business Unit Director: Alar Elken
Executive Editor: Sandy Clark
Acquisitions Editor: Vern Anthony
Editorial Assistant: Bridget Morrison
Executive Marketing Manager: Maura Theriault
Channel Manager: Mona Caron
Marketing Coordinator: Kasey Young
Executive Production Manager: Mary Ellen Black
Production Coordinator: Karen Smith
Senior Project Editor: Christopher Chien
Art Director: Cheri Plasse
Technology Project Manager: Tom Smith

Printed in Canada
1 2 3 4 5 6 7 8 9 10 XXX 05 04 03 02 01 00

For more information, contact Delmar, 3 Columbia Circle, PO Box 15015, Albany, NY 12212-0515; or find us on the World Wide Web at http://www.delmar.com

Asia:
Thomson Learning
60 Albert Street, #15-01
Albert Complex
Singapore 189969
Tel: 65 336 6411
Fax: 65 336 7411

Japan:
Thomson Learning
Palaceside Building 5F
I - I - I Hitotsubashi, Chiyoda-ku
Tokyo 100 0003 Japan
Tel: 813 5218 6544
Fax: 813 5218 6551

Australia/New Zealand:
Nelson/Thomson Learning
102 Dodds Street
South Melbourne, Victoria 3205
Australia
Tel: 61 39685 4111
Fax: 61 39 685 4199

UK/Europe/Middle East
Thomson Learning
Berkshire House
168-173 High Holborn
London
WC IV 7AA United Kingdom
Tel: 44 171497 1422
Fax: 44 171497 1426
Thomas Nelson & Sons LTD
Nelson House
Mayfield Road
Walton-on-Thames
KT 12 5PL United Kingdom
Tel: 44 1932 2522111
Fax: 44 1932 246574

Latin America:
Thomson Learning
Seneca, 53
Colonia Polanco
11560 Mexico D.F. Mexico
Tel: 525-281-2906
Fax: 525-281-2656

Canada:
Nelson/Thomson Learning
1120 Birchmount Road
Scarborough, Ontario
Canada MIK 5G4
Tel: 416-752-9100
Fax: 416-752-8102

Spain:
Thomson Learning
Calle Magallanes, 25
28015-Madrid
España
Tel: 34 91446 33 50
Fax: 34 91445 62 18

International Headquarters:
Thomson Learning
International Division
290 Harbor Drive, 2nd Floor
Stamford, CT 06902-7477
Tel: 203-969-8700
Fax: 203-969-8751

Library of Congress Cataloging-in-Publication Data
ISBN: 0-8273-7258-2

Contents

Introduction

Part 1 The History of ASE

History

Originally known as the National Institute for Automotive Service Excellence (NIASE), today's ASE was founded in 1972 as a nonprofit, independent entity dedicated to improving the quality of automotive service and repair through the voluntary testing and certification of automotive technicians. Until that time, consumers had no way of distinguishing between competent and incompetent automotive technicians. In the mid-1960s and early 1970s, efforts were made by several automotive industry affiliated associations to respond to this need. Though the associations were nonprofit, many regarded certification test fees merely as a means of raising additional operating capital. Also, some associations, having a vested interest, produced test scores heavily weighted in the favor of its members.

NIASE

From these efforts a new independent, nonprofit association, the National Institute for Automotive Service Excellence (NIASE), was established much to the credit of two educators, George R. Kinsler, Director of Program Development for the Wisconsin Board of Vocational and Adult Education in Madison, WI, and Myron H. Appel, Division Chairman at Cypress College in Cypress, CA.

Early efforts were to encourage voluntary certification in four general areas:

I. Engine: Engines, Engine Tune Up, Block Assembly, Cooling and Lube Systems, Induction, Ignition, and Exhaust

II. Transmission: Manual Transmissions, Driveline and Rear Axles, and Automatic Transmissions

III. Brakes and Suspension: Brakes, Steering, Suspension, and Wheels

IV. Electrical/Air Conditioning: Body/Chassis, Electrical Systems, Heating, and Air Conditioning

In early NIASE tests, Mechanic A–Mechanic B-type questions were used. Over the years the trend has not changed, but in mid-1984 the term was changed to Technician A–Technician B to better emphasize the sophistication of the skills needed to perform successfully in the modern motor vehicle industry. In certain tests the term used is Estimator A/B, Painter A/B, or Parts Specialist A/B. At about that same time, the logo was changed from "The Gear" to "The Blue Seal," and the organization adopted the acronym ASE for Automotive Service Excellence.

Since those early beginnings, several other related trades have been added. ASE now administers a comprehensive series of certification exams for automotive and light truck repair technicians, medium and heavy truck repair technicians, alternate fuels technicians, engine machinists, collision repair technicians, school bus repair technicians, and parts specialists.

The Series and Individual Tests

- Automotive and Light Truck Technician, consisting of: Engine Repair—Automatic Transmission/Transaxle—Manual Drivetrain and Axles—Suspension and Steering—Brakes—Electrical/Electronic Systems—Heating and Air Conditioning—Engine Performance
- Medium and Heavy Truck Technician, consisting of: Gasoline Engines—Diesel Engines—Drivetrain—Brakes—Suspension and Steering—Electrical/Electronic Systems—HVAC—Preventive Maintenance Inspection

- Alternate Fuels Technician, consisting of: Compressed Natural Gas Light Vehicles
- Advanced Series, consisting of: Automobile Advanced Engine Performance and Advanced Diesel Engine Electronic Diesel Engine Specialty
- Collision Repair Technician, consisting of: Painting and Refinishing—Nonstructural Analysis and Damage Repair—Structural Analysis and Damage Repair—Mechanical and Electrical Components—Damage Analysis and Estimating
- Engine Machinist Technician, consisting of: Cylinder Head Specialist—Cylinder Block Specialist—Assembly Specialist
- School Bus Repair Technician, consisting of: Body Systems and Special Equipment—Drivetrain—Brakes—Suspension and Steering—Electrical/Electronic Systems—Heating and Air Conditioning
- Parts Specialist, consisting of: Automobile Parts Specialist—Medium/Heavy Truck Parts Specialist

A Brief Chronology

1970–1971 Original questions were prepared by a group of forty auto mechanics teachers from public secondary schools, technical institutes, community colleges, and private vocational schools. These questions were then professionally edited by testing specialists at the Educational Testing Service (ETS) in Princeton, New Jersey, and thoroughly reviewed by training specialists associated with domestic and import automotive companies.

1971 July: About 800 mechanics tried out the original test questions at experimental test administrations.

1972 November and December: Initial NIASE tests were administered at 163 test centers. The original automotive test series consisted of four tests containing eighty questions each. Three hours were allotted for each test. Those who passed all four tests were designated Certified General Auto Mechanic (GAM).

1973 April and May: Test 4 was increased to 120 questions. Time was extended to four hours for this test. There were now 182 test centers. Shoulder patch insignias were made available.

November: Automotive series was expanded to five tests. Heavy-Duty Truck series of six tests was introduced.

1974 November: Automatic Transmission (Light Repair) test was modified. Name was changed to Automatic Transmission.

1975 May: Collision Repair series of two tests was introduced.

1978 May: Automotive recertification testing was introduced.

1979 May: Heavy-Duty Truck recertification testing was introduced.

1980 May: Collision Repair recertification testing was introduced.

1982 May: Test administration providers switched from Educational Testing Service (ETS) to American College Testing (ACT). Name of Automobile Engine Tune-Up test was changed to Engine Performance test.

1984 May: New logo was introduced. ASE's "The Blue Seal" replaced NIASE's "The Gear." All reference to Mechanic A–Mechanic B was changed to Technician A–Technician B.

1990 November: The first test of the Engine Machinist test series was introduced.

1991 May: The second test of the Engine Machinist test series was introduced.

November: The third and final Engine Machinist test was introduced.

1992 May: Name of Heavy-Duty Truck Test series was changed to Medium/Heavy Truck test series.

1993 May: Automotive Parts Specialist test was introduced. Collision Repair was expanded to six tests. Light Vehicle Compressed Natural Gas test was introduced. Limited testing began in English-speaking provinces of Canada.

1994 May: Advanced Engine Performance Specialist test was introduced.

1996 May: First three tests for School Bus Technician test series was introduced.

November: A Collision Repair test was added.

1997 May: A Medium/Heavy Truck test was added.

1998 May: A diesel advanced engine test was introduced: Electronic Diesel Engine Diagnosis Specialist. A test was added to the School Bus test series.

By the Numbers

Following are the approximate number of ASE technicians currently certified by category. The numbers may vary from time to time but are reasonably accurate for any given period. More accurate data may be obtained from ASE, which provides updates twice each year, in May and November after the Spring and Fall test series.

There are more than 338,000 Automotive Technicians with over 87,000 at Master Technician (MA) status. There are 47,000 Truck Technicians with over 19,000 at Master Technician (MT) status. There are 46,000 Collision Repair/Refinish Technicians with 7,300 at Master Technician (MB) status. There are 1,200 Estimators. There are 6,700 Engine Machinists with over 2,800 at Master Machinist Technician (MM) status. There are also 28,500 Automobile Advanced Engine Performance Technicians and over 2,700 School Bus Technicians for a combined total of more than 403,000 Repair Technicians. To this number, add over 22,000 Automobile Parts Specialists, and over 2,000 Truck Parts Specialists for a combined total of over 24,000 Parts Specialists.

There are over 6,400 ASE Technicians holding both Master Automotive Technician and Master Truck Technician status, of which 350 also hold Master Body Repair status. Almost 200 of these Master Technicians also hold Master Machinist status and five Technicians are certified in all ASE specialty areas.

Almost half of ASE certified technicians work in new vehicle dealerships (45.3 percent). The next greatest number work in independent garages with 19.8 percent. Next are tire dealerships with 9 percent, service stations at 6.3 percent, fleet shops at 5.7 percent, franchised volume retailers at 5.4 percent, paint and body shops at 4.3 percent, and specialty shops at 3.9 percent.

Of over 400,000 automotive technicians on ASE's certification rosters, almost 2,000 are female. The number of female technicians is increasing at a rate of about 20 percent each year. Women's increasing interest in the automotive industry is further evidenced by the fact that, according to the National Automobile Dealers Association (NADA), they influence 80 percent of the decisions of the purchase of a new automobile and represent 50 percent of all new car purchasers. Also, it is interesting to note that 65 percent of all repair and maintenance service customers are female.

The typical ASE-certified technician is 36.5 years of age, is computer literate, deciphers a half-million pages of technical manuals, spends 100 hours per year in training, holds four ASE certificates, and spends about $27,000 for tools and equipment. Twenty-seven percent of today's skilled ASE-certified technicians attended college, many having earned an Associate of Science degree in Automotive Technology.

ASE

ASE's mission is to improve the quality of vehicle repair and service in the United States through the testing and certification of automotive repair technicians. Prospective candidates register for and take one or more of ASE's thirty-three exams. The tests are grouped into specialties for automobile, medium/heavy truck, school bus, and collision repair technicians as well as engine machinists, alternate fuels technicians, and parts specialists.

Upon passing at least one exam and providing proof of two years of related work experience, the technician becomes ASE certified. A technician who passes a series of exams earns ASE Master Technician status. An automobile technician, for example, must pass eight exams for this recognition.

The tests, conducted twice a year at over 700 locations around the country, are administered by American College Testing (ACT). They stress real world diagnostic and repair problems. Though a good knowledge of theory is helpful to the technician in answering many of the questions, there are no questions specifically on theory. Certification is valid for five years. To retain certification, the technician must be retested to renew his or her certificate.

The automotive consumer benefits because ASE certification is a valuable yardstick by which to measure the knowledge and skills of individual technicians, as well as their commitment to their chosen profession. It is also a tribute to the repair facility employing ASE-certified technicians. ASE-certified technicians are permitted to wear blue and white ASE shoulder insignia, referred to as the "Blue Seal of Excellence," and carry credentials listing their areas of expertise. Often employers display their technicians' credentials in the customer waiting area. Customers look for facilities that display ASE's Blue Seal of Excellence logo on outdoor signs, in the customer waiting area, in the telephone book (Yellow Pages), and in newspaper advertisements.

The tests stress repair knowledge and skill. All test takers are issued a score report. In order to earn ASE certification, a technician must pass one or more of the exams and present proof of two years of relevant hands-on work experience. ASE certifications are valid for five years, after which time technicians must retest in order to keep up with changing technology and to remain in the ASE program. A nominal registration and test fee is charged.

To become part of the team that wears ASE's Blue Seal of Excellence®, please contact:

National Institute for Automotive Service Excellence
13505 Dulles Technology Drive
Herndon, VA 20171–3421

Part 2 Take and Pass Every ASE Test

ASE Testing

Participating in an Automotive Service Excellence (ASE) voluntary certification program gives you a chance to show your customers that you have the "know-how" needed to work on today's modern vehicles. The ASE certification tests allow you to compare your skills and knowledge to the automotive service industry's standards for each specialty area.

If you are the "average" automotive technician taking this test, you are in your mid-thirties and have not attended school for about fifteen years. That means you probably have not taken a test in many years. Some of you, on the other hand, have attended college or taken postsecondary education courses and may be more familiar with taking tests and with test-taking strategies. There is, however, a difference in the ASE test you are preparing to take and the educational tests you may be accustomed to.

Who Writes the Questions?

The questions on an educational test are generally written, administered, and graded by an educator who may have little or no practical hands-on experience in the test area. The questions on all ASE tests are written by service industry experts familiar with all aspects of the subject area. ASE questions are entirely job-related and designed to test the skills that you need to know on the job.

The questions originate in an ASE "item-writing" workshop where service representatives from domestic and import automobile manufacturers, parts and equipment manufacturers, and vocational educators meet in a workshop setting to share their ideas and translate them into test questions. Each test question written by these experts is reviewed by all of the members of the group. The questions deal with the practical problems of diagnosis and repair that are experienced by technicians in their day-to-day hands-on work experiences.

All of the questions are pretested and quality-checked in a nonscoring section of tests by a national sample of certifying technicians. The questions that meet ASE's high standards of accuracy and quality are then included in the scoring sections of future tests. Those questions that do not pass ASE's stringent tests are sent back to the workshop or are discarded. ASE's tests are monitored by an independent proctor and are administered and machine-scored by an independent provider, American College Testing (ACT). All ASE tests have a three-year revision cycle.

Testing

If you think about it, we are actually tested on just about everything we do. As infants, we were tested to see when we could turn over and crawl, later when we could walk or talk. As adolescents, we were tested to determine how well we learned the material presented in school and how we demonstrated our accomplishments on the athletic field. As working adults, we are tested by our supervisors on how well we have completed an assignment or project. As nonworking adults, we are tested by our families on everyday activities, such as housekeeping or preparing a meal. Testing, then, is one of those facts of life that begins in the cradle and follows us to the grave.

Testing is an important fact of life that helps us to determine how well we have learned our trade. Also, tests often help us to determine what opportunities will be available to us in the future. To become ASE certified, we are required to take a test in every subject in which we wish to be recognized.

Be Test-Wise

In spite of the widespread use of tests, most technicians are not very test-wise. An ability to take tests and score well is a skill that must be acquired. Without this knowledge, the most intelligent and prepared technician may not do well on a test. We will discuss some of the basic procedures necessary to follow in order to become a test-wise technician. Assume, if you will, that you have done the necessary study and preparation to score well on the ASE test.

Different approaches should be used for taking different types of tests. The different basic types of tests include: essay, objective, multiple-choice, fill-in-the-blank, true-false, problem solving, and open book. All ASE tests are of the four-part, multiple-choice-type. Before discussing the multiple-choice type test questions, however, there are a few basic principles that should be followed before taking any test.

Before the Test

- Do not arrive late. Always arrive well before your test is scheduled to begin. Allow ample time for the unexpected, such as traffic problems, so you will arrive on time and avoid the unnecessary anxiety of being late.
- Always be certain to have plenty of supplies with you. For an ASE test, three or four sharpened soft lead (#2) pencils, a pocket pencil sharpener, erasers, and a watch are all that are required.
- Do not listen to pretest chatter. When you arrive early, you may hear other technicians testing each other on various topics or making their best guess as to the probable test questions. At this time, it is too late to add to your knowledge. Also, the rhetoric may only confuse you. If you find it bothersome, take a walk outside the test room to relax and loosen up.
- Read and listen to all instructions. It is important to read and listen to the instructions. Make certain that you know what is expected of you. Listen carefully to verbal instructions and pay particular attention to any written instructions on the test paper. Do not dive into answering questions only to find out that you have answered the wrong question by not following instructions carefully. It is difficult to obtain a high score on a test if you answer the wrong questions.

These basic principles have been violated in almost every test ever given. Try to remember them. They are essential for success.

Objective Tests

A test is called an objective test if the same standards and conditions apply to everyone taking the test and there is only one correct answer to each question. Objective tests primarily measure your ability to recall information. A well-designed objective test can also test your ability to understand, analyze, interpret, and apply your knowledge. Objective tests include true-false, multiple-choice, fill-in-the-blank, and matching questions.

Objective questions, although not generally encountered in a classroom setting, are frequently used in standardized examinations. Objective tests are easy to grade and also reduce the amount of paperwork necessary to administer a test. The objective tests are used in entry-level programs or when very large numbers of people are being tested. ASE's tests consist exclusively of four-part, multiple-choice objective questions.

Taking an Objective Test

The principles of taking an objective test are somewhat different from those used in other types of tests. You should first quickly look over the test to determine the number of questions, but do not try to read through all of the questions. In an ASE test, there are usually between forty and eighty questions, depending on the subject matter. Read through each question before marking your answer. Answer the questions in the order they appear on the test. Leave the questions that you are not sure of blank and move on to the next question. You can return to those unanswered questions after you have finished the others. They may be easier to answer after your mind has had additional time to consider them on a subconscious level. In addition, you might find information in other questions that will help you to answer some of them. Do not be obsessed by the apparent pattern of responses. For example, do not be influenced by a pattern like d, c, b, a, d, c, b, a on an ASE test.

There is also a lot of folk wisdom about taking objective tests. For example, there are those who would advise you to avoid response options that use certain words such as all, none, always, never, must, and only, to name a few. This, they claim, is because nothing in life is exclusive. They would advise you to choose response options that use words that allow for some exception, such as sometimes, frequently, rarely, often, usually, seldom, and normally. They would also advise you to avoid the first and last option (A and D) because test writers, they feel, are more comfortable if they put the correct answer in the middle (B and C) of the choices. Another recommendation often offered is to select the option that is either shorter or longer than the other three choices because it is more likely to be correct. Some would advise you to never change an answer because your first intuition is usually correct.

Although there may be a grain of truth in this folk wisdom, ASE test writers try to avoid them and so should you. There are just as many A answers as there are B answers, just as many D answers as C answers. As a matter of fact, ASE tries to balance the answers at about 25 percent per choice A, B, C, and D. There is no intention to use "tricky" words, such as outlined above. Put no credence in the opposing words "sometimes" and "never," for example. When used in an ASE-type question, one or both may be correct; one or both may be incorrect.

There are some special principles to observe on multiple-choice tests. These tests are sometimes challenging because there are often several choices that may seem possible, and it may be difficult to decide on the correct choice. The best strategy, in this case, is to first determine the correct answer before looking at the options. If you see the answer you decided on, you should still examine the options to make sure that none seem more correct than yours. If you do not know or are not sure of the answer, read each option very carefully and try to eliminate those options that you know to be wrong. That way, you can often arrive at the correct choice through a process of elimination.

If you have gone through all of the test and you still do not know the answer to some of the questions, then guess. Yes, guess. You then have at least a 25 percent chance of being correct. If you leave the question blank, you have no chance. In ASE tests, there is no penalty for being wrong. As the late President Franklin D. Roosevelt once advised a group of students, "It is common sense to take a method and try it. If it fails, admit it frankly and try another. But above all, try something."

During the Test

Mark your bubble sheet clearly and accurately. One of the biggest problems an adult faces in test-taking, it seems, is in placing an answer in the correct spot on a bubble sheet. Make certain that you mark your answer for, say, question 21, in the space on the bubble sheet designated for the answer for question 21. A correct response in the wrong bubble will probably be wrong. Remember, the answer sheet is scored by machine and can only "read" what you have bubbled in. Also, do not bubble in two answers for the same question. For example, if you feel the answer to a particular question is A but think it may be C, do not bubble in both choices. Even if either A or C is correct, a double answer will score as an incorrect answer. It is better to take a chance with your best guess.

Review Your Answers

If you finish answering all of the questions on a test ahead of time, go back and review the answers of those questions that you were not sure of. You can often catch careless errors by using the remaining time to review your answers.

Do Not Be Distracted

At practically every test, some technicians will invariably finish ahead of time and turn their papers in long before the final call. Do not let them distract or intimidate you. Either they knew too little and could not finish the test, or they were very self-confident and thought they knew it all. Perhaps they were trying to impress the proctor or other technicians about how much they know. Often you may hear them later talking about the information they knew all the while but forgot to respond on their answer sheet.

Use Your Time Wisely

It is not wise to use less than the total amount of time that you are allotted for a test. If there are any doubts, take the time for review. Any product can usually be made better with some additional effort. A test is no exception. It is not necessary to turn in your test paper until you are told to do so.

Do Not Cheat

Some technicians may try to use a "crib sheet" during a test. Others may attempt to read answers from another technician's paper. If you do that, you are unquestionably assuming that someone else has a correct answer. You probably know as much, maybe more, than anyone else in the test room. Trust yourself. If you are still not convinced, think of the consequences of being caught. Cheating is foolish. If you are caught, you have failed the test.

Be Confident

The first and foremost principle in taking a test is that you need to know what you are doing in order to be test-wise. It will now be presumed that you are a test-wise technician and are now ready for some of the more obscure aspects of test-taking.

An ASE-style test requires that you use the information and knowledge at your command to solve a problem. This generally requires a combination of information similar to the way you approach problems in the real world. Most problems, it seems, typically do not fall into neat textbook cases. New problems are often difficult to handle, whether they are encountered inside or outside the test room.

An ASE test also requires that you apply methods taught in class as well as those learned on the job to solve problems. These methods are akin to a well-equipped toolbox in the hands of a skilled technician. You have to know what tools to use in a particular situation, and you must also know how to use them. In an ASE test, you will need to be able to demonstrate that you are familiar with and know how to use the tools.

You should begin a test with a completely open mind. At times, however, you may have to move out of your normal way of thinking and be creative to arrive at a correct answer. If you have diligently studied for at least one week before the test, you have bombarded your mind with a wide assortment of information. Your mind will be working with this information on a subconscious level, exploring the interrelationships among various facts, principles, and ideas. This prior preparation should put you in a creative mood for the test.

In order to reach your full potential, you should begin a test with the proper mental attitude and a high degree of self-confidence. You should think of a test as an opportunity to document how much you know about the various tasks in your chosen profession. If you have been diligently studying the subject matter, you will be able to take your test in serenity because your mind will be well organized. If you are confident, you are more likely to do well because you have the proper mental attitude. If, on the other hand, your confidence is low, you are bound to do poorly. It is a self-fulfilling prophecy. Perhaps you have heard athletic coaches talk about the importance of confidence when competing in sports. Mental confidence helps an athlete to perform at the highest level and gain an advantage over competitors. Taking a test is much like an athletic event. You are competing against yourself, in a certain sense, because you will be trying to approach perfection in determining your answers. As in any competition, you should aim your sights high and be confident that you can reach the apex.

Anxiety and Fear

Many technicians experience anxiety and fear at the very thought of taking a test. Many worry, become nervous, and even become ill at test time because of the fear of failure. Many often worry about the criticism and ridicule that may come from their employer, relatives, and peers. Some worry about taking a test because they feel that the stakes are very high. Those who spent a great amount of time studying may feel they must get a high grade to justify their efforts. The thought of not doing well can result in unnecessary worry. They become so worried, in fact, that their reasoning and thinking ability is impaired, actually bringing about the problem they wanted to avoid.

The fear of failure should not be confused with the desire for success. It is natural to become "psyched-up" for a test in contemplation of what is to come. A little emotion can provide a healthy flow of adrenaline to peak your senses and hone your mental ability. This improves your performance on the test and is a very different reaction from fear. Most technician's fears and insecurities experienced before a test are due to a lack of self-confidence. Those who have not scored well on previous tests or have no confidence in their preparation are those most likely to fail. Be confident that you will do well on your test and your fears should vanish. You will know that you have done everything possible to realize your potential.

Getting Rid of Fear

If you have previously experienced fear of taking a test, it may be difficult to change your attitude immediately. It may be easier to cope with fear if you have a better understanding of what the test is about. A test is merely an assessment of how much the technician knows about a particular task area. Tests, then, are much less threatening when thought of in this manner. This does not mean, however, that you should lower your self-esteem simply because you performed poorly on a test.

You can consider the test essentially as a learning device, providing you with valuable information to evaluate your performance and knowledge. Recognize that no one is perfect. All humans make mistakes. The idea, then, is to make mistakes before the test, learn from them, and avoid repeating them on the test. Fortunately, this is not as difficult as it seems. Practical questions in this study guide include the correct answers to consider if you have made mistakes on the practice test. You should learn where you went wrong so you will not repeat them in the ASE test. If you learn from your mistakes, the stage is set for future growth.

If you understood everything presented up until now, you have the knowledge to become a test-wise technician, but more is required. To be a test-wise technician, you not only have to practice these principles, you have to diligently study in your task area.

Effective Study

The fundamental and vital requirement to induce effective study is a genuine and intense desire to achieve. This is more basic than any rule or technique that will be given here. The key requirement, then, is a driving motivation to learn and to achieve. If you wish to study effectively, first develop a desire to master your studies and sincerely believe that you will master them. Everything else is secondary to such a desire. First, build up definite ambitions and ideals toward which your studies can lead. Picture the satisfaction of success. The attitude of the technician may be transformed from merely getting by to an earnest and energetic effort. The best direct stimulus to change may involve nothing more than the deliberate planning of your time. Plan time to study. Another drive that creates positive study is an interest in the subject studied. As an automotive technician, you can develop an interest in studying particular subjects if you follow these four rules:

1. Acquire information from a variety of sources. The greater your interest in a subject, the easier it is to learn about it. Visit your local library and seek books on the subject you are studying. When you find something new or of interest, make inexpensive photocopies for future study.
2. Merge new information with your previous knowledge. Discover the relationship of new facts to old known facts. Modern developments in automotive technology take on new interest when they are seen in relation to present knowledge.
3. Make new information personal. Relate the new information to matters that are of concern to you. The information you are now reading, for example, has interest to you as you think about how it can help.
4. Use your new knowledge. Raise questions about the points made by the book. Try to anticipate what the next steps and conclusions will be. Discuss this new knowledge, particularly the difficult and questionable points, with your peers.

You will find that when you study with eager interest, you will discover it is no longer work. It is pleasurable and you will be fascinated by what you study. Studying can be like reading a novel or seeing a movie that overcomes distractions and requires no effort or willpower. You will discover that the positive relationship between interest and effort works both ways. Even though you perhaps began your studies with little or no interest, simply staying with it helped you to develop an interest in your studies. Obviously, certain subject matter studies are bound to be of little or no interest, particularly in the beginning. Parts of certain studies may continue to be uninteresting. An honest effort to master those subjects, however, nearly always brings about some level of interest. If you appreciate the necessity and reward of effective studying, you will rarely be disappointed. Here are a few important hints for gaining the determination that is essential to carrying good conclusions into actual practice.

Make Study Definite

Decide what is to be studied and when it is to be studied. If the unit is discouragingly long, break it into two or more parts. Determine exactly what is involved in the first part and learn that. Only then should you proceed to the next part. Stick to a schedule.

The Urge to Learn

Make clear to yourself the relation of your present knowledge to your study materials. Determine the relevance with regard to your long-range goals and ambitions. Turn your attention away from real or imagined difficulties as well as other things that you would rather be doing. Some major distractions are thoughts of other duties and of disturbing problems. These distractions can usually be put aside, simply shunted off, by listing them in a notebook. Most technicians have found that by writing interfering thoughts down, their minds are freed from annoying tensions.

Adopt the most reasonable solution you can find or seek objective help from someone else for personal problems. Personal problems and worry are often causes of ineffective study. Sometimes there are no satisfactory solutions. Some manage to avoid the problems or to meet them without great worry. For those who may wish to find better ways of meeting their personal problems, the following suggestions are offered:

1. Determine as objectively and as definitely as possible where the problem lies. What changes are needed to remove the problem, and which changes, if any, can be made? Sometimes it is wiser to alter your goals than external conditions. If there is no perfect solution, explore the others. Some solutions may be better than others.
2. Seek an understanding confidant who may be able to help analyze your problems. Very often, talking over your problems with someone in whom you have confidence and trust will help you to arrive at a solution.
3. Do not betray yourself by trying to evade the problem or by pretending that it has been solved. If

social problem distractions prevent you from studying or doing satisfactory work, it is better to admit this to yourself. You can then decide what can be done about it.

Once you are free of interferences and irritations, it is much easier to stay focused on your studies.

Concentrate

To study effectively, you must concentrate. Your ability to concentrate is governed, to a great extent, by your surroundings as well as your physical condition. When absorbed in study, you must be oblivious to everything else around you. As you learn to concentrate and study, you must also learn to overcome all distractions. There are three kinds of distractions you may face:

1. Distractions in the surrounding area, such as motion, noise, and the glare of lights. The sun shining through a window on your study area, for example, can be very distracting. Some technicians find that, for effective study, it is necessary to eliminate visual distractions as well as noises. Others find that they are able to tolerate moderate levels of auditory or visual distraction.

   Improper ambient conditions. Make sure your study area is properly lighted and ventilated. The lighting should be adequate but should not shine directly into your eyes or be visible out of the corner of your eye. Also, try to avoid a reflection of the lighting on the pages of your book. Whether heated or cooled, the environment should be at a comfortable level. For most, this means a temperature of 78°F–80°F (25.6°C–26.7°C) with a relative humidity of 45 to 50 percent.

2. Distractions arising from your body, such as a headache, fatigue, and hunger. Be in good physical condition. Eat wholesome meals at regular times. Try to eat with your family or friends whenever possible. Mealtime should be your recreational period. Do not eat a heavy meal for lunch, and do not resume studies immediately after eating lunch. Just after lunch, try to get some regular exercise, relaxation, and recreation. A little exercise on a regular basis is much more valuable than a lot of exercise only on occasion.

3. Distractions of irrelevant ideas, such as how to repair the garden gate, when you are studying for an automotive-related test.

The problems associated with study are no small matter. These problems of distractions are generally best dealt with by a process of elimination. A few important rules for eliminating distractions follow.

Get Sufficient Sleep

You must get plenty of rest even if it means dropping certain outside activities. Avoid cutting in on your sleep time; you will be rewarded in the long run. If you experience difficulty going to sleep, do something to take your mind off your work and try to relax before going to bed. Some suggestions that may help include a little reading, a warm bath, a short walk, a conversation with a friend, or writing that overdue letter to a distant relative. If sleeplessness is an ongoing problem, consult a physician. Do not try any of the sleep remedies on the market, particularly if you are on medication, without approval of your physician.

If you still have difficulty studying, a final rule may help. Sit down in a favorable place for studying, open your books, and take out your pencil and paper. In a word, go through the motions.

Arrange Your Area

Arrange your chair and work area. To avoid strain and fatigue, whenever possible, shift your position occasionally. Try to be comfortable; however, avoid being too comfortable. It is nearly impossible to study rigorously when settled back in a large easy chair or reclining leisurely on a sofa.

When studying, it is essential to have a plan of action, a time to work, a time to study, and a time for pleasure. If you schedule your day and adhere to the schedule, you will eliminate most of your efforts and worries. A plan that is followed, then, soon becomes the easy and natural routine of the day. Most technicians find it useful to have a definite place and time to study. A particular table and chair should always be used for study and intellectual work. This place will then come to mean study. To be seated in that particular location at a regularly scheduled time will automatically lead you to assume a readiness for study.

Do Not Daydream

Daydreaming or mind-wandering is an enemy of effective study. Daydreaming is frequently due to an inadequate understanding of words. Use the glossary or a dictionary to look up the troublesome word. Another frequent cause of daydreaming is a deficient background in the present subject matter. When this is the problem, go back and review the subject matter to obtain the necessary foundation. Just one hour of concentrated study is equivalent to ten hours with frequent lapses of day-dreaming. Be on guard against mind-wandering, and pull yourself back into focus on every occasion.

Study Regularly

A system of regularity in study is believed by many scholars to be the secret of success. The daily time schedule must, however, be determined on an individual basis. You must decide how many hours of each day you can devote to your studies. Few technicians really are aware of where their leisure time is spent. An accurate account of how your days are presently being spent is an important first step toward creating an effective daily schedule.

WEEKLY SCHEDULE							
	SUN	MON	TUES	WED	THU	FRI	SAT
6:00							
6:30							
7:00							
7:30							
8:00							
8:30							
9:00							
9:30							
10:00							
10:30							
11:00							
11:30							
NOON							
12:30							
1:00							
1:30							
2:00							
2:30							
3:00							
3:30							
4:00							
4:30							
5:00							
5:30							
6:00							
6:30							
7:00							
7:30							
8:00							
8:30							
9:00							
9:30							
10:00							
10:30							
11:00							
11:30							

This convenient form is for keeping an hourly record of your week's activities. If you fill in the schedule each evening before bedtime, you will soon gain some interesting and useful facts about yourself and your use of your time. If you think over the causes of wasted time, you can determine how you might better spend your time. A practical schedule can be set up by using the following steps.

1. Mark your fixed commitments, such as work, on your schedule. Be sure to include classes and clubs. Do you have sufficient time left? You can arrive at an estimate of the time you need for studying by counting the hours used during the present week. An often-used formula, if you are taking classes, is to multiply the number of hours you spend in class by two. This provides time for class studies. This is then added to your work hours. Do not forget time allocation for travel.
2. Fill in your schedule for meals and studying. Use as much time as you have available during the normal workday hours. Do not plan, for example, to do all of your studying between 11:00 p.m. and 1:00 a.m. Try to select a time for study that you can use every day without interruption. You may have to use two or perhaps three different study periods during the day.
3. List the things you need to do within a time period. A one-week time frame seems to work well for most technicians. The question you may ask yourself is: "What do I need to do to be able to walk into the test next week, or next month, prepared to pass?"
4. Break down each task into smaller tasks. The amount of time given to each area must also be settled. In what order will you tackle your schedule? It is best to plan the approximate time for your assignments and the order in which you will do them. In this way, you can avoid the difficulties of not knowing what to do first and of worrying about the other things you should be doing.
5. List your tasks in the empty spaces on your schedule. Keep some free time unscheduled so you can deal with any unexpected events, such as a dental appointment. You will then have a tentative schedule for the following week. It should be flexible enough to allow some units to be rearranged if necessary. Your schedule should allow time off from your studies. Some

use the promise of a planned recreational period as a reward for motivating faithfulness to a schedule. You will more likely lose control of your schedule if it is packed too tightly.

Keep a Record

Keep a record of what you actually do. Use the knowledge you gain by keeping a record of what you are actually doing so you can create or modify a schedule for the following week. Be sure to give yourself credit for movement toward your goals and objectives. If you find that you cannot study productively at a particular hour, modify your schedule so as to correct that problem.

Scoring the ASE Test

You can gain a better perspective about tests if you know and understand how they are scored. ASE's tests are scored by American College Testing (ACT), a nonpartial, nonbiased organization with no vested interest in ASE or in the automotive industry. Each question carries the same weight as any other question. For example, if there are fifty questions, each is worth 2 percent of the total score. The passing grade is 70 percent. That means you must correctly answer thirty-five of the fifty questions to pass the test.

Understand the Test Results

The test results can tell you:

- where your knowledge equals or exceeds that needed for competent performance, or
- where you might need more preparation.

The test results cannot tell you:

- how you compare with other technicians, or
- how many questions you answered correctly.

Your ASE test score report will show the number of correct answers you got in each of the content areas. These numbers provide information about your performance in each area of the test. However, because there may be a different number of questions in each area of the test, a high percentage of correct answers in an area with few questions may not offset a low percentage in an area with many questions.

It may be noted that one does not "fail" an ASE test. The technician who does not pass is simply told "More Preparation Needed." Though large differences in percentages may indicate problem areas, it is important to consider how many questions were asked in each area. Because each test evaluates all phases of the work involved in a service speciality, you should be prepared in each area. A low score in one area could keep you from passing an entire test.

Note that a typical test will contain the number of questions indicated above each content area's description. For example:

Drive Train (Test T3)

Content Area	Questions	% of Test
A. Clutch Diagnosis and Repair	15	25%
B. Transmission Diagnosis and Repair	17	28%
C. Drive Shaft and Universal Joint Diagnosis and Repair	11	19%
D. Drive Axle Diagnosis and Repair	17	28%
Total	60*	100%

**Note: The test could contain up to ten additional questions that are included for statistical research purposes only. Your answers to these questions will not affect your score, but since you do not know which ones they are, you should answer all questions in the test. The five-year Recertification Test will cover the same content areas as those listed above. However, the number of questions in each content area of the Recertification Test will be reduced by about one-half.*

"Average"

There is no such thing as average. You cannot determine your overall test score by adding the percentages given for each task area and dividing by the number of areas. It doesn't work that way because there generally are not the same number of questions in each task area. A task area with twenty questions, for example, counts more toward your total score than a task with ten questions.

So, How Did You Do?

Your test score should give you a good picture of your results and a better understanding of your task areas of strength and weakness.

If you fail to pass the test, you may take it again at any time it is scheduled to be administered. You are the only one who will receive your test score. Test scores will not be given over the telephone by ASE nor will they be released to anyone without your written permission.

Gasoline Engines

Pretest

The purpose of this pretest is to determine the amount of review you may require before taking the ASE Medium/Heavy Truck Gasoline Engines Test. If you answer all the pretest questions correctly, complete the questions and study the information in this chapter to prepare for this test.

If two or more of your answers to the pretest questions are incorrect, complete a study of all the chapters in *Today's Technician Medium/Heavy-Duty Truck Engines* Classroom and Shop Manuals published by Delmar Publishers, plus a study of the questions and information in this chapter.

The pretest answers are located at the end of the pretest; these answers also are in the answer sheets supplied with this book.

1. A port fuel-injected engine has a steady "puff" noise in the exhaust with the engine idling. The MOST likely cause of this problem could be:
   A. a burned exhaust valve.
   B. excessive fuel pressure.
   C. a restricted fuel return line.
   D. a sticking fuel pump check valve.
2. During a cylinder power balance test on an engine with electronic fuel injection, cylinder number 3 provides very little rpm drop. *Technician A* says the ignition system may be misfiring. *Technician B* says the engine may have an intake manifold vacuum leak. Who is right?
   A. A only
   B. B only
   C. Both A and B
   D. Neither A nor B
3. The measurement being performed in Figure 1–1 is:
   A. valve concentricity.
   B. valve seat concentricity.
   C. valve face-to-seat contact.
   D. valve stem-to-guide clearance.
4. A bent connecting rod may cause:
   A. uneven connecting rod bearing wear.
   B. uneven main bearing wear.
   C. uneven piston pin wear.
   D. excessive cylinder wall wear.
5. While discussing cylinder measurement, *Technician A* says the cylinder taper is the difference between the cylinder diameter at the top of the ring travel compared to the cylinder diameter at the center of the ring travel. *Technician B* says cylinder out-of-round is the difference between the axial cylinder bore diameter at the top of the ring travel compared to the thrust cylinder bore diameter at the bottom of the ring travel. Who is right?
   A. A only
   B. B only
   C. Both A and B
   D. Neither A nor B

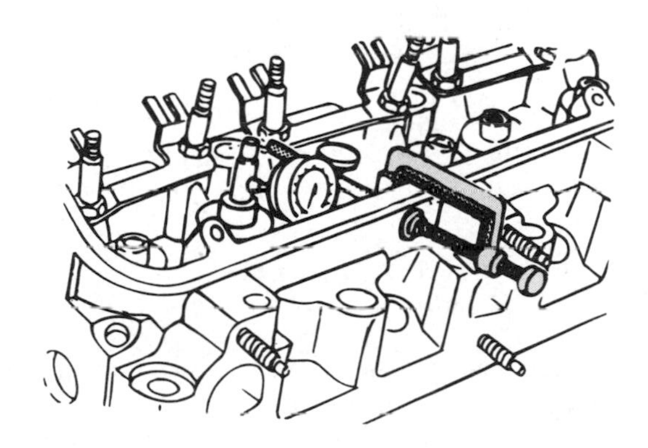

Figure 1–1 Valve measurement. *(Courtesy of General Motors Corporation, Service Technology Group)*

6. Valve seats are typically ground to an angle of:
   A. 15 or 20.
   B. 20 or 30.
   C. 30 or 45.
   D. 45 or 60.

7. The following are normal oil pump component measurements EXCEPT:
   A. outer rotor-to-housing clearance.
   B. clearance between the rotors.
   C. inner and outer rotor thickness.
   D. inner rotor diameter.

8. *Technician A* says the vibration damper counterbalances the back-and-forth twisting motion of the crankshaft each time a cylinder fires. *Technician B* says if the seal contact area on the vibration damper hub is scored, the damper assembly must be replaced. Who is right?
   A. A only
   B. B only
   C. Both A and B
   D. Neither A nor B

9. The thermostat is stuck open on a port fuel-injected engine. The MOST likely result of this problem is:
   A. a rich air-fuel ratio.
   B. a lean air-fuel ratio.
   C. excessive fuel pressure.
   D. engine overheating.

10. An excessively high coolant level in the recovery reservoir may be caused by any of these problems EXCEPT:
    A. restricted radiator tubes.
    B. a thermostat that is stuck open.
    C. a loose water pump impeller.
    D. An inoperative electric-drive cooling fan.

11. If ohmmeter 1 in Figure 1–2 shows an infinity reading, the pickup coil is:
    A. shorted to ground.
    B. shorted.
    C. open.
    D. not grounded.

12. *Technician A* says an intake manifold vacuum leak may cause a cylinder misfire with the engine idling. *Technician B* says an intake manifold vacuum leak may cause a cylinder misfire during hard acceleration. Who is right?
    A. A only
    B. B only
    C. Both A and B
    D. Neither A nor B

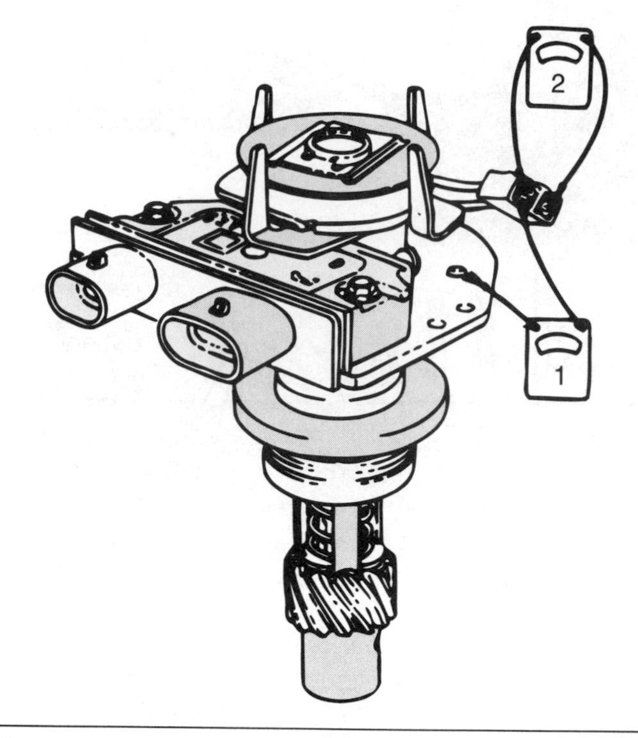

Figure 1–2 Distributor test. *(Courtesy of General Motors Corporation, Service Technology Group)*

13. *Technician A* says the PCV valve moves toward the closed position when the throttle is opened from idle to half throttle. *Technician B* says the PCV valve is moved toward the closed position by spring tension. Who is right?
    A. A only
    B. B only
    C. Both A and B
    D. Neither A nor B

14. During the battery test in Figure 1–3: *Technician A* says at the end of the test the voltage remains above 9.6V at 70°F (21°C) if the battery is satisfactory. *Technician B* says the battery should be discharged at half the cold cranking rating. Who is right?
    A. A only
    B. B only
    C. Both A and B
    D. Neither A nor B

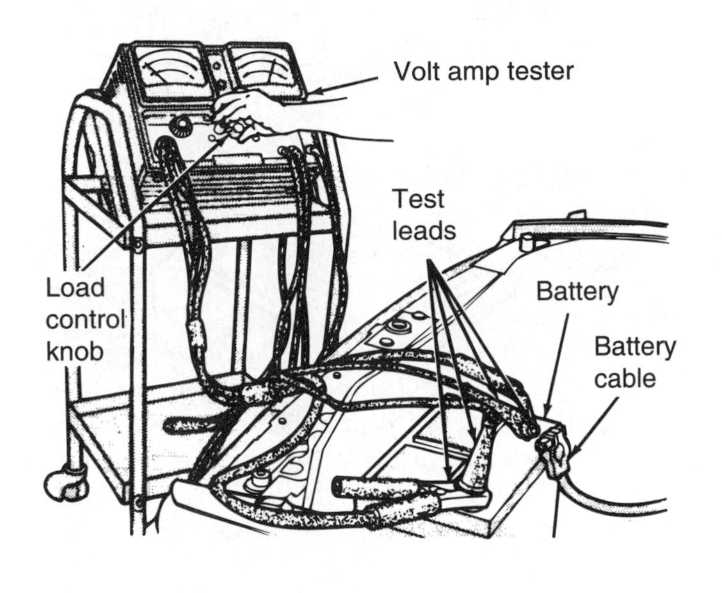

Figure 1–3 Battery test. *(Courtesy of Chrysler Corporation)*

Answers to Pretest

1. A, 2. C, 3. D, 4. A, 5. D, 6. C, 7. D, 8. A, 9. A, 10. B, 11. D, 12. A, 13. D, 14. C

General Engine Diagnosis

Task 1

Listen to driver's complaint and road test vehicle; determine needed repairs.

1. *Technician A* says that customer complaints must be listened to closely. *Technician B* says that when diagnosing a vehicle problem, always start with the easiest test to perform. Who is right?
   A. A only
   B. B only
   C. Both A and B
   D. Neither A nor B

2. *Technician A* says the customer complaint may be identified as the repairs progress on the customer's vehicle. *Technician B* says if the repairs are done carefully on the customer's car, there is no need to road test the vehicle when repairs are completed. Who is right?
   A. A only
   B. B only
   C. Both A and B
   D. Neither A nor B

Hint

The Technician must be familiar with a basic diagnostic procedure such as the following: Listen carefully to the customer's complaint, and question the customer to obtain more information regarding the problem. Identify the complaint, and road test the vehicle, if necessary. Think of the possible causes of the problem. Perform diagnostic tests to locate the exact cause of the problem. Always start with the easiest, quickest test. After the repair has been made, be sure the customer's complaint is eliminated; road test the vehicle again, if necessary.

Task 2

Inspect engine assembly for fuel, oil, coolant, and other leaks; determine needed repairs.

3. A customer has brought in a vehicle that has an engine oil leak. Upon inspection, the lower half of the engine is covered in oil. *Technician A* says that the engine must be steam cleaned before an accurate diagnosis can be given. *Technician B* says that if the lower portion of the engine is leaking, it is most likely the oil pan gasket. Who is right?
   A. A only
   B. B only
   C. Both A and B
   D. Neither A nor B

4. A cooling system is pressurized with a pressure tester to locate a coolant leak. After 15 minutes, the tester gauge has dropped from 15 to 5 psi, and there are no visible signs of coolant leaks in the engine compartment. *Technician A* says the engine may have a leaking head gasket. *Technician B* says that the heater core may be leaking. Who is right?
   A. A only
   B. B only
   C. Both A and B
   D. Neither A nor B

Hint

A Technician must understand the basic fuel, lubricating, and cooling systems and components. The location of all possible leaks in these systems must be identified. Coolant leaks may be internal or external in relation to the engine. If a vacuum leak is sufficient to cause the engine to stall while idling, it is possible to locate the leak by listening for the whistle generated as airflow passes the hole. Another method of vacuum leak detection is to spray carburetor cleaner around suspected areas while observing engine rpm with a tachometer. Because the carburetor cleaner is combustible, it will aid in enriching the mixture, which will increase engine speed.

Many times when trying to locate an engine oil leak, the engine has been leaking for an indefinite period and the engine oil has accumulated dirt. In this case, the engine must be steam-cleaned to make a leak more visible. Another way to find an oil leak is to add a dye to the engine oil that reacts with an ultraviolet lamp.

Task 3

Check the level and the condition of fuel, oil, and coolant; determine needed repairs.

5. What two things can engine coolant be tested for?
   A. Carbon and sulfur content
   B. Lead content and viscosity
   C. Acidity and freezing point
   D. Age and water pump condition.

6. What can be determined with the engine coolant by checking it with pH strips?
   A. Freezing point
   B. Head gasket failure
   C. Acidity
   D. Coolant age

Hint

When checking the condition of engine oil, one observes the level on the dipstick. The color or the darkness of the oil is a key to determining whether you should change the oil. The darker in color, the greater amount of carbon particulate in the oil.

Always use caution when working with pressurized cooling systems. Engine coolant should be tested for freezing point and for acidity. You use a cooling system hydrometer with a graduated temperature scale to check the freezing point of the coolant. Acidity can be checked using pH strips that are dipped into the coolant.

Task 4

Listen to engine noises; determine needed repairs.

7. A heavy thumping noise occurs with the engine idling, but the oil pressure is normal. This noise may be caused by:
   A. worn pistons.
   B. loose flywheel bolts.
   C. worn main bearings.
   D. loose camshaft bearings.

8. An engine produces a clacking noise during acceleration at low speeds. This noise is worse when the engine coolant is cold, but the noise is still present with the engine at normal operating temperature. The cause of this noise could be:
   A. loose connecting rod bearings.
   B. worn main bearings.
   C. a collapsed piston.
   D. sticking valve lifters.

Hint

Worn pistons and cylinders cause a rapping noise while accelerating, and worn main bearings cause a thumping noise when you accelerate the engine. Loose camshaft bearings usually do not cause a noise unless severely worn.

The most common noise complaint is valvetrain noise. The Technician identifies valve tappet noise by a light, regular clicking sound at twice the engine speed that comes from the upper portion of the engine. Dirty hydraulic lifters, lack of lubrication, and misadjusted valve clearance are some of the causes of valvetrain noise.

Ignition detonation and preignition can cause noises that can be mistaken for internal engine components. The sound is the result of a second flame that starts after the air-fuel mixture ignites. When the two flame fronts collide, one hears a loud explosion or knock. These conditions can be caused by internal engine components, low-quality fuel, or a faulty ignition system. Other components that can create noise include flywheels, harmonic balancers, any belt-driven component, torque converters, and engine mounts.

Task 5

Check color and quantity of the engine exhaust smoke; determine needed repairs.

9. Which of the following exhaust conditions are the MOST likely indication of an engine misfire?
   A. Blue colored smoke from the tailpipe
   B. Excessive rattling of the exhaust system
   C. Cold exhaust temperature
   D. Puffing or wheezing

10. Blue smoke that is coming out of the tailpipe can be caused by all of the following EXCEPT:
   A. worn valve guides.
   B. worn valve stem seals.
   C. fouled spark plugs.
   D. worn piston rings.

Hint

Blue colored exhaust smoke indicates that an excessive amount of oil is entering the combustion chamber. Blue smoke during acceleration indicates worn piston rings. Blue smoke after startup indicates worn valve guides or seals. Black smoke indicates excessive fuel consumption;

this is most likely a fuel system component concern. Gray or white smoke is an indication that coolant is entering the combustion chamber. The sound of the exhaust should be smooth and even. Puffing or wheezing exhaust or excessive rattling of the exhaust system itself indicates a misfire.

Task 6

Perform engine vacuum test; determine needed repairs.

11. With a vacuum gauge connected to manifold vacuum, the needle fluctuates between 15 and 20 in. Hg (Figure 1–4). What could be the cause?
    A. Late ignition timing
    B. Intake manifold gasket failure
    C. A restricted exhaust system
    D. Sticking valve stems and guides

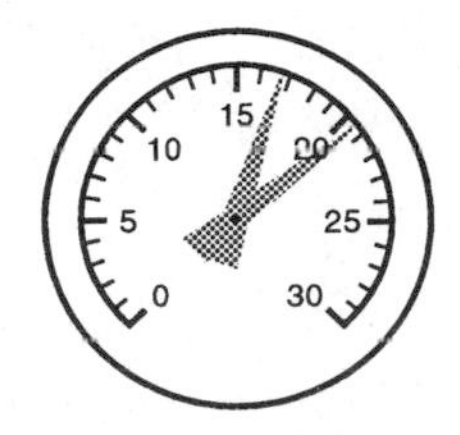

Figure 1–4 Vacuum gauge reading.

12. With a vacuum gauge connected to an engine, the vacuum gauge reading is a steady 18 in. Hg at idle. When the engine speed is increased to 1,500 rpm, the reading increases to a steady 25 in. Hg. *Technician A* says that the engine may have a blocked catalytic converter. *Technician B* says that the EGR valve may be stuck open. Who is right?
    A. A only
    B. B only
    C. Both A and B
    D. Neither A nor B

Hint

The Technician connects a vacuum gauge directly to the intake manifold to diagnose engine and related system conditions. When a vacuum gauge is connected to the intake manifold, the reading on the gauge should provide a steady reading between 17 and 22 inches of mercury (in. Hg) with the engine idling. Abnormal vacuum gauge readings indicate these problems:

- *A low, steady reading indicates late ignition timing.*
- *If the vacuum gauge reading is steady and much lower than normal, the intake manifold has a significant leak.*
- *When the vacuum gauge pointer fluctuates between approximately 11 and 16 in. Hg on a carbureted engine at idle speed, the carburetor idle mixture screws require adjusting. On a fuel-injected engine, the injectors require cleaning or replacing.*
- *Burned or leaking valves cause a vacuum gauge fluctuation between 12 and 18 in. Hg.*
- *Weak valve springs result in a vacuum gauge fluctuation between 10 and 25 in. Hg.*
- *A leaking head gasket may cause a vacuum gauge fluctuation between 7 and 20 in. Hg.*
- *If the valves are sticking, the vacuum gauge fluctuates between 14 and 18 in. Hg.*

If the vacuum gauge pointer drops to a very low reading when the technician accelerates the engine and holds a steady higher rpm, the catalytic converter or other exhaust system components are restricted.

Task 7

Perform cylinder power balance test; determine needed repairs.

13. While performing a cylinder power balance test, *Technician A* says that when all the cylinders are contributing equally to the engine power, all the cylinders will provide the specified rpm drop. *Technician B* says that often the reason for a cylinder not contributing to the engine power can be traced to the fuel or the ignition system. Who is right?
    A. A only
    B. B only
    C. Both A and B
    D. Neither A nor B

14. When testing cylinder balance on a V6 port fuel-injected engine, number 2 cylinder provides 20 rpm decrease and all the other cylinders provide a 60 rpm decrease. The cylinder compression is tested and all cylinders have the specified compression. The cause of this problem could be:
    A. excessive carbon buildup in number 2 cylinder.
    B. a defective fuel injector in number 2 cylinder.
    C. worn compression rings on number 2 cylinder.
    D. a defective head gasket on number 2 cylinder.

Hint

If the cylinder is working normally, a noticeable rpm decrease occurs when the cylinder misfires. If there is very little rpm decrease when the analyzer causes a cylinder to misfire, the cylinder is not contributing to engine power. Under this condition the engine compression, ignition system, and fuel system should be checked to locate the cause of the problem. An intake manifold vacuum leak may cause a cylinder misfire with the engine idling or operating at low speed. If this problem exists, the misfire will disappear at a higher speed when the manifold vacuum decreases. When all the cylinders provide the specified rpm drop, are all the cylinders contributing equally to the engine power?

Task 8

Perform cylinder compression tests; determine needed repairs.

15. *Technician A* says that if an engine has one cylinder that is oversized, it will have a higher compression reading than the others. *Technician B* says if the compression pressure increases after adding a small amount of oil to the cylinder, the valve guides are worn. Who is right?
    A. A only
    B. B only
    C. Both A and B
    D. Neither A nor B

16. While performing a compression test, a gradual buildup of compression occurs with each stroke. *Technician A* says this is an indication of a burned exhaust valve. *Technician B* says that it indicates a cracked cylinder head. Who is right?
    A. A only
    B. B only
    C. Both A and B
    D. Neither A nor B

Hint

You disable the ignition and fuel injection system before proceeding with the compression test. During the compression test the engine is cranked through four compression strokes on each cylinder and the compression readings recorded. One interprets lower-than-specified compression readings as follows:

- *Low compression readings on one or more cylinders indicate worn rings, valves, a blown head gasket, or a cracked cylinder head.*
- *A gradual buildup on the four compression readings on each stroke indicates worn rings, whereas little buildup on the four strokes usually is the result of a burned exhaust valve.*
- *When the compression readings on all the cylinders are even but lower than the specified compression, you can suspect worn rings and cylinders.*
- *A leaking head gasket or cracked cylinder head causes low compression on two adjacent cylinders.*
- *Higher-than-specified compression usually indicates carbon deposits in the combustion chamber.*
- *A hole in the piston or a severely burned exhaust valve will cause zero compression in a cylinder. If the zero compression reading is caused by a hole in the piston, the engine will have excessive blowby.*

When the engine spins freely and compression in all cylinders is low, check the valve timing. If a cylinder compression reading is below specifications, perform a wet test to determine if the valves, or rings, is the cause of the problem. Squirt approximately 2 or 3 teaspoons of engine oil through the spark plug opening into the cylinder, with the low compression reading. Crank the engine to distribute the oil around the cylinder wall and then retest the compression. If the compression reading improves considerably, the rings (or cylinders) are worn. When there is little change in the compression reading, one of the valves is leaking.

Task 9

Perform cylinder leakage tests; determine needed repairs.

17. When performing a cylinder leakage test, which of the following is considered to be the maximum tolerable percentage?
    A. 5 to 10 percent
    B. 30 to 40 percent
    C. 40 to 45 percent
    D. 15 to 20 percent.

18. During a cylinder leakage test, cylinder number 4 has 50 percent leakage and the air can be heard escaping from the PCV valve opening. *Technician A* says that the intake valve in that cylinder is leaking. *Technician B* says that the piston rings in that cylinder are worn. Who is right?
    A. A only
    B. B only
    C. Both A and B
    D. Neither A nor B

Hint

During the leakage test a regulated amount of air from the shop air supply is forced into the cylinder with both exhaust and intake valves closed. The gauge on the leak-

age tester indicates the percentage of air escaping the cylinder. A gauge reading of 0 percent indicates there is no cylinder leakage; if the reading is 100 percent, the cylinder is not holding any air.

An excessive reading is one that exceeds 20 percent. Check for air escaping from the tailpipe, positive crankcase valve (PCV) opening, or the top of the throttle body or carburetor. Air escaping from the tailpipe indicates an exhaust valve leak. When the air is escaping out of the PCV valve opening, the piston rings are leaking. An intake valve is leaking if air is escaping from the top of the throttle body, or carburetor. Remove the radiator cap and check the coolant for bubbles, which indicate a leaking head gasket or cracked cylinder head.

Task 10

Diagnose engine mechanical, ignition, or fuel problems with an engine oscilloscope and or analyzer (scan tool); determine needed repairs.

19. What is the purpose of the snap test function when using an engine analyzer?
    A. To check the voltage increase for each cylinder under load
    B. To check if all the spark plugs have the same air gap
    C. To indicate the condition of the primary circuit switching device
    D. To observe the ratio between engine speed and the firing kV

20. *Technician A* says that a fouled spark plug will cause a higher firing voltage. *Technician B* says the top of the firing line is the secondary voltage required to start firing the spark plug in the secondary circuit. Who is right?
    A. A only
    B. B only
    C. Both A and B
    D. Neither A nor B

Hint

On many engine analyzers, ignition performance tests include primary circuit tests, secondary kilovolt (kV) tests, acceleration tests, scope patterns, and cylinder miss recall. Primary circuit tests include coil-input voltage, coil primary resistance, dwell, curb idle speed, and idle vacuum. The kV test measures the voltage required to start firing a spark plug. A high resistance problem in a spark plug or a spark plug wire causes higher firing kV, whereas a fouled spark plug or a cylinder with low compression results in a lower firing kV.

Some secondary kV tests include a snap kV test in which the analyzer directs the Technician to accelerate the engine suddenly. When this action is taken, the firing kV should increase evenly on each cylinder. Some engine analyzers also display circuit gap for each cylinder. The circuit gap is the voltage to fire all the gaps in the secondary circuit, such as the rotor gap, but the spark plug gap is excluded. Some analyzers display the burn time for each cylinder with the secondary kV test. The burn time is the length of the spark line in milliseconds (ms). The average burn time should be 1 to 1.5 milliseconds.

Cylinder Head and Valve-train Diagnosis and Repair

Task 1

Inspect cylinder heads for cracks and gasket surface areas for warpage; check passage condition.

21. When measuring the cylinder head for warpage with a straightedge, the feeler gauge measurement is 0.025 in. (0.63 mm) (Figure 1–5). *Technician A* says the cylinder head needs resurfacing. *Technician B* says the top surface of the block should be checked for warpage. Who is right?
    A. A only
    B. B only
    C. Both A and B
    D. Neither A nor B

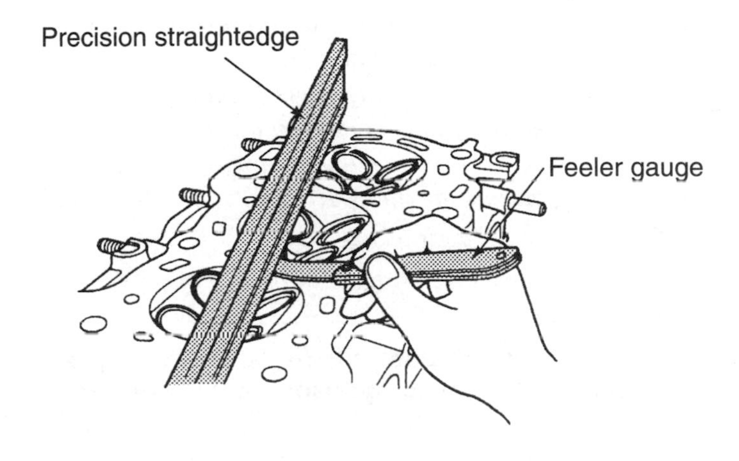

Figure 1–5 Measuring cylinder head warpage.

22. An engine has experienced repeated head gasket failures. The same head bolts were reused each time the head was installed. *Technician A* says the torque procedure on some head bolts requires the use of an angle gauge or torque meter. *Technician B* says some engines require head bolt replacement each time the cylinder head is removed. Who is right?
    A. A only
    B. B only
    C. Both A and B
    D. Neither A nor B

Hint

When inspecting the cylinder head, it is normal if carbon buildup is a light, even layer across the entire combustion chamber. If the carbon is excessive, the cylinder head should be thoroughly cleaned. This carbon buildup could be caused by worn valve guides, valve seals, or rings.

Once the initial cleaning has been completed, inspect the cylinder head for cracks and other obvious damage. Remember, not all cracks are visible to the eye. Therefore, it may be necessary to perform additional tests if a crack is suspected. Common locations for cracks are between the valve seats, or around the spark plug hole. If the cylinder head has excessive damage, it may be more cost efficient to replace it instead of repairing it.

Check the cylinder head mating surface for texture and warpage. Use a straightedge with a feeler gauge to measure deck warpage. Warpage can occur in any direction on the head surface. Measure for warpage in three areas along the edges and three areas across the center. Compare the measurements with the manufacturer's specifications. If there are no specifications available, a rule of thumb is 0.003 in. (0.08 mm) for any 6-inch length. Use the feeler gauge to determine whether the cylinder head needs to be resurfaced.

Use the straightedge in the same matter to check for warpage of the intake and exhaust manifold mating surfaces. The general rule for maximum warpage limit for the manifold mating surfaces allowed is 0.004 in. (0.1 mm).

Task 2

Inspect and test valve springs for squareness, pressure, and free height comparison; replace as necessary.

23. Engine valve springs should be inspected for all of the following EXCEPT:
    A. cracks.
    B. spring tension.
    C. pitting and nicks.
    D. spring coil gap.

24. How is valve stem height corrected?
    A. By removing material from the valve tip
    B. Adding shims
    C. Installing a valve spring that is the correct height
    D. Removing material from the valve seat

25. While discussing the measurement in Figure 1–6, *Technician A* says the valve spring is being measured for installed height. *Technician B* says the valve spring must be held in one position. Who is right?
    A. A only
    B. B only
    C. Both A and B
    D. Neither A nor B

Hint

Begin valve spring inspection with a visual check for obvious signs of wear, cracks, corrosion, pitting, and nicks. If a valve has any of these defects, discard it and replace it with a new one. After a visual inspection, the spring must be checked for squareness, free length, and spring pressure.

If a spring is not square, it can side load the valve stem and the valve guide, causing excessive wear. The free length of the spring must be measured and compared with the specifications. Some springs that have a free length below specification can be corrected with the use of shims. To check a spring for squareness and free length, place the spring next to a square and rotate it while watching for warpage. Valve spring pressure is

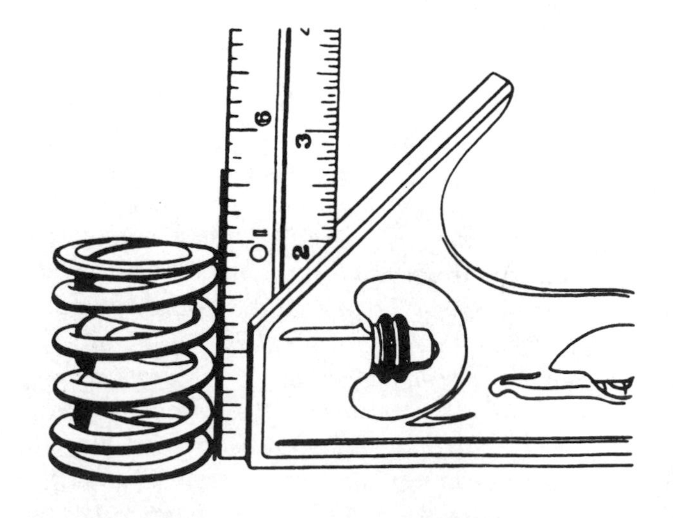

Figure 1–6 Valve spring measurement. *(Courtesy of General Motors Corporation, Service Technology Group)*

tested using a special spring tension gauge. The tension gauge works by compressing the spring to the specified height and observing the spring pressure reading on the gauge. Spring tension should be within 10 percent of the manufacturer's specification, and no more than 10 pounds difference between springs.

Task 3

Inspect and replace valve springs retainers, rotators, and locks; replace valve stem seals.

26. What preparation must be done to the valve guides before measuring valve seat runout?
    A. Use 30 weight engine oil to lubricate the guide.
    B. Clean the valve guide with a bore brush.
    C. Ream the valve guide to its original diameter.
    D. Use a pilot to hold the guide in place.

27. The valves in Figure 1–7 are from an engine with valve rotators. As indicated in the figure the valve stem tip wear patterns indicate:
    A. valve 1 indicates a normal wear pattern.
    B. valve 1 indicates a worn-out rotator.
    C. valve 2 indicates a worn valve guide.
    D. valve 3 indicates a weak valve spring.

Hint

Valve spring retainers and locks must be checked for wear and scoring. Check for signs of cracks and areas of discoloration. When any of these conditions are present, replace the components. The valve lock grooves on the valve stems must be inspected for wear, particularly round shoulders. If these shoulders are uneven or rounded, replace the valve. It is a good practice to replace the valve stem seals any time the cylinder head is disassembled.

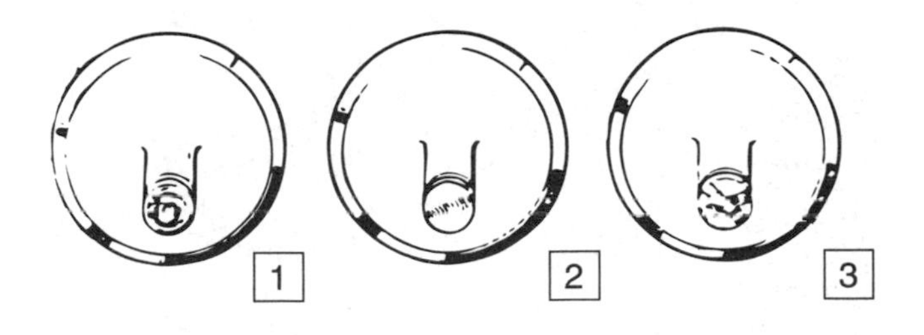

Figure 1–7 Valve stem tip conditions. *(Courtesy of General Motors Corporation, Service Technology Group)*

Task 4

Inspect valve guides for wear, and check valve guide-to-stem fit; determine needed repair.

28. *Technician A* says the valve stem must be measured at the top, middle, and near the fillet. *Technician B* says many modern engines use tapered stems to provide additional clearance between the stem and guide close to the head of the valve. Who is right?
    A. A only
    B. B only
    C. Both A and B
    D. Neither A nor B

29. *Technician A* says that a worn-out valve guide will provide an inaccurate valve seat runout measurement. *Technician B* says that the valve seat-to-valve face contact area provides a path for heat from the valve head to dissipate. Who is right?
    A. A only
    B. B only
    C. Both A and B
    D. Neither A nor B

Hint

Before attempting to measure valve guide wear and valve seat runout, it is important to use a valve guide cleaner or a bore brush. After preparing the guide, use a small bore gauge and measure the valve guide at three different locations from top to bottom. Measure the fingers of the bore gauge with an outside micrometer. Measure the diameter of the valve stem, and subtract the difference to find the clearance. Another method is to insert the correct valve into the guide and install the special tool that maintains the height of the valve during the inspection. Place a dial indicator magnetic base on the cylinder head with the tip of the dial indicator at a right angle to the valve head. Zero the dial indicator and rock the valve while observing the clearance indicated on the dial.

The maximum amount of wear that most manufacturers recommend is 0.005 in. (0.12 mm) or less. The desirable clearances are 0.001 to 0.03 in. (0.025 to 0.080 mm) for intake and 0.0015 to 0.0035 in. (0.04 to 0.009 mm) for the exhaust valves. If any of the valve guides are out of specification, replace the guide.

Task 5

Inspect, replace, and/or grind valves.

30. Reconditioning valves is being discussed. *Technician A* says the valve is stroked across the stone as the stone is being fed in. *Technician B*

says when the last pass is completed, the stone should be backed away from the valve. Who is right?

A. A only
B. B only
C. Both A and B
D. Neither A nor B

31. *Technician A* says the amount removed from the valve face is the size of the required valve spring shim. *Technician B* says valve stem tip height must be correct for proper rocker arm geometry. Who is right?

A. A only
B. B only
C. Both A and B
D. Neither A nor B

32. When resurfacing valves, *Technician A* says that valve stem tips usually do not require resurfacing. *Technician B* says that there is no need for an interference angle between the valve and the seat. Who is right?

A. A only
B. B only
C. Both A and B
D. Neither A nor B

Hint

There are two valve areas that require reconditioning: the stem tip and, more commonly, the valve face. If a stone is to be used, it must be dressed before any grinding is performed. Usually, after conditioning eight to ten valves, it is time to dress the stone again. Watch the valve face surface while grinding. If the surface is rough or the valve chatters during the process, the stone needs to be dressed. Refer to the valve grinding manufacturer's procedure. Refer to the following tips when resurfacing the valve.

- *Locate the valve into the bottom of the chuck each time a new one is inserted.*
- *Use a full stroke, using the entire width of the stone.*
- *Keep cooling oil flowing over the valve face.*
- *Never remove more material from the valve than is needed to provide a fresh surface.*
- *An interference fit of one degree is required for installation. You accomplished this by cutting the valve face one-degree different from the valve seat.*

Task 6

Inspect and grind valve seats, and/or determine need for replacement.

33. All of the following are considerations when grinding valve seats EXCEPT:

A. dress the stone before cutting any seat.
B. remove only enough material to provide a new surface.
C. use transmission fluid to lubricate the stone only.
D. do not apply pressure to the grinding stone.

34. In Figure 1–8 the measurement being performed is:

A. valve guide wear.
B. valve guide-to-stem clearance.
C. valve seat concentricity.
D. valve guide installed height.

Hint

Valve seats are usually ground at three different angles: the seat angle, the topping angle, and the throat angle. Any valve seat that is excessively worn or deeply scored must be replaced with a new one. Refer to the manufacturer's specifications for proper grinding angles. The following are some considerations when grinding valve seats:

- *The seat contact pattern should be 1/16 to 3/32 in. (1.60 to 2.39 mm) wide.*

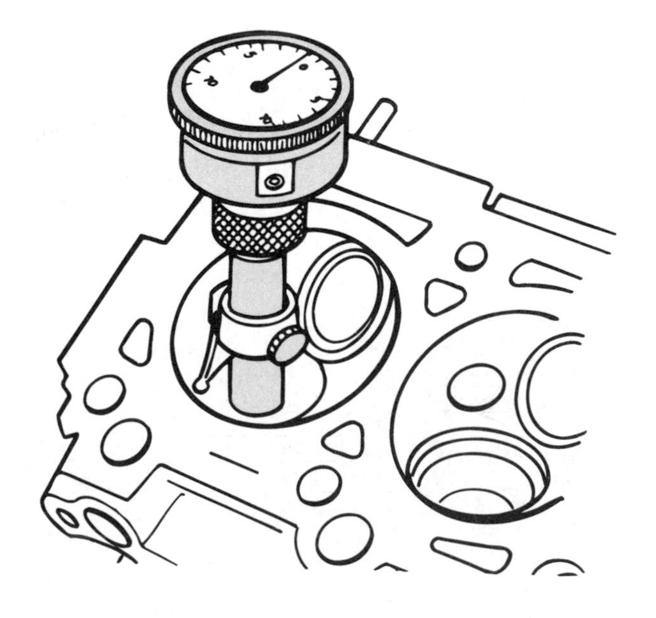

Figure 1–8 Valve measurement. *(Courtesy of Ford Motor Company)*

- *Dress each stone with a diamond bit-dressing tool before cutting any seat.*
- *Worn valve guides must be replaced before attempting to grind any seats.*
- *When grinding the seat, only remove enough material to provide a new surface.*
- *Never apply pressure to the grinding stone because it will remove too much material and cause the stone to require dressing sooner.*
- *It is critical that the stone is dressed to the proper angle.*
- *Be sure to use the correct size pilot as a guide for the grinding seat assembly.*

Task 7

Check valve face-to-seat contact and valve seat runout; service seats and valves as necessary.

35. When grinding valves and seats, the valve seat contact area on the valve face is shown in Figure 1–9. To correct this improper valve seat contact area it is necessary to use:
    A. a 45-degree and a 60-degree stone.
    B. a 30-degree and a 45-degree stone.
    C. a 45-degree and a 90-degree stone.
    D. a 60-degree and a 90-degree stone.

36. An acceptable valve seat width is:
    A. 1/16 in. (1.58 mm).
    B. 3/16 in. (4.76 mm).
    C. 7/32 in. (5.56 mm).
    D. 1/4 in. (6.35 mm).

Hint

When a valve is closed, it must make a pressure-tight seal. The valve face-to-valve seat contact provides the seal. In addition, this contact provides a path for heat to dissipate to the cylinder head.

First, measure the width of the seat with a machinist's ruler. If out of specification, remove material from the seat accordingly to obtain proper seat width. Proper seat concentricity is critical for a proper seal. Use a valve runout gauge with an arbor installed in the valve guide bore. Slowly rotate the gauge around the seat while observing the readings. Compare the results with specifications; generally 0.002 in. (0.050 mm) is a usual tolerance. Remember, a worn valve guide will give an improper measurement.

Task 8

Check valve seat assembled height and valve stem height; service valve spring assembles as necessary.

37. Measurement B in Figure 1–10 is more than specified. *Technician A* says this problem may bottom the lifter plunger. *Technician B* says a shim should be installed under the valve spring. Who is right?
    A. A only
    B. B only
    C. Both A and B
    D. Neither A nor B

38. Refer to Figure 1–10. Measurement A is more than specified with a new valve installed. The most likely correction for this problem is:
    A. grind the valve seat to move the seat contact area downward on the valve face.
    B. grind the valve seat to move the seat contact area upward on the valve face.
    C. install a new valve seat.
    D. grind some metal off the valve face.

Hint

The cylinder head is now ready for assembly. Make sure that cylinder head has been cleaned. Before installing the valves, polish the valve stems with fine crocus cloth

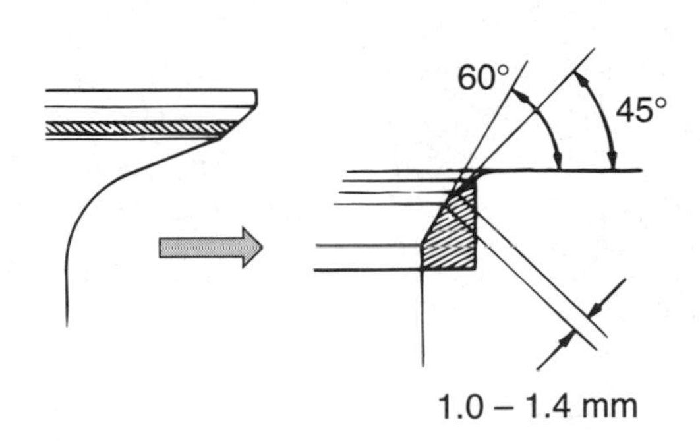

Figure 1–9 Valve seat contact area.

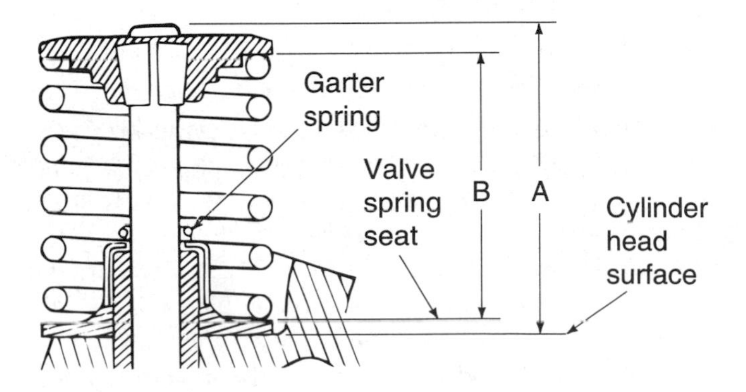

Figure 1–10 Valve spring and stem measurements. *(Courtesy of Chrysler Corporation)*

and solvent. Lubricate the valve stems with assembly grease or, at the least, engine oil. Bottom the valve against the seat and assemble the valve spring retainer with the locks onto the valve. While holding the valve seated, use a machinist's ruler and measure from the spring seat to the underside of the spring retainer. Compare this measurement to the specification and add shims, if necessary.

Stem height increases due to resurfacing the valve. Stem height is corrected by removing material from the valve tip. This measurement is from the valve tip to the spring seat. Again, use a machinist's ruler.

Task 9

Inspect pushrods, rocker arms, rocker arm pivots, and shafts for wear, bending, cracks, looseness, and blocked oil passages; repair or replace.

39. *Technician A* says that the pushrods can be checked for straightness by rolling them on a known flat surface. *Technician B* says a bent pushrod may be caused by a sticking valve lifter. Who is right?
    A. A only
    B. B only
    C. Both A and B
    D. Neither A nor B

40. All of these statements about pushrods are true EXCEPT:
    A. if the drilled passage through the center of a pushrod is plugged, rocker arm lubrication is insufficient.
    B. pushrods are designed to rotate when the engine is running.
    C. replacement pushrods must be the same length as the original pushrods.
    D. a shorter pushrod may be installed if the installed valve stem height is more than specified.

Hint

Before assembling the rocker arm and the related components, some inspection of the components should be performed. The rocker arms must be inspected for wear and damage. Rocker arms are constructed of cast iron, stamped steel, or aluminum. There are three areas on the rocker arm that receives high stress. These areas are the pushrod contact, the pivot, and the valve stem contact surface. The pushrod and the valve stem contact areas should be round and show signs of even wear. The oil passages in the rocker arms should be clean and free from debris.

Inspect rocker arm shafts (if applicable) for straightness and for excessive wear at the points where the rocker arm rides. The shafts should be free of scoring and galling. Inspect the shaft for wear at the places where the rocker arm pivots. A slight polishing with no ridges indicates normal wear. Wear in this location usually indicates a lack of lubrication, which in turn generates excessive heat. The shaft itself can be checked for straightness by rolling it on a smooth surface.

Pushrods should be inspected for signs of wear and bending. Inspect the tips for normal wear patterns. Check the pushrod runout by rolling it on a known true surface, a piece of glass works well. Runout should not exceed 0.003 in. (0.08 mm).

Task 10

Inspect and replace hydraulic or mechanical lifters.

41. Which of the following types of wear patterns is MOST likely desired for the lifter mating surface?
    A. Convex countermachined face
    B. Smooth
    C. Centered circular wear pattern
    D. Cross-hatched

42. *Technician A* says that the tool in Figure 1–11 is used to prime the lifters. *Technician B* says this tool is used to determine lifter leak-down. Who is right?
    A. A only
    B. B only
    C. Both A and B
    D. Neither A nor B

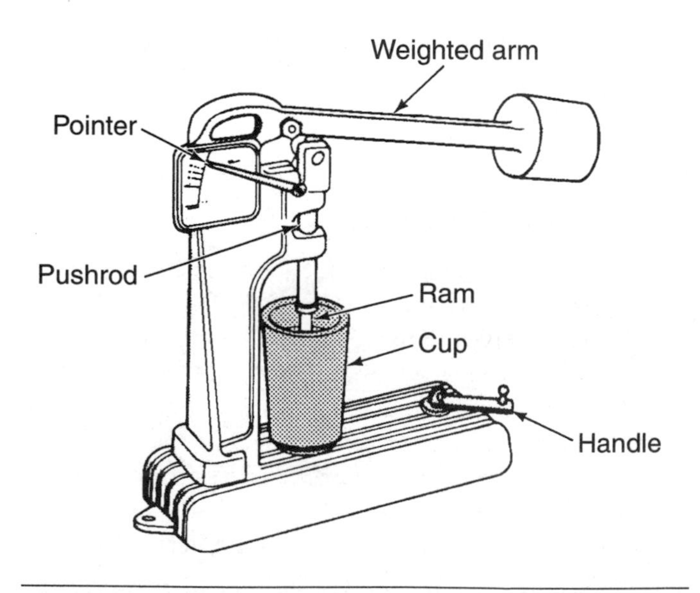

Figure 1–11 Valve component tester. *(Courtesy of Chrysler Corporation)*

Hint

When inspecting valve lifters, the surface face of the lifter must be smooth with a centered circular wear pattern. The surface should be a convex countermachined face. If wear extends to the edge of the lifter, the convex shape is worn away and the lifter must be replaced. The lifter body should be polished and smooth, free from any ridges, scarring, and signs of scuffing.

Hydraulic lifters that pass the visual inspection should be tested for leak-down. This is done using a special tool that applies weight to a primed lifter submerged in engine oil. While the lifter is bleeding down, observe the scale and compare the rate of bleed-down over a certain length of time. Leak-down range is between 20 and 90 seconds.

Sometimes it might be necessary to disassemble a lifter. Only disassemble one lifter at a time. Keep track of the order in which it comes apart.

Flat tappet lifters require a break-in period. A normal break-in involves applying break-in lube on the camshaft and the lifter face. Immediately following startup, the engine must be run at varying speeds between 1,500 to 2,500 rpm over a 20-minute period. The engine is run at off idle speeds to provide adequate oil splashed onto the machined surfaces that are in contact with the camshaft.

Task 11

Adjust valves.

43. While adjusting mechanical lifters, *Technician A* says when the valve clearance is checked on a cylinder, that cylinder should be positioned at TDC on the exhaust stroke. *Technician B* says some mechanical lifters have removable shim pads, available in various thickness to provide proper valve clearancces. Who is right?
    A. A only
    B. B only
    C. Both A and B
    D. Neither A nor B

44. All the statements about the valve adjustment in Figure 1–12 are true EXCEPT:
    A. both valves in the cylinder should be closed.
    B. the valve has a hydraulic valve lifter.
    C. turn the adjusting screw clockwise to decrease the valve clearance.
    D. excessive clearance may cause a clicking noise at idle speed.

Hint

Valve clearance usually needs to be adjusted after cylinder head reconditioning. The following are typical methods of adjusting valve lash:

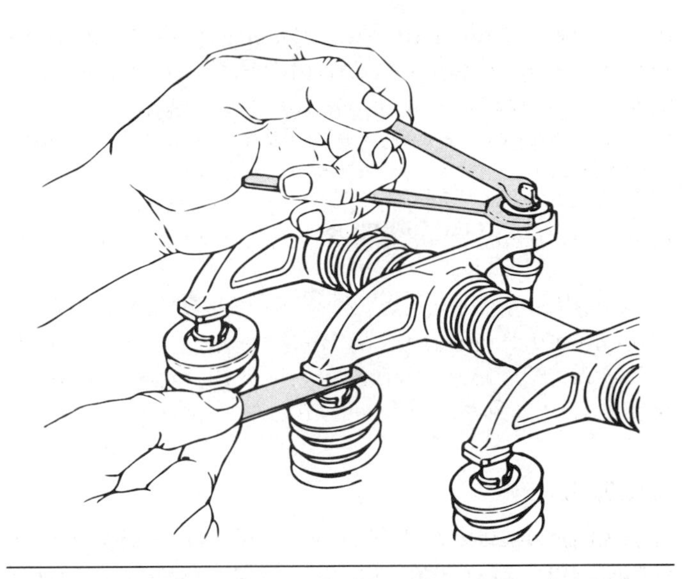

Figure 1–12 Valve adjustment. *(Courtesy of Chrysler Corporation)*

- *Adjustable nut attaching the rocker arm to the stud*
- *Adjustable screw located in the end of the rocker*
- *Selectable shims that are positioned between the camshaft lobes and the followers*
- *Selective pushrod length*

Typical adjustments require that number 1 cylinder to be positioned at top dead center (TDC) on the compression stroke. Adjust the hydraulic lifters to the specified preload, while setting the specified clearance for solid lifter engines. After the number 1 cylinder is adjusted, the crankshaft must be rotated to the next designated location and the specified valve clearance adjusted. Depending on the manufacturer, all the valves can be adjusted at two different crankshaft locations, while others require locating each piston to TDC.

Engine Block Diagnosis and Repair

Task 1

Inspect and replace pans, covers, gaskets, and seals.

45. When using room temperature vulcanizing (RTV) for gasket applications:
    A. the gasket surfaces should be cleaned with carburetor cleaner.
    B. the RTV should be placed on one side of the bolt holes only.

C. the RTV should be allowed to cure for 25 minutes before assembly.
D. the RTV bead should be 1/8 in. (3 mm) in width.

46. *Technician A* says a springless seal may be installed in the timing gear cover. *Technician B* says when a seal is installed, the seal lip must face in the same direction as the oil flow. Who is right?
A. A only
B. B only
C. Both A and B
D. Neither A nor B

Hint

Inspect the oil pan for cracks or dents. Also, inspect the pan for extensive areas of rust. If the oil pan is severely rusted, it should be replaced. Check the gasket-mating surface of the pan for straightness, using a straightedge and a feeler gauge. If the gasket surface is warped, it can be straightened by striking it with a ball peen hammer on a flat, true surface. Stamped steel valve cover gasket-mating surfaces can be straightened using the same method.

When installing new gaskets, make sure that the gasket and the surface that it is going to seal are free from debris and oil or grease. It is imperative that all of the old gasket material is removed for proper sealing.

Task 2

Inspect engine block for cracks, passage condition, core and gallery plug condition, and surface warpage; service block or determine repairs.

47. *Technician A* says a warped cylinder head mounting surface on an engine block may cause valve seat distortion. *Technician B* says a warped cylinder head mounting surface on an engine block may cause coolant and combustion leaks. Who is right?
A. A only
B. B only
C. Both A and B
D. Neither A nor B

48. The LEAST likely cause of a cracked block is:
A. metal fatigue.
B. impact damage from a vehicle collision.
C. engine overheating from low coolant.
D. not enough antifreeze in the coolant in cold climates.

Hint

Cracks in the cylinder block are usually found during a visual inspection. When cracks are found, you should attempt to determine the cause. The following usually causes cracks:

- *Engine overheating from low coolant*
- *Insufficient antifreeze in the coolant in cold climates*
- *Impact damage*
- *Detonation*
- *Fatigue*

Inspect all the oil passages with a shop light. The oil passages should be free from any gasket material, metal shavings, and other foreign objects. An engine brush kit can be used to clean all the oil passages.

Visually inspect the cylinder block deck for scoring, corrosion, cracks, and nicks. If a scratch in the deck is deep enough to catch on a fingernail, the deck needs to be resurfaced. Measure deck warpage with a precision straightedge and feeler gauge. To obtain proper results, the deck must be perfectly clean. Check for warpage across the four edges of the deck. The thickest feeler gauge that will fit between the straightedge and the deck determines the amount of warpage. If a deck is warped, the amount of material to be removed must be determined. On V-type engine blocks, when one deck is warped and needs to be machined, both sides have to be machined at the same time.

Task 3

Inspect and measure cylinder walls for wear and damage; determine needed service.

49. While discussing cylinder taper, *Technician A* says taper in the bore causes the ring end gaps to change while the piston moves in the bore. *Technician B* says a cylinder bore gauge can be used to determine taper. Who is right?
A. A only
B. B only
C. Both A and B
D. Neither A nor B

50. If new rings are installed without removing the piston ring ridge, which of the following may result?
A. The piston skirt may be damaged.
B. The piston pin may be broken.
C. The connecting rod bearings may be damaged.
D. The piston ring lands may be broken.

Hint

After visually inspecting the cylinder bores, use a dial bore gauge, an inside micrometer, or a telescoping gauge to measure the bore diameter. Piston movement in the cylinder bores produces uneven wear throughout the cylinder. The cylinder wears the most—90 degrees to the piston pin and in the area of the upper ring contact at TDC. This is because the cylinder receives less lubrication while being subjected to the greatest amount of pressure and heat at the top of the cylinder.

Taper in the cylinder bore causes the piston ring gaps to change as the piston travels in the bore. To measure taper and out-of-round using a dial bore gauge, simply rotate and move the gauge up and down in the bore. If you use a telescoping gauge or inside micrometer, measure the top of the bore just below the deck and at the bottom of the ring travel. As a general rule of thumb, the maximum taper allowed is 0.005 in. (0.125 mm) and 0.001 in. for out-of-round. If any cylinders are out of specification, all the cylinders should be bored to a standard oversize.

Task 4

Remove cylinder wall ridges; hone and clean cylinder walls.

51. While removing the cylinder wall ring ridge, which of the following should be performed?
    A. Place an oiled rag in the cylinder to catch any metal shavings.
    B. After removing the ridge, smooth the surface with 80 grit sandpaper.
    C. If the ridge has a step less than 0.050 in. (1.2 mm), there is no need for it to be removed.
    D. Wash the cylinder walls with soap and water solution before removing the ridge.

52. The tool shown in Figure 1–13 is used to:
    A. hone the cylinder wall.
    B. deglaze the cylinder wall.
    C. remove the ring ridge.
    D. measure the cylinder out-of-round.

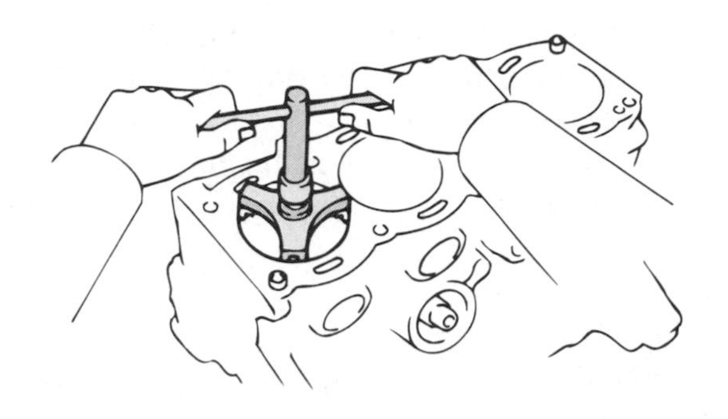

Figure 1–13 Cylinder service tool.

Hint

The ridge at the top of the cylinder walls should be removed with a ridge reamer before attempting to remove the pistons. When using a ridge reamer, do not remove any more material than necessary. Begin the procedure by turning the crankshaft to lower the piston in the cylinder. Place an oiled rag on top of the piston to catch any metal shavings that might fall into the cylinder. Always follow the tool manufacturer's instructions for the ridge reamer you are using. Rotate the tool in a clockwise rotation until the ridge is removed. Remove the rag from the cylinder and clean out any remaining metal shavings.

If the cylinder needs to be bored, the cylinder oversize must be determined. Use the cylinder that has the worst wear as a reference for overbore size. It is a good practice to match the pistons to the bore. The pistons usually have some differences in size, due to manufacturing tolerances. Measure the exact size of the replacement pistons. Then determine the desired finished size of the cylinder and how much it will be bored. Leave 0.003 to 0.005 in. (0.075 to 0.125 mm) for finish honing.

Task 5

Inspect camshaft and bearings for wear and damage; determine bearing clearance and replace as necessary.

53. The tool shown in Figure 1–14 is used to:
    A. remove camshaft bearings.
    B. install camshaft bearings.
    C. remove and install camshaft bearings.
    D. measure camshaft bearing alignment.

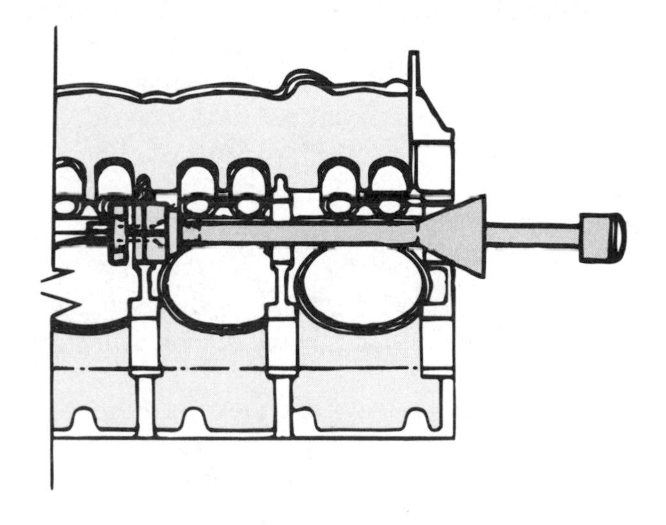

Figure 1–14 Block service tool. *(Courtesy of General Motors Corporation, Service Technology Group)*

54. One camshaft bearing is burned and scored in an engine with the camshaft in the block. *Technician A* says the oil hole in the bearing may not have been properly aligned with the oil hole in the block. *Technician B* says the oil pump components should be inspected and measured for wear. Who is right?
    A. A only
    B. B only
    C. Both A and B
    D. Neither A nor B

Hint

Begin the camshaft inspection with a visual inspection of the lobes and journals. Both of these must be free of scoring and galling. Normal lobe wear pattern is slightly off center with a wider wear pattern at the nose than at the heel. The off center wear is a result of the slight taper of the lobe used in conjunction with the convex shape of the lifter to rotate the lifter. If the wear pattern extends to the edges of the lobe, the lifter will not rotate. If this condition exists, the camshaft must be replaced.

To measure camshaft bearing oil clearance, use an expandable bore gauge and a micrometer to measure the inner diameter of the bearing. Then measure the diameter of the camshaft journal. Subtract the two measurements to obtain the bearing oil clearance. Refer to the manufacturer's specifications for the proper oil clearance.

Task 6

Inspect crankshaft for surface cracks and journal damage and wear; check oil passage condition; determine needed repair.

55. *Technician A* says metal burrs on the crankshaft flange may cause excessive wear on the flexplate gear teeth. *Technician B* says metal burrs may cause improper torque converter to transmission alignment. Who is right?
    A. A only
    B. B only
    C. Both A and B
    D. Neither A nor B

56. In Figure 1–15, what is being performed?
    A. Removing a scratch in the bearing with crocus cloth
    B. Using a special tool to remove the bearing insert
    C. Measuring the thickness of the crushed plastigage
    D. Determining if the crankshaft journal has been machined

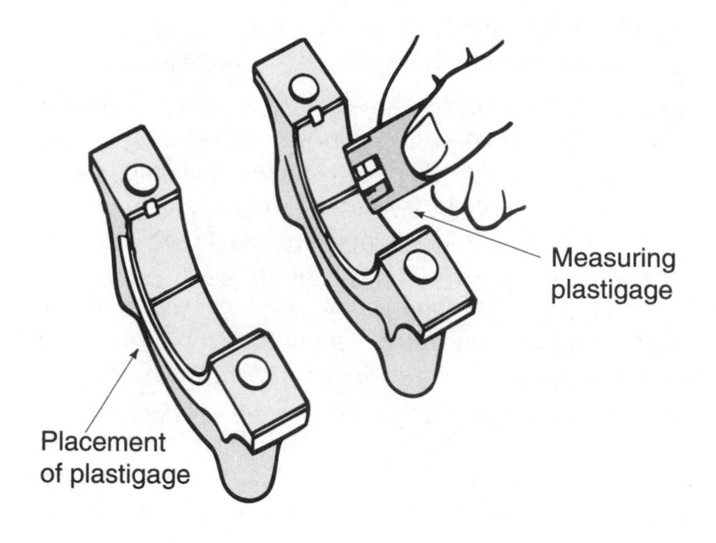

Figure 1–15 Crankshaft and bearing service. *(Courtesy of Ford Motor Company)*

Hint

The crankshaft transmits the torque from the connecting rods to the drivetrain. These pressures and rotational forces eventually cause wear and stress on the crankshaft. Before reusing the crankshaft, a thorough visual inspection of the subject areas is required. These areas include the main bearing journals, connecting rod journals, fillets, thrust surface, oil passages, counterweights, seal surfaces, and vibration damper journal. Next, inspect for warpage and measure the journals for excessive wear. An additional check is to inspect for stress cracks using Magnetic Particle Inspection (MPI).

First, visually inspect the crankshaft for obvious wear and damage. This includes inspecting the threads at the front of the crank, the keyways, and the pilot bushing bore. When inspecting the journals, run your fingernail across the surface to feel for nicks and scratches. If a journal is scored, it must be polished before an accurate measurement can be obtained. Remember to inspect the area around the fillet very closely. Stress cracks can develop in this area.

Clean the oil passages by running a length of wire or a small bore brush through them, followed by a spray cleaner. Inspect the passage openings.

Task 7

Inspect and replace main and rod bearings; check assembled clearances.

57. When inspecting the old crankshaft bearings, all of the following can be determined EXCEPT:
    A. mileage.
    B. crankshaft misalignment.

C. lack of lubrication.
D. metal-to-metal contact in the engine.

58. *Technician A* says the curvature of connecting rod bearings is slightly larger than the curvature of the bearing bores, and this feature is called bearing spread. *Technician B* says when a connecting rod bearing half is installed, its edges should be slightly below the surface of the bearing bore. Who is right?
A. A only
B. B only
C. Both A and B
D. Neither A nor B

59. When measuring the crankshaft journal in Figure 1–16, the difference between measurements:
A. A and B indicates horizontal taper.
B. C and D indicates vertical taper.
C. A and C indicates out-of-round.
D. A and D indicates vertical taper.

Hint

Inspect the main and rod bearings for wear patterns and record your determination in your notebook or on the work order. Note any unusual wear patterns indicating crankshaft or crankcase misalignment, lack of oil, and so forth. Inspect the backside of the bearings to determine if the crankshaft has been machined. Most oversized bearings are stamped to indicate the over- or under-size.

When an engine is reconditioned, the main and rod bearings are replaced. However, inspection of the old bearings provides clues as to the cause of an engine failure. For example, the soft material used to construct the bearings allows impurities to embed into it. Excessive metal flakes may alert the technician that there is metal-to-metal contact between moving parts in the engine.

To determine oil clearance with rod or main bearings, use plastigage. Place the bearing inserts into the cap and in the saddle. Next, install the crankshaft into the cylinder block. The crankshaft must be free from oil so that the plastigage will stay in position while the cap is torqued. Use a plastigage strip long enough to fit across the journal. Now, coat the bearing insert with oil to ensure that the plastigage will stick to the journal instead of the bearing. Install and torque the cap. Remove the cap and measure the width of the plastigage with the scale on the package. Use an oiled rag to remove all traces of the plastigage from the journal. After all clearances have been checked, remove the crankshaft and lubricate both the bearings and the crankshaft journals. If all clearances are within specifications, reinstall the crankshaft, caps, and the bolts.

Task 8

Recognize piston and bearing wear patterns that indicate connecting rod alignment and bearing bore problems; inspect rod alignment and bore condition.

60. The tool in Figure 1–17 is being used for what purpose?
A. To widen the piston ring grooves
B. To deepen the piston ring grooves
C. To remove and replace the piston rings
D. To remove carbon from the piston ring grooves

61. The main reason for seized or scored pistons is MOST likely:
A. a lack of lubrication.
B. overheating of the piston.
C. preignition.
D. excessive engine rpm.

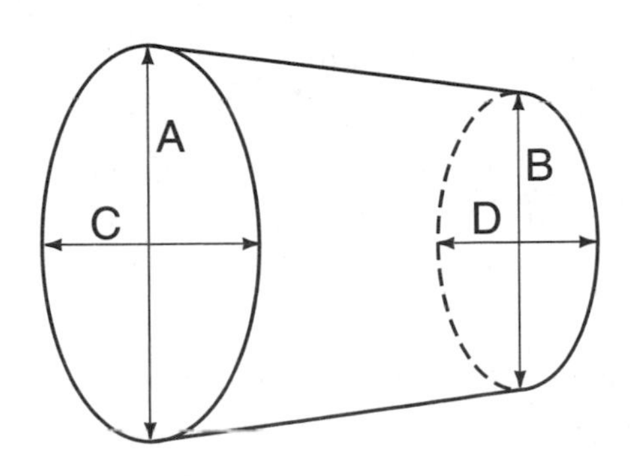

Figure 1–16 Crankshaft journal measurement. *(Courtesy of Ford Motor Company)*

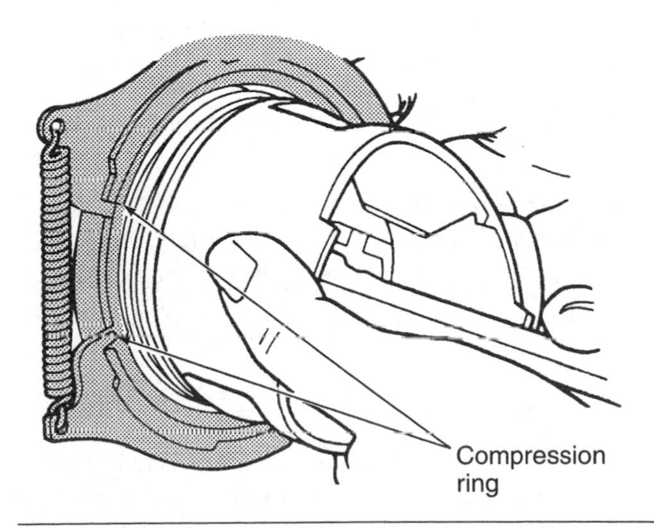

Figure 1–17 Piston service tool. *(Courtesy of Chrysler Corporation)*

62. Removing press-fit piston pins is being discussed. *Technician A* says the pin can be driven out with a punch and a hammer while securing the piston in a vise. *Technician B* says removing the pin requires a press and special adapters. Who is right?
    A. A only
    B. B only
    C. Both A and B
    D. Neither A nor B

Hint

Before a connecting rod is accepted for reuse, it must be carefully inspected and measured. Both the big end and the small end bores should be inspected for clearance, out-of-round, and taper. In addition, the rod should be checked for length.

When measuring the inside diameter of the big end, assemble the cap to the rod, leaving the nuts loose. Place the rod in a soft jaw vise, with the jaws covering the parting line, and torque the cap nuts to specifications. This procedure ensures that the cap and rod are properly aligned with each other. The easiest way to measure the bore is to use a dial bore gauge. If an inside micrometer is used, measure the diameter in two or three directions near each end of the bore to obtain out-of-round and taper measurements. The greatest amount of out-of-round will occur in the cap and in a vertical direction.

Inspect the small end bore in the same manner as the big end bore. Use a dial bore gauge to measure clearance, out-of-round, and taper. If the piston assembly is designed with a press-fit pin in the connecting rod, the bore must be the correct size to provide interference fit. In addition, any scuffing or nicks in the bore may inhibit the piston from rocking properly. Sometimes the bore can be honed to a larger size to accept an oversized piston pin.

Task 9

Inspect, measure, service, or replace pistons, pins, and bushings.

63. *Technician A* says pistons are cam ground. *Technician B* says that pistons should be measured across the center of the skirt thrust surface. Who is right?
    A. A only
    B. B only
    C. Both A and B
    D. Neither A nor B

64. When assembling a connecting rod, piston, and piston pin, *Technician A* says there is no need to center the connecting rod and piston pin in the piston bore. *Technician B* says the piston may be assembled on the connecting rod in either of two positions. Who is right?
    A. A only
    B. B only
    C. Both A and B
    D. Neither A nor B

Hint

The piston pin and the boss are subjected to severe operating conditions. Compounding this is the difference in expansion rates between the steel piston pin and the aluminum piston. A steel pin expands about 0.0003 in. (0.008 mm) for every 50°F increase in temperature. Consequently, proper clearance is critical.

Manufacturers vary concerning recommended procedures for checking pin clearance and how much clearance is allowed. For example, Chevrolet suggests measuring the pin diameter and the bore diameter; any clearance over 0.001 in. (0.025 mm) requires replacement of the pin and piston. Pontiac says the pin will fall through the bore with 0.0005 in. (0.01 mm) clearance, but not with 0.0003 in. (0.008 mm). Some import manufacturers check pin and bore wear by holding the piston and attempting to move the connecting rod up and down. Any movement felt indicates wear. Feel is not always the most reliable method. If you have any doubts, measure the clearance.

Measure the pin using the recommended procedure for the engine being serviced. With the pin removed, visually inspect it for wear. Use your hands to feel for scuffing or scoring. If there is any wear, the pin must be replaced.

Task 10

Measure piston-to-cylinder bore clearance; determine needed service.

65. An acceptable piston-to-cylinder bore clearance is:
    A. 0.0015 in. (0.038 mm).
    B. 0.004 in. (0.101 mm).
    C. 0.006 in. (0.152 mm).
    D. 0.008 in. (0.203 mm).

66. When performing the service operation in Figure 1–18: *Technician A* says the tool must be held squarely against the top of the block. *Technician B* says the nuts should be installed on the connecting rod bolts. Who is right?
    A. A only
    B. B only
    C. Both A and B
    D. Neither A nor B

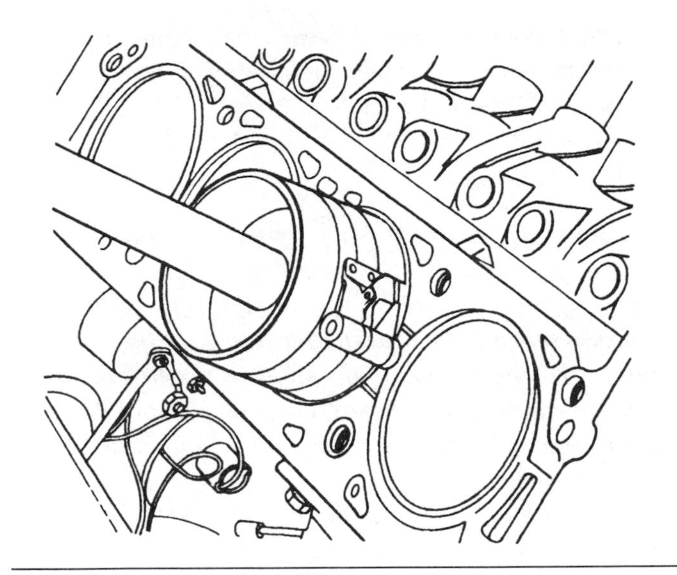

Figure 1–18 Engine service tool. *(Courtesy of Ford Motor Company)*

Hint

Overheating of the piston generally causes seized or scored pistons. Poor cooling system operation and improper combustion due to preignition or detonation can cause this. These overheating conditions cause lubrication failures of the pin.

Piston clearance is determined by measuring the size of the piston skirt at the manufacturer's sizing point. This measurement is subtracted from the size of the cylinder bore. If the piston clearance is not within specifications, it may be necessary to bore the cylinder to accept an oversized piston.

Since most pistons are cam ground, it is important to measure the piston diameter at the specified location. Some manufacturers require measurements across the thrust surface of the skirt centerline. Others require measuring a specified distance from the bottom of the oil ring groove. Always refer to the appropriate service manual for the engine you are servicing.

Task 11

Replace piston rings; check ring groove clearance and end gap.

67. *Technician A* says that if the piston ring groove clearance has exceeded specifications, a larger piston ring must be used. *Technician B* says excessive piston ring side clearance can cause a piston ring to break. Who is right?
    A. A only
    B. B only
    C. Both A and B
    D. Neither A nor B

68. When measuring the piston ring gap as shown in Figure 1–19, *Technician A* says that the ring gap should be measured with the ring positioned at the top of the ring travel in the cylinder. *Technician B* says the two compression rings are interchangeable on some pistons. Who is right?
    A. A only
    B. B only
    C. Both A and B
    D. Neither A nor B

Hint

Measure the ring groove for wear. Install a new ring backward in the groove and use a feeler gauge to measure the clearance. Check the groove at several locations around the piston. If the side clearance is excessive, replace the piston. Excessive side clearance can result in ring breakage.

The topmost ring groove wears the most. Normal ring-to-groove side clearance is between 0.002 and 0.004 in. (0.05 to 0.10 mm). Roll the ring around the entire groove while observing for binding. The ring depth should remain consistent. To check ring gap, place the ring into the appropriate cylinder bore. Use the piston to slide the ring to the specified depth, usually the bottom of ring travel in the stroke. The piston head will keep the ring square in the bore. Measure the gap at the ring ends. The value will be in the range of 0.004 in. (0.101 mm) per inch of diameter.

Task 12

Inspect, repair, or replace crankshaft vibration dampener (harmonic balancer) and flywheel.

69. If a harmonic balancer outer ring has slipped on the rubber mounting, what must be done to correct it?
    A. Twist the ring back to its original position.
    B. Secure the outer ring by welding it in place.

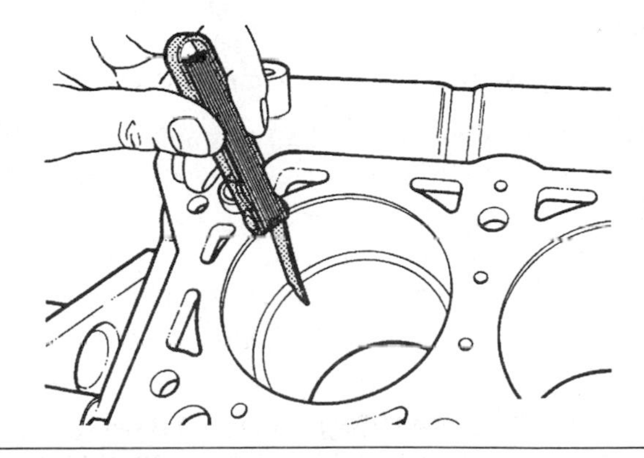

Figure 1–19 Piston ring measurement. *(Courtesy of Chrysler Corporation)*

C. Replace the harmonic balancer.
D. Press on a new outer ring.

70. When discussing flywheel service, *Technician A* says excessive flywheel runout may cause a grabbing clutch. *Technician B* says when reusing a pressure plate, it may be installed in any position on the flywheel. Who is right?
    A. A only
    B. B only
    C. Both A and B
    D. Neither A nor B

Hint

Inspect the harmonic balancer for signs of wear in its center bore. Also, inspect the rubber mounting for indications of twisting and deterioration. If a balancer has slipped or rotated, it must be replaced. If wear to the center bore is present in the form of a groove worn in from the front engine seal, a repair sleeve kit is available.

Inspect the flywheel for signs of stress or heat cracks. Flexplates usually crack around the mounting area. Overheating because of a slipping clutch can cause these cracks. If the cracks are deep or the flywheel is blue in color, replace the flywheel.

Flywheels with light scoring and small cracks can be resurfaced. During resurfacing, flywheel warpage is also eliminated. Flywheels and flexplates also contain the ring gear that the starter drive engages. Inspect the ring gear for missing or cracked teeth. If a flexplate ring gear is damaged, replace the flexplate. Most flywheels have replaceable ring gears.

Lubrication and Cooling Systems Diagnosis and Repair

Task 1

Perform oil pressure tests; determine needed repairs.

71. All of the following are causes of low engine oil pressure EXCEPT:
    A. worn camshaft bearings.
    B. worn crankshaft bearings.
    C. weak oil pressure regulator spring tension.
    D. restricted pushrod oil passages.

72. An engine has zero oil pressure. *Technician A* says the pin may be sheared off in the oil pump drive gear. *Technician B* says the oil pump inlet screen may be partially plugged. Who is right?
    A. A only
    B. B only
    C. Both A and B
    D. Neither A nor B

Hint

When low oil pressure is evident, first check the oil level. A low oil level causes the oil pump to aerate and lose volume. If the oil level is too high, it may be caused by gasoline entering the crankcase because of a faulty fuel pump, improperly adjusted carburetor, sticking choke plate, or leaking fuel injector. If the level and the condition are not in question, use a mechanical gauge to check oil pressure.

When performing oil pressure tests, remove the oil gauge sending unit from the engine. Use the appropriate adapters and connect the gauge to the engine. With the engine at normal operating temperature observe the pressure at idle. Increase the engine speed to 2,000 rpm while observing the gauge. Compare the test results with the manufacturer's specifications.

No oil pressure at all indicates that there is a problem with the oil pump drive mechanism or the oil pump itself. Other possible causes include:

- *Oil pump pickup is plugged.*
- *Gallery plugs are leaking.*
- *A hole is present in the pickup tube.*
- *Lower than specified oil level.*
- *Improper oil viscosity.*
- *Sticking or weak oil pressure relief valve.*
- *Worn crankshaft, connecting rod, or camshaft bearings.*

Task 2

Inspect, repair, or replace oil pump and drives.

73. *Technician A* says that it is more cost effective to rebuild an oil pump than to buy a new one. *Technician B* says that it is a good practice to replace an oil pump when rebuilding a engine. Who is right?
    A. A only
    B. B only
    C. Both A and B
    D. Neither A nor B

74. The LEAST likely cause of oil pump drive failure is:
    A. oil pump gear lockup.
    B. oil pump drive rod failure.
    C. excessive engine speed.
    D. a defective oil pressure relief valve.

Hint

The oil pump is usually replaced whenever the engine is rebuilt. If they are reused, they must pass a visual inspection. If it fails an inspection, some manufacturers provide a rebuild kit. Proper inspection requires the oil pump to be cleaned and disassembled. Most often the pump is replaced rather than rebuilt because of the low cost of a new pump.

Oil pump drives can often be the cause of no oil pressure. The oil pump may ingest foreign particles, which are too large to pass through the gear assembly. This condition causes the pump to lock up, which causes the drive shaft to break.

Task 3

Inspect, adjust, and replace belts and pulleys.

75. All of these are methods of measuring engine drive V-belt tension EXCEPT:
    A. use a belt tension gauge.
    B. measure the amount of belt deflection.
    C. visually see if the belt is contacting the bottom of the pulley.
    D. measure the length of the belt compared to a new one.

76. A loose alternator belt may cause:
    A. a discharged battery.
    B. a squealing noise while decelerating.
    C. a damaged alternator bearing.
    D. engine overheating.

Hint

Since the friction surfaces are the sides of a V-belt, the belt must be replaced if the sides are worn and the belt is contacting the bottom of the pulley. The belt tension may be checked with the engine shut off, and a belt tension gauge placed over the belt at the center of the belt span. A loose or worn belt may cause a squealing noise when the engine is accelerated. Measuring the amount of belt deflection with the engine shut off also may check the belt tension. Use your thumb to depress the belt at the center of the belt span. If the belt tension is correct, the belt should have 1/2-in. deflection per foot of belt span.

Ribbed V-belts usually have a spring-loaded belt tensioner, with a belt wear indicator scale on the tensioner housing. If a power steering pump belt requires tightening, always pry on the pump ear, not on the housing.

Task 4

Perform cooling system pressure tests; determine needed repairs.

77. While discussing cooling systems, *Technician A* says that radiator pressure caps can be tested with a special adapter. *Technician B* says that pressure testing the cooling system can determine the condition of the thermostat. Who is right?
    A. A only
    B. B only
    C. Both A and B
    D. Neither A nor B

78. When filling a cooling system, *Technician A* says to fill half the system capacity with pure antifreeze, and the remaining amount with water. *Technician B* says to use a 50/50 mixture of water-to-coolant mixture to fill the cooling system. Who is right?
    A. A only
    B. B only
    C. Both A and B
    D. Neither A nor B

79. The tester in the Figure 1–20 may be used to test the following items EXCEPT:
    A. cooling system leaks.
    B. radiator cap pressure relief valve.
    C. coolant specific gravity.
    D. heater core leaks.

Hint

A pressure tester may be connected to the radiator filler neck to check for cooling system leaks. Operate the tester pump and apply 15 psi to the cooling system. Inspect the cooling system for external leaks with the system pres-

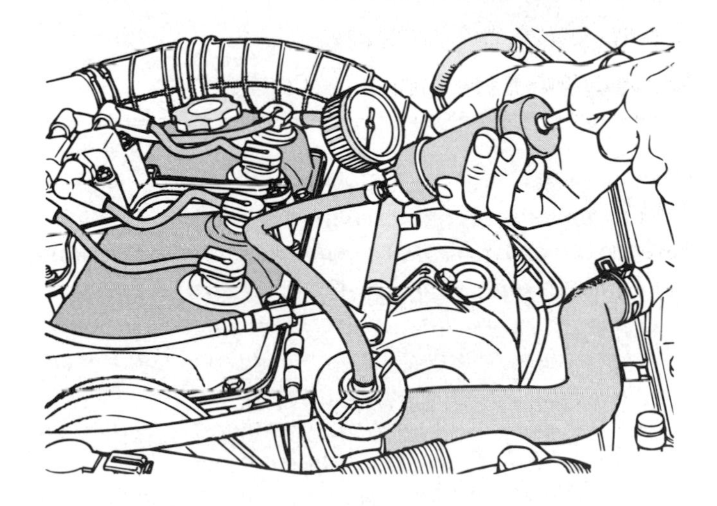

Figure 1–20 Cooling system service tool. *(Courtesy of Chrysler Corporation)*

surized. If the gauge pressure drops more than specified by the vehicle manufacturer, the cooling system has a leak. If there are no visible external leaks, check the front floor mat for coolant dripping out of the heater core. When there are no external leaks, check the engine for combustion chamber leaks.

The radiator pressure cap may be tested with the pressure tester. When the tester pump is operated, the cap should hold the rated pressure. Always relieve the pressure before removing the tester.

Task 5

Inspect and replace thermostat, bypass, and housing.

80. A vehicle has an overheating problem. The vehicle sometimes overheats within the first five minutes after the engine is started. Other times, the cooling system operates normally. *Technician A* says the water pump can be working intermittently. *Technician B* says the thermostat is sometimes sticking and needs replacement. Who is right?
    A. A only
    B. B only
    C. Both A and B
    D. Neither A nor B

81. While discussing thermostat service, *Technician A* says a thermostat with a lower temperature rating may be installed in a fuel-injected engine without adversely affecting fuel system operation. *Technician B* says a thermostat that sticks open may cause the powertrain control module (PCM) to enter closed loop too quickly as the engine warms up. Who is right?
    A. A only
    B. B only
    C. Both A and B
    D. Neither A nor B

Hint

If a customer brings in a vehicle with an overheating problem, it is possible that the thermostat is not opening. A thermostat that is stuck in the open position prevents an engine from reaching operating temperature. Check the temperature rating of the thermostat and confirm that it is the proper one for the engine application. Visually inspect the thermostat for rust and other contamination. Make sure the thermostat is installed properly.

To test the thermostat, submerge it into a container of water. Use a thermometer so the temperature when the thermostat opens can be determined. Heat the water while observing the thermostat. At the rated temperature of the thermostat, it should begin to open.

Task 6

Drain, flush, refill, and bleed cooling system in accordance with accepted procedures.

82. While discussing cooling system flushing, *Technician A* says that to properly flush the radiator, reverse flushing should be performed to dislodge deposits. *Technician B* says that a flush involves draining the coolant and refilling the system with new antifreeze. Who is right?
    A. A only
    B. B only
    C. Both A and B
    D. Neither A nor B

83. While discussing cooling system service, *Technician A* says if the cooling system pressure is reduced the coolant boiling point is increased. *Technician B* says when more antifreeze is added to the coolant the coolant boiling point is increased. Who is right?
    A. A only
    B. B only
    C. Both A and B
    D. Neither A nor B

Hint

Flushing of the cooling system is accomplished by using pressurized water through the cooling system in a reverse direction to normal coolant flow. A special flushing gun mixes low-pressure air with tap water. Reverse flushing causes the deposits to dislodge from the various components. They can then be removed from the system. The engine block and radiator should be flushed separately.

To flush the radiator, drain the system and disconnect the upper and lower hoses. Attach a long hose to the upper hose outlet to deflect the water. Disconnect and plug the heater hoses that are attached to the radiator. Fit the flush gun to the lower hose opening. This causes the radiator to be flushed in the reverse direction. Fill the radiator with water and turn on the gun in short bursts. Continue flushing until the water exiting the radiator is clean.

The cooling system is filled with a 50/50 mix of antifreeze and water. Before filling the cooling system, make sure that all hose clamps are tight and the drain plug is tight. Fill half of the system capacity with 100 percent antifreeze, the rest of the capacity with pure water. Continue to fill the cooling system as needed as the engine warms up.

Task 7

Inspect and replace water pump and hoses.

84. *Technician A* says a defective water pump bearing may cause a growling noise when the engine is idling. *Technician B* says the water pump bearing may be ruined by coolant leaking past the pump seal. Who is right?
    A. A only
    B. B only
    C. Both A and B
    D. Neither A nor B

85. Referring to Figure 1–21, what is the most common reason for replacing the water pump?
    A. The pump impeller becomes deteriorated from corrosion.
    B. The water pump-to-engine block gasket fails.
    C. The hub separates from the water pump shaft.
    D. Failure of the lip seal.

Hint

The cooling system water pump must be replaced when the bearing has failed, or more commonly when the front lip seal has failed and has caused a coolant leak. Most water pumps can be removed after disconnecting the lower or upper radiator hoses at the pump. Remove any by-pass or heater hoses that are attached to the pump. Remove the bolts attaching the pump to the block. When removing the bolts, take time to keep them in order because they are usually different lengths. The water pump might need a tap with a hammer to separate it from the block. Remove all traces of gasket material from the block and be sure to use approved sealer with the new gasket.

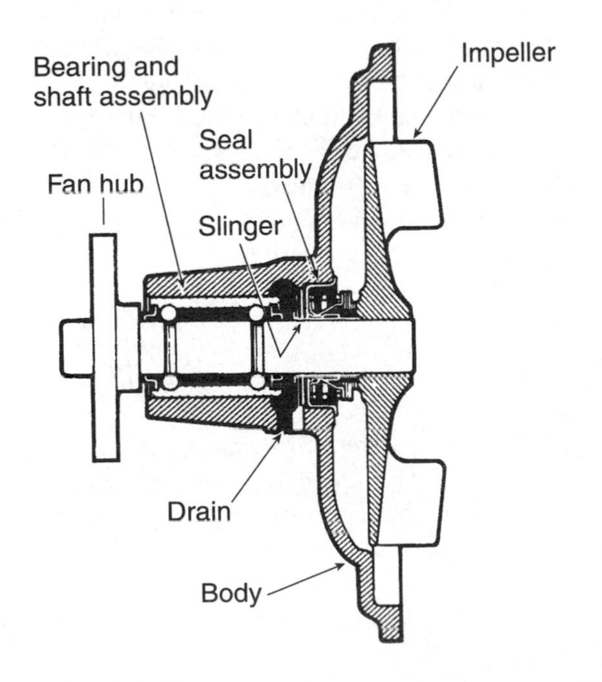

Figure 1–21 Water pump components. *(Courtesy of General Motors Corporation, Service Technology Group)*

Task 8

Inspect and replace radiator, pressure cap, expansion tank, and coolant recovery system.

86. In Figure 1–22, what type of radiator is shown?
    A. Downdraft
    B. Updraft
    C. Crossflow
    D. Downflow

87. The vacuum valve in the radiator cap is stuck closed. The result could be:
    A. collapsed upper radiator hose after the engine is shut off.
    B. excessive cooling system pressure at normal engine temperature.
    C. engine overheating when operating under heavy load.
    D. engine overheating during extended idle periods.

Hint

The radiator should be inspected for coolant leaks and debris in the air passages through the core. With the radiator cap removed some of the coolant passages can be inspected for restriction. The radiator and the cap may be tested for leaks with a pressure tester. If the vacuum valve in the radiator cap is sticking, the top radiator hose usually collapses as the engine cools down. When the pressure valve in the cap is leaking, there is excessive coolant in the coolant recovery container.

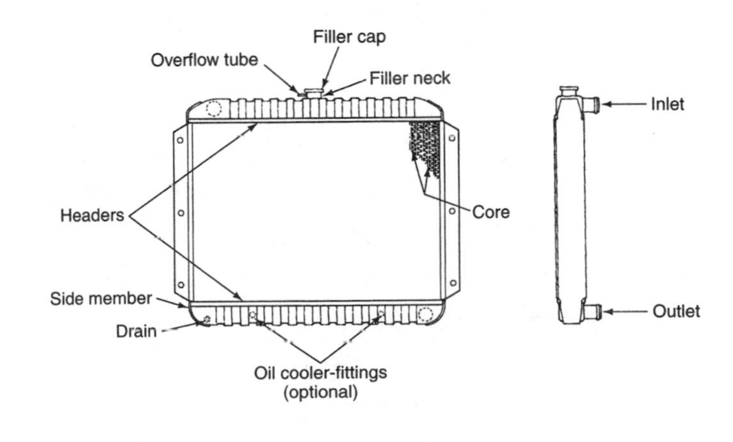

Figure 1–22 Radiator. *(Reprinted with permission from SAE Handbook 1980, Society of Automotive Engineers, Inc.)*

Task 9

Clean, inspect, and replace fan, fan clutch, and fan shroud.

88. *Technician A* says that when a viscous fan clutch has oily streaks radiating out from the hub shaft, the clutch should be replaced. *Technician B* says that all types of fan blades should be inspected for stress fractures. Who is right?
    A. A only
    B. B only
    C. Both A and B
    D. Neither A nor B

89. An engine with a viscous cooling fan clutch is overheating. All of these defects may be the cause of the problem EXCEPT:
    A. slipping viscous clutch.
    B. a seized viscous clutch.
    C. missing fan shroud.
    D. a slipping water pump and cooling fan drive belt.

Hint

The cooling fan can be driven from an accessory belt or be electrically operated. Regardless of how it is powered, the fan blades must be inspected for stress cracks. Because the fan blades are balanced to prevent vibration to the water pump, if any of these blades are damaged, the fan must be replaced. Belt-driven fans use either flex fans or a viscous fan clutch. Flex fans should be inspected for stress cracks, while viscous fan clutches should be inspected for indications of leakage. Also, inspect the thermostatic spring for free movement and accumulations of dirt and debris. If any free movement exists, replace the clutch.

Electric fans are also inspected for damage and looseness. If the fan fails to turn on at the specific temperature, the temperature-sending unit is most likely the cause. If the fan motor is in question, the easiest method to check the operation is to apply direct battery voltage to the fan motor. If the fan operates, it is functioning properly. If the fan does not operate, check the motor ground before faulting the motor.

Ignition System Diagnosis and Repair

Task 1

Diagnose no-starting, hard starting, engine missing, power loss, and/or poor mileage problems on trucks with point-type and electronic ignition systems; determine needed repairs.

90. A vehicle that is experiencing poor fuel economy has entered the shop. *Technician A* says that if there is no fuel leaking from the fuel pump, then there is no problem with the pump. *Technician B* says that the fuel pump could be leaking internally and the engine oil level should be checked to help indicate a problem. Who is right?
    A. A only
    B. B only
    C. Both A and B
    D. Neither A nor B

91. While diagnosing a fuel-related no-start condition on a port fuel-injected engine, *Technician A* says that one terminal on each fuel injector is supplied with 12 volts when the ignition switch is turned on. *Technician B* says the other terminal on each injector is connected directly to ground. Who is right?
    A. A only
    B. B only
    C. Both A and B
    D. Neither A nor B

92. When diagnosing a no-start condition on a throttle body injected engine with a distributor ignition (DI), a 12V test light is connected from the coil tach terminal to ground and a test spark plug is connected from one of the spark plug wires to ground. When the engine is cranked, the test light flutters on and off, but the spark plug does not fire. The cause of this problem could be:
    A. a defective distributor pickup coil.
    B. a defective ignition module.
    C. a defective ignition coil.
    D. an open circuit in one of the pickup coil leads.

Hint

When diagnosing a no-start condition, many components could be the cause. For this instance, we will assume that the problem is with the ignition system. The ignition defects that may cause a no-start condition could be some of the following: defective coil, defective cap and rotor, defective pickup coil, open secondary wire, no primary voltage at the coil, fouled spark plugs, or defective ignition module.

If an engine has misfiring problems, the following components could be the cause: engine compression, intake manifold leaks, high resistance in spark plug wires or secondary coil wire, electrical leakage in the distributor cap, defective coil, defective spark plugs, low primary voltage and current, improperly routed spark plug wires, insufficient dwell, or worn distributor bushings.

If a power loss condition exists, check the following: engine compression, restricted exhaust or air intake, insufficient timing advance, late ignition timing, or cylinder misfire.

The following can cause engine detonation: higher than specified compression, ignition timing too far advanced, excessive timing advance, spark plug heat range too hot, or improperly routed spark plug wires.

Task 2

Inspect, test, repair, or replace ignition primary circuit wiring and components.

93. *Technician A* says that with a 12V test lamp connected from the tach terminal to ground, if the light flutters while the engine is cranking, the primary ignition system is functioning properly. *Technician B* says that a magnetic-type pickup coil cannot be tested with an ohmmeter. Who is right?
    A. A only
    B. B only
    C. Both A and B
    D. Neither A nor B

94. *Technician A* says the secondary ignition circuit starts at the ignition switch and follows through to the spark plug. *Technician B* says the primary circuit is from the coil terminal through the distributor cap to the spark plugs. Who is right?
    A. A only
    B. B only
    C. Both A and B
    D. Neither A nor B

Hint

When diagnosing a no-start condition in the primary ignition circuit, connect a 12-volt test lamp from the coil tachometer terminal to ground, and turn on the ignition switch. Crank the engine and observe the test light. If the light flutters while the engine is cranked, the pickup coil signal and the module are satisfactory. When the test lamp does not flutter, one of these components is defective. The pickup coil can be tested with an ohmmeter. If the pickup is satisfactory, the module is defective.

Task 3

Inspect, test, and service distributor.

95. While discussing ignition point adjustment, *Technician A* says an increase in point gap increases the cam dwell reading. *Technician B* says point dwell must be long enough to allow the magnetic field to build up in the coil. Who is right?
    A. A only
    B. B only
    C. Both A and B
    D. Neither A nor B

96. The MOST likely result of a cracked distributor cap is:
    A. cylinder misfiring during engine acceleration.
    B. stalling during deceleration.
    C. detonation at a constant engine speed.
    D. surging at idle.

Hint

When inspecting a distributor, start with a visual inspection. Check for signs of cracks on the cap, damaged wires, or terminals. Remove the cap and inspect the rotors for signs of arcing. Inspect all lead wires for worn insulation and loose terminals. Check the centrifugal advance mechanism for wear, particularly the weights for wear on the pivot holes. Also, inspect the pickup plate for wear and make sure that it moves freely. Test the vacuum advance unit using a hand-held vacuum pump. Apply 20 inches of vacuum to the advance unit and make sure that it moves the advance plate and holds vacuum. Inspect the drive gear for worn or chipped teeth. On point-type ignitions, check the points for burning or pitting. Replace the points if either one of these signs are visible.

Task 4

Inspect, test, service, repair, or replace ignition system secondary circuit wiring and components.

97. If certain spark plugs fire and others do not, and a secondary circuit problem is suspected, which of the following could be the cause?
    A. Ignition switch
    B. Hall effect switch
    C. Ignition coil
    D. Spark plug wires

98. What type of tool is most commonly used when testing ignition coil windings?
    A. OEM specific scan tool
    B. Digital multimeter (DMM)
    C. Oscilloscope
    D. Test light

Hint

If the primary ignition circuit is ruled out as the cause of a no-start condition, the secondary circuit must be inspected. Connect a test spark plug to a secondary ignition wire and ground the case. Be sure to use the proper spark plug for the vehicle application. Crank the engine and observe the plug. If the spark plug fires, it will indicate the coil is not the problem. Connect the spark plug to several different spark plug wires and crank the engine while observing the plug for each individual cylinder. If one or more plugs fire while others do not, this indicates that current is leaking through a defective distributor cap, rotor, or spark plug wires. If the test spark plug fires at all the individual wires, the system is functioning properly.

Task 5

Inspect, test, and replace ignition coils.

99. The primary circuit current flow is 7 amperes, and the specified current flow is 4 amperes. *Technician A* says this high current flow may be caused by a shorted primary coil winding. *Technician B* says this high current flow may damage the pickup coil. Who is right?
    A. A only
    B. B only
    C. Both A and B
    D. Neither A nor B

100. An ignition system that is sensitive to moisture could be caused by all of the following EXCEPT:
    A. a cracked distributor cap.
    B. deteriorated ignition wires.
    C. a cracked ignition coil tower.
    D. loose distributor mounting.

Hint

The ignition coil should be inspected for cracks and signs of electrical leakage. The cylinder-type coil should be inspected for oil leaks. If oil is leaking from the coil, air space is present in the coil, which allows condensation to collect inside the coil. Condensation in an ignition coil causes high voltage leaks and engine misfiring.

Use an ohmmeter to test the primary coil windings. When a pair of ohmmeter leads are connected to the primary coil terminals, an infinity reading indicates an open winding. The coil is shorted if the meter reading is below the specified resistance. Most primary windings have a resistance of 0.5 to 2 ohms.

Use an ohmmeter to test the secondary windings of the coil. Attach one test lead to the primary terminal and one to the secondary terminal. As before, an infinity reading indicates open in the circuit. Any reading below the specified resistance indicates a shorted secondary winding.

Task 6

Check and adjust ignition timing and timing advance/retard.

101. The timing light like the one in Figure 1–23 has what capability to do which of the following?
    A. Be used as a trouble light
    B. To check spark advance
    C. Be used as an oscilloscope on the secondary ignition circuit
    D. Check the primary circuit ignition timing

102. When adjusting the ignition timing on a vehicle, *Technician A* says that the underhood emission label has specific instructions for setting the timing. *Technician B* says that on non-feedback style carburetors, the engine idle speed must be verified before the timing can be adjusted. Who is right?
    A. A only
    B. B only
    C. Both A and B
    D. Neither A nor B

Hint

Ignition timing procedures vary from vehicle to vehicle. The procedure and the specifications for each vehicle are located on the underhood emission label. On distributors with advance mechanisms, manufacturers usually recommend disconnecting and plugging the vacuum advance hose while checking the timing. On carburetor engines, the engine speed must be set beforehand to the manufacturer's specifications. The timing light pickup is placed on the number 1 spark plug wire, and the power supply

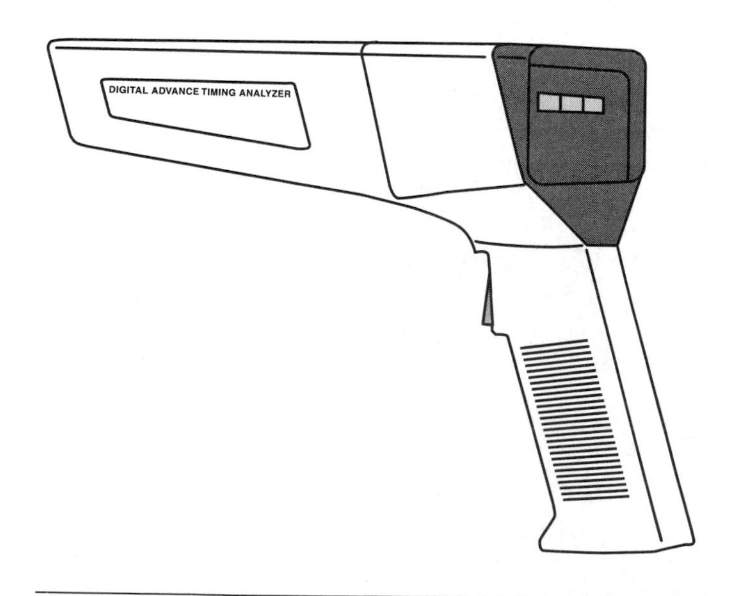

Figure 1–23 Timing light.

wires are connected to the battery terminals. On many fuel-injected engines an in-line timing connector must be disconnected to eliminate any spark advance provided by the PCM. On distributor ignition (DI) systems the distributor clamp bolt may be loosened, and the distributor is rotated until the timing mark appears at the specified degrees of advance on the timing indicator.

Many timing lights have the capability to check the spark advance. An advance control on the light slows the flashes of the light as the advance knob is rotated. When the light flashes are slowed with the engine running at higher speed, the timing marks move back to the basic setting.

Task 7

Inspect, test, and replace ignition system pickup sensor or triggering devices.

103. In Figure 1–24, if ohmmeter provides an infinity reading in the auto range setting, the pickup coil has which of these conditions?
    A. Short to ground
    B. Shorted
    C. Grounded
    D. Open

104. *Technician A* says a Hall effect distributor pickup produces an analog voltage signal. *Technician B* says a grounded pickup coil may cause a no-start condition. Who is right?
    A. A only
    B. B only
    C. Both A and B
    D. Neither A nor B

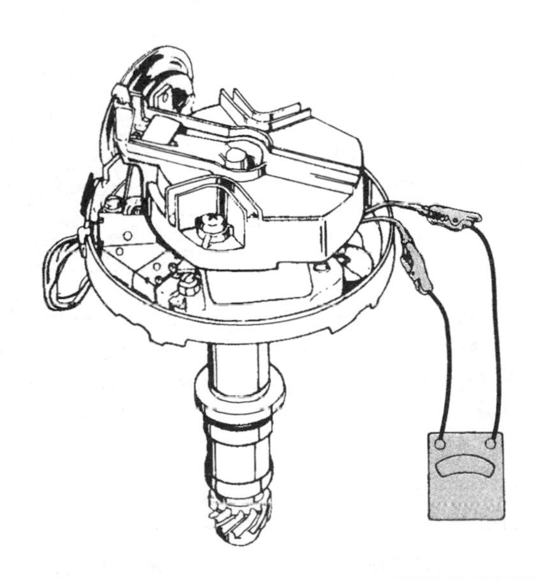

Figure 1–24 Distributor pickup tests. *(Courtesy of General Motors Corporation, Service Technology Group)*

Hint

The distributor pickup gap is adjustable on some engines. To adjust the gap, use a nonmagnetic feeler gauge positioned between the reluctor high points and the pickup coil. If a pickup gap adjustment is required, loosen the bolts, insert the correct feeler gauge, close the gap, and tighten the mounting bolts. Pickup coils that are riveted to the advance plate do not require adjustment.

Connect an ohmmeter to the pickup coil leads to test the pickup coil for an open or shorted circuit. While the leads are still connected to the ohmmeter, pull on the leads that are connected to the pickup coil and watch for an erratic reading. This would indicate an intermittent open in the wires. Most pickup coils have a resistance of 150 to 900 ohms. If the pickup coil has an open circuit it would be indicated by an infinity meter reading, whereas a meter reading below specifications would indicate a shorted coil.

Task 8

Inspect, test, and replace ignition control module.

105. What does an ignition module tester analyze?
    A. It tests the secondary ignition circuit.
    B. It tests the primary ignition circuit.
    C. It tests the available voltage to the ignition module.
    D. It checks the ability of the module to turn on and off.

106. All these statements about ignition modules are true EXCEPT:
    A. ignition modules mounted in the distributor require a heat dissipating grease on the module mounting surface.
    B. some ignition modules mounted externally from the distributor must be grounded where they are mounted.
    C. an ignition module may be damaged by a shorted primary coil winding.
    D. ignition modules are protected from voltage spikes by an external zener diode.

Hint

Each vehicle manufacturer has its own ignition module tester. These testers check the module's ability to switch the primary ignition circuit on and off. Follow the instructions of the test equipment manufacturer. The module leads are connected to the test equipment while the power supply connections for the module are connected to the battery.

Task 9

Inspect, test, and replace electronic and vacuum-type governor assemblies.

107. In the electronic governor system shown in Figure 1–25:
    A. the ECM directly controls the governor motor.
    B. the governor motor is connected directly to the throttle shafts.
    C. the ECM sends voltage signals to the governor module to provide the proper governor motor control.
    D. the ECM does not require any input information to provide proper governor operation.

108. The vacuum-type governor shown in Figure 1–26 provides a governed engine speed of 2,100 rpm, but the specified governed engine rpm is 3,800. This problem could be caused by:
    A. a defective governor solenoid valve that is sticking closed.
    B. a plugged vacuum passage in the vacuum regulator.
    C. an inoperative electric vacuum pump.
    D. a disconnected vacuum hose at the governor assembly.

Hint

Governors protect the engine by limiting the maximum engine rpm. Some engines are equipped with vacuum governors. In this type of governor a regulated vacuum is supplied from an electric vacuum pump to the governor diaphragm on the throttle body. When the engine is operating at the governed rpm, a module-operated solenoid on the governor assembly controls the amount of vacuum supplied to the governor diaphragm. This diaphragm is linked to the throttles, and the vacuum supplied to the diaphragm holds the throttles in the proper position to provide the governed rpm.

Other truck engines have electronic governors connected to the ignition system. In this type of governor a governor motor on the throttle body holds the throttles in the governed rpm position. The governor motor is driven by a governor module, and the PCM sends control signals to the governor module. The PCM uses input signals such as engine rpm, manifold absolute pressure (MAP) sensor, and throttle position sensor (TPS) to calculate the control signal that the PCM sends to the governor module.

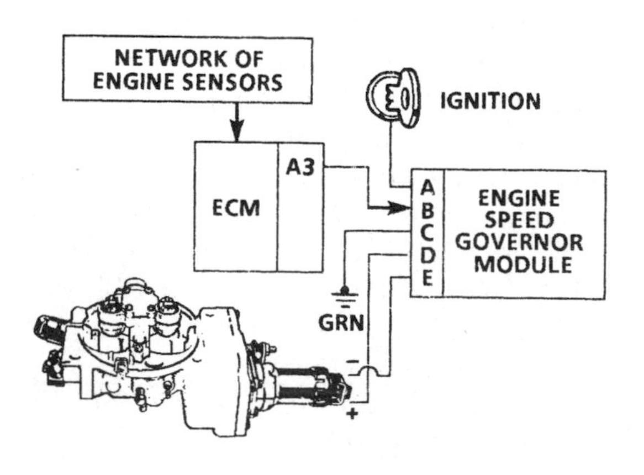

Figure 1–25 Electronic governor. *(Courtesy of Service Technology Group, General Motors Corporation)*

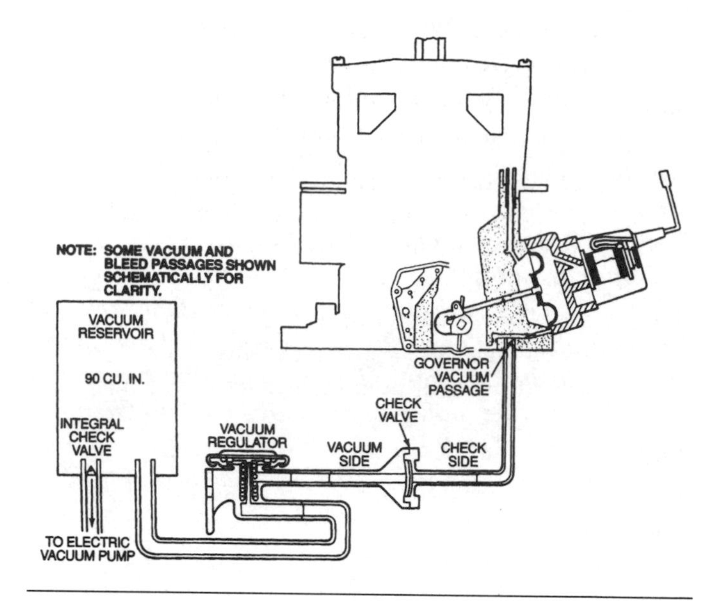

Figure 1–26 Vacuum-type governor. *(Courtesy of Ford Motor Company)*

Fuel and Exhaust Diagnosis and Repair

Task 1

Diagnose no-starting, hard starting, poor idle, flooding, hesitation, engine missing, power loss, poor mileage, and/or dieseling problems on trucks with carburetor and fuel injection systems; determine needed repairs.

109. A throttle body injected truck engine hesitates during acceleration with the engine at normal operating temperature. The engine idles smoothly and performs properly in all other operating modes. The cause of this problem could be:

A. a defective throttle position sensor (TPS).
B. a burned exhaust valve.
C. a partially restricted fuel injector.
D. excessive fuel pump pressure.

110. The driver complains about excessive fuel consumption on a throttle body injected truck engine. The cause of this problem could be:
A. a restricted fuel filter.
B. a restricted fuel return line.
C. a weak pressure regulator spring.
D. a defective fuel pump relay.

Hint

For fuel-injected vehicles, it must be verified that the fuel pump has the specified pressure and flow. Next, the injectors must be inspected for operation. Twelve volts must be available at each fuel injector with the ignition switch on. The powertrain control module (PCM) controls the ground path for the injectors. If it is suspected that the ground path is faulty, further diagnosis with a scan tool should take place to determine if there are any fault codes or problems with the injector drivers located in the PCM.

The carburetor must be inspected before it is determined that it needs a rebuild. Check that the choke plate is functioning properly along with the accelerator linkage. If possible, check the float level. A faulty float level can cause flooding, as well as fuel starvation.

Task 2

Inspect, test, and/or replace fuel pumps (mechanical and electrical); inspect, service, and replace fuel filter, gas cap, fuel lines, fuel tank, hoses, and fittings.

111. A throttle body injected truck engine has 6 psi (41 kPa) fuel pressure, and the specified fuel pressure is 9 to 13 psi (62 to 90 kPa). This low fuel pressure could cause all of the following problems EXCEPT:
A. a hesitation during acceleration.
B. engine cutout during high speed and heavy load.
C. rough idle operation.
D. surging at medium and high speeds.

112. When discussing fuel pump testing on throttle body injected truck engines, *Technician A* says some engines have an electric fuel pump in the fuel tank(s), and another electric fuel pump mounted on the frame rail. *Technician B* says if the fuel pump has the specified pressure the fuel pump is satisfactory. Who is right?
A. A only
B. B only
C. Both A and B
D. Neither A nor B

Hint

Electric fuel pumps operate for two to ten seconds during ignition key-up. The reason for this is to run the pump so that the fuel rail can be pressurized at startup. The PCM controls the relay that supplies power to the electric fuel pump.

A cam that is located on front end of the camshaft drives most mechanical fuel pumps. These fuel pumps are bolted to the engine block. The pump must be observed for signs of external or internal leakage. Internal leakage will usually contaminate the engine oil with raw fuel. This can be a dangerous situation if not attended to. Mechanical fuel pumps have two main lines, one from the fuel tank, and one to the carburetor.

Task 3

Remove and replace carburetor; adjust linkages.

113. What is the function of the automatic choke?
A. To limit the amount air entering the engine
B. To provide blowby vapor for the PCV system
C. To provide the best engine performance in cold weather
D. To add raw fuel to the engine during cold startup

114. An improperly adjusted throttle linkage on a throttle body injected engine could cause:
A. stalling on deceleration.
B. loss of engine power at high speeds and heavy loads.
C. rough idle operation.
D. engine surging at low speeds.

Hint

To remove the carburetor, all linkages, vacuum hoses, and fuel lines must be disconnected. Take your time and, if necessary, label all the connections so that reassembly will not be confusing. After disconnecting the vacuum hoses and linkage connections, loosen the fuel supply line with a flare nut wrench. Loosen the fuel line fitting one or two turns only. Now, remove the retaining hardware that holds the carburetor to the intake manifold. After the bolt or nuts have been removed, disconnect the fuel line from the carburetor. Take the proper precautions, because there will be liquid fuel lost from the carburetor and the fuel supply line.

Task 4

Rebuild carburetor including: disassembling, cleaning, replacing faulty parts, assembling, and adjusting.

115. The accelerator pump height is adjusted so it is less than specified. *Technician A* says this condition may cause a stalling at idle speed. *Technician B* says this condition may cause a hesitation during acceleration. Who is right?
  A. A only
  B. B only
  C. Both A and B
  D. Neither A nor B

116. A carbureted truck engine has a hesitation during acceleration only when the engine coolant is cold. The cause of this problem could be:
  A. the vacuum break adjustment on the choke is too wide.
  B. the choke spring adjustment is too rich.
  C. the fast idle cam adjustment is too wide.
  D. the accelerator pump height is less than specified.

Hint

After removing the carburetor, remove any plastic parts that can be dissolved in cleaning solution. If the carburetor has a large amount of varnish and grime, it may be soaked in cleaning solution before disassembly, otherwise disassemble the carburetor and soak the parts in a cleaning solution for half an hour to an hour. During disassembly, check for obvious signs of damage or worn-out parts. Check springs for signs of binding or broken coils. Inspect seals for tears or signs of swelling. Observe where all the check balls are placed so that they can be replaced. Inspect all the metering passages for obstructions. Check the subassemblies for proper operation and broken parts. The subassemblies include the choke, float and linkage, dashpots, and accelerator pump.

Task 5

Remove, clean, and replace throttle body; adjust related linkages.

117. For what reason do some throttle bodies such as the one shown in Figure 1–27 have a special coating?
  A. To prevent the engine from backfiring during deceleration
  B. So that the airflow is evenly distributed to all the cylinders
  C. To prevent carbon buildup on the throttle plate
  D. To increase the velocity of the incoming air

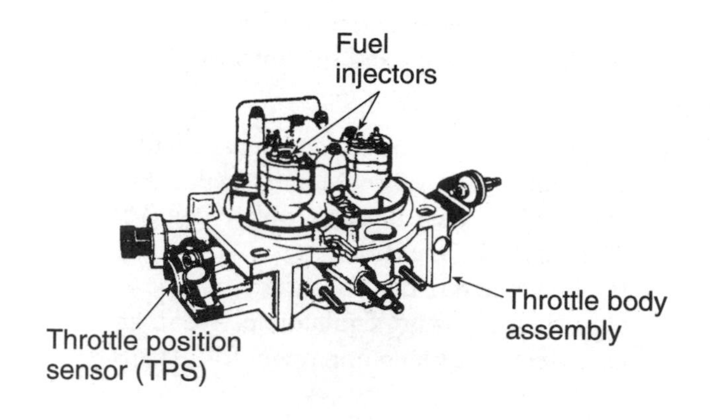

Figure 1–27 Throttle body assembly. *(Courtesy of General Motors Corporation, Service Technology Group)*

118. When replacing a throttle body on a vehicle, *Technician A* says the PCV valve supply hose is connected above the throttles in the throttle body. *Technician B* says that the throttle body gasket can be reused. Who is right?
  A. A only
  B. B only
  C. Both A and B
  D. Neither A nor B

Hint

To remove the throttle body, first remove the air intake assembly that is connected to the throttle body. Disconnect the throttle linkage and, if equipped, the transmission kick-down cable. Disconnect the throttle position sensor and remove the throttle body. Remove the old gasket, because a new one will be installed during installation. Be sure to place a rag in the intake manifold opening to ensure that no foreign objects enter the engine. Some throttle bodies have a special coating that does not allow carbon to build up on the surface of the throttle plate. If the throttle body has a special coating, no chemical cleaners can be used to clean the throttle body.

Task 6

Inspect and replace carburetor/fuel injection mounting plates, intake manifold, and gaskets.

119. All of the following are true about intake manifold gaskets EXCEPT:
  A. it is a good practice to replace all the intake manifold gaskets at the same time.
  B. the replacement gasket should be made of the same material as the original.

C. conveniently, all the gaskets in the intake system come together in a kit.
D. cracked intake manifold gaskets can be repaired with silicone.

120. A throttle body injected truck engine has a cylinder misfire during idle operation. As the engine speed is increased the cylinder misfire disappears. *Technician A* says one of the intake manifold gaskets may be leaking. *Technician B* says the spark plug may be defective in the misfiring cylinder. Who is right?
A. A only
B. B only
C. Both A and B
D. Neither A nor B

Hint

Faulty intake manifold gaskets can cause driveability problems that can commonly be confused with other fuel system components. A defective gasket in the intake system can easily be overlooked. Perform specific tests to determine if any intake gaskets have failed. It is a good practice when replacing the intake manifold gaskets to replace all the gaskets in the intake system. Conveniently, they usually come together in a kit. Be sure that the gasket(s) you are replacing are made from the same material as the Original Equipment Manufacturer (OEM).

Task 7

Inspect, clean, and adjust carburetor choke (automatic and manual).

121. On a carburetor with an automatic choke housing connected to a heat pipe, the choke housing contains an accumulation of carbon. The cause of this problem could be:
A. engine backfiring through the carburetor.
B. engine overheating.
C. a burned-out heat pipe in the exhaust manifold.
D. excessive carbon deposits in the exhaust manifold.

122. On a carburetor with an automatic choke housing connected to a heat pipe, the choke remains partially closed after the engine is at normal operating temperature. *Technician A* says the choke housing may be warped causing a vacuum leak. *Technician B* says the heat pipe may be restricted. Who is right?
A. A only
B. B only
C. Both A and B
D. Neither A nor B

Hint

The function of the automatic choke is to provide the best possible engine performance in cold temperatures. A properly operating choke should open and close freely. The choke should open as soon as the engine can run without its help. Automatic chokes that use a thermostatic coil have index markings on the plastic cover. The index marks indicate richer or leaner adjustment. Check for proper operation of the choke plate before attempting to adjust the choke. Make sure the choke linkage is free from dirt and debris. In addition, the choke spring must be "cold" before attempting an adjustment.

Task 8

Inspect, test, adjust, and replace cold enrichment systems components.

123. When the engine coolant is cold:
A. the engine coolant temperature (ECT) sensor resistance decreases and the injector pulse width decreases.
B. the ECT sensor resistance increases and the injector pulse width also increases.
C. the ECT sensor voltage drop decreases.
D. the PCM should be operating in closed loop.

124. When the engine coolant and the ECT sensor are 0°F (-18°C), the ECT sensor has 5,000 ohms resistance. The specified ECT sensor resistance at this temperature is 24,000 ohms. *Technician A* says the engine may provide reduced fuel economy. *Technician B* says the engine may be hard to start when cold. Who is right?
A. A only
B. B only
C. Both A and B
D. Neither A nor B

Hint

The fuel injection system has to enrich the air-fuel mixture during cold temperatures. This is accomplished by means of electronically controlling the pulse width of the injectors. The pulse width of the injector is the amount of time that the injector is energized. This time is rated in milliseconds. The PCM monitors the engine coolant temperature sensor. If the engine temperature is below a specified level, the PCM increases the pulse width to provide a richer air-fuel ratio. When the engine coolant reaches a specified temperature, anywhere from 130° to 170°F (54° to 76°C) and the oxygen sensor reaches 600°F (315°C), the system enters closed loop. In this mode, the fuel mixture is adjusted according to the input sensor signals.

Task 9

Inspect, test, and replace deceleration fuel reduction or shut off systems components.

125. The PCM can determine the vehicle is in the deceleration mode by which of the following sensors?
   A. Antilock brake sensor
   B. Throttle position sensor
   C. Canister purge solenoid
   D. Camshaft position sensor

126. A throttle body injected truck engine stalls frequently during deceleration with the engine at normal operating temperature. The engine idles smoothly and operates normally under all other driving conditions. All of these defects may be the cause of the problem EXCEPT:
   A. carbon accumulation around the throttles.
   B. carbon accumulation in the idle air control (IAC) motor.
   C. low fuel pressure.
   D. a badly worn IAC motor pintle.

Hint

The PCM determines deceleration by observing the following sensors (in order of importance):

- *Throttle position sensor*
- *Engine vacuum*
- *Engine speed*
- *Vehicle speed*

When the driver releases the accelerator pedal, the throttle position sensor value immediately changes, indicating to the PCM a change has occurred and the driver is decelerating. Next, the engine vacuum will increase as the throttle plates close. This is the second indication of the driver's desire to decelerate, and is measured by the manifold absolute pressure (MAP) sensor.

Task 10

Inspect, service, and repair or replace air filtration system components.

127. *Technician A* says that air cleaner elements on newer fuel-injected vehicles can last for 75,000 miles before needing replacement. *Technician B* says that a shop light can be used to determine how dirty the element is. Who is right?
   A. A only
   B. B only
   C. Both A and B
   D. Neither A nor B

128. A port fuel injected truck engine with a mass air flow (MAF) sensor mounted in the air intake hose has a hesitation during acceleration. The MOST likely cause of this problem is:
   A. higher than specified fuel pressure.
   B. a leak in the air intake system between the MAF sensor and the intake manifold.
   C. a restricted fuel return line.
   D. a vacuum line disconnected from the fuel pressure regulator.

Hint

The automotive engine burns about 9,000 gallons (34,065 liters) of air for every gallon of gasoline at an air-fuel ratio to 14.7:1. With today's engines, which run much leaner to meet emissions standards, the ratio is closer to 10,000 gallons (37,850 liters) of air. This is a large amount of air that needs to be filtered. This is accomplished by using a paper element enclosed in a sealed box that is attached to the throttle body through intake plumbing. The paper element is disposable and should be replaced at the vehicle manufacturer's specified intervals. A good practice to follow is to remove the air cleaner element and hold a shop light directly behind it while trying to see any light that passes through the element. If very little light passes through the element, the filter should be replaced.

Task 11

Remove, clean, inspect/test, and repair or replace fuel system vacuum and electrical components and connections.

129. *Technician A* says that the optimum air fuel ratio is 18:1. *Technician B* says that for every gallon of gasoline that the engine burns, the engine uses 9,000 gallons of air. Who is right?
   A. A only
   B. B only
   C. Both A and B
   D. Neither A nor B

130. *Technician A* says that the fuel pressure regulator on a port fuel-injected engine senses engine vacuum. *Technician B* says the fuel pressure regulator controls the amount of fuel that is returned to the tank. Who is right?
   A. A only
   B. B only
   C. Both A and B
   D. Neither A nor B

Hint

The fuel pressure regulator controls the amount of fuel that is returned to the fuel tank. This is also how the proper fuel pressure is maintained in the fuel rail. On a port fuel-injected engine, the pressure regulator valve is controlled by spring tension and intake manifold vacuum. As engine manifold vacuum decreases, which indicates the engine is under a load, the spring in the regulator closes the orifice and raises the fuel pressure. The higher fuel pressure is needed to enrich the air-fuel mixture under load.

The fuel pressure regulator is bolted to the fuel rail near the fuel return line connection. To remove the fuel pressure regulator, first relieve the fuel rail pressure to prevent fuel from being sprayed in the engine compartment. Always replace the old regulator gasket with a new one.

Task 12

Inspect exhaust manifold, exhaust pipe, muffler, and other components of the exhaust system; repair/replace as needed.

131. In Figure 1–28, as engine manifold vacuum decreases, what does the spring in the fuel pressure regulator do?
   A. Open the valve to allow more fuel to the rail
   B. Open the valve to allow less fuel to return to the tank
   C. Close the valve to allow fuel pressure to build in the rail
   D. Close the valve so that the fuel pump will operate under a load

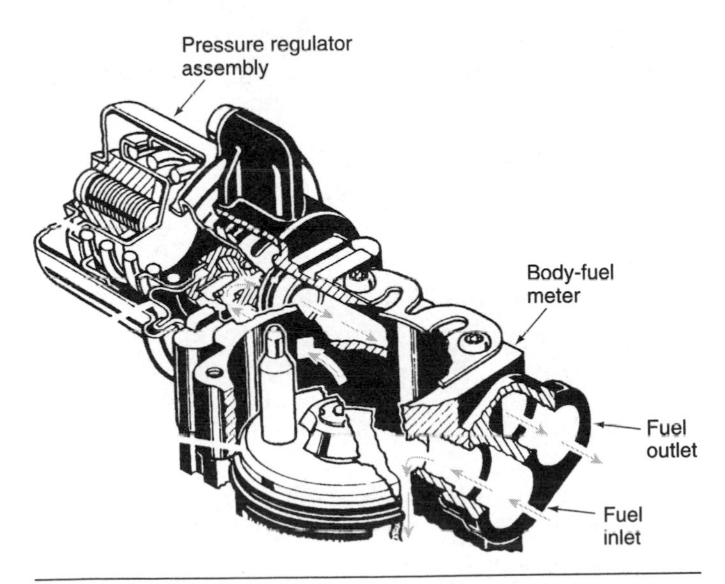

Figure 1–28 Pressure regulator and central port injector assembly. *(Courtesy of General Motors Corporation, Service Technology Group)*

132. In Figure 1–29, what advantages do stainless steel exhaust manifolds have over conventional cast iron manifolds?
   A. Lower in cost
   B. Heavier in weight
   C. Higher emissions
   D. Lighter in weight

Hint

Exhaust manifolds are made of cast iron, which offer good resistance to change from temperature. The exhaust manifold bolts to the cylinder head. All the corrosive exhaust gas flows through the manifold to the catalytic converter and then out through the muffler and tailpipe. Vehicle manufacturers now use stainless steel for exhaust manifolds. Stainless steel offers greater resistance to corrosion, and quicker warmup, which helps the catalytic converter to reach operating temperature sooner. Stainless steel is considerably lighter than cast iron.

Exhaust manifolds should be inspected for cracks and for extreme corrosion that can cause exhaust leakage. Exhaust manifold gaskets can sometimes deteriorate and blow out from between the manifold and the cylinder head.

Battery and Starting Systems Diagnosis and Repair

Task 1

Inspect, clean, fill, or replace battery.

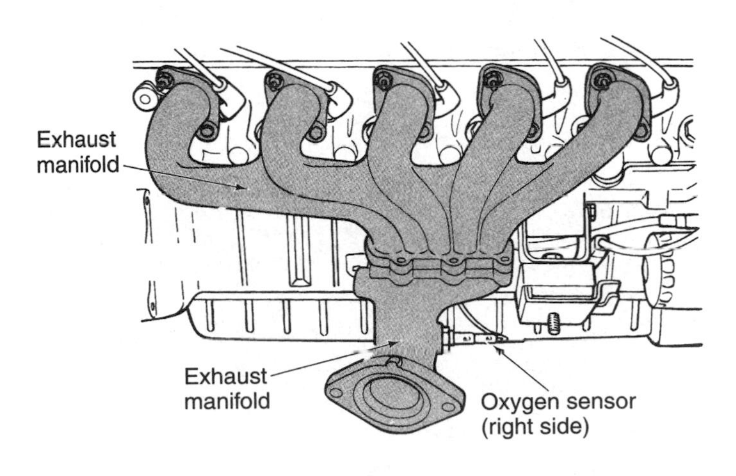

Figure 1–29 Exhaust manifold. *(Courtesy of Chrysler Corporation)*

133. When discussing cleaning the battery as shown in Figure 1–30: *Technician A* says that a solution of baking soda and water helps prevent future corrosion. *Technician B* says the leftover solution of water and baking soda can be used to fill the cells in the battery. Who is right?
   A. A only
   B. B only
   C. Both A and B
   D. Neither A nor B

134. While discussing battery post service, *Technician A* says use a wire brush to remove corrosion. *Technician B* says use the electrolyte solution to remove the oxidation from the posts. Who is right?
   A. A only
   B. B only
   C. Both A and B
   D. Neither A nor B

Hint

Inspect the battery for cracks and signs of electrolyte leakage. Check that the terminals are tight and not heavily corroded. Also, check to make sure that the battery holddown is in place and holding the battery securely. Use a solution of one part baking soda and one part water to clean the battery. Pour the solution over the entire battery, allowing it to run down the side and onto the battery tray. After the solution has been used up, rinse the battery with a generous amount of water. Some batteries allow you to add water to the cells. Use only distilled water to top off the cells. When replacing the battery, always disconnect the negative battery cable first. This is to prevent a short if the wrench touches a ground while disconnecting the positive cable. When installing the battery, connect the positive battery first, for the same reason.

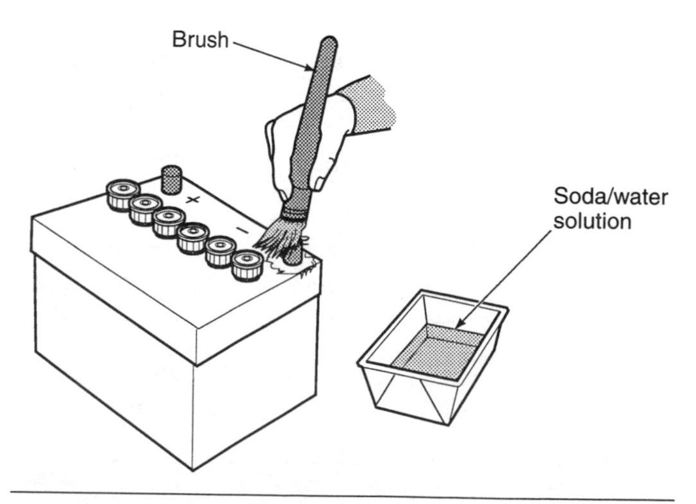

Figure 1–30 Battery cleaning. *(Courtesy of Chrysler Corporation)*

Task 2

Slow and fast charge a battery.

135. *Technician A* says that a battery can be charged at any rate as long as the electrolyte does not boil and the temperature does not exceed 120°F (50°C). *Technician B* says fast charging is just as complete as charging a battery at a slower rate. Who is right?
   A. A only
   B. B only
   C. Both A and B
   D. Neither A nor B

136. All of these statements about battery charging are true EXCEPT:
   A. the battery is completely charged when the specific gravity reaches 1,225.
   B. explosive hydrogen gas is produced in the battery cells during the charging process.
   C. oxygen gas is produced in the battery cells during the charging process.
   D. do not charge a maintenance-free battery if the hydrometer in the battery top indicates yellow.

Hint

Slow charging a battery is charging the battery at 1.5 to 7 amps per hour. The slower a battery is charged, the more complete the charge. A battery can be fast charged at any rate (within reason) as long as the electrolyte does not boil over and the temperature does not exceed 120°F (50°C).

Task 3

Perform a battery capacity (load, high rate discharge) test; determine needed service.

137. In Figure 1–31, what is being tested?
   A. Alternator current draw test
   B. Starter current draw test
   C. Battery load test
   D. Ground circuit voltage drop test

138. When load testing a battery, how many amps should be placed on a 525 cold cranking amp battery?
   A. 125
   B. 325
   C. 262
   D. 425

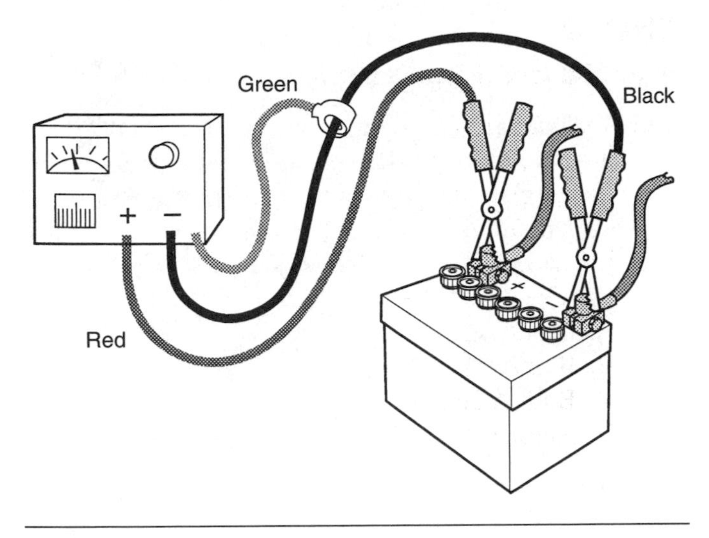

Figure 1–31 Electrical system test.

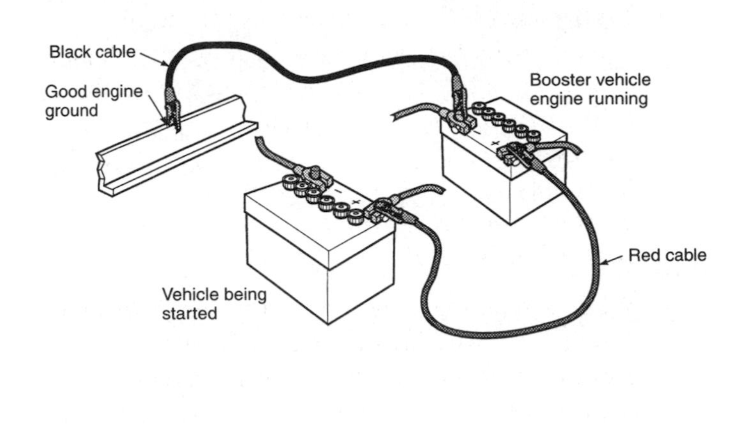

Figure 1–32 Vehicle boosting.

Hint

A battery load test is performed to determine the electrical storage capacity of the battery. The test is performed by applying an ampere load that is half of the cold cranking amperage of the battery. For example, if a battery has a cold cranking rating of 800 amps, to load test apply a load of 400 amps for 15 seconds. The battery voltage should not drop below 9.6 volts for the 15 seconds of the load test.

Task 4

Jump-start a vehicle with jumper cables and a booster battery or auxiliary power supply in accordance with manufacturers.

139. *Technician A* says when jump-starting a vehicle, connect the negative booster cable to the negative battery terminal in the vehicle being boosted. *Technician B* says to connect the negative cables first to prevent sparks. Who is right?
    A. A only
    B. B only
    C. Both A and B
    D. Neither A nor B

140. While jump-starting a vehicle as shown in Figure 1–32: *Technician A* says the negative booster cable on the boost vehicle should be removed first. *Technician B* says that all the accessories must be turned off to prevent electrical damage. Who is right?
    A. A only
    B. B only
    C. Both A and B
    D. Neither A nor B

Hint

The accessories must be off in both vehicles during the boost procedure. The negative booster cable must be connected to an engine ground in the vehicle being boosted. Always connect the positive booster cable, followed by the negative booster cable, and complete the negative cable connection last on the vehicle being boosted. When disconnecting the booster cables, remove the negative booster cable first on the vehicle being boosted.

Task 5

Inspect, clean, repair, and replace battery cables and clamps.

141. All these statements about battery cable service are true EXCEPT:
    A. the terminals should be removed from the battery with a puller.
    B. remove the negative battery cable first.
    C. remove the positive battery cable before the negative cable.
    D. cleanse the battery with a baking soda and water solution.

142. *Technician A* says that battery terminals can be treated with petroleum jelly to help prevent corrosion. *Technician B* says that protective pads can be placed on the terminals to prevent corrosion. Who is right?
    A. A only
    B. B only
    C. Both A and B
    D. Neither A nor B

Hint

After removing the battery cable from the post, clean the post and the terminals with a wire brush. It is always wise to spray the cable clamps with a protective coating to prevent corrosion. A little grease or petroleum jelly will also prevent corrosion. In addition, protective pads are available that go under the clamp and around the terminal to prevent corrosion.

Task 6

Inspect, test, and replace starter relays and solenoids.

143. In Figure 1–33 what procedure is being performed?
    - A. Starter solenoid current draw
    - B. Starter motor current draw
    - C. Battery load test
    - D. Resistance check

144. *Technician A* says a starting and charging system tester can be used to test free spinning current draw tests. *Technician B* says that free spinning current draw tests must be performed with the starter solenoid removed. Who is right?
    - A. A only
    - B. B only
    - C. Both A and B
    - D. Neither A nor B

145. In Figure 1–34, which of the following is being performed?
    - A. Pinion gear clearance
    - B. Checking the voltage drop across the ground circuit
    - C. Battery load test
    - D. Starter free speed test

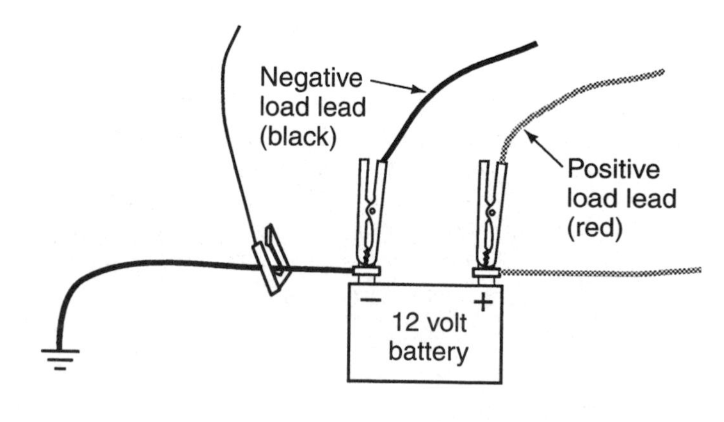

Figure 1–33 Electrical system test.

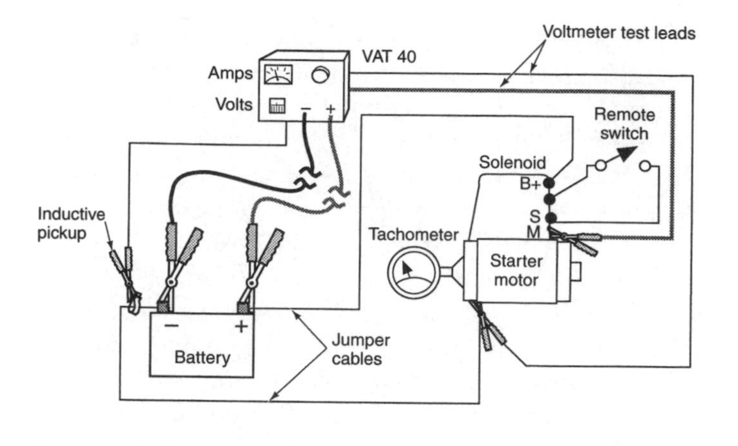

Figure 1–34 Electrical system test.

Hint

Conditions of high current draw, low cranking speed, and low cranking voltage usually indicate a defective starter. This condition also may be caused by internal engine problems, such as partially seized bearings. Low current draw, low cranking speed, and high cranking voltage indicate excessive resistance in the starter circuit. A defective starter relay or solenoid also can cause this resistance. Relays and solenoids have disks that transmit the current; these disks become pitted and burned. This is what causes high voltage drops.

Task 7

Remove and replace the starter.

146. *Technician A* says that high resistance in the starter motor circuit can cause low cranking speed. *Technician B* says that internal engine problems can be the cause of the starter motor not functioning properly. Who is right?
    - A. A only
    - B. B only
    - C. Both A and B
    - D. Neither A nor B

147. A starting motor fails to disengage properly and the armature rotates with the engine for a short time after the engine starts. The cause of this problem could be:
    - A. high resistance in the starting motor ground.
    - B. misalignment of the starting motor in relation to the flywheel housing.
    - C. high resistance between the battery positive terminal and the starter solenoid.
    - D. a shorted condition in the starting motor armature.

Hint

When reinstalling the starter motor you should perform a free spin test or a current draw test, and also test the pinion gear clearance. Disconnect the M-terminal so the solenoid can be energized to shift the pinion gear into the cranking position without spinning the starting motor armature and pinion gear. Then you will be able to check the clearance with a feeler gauge. Normally, specifications call for a clearance of 0.010 to 0.140 in. (0.25 to 0.35 mm). Some starting motors have shims between the starting motor and the flywheel housing mating surfaces. When removing and replacing a starting motor, these shims must be reinstalled in their original position. If these shims are not properly installed, the starting motor may be misaligned in relation to the flywheel housing. This condition may cause failure of the starting motor to disengage properly when the engine starts.

Emissions Control Systems Diagnosis and Repair

Task 1

Test, inspect, clean, and replace exhaust gas recirculation EGR valves, valve manifolds, and passages of the EGR system.

148. Refer to Figure 1–35. *Technician A* says that some EGR valves with heavy carbon buildup must be cleaned with a sandblaster. *Technician B* says to use fuel injection cleaner to clean the passages in the EGR valve. Who is right?
    A. A only
    B. B only
    C. Both A and B
    D. Neither A nor B

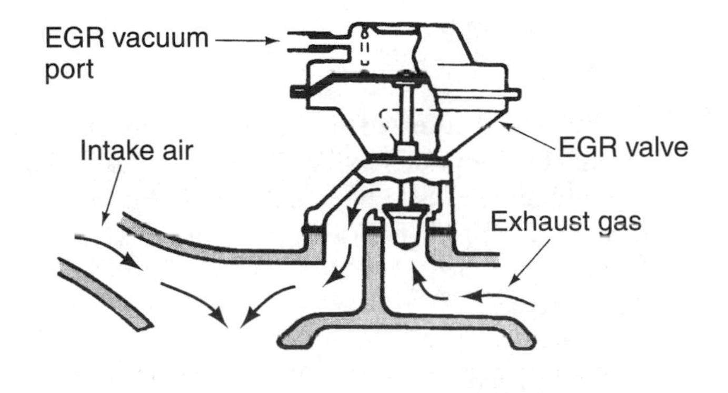

Figure 1–35 EGR valve. *(Courtesy of Cadillac Motor Car Division)*

149. All of the following statements about EGR valves are true EXCEPT:
    A. they operate on ported venturi vacuum systems.
    B. they operate on ported vacuum.
    C. they may be mounted on the intake.
    D. they operate at wide-open throttle.

Hint

Exhaust gas recirculation (EGR) valves redirect a metered amount of exhaust gas into the intake manifold. The reason for this is to help reduce exhaust emissions. Most EGR valves are vacuum operated. Some EGR valves are controlled by an electric actuator. To test an EGR and to tell if it functioning properly, use a hand-operated vacuum pump and apply 10 to 15 inches of vacuum to the port on the EGR. Within this range, the valve should open fully. When an EGR valve is stuck in the open position, the engine will run rough at idle and experience power loss and poor fuel economy. Depending on the application, the EGR passages in the manifold become blocked and need to be cleaned. Use special fuel injection spray cleaner, or any cleaner that is safe to use on vehicles with oxygen sensors. Spray the cleaner into the intake manifold passages to dissolve the carbon.

Task 2

Inspect, repair, and replace controls and hoses of the EGR system.

150. While discussing EGR valves, *Technician A* says that the vacuum hoses usually do not cause failures with the EGR system. *Technician B* says that the plastic used for the EPT and EVR solenoids is very durable and will not crack around the vacuum hose connection. Who is right?
    A. A only
    B. B only
    C. Both A and B
    D. Neither A nor B

151. *Technician A* says that the coolant temperature override switch prevents EGR valve operation until the engine warms up to operating temperature. *Technician B* says the EVR solenoid converts the electrical signals from the PCM into a mechanical action, which directs vacuum to the EGR valve. Who is right?
    A. A only
    B. B only
    C. Both A and B
    D. Neither A nor B

Hint

The EGR controls include the following:

- *Coolant temperature override switch*
- *EGR pressure transducer (EPT)*
- *EGR vacuum regulator (EVR) solenoid*
- *Vacuum hoses*
- *PCM*

Inspect all the hoses for cracking or any areas that could cause a vacuum leak. Most EPT and EVR solenoids are made of plastic and should be inspected where the vacuum lines attach.

Task 3

Test the operation of the air injection system.

152. In Figure 1–36, which of the following conditions would best describe what should happen when the two hoses indicated are disconnected while the engine is idling?
    A. A strong suction should be felt at the end of the disconnected hose.
    B. Either of the hoses should discharge airflow out the end of the hose.
    C. There should be no positive air flow under 4,000 rpm.
    D. There should be no suction at the hoses until the air pump clutch is engaged.

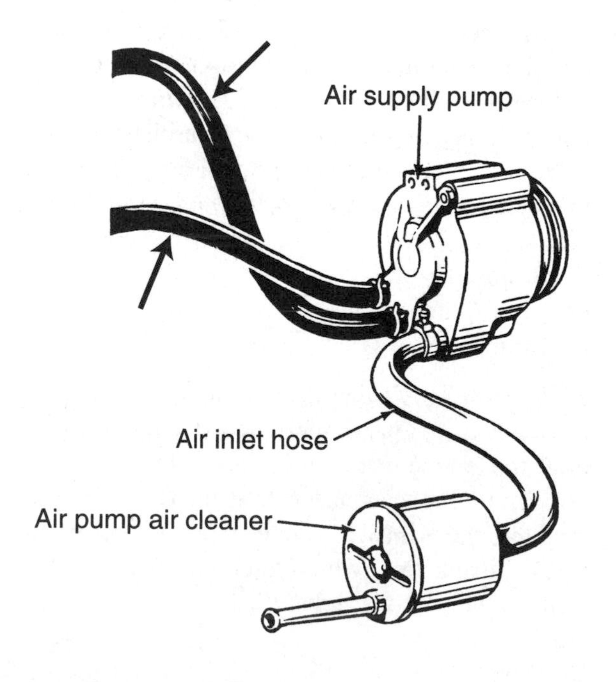

Figure 1–36 Air pump system.

153. When troubleshooting the air injection system, the first thing to be inspected should be:
    A. the air pump drive belt.
    B. the air injection hoses.
    C. the vacuum-operated valves in the air injection system.
    D. checking for air leaks at the three-way catalytic converter.

Hint

To check the operation of the air injection system, your inspection should include the following:

- *Check the condition of the drive belt and the tension on the belt.*
- *Inspect all the air system hoses and vacuum lines. Look for signs of cracking, brittleness, or burning.*
- *Inspect the check valves for exhaust leakage.*
- *Disconnect the hose at the pump side of the check valve, start the engine, and check for airflow from the hose. There should be adequate airflow at the hose opening. Increase the engine speed. As the engine speed increases, so should the airflow. Next, pinch off the pump outlet hose. A popping noise should be heard as the pressure valve opens.*

Task 4

Inspect, repair, and replace pumps, pressure relief valves, filters, pulleys, and belts of the air injection system.

154. When removing an air pump from a vehicle, *Technician A* says new air pumps will come with a new diverter valve. *Technician B* says that the new air pump may not come with a new pulley. Who is right?
    A. A only
    B. B only
    C. Both A and B
    D. Neither A nor B

155. A vehicle with a seized air injection pump has entered the shop for service. *Technician A* says that the air pump needs to be rebuilt. *Technician B* says that you should never pry on the housing while tightening the drive belt. Who is right?
    A. A only
    B. B only
    C. Both A and B
    D. Neither A nor B

Hint

Air pumps do not require periodic maintenance, and one cannot rebuild an EGR valve. When removing or servicing the air pump, never pry against the pump housing, because it is made of aluminum and can be cracked easily. Removing the air pump is similar to removing an alternator or any engine-driven accessory. The following steps are for general air pump removal.

- *Disconnect the vacuum and output hoses from the pump.*
- *Loosen the pump mounting bolts and remove the belt. Depending on the vehicle application, some of the other belts may have to be removed.*
- *Remove the pump bolts and remove the pump.*

Task 5

Inspect, and replace by-pass (anti-backfire) valves and vacuum hoses of air injection systems.

156. After disconnecting the vacuum supply hose to the by-pass valve, what should be present at the hose?
    A. No vacuum
    B. Positive air displacement
    C. Exhaust gas
    D. Vacuum

157. *Technician A* says that the gulp valve is used to prevent engine backfires. *Technician B* says the gulp valve has largely been replaced by the diverter valve. Who is correct?
    A. A only
    B. B only
    C. Both A and B
    D. Neither A nor B

Hint

The main purpose of the by-pass valve is to redirect the air to the intake manifold during deceleration, to prevent backfire. Check the by-pass valve as follows:

- *Check the condition of all the hoses and the hose connections.*
- *Start the engine and chock the wheels.*
- *Disconnect the vacuum line at the by-pass valve and place a finger over the end of the hose. If there is no suction, the hose is plugged or broken.*
- *Reconnect the vacuum hose and increase the engine speed to 2,000 rpm, then quickly close the throttle. If the engine backfires, the valve is defective and needs replacement.*

Task 6

Inspect, service, and replace hoses, check valves, air manifolds, and injectors of the air injection systems.

158. While replacing an air injection manifold, which of the following should be done?
    A. Remove the exhaust manifold from the vehicle then remove the air injection manifold.
    B. Remove the air pump from the vehicle and then remove the air injection manifold.
    C. Use high temperature silicone in place of the gasket.
    D. Apply penetrating oil to the nuts before removal.

159. What type of hose can be used to replace air injection hoses?
    A. Preformed air injection hose
    B. Preformed heater hose
    C. Fuel supply hose
    D. Heater hose

Hint

The air injection hoses should be inspected for cracks, deterioration, and wear. Check the hoses for air leaks. Any hoses that are replaced must be the same preformed shape as the original. Any hose that is not suited for air injection systems may deteriorate prematurely. Most manifolds are one-piece assemblies and are connected to the exhaust manifold with nozzle fittings. Before attempting to loosen the nuts at the manifold, apply penetrating oil to the nuts so that the nuts can be loosened with fewer or no problems. Use a tube wrench to loosen the nuts. If the nuts still are a problem, heat from a torch can be used to loosen the nuts. When installing the new air injection manifold, use antiseize compound on the threads to help start the threads and to prevent rusting between the threads.

Task 7

Inspect, test, service, and replace positive crankcase ventilation (PCV) valve and system components.

160. *Technician A* says that regular oil change intervals helps to prevent sludge in the PCV system. *Technician B* says to check the air filter for oil deposits and oil puddling. Who is right?
    A. A only
    B. B only
    C. Both A and B
    D. Neither A nor B

161. Which of the following are indications that the PCV system is not functioning properly?
 A. No oil consumption
 B. Normal HC and CO emission readings
 C. Normal fuel consumption
 D. Excessive crankcase vapors escaping from the oil filler cap opening

Hint

The PCV system requires service at regular intervals to ensure that the system operates properly. Sludge and carbon can plug the PCV valve. Service the PCV valve by inspecting it, making sure that the valve is not clogged and it moves freely. Some symptoms of a PCV system that is not functioning properly include the following:

- *Increased oil consumption*
- *Diluted or dirty oil, or oil that has sludge or is acidic*
- *Excessive blowby, or an oil level dipstick that is blown out of its seat*
- *Rough idling or stalling at idle or low engine speeds*

Inspect all PCV system hoses for cracks and deterioration. Check the air filter for oil deposits and for oil puddling in the oil cleaner housing. In addition, the system should be inspected for clogs. Clogs are usually caused by not changing the oil at regular intervals, and by not maintaining the PCV system. Regular maintenance should include cleaning the PCV valve with a carburetor or fuel injector cleaner.

Task 8

Inspect, test, and replace components of the catalytic converter systems.

162. *Technician A* says that if the catalytic converter is operating normally, the converter inlet should be hotter than the outlet. *Technician B* says that the heat shields should be reinstalled after the catalytic converter is replaced. Who is right?
 A. A only
 B. B only
 C. Both A and B
 D. Neither A nor B

163. *Technician A* says that all three-way catalytic (TWC) converters are interchangeable. *Technician B* says that an overly rich air-fuel mixture will not affect the catalytic converter. Who is right?
 A. A only
 B. B only
 C. Both A and B
 D. Neither A nor B

Hint

Typical catalytic converter replacement is essentially the same as replacing a muffler or a tailpipe. Always wait until the exhaust system has cooled sufficiently. Use heat from a torch and a hammer to remove the converter from the converter inlet pipe. Use the correct application converter for the vehicle. Remember to install all the heat shields back onto the converter.

Computerized Engine Controls Diagnosis and Repair

Task 1

Diagnose the causes of emission problems resulting from failure of computerized engine controls.

164. A fuel-injected engine has a severe surging problem only at 55 mph (88 km/h) or faster. Engine operation is normal at idle and low speeds. *Technician A* says there may be low voltage at the fuel pump. *Technician B* says that the inertia switch may have high resistance. Who is right?
 A. A only
 B. B only
 C. Both A and B
 D. Neither A nor B

165. An engine has high hydrocarbon (HC) and carbon monoxide (CO) emissions. All of these defects could cause this problem EXCEPT:
 A. a defective oxygen (O_2) sensor.
 B. an engine coolant temperature (ECT) sensor with less than the specified resistance.
 C. a defective fuel pressure regulator causing excessive fuel pressure.
 D. a defective mass airflow (MAF) sensor.

Hint

Depending on what type of sensor or computerized fuel injection component has failed, the results can range from poor fuel economy to the system operating in failure or limp-in mode. When a sensor fails, the PCM will recognize there is a fault and illuminate the malfunction indicator light (MIL). If the component that has failed is one that is critical to engine driveability, the PCM will substitute a known good value so that the engine can still operate. This is what is known as a limp-in mode. For instance, with a faulty oxygen sensor sending information to the PCM, the PCM may never think the engine is at operating temperature. With the PCM thinking that the engine is always cold, it will always provide a rich fuel mixture. A rich air-fuel mixture can cause the catalytic converters to overheat and melt down, and also cause the engine oil to become sludged or diluted with liquid fuel. Any type of computerized fuel injection that has failed should be attended to as soon as possible to prevent other problems.

Task 2

Perform analytic/diagnostic procedures on vehicles with onboard diagnostic computer systems; determine needed action.

166. When diagnosing an EVAP system with a scan tool, the PCM never provides the on command to the EVAP solenoid at any engine or vehicle speed. *Technician A* says to check the ECT sensor signal to the PCM. *Technician B* says check the vacuum hoses from the intake to the EVAP canister. Who is right?
    A. A only
    B. B only
    C. Both A and B
    D. Neither A nor B

167. All of the following are true regarding the use of a scan tool EXCEPT:
    A. warm the engine to normal operating temperature.
    B. connect the power adapter to the cigar lighter.
    C. disconnect the negative battery cable before connecting the DLC connector.
    D. enter the model year, engine size, and vehicle model.

Hint

Trouble codes and other diagnostic tests can be performed with a hand-held scan tool. Scan tool operation varies depending on the make of the tester, but the following is a typical example:

1. *Warm the engine until the engine is at normal operating temperature. Turn the ignition off. Connect the power adapter to the cigar lighter, and connect the other lead of the scan tool to the data link connector (DLC).*
2. *Follow the procedure that the scan tool provides on the display. These include entering the model year, engine size, and vehicle model. These are all listed in the tester operation manual. Now that the initial entries have been selected, different entries appear. A typical list of the next choice of entry are:*

- *Engine*
- *Antilock brake system*
- *Suspension*
- *Transmission*

The Technician must make the selection of the system they want to troubleshoot. Next, they have to decide if they want to retrieve stored trouble codes or see a live data screen.

Task 3

Inspect, test, adjust, and replace sensor, control, and actuator components and circuits of computerized engine control systems.

168. While discussing ECT sensor diagnosis, *Technician A* says that a defective ECT sensor may cause hard cold starting. *Technician B* says that a defective ECT sensor may cause improper emissions. Who is right?
    A. A only
    B. B only
    C. Both A and B
    D. Neither A nor B

169. Which of the following can occur if the oxygen sensor fails?
    A. The vehicle's cruise control will not engage.
    B. The engine coolant temperature will increase.
    C. The system will never go to closed loop.
    D. Excessive oxides of nitrogen will be produced.

Hint

A defective engine coolant temperature (ECT) sensor may cause the following problems:

- *Hard starting*
- *Rich or lean air-fuel ratio*
- *Improper operation of emissions devices*
- *Reduced fuel economy*
- *Driveability problems*
- *Engine stalling*

An ECT sensor can be diagnosed easily with a scan tool or a digital voltmeter. An ECT sensor can be removed and placed in a container of water that is heated. Connect an ohmmeter across the terminals, and place a thermometer in the container. When the water is heated, the sensor should have the specified resistance at any temperature. If the sensor does not have the proper readings, replace it.

Task 4

Use and interpret digital multimeter, digital volt ohm-meter (DMM, DVOM) readings.

170. Digital multimeters (DMM) can be used on all of the following scales EXCEPT:
    A. kiloamps.
    B. DC volts.
    C. AC volts
    D. ohms.

171. While using a DMM to diagnose a throttle position sensor, *Technician A* says that the sensor can be checked with the meter on the resistance scale or the voltage scale. *Technician B* says that if the PCM does not set a throttle position sensor fault code, there is no problem with the sensor. Who is right?
    A. A only
    B. B only
    C. Both A and B
    D. Neither A nor B

Hint

The use of digital multimeters can be very helpful when diagnosing sensors or circuits of the fuel injection system. Multimeters provide several different types of readings: DC volts, AC volts, ohms, amperes, and milliamperes. With an ECT sensor installed in the engine, the sensor may be back-probed to connect the multimeter (set on DC volt scale) to the sensor terminals. The sensor should provide the proper voltage drop at the given coolant temperature.

To diagnose a Throttle Position Sensor (TPS) with the ignition switch on, connect the voltmeter to the 5-volt reference wire to ground. If there is no voltage to the reference wire, the problem is between the electrical connector and the PCM.

Task 5

Read and interpret technical literature (service publications and information).

172. *Technician A* says that most manufacturers provide technical service information on compact discs (CD). *Technician B* says that if available, service manuals from the vehicle's manufacturer have more detailed information regarding procedures and diagnostic information compared to generic manuals. Who is right?
    A. A only
    B. B only
    C. Both A and B
    D. Neither A nor B

173. *Technician A* says that service manuals cannot anticipate all types of situations that may occur to a technician performing work on a vehicle. *Technician B* says OEM service manuals cover several model years on a specific vehicle. Who is right?
    A. A only
    B. B only
    C. Both A and B
    D. Neither A nor B

Hint

Technical service manuals and other published literature, such as wiring diagrams and powertrain and emission control diagnostic books, can be very beneficial. One should not overlook these diagnostic tools. The vehicle manufacturers publish the material that is contained in the books. These books have instructions in the front of the book that are easy to use.

Appropriate service repair procedures are essential for the safe, reliable operation of all motor vehicles as well as the personal safety of the Technician performing the work. The service manuals provide general directions for accomplishing repair work with tested techniques. There are numerous differences in procedures, techniques, tools, and the parts used to service the vehicle.

The skill of the Technician plays a large role in the quality of the service done. A service manual cannot anticipate all types of situations and variations that will occur to the Technician performing the work. Any Technician who deviates from the written procedure provided in the service manual must understand that they are compromising personal safety and risking damage to the vehicle that is being serviced.

Task 6

Test, remove, inspect, clean, service, and repair or replace electrical distribution circuits and connections.

174. *Technician A* says that circuit protection devices open when the current in the circuit falls below a predetermined level. *Technician B* says excessive current results from a decrease in the circuit's resistance. Who is right?
    A. A only
    B. B only
    C. Both A and B
    D. Neither A nor B

175. How are circuits designed that are susceptible to overloads on a routine basis?
    A. Maxi fuses
    B. A fusible link

C. A circuit breaker
D. Fuse

Hint

Most automotive circuits are protected from high current flow that would exceed the capacity of the circuit's conductors and/or loads. Excessive current results from a decrease in the circuit's resistance. When the current flow reaches a predetermined level, most circuit protection devices open and stop current flow in the circuit. This action prevents damage to the circuit and its components.

The most commonly used circuit protection devices are the fuses. A fuse contains a metal strip that melts when the current flowing through it exceeds its rating. The thickness of the metal strip depends on the ampere rating of the fuse.

A vehicle may have one or more fusible links to provide protection for the main power wires before they are divided into smaller circuits.

A circuit that is susceptible to an overload on a routine basis is usually protected by a circuit breaker. Some circuit breakers require manual resetting by pressing a button. Others must be removed from the circuit to reset themselves. Some circuit breakers are self-resetting. This type of circuit breaker uses a bimetallic strip that reacts to excessive current. When an overload or a circuit defect occurs that causes an excessive amount of current draw, the current flowing through the bimetallic strip causes it to heat. As the strip heats, it bends and opens the contacts. Once the contacts are open, the current can no longer flow.

Task 7

Practice recommended precautions when handling static sensitive devices.

176. A powertrain control module (PCM) must be replaced in a vehicle. *Technician A* says a ground strap and conductive mat must be used to ensure that there is no static electricity damage to the vehicle. *Technician B* says to place the programmable read only memory (PROM) chip in your pocket until ready to install it in the PCM. Who is right?
   A. A only
   B. B only
   C. Both A and B
   D. Neither A or B

177. When should a ground strap and conductive mat be used?
   A. When performing battery service
   B. When working on the secondary ignition system
   C. When replacing the ignition coil
   D. When working with electronic control modules

Hint

When working with electronic components, steps must be taken to prevent damage to the components by static discharge. The human body can generate voltages as great as 35,000 volts by simply walking across the carpet on a dry day. As little as 300 volts can severely damage the sensitive electronic components of the vehicle. Due to the sensitive nature of these components, steps must be taken to protect them from static discharge. The Technician must be grounded to safely drain off any static charge, and the electronic parts should be placed on a grounded conductive mat, rather than the vehicle's carpet or upholstery. Any time there is contact with sensitive electronics, the technician must remain grounded. Touching a metal object is not sufficient. If a replacement part is sensitive to static discharge, it will normally be shipped in a static bag. The part should not be removed from the static bag until the Technician is properly grounded and the part is ready to be installed.

Task 8

Diagnose driveability and emissions problems resulting from failures of interdependent systems (engine alarm, air conditioning, and similar systems).

178. The PCM on a truck controls both the engine electronic functions and the transmission shifting. The transmission upshifting is erratic, but all hydraulic pressure tests on the transmission are within specifications, and all input sensors in the transmission are also within specifications. The cause of this problem could be:
   A. a defective throttle position sensor (TPS).
   B. an open circuit in one of the shift solenoids.
   C. a shorted circuit in the converter clutch lockup solenoid.
   D. an open circuit in the voltage supply wire to the shift solenoids.

179. Refer to Figure 1–37. On a truck with air conditioning and throttle body injection, the engine speed slows down when the A/C compressor is engaged. The engine speed is normal with the engine hot or cold and the A/C off. The cause of this problem could be:
   A. a defective idle air control (IAC) motor.
   B. an open circuit in one of the IAC motor lead wires.
   C. a defective ECT sensor.
   D. an open wire from the A/C clutch circuit to the PCM.

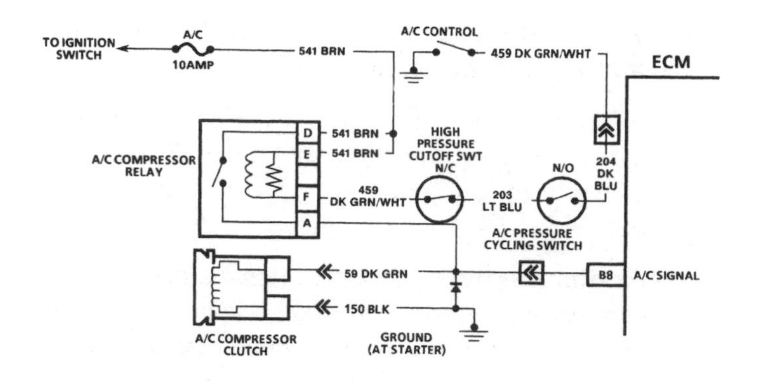

Figure 1–37 A/C compressor clutch circuit. *(Courtesy of General Motors Corporation, Service Technology Group)*

Answers to Questions

1. Answer A is wrong because customer complaints must be listened to closely, but *Technician B* is also right.

   Answer B is wrong because when diagnosing a vehicle problem, always start with the easiest test to perform, but *Technician A* is also right.

   Answer C is right because both *Technician A* and *Technician B* are right.

   Answer D is wrong because both *Technician A* and *Technician B* are right.

2. Answer A is wrong because the customer's complaint must be identified to diagnose the vehicle properly before repairs begin.

   Answer B is wrong because a road test may be necessary when repairs are completed to be sure the customer's complaint is eliminated.

   Answer C is wrong because both *Technician A* and *Technician B* are wrong.

   Answer D is right because both *Technician A* and *Technician B* are wrong.

3. **Answer A is right** because the engine must be steam cleaned before an accurate diagnosis for an oil leak can be given.

   Answer B is wrong because even if the lower portion of the engine is leaking, it can be areas on the top of the engine, rear main seal, or cam galley plugs as well as the oil pan gasket.

   Answer C is wrong because *Technician A* is right and *Technician B* is wrong.

   Answer D is wrong because *Technician A* is right and *Technician B* is wrong.

4. Answer A is wrong because with no visible external signs, the engine may have a leaking head gasket, but *Technician B* is also right.

   Answer B is wrong because the heater core may be leaking, but *Technician A* is also right.

   Answer C is right because both *Technician A* and *Technician B* are right.

   Answer D is wrong because both *Technician A* and *Technician B* are right.

5. Answer A is wrong because engine coolant cannot be tested for these two things.

   Answer B is wrong because lead content and the viscosity are not of concern when discussing coolant.

   Answer C is right because you can test coolant for acidity and freezing point.

   Answer D is wrong because water pump condition cannot be determined by the condition of the coolant.

6. Answer A is wrong because freezing point cannot be determined by the pH.

   Answer B is wrong because the condition of the head gasket cannot be determined without disassembly.

   Answer C is right because coolant acidity can be checked with pH strips.

   Answer D is wrong because there is no way to determine the age of the coolant.

7. Answer A is wrong because piston noise would diminish after the engine was at operating temperature.

   Answer B is right because loose flywheel bolts can cause this thumping noise.

   Answer C is wrong because if the main bearings were in question, the oil pressure would indicate there was a problem.

   Answer D is wrong because again, oil pressure would indicate there was a problem.

8. Answer A is wrong because loose connecting rod bearings cause a clattering noise during acceleration and deceleration.

   Answer B is wrong because worn main bearings cause a thumping noise when the engine is started after sitting for several hours.

   Answer C is right because a collapsed piston causes a clacking noise during acceleration, and this noise is more pronounced when the engine is cold.

   Answer D is wrong because sticking valve lifters cause a clicking noise at idle speed, and this noise occurs more frequently on a cold engine.

9. Answer A is wrong because blue colored smoke from the tailpipe usually indicates an internal oil leak.

   Answer B is wrong because excessive rattling of the exhaust system would only indicate a very severe misfire.

   Answer C is wrong because misfires cannot be determined by exhaust temperature.

   Answer D is right because puffing or wheezing indicates an engine misfire.

10. Answer A is wrong because worn valve guides makes blue smoke.

    Answer B is wrong because worn valve stem seals cause blue smoke in the exhaust.

    Answer C is right because a fouled spark plug will not allow the fuel to burn resulting in black smoke, therefore, the exception.

    Answer D is wrong because worn piston rings will cause blue smoke.

11. Answer A is wrong because late ignition timing causes a low, steady vacuum reading.

    Answer B is wrong because intake manifold leaks will cause a low vacuum reading.

    Answer C is wrong because a restricted exhaust system will cause a low vacuum reading.

    Answer D is right because sticking valve stems and guides causes a vacuum gauge needle fluctuation in the 15 to 20 in. Hg range.

12. Answer A is wrong because the vacuum gauge readings indicate normal engine operation.

    Answer B is wrong because the vacuum gauge readings indicate normal engine operation.

    Answer C is wrong because both *Technician A* and *Technician B* are wrong.

    Answer D is right because both *Technician A* and *Technician B* are wrong.

13. Answer A is wrong because when all the cylinders are contributing equally to the engine power, all the cylinders will provide the specified rpm drop, but *Technician B* is also right.

    Answer B is wrong because often the reason for a cylinder not contributing to the engine power can be traced to the fuel or the ignition system, but *Technician A* is also right.

    Answer C is right because both *Technician A* and *Technician B* are right.

    Answer D is wrong because both *Technician A* and *Technician B* are right.

14. Answer A is wrong because excessive carbon buildup in number 2 cylinder would not cause a low rpm decrease. This problem causes detonation and excessive compression pressure.

    Answer B is right because a defective fuel injector in number 2 cylinder could result in less rpm decrease on this cylinder.

    Answer C is wrong because worn compression rings on number 2 cylinder results in reduced cylinder compression, and the question informs us the compression is satisfactory.

    Answer D is wrong because a defective head gasket on number 2 cylinder results in reduced cylinder compression and possible coolant leaks into the cylinder, and the question informs us the compression is satisfactory.

15. Answer A is wrong because an oversized cylinder will have the same compression as the others.

    Answer B is wrong because oil in the cylinder can only seal the piston rings.

    Answer C is wrong because both *Technician A* and *Technician B* are wrong.

    Answer D is right because both *Technician A* and *Technician B* are wrong.

16. Answer A is wrong because while performing a compression test, a gradual buildup of compression occurs with each stroke is normal and nothing is wrong. A burned exhaust valve or certain cylinder head cracks would cause very low compression in that particular cylinder.

Answer B is wrong because while performing a compression test, a gradual buildup of compression occurs with each stroke is normal and nothing is wrong. A burned exhaust valve or certain cylinder head cracks would cause very low compression in that particular cylinder.

Answer C is wrong because both *Technician A* and *Technician B* are wrong.

Answer D is right because both *Technician A* and *Technician B* are wrong.

17. Answer A is wrong because 5 to 10 percent is too low.

Answer B is wrong because 30 to 40 percent is too high.

Answer C is wrong because 40 to 45 percent is too high.

Answer D is right because 15 to 20 percent is the correct tolerance.

18. Answer A is wrong because a leaking intake valve would not pass air into the crankcase only through the throttle.

Answer B is right because if the piston rings in that cylinder are worn, air would pass into the crankcase and out the PCV.

Answer C is wrong because *Technician A* is wrong and *Technician B* is right.

Answer D is wrong because *Technician A* is wrong and *Technician B* is right.

19. **Answer A is right** because the snap test is used to check the voltage increase for each cylinder under load.

Answer B is wrong because the spark plug gap has to be checked with a feeler gauge.

Answer C is wrong because snap tests are only used on the secondary ignition circuit.

Answer D is wrong because this is not the reason a snap test is performed.

20. Answer A is wrong because a fouled spark plug would not fire at all and on a scope shows a very low firing line.

Answer B is right because the top of the firing line is the voltage required in the spark plugs in the secondary circuit.

Answer C is wrong because *Technician A* is wrong and *Technician B* is right.

Answer D is wrong because *Technician A* is wrong and *Technician B* is right.

21. Answer A is wrong because the head needs to be resurfaced. The feeler gauge measurement is 0.025 in. (0.63 mm), and the standard is .003 inch for every 6 inches, but *Technician B* is also right.

Answer B is wrong because a warped block surface may have caused the warped cylinder head, and so the block surface should be checked, but *Technician A* is also right.

Answer C is right because both *Technician A* and *Technician B* are right.

Answer D is wrong because both *Technician A* and *Technician B* are right.

22. Answer A is wrong because the torque sequence on some head bolts does require the use of an angle gauge or torque meter, but *Technician B* is also right.

Answer B is wrong because the head bolts must be replaced each time the cylinder head is removed on some engines, but *Technician A* is also right.

Answer C is right because both *Technician A* and *Technician B* are right.

Answer D is wrong because both *Technician A* and *Technician B* are right.

23. Answer A is wrong because springs should be inspected for cracks.

Answer B is wrong because the valve springs' spring tension should be inspected.

Answer C is wrong because pitting and nicks are unwanted.

Answer D is right because spring coil gap is the exception and not a standard valve spring measurement.

24. **Answer A is right** because you correct valve stem height by removing material from the valve tip.

Answer B is wrong because shims are used for assembled spring height.

Answer C is wrong because this will not correct the length of the valve stem itself.

Answer D is wrong because this will not correct the length of the valve stem itself.

25. Answer A is wrong because the valve spring is being measured for squareness.

Answer B is wrong because the valve spring must be rotated when checking for squareness.

Answer C is wrong because both *Technician A* and *Technician B* are wrong.

Answer D is right because both *Technician A* and *Technician B* are wrong.

26. Answer A is wrong because the guides do not need to be lubricated.

    Answer B is right because you do clean the valve guide with a bore brush.

    Answer C is wrong because many valve guides cannot be reamed.

    Answer D is wrong because the valve guide does not need to be held in place.

27. **Answer A is right** because valve 1 indicates normal wear.

    Answer B is wrong because valve 1 indicates normal wear.

    Answer C is wrong because valve 2 indicates partial valve rotation.

    Answer D is wrong because valve 3 indicates complete lack of valve rotation.

28. **Answer A is right** because the valve stem should be measured at the top, middle, and near the fillet, but *Technician B* is also right

    Answer B is wrong because engines do not use tapered valve stems.

    Answer C is wrong because *Technician A* is right and *Technician B* is wrong.

    Answer D is wrong because *Technician A* is right and *Technician B* is wrong.

29. Answer A is wrong because a worn-out valve guide will provide an inaccurate valve seat runout measurement, but *Technician B* is also right.

    Answer B is wrong because the valve seat-to-valve face contact area provides a path for heat from the valve head to dissipate, but *Technician A* is also right.

    Answer C is right because both *Technician A* and *Technician B* are right.

    Answer D is wrong because both *Technician A* and *Technician B* are right.

30. Answer A is wrong because the valve is stroked across the stone as the stone is being fed in, but *Technician B* is also right.

    Answer B is wrong because when the last pass is completed the stone should be backed away from the valve, but *Technician A* is also right.

    Answer C is right because both *Technician A* and *Technician B* are right.

    Answer D is wrong because both *Technician A* and *Technician B* are right.

31. Answer A is wrong because the amount of material removed from the valve face and the shim size has nothing to do with each other.

    Answer B is right because valve stem tip height must be correct for proper rocker arm geometry.

    Answer C is wrong because *Technician A* is wrong and *Technician B* is right.

    Answer D is wrong because *Technician A* is wrong and *Technician B* is right.

32. **Answer A is right** because on occasion the valve stem requires resurfacing.

    Answer B is wrong because many engines use an interference fit that comes from having a degree difference between the seat and valve face angles.

    Answer C is wrong because *Technician A* is right and *Technician B* is wrong.

    Answer D is wrong because *Technician A* is right and *Technician B* is wrong.

33. Answer A is wrong because you do dress the stone before cutting any seat.

    Answer B is also wrong because you do remove only enough material to provide a new surface.

    Answer C is right because you do not use transmission fluid to lubricate the stone, only special valve cutting oil.

    Answer D is wrong because you do not apply pressure to the grinding stone.

34. Answer A is wrong because the measurement being performed in Figure 1–8 is valve seat concentricity.

    Answer B is wrong because the measurement being performed in Figure 1–8 is valve seat concentricity.

    Answer C is right because the measurement being performed in Figure 1–8 is valve seat concentricity.

Answer D is wrong because the measurement being performed in Figure 1–8 is valve seat concentricity.

35. **Answer A is right** because it is necessary to use a 45-degree and a 60-degree stone to raise the valve seat contact area.

    Answer B is wrong because it is necessary to use a 45-degree and a 60-degree stone to raise the valve seat contact area.

    Answer C is wrong because it is necessary to use a 45-degree and a 60-degree stone to raise the valve seat contact area.

    Answer D is wrong because it is necessary to use a 45-degree and a 60-degree stone to raise the valve seat contact area.

36. **Answer A is right** because 1/16 in. (1.58 mm) to 3/32 in. (2.38 mm) is an acceptable valve seat width.

    Answer B is wrong because 1/16 in. (1.58 mm) to 3/32 in. (2.38 mm) is an acceptable valve seat width.

    Answer C is wrong because 1/16 in. (1.58 mm) to 3/32 in. (2.38 mm) is an acceptable valve seat width.

    Answer D is wrong because 1/16 in. (1.58 mm) to 3/32 in. (2.38 mm) is an acceptable valve seat width.

37. Answer A is wrong because excessive installed valve spring height does not bottom the lifter plunger unless the installed valve stem height is excessive.

    Answer B is right because a shim should be installed under the valve spring to correct excessive installed valve spring height.

    Answer C is wrong because *Technician A* is wrong, and *Technician B* is right.

    Answer D is wrong because *Technician A* is wrong, and *Technician B* is right.

38. Answer A is wrong because grinding the valve seat will increase the installed valve stem height.

    Answer B is wrong because grinding the valve seat will increase the installed valve stem height.

    Answer C is right because installing a new valve seat decreases the installed valve stem height.

    Answer D is wrong because grinding metal off the valve face will increase the installed valve stem height.

39. **Answer A is right** because the pushrods can be checked for straightness by rolling them on a known flat surface.

    Answer B is wrong because a bent pushrod may be caused by a sticking valve, but this problem is not caused by a sticking valve lifter.

    Answer C is wrong because *Technician A* is right and *Technician B* is wrong.

    Answer D is wrong because *Technician A* is right and *Technician B* is wrong.

40. Answer A is wrong because if the drilled passage through the center of a pushrod is plugged, rocker arm lubrication is insufficient.

    Answer B is wrong because pushrods are designed to rotate when the engine is running.

    Answer C is wrong because replacement pushrods must be the same length as the original pushrods.

    Answer D is right because it is not true that a shorter pushrod may be installed if the installed valve stem height is more than specified.

41. Answer A is wrong because a convex countermachined face is not appropriate for a valve lifter.

    Answer B is wrong because a smooth will not retain oil for lubrication.

    Answer C is right because a centered circular wear pattern is MOST desirable.

    Answer D is wrong because a crosshatched pattern is mostly seen in cylinder walls and would be too coarse for a valve lifter.

42. Answer A is wrong because priming new lifters only requires soaking them in oil.

    Answer B is right because the tool is used to determine lifter leak-down.

    Answer C is wrong because *Technician A* is wrong and *Technician B* is right.

    Answer D is wrong because *Technician A* is wrong and *Technician B* is right.

43. Answer A is wrong because when the valve clearance is checked on a cylinder, that cylinder should be positioned at TDC on the compression stroke.

Answer B is right because some mechanical lifters have removable shim pads, available in various thickness to provide proper valve clearances.

Answer C is wrong because *Technician A* is wrong and *Technician B* is right.

Answer D is wrong because *Technician A* is wrong and *Technician B* is right.

44. Answer A is wrong because both valves in the cylinder should be closed.

    Answer B is right because this statement is not true. A feeler gauge is not used to adjust hydraulic valve lifters.

    Answer C is wrong because the adjusting screw is turned clockwise to decrease the valve clearance.

    Answer D is wrong because excessive clearance may cause a clicking noise at idle speed.

45. Answer A is wrong because the leftover residue from the carburetor cleaner can affect the gasket surface.

    Answer B is wrong because RTV should be placed completely around the bolt hole.

    Answer C is wrong because RTV will set up after 10 to 15 minutes.

    Answer D is right because the RTV bead should be 1/8 in. (3 mm) in width.

46. Answer A is wrong because the timing gear cover seal has a garter spring.

    Answer B is wrong because the seal lip must face toward the oil flow.

    Answer C is wrong because both *Technician A* and *Technician B* are wrong.

    Answer D is right because both *Technician A* and *Technician B* are wrong.

47. Answer A is wrong because a warped cylinder head mounting surface on an engine block may cause valve seat distortion, but *Technician B* is also right.

    Answer B is wrong because a warped cylinder head mounting surface on an engine block may cause coolant and combustion leaks, but *Technician A* is also right.

    Answer C is right because both *Technician A* and *Technician B* are right.

    Answer D is wrong because both *Technician A* and *Technician B* are right.

48. **Answer A is right** because metal fatigue in engine blocks is rare.

    Answer B is wrong because impact damage from a vehicle collision is a more likely cause of a cracked block.

    Answer C is wrong because engine overheating from low coolant is a more likely cause of a cracked block.

    Answer D is wrong because not enough antifreeze in the coolant is a more likely cause of a cracked block in cold climates.

49. Answer A is wrong because taper in the bore causes the ring end gaps to change while the piston moves in the bore, but *Technician B* is also right.

    Answer B is wrong because a cylinder bore gauge can be used to determine taper, but *Technician A* is also right.

    Answer C is right because both *Technician A* and *Technician B* are right.

    Answer D is wrong because both *Technician A* and *Technician B* are right.

50. Answer A is wrong because the skirt never meets the top of the cylinder.

    Answer B is wrong because the cylinder ridge would not affect the piston pin.

    Answer C is wrong because the connecting rod bearings would not be affected.

    Answer D is right because the piston ring lands may be broken when the new rings strike the ring ridge.

51. **Answer A is right** because you place an oiled rag in the cylinder to catch any metal shavings.

    Answer B is wrong because you never use sandpaper on the cylinder walls.

    Answer C is wrong because a step of 0.050 in. (1.2 mm) is excessive and the ridge needs to be removed.

    Answer D is wrong because you only wash the cylinders after the ridges have been removed.

52. Answer A is wrong because the tool in Figure 1–13 is used to remove the ring ridge.

Answer B is wrong because the tool in Figure 1–13 is used to remove the ring ridge.

Answer C is right because the tool in Figure 1–13 is used to remove the ring ridge.

Answer D is wrong because the tool in Figure 1–13 is used to remove the ring ridge.

53. Answer A is wrong because the tool in Figure 1–14 is used to remove and install camshaft bearings.

Answer B is wrong because the tool in Figure 1–14 is used to remove and install camshaft bearings.

Answer C is right because the tool in Figure 1–14 is used to remove and install camshaft bearings.

Answer D is wrong because the tool in Figure 1–14 is used to remove and install camshaft bearings.

54. Answer A is wrong because the oil hole in the bearing may not have been properly aligned with the oil hole in the block, but *Technician B* is also right.

Answer B is wrong because the oil pump components should be inspected and measured for wear, but *Technician A* is also right.

Answer C is right because both *Technician A* and *Technician B* are right.

Answer D is wrong because both *Technician A* and *Technician B* are right.

55. Answer A is wrong because metal burrs on the crankshaft flange may cause excessive wear on the flexplate gear teeth, but *Technician B* is also right.

Answer B is wrong because metal burrs may cause improper torque converter to transmission alignment, but *Technician A* is also right.

Answer C is right because both *Technician A* and *Technician B* are right.

Answer D is wrong because both *Technician A* and *Technician B* are right.

56. Answer A is wrong because if a scratch is deep enough to catch with a fingernail, the crankshaft needs machining.

Answer B is wrong because there is no special tool shown in the figure.

Answer C is right because the Technician is measuring the thickness of the crushed plastigage in the figure.

Answer D is wrong because to determine if the crankshaft has been machined, the bearing insert will indicate that it is oversized.

57. **Answer A is right** because it is nearly impossible to determine mileage on bearing inspection.

Answer B is wrong because you can determine crankshaft misalignment from bearing inspection.

Answer C is wrong because you can determine lack of lubrication during bearing inspection.

Answer D is wrong because you can determine metal-to-metal contact during bearing inspection.

58. **Answer A is right** because the curvature of connecting rod bearings is slightly larger than the curvature of the bearing bores, and this feature is called bearing spread.

Answer B is wrong because when a connecting rod bearing half is installed, its edges should be slightly above the surface of the bearing bore.

Answer C is wrong because *Technician A* is right and *Technician B* is wrong.

Answer D is wrong because *Technician A* is right and *Technician B* is wrong.

59. Answer A is wrong because A and B measures vertical taper.

Answer B is wrong because C and D measures horizontal taper.

Answer C is right because A and C measures out-of-round.

Answer D is wrong because A and D is not a valid measurement.

60. Answer A is wrong because one does NOT use the tool shown in the figure to widen the piston ring grooves.

Answer B is wrong because one does NOT use the tool shown in the figure to deepen the piston ring grooves.

Answer C is right because one uses the tool shown in the figure to remove and replace the piston rings.

Answer D is wrong because one does NOT use the tool shown in the figure to remove carbon from the piston ring grooves.

61. Answer A is wrong because a lack of lubrication is also a cause for scoring and seizure but not the MOST likely.

Answer B is right because overheating of the piston is the MOST likely reason for piston scoring.

Answer C is wrong because preignition is also a cause for scoring and seizure but not the MOST likely.

Answer D is wrong because excessive engine rpm is a potential cause for scoring and seizure but not the MOST likely.

62. Answer A is wrong because you never drive out a pressed-in piston pin with a hammer and drift.

    Answer B is right because removing a pressed-in pin requires a press and special adapters.

    Answer C is wrong because *Technician A* is wrong and *Technician B* is right.

    Answer D is wrong because *Technician A* is wrong and *Technician B* is right.

63. Answer A is wrong because pistons are cam ground, but *Technician B* is also right.

    Answer B is wrong because pistons should be measured across the center of the skirt thrust surface, but *Technician B* is also right.

    Answer C is right because both *Technician A* and *Technician B* are right.

    Answer D is wrong because both *Technician A* and *Technician B* are right.

64. Answer A is wrong because the connecting rod and piston pin must be centered in the piston bore.

    Answer B is wrong because the marks on the piston and connecting rod must be properly aligned before assembling the piston and connecting rod.

    Answer C is wrong because both *Technician A* and *Technician B* are wrong.

    Answer D is right because both *Technician A* and *Technician B* are wrong.

65. **Answer A is right** because an acceptable piston-to-cylinder bore clearance is 0.0015 in. (0.038 mm).

    Answer B is wrong because an acceptable piston-to-cylinder bore clearance is 0.0015 in. (0.038 mm).

    Answer C is wrong because an acceptable piston-to-cylinder bore clearance is 0.0015 in. (0.038 mm).

    Answer D is wrong because an acceptable piston-to-cylinder bore clearance is 0.0015 in. (0.038 mm).

66. **Answer A is right** because the tool must be held squarely against the top of the block.

    Answer B is wrong because crankshaft protecting sleeves should be installed on the connecting rod bolts.

    Answer C is wrong because *Technician A* is right and *Technician B* is wrong.

    Answer D is wrong because *Technician A* is right and *Technician B* is wrong.

67. Answer A is wrong because if the piston ring groove gap has exceeded specifications, you replace the piston, not the piston ring.

    Answer B is right because excessive piston ring clearance can cause a piston ring to break.

    Answer C is wrong because *Technician A* is wrong and *Technician B* is right.

    Answer D is wrong because *Technician A* is wrong and *Technician B* is right.

68. Answer A is wrong because you position the ring at the bottom of ring travel, not the top.

    Answer B is right because the two compression rings are interchangeable on some pistons.

    Answer C is wrong because *Technician A* is wrong and *Technician B* is right.

    Answer D is wrong because *Technician A* is wrong and *Technician B* is right.

69. Answer A is wrong because this is not an approved method for fixing a balancer.

    Answer B is wrong because this will melt the rubber insulator.

    Answer C is right because if the outer ring has slipped you replace the harmonic balancer.

    Answer D is wrong because outer rings are not available separately.

70. **Answer A is right** because excessive flywheel runout may cause a grabbing clutch.

    Answer B is wrong because before a pressure plate is removed it should be marked in relation to the flywheel, and then reinstalled in the same position.

    Answer C is wrong because *Technician A* is right and *Technician B* is wrong.

Answer D is wrong because *Technician A* is right and *Technician B* is wrong.

71. Answer A is wrong because worn camshaft bearings will cause low oil pressure.

    Answer B is wrong because worn crankshaft bearings will cause low oil pressure.

    Answer C is wrong because weak oil pressure regulator spring tension can cause low oil pressure.

    Answer D is right because restricted pushrod oil passages WILL NOT cause low oil pressure.

72. **Answer A is right** because a sheared pin in the oil pump drive gear causes zero oil pressure.

    Answer B is wrong because if the oil pump inlet screen is partially plugged, the engine still has some oil pressure.

    Answer C is wrong because *Technician A* is right and *Technician B* is wrong.

    Answer D is wrong because *Technician A* is right and *Technician B* is wrong.

73. Answer A is wrong because oil pumps are cheap compared to the labor it would take to rebuild them.

    Answer B is right because it is a good practice to replace an oil pump when rebuilding an engine.

    Answer C is wrong because *Technician A* is wrong and *Technician B* is right.

    Answer D is wrong because *Technician A* is wrong and *Technician B* is right.

74. Answer A is wrong because oil pump gear lockup will cause drive failure.

    Answer B is wrong because if the oil pump drive rod fails, the drive fails.

    Answer C is right because excessive engine speed will not cause an oil pump drive failure and therefore the LEAST likely cause.

    Answer D is wrong because a defective oil pressure relief valve stuck will cause excessive stress on the drive shaft and it could fail.

75. Answer A is wrong because you can use a belt tension gauge.

    Answer B is wrong because you can measure the amount of belt deflection.

    Answer C is wrong because you can visually see if the belt is contacting the bottom of the pulley.

    Answer D is right because you do NOT measure the length of the belt compared to a new one.

76. **Answer A is right** because a loose alternator belt may cause a discharged battery.

    Answer B is wrong because a loose alternator belt will not cause a squealing noise while decelerating only on acceleration.

    Answer C is wrong because a loose alternator belt will not cause a damaged alternator bearing.

    Answer D is wrong because a loose alternator belt will not cause engine overheating.

77. **Answer A is right** because radiator pressure caps can be tested with a special adapter.

    Answer B is wrong because pressure testing the coolant system only measures the system integrity. The thermostat must be tested for opening and closing.

    Answer C is wrong because *Technician A* is right and *Technician B* is wrong.

    Answer D is wrong because *Technician A* is right and *Technician B* is wrong.

78. Answer A is wrong because you fill half the system capacity with pure antifreeze, and the remaining amount with water, but *Technician B* is also right.

    Answer B is wrong because you use a 50/50 mixture of water-to-coolant mixture to fill the cooling system, but *Technician A* is also right.

    Answer C is correct because both *Technician A* and *Technician B* are right.

    Answer D is wrong because both *Technician A* and *Technician B* are right.

79. Answer A is wrong because the tool in Figure 1–20 may be used to check cooling system leaks.

    Answer B is wrong because the tool in Figure 1–20 may be used to check the radiator pressure cap.

    Answer C is right because the tool in Figure 1–20 is not used to check coolant specific gravity.

    Answer D is wrong because the tool in Figure 1–20 is used to check for heater core leaks.

80. Answer A is wrong because water pumps do not work intermittently. They either pump or do not pump.

Answer B is right because if the thermostat sometimes sticks, it could cause this condition.

Answer C is wrong because *Technician A* is wrong and *Technician B* is right.

Answer D is wrong because *Technician A* is wrong and *Technician B* is right.

81. Answer A is wrong because a thermostat with a lower temperature rating causes a rich air-fuel ratio on a fuel-injected engine.

    Answer B is wrong because a thermostat that sticks open may cause the powertrain control module (PCM) to enter closed loop too slowly as the engine warms up.

    Answer C is wrong because both *Technician A* and *Technician B* are wrong.

    Answer D is right because both *Technician A* and *Technician B* are wrong.

82. Answer A is wrong because to properly flush the radiator, reverse flushing should be performed to dislodge deposits, but *Technician B* is also right.

    Answer B is wrong because a flush involves draining the coolant and refilling the system with new antifreeze, but *Technician A* is also right.

    Answer C is right because both *Technician A* and *Technician B* are right.

    Answer D is wrong because both *Technician A* and *Technician B* are right.

83. Answer A is wrong because if the cooling system pressure is reduced, the coolant boiling point is decreased.

    Answer B is right because when more antifreeze is added to the coolant, the coolant boiling point is increased.

    Answer C is wrong because *Technician A* is wrong and *Technician B* is right.

    Answer D is wrong because *Technician A* is wrong and *Technician B* is right.

84. Answer A is wrong because a defective water pump bearing may cause a growling noise when the engine is idling, but *Technician B* is also right.

    Answer B is wrong because the water pump bearing may be ruined by coolant leaking past the pump seal, but *Technician A* is also right.

    Answer C is right because both *Technician A* and *Technician B* are right.

    Answer D is wrong because both *Technician A* and *Technician B* are right.

85. Answer A is wrong because the pump impeller becoming deteriorated from corrosion is a rare occurrence.

    Answer B is wrong because a water pump-to-engine block gasket failure is not that common of a failure.

    Answer C is wrong because hub separation is rare.

    Answer D is right because lip seal failure is the most common reason for pump replacement.

86. Answer A is wrong because there is no such radiator as a downdraft.

    Answer B is wrong because there is no such radiator as an updraft.

    Answer C is wrong because this is not the type of radiator shown.

    Answer D is right because a downflow radiator is shown.

87. **Answer A is right** because a stuck-closed vacuum valve would cause a collapsed upper radiator hose after the engine is shut off.

    Answer B is wrong because a faulty vacuum valve would not cause the system to overpressurize.

    Answer C is wrong because if the engine is overheating under a load, it is more likely that the ignition timing is incorrect.

    Answer D is wrong because this would indicate cooling fan problems.

88. Answer A is wrong because if oily streaks are radiating outward from the viscous clutch hub shaft, all or most of the liquid has leaked out of the hub, and replacement is necessary, but *Technician B* is also right.

    Answer B is wrong because all types of fan blades should be inspected for stress fractures, but *Technician A* is also right.

    Answer C is right because both *Technician A* and *Technician B* are right.

    Answer D is wrong because both *Technician A* and *Technician B* are right.

89. Answer A is wrong because a slipping viscous clutch may cause engine overheating.

    Answer B is right because a seized viscous clutch does not cause engine overheating.

Answer C is wrong because a missing fan shroud may cause engine overheating.

Answer D is wrong because a slipping water pump and cooling fan drive belt may cause engine overheating.

90. Answer A is wrong because if there is no fuel leaking from the fuel pump, then you still need to test it for internal leaks.

    Answer B is right because the fuel pump could be leaking internally and the engine oil level should be checked to help indicate a problem.

    Answer C is wrong because *Technician A* is wrong and *Technician B* is right.

    Answer C is wrong because *Technician A* is wrong and *Technician B* is right.

91. **Answer A is right** because one terminal on each fuel injector is supplied with 12 volts when the ignition switch is turned on.

    Answer B is wrong because the other terminal on each injector is connected to the power train control module (PCM), and this module switches the injectors on and off.

    Answer C is wrong because *Technician A* is right and *Technician B* is wrong.

    Answer D is wrong because *Technician A* is right and *Technician B* is wrong.

92. Answer A is wrong because if the distributor pickup coil is defective, the 12V test light would not flutter.

    Answer B is wrong because if the ignition module is defective, the 12V test light would not flutter.

    Answer C is right because a defective ignition coil would cause the 12V test light to flutter, but the spark plug would not fire.

    Answer D is wrong because if a distributor pickup lead was open, the 12V test light would not flutter.

93. **Answer A is right** because with a 12-volt test lamp connected to the tach terminal to ground, if the light flutters while cranking the engine, the primary ignition system is functioning properly.

    Answer B is wrong because a magnetic-type pickup coil can be tested with an ohmmeter.

    Answer C is wrong because *Technician A* is right and *Technician B* is wrong.

    Answer D is wrong because *Technician A* is right and *Technician B* is wrong.

94. Answer A is wrong because secondary ignition circuit DOES NOT start at the ignition switch and follow through to the spark plug.

    Answer B is wrong because the primary circuit is NOT from the coil terminal through the distributor cap to the spark plugs.

    Answer C is wrong because both *Technician A* and B are wrong.

    Answer D is right because both *Technician A* and B are wrong.

95. Answer A is wrong because an increase in point gap decreases, not increases, the cam dwell reading.

    Answer B is right because point dwell must be long enough to allow the magnetic field to build up in the coil.

    Answer C is wrong because *Technician A* is wrong and *Technician B* is right.

    Answer D is wrong because *Technician A* is wrong and *Technician B* is right.

96. **Answer A is right** because the MOST likely result of a cracked distributor cap is cylinder misfiring during engine acceleration.

    Answer B is wrong because the MOST likely result of a cracked distributor cap is cylinder misfiring during engine acceleration.

    Answer C is wrong because the MOST likely result of a cracked distributor cap is cylinder misfiring during engine acceleration.

    Answer D is wrong because the MOST likely result of a cracked distributor cap is cylinder misfiring during engine acceleration.

97. Answer A is wrong because an ignition switch failure would affect all of the plugs, not just one.

    Answer B is wrong because an ignition coil failure would affect all of the plugs, not just one.

    Answer C is wrong because a Hall effect switch failure would affect all of the plugs, not just one.

    Answer D is right because a failure of one or more of the spark plug secondary wires is the cause.

98. Answer A is wrong because there are no scan tools that specifically check the coil, unless it is included in the OBD II diagnostic parameters.

    Answer B is right because you use the DMM to make ohmmeter tests on the coil windings.

    Answer C is wrong because you can use a scope to test normal and maximum coil voltage, but this equipment is not specifically used to test the coil windings.

    Answer D is wrong because a test light cannot be used to test coil windings accurately.

99. **Answer A is right** because this high current flow may be caused by a shorted primary coil winding.

    Answer B is wrong because this high current flow does not flow through the pickup coil, and therefore, this component will not be damaged. The high primary current may damage the ignition module.

    Answer C is wrong because *Technician B* is wrong and *Technician A* is right.

    Answer D is wrong because *Technician B* is wrong and *Technician A* is right.

100. Answer A is wrong because a cracked distributor cap is sensitive to moisture.

    Answer B is wrong because deteriorated ignition wires are sensitive to moisture.

    Answer C is wrong because a cracked coil tower is sensitive to moisture.

    Answer D is right because a loose distributor mounting has no effect on moisture in the system. Therefore it is the exception.

101. Answer A is wrong because timing lights cannot be used as trouble lights.

    Answer B is right because this timing light can check spark advance.

    Answer C is wrong because timing lights cannot be used as oscilloscopes.

    Answer D is wrong because all timing checks occur on the secondary side.

102. Answer A is wrong because the underhood emission label has specific instructions for setting the timing, but *Technician B* is also right.

    Answer B is wrong because on non-feedback style carburetors, the engine idle speed must be verified before the timing can be adjusted, but *Technician A* is also right.

    Answer C is right because both *Technician A* and *Technician B* are right.

    Answer D is wrong because both *Technician A* and *Technician B* are right.

103. Answer A is wrong because ohmmeter number 2 is connected to test the pickup for an open or shorted condition.

    Answer B is wrong because a low resistance, not infinity, indicates a short in the pickup coil winding.

    Answer C is wrong because ohmmeter number 2 is connected to test the pickup for an open or shorted condition.

    Answer D is right because resistance infinity on the two pickup coil leads indicates an open circuit in the pickup coil.

104. Answer A is wrong because a Hall effect distributor pickup produces a digital voltage signal.

    Answer B is right because a grounded pickup coil may cause a no-start condition.

    Answer C is wrong because *Technician A* is wrong and *Technician B* is right.

    Answer D is wrong because *Technician A* is wrong and *Technician B* is right.

105. Answer A is wrong because the tester has no capability to analyze the primary circuit.

    Answer B is wrong because the tester is not a voltmeter.

    Answer C is wrong because the tester has no capability to analyze the secondary circuit.

    Answer D is right because the ignition module tester checks the ability of the module to turn on and off.

106. Answer A is wrong because ignition modules mounted in the distributor require a heat dissipating grease on the module mounting surface.

    Answer B is wrong because some ignition modules mounted externally from the distributor must be grounded where they are mounted.

    Answer C is wrong because an ignition module may be damaged by a shorted primary coil winding, which results in high current flow through the module.

Answer D is right because ignition modules are NOT protected from voltage spikes by an external zener diode.

107. Answer A is wrong because the ECM sends voltage signals to the governor module and this module directly controls the governor motor.

Answer B is wrong because the governor motor is NOT connected directly to the throttle shafts. The governor motor only provides an override function on the throttle shafts to limit engine rpm.

Answer C is right because the ECM sends voltage signals to the governor module to provide the proper governor motor control.

Answer D is wrong because the ECM uses engine rpm, throttle position sensor (TPS), and manifold absolute pressure (MAP) sensor inputs to provide proper governor operation.

108. **Answer A is right** because a defective governor solenoid valve that is sticking closed causes excessive vacuum applied to the governor diaphragm and this results in a lower than specified governed engine rpm.

Answer B is wrong because the lower than specified governed engine rpm is caused by excessive vacuum at the governor diaphragm, and a plugged vacuum passage in the vacuum regulator causes zero vacuum at the governor diaphragm.

Answer C is wrong because the lower than specified governed engine rpm is caused by excessive vacuum at the governor diaphragm, and an inoperative electric vacuum pump causes zero vacuum at the governor diaphragm.

Answer D is wrong because the lower than specified governed engine rpm is caused by excessive vacuum at the governor diaphragm, and a disconnected vacuum hose at the governor assembly causes zero vacuum at the governor diaphragm.

109. **Answer A is right** because a defective TPS may cause a hesitation during acceleration.

Answer B is wrong because a burned exhaust valve causes rough idle operation, and the question states that idle operation is normal.

Answer C is wrong because a partially restricted injector causes rough idle operation, and the question states that idle operation is normal.

Answer D is wrong because excessive fuel pressure causes a rich air-fuel ratio, and this problem does not result in a hesitation during acceleration.

110. Answer A is wrong because a restricted fuel filter may cause low fuel pressure and a lean air-fuel ratio.

Answer B is right because a restricted fuel return line causes excessive fuel pressure, a rich air-fuel ratio, and excessive fuel consumption.

Answer C is wrong because a weak pressure regulator spring causes low fuel pressure and a lean air-fuel ratio.

Answer D is wrong because a defective fuel pump relay usually causes an inoperative fuel pump, and this causes the engine to stop running.

111. Answer A is wrong because low fuel pressure may cause a hesitation during acceleration.

Answer B is wrong because low fuel pressure may cause engine cutout during high speed and heavy load.

Answer C is right because the low fuel pressure would NOT cause rough idle operation.

Answer D is wrong because low fuel pressure may cause surging at medium and high speeds.

112. **Answer A is right** because some engines have an electric fuel pump in the fuel tank(s), and another electric fuel pump mounted on the frame rail.

Answer B is wrong because a fuel pump may have the specified pressure but inadequate flow because of a restricted fuel line or filter or low input voltage to the pump.

Answer C is wrong because *Technician A* is right and *Technician B* is wrong.

Answer D is wrong because *Technician A* is right and *Technician B* is wrong.

113. Answer A is wrong because there is no device that limits the amount of air into the engine.

Answer B is wrong because the PCV system does not need blowby vapor to operate.

Answer C is right because the automatic choke lowers the pressure in the carburetor venturi and richens the mixture, thus providing the best engine performance in cold weather.

Answer D is wrong because providing raw fuel is not the function of the choke.

114. Answer A is wrong because a misadjusted throttle linkage does not cause stalling on deceleration.

Answer B is right because a misadjusted throttle linkage may cause a loss of engine power at high speeds and heavy loads, because the linkage is not pulling the throttles open far enough in relation to accelerator pedal travel.

Answer C is wrong because a misadjusted throttle linkage does not cause rough idle operation.

Answer D is wrong because a misadjusted throttle linkage does not cause engine surging a low speeds.

115. Answer A is wrong because a lower than specified accelerator pump height does not cause stalling at idle speed.

Answer B is right because a lower than specified accelerator pump height may cause a hesitation during acceleration.

Answer C is wrong because *Technician A* is wrong and *Technician B* is right.

Answer D is wrong because *Technician A* is wrong and *Technician B* is right.

116. **Answer A is right** because a wider than specified vacuum break adjustment may cause a hesitation only when the engine coolant is cold.

Answer B is wrong because a rich choke spring adjustment causes a rich air-fuel ratio during engine warmup.

Answer C is wrong because a wider than specified fast idle cam adjustment does not cause a hesitation when the engine is cold.

Answer D is wrong because if the accelerator pump height is less than specified, the engine has a hesitation when the engine is cold or hot.

117. Answer A is wrong because the coating has nothing to do with engine backfiring.

Answer B is wrong because the coating does not affect the airflow.

Answer C is right because the coating prevents carbon buildup on the throttle plate.

Answer D is wrong because the special coating cannot increase the incoming airflow.

118. Answer A is wrong because the PCV valve supply hose is connected below the throttles to the throttle body base plate or intake manifold vacuum.

Answer B is wrong because you cannot reuse the throttle body gasket.

Answer C is wrong because both *Technician A* and *Technician B* are wrong.

Answer D is right because both *Technician A* and *Technician B* are wrong.

119. Answer A is wrong because it is a good practice to replace all the intake manifold gaskets at the same time.

Answer B is wrong because the replacement gasket should be made of the same material as the original.

Answer C is wrong because all the gaskets in the intake system do come together in a kit.

Answer D is right because cracked intake manifold gaskets cannot be repaired with silicone.

120. **Answer A is right** because a leaking intake manifold gasket may cause a cylinder misfire at idle that disappears when the engine speed is increased.

Answer B is wrong because a defective spark plug causes cylinder misfiring at idle and also under other operating conditions such as acceleration.

Answer C is wrong because *Technician A* is right and *Technician B* is wrong.

Answer D is wrong because *Technician A* is right and *Technician B* is wrong.

121. Answer A is wrong because engine backfiring through the carburetor does not cause carbon accumulation in the choke housing.

Answer B is wrong because engine overheating does not cause carbon accumulation in the choke housing.

Answer C is right because a burned-out heat pipe in the exhaust manifold causes carbon accumulation in the choke housing.

Answer D is wrong because excessive carbon deposits in the exhaust manifold do not cause carbon accumulation in the choke housing.

122. Answer A is wrong because the choke housing may be warped, causing a vacuum leak, but *Technician B* is also right.

Answer B is wrong because the heat pipe may be restricted, but *Technician A* is also right.

Answer C is right because both *Technician A* and *Technician B* are right.

Answer D is wrong because both *Technician A* and *Technician B* are right.

123. Answer A is wrong because the engine coolant temperature (ECT) sensor resistance increases and the injector pulse width also increases.

Answer B is right because the ECT sensor resistance increases and the injector pulse width also increases.

Answer C is wrong because the ECT sensor voltage drop increases.

Answer D is wrong because the PCM should be operating in open loop.

124. Answer A is wrong because with less than the specified resistance in the ECT sensor, the air-fuel ratio is lean and the engine should provide good fuel economy.

Answer B is right because the engine may be hard to start when cold because with less than the specified resistance in the ECT sensor, the air-fuel ratio is lean.

Answer C is wrong because *Technician A* is wrong and *Technician B* is right.

Answer D is wrong because *Technician A* is wrong and *Technician B* is right.

125. Answer A is wrong because the PCM does not use antilock information to determine if the vehicle is decelerating.

Answer B is right because the throttle position sensor determines deceleration mode.

Answer C is wrong because the purge solenoid can give no indication of deceleration mode.

Answer D is wrong because the camshaft position sensor will not indicate deceleration mode.

126. Answer A is wrong because carbon accumulation around the throttles may cause stalling during deceleration.

Answer B is wrong because carbon accumulation in the idle air control (IAC) motor may cause stalling during deceleration.

Answer C is right because low fuel pressure causes surging and cutting out at high speed and under heavy load.

Answer D is wrong because a badly worn IAC motor pintle may cause stalling during deceleration.

127. Answer A is wrong because fuel injected vehicles need air filter replacement at the manufacturer's specified intervals.

Answer B is right because a shop light can be used to determine how dirty the element is.

Answer C is wrong because *Technician A* is wrong and *Technician B* is right.

Answer D is wrong because *Technician A* is wrong and *Technician B* is right.

128. Answer A is wrong because higher than specified fuel pressure causes a rich air-fuel ratio and this does not result in a hesitation during acceleration.

Answer B is right because a leak in the air intake system between the MAF sensor and the intake manifold causes a lean air-fuel ratio and a hesitation during acceleration.

Answer C is wrong because a restricted fuel return line causes a rich air-fuel ratio and this does not result in a hesitation during acceleration.

Answer D is wrong because a vacuum line disconnected from the fuel pressure regulator causes a rich air-fuel ratio and this does not result in a hesitation during acceleration.

129. Answer A is wrong because the optimal air-fuel-ratio is 14.7:1.

Answer B is right because for every gallon of gasoline that the engine burns, the engine uses 9,000 gallons of air.

Answer C is wrong because *Technician A* is wrong and *Technician B* is right.

Answer D is wrong because *Technician A* is wrong and *Technician B* is right.

130. Answer A is wrong because the fuel pressure regulator senses engine vacuum, but *Technician B* is also right.

Answer B is wrong because the fuel pressure regulator controls the amount of fuel that is returned to the tank, but *Technician A* is also right.

Answer C is right because both *Technician A* and *Technician B* are right.

Answer D is wrong because both *Technician A* and *Technician B* are right.

131. Answer A is wrong because a decrease in engine vacuum indicates an increase in load, which requires more fuel pressure.

Answer B is wrong because the valve closes with an increase in engine load.

Answer C is right because when the engine vacuum decreases, the fuel pressure regulator spring pushes the valve closed to allow fuel pressure to build in the rail.

Answer D is wrong because the fuel pump is always operating, no matter what the load is.

132. Answer A is wrong because stainless steel manifolds are usually more costly.

Answer B is wrong because they are lighter in weight.

Answer C is wrong because they do not produce higher emissions.

Answer D is right because they are lighter in weight.

133. **Answer A is right** because the solution of baking soda and water helps to prevent future corrosion.

Answer B is wrong because you never use anything but water or distilled water to fill the cells of the battery.

Answer C is wrong because *Technician A* is right and *Technician B* is wrong.

Answer D is wrong because *Technician A* is right and *Technician B* is wrong.

134. **Answer A is right** because you use a wire brush to remove corrosion on the battery terminals.

Answer B is wrong because the post should be cleaned using a battery terminal brush.

Answer C is wrong because *Technician A* is right and *Technician B* is wrong.

Answer D is wrong because *Technician A* is right and *Technician B* is wrong.

135. **Answer A is right** because a battery can be charged at any rate as long as the electrolyte does not boil and the temperature does not exceed 120°F (50°C).

Answer B is wrong because fast charging is NOT as complete as charging a battery at a slower rate.

Answer C is wrong because *Technician A* is right and *Technician B* is wrong.

Answer D is wrong because *Technician A* is right and *Technician B* is wrong.

136. **Answer A is right** because the battery is not completely charged until the specific gravity reaches 1,265.

Answer B is wrong because explosive hydrogen gas is produced in the battery cells during the charging process.

Answer C is wrong because oxygen gas is produced in the battery cells during the charging process.

Answer D is wrong because you do not charge a maintenance-free battery if the hydrometer in the battery top indicates yellow.

137. Answer A is wrong because there is no alternator current test to perform.

Answer B is wrong because the amp pickup lead would have to be placed on the starter cable.

Answer C is right because the battery load test is shown in the figure.

Answer D is wrong because the figure does not illustrate a ground circuit voltage drop test.

138. Answer A is wrong because the load should be half that of the cold cranking amps.

Answer B is wrong because the load should be half that of the cold cranking amps.

Answer C is right because the load should be half of the cold cranking amps and 262 is half of the cold cranking amps.

Answer D is wrong because the load should be half that of the cold cranking amps.

139. Answer A is wrong because the negative booster cable should be connected to a chassis ground connection on the vehicle being boosted.

Answer B is wrong because the positive booster cable should be connected first.

Answer C is wrong because both *Technician A* and *Technician B* are wrong.

Answer D is right because both *Technician A* and *Technician B* are wrong.

140. Answer A is wrong because you remove the negative booster cable on the vehicle being boosted first.

Answer B is right because all the accessories must be turned off to prevent electrical damage.

Answer C is wrong because *Technician A* is wrong and *Technician B* is right.

Answer D is wrong because *Technician A* is wrong and *Technician B* is right.

141. Answer A is wrong because the terminal should be removed with a puller.

Answer B is wrong because you always remove the battery cable first.

Answer C is right because you never remove the positive battery cable before the negative cable.

Answer D is wrong because a baking soda and water solution can be used to clean the battery.

142. Answer A is wrong because battery terminals can be treated with petroleum jelly to help prevent corrosion, but *Technician B* is also right.

Answer B is wrong because protective pads can be placed on the terminals to prevent corrosion, but *Technician A* is also right.

Answer C is right because both *Technician A* and *Technician B* are right.

Answer D is wrong because both *Technician A* and *Technician B* are right.

143. Answer A is wrong because the starter solenoid current draw is not being tested.

Answer B is right because the starter motor current draw test is shown.

Answer C is wrong because the amp pickup lead is on the starter cable.

Answer D is wrong because the resistance in the circuit is not being checked.

144. **Answer A is right** because a starting and charging system tester can be used to test free spinning current draw tests.

Answer B is wrong because you do not have to remove the solenoid to do the free spinning current draw test.

Answer C is wrong because *Technician A* is right and *Technician B* is wrong.

Answer D is wrong because *Technician A* is right and *Technician B* is wrong.

145. Answer A is wrong because pinion gear clearance is checked using a feeler gauge.

Answer B is wrong because the starter would have to be installed in the vehicle to perform this test.

Answer C is wrong because the battery is not shown in the figure.

Answer D is right because the starter free speed test is shown in the figure.

146. Answer A is wrong because high resistance in the starter motor circuit can cause low cranking speed, but *Technician B* is also right.

Answer B is wrong because internal engine problems can be the cause of the starter motor not functioning properly, but *Technician A* is also right.

Answer C is right because both *Technician A* and *Technician B* are right.

Answer D is wrong because both *Technician A* and *Technician B* are right.

147. Answer A is wrong because high resistance in the starting motor ground does not cause failure of the starting motor to disengage when the engine starts.

Answer B is right because misalignment of the starting motor in relation to the flywheel housing may cause failure of the starting motor to disengage when the engine starts.

Answer C is wrong because high resistance between the battery positive terminal and the starter solenoid does not cause failure of the starting motor to disengage when the engine starts.

Answer D is wrong because a shorted condition in the starting motor armature does not cause failure of the starting motor to disengage when the engine starts.

148. Answer A is wrong because if the carbon buildup on the EGR valve cannot be cleaned with solvent, the valve should be replaced.

Answer B is right because you do use fuel injection cleaner to clean the passages in the EGR valve.

Answer C is wrong because *Technician A* is wrong and *Technician B* is right.

Answer D is wrong because *Technician A* is wrong and *Technician B* is right.

149. Answer A is wrong because some EGR systems operate on ported venturi vacuum systems.

Answer B is wrong because EGR operates on ported vacuum.

Answer C is wrong because EGR valves can be mounted on the intake.

Answer D is right because they DO NOT operate at wide-open throttle.

150. Answer A is wrong because vacuum hoses usually do cause failures with the EGR system.

Answer B is wrong because the plastic used for the EPT and EVR solenoids is NOT very durable and will not crack around the vacuum hose connection.

Answer C is wrong because both *Technician A* and *Technician B* are wrong.

Answer D is right because both *Technician A* and *Technician B* are wrong.

151. Answer A is wrong because the coolant temperature override switch prevents EGR valve operation until the engine warms up to operating temperature, but *Technician B* is also right.

Answer B is wrong because the EVR solenoid converts the electrical signals from the PCM into a mechanical action, which directs vacuum to the EGR valve, but *Technician A* is also right.

Answer C is right because both *Technician A* and *Technician B* are right.

Answer D is wrong because both *Technician A* and *Technician B* are right.

152. Answer A is wrong because the hoses indicated are outlet hoses.

Answer B is right because either hose should discharge airflow out the end of the hose.

Answer C is wrong because the air pump is always positive displacement no matter what the rpm is.

Answer D is wrong because there is no air pump clutch.

153. **Answer A is right** because the air pump drive belt is the first area you check.

Answer B is wrong because the air injection hoses are not the most logical component to be inspected first.

Answer C is wrong because this is not the easiest thing to perform first.

Answer D is wrong because it is easier to inspect the belt first.

154. **Answer A is right** because new air pumps will come with a new diverter valve.

Answer B is wrong because new air pumps come with a new pulley.

Answer C is wrong because *Technician A* is right and *Technician B* is wrong.

Answer D is wrong because *Technician A* is right and *Technician B* is wrong.

155. Answer A is wrong because you cannot rebuild the air pump.

Answer B is right because you should never pry on the housing while tightening the drive belt.

Answer C is wrong because *Technician A* is wrong and *Technician B* is right.

Answer D is wrong because *Technician A* is wrong and *Technician B* is right.

156. Answer A is wrong because if there were no vacuum present, the engine would be off or the hose clogged or restricted.

Answer B is wrong because you would not feel positive air pressure at this hose, only vacuum or nothing.

Answer C is wrong because you would never feel the presence of exhaust gas at this hose.

Answer D is right because after disconnecting the vacuum supply hose to the by-pass valve, vacuum should be present at the hose end.

157. Answer A is wrong because the gulp valve is used to prevent engine backfire, but *Technician B* is also right.

Answer B is wrong because the gulp valve has largely been replaced by the diverter valve, but *Technician A* is also right.

Answer C is right because both *Technician A* and *Technician B* are right.

Answer D is wrong because both *Technician A* and *Technician B* are right.

158. Answer A is wrong because you remove the exhaust manifold from the vehicle, after you remove the air injection manifold.

Answer B is wrong because you do not have to remove the air pump from the vehicle to remove the air injection manifold.

Answer C is wrong because you do not use high temperature silicone in place of the gasket.

Answer D is right because you do apply penetrating oil to the nuts before removal.

159. **Answer A is right** because preformed air injection hose is the correct material for the repair.

Answer B is wrong because heater hose is not suitable for air injection hose.

Answer C is wrong because fuel supply hose is too small in diameter.

Answer D is wrong because heater hose is not the correct diameter.

160. Answer A is wrong because regular oil change intervals prevents sludge formation in the PCV system, but *Technician B* is also right.

Answer B is wrong because you do check the air filter for oil deposits and oil puddling, but *Technician A* is also right.

Answer C is right because both *Technician A* and *Technician B* are right.

Answer D is wrong because both *Technician A* and *Technician B* are right.

161. Answer A is wrong because oil consumption is not a direct indication of PCV faults.

Answer B is wrong because normal HC and CO emission readings would indicate that the PCV is operating normally.

Answer C is wrong because normal fuel consumption would indicate that the PCV is operating normally.

Answer D is right because excessive crankcase vapors escaping from the oil filler cap opening may indicate a PCV failure.

162. Answer A is wrong because the converter outlet should be hotter than the inlet if the converter is operating normally.

Answer B is right because the heat shields should be reinstalled after the catalytic converter is replaced.

Answer C is wrong because *Technician A* is wrong and *Technician B* is right.

Answer D is wrong because *Technician A* is wrong and *Technician B* is right.

163. Answer A is wrong because all three-way catalytic (TWC) converters are NOT interchangeable.

Answer B is wrong because that an overly rich air-fuel mixture will overheat the catalytic converter.

Answer C is wrong because both *Technician A* and *Technician B* are wrong.

Answer D is right because both *Technician A* and *Technician B* are wrong.

164. Answer A is wrong because a low voltage at the fuel pump may cause a severe surging problem only at 55 mph, but *Technician B* is also right.

Answer B is wrong because the inertia switch with high resistance may cause a severe surging problem only at 55 mph, but *Technician A* is also right.

Answer C is right because both *Technician A* and *Technician B* are right.

Answer D is wrong because both *Technician A* and *Technician B* are right.

165. Answer A is wrong because a defective oxygen (O_2) sensor may cause a rich air-fuel ratio and excessive HC and CO emissions.

Answer B is right because an engine coolant temperature (ECT) sensor with less than the specified resistance causes a lean air-fuel ratio and this does NOT cause high HC and CO emissions.

Answer C is wrong because a defective fuel pressure regulator causing excessive fuel pressure may cause a rich air-fuel ratio and excessive HC and CO emissions.

Answer D is wrong because a defective mass airflow (MAF) sensor may cause a rich air-fuel ratio and excessive HC and CO emissions.

166. **Answer A is right** because you check the ECT sensor signal to the PCM when diagnosing an EVAP system with a scan tool. If the PCM never provides the on command to the EVAP solenoid, the ECT sensor may be defective.

Answer B is wrong because the vacuum hoses from the intake to the EVAP canister will have no effect on the PCM command to the EVAP solenoid.

Answer C is wrong because *Technician A* is right and *Technician B* is wrong.

Answer D is wrong because *Technician A* is right and *Technician B* is wrong.

167. Answer A is wrong because you do warm the engine to normal operating temperature when using a scan tool.

Answer B is wrong because you do connect the power adapter to the cigar lighter.

Answer C is right because you DO NOT disconnect the negative battery cable before connecting the DLC connector.

Answer D is wrong because you do enter the model year, engine size, and vehicle model.

168. Answer A is wrong because a defective ECT sensor may cause hard cold starting, but *Technician B* is also right.

Answer B is wrong because a defective ECT sensor may cause improper emissions, but *Technician A* is also right.

Answer C is right because both *Technician A* and *Technician B* are right.

Answer D is wrong because both *Technician A* and *Technician B* are right.

169. Answer A is wrong because on most OBD systems the oxygen sensor failure has no effect on cruise control operation.

Answer B is wrong because on most OBD systems, oxygen sensor failure has no effect on engine temperature.

Answer C is right because oxygen sensor failure will prevent the OBD system from going to closed loop.

Answer D is wrong because oxygen sensor failure has no effect on the production of oxides of nitrogen (NOx).

170. **Answer A is right** because kiloamps is not used on a DMM.

Answer B is wrong because DC volts are found on a DMM.

Answer C is wrong because AC volts are found on a DMM.

Answer D is wrong because ohms are found on a DMM.

171. **Answer A is right** because the sensor can be checked with the meter on the resistance scale or the voltage scale.

Answer B is wrong because if the PCM does not set a throttle position sensor fault code, there may still be a problem with the sensor.

Answer C is wrong because *Technician A* is right and *Technician B* is wrong.

Answer D is wrong because *Technician A* is right and *Technician B* is wrong.

172. Answer A is wrong because most manufacturers provide technical service information on compact discs (CD), but *Technician B* is also right.

Answer B is wrong because if available, service manuals from the vehicle's manufacturer have more detailed information regarding procedures and diagnostic information, but *Technician A* is also right.

Answer C is right because both *Technician A* and *Technician B* are right.

Answer D is wrong because both *Technician A* and *Technician B* are right.

173. **Answer A is right** because service manuals cannot anticipate all types of situations that may occur to a technician performing work on a vehicle.

Answer B is wrong because OEM service manuals only cover one model year on a specific vehicle.

Answer C is wrong because *Technician A* is right and *Technician B* is right.

Answer D is wrong because *Technician A* is right and *Technician B* is right.

174. Answer A is wrong because circuit protection devices open when the current in the circuit falls above a predetermined level.

Answer B is right because excessive current results from a decrease in the circuit's resistance.

Answer C is wrong because *Technician A* is wrong and *Technician B* is right.

Answer D is wrong because *Technician A* is wrong and *Technician B* is right.

175. Answer A is wrong because maxifuses are for one time only.

Answer B is wrong because fusible links are good one time only.

Answer C is right because a circuit breaker can be used many times.

Answer D is wrong because a fuse is good one time only.

176. **Answer A is right** because when replacing a ECM/PCM, a ground strap and conductive mat must be used to ensure that there is no static electricity damage to the electronic components on the vehicle.

Answer B is wrong because the PROM chip should be left in the static protective envelope until you are ready to install it in the PCM. This chip should never be placed in your pocket.

Answer C is wrong because *Technician A* is right and *Technician B* is wrong.

Answer D is wrong because *Technician A* is right and *Technician B* is wrong.

177. Answer A is wrong because a ground strap and conductive mat is not used to perform battery service.

Answer B is wrong because you do not use a ground strap and conductive mat when working on the secondary ignition system.

Answer C is wrong because when replacing the ignition coil, you do not use a ground strap and conductive bus.

Answer D is right because when working with electronic control modules, you do use a grounding strap and conductive material.

178. **Answer A is right** because a defective throttle position sensor (TPS) may cause erratic transmission upshifting.

Answer B is wrong because an open circuit in one of the shift solenoid causes the transmission to completely miss one or two shifts.

Answer C is wrong because an open circuit in the converter clutch lockup solenoid only affects converter clutch lockup.

Answer D is wrong because an open circuit in the voltage supply wire to the shift solenoids prevents all transmission shifts. Under this condition the transmission will operate in reverse and one forward gear, which is usually second gear in many transmissions.

179. Answer A is wrong because a defective idle air control (IAC) motor results in improper idle speed under all conditions.

Answer B is wrong because an open circuit in one of the IAC motor lead wires causes improper idle speed under all conditions.

Answer C is wrong because a defective ECT sensor may cause improper idle speed when the engine is cold.

Answer D is right because an open wire from the AC clutch circuit to the PCM causes the PCM not to increase the idle speed when the clutch is engaged, and this results in low idle speed.

2 Diesel Engines

Pretest

The purpose of this pretest is to determine the amount of review you may require before taking the ASE Medium/Heavy Truck Diesel Engine Test. If you answer all the pretest questions correctly, complete the questions and study the information in this chapter to prepare for this test.

If two or more of your answers to the pretest questions are incorrect, study the required information in *Today's Technician Medium/Heavy-Duty Truck Diesel Engines* published by Delmar Publishers, plus a study of the questions and information in this chapter.

The pretest answers are located at the end of the pretest; these answers also are in the answer sheets supplied with this book.

1. A diesel engine is idling lower than specifications. *Technician A* says this could be caused by worn governor parts. *Technician B* says that a clogged fuel filter could be the cause. Who is right?
   A. A only
   B. B only
   C. Both A and B
   D. Neither A nor B

2. *Technician A* says excessive black smoke in a diesel engine may be caused by a partially restricted air intake. *Technician B* says this problem may be caused by a defective low-pressure pump with low fuel pressure. Who is right?
   A. A only
   B. B only
   C. Both A and B
   D. Neither A nor B

3. Which one of the following diesel engine types uses an O-ring to prevent coolant from leaking into the crankcase?
   A. Dry sleeve
   B. Wet sleeve
   C. Two cycle
   D. Direct injection

4. The valves of a diesel engine must be replaced if:
   A. there is pitting on the valve face.
   B. the engine is undergoing an overhaul.
   C. the seats are to be resurfaced.
   D. they show signs of cracking or warpage.

5. The MOST likely cause of bearing failure is:
   A. overloading.
   B. corrosion.
   C. contaminated oil.
   D. misalignment.

6. The LEAST likely cause for a diesel engine misfire is:
   A. a leak in the exhaust manifold.
   B. bad injector.
   C. valve clearance out of specification.
   D. cracked wet cylinder sleeve.

7. A diesel engine will not crank when the key is moved to the start position. The LEAST likely cause of this problem is :
   A. a high-resistance battery connection.
   B. a defective starter switch.
   C. a discharged battery.
   D. contaminated fuel.

8. To check a cylinder head for cracks after it is removed from the engine, all of the following may be necessary EXCEPT:
   A. use a putty knife to remove excess gasket material from the head and inspect for visible cracks.
   B. fill the coolant passages with coolant and pressure test the head.
   C. use magnetic or dye crack detection to find small cracks that are not otherwise visible.
   D. remove carbon buildup from the combustion chambers.

9. Excessive surges in the coolant level could be caused by all of the following EXCEPT:
   A. a loose fan belt.
   B. a faulty fuel injector.
   C. a restricted radiator.
   D. a blown head gasket.

10. In Figure 2–1 an acceptable minimum valve margin is:
    A. 0.010 in. (0.254 mm).
    B. 0.020 in. (0.508 mm).
    C. 0.025 in. (0.635 mm).
    D. 0.031 in. (0.787 mm).

11. The oil pump measurement shown in Figure 2–2 is:
    A. rotor tip clearance.
    B. rotor end play.
    C. rotor-to-pump body clearance.
    D. drive gear backlash.

12. The average compression ratio in a heavy-duty truck diesel engine is:
    A. 10.25:1
    B. 14:1
    C. 18:1
    D. 24:1

13. While discussing the air-operated cooling fan clutch in Figure 2–3, *Technician A* says if air pressure is supplied to the fan clutch, the clutch is engaged. *Technician B* says the air pressure is supplied through an electrically controlled air solenoid to the fan clutch. Who is right?
    A. A only
    B. B only
    C. Both A and B
    D. Neither A nor B

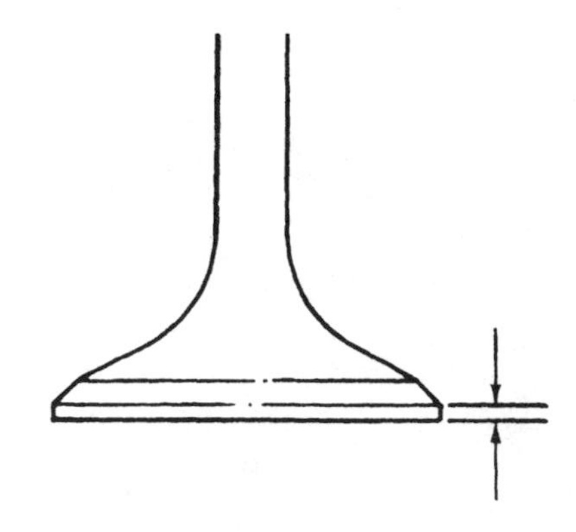

Figure 2–1 Valve measurement. *(Courtesy of Ford Motor Company)*

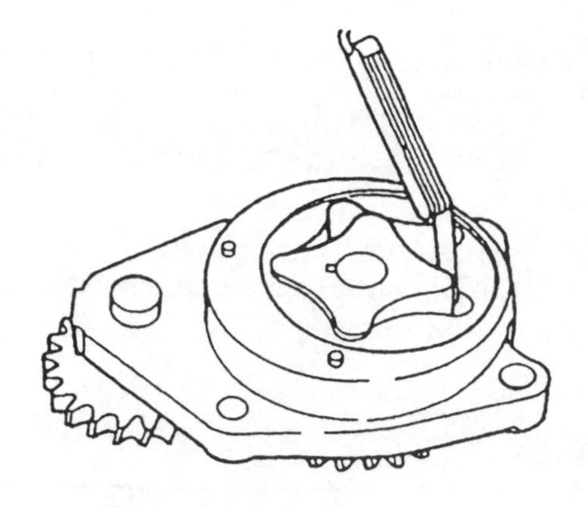

Figure 2–2 Oil pump measurement. *(Courtesy of Ford Motor Company)*

14. The starting motor on a diesel engine has low current draw, high cranking voltage, and slow cranking speed. The batteries are fully charged and pass a load test. *Technician A* says the battery and starting motor cables should be tested for excessive resistance. *Technician B* says the starting motor bushings may be worn. Who is right?
    A. A only
    B. B only
    C. Both A and B
    D. Neither A nor B

Answers to Pretest

1. C, 2. A, 3. B, 4. D, 5. C, 6. A, 7. D, 8. B, 9. B, 10. D, 11. A, 12. C, 13. B, 14. A

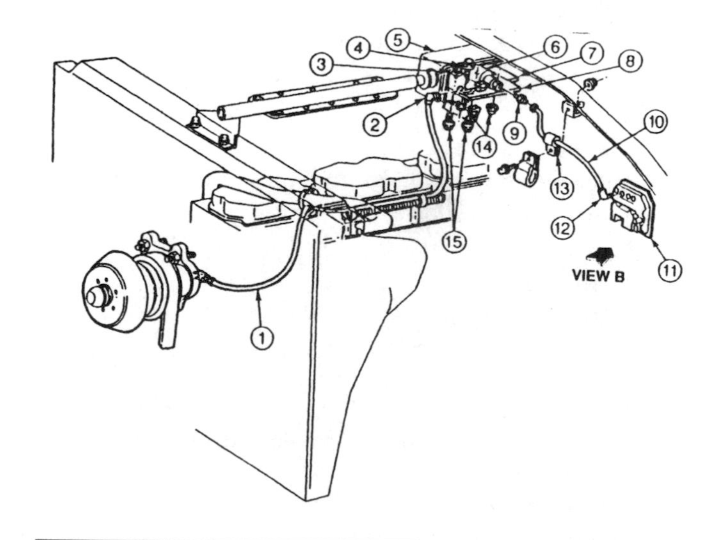

Figure 2–3 Air-operated cooling fan clutch. *(Courtesy of Ford Motor Company)*

General Engine Diagnosis

Task 1

Verify the complaint, and road/dyno test vehicle; review past maintenance documents (if available); determine further diagnosis.

1. Which of the following is LEAST likely to produce a loss of power in a diesel at full rated loads?
   A. High exhaust backpressure
   B. High-density air supplied to the cylinder
   C. Oil and dirt in the air tubes or cooling fins of the aftercooler
   D. Small leak in the fuel line from the tank to the pump, or a crimped line to an injector.

2. A diesel engine that has a power loss or rough operating problem can be caused by all of the following EXCEPT:
   A. a plugged air filter.
   B. a leaking fuel line.
   C. a plugged fuel filter.
   D. a low coolant level.

Hint

The service technician must be aware of normal diesel engine operating noises in order to determine whether a system requires service. Normal noises include the sounds of diesel combustion, fuel injectors, etc.

The diesel engine technician needs to apply a basic diagnostic procedure such as a scientific process of elimination. This technician should perform the following diagnostic tasks:

1. *Listen carefully to the customer's complaint, and question the customer to obtain more information concerning this complaint.*
2. *Identify the complaint and road test or dyno the truck if necessary.*
3. *Think of possible causes for this complaint.*
4. *Perform diagnostic tests to locate the exact cause of the complaint. Always start with the easiest, quickest test, and then work toward the more difficult test.*
5. *Be sure the customer's complaint is eliminated; road test or dyno the truck to verify the repair that eliminated the cause.*

Task 2

Inspect engine assembly and compartment for fuel, oil, coolant, and other leaks; determine needed repairs.

3. *Technician A* says coolant may form a pool of green or blue liquid below the component that is leaking. *Technician B* says that a coolant leak must leave some visible residue in the engine compartment or under the engine. Who is right?
   A. A only
   B. B only
   C. Both A and B
   D. Neither A nor B

4. *Technician A* says small leaks can easily be detected, because antifreeze does not evaporate and generally leaves a red or green colored buildup or stain nearby. *Technician B* says a defective thermostat is most frequently caused by repeated overheating of the engine. Who is right?
   A. A only
   B. B only
   C. Both A and B
   D. Neither A nor B

Hint

Engine manufacturers design engines to reduce exhaust and evaporative emissions. Careful inspection of the engine and engine compartment can reveal potential problems.

For example, oil residue on the bottom side of the hood may indicate excess crankcase pressure caused by blowby or an overfilled crankcase. Oil on the exhaust could be caused by a leaking head gasket or a bad valve cover gasket.

Stains on the engines that appear discolored or lighter than the surrounding area could be from a small coolant leak from the hose or the head. A light, slippery, almost invisible coating on some of the engine compartment surfaces could be caused by a small leak in the radiator. This could be from excess venting of radiator pressure or a very small leak in a high-pressure fuel or hydraulic line.

Deposits of liquid in cracks and crevices or low spots can be indicators of leaks from the surrounding components, lines, and/or hoses.

Task 3

Inspect engine compartment wiring harness, connectors, seals, and locks; determine needed repairs.

5. A connector is found to be distorted from heat but when checked with an ohmmeter the electrical connection is still good. *Technician A* says to replace the connector. *Technician B* says to apply electrical tape around the connector. Who is right?
   A. A only
   B. B only
   C. Both A and B
   D. Neither A nor B

6. The type of connector being replaced in Figure 2–4 is a:
   A. micro.
   B. weather pack.
   C. pull-to-seal.
   D. molded plug.

Hint

Modern engine electrical/electronic wiring is more complex and critical. If connectors or wires are defective, the circuit sends faulty inputs from the monitoring sensors that operate on milliamps of current. This, in turn, causes computerized circuits using these inputs to make faulty adjustments to the engine control. Some of the common problems associated with circuits are corrosion and other foreign substances in connectors, loose or improperly mated connectors, and improper grounding. If a visual inspection of a connector reveals corrosion, the connector should be cleaned or replaced. Wiring should be replaced if it is visibly lumpy or has nicked insulation to prevent faults from corrosion. Loose or corroded grounds should be removed, cleaned, treated for corrosion, reinstalled, and coated with a corrosion inhibitor. All required connections to a ground must be free of all rust and/or paint to ensure a good electrical connection.

Task 4

Listen to engine noises; determine needed repairs.

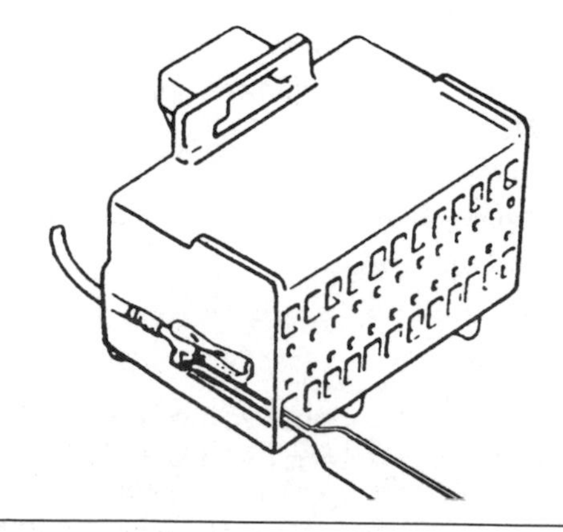

Figure 2–4 Electrical connector.

7. A knocking noise is heard in the engine. The noise increases at the same rate as engine rpm. Which of the following is LEAST likely to cause this?
   A. Fan belt
   B. Camshaft
   C. Crankshaft
   D. Rocker arms

8. When an engine is started after being shut down for several hours a heavy thumping noise occurs for 30 seconds and then disappears. The cause of this problem could be:
   A. worn camshaft bearings.
   B. worn rocker arms.
   C. loose main bearings.
   D. excessive crankshaft endplay.

Hint

A technician who develops a discriminating ear for engine noises can simplify troubleshooting or find problems before they are indicated in diagnostics. A stethoscope may be used to listen to engine noises. If one of the cylinders appears to be misfiring, manually disable each injector. Every time you disable a good cylinder, the sound of the engine should change. On the other hand, a missing cylinder will have no noticeable change in sound. Often the frequency of an unusual noise is relevant to the defective component's operating speed. A loud knocking that increases at the same rate as the engine would probably be associated with the crank, whereas a sound that is half the engine speed might be created by a cam or something directly attached to it. Other components may produce high-frequency sounds like a squeal. Squeals may also be produced when two surfaces are rubbing against each other, like belts slipping.

Task 5

Check engine exhaust emissions, odor, smoke color, opacity (density), and quantity; determine needed repairs.

9. Which of the following is LEAST likely to happen?
   A. A faulty injector may cause an engine to produce a white smoke.
   B. Faulty or improperly installed oil control rings may result in the engine producing a white smoke.
   C. A blown head gasket may result in the engine producing either a blue or white smoke, depending on where the gasket is blown.
   D. Operating a diesel engine for extended periods in a low-load or no-load condition may result in the engine producing a black smoke or soot.

10. *Technician A* says overfueling can cause very black smoke to come out of the exhaust pipe. *Technician B* says that low coolant level can cause this problem. Who is right?
    A. A only
    B. B only
    C. Both A and B
    D. Neither A nor B

Hint

The color and consistency of the exhaust can reveal problems associated with the combustion of the engine. When a diesel engine is running properly with a heavy load, its exhaust is almost clear. The only indication of the exhaust is a slight distortion in the objects viewed through the exhaust. As a rule, when chemicals are burnt or suspended in the exhaust they produce a discoloration or smoke. Oil, for instance, adds a blue haze to the exhaust, causing a blue smoke.

White smoke results from incomplete combustion in lean combustion chamber areas, by fuel spray on metal surfaces, and with low temperatures in the cylinder, such as when starting an engine in cold weather. Fuel with too high a cetane rating can also cause white smoke when used in high-ambient-temperature conditions. White smoke occurs more often in an indirect injected (IDI) engine from retarded timing. Very late timing on a direct injected (DI) engine causes white smoke. Gray or black smoke is the result of incomplete combustion in rich combustion chamber areas, caused by such conditions as engine overload, insufficient fuel injector spray penetration, retarded injection timing, or poor fuel evaporation and mixing due to advanced timing. Air starvation is the major cause of black smoke.

Task 6

Perform fuel system tests; check fuel consumption; determine needed repairs.

11. Which of the following is LEAST likely to cause low fuel pressure?
    A. A partially plugged fuel filter
    B. Low engine idle
    C. A sticking injector
    D. Low coolant level.

12. *Technician A* says that high-pressure fuel lines should be bled of any air that might be trapped after replacement. *Technician B* says you can check for air in the fuel system by installing a clear plastic hose in the return line and looking for bubbles. Who is right?
    A. A only
    B. B only
    C. Both A and B
    D. Neither A nor B

Hint

If the engine runs but is low on horsepower, examine the following items to detect the problem:

Check the fuel filter and change it if necessary. Check the snap pressure by attaching a pressure gauge to an extra outlet fitting on the shutdown solenoid and then accelerating the engine quickly from idle to full throttle. Read the maximum pressure on the gauge during acceleration. It should be very near the pressure listed on the pump calibration sheet under engine fuel pressure. Snap pressure checks will not be accurate on air-fuel ratio control (AFC) pumps.

Maximum delivery check: Before making this check on a pump test stand, back out the torque screw and high idle screw approximately five turns. Run the pump at speed specified for full (rated) fuel and divert fuel flow to a graduated container. Collect fuel for 1,000 strokes. Run at least three tests before recording the delivery. This wets the glass of the graduated containers and removes air from the lines. If the delivery is incorrect, it must be adjusted by removing the governor cover, main governor spring, and guide stud.

Rotate the pump until the leaf spring adjusting screw can be seen through the cam ring. With an Allen wrench, adjust the leaf spring. Replace the parts removed and recheck the fuel delivery.

Task 7

Perform air intake system restriction and leakage tests; determine needed repairs.

13. An air intake is being check for a restriction using a water manometer. *Technician A* says that if the air restriction is excessive the filter or intake tubes need to be inspected. *Technician B* says that the engine speed should be the same for the two readings taken on both sides of the air cleaner. Who is right?
    A. A only
    B. B only
    C. Both A and B
    D. Neither A nor B

14. *Technician A* says that manufacturers allow for movement in the intake system components by using durable rubber couplings, which are clamped at both ends. *Technician B* says that if an air intake pipe, tube, or hose is loose or missing a clamp, an inspection should be made of the turbocharger compressor vanes. Who is right?

A. A only
B. B only
C. Both A and B
D. Neither A nor B

Hint

When an engine suffers from a loss of power or rough operation, you should go back to the basics. To obtain the optimum performance from an engine, the proper quantity and quality of fuel must be supplied to the cylinder at the right time with the right amount of air and heat.

Air starvation will generally result in incomplete combustion, increasing black smoke, a decrease in power, and heat buildup in the cylinder due to insufficient scavenging. Some of the reasons for air starvation are restricted intake hoses and pipes, dirty air filter, excessive exhaust backpressure, or high altitude operation. When the diesel engine is equipped with a blower or turbocharger; check for leaks in the output tubes and the aftercooler. Verify that the aftercooler is functioning properly. Ultimately, the denser the air, the better the engine runs. Fuel starvation will generally result in total combustion of fuel supplied with a loss of power, no noticeable change in exhaust color, and low cylinder temperatures. Some of the reasons for fuel starvation are a defective fuel pump, leaks in fuel lines from the tank, restricted filters or strainers, and crimped fuel lines between the pump and injectors. A temperature spread of less than 10°F (6°C) between the fuel's pour point and temperature can restrict fuel flow enough to cause starvation. Faulty injectors or improper spray patterns can also reduce power and cause erratic operation; this is usually accompanied with black or gray smoke.

Task 8

Perform manifold pressure and/or air box pressure tests; determine needed repairs.

15. A blower air box is being tested for pressure. *Technician A* says to use a manometer to check pressure. *Technician B* says the engine must not be operated at speeds above idle, to prevent damage to the camshaft lobes. Who is right?
    A. A only
    B. B only
    C. Both A and B
    D. Neither A nor B

16. *Technician A* says that low air box pressure can cause an air inlet restriction. *Technician B* says that you can check air box pressure with an inches of mercury vacuum gauge. Who is right?
    A. A only
    B. B only
    C. Both A and B
    D. Neither A nor B

Hint

Excessive air box pressure may be caused by exhaust restriction, or restricted air inlet ports in the cylinders. Low air box pressure may be caused by air leaks between the turbocharger and the air box, or a defective turbocharger. Use a manometer to test the air box pressure at the specified engine rpm.

Task 9

Perform exhaust backpressure and temperature tests; determine needed repairs.

17. When checking exhaust backpressure, *Technician A* says to connect a manometer to the exhaust manifold and check against the manufacturer's specifications. *Technician B* says if a plug has not been provided for this test, drill and tap the exhaust manifold pipe. Who is right?
    A. A only
    B. B only
    C. Both A and B
    D. Neither A nor B

18. While inspecting the exhaust system, the technician notices that the vehicle has an excess amount of exhaust pressure. Which of the following is LEAST likely to cause this?
    A. Improper muffler type
    B. Excessive carbon buildup in the exhaust system
    C. Collapsed exhaust pipe
    D. Low oil pressure

Hint

A slight pressure in the exhaust system is normal. However, excessive exhaust backpressure seriously affects engine operation. It may cause an increase in the air box pressure with a resultant loss of efficiency of the blower. This means less air for scavenging, which results in poor combustion and higher temperatures. Causes of high exhaust backpressure are usually a result of an inadequate or improper type of muffler, an exhaust pipe that is too long or too small in diameter, an excessive number of sharp bends in the exhaust system, or obstructions such as excessive carbon formation or foreign matter in the exhaust system. Check the exhaust backpressure, measured in inches of mercury, with a manometer. Connect the manometer to the exhaust manifold (except on turbocharged engines) by removing a pipe plug, which is provided for that purpose. If no opening is provided, drill an 11/32-inch hole in the exhaust manifold companion flange and tap the hole to accommodate a pipe plug. On turbocharged engines, check the exhaust backpressure in the exhaust piping 6 to 12 inches from the turbine outlet. The tapped hole must be in a comparatively straight pipe area for an accurate measurement.

Task 10

Perform crankcase pressure test; determine needed repairs.

19. *Technician A* says to test an engine for excess crankcase pressure the engine must be at an idle. *Technician B* says worn piston rings may cause excessive crankcase pressure. Who is right?
    A. A only
    B. B only
    C. Both A and B
    D. Neither A nor B

20. A diesel engine has excessive crankcase pressure. The cause of this problem could be:
    A. a leaking oil pan gasket.
    B. a collapsed piston skirt.
    C. a restricted crankcase ventilator.
    D. a burned exhaust valve.

Hint

The crankcase pressure indicates the amount of air passing between the rings and the cylinder liners into the crankcase, most of which is clean air from the air box. A slight pressure in the crankcase is desirable to prevent the entrance of dust. A loss of engine lubricating oil through the breather tube, crankcase ventilator, or dipstick hole in the cylinder block is indicative of excessive crankcase pressure. The causes of high crankcase pressure may be traced to excessive blowby due to worn piston rings, a hole or crack in a piston crown, loose piston pin retainers, worn blower oil seals, defective blower, cylinder head gaskets, or excessive exhaust backpressure. In addition, the breather tube or crankcase ventilator should be checked for obstructions. Check the crankcase pressure with a manometer connected to the oil level dipstick opening in the cylinder block. Check the readings obtained at various engine speeds with the manufacturer manual.

Task 11

Diagnose no cranking, cranks but fails to start, hard starting, and starts but does not continue to run problems; determine needed repairs.

21. *Technician A* says hard starting may be caused by air in the fuel system. *Technician B* says low cylinder compression does not affect engine starting. Who is right?
    A. A only
    B. B only
    C. Both A and B
    D. Neither A nor B

22. An engine with a distributor-type injector pump cranks but fails to start. What is the probable cause?
    A. A faulty injector
    B. A defective starter cutout relay
    C. A defective shutdown solenoid
    D. A leaky intake hose

Hint

Often when an engine cranks but fails to start it is an indication of air in the fuel system. To determine if air is causing the problems pressurize the system using a hand primer and bleed the lines to each injector while cranking the engine. If air is found, examine the area around the fuel lines between the primer and injectors for an accumulation of fuel. If the primer pump fails to build up pressure within the required time, check fuel filter lines for leaks and the tank for adequate fuel level.

A diesel engine may not start or be hard to start because the temperature of the air in the cylinder is too low for full combustion of the atomized fuel. If the starter does not rotate the engine fast enough to allow the air in the cylinders to reach combustion temperature, combustion will not occur.

Electrically, the starter may turn slowly if it is defective, has a low battery voltage, corroded cables, or high-resistance electrical connections. Inspect the starter, battery, and connections to ensure proper operation of the starter motor.

Mechanically, the engine may offer too much load on the starter. This happens when the oil is too thick. Check oil specifications for that climate zone, check immersion heaters if installed.

1. *Low combustion temperature may also result if the outside temperature is too low and the engine does not use preheating.*
2. *To assist in starting the engine at low temperatures use ether, which will reduce the flash point of the fuel/air mixture.*
3. *To maintain a warm engine for starting at low temperatures, the engine may use an immersion heater in the cooling system or submerged in the crankcase oil.*

Task 12

Diagnose surging, rough operation, misfiring, low power, slow deceleration, slow acceleration, and shutdown problems; determine needed repairs.

23. Low engine power is a result of insufficient fuel delivery to a unit injector. Which of the following would be LEAST likely cause?
    A. Suction leaks in fuel supply system
    B. Incorrect throttle travel
    C. Clogged fuel filter
    D. A leak in the injector return line

24. When servicing the emergency shutdown device shown in Figure 2–5, perform all of these service operations EXCEPT:
    A. replace the blower.
    B. resurface face areas that are not flat.
    C. replace warped valve plates.
    D. replace blower screens.

Hint

Engine surging may be caused by a defective governor. Low cylinder compression or a defective injector may cause rough low-speed operation and cylinder missing. Low engine power may be caused by an improper throttle linkage adjustment. An improper throttle linkage adjustment or a defective governor causes slow acceleration. Shutdown problems are often caused by a defective shutdown solenoid or related circuit.

Task 13

Isolate and diagnose engine-related vibration problems; determine needed repairs.

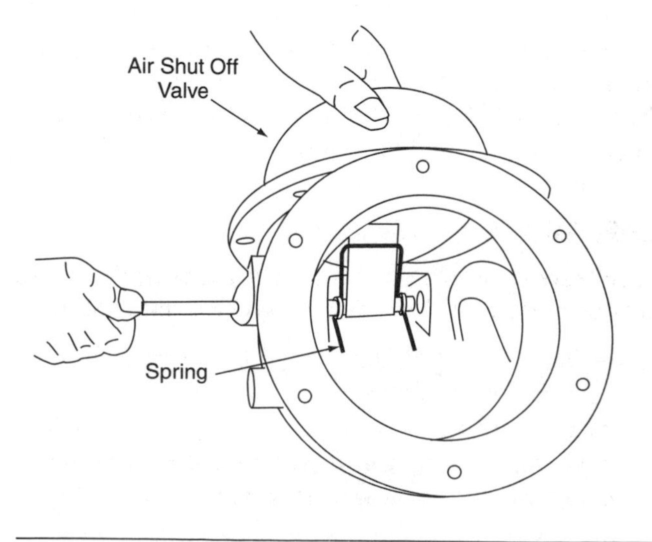

Figure 2–5 Air shut off valve.

25. An engine has a vibration occurrence rate that is equal to half the engine rpm. *Technician A* says that this might be caused by a bent cam. *Technician B* says that this could be caused by a bent connecting rod. Who is right?
    A. A only
    B. B only
    C. Both A and B
    D. Neither A nor B

26. The engine component shown in Figure 2–6 is the:
    A. flywheel assembly.
    B. vibration damper assembly.
    C. timing gear assembly.
    D. water pump assembly.

Hint

Often the frequency of occurrence of an unusual vibration is a direct or multiple of the operational speed of the device creating the vibrations. A strong vibration that increases with engine speed would probably be the crank or something directly attached to it, whereas a vibration that has an occurrence rate half the engine speed might be created by a cam or something directly attached to it. Other components, like an out-of-balance turbocharger or alternator, may produce high-frequency vibrations that seem to have a constant occurrence rate. In these cases use a stethoscope to find the strongest concentration of that vibration. High-frequency vibrations can also be produced by bearings that are failing, or two surfaces that are contacting each other. Some vibrations may be traced to specific components. If a vibration only occurs when the A/C compressor clutch is engaged, the cause of the vibration is in the clutch or compressor.

Task 14

Check cooling system for protection level, contamination, coolant level, temperature, pressure, conditioner concentration, filtration, and fan operation; determine needed repairs.

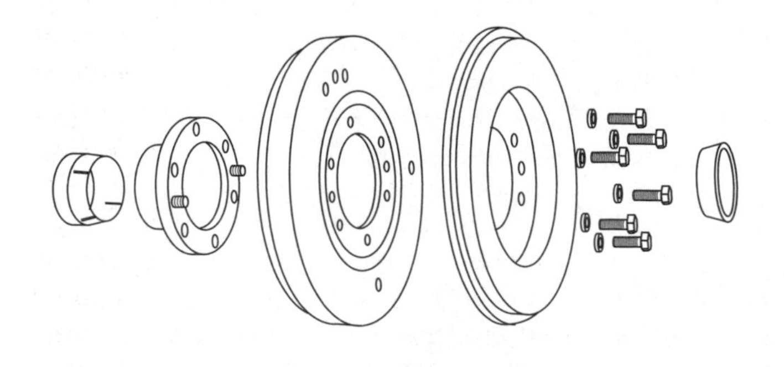

Figure 2–6 Engine-related component.

27. *Technician A* says a radiator cap in a cooling system that cannot maintain a system pressure of its rating of 15 psi (103 kPa) may cause an overheat condition. *Technician B* says when the pressure exceeds the rated amount, the cap releases the pressure. Who is right?
    A. A only
    B. B only
    C. Both A and B
    D. Neither A nor B

28. All of the following are true if a pressure cap is found to be defective EXCEPT:
    A. replace the cap.
    B. excess pressure could cause damage to the radiator tubes.
    C. excess vacuum pressure could collapse radiator hoses.
    D. adjust or replace the spring in the cap.

Hint

The radiator cap will maintain the pressure within the cooling system at or below approximately 15 psi (103 kPa). During heavy loads and prolonged periods of idling, this pressure may exceed the rating of the radiator cap and fluid will be vented to an expansion tank or bottle. Once enough fluid has been vented to reduce the pressure within limits, the spring tension of the radiator cap will close the vent line. At normal temperatures, with the pressure cap removed, small fluctuations in coolant level can be expected. Inspection of this fluid level may indicate deeper problems. A constant overflowing of the fluid from the radiator overflow tube may indicate air pockets, uneven heating, or hot spots in the engine. Repetitious and rhythmic excessive surges in this level usually indicate a leak to the combustion cylinder, such as a blown head gasket.

If the cooling system overheats, check tightness on hose clamps, check for soft or sometimes mushy feeling hoses and check for cracks. Check for leaks in the radiator, water pump, cooling filter, head gaskets, and water manifold. An external coolant leak is usually indicated by a red or greenish color, or a residue that builds up at the leak. All belts should be checked for proper condition and tension. Pressurize the system and watch hoses, radiator tank joints, and head gaskets for leaks. Test the lubrication oil for water and inspect the radiator and fins for obstructions. Remove and test the thermostat for proper opening and closing temperature, and replace as necessary. If the thermostat is found defective, usually it is an indication of something else causing an overheating condition.

Task 15

Check lubrication system for contamination, oil level, temperature, pressure, filtration, and oil consumption; determine needed repairs.

29. *Technician A* says canister oil filters should be prefilled to protect the bearings. *Technician B* says that the filter housing mounted by-pass valve needs to be removed and inspected when replacing the filter. Who is right?
    A. A only
    B. B only
    C. Both A and B
    D. Neither A nor B

30. The LEAST likely cause of water droplets or gray oil on the dipstick would be:
    A. cracked block or cylinder head.
    B. blown head gasket.
    C. leaky oil cooler.
    D. worn oil pump.

Hint

Examination of lubrication oil can reveal problems associated with engine performance. An indication of oil dilution from coolant is water droplets forming on the dipstick while the engine is off, or if the oil appears gray in color after the engine has run. This can occur for any of the following; a cracked block or cylinder head, blown head gasket, leaky oil cooler, or a leaky sleeve O-ring. Fuel in the oil will cause the oil to thin out when cold and may appear on the dipstick as clearer oil at the top of the level indicator if the engine has not been run for a while. This problem may be caused by leaky injectors or a leaky injector pump.

Excessive oil consumption can be a good indicator of improperly seated or defective rings, a damaged piston, or defective valve stem seals. These problems are usually accompanied by blue smoke in the exhaust. Low oil pressure may indicate a worn oil pump, restricted suction tube or screen open, or improperly adjusted relief valve, clogged oil filter, or worn bearings.

Task 16

Check, record, and clear electronic diagnostic codes; monitor electronic data; determine needed repairs.

31. *Technician A* says that only a truck manufacturer's scan tool is compatible with the truck manufacturer's electronic systems. *Technician B* says that each truck manufacturer uses a different data link connector. Who is right?
    A. A only
    B. B only

C. Both A and B
D. Neither A nor B

32. What is the meaning of the abbreviation DTC as it pertains to the error codes on the modern diesel engine?
A. Drivetrain codes
B. Diagnostic trouble codes
C. Direct trouble codes
D. Direct transmission codes

Hint

Today's diesel engines contain a variety of computerized engine control systems depending on the engine manufacturer. According To SAE Standards J1930, diagnostic codes are known as diagnostic trouble codes (DTCs). The DTC extraction method varies with the engine and truck manufacturer. The typical process is to use a scan tool, diagnostic reader, or laptop computer to check for and clear DTCs.

Some vehicles display limited readout DTCs by blinking an instrument panel indicator in two-digit codes called a flash codes.

Cylinder Head and Valvetrain Diagnosis and Repair

Task 1

Remove, inspect, disassemble, and clean cylinder head assembly(ies).

33. You suspect a cylinder head is warped or cracked, or has a blown head gasket. What would you do first?
A. Check head bolt torque.
B. Apply a dye crack detector to reveal the presence of small cracks.
C. Remove all residue oil, carbon, and gasket material from surfaces and visually inspect for cracks.
D. Remove excess scale deposits from the passage and perform a pressure test of the head to check for internal cracks.

34. When discussing cylinder head inspection, *Technician A* says the exhaust manifold mating surface on the cylinder head should be checked for warpage with a dial indicator. *Technician B* says there is a limit to the amount of metal that can be machined from the cylinder head to block mating surface. Who is right?
A. A only
B. B only
C. Both A and B
D. Neither A nor B

Hint

Inspection of the cylinder head begins by checking the torque of cylinder head bolts before removing the head. This can indicate the potential for a warped or cracked head. Inspect the head gasket and head surface for signs of carbon or coolant stains between cylinders and passages before cleaning the head. Use a parts cleaner to remove oil, coolant, and carbon residue. When cracks are suspect, but not apparent, magnetic or dye crack detection may be required to highlight cracks. Visually reinspect the head for cracks in the casting, scale buildup in the coolant passages, and pitting of machined surfaces. Excess accumulation of scale in water passages can usually be removed by placing the head in a noncorrosive dip for about two hours. Pressure testing the head can be used to locate a crack in coolant and oil passages.

Task 2

Inspect threaded holes, studs, and bolts for serviceability; service/replace as needed.

35. *Technician A* says that before installation head bolts must be cleaned and inspected for erosion or pitting. *Technician B* says the cylinder needs to be checked for foreign objects before installing the head. Who is right?
A. A only
B. B only
C. Both A and B
D. Neither A nor B

36. *Technician A* says that oil in a cylinder head holddown bolt hole may cause an inaccurate torque reading. *Technician B* says that a head bolt with stripped threads should be redressed before installing. Who is right?
A. A only
B. B only
C. Both A and B
D. Neither A nor B

Hint

When removing or installing a component, carefully inspect all bolts, studs, holes, and nuts for damage to threads. Stripped, cross-threaded, nicked, rolled, and rusty threads can produce errors in the torque values. Use compressed air to remove foreign matter from holes. Obstructions lodged in holes, such as metal and liquids, can cause a bolt to bottom out and reach torque value without obtaining the desired results. Examine the shank of bolts and studs for signs of twisting or overtorque fracturing. Replace defective bolts and studs. Ensure that nuts are not distorted or cracked, replace self-locking and defective nuts. Clean and redress threads in holes. If threads in a hole are damaged beyond use, drill and tap the hole and install an approved helicoil. Use an approved stud extractor to remove broken studs and bolts.

Task 3

Measure cylinder head deck-to-deck thickness, and check mating surfaces for warpage; inspect for cracks/damage; check condition of passages; inspect core, gallery, and plugs; service as needed.

37. *Technician A* says the cylinder head warpage should be measured transversely across the cylinder head at each end and between each cylinder. *Technician B* says a blown head gasket between two cylinders may indicate a transversely warped cylinder head. Who is right?
    A. A only
    B. B only
    C. Both A and B
    D. Neither A nor B

38. To check a cylinder head to block mating surface, *Technician A* says use a straightedge and an inside micrometer for detection of warpage. *Technician B* says to measure the cylinder head mating surface at 3, 6, 9, and 12 o'clock positions around each cylinder. Who is right?
    A. A only
    B. B only
    C. Both A and B
    D. Neither A nor B

Hint

The cylinder head should be checked for warping. Using a straightedge and feeler gauge, measure the face for flatness. Check for transverse (across the width of a cylinder head) warpage at each end and between each cylinder. Check for longitudinal (across the length of the cylinder head) warpage at points above and below each cylinder, between centerlines of valves, and the outer edge of the head surface. Compare measurements to manufacturer's specifications to determine the need for resurfacing. Resurfacing may necessitate removing inserts, such as valve guides, and/or deburring of water ports. Check cylinder head thickness, called deck-to-deck thickness, to determine the maximum resurfacing allowable.

Task 4

Test cylinder head for leakage using approved methods; service as needed.

39. When testing a cylinder head for potential coolant leakage, *Technician A* says the coolant needs to be heated to operating temperature. *Technician B* says the coolant needs to be under pressure. Who is right?
    A. A only
    B. B only
    C. Both A and B
    D. Neither A nor B

40. When performing a cylinder head leakage test of fuel passages, which of the following is LEAST likely to be used?
    A. Discarded injectors
    B. Compressed air
    C. Threaded plug
    D. Injection pump

Hint

When manufacturers incorporate fuel delivery passages into a cylinder head, use their procedures to pressurize with air and test for fuel passage leaks. Often this procedure requires installing the discarded injectors or a jig, sealing the return port with a threaded plug, and using a regulated air supply. Water passages are checked using a combination of air pressure and water heated to operating temperature. Usually this requires installation of specialized jigs to seal coolant ports and provide connections for coolant and air supply.

Task 5

Inspect and test valve springs for squareness, pressure, and free height comparison; replace as needed.

41. *Technician A* says to replace springs that are fractured, broken, out of square, or do not meet the specifications. *Technician B* says companion springs under the same valve bridge as defective springs must also be replaced. Who is right?
    A. A only
    B. B only
    C. Both A and B
    D. Neither A nor B

42. Which of the following is LEAST likely to be revealed during inspection and testing of valve springs?
    A. Springs that produce harmonic distortion
    B. Springs with uneven free height
    C. Cracks or breaks in springs
    D. Springs that are not square

Hint

Check valve spring straightness with a T-square. If valve springs are not square they cause uneven side pressure, which results in faster guide wear. Also check the free length, loaded length, and tension of the spring using a spring tension gauge. Replace valve springs that do not meet specifications. Valve springs using the same valve bridge must be replaced together to prevent unbalanced valve operation. This can cause damage to the bridge from uneven stress, valve guides and stems to wear faster, and insufficient scavenging of the cylinder.

Task 6

Inspect valve spring retainers and/or rotators, locks, and seals; replace as needed.

43. *Technician A* says that mechanical rotators on a valve help eliminate carbon deposits on the valve seat. *Technician B* says that they can prevent oil leakage from the valve guide. Who is right?
    A. A only
    B. B only
    C. Both A and B
    D. Neither A nor B

44. An engine drops a valve 20 hours after an overhaul. *Technician A* says the probable cause was missing or defective valve rotators. *Technician B* says the probable cause was improperly installed keepers. Who is right?
    A. A only
    B. B only
    C. Both A and B
    D. Neither A nor B

Hint

Valve spring retainers keep the valve spring on the valve, and valve rotators are used to rotate the valve each time it is opened to reduce carbon deposits on the valve face and seat. Some engines also have seals. Careful inspection of these components can prevent serious damage and extend the life of the engine. Retainers that become worn or damaged may allow the spring to release the valve. If this happens, the valve drops into the cylinder and damages the piston and/or cylinder head. Seals between the valve guide and valve stem prevent oil from entering the cylinder on the intake stroke. Defective seals can cause oil consumption, leading to excess smoke in the exhaust, carbon buildup, fouling of injectors, formation of hot spots, and precombustion or dieseling.

Task 7

Measure valve guides for wear, check valve guide-to-stem clearance, and measure valve guide height; recondition/replace as needed.

45. When replacing valve guides, *Technician A* says that when the manufacturer does not supply specifications, press the new guide in head to the same position the old guide was in. *Technician B* says excess scoring of the guide bore will require the head to be replaced. Who is right?
    A. A only
    B. B only
    C. Both A and B
    D. Neither A nor B

46. Which of the following is LEAST likely to be used to check the inside diameter of a valve guide?
    A. Snap gauge
    B. Feeler gauge
    C. Ball gauge
    D. Dial indicator

Hint

After the cylinder head has been checked or resurfaced and is considered usable, the valve guides should be checked. Measure the guide inside diameter with a snap, ball, or dial gauge in three different locations throughout the length of the guide. If guides are worn excessively, they should be replaced if they are the replaceable types. In most cylinder heads, the valve guides can be removed by using a driver. Select a driver that fits the valve guide. Insert the guide driver and drive or press the new guide into the cylinder head until the correct guide position is reached. If the manufacturer's specifications are not available, position the guide in the same position of the old guide.

Task 8

Inspect, recondition, or replace valves.

47. *Technician A* says that if valves are to be reused, do not grind valve stem ends as this will cause a change in tappet clearance. *Technician B* says existing valves should only be resurfaced if the valve seat was replaced. Who is right?
    A. A only
    B. B only
    C. Both A and B
    D. Neither A nor B

48. Which of the following valve conditions are shown in Figure 2–7?
    A. Cracks in the valve stem
    B. Carbon deposits on the face of the valve

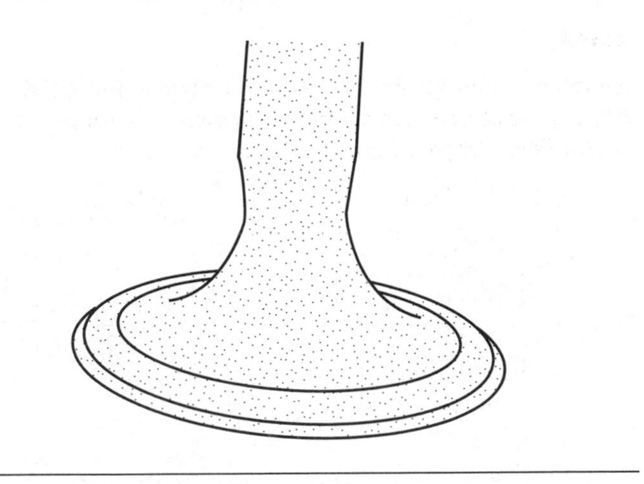

Figure 2–7 Defective valve condition.

C. Cracks on the face of the valve
D. A stretched valve stem

Hint

A common practice during major maintenance is to replace all valves, if they have been in the head for several thousand operating hours. Valves with scratched, scored, pitted, or worn stems or stems with worn keeper grooves should be replaced. If the valve head is cupped, nicked, or marred it should be replaced. If a valve passes the visual inspection, it must be cleaned using a wire buffing wheel or glass bead blaster, if available. Use a valve grinder to obtain the correct face angle. Inspect the valve to ensure that the valve margin is within specifications. Insufficient valve margin may result in premature valve burning and failure. Finally, valve stem ends are ground flat and a small chamfer, usually about 1/32 to 1/16 in. (0.79 to 1.59 mm) is applied.

Task 9

Inspect, recondition, or install/replace valve seats.

49. Which procedure is LEAST likely to be used when performing valve seat maintenance?
    A. Testing seats for looseness with a ball peen hammer
    B. Sharpening a chisel to remove defective seats
    C. Checking seats with a concentricity indicator
    D. Dressing a valve seat grinding stone

50. All of the following will indicate valve seat leaks EXCEPT:
    A. visible cracks.
    B. oblong valve seats.
    C. discoloration of the back of the valve head.
    D. worn piston rings.

Hint

It is important to properly inspect, clean, and recondition or replace valve seats to maintain a good valve-to-head seal. Loose valve seats are located by lightly tapping each seat with the peen end of a ball peen hammer. A loose seat will produce a sound that differs from the ring sound created by tapping the head adjacent to the seat. The seat may even move slightly. If the seat is solid, check it for cracks and excessive width. Normal valve seat width is 0.0625 to 0.125 in. (1.59 to 3.175 mm). Leaks may occur if the seat has visible cracks, uneven burns or discoloration, or an opening that is oblong or excessive in diameter. Replace any seats that cannot be reconditioned. Valve seat checking and reconditioning should be done after guides have been replaced or reconditioned.

To recondition a valve seat, a mandrel pilot is installed in the valve guide to keep a specially dressed grinding wheel centered in the seat. The grinding wheel may be dressed with a one-degree steeper angle than the valve. This allows the valve to cut through future carbon buildup. The one-degree difference is not necessary if valves have rotators. New seats must be reconditioned after installation. Also, check the concentricity of the seat opening. If the seat is not round, pressure may escape during compression and power strokes. Verify that the valve head height is correct. If the head height is out of specification seats may need to be replaced, as this may alter the compression ratios or injector spray pattern. It is important not to remove any more material than is necessary to refinish the valve seat.

Task 10

Measure valve head height relative to deck, valve face-to-seat contact, and valve seat concentricity.

51. Which tool combination is needed to check valve head height?
    A. Straightedge and feeler gauge
    B. Dial indicator, step block, steel ruler and outside caliper
    C. Vernier micrometer and point gauge
    D. Steel ruler and T-square

52. Before installing the valves into the cylinder head, you should check which one of the following first?
    A. Valve-to-seat contact
    B. Rocker arm adjustment
    C. Lift on the camshaft
    D. Head gasket

Hint

Before installing valves, check the valve-to-seat contact. One method of doing this is to coat the face of the valve with Prussian Blue and lightly snap the valve into its seat. Remove the valve and check its face. If the valve face does not have an even seating mark, the seat is not concentric and must be reground. When the valve face-to-seat contact has been checked, measure the valve head height. This checks the distance the valve protrudes above or below the machined surface of the head. Place a straightedge across the cylinder head and use a feeler gauge to measure distance between the valve head and straightedge. If the valve is designed to protrude above the head, place the straightedge on the valve. Use a feeler gauge between the valve head and the cylinder head machined surface.

Task 11

Inspect and replace injector sleeves and seals; measure injector tip or nozzle protrusion where specified by manufacturer.

53. After replacing injector sleeves and seals as shown in Figure 2–8, *Technician A* says injector tip height must be checked to confirm proper clearance. *Technician B* says to perform a cylinder head leak test. Who is right?
    A. A only
    B. B only
    C. Both A and B
    D. Neither A nor B

54. If an injector sleeve is found to be leaking, which of following procedures is LEAST likely to be required?
    A. Using the OEM process to remove and install the sleeve
    B. Perfoming a cylinder head pressure test
    C. Checking injector tip protrusion
    D. Replacing the injector nozzle

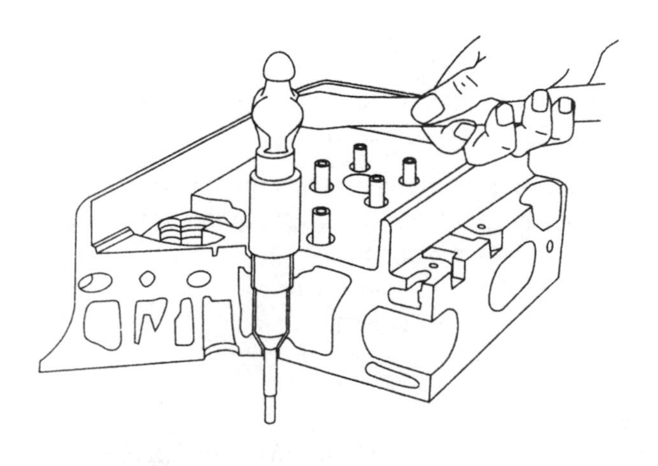

Figure 2–8 Cylinder head service.

Hint

If a leak is detected in an injector sleeve seal, the seal must be replaced to prevent further damage to the engine. After performing the manufacturer's procedures for removing and reinstalling the injector sleeve, perform a cylinder head pressure test to ensure the sleeve is sealed. Check the injector tip height to ensure it meets manufacturer specifications. Failure to do so may result in improper combustion due to the injector spray pattern striking the cylinder walls or valves instead of the piston.

Task 12

Inspect, clean, and/or replace precombustion chambers where specified by manufacturer.

55. A precombustion chamber is being replaced on a medium-duty diesel engine. Which of these processes apply in this repair?
    A. Press the prechamber in the cylinder head with an arbor press.
    B. Adjust the clearance using a grinding wheel.
    C. Check the clearance with a feeler gauge flush to plus 0.002 in. (0.050 mm).
    D. Use 400 grit sandpaper to grind the prechamber to fit.

56. If a precombustion chamber is sticking too far out of the cylinder head, all of these components may be damaged EXCEPT:
    A. the piston.
    B. the intake valve.
    C. the block.
    D. the exhaust valve.

Hint

On IDI applications, precombustion chambers are either integral or installed as units in the head. Install a precombustion chamber into the cylinder head with a plastic hammer flush to plus .002 in. (.050 mm).

Task 13

Inspect, and/or replace valve bridges (crossheads) and guides; adjust bridges (crossheads).

57. *Technician A* says that crosshead guide pins should be at right angles to the heads' milled surface. *Technician B* says to check crosshead guide pin diameter with a micrometer, and compare to manufacturer's specifications. Who is right?
    A. A only
    B. B only
    C. Both A and B
    D. Neither A nor B

58. *Technician A* says that the four valve heads on a four-stroke engine use four exhaust valves in each cylinder. *Technician B* says that a four-stroke diesel engine with valve crossheads will use a four-valve head. Who is right?
    A. A only
    B. B only
    C. Both A and B
    D. Neither A nor B

Hint

Valve crossheads are used on some types of diesel engines that use "four valve heads." Four-valve heads used on four-stroke cycle engines have two intake and two exhaust valves per cylinder. Two-stroke cycle engines, such as Detroit Diesel, use four exhaust valves in each cylinder. The crosshead is a bracket or bridge-like device that allows a single rocker arm to open two valves at the same time. The crossheads must be checked for wear as follows:

1. *Visually check crosshead for cracks with magnetic crack detector if available.*
2. *Check crosshead inside diameter for out of roundness and excessive diameter.*
3. *Visually check for wear at the point of contact between the rocker lever and the crosshead.*
4. *Check the adjusting screw threads for broken or worn threads.*
5. *Check the crosshead guide pin for diameter with a micrometer.*
6. *Check the crosshead guide pin to ensure that it is at right angles to the head-milled surface.*

Task 14

Clean components; reassemble, check, and install cylinder head assembly as specified by the manufacturer.

59. Which of the following should the technician check first when reinstalling a cylinder head to the engine block?
    A. Cylinder bores for foreign matter and cross threading
    B. The water pump bearings
    C. The front crank seal
    D. The rear crank seal

60. *Technician A* says cylinder heads are always torqued in sequence according to specifications set by the manufacturer. *Technician B* says that bent pushrods must be straightened. Who is right?
    A. A only
    B. B only
    C. Both A and B
    D. Neither A nor B

Hint

The assembly of the cylinder head or heads onto the engine block will involve many different procedures that are peculiar to a given engine. The following procedures are general in nature. Before attempting to install the cylinder head, make sure the cylinder head and block surface are free from all rust, dirt, old gasket materials, and grease or oil. Before placing the head gasket on the block make sure all bolt holes in the block are free of oil and dirt by blowing them out with compressed air.

Select the correct head gasket and place it on the cylinder block, checking it closely for an "up" or "top" mark that some head gaskets may have. In most cases head gaskets are installed dry with no sealer, although in some situations an engine manufacturer may recommend applying sealer to the gasket before cylinder head installation.

Clean and inspect all head bolts or cap screws for erosion or pitting. To clean bolts that are very rusty and dirty use a wire wheel. Coat bolt threads with light oil or diesel fuel and place the bolts or cap screws into the bolt holes in the head and the block.

Using a speed handle wrench and the appropriate size socket, start at the center of the cylinder head and turn the bolts down snug, which is until the bolt touches the head and increased torque is required to turn it. Gradually tighten the head bolts in the specified sequence.

When all bolts have been tightened this far, continue to tighten each bolt one-quarter to one-half turn at a time with a torque wrench until the recommended torque is reached.

Task 15

Inspect pushrods, rocker arms, rocker arm shafts, electronic wiring harness, and brackets for wear, bending, cracks, looseness, and blocked oil passages; repair/replace as needed.

61. All of the following must be checked before installing rocker arms, EXCEPT:
    A. rocker arm bushings for wear.
    B. rocker arm shaft for wear.
    C. rocker arm shaft for pitting or scoring.
    D. rocker arm tappet screws are tightened down.

62. *Technician A* says a sticking valve may cause a bent pushrod. *Technician B* says pushrods must be checked for proper fit to cam followers. Who is right?

A. A only
B. B only
C. Both A and B
D. Neither A nor B

Hint

After the cylinder head bolts have been tightened to specifications, the pushrods and rocker arm assemblies may be installed. Bent pushrods must be replaced. Pushrods must be checked for breaks or cracks where the ball socket on either end is fitted into the rod. Place the pushrods in the engine, making sure that they fit into the cam followers or tappets. Replace all the rocker arms, shafts, and retainers that are worn, cracked, or scored.

Task 16

Inspect, adjust/replace cam followers.

63. *Technician A* says a lifter bottom should be smooth and slightly concave. *Technician B* says if a lifter bottom is pitted, the camshaft mating lobe should be inspected. Who is right?
    A. A only
    B. B only
    C. Both A and B
    D. Neither A nor B

64. A misfiring cylinder is diagnosed on a diesel engine with unit injectors. The Technician replaces one injector. However, the engine continues to miss. The cause of this problem could be:
    A. a worn cam follower roller.
    B. a restricted fuel filter.
    C. high fuel pump pressure.
    D. a leaking injector tube.

Hint

The function of cam followers is to reduce friction and evenly distribute the force imparted from the cam's profile to the valvetrain. Diesel engines using cylinder block mounted camshafts use two types of followers while those using overhead camshafts have either direct actuated rockers or roller-type cam followers:

Hydraulic lifters should be inspected during any head maintenance for:

1. *Hydraulic lifter body for scuffing and scoring. If the lifter body wall is worn or damaged, the mating bore in the block should also be checked.*
2. *Check the fit of each hydraulic lifter in its mating bore in the block. If the clearance is excessive, try a new lifter.*
3. *The hydraulic lifter bottom must be smooth and slightly convex. If worn, pitted, or damaged, the mating camshaft lobe should also be checked.*

Task 17

Adjust valve clearance(s).

65. The valve adjustment in Figure 2–9 is being performed on:
    A. an exhaust valve with a spring-loaded bridge.
    B. an exhaust valve with a non-spring-loaded bridge.
    C. an intake valve in a four-cycle diesel engine.
    D. an exhaust valve bridge in a two-cycle engine.

66. While discussing the adjustment in Figure 2–10, *Technician A* says the valve bridge is being adjusted. *Technician B* says this adjustment must be performed after the valve clearance is adjusted. Who is right?
    A. A only
    B. B only
    C. Both A and B
    D. Neither A nor B

Hint

When valves are properly adjusted, there should be clearance between the end of the rocker arm and the top of the valve stem with the valve closed. Valve lash is required because as the moving parts become heated, they expand. If valve clearance is insufficient, the valves may remain open when an engine reaches operating temperature. A loose valve adjustment retards valve opening and advances valve closing, decreasing the cylinder breathing time. Exhaust valves are subject to more heat

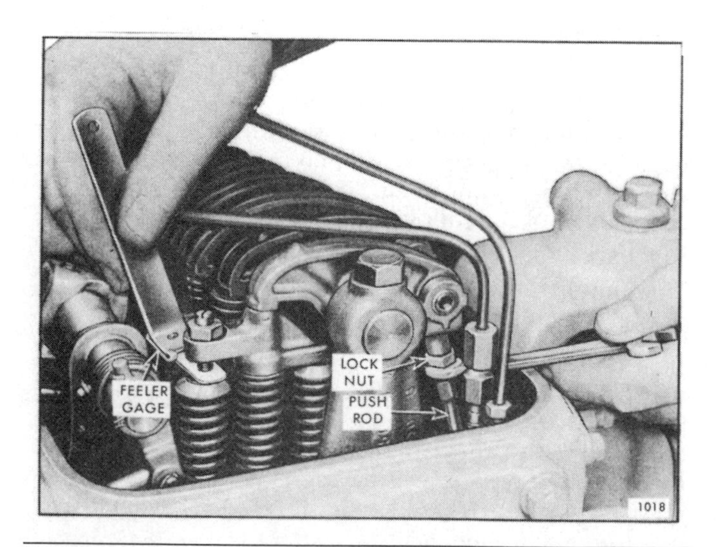

Figure 2–9 Valvetrain adjustment. *(Courtesy Detroit Diesel Corporation)*

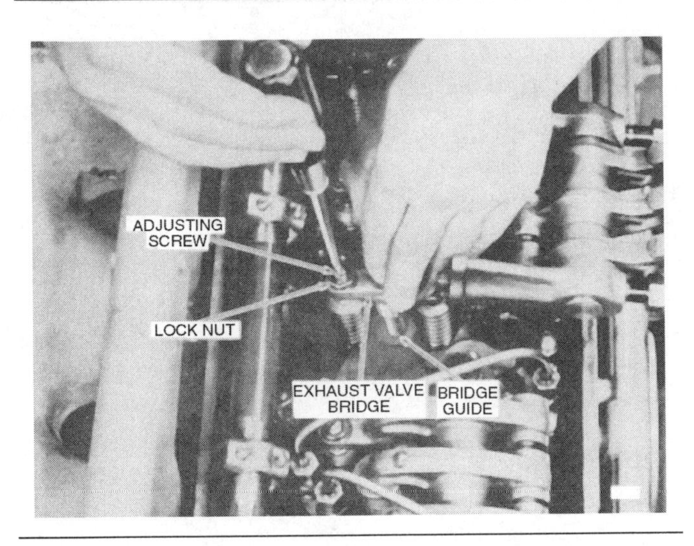

Figure 2–10 Valvetrain adjustment. *(Courtesy Detroit Diesel Corporation)*

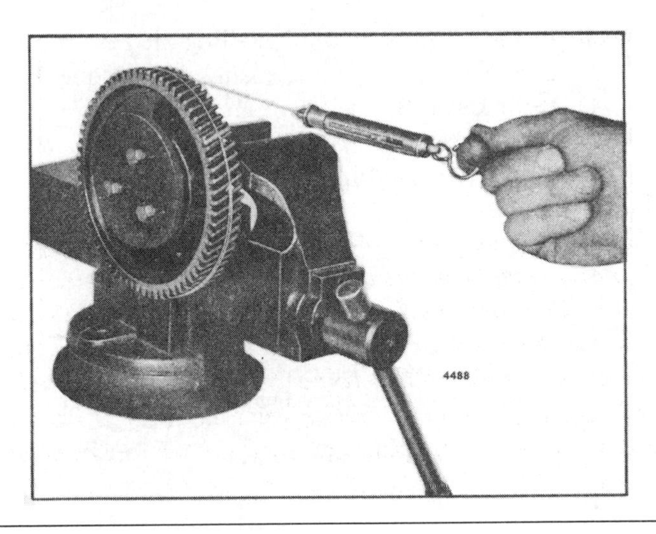

Figure 2–11 Valvetrain gear service. *(Courtesy Detroit Diesel Corporation)*

and as a consequence, OEMs require a valve lash setting for exhaust valves that is usually greater than the intake valve lash setting.

Some diesel engine manufacturers recommend valve adjustment under static conditions and with the engine coolant 100°F (37°C) or less. The valve clearance must be adjusted to specifications with the valves closed. The fuel injection system must be disabled during the valve adjustment. If the engine has valve bridges, these components must be adjusted before the valve clearance adjustment.

Task 18

Inspect, measure, and replace/reinstall overhead camshaft and bearings; measure and adjust end play and backlash.

67. Which of the following is LEAST likely to be used to measure the backlash in the timing geartrain?
    A. Dial indicator.
    B. Feeler gauge.
    C. Depth gauge.
    D. Ruler.

68. *Technician A* says Figure 2–11 illustrates checking the freeplay between the camshaft gear and the camshaft. *Technician B* says this figure shows checking the preload on an idler gear that transfers the drive from the crankshaft gear to the camshaft gear. Who is right?
    A. A only
    B. B only
    C. Both A and B
    D. Neither A nor B

Hint

Visually inspect the camshaft for worn, pitted, or scored lobes or journals. Be sure the oil supply passages that supply oil to the camshaft bearings are clean. Discard any camshaft that does not pass a visual inspection. Heel-to-toe camshaft lobe measurement is equal to the diameter of the circular portion of the lobe and the maximum amount of lift created by the rise at the top of the lobe. The cam follower rides in the center of lobe, causing a groove to form. This groove should not be greater than 0.003 in. (0.076 mm) deep. If either the top of the lobe or the circular portion is worn, the amount of lift is reduced. If an intake or exhaust cam lobe fails to produce enough lift, the valve it controls may not open enough to supply adequate air for combustion to occur. If the lobe controls an injector, the injector may not fire. Visually inspect bearing journals for bluing, scoring, or wear. Measure the diameter of journals with an outside micrometer, and check for out-of-round condition by checking around the journal in several places. Inspect the shaft and gear keyway for cracks or distortion. Ensure the keyway in the gear and shaft matches the woodruff key. Any movement between these three can randomly change the timing of valves and injectors.

Engine Block Diagnosis and Repair

Task 1

Remove, inspect, service, and install pans, covers, vents, gaskets, seals, and wear rings.

69. *Technician A* says before installing die cast rocker covers on a cylinder head, ensure the silicone gasket is secure in the cover groove to ensure a good seal. *Technician B* says to always use new gaskets when installing valve covers. Who is right?
    A. A only
    B. B only
    C. Both A and B
    D. Neither A nor B

70. All of the following are parts of a valve cover EXCEPT:
    A. the head rail.
    B. the breather housing.
    C. the filler cap.
    D. the strainer.

Hint

Oil pans are usually located in the airflow under the frame rails and are vulnerable to damage from objects on the road. Most highway diesel engines have oil pans that can be removed from the engine while it is in chassis. It is advisable to drain the oil sump before removing it. After removal the oil pan should be cleaned of gasket residues and washed with a pressure washer. Inspect the oil pan mating flanges, check for cracks and inspect the drain plug threads. Cast aluminum oil pans that fasten both to the cylinder block and to the flywheel housing are prone to stress cracking in the rear. With these oil pans torque sequences and values are critical. After removing and cleaning, cast aluminum oil pans can be successfully repair welded without distortion using the tungsten inert gas (TIG) process providing the cracks are not too large and the oil saturated area of the crack is ground clean. Whenever an oil pan has indications of porosity resulting from corrosion, it should always be replaced. The Technician should always be aware of the fact that a failure of the oil sump, or its drain plug, on the road may cause severe engine damage.

Task 2

Disassemble, clean, and inspect engine block for cracks; check mating surfaces for damage or warpage; check condition of passages, core, and gallery plugs; inspect threaded holes, studs, dowel pins, and bolts for serviceability; service/replace as needed.

71. After the Technician cleans the engine block, what should he or she do first to the block?
    A. Paint the engine block.
    B. Test the engine block for cracks.
    C. Remove all cylinder sleeves.
    D. Plug all oil ports.

72. A recently overhauled engine develops cylinder liquid lock. After clearing the liquid from the cylinder, inspecting the heads, and replacing the head gaskets the problem reoccurs. *Technician A* says the intercooler may be defective. *Technician B* says the head may have a hairline crack. Who is right?
    A. A only
    B. B only
    C. Both A and B
    D. Neither A nor B

Hint

Electromagnetic flux test the block for cracks at each out-of-chassis overhaul. The block counterbore and packing ring area must be completely cleaned of all rust, scale, and grease. Most scale and/or rust can be removed using a piece of crocus cloth or 100 to 120 grit emery paper or wet-dry sandpaper. Inspect the counterbore closely for cracks, and ensure that no rough or eroded areas exist to ruin a sleeve or cut its O-ring. Damage of this nature may require resleeving the counterbore by an experienced mechanic with special tools. If the counterbore is in good shape, use a dial indicator mounted on a fixture to measure the counterbore top depth. A depth micrometer can be used in place of the dial indicator if firm contact is maintained with the block surface.

Check for deck warpage using a straightedge and feeler gauge. Check the main bearing bore alignment. Check the engine service history to ensure that the engine has not previously line bored. Select the correct master bar for the engine being tested, and then clamp it in position by torquing down the main bearing caps minus the main bearings. The master bar should rotate in the cylinder block main bearing line bore without binding. If the master bar binds, the main bearing bores should be line bored.

Check the cam bore dimensions and install the cam bushings with the correct cam bushing installation equipment Be sure the oil holes are aligned between the camshaft bearing and the oil hole in the bearing bore before installing the bushing. Install sealant on the gallery and expansion plugs before installation.

Task 3

Pressure test engine block for coolant leakage; service as needed.

73. When pressure testing an engine block, *Technician A* says the block need only be submerged in water

for about 20 to 30 minutes and the water checked for bubbles. *Technician B* says cracked cylinder blocks must be replaced. Who is right?
A. A only
B. B only
C. Both A and B
D. Neither A nor B

74. *Technician A* says to properly test an engine block for cracks, it should be submerged in a tank filled with cold water. *Technician B* says an engine block can be checked for water leaks by pressurizing it for about two hours without loss of pressure. Who is right?
A. A only
B. B only
C. Both A and B
D. Neither A nor B

Hint

After the cylinder block has been cleaned, it must be pressure tested for leaks. The following method may be used when a large enough water tank is available and the cylinder block is completely stripped of all parts.

1. *Seal off the water inlet and outlet holes airtight. This can be done by using steel plates and suitable rubber gaskets held in place by bolts. Drill and tap one cover plate to provide a connection for an airline.*
2. *Immerse the block for 20 minutes in a tank of water heated to 180 to 200°F (82–93°C).*
3. *Apply 40 psi (276 kPa) air pressure to the water jacket and observe the water in the tank for bubbles, which indicate the presence of cracks, or leaks in the block. A cracked cylinder block must be replaced.*
4. *After the pressure test is completed, remove the block from the water tank. Then, remove the plates and gaskets and dry the block with compressed air.*

Task 4

Inspect cylinder sleeve counterbore and lower bore; check bore distortion; determine needed service.

75. A small crack is detected in a wet sleeve counterbore. *Technician A* says to use a heliarc to seal the crack and recut the counterbore. *Technician B* says the block needs to be sent out to have the counterbore resleeved. Who is right?
A. A only
B. B only
C. Both A and B
D. Neither A nor B

76. *Technician A* says that with a wet sleeve type cylinder, the cylinder only seals at the top. *Technician B* says in this type of sleeve the coolant can pass around the sleeve. Who is right?
A. A only
B. B only
C. Both A and B
D. Neither A nor B

Hint

Check the cylinder sleeve counterbore for the correct depth and circumference. Counterbore depth should typically not vary by more than 0.001 in. (0.025 mm). Counterbore depth can be subtracted from the sleeve flange dimension to calculate sleeve protrusion. Recut and shim the counterbore using the OEM recommended tools and specifications.

Task 5

Inspect and measure cylinder walls or liners for wear and damage; determine needed service.

77. *Technician A* says that when replacing a wet sleeve you should always replace the seals on the sleeve. *Technician B* says that to remove the sleeve you should tap it out with a hammer. Who is right?
A. A only
B. B only
C. Both A and B
D. Neither A nor B

78. Which of the following tools is LEAST likely to reveal a defective cylinder sleeve?
A. Inside micrometer
B. Cylinder gauge
C. Dial indicator
D. Snap gauge or bore gauge

Hint

Before reinstalling, old wet cylinder sleeves should be cleaned by immersion or glass bead blasting and closely inspected. Sleeves that fail the inspection should be discarded to ensure maximum overhaul life. The inspection should include checking for tapers using an inside micrometer, snap gauge, or cylinder gauge. An out-of-round condition can be detected by measuring the diameter at one point inside the cylinder and rotating the inside micrometer around the cylinder 180 degrees. To check liner-to-block height, first install and leave a liner clamping tool in place, holding the liner down. Next measure the liner protrusion with a sled gauge. The liner protrusion should be less than 0.003 in. (0.0762 mm) with no more than a 0.002 in. (0.050 mm) variation between adjacent liners.

Task 6

Deglaze and clean cylinder walls or liners.

79. After honing a cylinder liner, *Technician A* says the liner should have a 60-degree crossover angle pattern with a 15 to 20-microinch crosshatch. *Technician B* says to wash the liner surface with solvent. Who is right?
    A. A only
    B. B only
    C. Both A and B
    D. Neither A nor B

80. After wet-type cylinder sleeves have been removed, the packing ring area and counterbore lip show signs of light rust and scale. *Technician A* says this is normal and no additional action is required except ensuring antifreeze protection includes rust inhibitors. *Technician B* says to use 100 to 120 grit emery paper to remove rust and scale, and check for erosion and pitting. Who is right?
    A. A only
    B. B only
    C. Both A and B
    D. Neither A nor B

Hint

After cleaning and pressure testing, inspect the cylinder block. Since most of the engine cooling is accomplished by heat transfer through the cylinder liners to the water jacket, a good liner to block contact must exist when the engine is operating. Whenever the cylinder liners are removed from an engine, the block bores must be inspected. Before attempting to check the block bores, hone them throughout their entire length until about 75 percent of the area above the ports has been cleaned up.

When honing a sleeve, use a hone in which the cutting radius of the stones can be set in a fixed position to remove irregularities in the bore rather than following the irregularities as with a spring-loaded hone. Clean the stones frequently with a wire brush to prevent stone loading. Follow the hone manufacturer instructions regarding the use of oil or kerosene on the stones. Do not use such cutting agents with a dry hone.

Insert the hone in the bore and adjust the stones snugly to the narrowest section. When correctly adjusted, the hone will not shake in the bore, but will drag freely up and down the bore when the hone is not running.

When honing is completed, the bore should have a 60-degree crossover angle pattern 15 to 20 microinch crosshatch. It should be clearly visible by eye. Wash the cylinder block thoroughly with soap and water after the honing operation is completed.

Task 7

Replace cylinder liners and seals; check and adjust liner height.

81. When installing the cylinder insert and liner shown in Figure 2–12, all of these statements are true EXCEPT:
    A. an aluminum block should be heated by submerging it in a tank of hot water for 20 minutes.
    B. the block bore and counterbore must be clean before insert and liner installation.
    C. a hydraulic press must be used to install the liner in the block.
    D. the liner must be pushed downward until the liner flange contacts the insert.

82. When performing the adjustment in Figure 2–13, *Technician A* says the liner flange must extend the specified distance above the top of the block surface. *Technician B* says a 0.010 in. (0.254 mm) variation in cylinder liner heights is acceptable. Who is right?
    A. A only
    B. B only

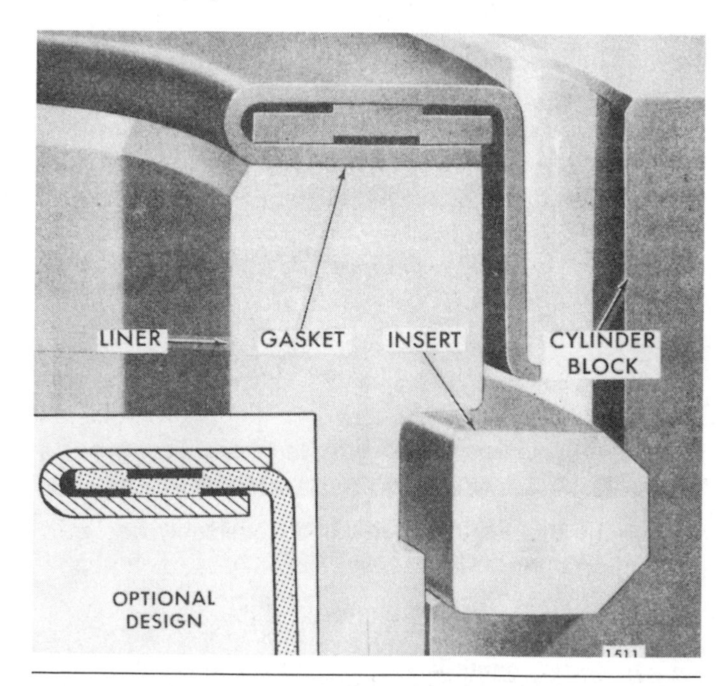

Figure 2–12 Cylinder insert and liner installation. *(Courtesy Detroit Diesel Corporation)*

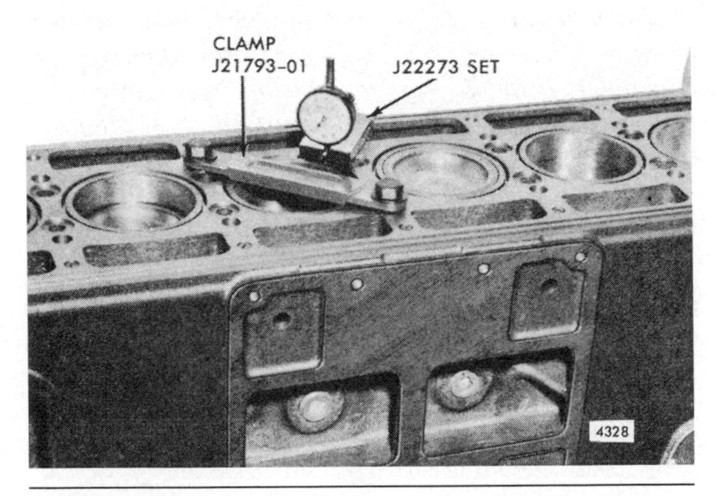

Figure 2–13 Cylinder liner measurement. *(Courtesy Detroit Diesel Corporation)*

C. Both A and B
D. Neither A nor B

Hint

Before liner removal from an aluminum block, the block should be heated by submerging it in a tank of hot water or circulating steam through the coolant passages. A puller with the proper size adapters is used to remove the liners from the block. Selective fitting of liners to block bores is good practice whether or not the block bore has been machined. To selective fit a set of dry liners to a block, measure the inside diameter of each block bore across the North-South and East-West faces and grade in order of size. Next, get the set of new liners and measure the outside diameter of each. Once again grade in order of size. Ensure that every measurement falls within the OEM specifications. Then fit the liner with the largest OD to the block bore with the largest ID. Loose fits of dry liners around 0.0015 in. (0.035 mm) are common. An aluminum block should be heated before installing liners and inserts. On these installations the liners are installed by hand. After installation the distance must be measured from the top of the block to the top of the liner. If this distance is not within specifications, shims may be placed under the insert.

Task 8

Inspect in-block camshaft bearings for wear and damage; replace as necessary.

83. In Figure 2–14 the Technician is measuring:
    A. cam lobe wear.
    B. cam bearing journal wear.
    C. cam lift.
    D. cam duration.

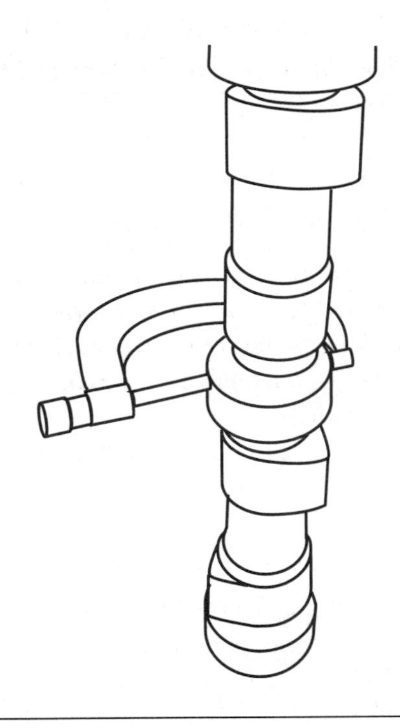

Figure 2–14 Camshaft measurement.

84. When servicing the camshaft bearing in Figure 2–15:
    A. the bearing clearance is the difference between the camshaft journal diameter and the camshaft bearing diameter.
    B. the bearing may be installed in any position in the bearing bore.
    C. the bearing must extend 0.005 in. (0.127 mm) from the front block surface.
    D. the circular bearing may be installed in a scored bearing bore.

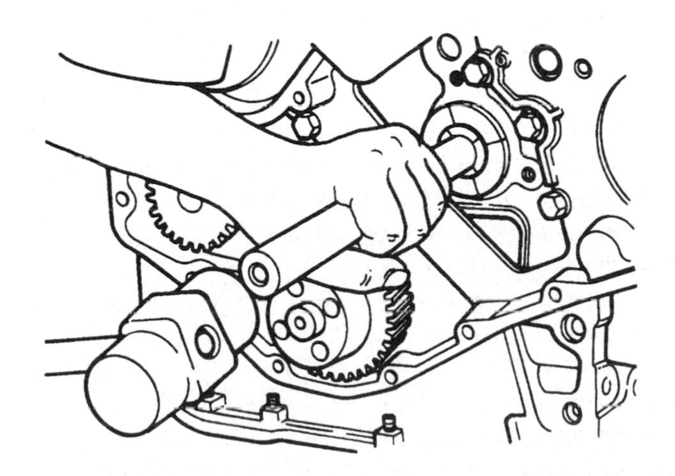

Figure 2–15 Camshaft bearing installation. *(Courtesy of Ford Motor Company)*

Hint

The camshaft journals are supported by pressure-lubricated bearings in the block. The camshaft is subjected to loading whenever a cam is actuating the valvetrain. Camshaft bearings are normally replaced during engine overhaul. If they are to be reused they must be measured with a dial bore gauge or telescoping gauge and micrometer to ensure that they are within the OEM reuse parameters. Camshaft bushings in the block are removed sequentially starting usually from the front to the back of the cylinder block using a correctly sized bushing driver (mandrel) and slide hammer. Cam bearing split shells are retained by camshaft bearing cap crush on overhead camshafts or lock rings on Detroit Diesel Corporation (DDC) 53, 71, and 92 Series. Camshaft bearings in the block are installed with the same tools used to remove them. When installing bushings to a cylinder block, care should be taken to properly align the oil holes in the bearing and block. Bearing clearance is the camshaft bushing dimension minus the camshaft journal dimension measured with a micrometer.

Task 9

Inspect, measure, and replace/reinstall in-block camshaft; measure/adjust end play.

85. When inspecting and measuring the camshaft lobes as shown in Figure 2–16 all of these statements are true EXCEPT:
    A. if the camshaft lobes are pitted, camshaft replacement is necessary.
    B. if the camshaft has edge deterioration for 20 degrees on either side of the cam lobe tip, camshaft replacement is required.
    C. if the transfer pump cam lobe measures 1.63 in. (41.4 mm) the camshaft must be replaced.
    D. if one intake camshaft lobe measures 1.95 in. (49.53 mm) the camshaft must be replaced.

86. After the camshaft gear in Figure 2–17 has been installed, a 0.010 in. feeler gauge can be inserted between the gear and the camshaft shoulder. *Technician A* says the gear is not properly seated. *Technician B* says the camshaft should be replaced. Who is right?
    A. A only
    B. B only
    C. Both A and B
    D. Neither A nor B

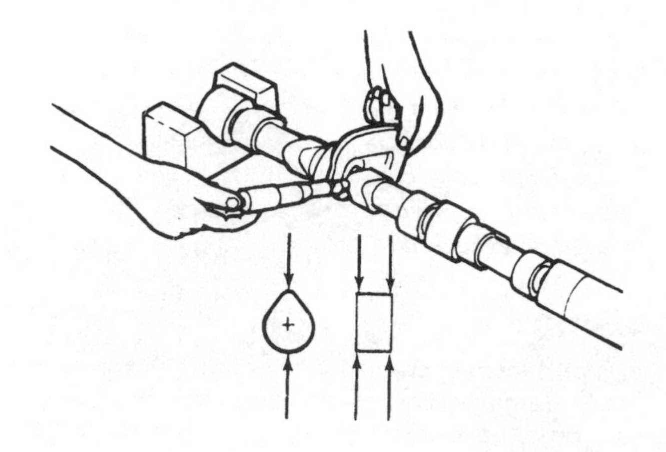

DIAMETER AT PEAK OF CAM LOBE

Lobe	Min.	Max.
Intake	51.774mm (2.0383 In.)	52.251mm (2.0571 In.)
Exhaust	51.569mm (2.0313 In.)	52.073mm (2.0501 In.)
Lift Pump	41.310mm (1.6264 In.)	41.829mm (1.6468 In.)

Figure 2–16 Camshaft inspection and measurement. *(Courtesy of Ford Motor Company)*

Hint

Some camshaft gears must be pressed onto the camshaft with a hydraulic press. Be sure the camshaft is properly supported on the press to avoid camshaft damage. If a key is positioned between the camshaft and the gear, be sure the key is properly positioned. The gear must be pressed onto the shaft until it contacts the front journal shoulder.

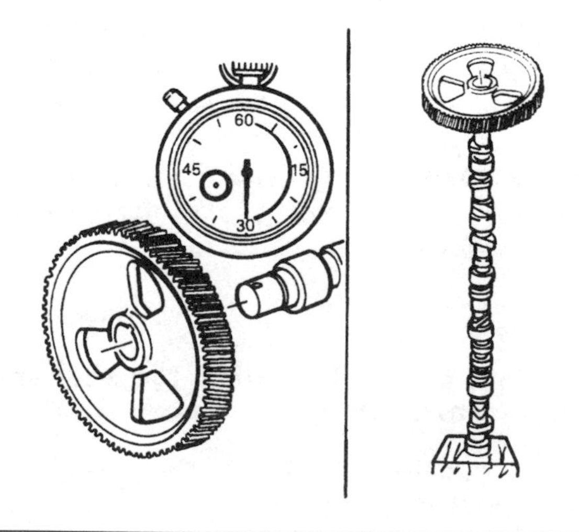

Figure 2–17 Camshaft and gear. *(Courtesy of Ford Motor Company)*

Camshaft end play is defined either by free or captured thrust washers/plates. Thrust loads are not normally excessive unless the camshaft is driven by a helical toothed gear in which instance there is more likely to be wear at the thrust faces. End play is best measured with a dial indicator: the camshaft should be gently levered longitudinally rearward then forward and the travel measured and checked to specifications.

Task 10

Clean and inspect crankshaft for surface cracks and journal damage; check condition of oil passages; check passage plugs; measure journal diameters; determine needed service.

87. Multiple fractures are discovered in the crankshaft. *Technician A* says the cause may be found by carefully inspecting the vibration damper. *Technician B* says the flywheel or torque converter needs to be closely inspected as a possible cause. Who is right?
    A. A only
    B. B only
    C. Both A and B
    D. Neither A nor B

88. When measuring the crankshaft journal in Figure 2–18 the difference between measurements:
    A. A and B indicates horizontal taper.
    B. C and D indicates vertical taper.
    C. A and C indicates out-of-round.
    D. A and D indicates vertical taper.

Hint

An inspection of the crankshaft begins with the connecting rod and main bearing shells. The bearing shells will usually show signs of damage before the crank. If the bearings indicate damage, inspect the crankshaft journal associated with that shell before it is cleaned. Clean the crankshaft and visually inspect for damage. This includes cracked or worn front hub key slots, the main bearing journals for scoring or bluing, dowel pins and holes in the flange for cracks and wear, oil supply holes for cracks, and grooving in the seal contact areas. An out-of-round journal can be detected by using an outside micrometer to measure in at least two places around the journal's diameter. Generally, this measurement should not exceed 0.002 in. (0.051 mm). To find a tapered journal, measure the diameter nearest each side and the middle of the journal. Check thrust surfaces visually for scoring, nicks, and gouges.

Torsional crankshaft failures result in cracks in the crankshaft. Torsional failures may be caused by slipped vibration damper, or improper flywheel balance. A bent crankshaft may be caused by improper main bearing bore alignment.

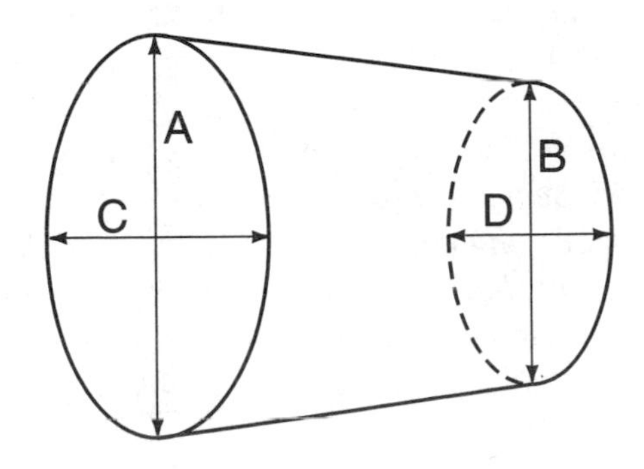

Figure 2–18 Crankshaft measurements. *(Courtesy of Ford Motor Company)*

Task 11

Check main bearing bore alignment and cap fit; determine needed service.

89. When replacing the existing main bearings with undersize bearings, *Technician A* says the crankshaft journals should be machined to fit the bearings. *Technician B* says the main bores should also be line bored. Who is right?
    A. A only
    B. B only
    C. Both A and B
    D. Neither A nor B

90. On the main bearing cap in Figure 2–19, *Technician A* says the numbered side of the cap goes toward the blower side of the engine. *Technician B* says this bearing cap may be installed on any of the main bearings. Who is right?
    A. A only
    B. B only
    C. Both A and B
    D. Neither A nor B

Hint

Inspection of the main bearing bore alignment and out-of-roundness can prevent possible damage to the crankshaft or premature bearing failure. If the bearing bore is out of alignment uneven pressure is applied to each of the bearings, causing the crank to flex as it is rotated. This flexing may cause the crank to fracture and result in bearings with unusual wear patterns. Verify the alignment using a master bar placed across the main bearing bores with the bearings removed. Use an inside microm-

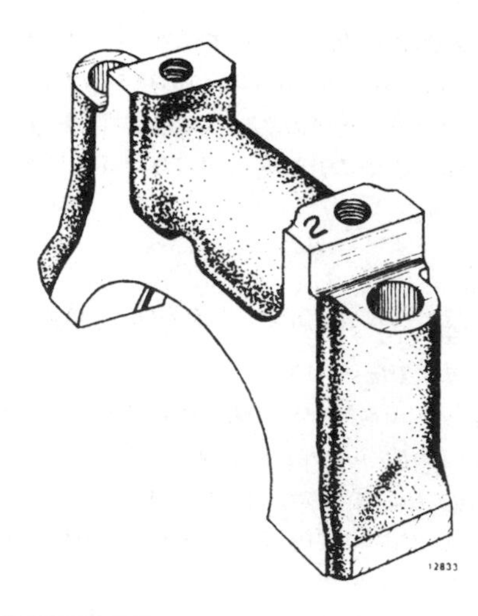

Figure 2–19 Main bearing cap. *(Courtesy Detroit Diesel Corporation)*

eter to check the bearing bore for correct diameter and an out-of-round condition. If the bearing bore is too large, bearing shells will not seat properly against the bore. This may cause inadequate heat transfer from the shells to the bore, turning, or movement of shells in the bore, and misalignment of the crank when load is applied. An out-of-round bore will cause uneven torque on the crank, which could lead to fatigue fracturing, or a broken crankshaft.

Task 12

Inspect and replace main bearings; check bearing clearances; check and adjust crankshaft end play.

91. *Technician A* says that main bearing shells should receive a light coat of oil before final installation. *Technician B* says that when checking the bearing, install plastigage in all the bearings, torque to specifications, and rotate the crankshaft. Who is right?
    A. A only
    B. B only
    C. Both A and B
    D. Neither A nor B

92. The measurement in Figure 2–20 is more than specified. To correct this problem it is necessary to:
    A. install thicker thrust washers on each side of the rear main bearing cap.
    B. install number 3 main bearing inserts with thicker thrust surfaces.
    C. install thicker spacers between the front of the crankshaft and the front main bearing.

Figure 2–20 Crankshaft end play measurement. *(Courtesy Detroit Diesel Corporation)*

    D. install thicker spacers between the vibration damper and the crankshaft gear.

Hint

Correct installation of bearing shells is essential to any successful overhaul. Many premature bearing failures are caused by improper assembly of bearings. After reconditioning the block and selecting proper-sized bearings, carefully install the top half of each bearing into the cylinder block. Ensure the locating lugs are fitted into the matching slot in the cylinder block bore. Look through the oil passage in the shell to ensure that it is aligned to the oil passage in the bore. Failure to align the oil passages will result in oil starvation, bearing fatigue, scoring, metal in oil, and bearing failure. Carefully install the upper half of the rear main seal, and then install thecrank in the block. Align the timing marks on the crank gear and cam gear. Use plastigage to check all main bearing clearances. If clearances are not within specification, check for improperly sized bearings, dirt, or metal under shells, or misalignment of shells. If clearances are correct, remove the plastigage, lubricate bearings, reinstall bearing caps, and torque to specifications. After all the bearing caps are torqued to specifications, rotate the crankshaft by hand to check for binding. If binding occurs, loosen all caps and tighten individually to determine which bearing is at fault. Check and adjust crankshaft end play.

Task 13

Inspect, reinstall, and time the drive geartrain. This includes checking timing sensors, gear wear, and backlash of crankshaft, camshaft, auxiliary, drive, and idler gears; service shafts, bushings, and bearings.

93. *Technician A* says that all gears in the timing geartrain must be inspected for tooth wear, including the crankshaft gear. *Technician B* says a slight roll or lip on each gear tooth is acceptable, because gears will normally mate this way during the first 500 miles of operation. Who is right?
    A. A only
    B. B only
    C. Both A and B
    D. Neither A nor B

94. Figure 2–21 indicates the condition of a key used on the camshaft gear. *Technician A* says the manufacturer used this to offset the timing and it must be replaced with the same type. *Technician B* says that the camshaft, gear, and key must be replaced because the camshaft became jammed and twisted and started to shear the key. Who is right?
    A. A only
    B. B only
    C. Both A and B
    D. Neither A nor B

Hint

The timing geartrain including all gears that drive the camshaft must be inspected for damage and wear. The gear teeth must be in satisfactory condition to maintain the timing of intake and exhaust valves and on some engines, the injectors. Excess tooth wear can seriously affect engine performance and long-term operation and may cause damage to the piston, cylinder, and crank. Carefully inspect for a slight roll or lip on each gear tooth, in addition to looking for chipping, pitting, and burring. If these are present on any gear replace it and examine the mating gears for similar damage. Installation of these gears is very critical as it affects the engine timing. The camshaft is usually keyed between the shaft and gear, although some manufacturers use offset keys to shift the keyway and thereby advance the timing.

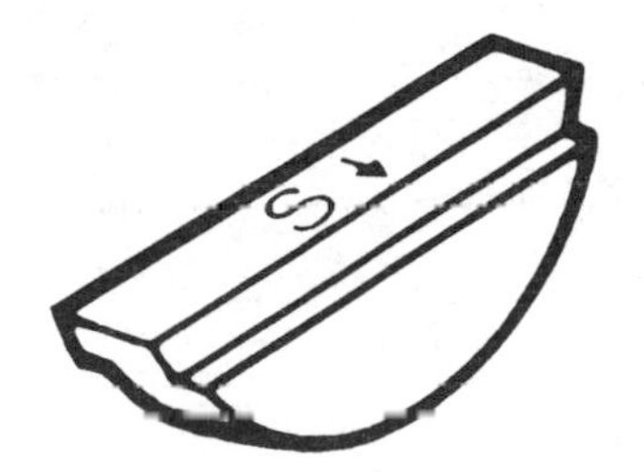

Figure 2–21 Camshaft key. *(Courtesy Detroit Diesel Corporation)*

Task 14

Clean, inspect, measure, or replace pistons, pins, and retainers.

95. *Technician A* says pistons should be laid out in the order in which they came out. *Technician B* says installing an improper injector nozzle could cause scoring and cracking of pistons skirts. Who is right?
    A. A only
    B. B only
    C. Both A and B
    D. Neither A nor B

96. The LEAST likely cause of cracks and scoring in the piston skirts is:
    A. excessive lubrication levels in the crankcase.
    B. improper piston clearance.
    C. excessive fuel setting.
    D. engine overheating.

Hint

After removing each piston, carefully inspect it and its matching cylinder for possible signs of damage and try to determine probable cause. Check for scoring on the piston skirt and compare it to the matching side of the cylinder. Scoring may have been caused by engine overheating, excessive fuel setting, improper piston clearance, insufficient lubrication, or an improper injector or nozzle. A crack in the skirt or ring lands may be accompanied by a bluing or gouges in the skirt and the matching cylinder. Cracked pistons may be caused by excessive use of starting fluid or introduction of starter fluid at operating temperatures, excessive piston clearance, or foreign objects in the cylinder such as a dropped valve. A nick in the bottom of the skirt and connecting rod may be caused by improperly installed connecting rods. Uneven wear in the skirt matched with bluing or scuffing on the cylinder could be caused by normal operation at low temperatures, dirty lubricating oil, too little piston clearance, or dirty intake air. The effects of stuck or broken rings may appear on the cylinder as scrapes and gouges, scratches and carbon buildup, or glazing. The leading causes are overheating, insufficient lubrication, and excessive fuel settings. Worn piston pin bores can also be caused by normal wear, insufficient lubrication, or dirty lubricating oil. During an overhaul the piston and rings are usually replaced.

Retarded injection timing causes excessive cylinder temperatures and burning/erosion damage through the central crown area of the piston. An indicator of retarded injection timing is piston crown scorching under the injector nozzle orifice.

Task 15

Measure piston-to-cylinder wall clearance.

97. While performing the measurement in Figure 2–22, *Technician A* says if a 6 lb. (2.72 kg) pull is required to remove a 0.004 in. (0.101 mm) feeler gauge, the piston to cylinder clearance is 0.005 in. (0.127 mm). *Technician B* says the piston should be placed in hot water before the measurement. Who is right?
    A. A only
    B. B only
    C. Both A and B
    D. Neither A nor B

98. All of these statements about piston-to-cylinder clearance are true EXCEPT:
    A. piston scoring may be caused by less than specified piston-to-cylinder clearance.
    B. excessive piston-to-cylinder clearance may cause piston slapping.
    C. the piston-to-cylinder clearance on cam ground pistons must be measured at 90 degrees to the piston pin.
    D. a scored piston may cause lower than specified oil pressure.

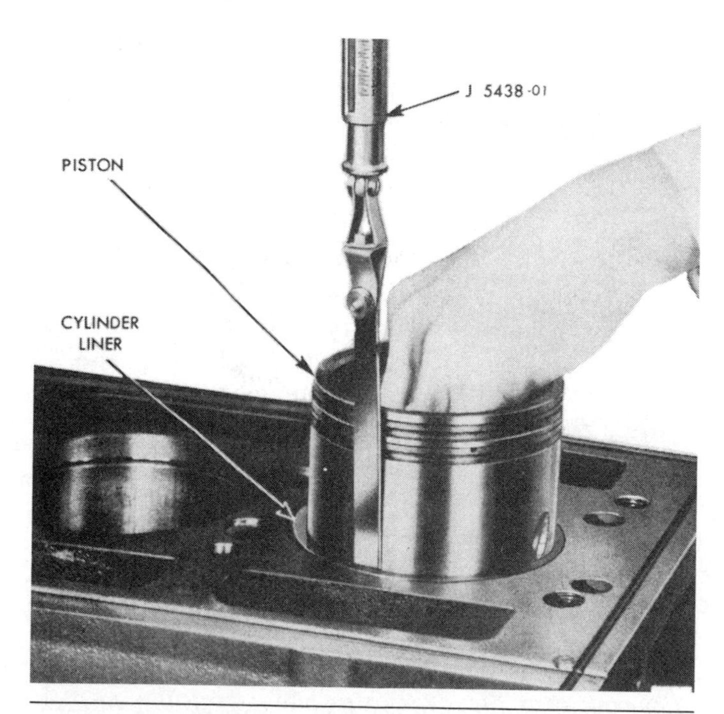

Figure 2–22 Piston-to-cylinder clearance measurement. *(Courtesy Detroit Diesel Corporation)*

Hint

Use a micrometer to measure the piston diameter at right angles to the piston pin bore and one inch below the bottom edge of the lowest ring groove. Compare this measurement to the parent bore diameter. The difference between these two values becomes the piston-to-cylinder wall clearance. The average running clearance is 0.006 in. (0.152 mm), but not less than 0.002 in. (0.050 mm). Insufficient clearance will cause premature piston or cylinder/liner failure. Measure cam ground pistons at right angles (90 degrees) to the piston pin. The running clearance of all pistons can be measured using a spring scale with a feeler gauge attached to its end. Insert the specified feeler gauge in the cylinder. Lubricate the piston with oil and install it bottom up with no rings, and the feeler gauge between the cylinder and the piston. Position the piston about 2 inches (50 mm) below the fire deck with the piston pin bore in line with the crankshaft. When the spring scale indicates the specified force in pounds as the feeler gauge is being withdrawn the running clearance is correct.

Task 16

Check ring-to-groove clearance and end gaps; install rings on pistons.

99. All of these statements about the measurement in Figure 2–23 are true EXCEPT:
    A. if a compression ring gap is less than specified, the ends of the ring may be filed or stoned.

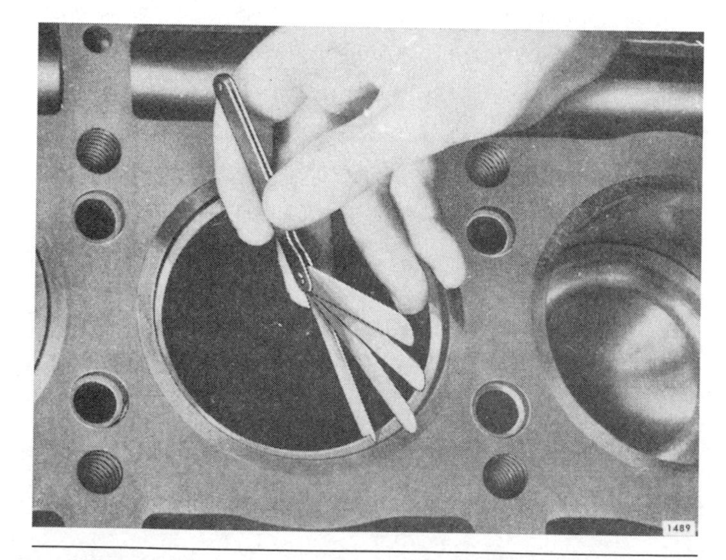

Figure 2–23 Piston ring gap measurement. *(Courtesy Detroit Diesel Corporation)*

B. the piston ring should be positioned below the normal ring travel in the cylinder.
C. the ends of the piston ring must remain square if they are filed or stoned.
D. a piston should be used to push the ring down in the cylinder.

100. While discussing pistons and piston rings, *Technician A* says the second ring from the top of the piston may be called a fire ring. *Technician B* says all fire rings are keystone shaped. Who is right?
A. A only
B. B only
C. Both A and B
D. Neither A nor B

Hint

Ensure the rings have the proper end gap before installing them on the piston. This measurement is taken while the rings are in the cylinder. Slide each ring into the cylinder, using a piston with the rings removed, and measure the gap between the ends of the ring. Install the rings in the proper piston grooves, and stagger the ring gaps around the piston. Ring side clearance is the installed clearance between the ring and the ring groove. The dimension is measured using a feeler gauge.

Task 17

Identify piston and bearing wear patterns that indicate connecting rod alignment or bearing bore problems; check bearing bore and bushing condition; replace as needed.

101. Cracks are present in the piston skirt. *Technician A* says to check the connecting rod and bottom of the cylinder liner for damage caused by improperly installed connecting rods. *Technician B* says to check for insufficient piston-to-cylinder clearance. Who is right?
A. A only
B. B only
C. Both A and B
D. Neither A nor B

102. Wear on the edges of a connecting rod bearing indicates:
A. excessive bearing clearance.
B. connecting rod misalignment.
C. insufficient bearing clearance.
D. contaminated engine oil.

Hint

When connecting rod bearings show excessive wear on the sides of the bearing, the connecting rod may be misaligned. If a connecting rod bore is scored or damaged the rod must be replaced. Any evidence of cracks in the connecting rod also requires rod replacement. A connecting rod alignment tester may be used to check connecting rods for bends and twists. The connecting rod length must be measured to check for a stretched condition.

Task 18

Assemble pistons and connecting rods and install in block; replace rod bearings and check clearances; check condition, position, and clearance of piston cooling jets (nozzles).

103. When assembling pistons and connecting rods:
A. the markings on the connecting rod and bearing cap may be on opposite sides of the connecting rod.
B. the markings on the connecting rod and bearing cap may face either side of the engine.
C. on two-cycle diesel engines, the piston and connecting rod assembly should be installed in the liner before installation in the engine.
D. the piston ring gaps should be directly below each other on the piston before it is installed in the cylinder.

104. When servicing spray nozzles located in the top of the connecting rod:
A. the spray nozzle may be removed without removing the top bushing.
B. when installing the nozzle, the holes on the nozzle may be in any position.
C. a press, sleeve, and remover are required to remove the spray nozzle.
D. the spray nozzle should be installed so it is not bottomed in the counterbore.

Hint

Installation of piston pins into the aluminum piston can be made easier by preheating the piston to 200°F (93° C). Slide the piston pin through the piston bore and connecting rod, and then install the piston pin retaining rings. Clamp in a soft-jawed vise by the connecting rod to support the assembly while installing the rings. Ensure the rod number is correct for the cylinder in which it is being placed, and that number is facing the cam on six-cylinder engines or the outside on V8 engines. Protect the rod bearing and journal by covering rod bolts with plastic caps or rubber hose, and carefully guide the rod onto the journal until the bearing shell seats. Use plastigage to measure the connecting rod bearing clearances. The connecting rod caps must be installed so the number on the cap is on the same side as the number on the rod. Recheck the torque of all cap bolts.

Task 19

Inspect, measure, and service/replace crankshaft vibration damper.

105. While discussing viscous-type vibration dampers, *Technician A* says the vibration damper may be disassembled and repaired. *Technician B* says most engine manufacturers recommend damper replacement during a major engine overhaul. Who is right?
   A. A only
   B. B only
   C. Both A and B
   D. Neither A nor B

106. All of these statements about viscous-type vibration dampers are true EXCEPT:
   A. if a puller is not available, a hammer and punch may be used to remove the damper.
   B. if the viscous fluid is leaking from the damper, damper replacement is necessary.
   C. dents in the damper housing may render the damper ineffective.
   D. if the damper is bulged or split, the fluid in the damper has exploded.

Hint

The primary vibration damper function is to reduce crankshaft torsional vibration. The damper housing is coupled to the crankshaft and uses springs, rubber, or viscous fluid to drive the inertia ring. Viscous-type harmonic balancers have become almost universal in truck and bus diesels; the annular housing is hollow and bolted to the crankshaft. Within the hollow housing, the inertia ring is suspended in and driven by silicone gel. The shearing of the viscous fluid film between the drive ring and the inertia ring affect the damping action. A defective vibration damper may result in a failed crankshaft. Visually inspect the damper housing noting any dents or signs of warpage: evidence of either is reason to reject the component.

Check for indications of fluid leakage with the damper in place. Trace evidence of leakage justifies replacement of the damper. With the damper on the crankshaft, check for wobble using a dial indicator. If the wobble exceeds specifications, replace the damper. Viscous dampers should be checked for nicks, cracks, or bulges. Cracks or bulges may indicate that internal fluid has ignited and caused the case to expand.

Task 20

Inspect, install, and align flywheel housing.

107. When performing the adjustment in Figure 2–24, *Technician A* says the dial indicator stem must be positioned so it does not drop into a capscrew hole. *Technician B* says a flywheel housing face runout measurement of 0.005 in. (0.127 mm) is acceptable. Who is right?
   A. A only
   B. B only
   C. Both A and B
   D. Neither A nor B

108. The measurement shown in Figure 2–25 is the:
   A. flywheel bore alignment measurement.
   B. flywheel runout measurement.
   C. flywheel housing bore measurement.
   D. crankshaft end play measurement.

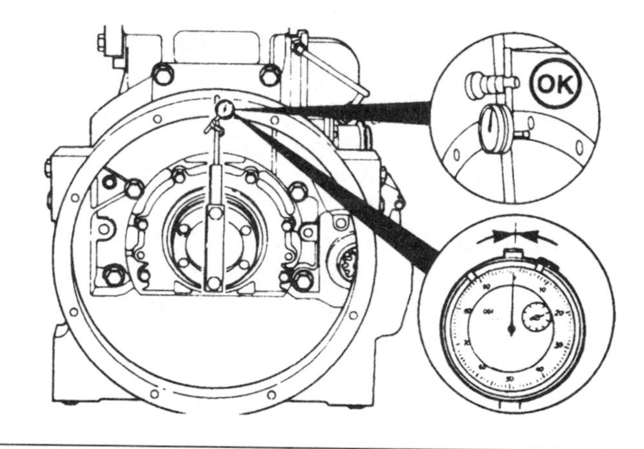

Figure 2–24 Flywheel housing face runout measurement. *(Courtesy of Ford Motor Company)*

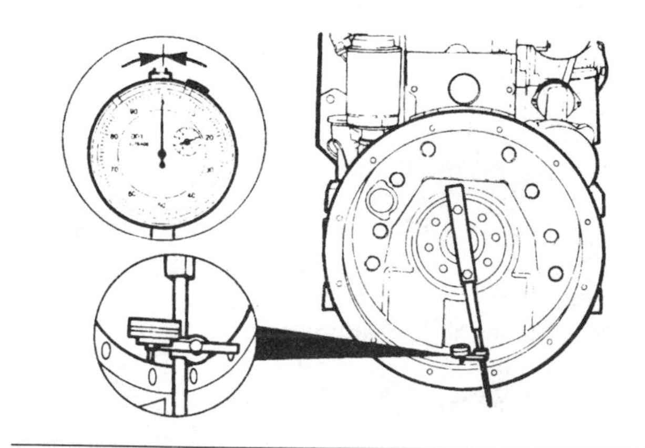

Figure 2–25 Dial indicator measurement. *(Courtesy of Ford Motor Company)*

Hint

Misalignment of the flywheel housing with a manual transmission may cause a grabbing clutch and excessive pilot bushing wear. With an automatic transmission, misalignment of the flywheel housing causes torque converter misalignment. This could result in excessive transmission pump wear and pump seal failure.

Task 21

Inspect crankshaft flange and flywheel/flexplate mating surfaces for burrs; measure runouts; repair as needed.

109. In Figure 2–26 the dial indicator reading is more than specified. *Technician A* says to check the flywheel-to-crankshaft mating surfaces for metal burrs. *Technician B* says this condition may cause the transmission to jump out of gear. Who is right?
   A. A only
   B. B only
   C. Both A and B
   D. Neither A nor B

110. When inspecting and measuring flywheels:
   A. on a pot-type flywheel, the flywheel bore runout may be measured with a dial indicator.
   B. on a pot-type flywheel, the flywheel face may be resurfaced without machining the pot face.
   C. when it is necessary to remove deep scoring from a conventional flywheel, 0.250 in. (6.35 mm) may be machined from the flywheel face.
   D. on a pot-type flywheel if the bore runout is excessive, the flywheel must be replaced.

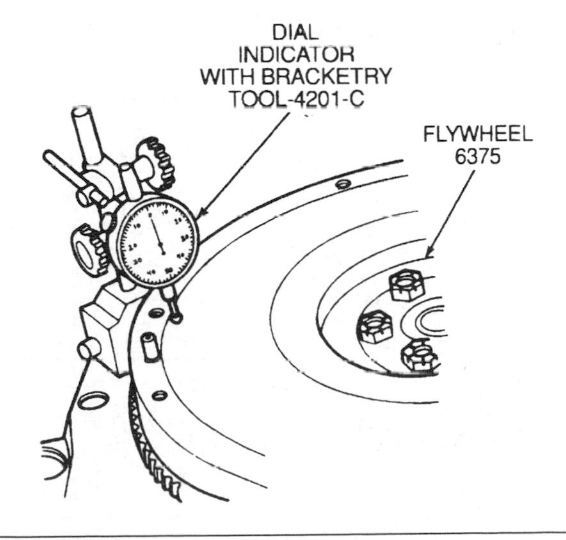

Figure 2–26 Flywheel measurement. *(Courtesy of Ford Motor Company)*

Hint

Flywheels are commonly removed from engines for reasons such as clutch damage, leaking rear main seals, leaking cam plugs, and so on, and care should be taken both when inspecting and reinstalling the flywheel and the flywheel housing. Flywheels should be inspected for face warpage, heat checks, scoring, intermediate drive lug alignment and condition, axial and radial runout using dial indicators, straightedges, and thickness gauges. Damaged flywheel faces may be machined using a flywheel resurfacing lathe to OEM tolerances. Typical maximum machining tolerances range from 0.060 to 0.090 in. (1.50 to 2.30 mm). It is important to note that when resurfacing pot-type flywheel faces, the pot face must have the same amount of material ground away as the flywheel face: the consequence of machining the clutch face only is to have an inoperable clutch.

Task 22

Inspect flywheel/flexplate (including ring gear) for cracks, wear, and runout; determine needed repairs.

111. *Technician A* says the starter ring gear is usually pinned or bolted to the flywheel assembly. *Technician B* says the ring gear may be installed in either upward or downward position on the flywheel. Who is right?
   A. A only
   B. B only
   C. Both A and B
   D. Neither A nor B

112. When trying to start a diesel engine, the starting motor is turning normally but the engine does not crank over. There are no unusual noises from the starting motor or starter drive. *Technician A* says the flywheel mounting bolts may be loose. *Technician B* says the ring gear may be loose and turning on the flywheel. Who is right?
   A. A only
   B. B only
   C. Both A and B
   D. Neither A nor B

Hint

Inspect the flywheel for cracks, missing or damaged ring gear teeth, distortion, or oblong mounting holes, or a pilot bearing with a loose or improper fit. Inspect the vibration damper, flywheel/flexplate, and clutch/torque converter for damage that may cause an unbalanced condition. If damage is detected there may also have been uneven wear on the main bearings and/or stress on the crankshaft.

The ring gear is a shrink fit on the flywheel. To install a new ring gear, place the flywheel on a flat, level surface and check that the ring gear seating surface is free from dirt, nicks, and burrs. Ensure that the new ring gear is the correct one and if its teeth are chamfered on one side, that they will face the cranking motor pinion after installation. Next, the ring gear must be expanded using heat so that it can be shrunk to the flywheel. When the ring gear has been heated evenly to the correct temperature it will usually drop into position and almost instantly contract to the flywheel.

Lubrication and Cooling Systems Diagnosis and Repair

Task 1

Check engine oil pressure, pressure gauge, and sending unit.

113. On a vehicle with an electrical oil pressure gauge, all of the following will cause the faulty readings EXCEPT:
    A. low voltage to the gauge.
    B. a short in the wire from the sending unit.
    C. low coolant level.
    D. a defective sending unit.

114. An electrical oil pressure gauge does not indicate any pressure and the engine operation is satisfactory. Which of the following is at fault?
    A. The passage to the sensor is open.
    B. The wrong engine oil was installed in the engine.
    C. The wiring to the sensor is open.
    D. The wrong sensor was installed in this engine.

Hint

Oil pressure gauges are either electrically or mechanically operated, and low oil pressure lights are electrically operated indicators. These devices must be considered any time incorrect or faulty oil pressure readings are obtained. Mechanical gauges use the engine oil pressure routed through a tube to operate the meter movements. If the tube becomes punctured, kinked, or clogged, the fluid is not able to operate the meter movements. In this case, the tube must be replaced if it is punctured or kinked or blown clean if clogged. Common problems associated with electrical gauges are open or shorted wiring, a clogged passage to the sensor, or a defective sensor. Use a voltmeter to check the resistance of the sensor and continuity of the wiring. Indicator lights use a pressure switch and a light bulb to indicate oil pressure. If the engine oil pressure is actually low, ensure that the oil is of the proper grade as specified by the manufacturer, and that the by-pass valve is properly set before considering the engine for overhaul.

Task 2

Inspect, measure, repair/replace oil pump, drives, inlet pipes, and screens.

115. The following are normal oil pump component measurements EXCEPT:
    A. inner rotor diameter.
    B. clearance between the rotors.
    C. inner and outer rotor thickness.
    D. outer rotor-to-housing clearance.

116. The oil pump measurement in Figure 2–27 is:
    A. rotor tooth tip clearance.
    B. outer rotor-to-housing clearance.
    C. rotor backlash.
    D. rotor end play.

Hint

Some engine manufacturers use external oil pumps that have oil lines routed out of the block to the pump. Other manufacturers use internal oil pumps that can be accessed from the crankcase by removing the oil pan. Remove the bolts that hold the oil pump to the engine and then carefully remove the oil pump. With external oil pumps, remove the cover from the pump body and inspect it for wear. Remove and inspect the idler and driven gears for pitting and their teeth for wear. Oil pump measurements are performed with a feeler gauge and a straightedge. In some cases an outside micrometer is used to measure the rotor thickness. If the pump includes a by-pass valve, it must also be checked for wear and cleaned.

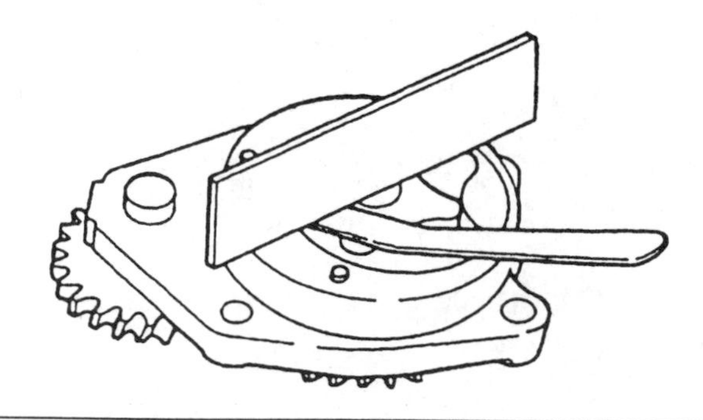

Figure 2–27 Oil pump measurement. *(Courtesy of Ford Motor Company)*

Task 3

Inspect, repair/replace oil pressure regulator valve(s), by-pass valve(s), and filters.

117. While inspecting an oil filter housing or its mounting, which of the following is LEAST likely to be a procedure?
   A. Visually checking for cracks
   B. Visually checking gasket surface for nicks
   C. Magna fluxing the housing and metal parts to detect small cracks
   D. Visually inspecting housing passageways for obstructions and racks

118. Two Technicians are discussing the lubrication system on a diesel engine. *Technician A* says there are two different types of filters used, the cartridge and the spin-on type. *Technician B* says they also use a partial type and a full-flow type. Who is right?
   A. A only
   B. B only
   C. Both A and B
   D. Neither A nor B

Hint

Oil filters trap particles that would otherwise cause damage to other critical components, like engine bearings. Regular replacement of filter elements and inspection of the housing and other component parts is necessary. Cracks anywhere in the housing and nicks on machined surfaces may cause major engine failures, and any cracked components must be replaced. Some filter assemblies incorporate a by-pass valve in their housing. These valve assemblies must be removed and inspected for corrosion, wear, and other signs of damage during overhaul or any time the filter assembly contains metal particles. Canister filters should be cut open and the element material examined for content. Suspended metal particles indicate a potential for bearing failure. A milky gray sludge indicates water in the oil caused by a possible head or cylinder block leak, which may already have caused bearing damage. Ensure all gasket surfaces are straight and free of nicks, which could cause an improper seal of the assembly. Reinstall the by-pass valve, and replace the element, gaskets, and seals. When reassembling filter assemblies, apply a small amount of oil to all seals to ensure a tight seal. Prefilling of canister-type filters is recommended to ensure adequate oil is present during startup to lubricate the bearings.

Task 4

Inspect, clean, test, reinstall/replace, and align oil cooler; test, reinstall/replace by-pass valve and oil thermostat; inspect and repair/replace lines and hoses.

119. Which of the following would indicate an oil cooler failure?
   A. Low oil pressure
   B. High oil pressure
   C. Milky gray sludge in oil filter
   D. Bubbles in the cooling system.

120. While discussing the components in Figure 2–28, *Technician A* says the figure shows an engine oil cooler with oil-to-air cooling. *Technician B* says component G is the inlet valve. Who is right?
   A. A only
   B. B only

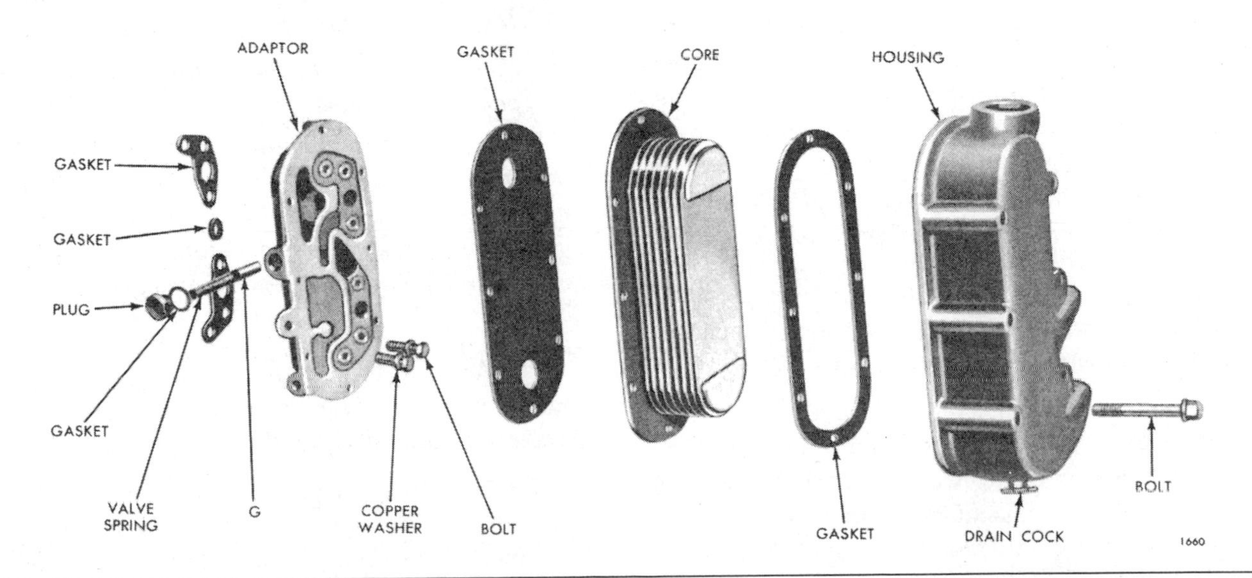

Figure 2–28 Engine oil cooler. *(Courtesy Detroit Diesel Corporation)*

C. Both A and B
D. Neither A nor B

Hint

Complete servicing of the oil cooler should be done as part of a major engine overhaul. Oil cooler failures may cause oil in the coolant, coolant in the oil pan, or milky gray sludge in the filter.

Task 5

Inspect and clean turbocharger lubrication system; repair/replace as needed.

121. A turbocharged diesel engine has blue smoke in the exhaust. *Technician A* says the turbocharger oil drain line should be checked. *Technician B* says the turbine wheel in the turbocharger may be pitted. Who is right?
A. A only
B. B only
C. Both A and B
D. Neither A nor B

122. The MOST likely cause of repeated turbocharger bearing failure is:
A. excessive turbocharger shaft end play.
B. contaminated engine oil.
C. a leak in the air intake system.
D. a leak in the exhaust system before the turbo.

Hint

Inspecting the turbocharger lubrication system begins by checking for signs of leaks. Inspect hoses and tubes for damage, such as frayed or cut hoses and tubes for kinks or nicks. Remove hoses and tubes and check for obstructions. Install the lines after replacing any damaged ones, and install a tee in one of the connections. Attach an oil pressure gauge to the tee, and tighten all connections. Start the engine and monitor the oil pressure gauge. Operate the engine over a range of operating speeds while monitoring the oil pressure. Stop the engine and ensure the pressure drops. All gauge readings must be the same as the instrument panel oil pressure gauge.

Task 6

Inspect, reinstall/replace, and adjust drive belts.

123. *Technician A* says misalignment between pulleys must not exceed 1/16 inch (1.59 mm) for each 12 inches (30.5 cm) of distance between pulley centers. *Technician B* says belts must not touch the bottom of grooves and protrude more than 3/32 inch (2.38 mm) above the top of the groove. Who is right?
A. A only
B. B only
C. Both A and B
D. Neither A nor B

124. *Technician A* says that when two or more identical belts are used on the same pulley, change only the belt that is defective. *Technician B* says belts can be installed by starting them on the edge of the pulley and cranking the engine. Who is right?
A. A only
B. B only
C. Both A and B
D. Neither A nor B

Hint

The following points summarize the most important points when servicing drive belts:

1. *When two or more identical belts are used on the same pulley, all of the belts must be replaced at the same time.*
2. *Make sure the distance between the pulley centers is as short as possible when you install the belts. Do not roll the belts over the pulley. Do not use a tool to pry the belts onto the pulley.*
3. *The pulleys must not be out of alignment more than 1/16 inch (1.59 mm) for each 12 inches (30.5 mm) of distance between the pulley center.*
4. *The belts must not touch the bottom of the pulley grooves. The belts must not protrude more than 3/32 inch (2.38 mm) above the outside diameter of the pulley.*
5. *When identical belts are installed on a pulley, the protrusion of the belts must not vary more than 1/16 inch (1.59 mm).*
6. *Make sure the belts do not touch or hit against any part of the engine.*

Drive belt tension should be measured with a belt tension gauge. After a belt has been adjusted and the engine has been running for at least one hour, stop the engine and check the belt tension. If the tension is less than specified, adjust the belt to the correct value.

Task 7

Check coolant temperature, temperature sensor, temperature gauge, and sending unit.

125. A Technician suspects the mechanical temperature gauge is giving false readings. To check the temperature gauge, all of the following could be used EXCEPT:

A. heat sensitive tape.
B. radiator thermometer.
C. heat sensitive gun.
D. voltmeter.

126. In Figure 2–29 the water temperature indicator sensor:
A. contains a thermostatic switch.
B. increases in resistance value as the coolant temperature decreases.
C. decreases in resistance value as the coolant temperature decreases.
D. decreases in voltage drop as the coolant temperature decreases.

Hint

Temperature gauges may be mechanical or electrical. Mechanical gauges use a liquid-filled sensor attached to the hydraulic meter movement by a tube. This type of gauge is self-contained and must be replaced as an assembly. Common reasons for failure of a mechanical gauge are corrosion on sensor, leaking sensor, a pinched tube, and binding of meter movement of needle. Electrical gauges use a sensor that changes its resistance as temperature changes. This resistance is used to vary the current applied to the meter movements. Failures of an electrical gauge can be caused by defective sensors, an open or short in wiring and open or bad grounds. While meter movements will fail, they are usually a symptom of other problems. Indicator lights typically use a sensor that acts like a switch to turn on a light circuit.

Task 8

Inspect and replace thermostat(s), bypasses, housing(s), and seals.

127. On a vehicle with an overcooling problem, *Technician A* says that a stuck thermostat could cause this problem. *Technician B* says that an overfilled radiator could cause this problem. Who is right?
A. A only
B. B only
C. Both A and B
D. Neither A nor B

128. In the cooling system shown in Figure 2–30, *Technician A* says failure to install one of the thermostats (item 5) causes engine overheating. *Technician B* says the thermostats may be installed in either direction. Who is right?
A. A only
B. B only
C. Both A and B
D. Neither A nor B

Hint

To function effectively, a thermostat must start to open at a specified temperature and be fully open at a set number of degrees above the start-to-open temperature. The cooling system thermostat is normally located either in the coolant manifold or in a housing attached to the coolant manifold. Its primary function is to permit a

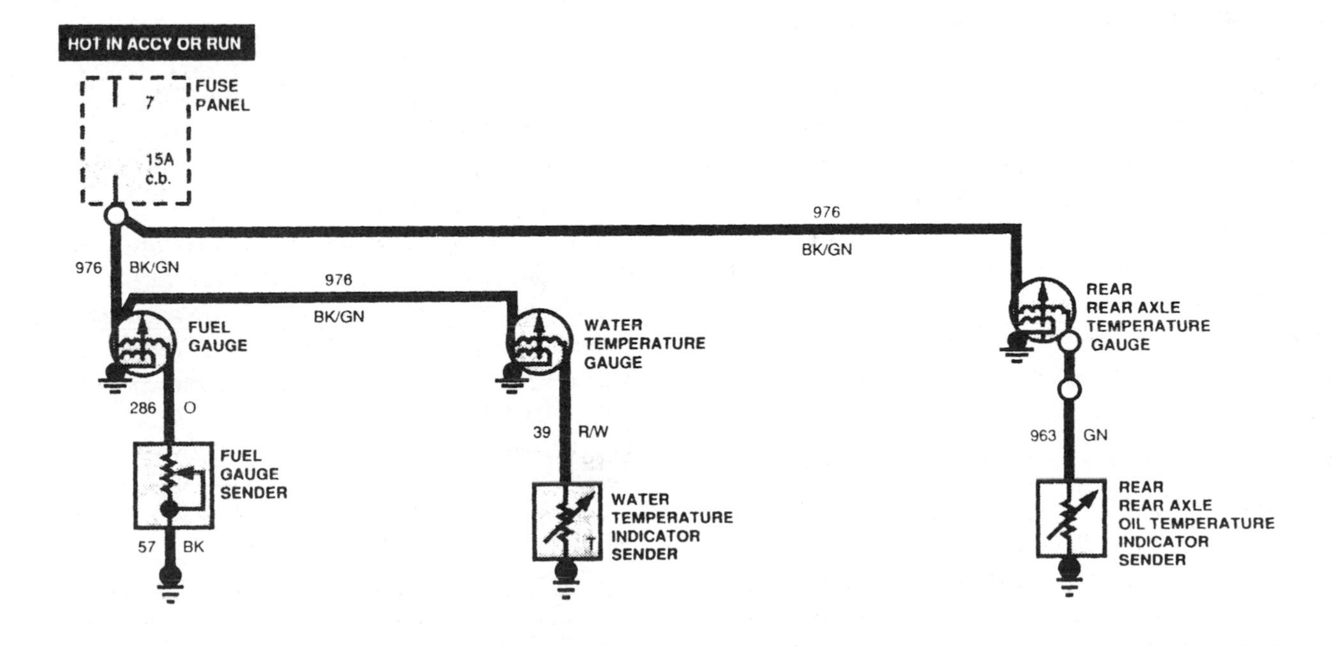

Figure 2–29 Coolant temperature gauge circuit. *(Courtesy of Ford Motor Company)*

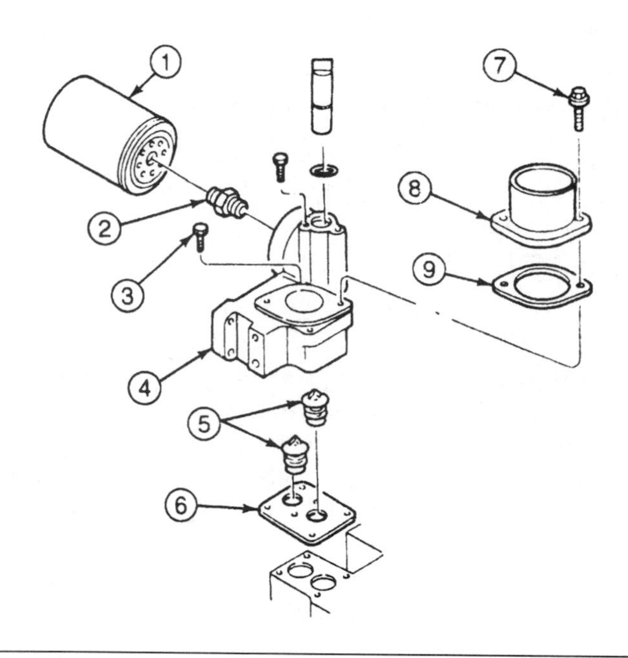

Figure 2–30 Engine thermostats. *(Courtesy of Ford Motor Company)*

rapid warmup of the engine. When the engine has attained its normal operating temperature, the thermostat opens and permits coolant circulation. The heat-sensing thermostat element consists of a wax pellet into which the actuating shaft of the thermostat is immersed. As the wax expands, the actuating shaft is forced outward in the pellet, opening the thermostat.

The term "by-pass circuit" describes the routing of the coolant through the engine cylinder block and head before the thermostat opens. The flow of by-pass coolant permits rapid engine warmup to the required operating temperature.

Running without a thermostat is not recommended. It also contravenes the Environmental Protection Agency (EPA) requirements regarding tampering with emission control components. Removing the thermostat invariably results in the engine running too cool. This can cause vaporized water in the crankcase to condense and cause corrosive acids and sludge in the crankcase. Additionally, low engine running temperatures will increase the emission of hydrocarbons (HC).

Task 9

Flush and refill cooling system; bleed air from system; recover coolant.

129. All these statements about cooling system service are true EXCEPT:
   A. when the cooling system pressure is decreased, the coolant boiling point increases.
   B. if more antifreeze is added to the coolant, the boiling point is increased.
   C. a good quality ethylene glycol antifreeze contains antirust and corrosion inhibitors.
   D. equipment is available to recover, recycle, and reuse coolant solutions.

130. While discussing cooling system service, *Technician A* says that when filling the cooling system the air bleeder valves should be closed. *Technician B* says that if the coolant solution contains over 67 percent antifreeze, heat transfer from the cylinders to the coolant is adversely affected. Who is right?
   A. A only
   B. B only
   C. Both A and B
   D. Neither A nor B

Hint

Engine coolant is a mixture of water, antifreeze, and supplemental cooling additives (SCA). To cool the engine, coolant must have an unobstructed flow throughout the cooling system. Often when refilling the cooling system, air becomes trapped inside the block. If this condition persists and the engine temperature exceeds 212°F (100°C) water will turn to steam. This steam pressure forces more water out of the block and thereby causes an air lock in the cooling system. To prevent this from happening, manufacturers usually install bleeder valves and provide instructions on their use. Generally, bleeder valves are located near the top of the cooling system and block. After a cooling system has been in operation for several years, its efficiency may be reduced due to formation of scale and sludge. If this is the case, the radiator may need to be removed and flushed. While this can be performed in the vehicle, it generally is not very effective. Drain the coolant system into a container by opening the lower drain valves and upper bleed valves. Test the antifreeze protection level. Once the radiator or some other component is replaced, refill the system ensuring that the bleeder valves are open. When all the air is out of the system, close the bleeder valves, start the engine, and check the coolant level. In modern engines the antifreeze will prevent corrosion and raise the temperature at which the coolant boils under pressure. This lengthens the life of the engine and cooling system components.

Task 10

Inspect, repair/replace coolant conditioner/filter, check valves, lines, and fittings.

131. When checking the coolant condition with an supplemental coolant additive (SCA) test strip, the Technician finds that the coolant pH condition is higher than specification limits. Which of these items should the Technician do?
   A. Add more antifreeze to increase the SCA.
   B. Continue to run the truck until the next PMI.
   C. Drain the entire coolant system and add the proper SCA mixture.
   D. Run the truck with no SCA additives until the next PMI.

132. While discussing coolant solutions, *Technician A* says a cooling system may be filled with water if the truck is operating in a climate where the temperature is always above freezing. *Technician B* says a 50/50 water and antifreeze solution has a lower boiling point compared to water. Who is right?
   A. A only
   B. B only
   C. Both A and B
   D. Neither A nor B

Hint

When freeze protection is required, an antifreeze meeting engine manufacturer specification must be used. An inhibitor system is included in most types of antifreeze and no additional inhibitors are required on initial fill if a minimum antifreeze concentration of 30 percent by volume is used. Solutions of less than 30 percent concentration do not provide sufficient corrosion protection. Concentrations over 67 percent adversely affect freeze protection and heat transfer rates. Some manufacturers suggest a 40 percent antifreeze to 60 percent water solution. Ethylene glycol base antifreeze is recommended, such as in all Detroit Diesel engines. Methyl alcohol base antifreeze is not recommended because of its effect on the nonmetallic components of the cooling system and because of its low boiling point. Methoxy propanol base antifreeze is not recommend for use by Detroit Diesel due to the presence of fluoroelastomer seals in the cooling system. Before installing ethylene glycol base antifreeze in a unit that has previously operated with methoxy propanol, the entire cooling system should be drained, flushed with clean water, and examined for rust and scale contaminants, etc. If deposits are present, the cooling system must be chemically cleaned with a commercial grade heavy-duty descaler. The inhibitors in antifreeze should be replenished at approximately 500-hour intervals or by testing with a non-chromate inhibitor system. Commercially available inhibitor systems may be used to reinhibit antifreeze solutions.

Most OEMs suggest testing the SCA levels in the coolant followed by adding SCA to adjust to the required values. Never dump unmeasured quantities of SCA into the cooling system at each PM. Generally, OEMs recommend that the coolant SCA level be tested at each oil change interval. Additionally, whenever there is a substantial loss of coolant and the system has to be replenished, the SCA level should be tested. The test kits usually consist of test strips, which must be stored in airtight containers and which have expiration dates that should be observed. The pH test is a litmus test in which a test strip is first inserted into sample of the coolant, then removed and the color of the test strip indexed to a color chart provided with the kit. The optimum pH window is defined by each OEM, but normally falls between 7.5 and 11.0 on the pH scale.

Task 11

Inspect, repair/replace water pump, hoses, and idler pulley.

133. While discussing cooling system hose replacement, *Technician A* says a collapsed upper radiator hose may be caused by a defective pressure release valve in the radiator cap. *Technician B* says a collapsed upper radiator hose may be caused by a plugged hose from the radiator filler neck to the coolant recovery reservoir. Who is right?
   A. A only
   B. B only
   C. Both A and B
   D. Neither A nor B

134. While discussing water pumps, *Technician A* says a defective water pump bearing may cause a growling noise with the engine idling. *Technician B* says the water pump bearing may be ruined by coolant leaking past the water pump seal. Who is right?
   A. A only
   B. B only
   C. Both A and B
   D. Neither A nor B

Hint

Most water pumps are belt driven using an idler pulley to maintain belt tension. As belts age, they stretch and the idler pulley automatically adjusts tension. If the belt is loose, check for binding in the idler pulley. The water pump is usually mounted on the front of the block with a gasket between the block and the water pump. A leaking water pump seal causes coolant to drip from the drain hole on the bottom side of the pump.

Task 12

Inspect and clean radiator, pressure cap, and tank(s); determine needed service.

135. *Technician A* says if the pressure in the radiator is too great, the pressure cap will release the excess pressure and fluid into an expansion tank.
Technician B says most radiator pressure caps have a release pressure of 15 psi or lower. Who is right?
A. A only
B. B only
C. Both A and B
D. Neither A nor B

136. *Technician A* says backflushing the radiator in the vehicle should be accomplished annually.
Technician B says air locks usually occur when the cooling system is filled with the bleeder valves open. Who is right?
A. A only
B. B only
C. Both A and B
D. Neither A nor B

Hint

The radiator has a pressure control cap with a normally closed valve. The cap, with a number 7 stamped on its top, is designed to permit a pressure of approximately 7 pounds (48 kPa) in the system before the valve opens. This pressure raises the boiling point of the cooling liquid and permits somewhat higher engine operating temperatures without loss of any coolant. To prevent the collapse of hoses and other parts, which are not internally supported, a second valve in the cap opens under vacuum when the system cools. Use extreme care when removing the coolant pressure control cap. Remove the cap slowly after the engine has cooled. The sudden release of pressure from a heated cooling system can result in loss of coolant and possible personal injury (scalding) from the hot liquid. To ensure against possible damage to the cooling system from either excessive pressure or vacuum, check both valves in the cap periodically for proper opening and closing pressures. If the pressure valve does not open at the rated pressure or the vacuum valve does not open at 0.625 psi (4.3 kPa) differential pressure, replace the pressure cap.

Task 13

Inspect, repair/replace fan hub, fan, fan clutch, mechanical and electronic fan controls, fan thermostat, and fan shroud.

137. Item 128 in Figure 2–31 is:
A. an air-applied fan clutch.
B. an electrically activated fan clutch.
C. a thermostatic fan clutch.
D. a hydraulic applied fan clutch.

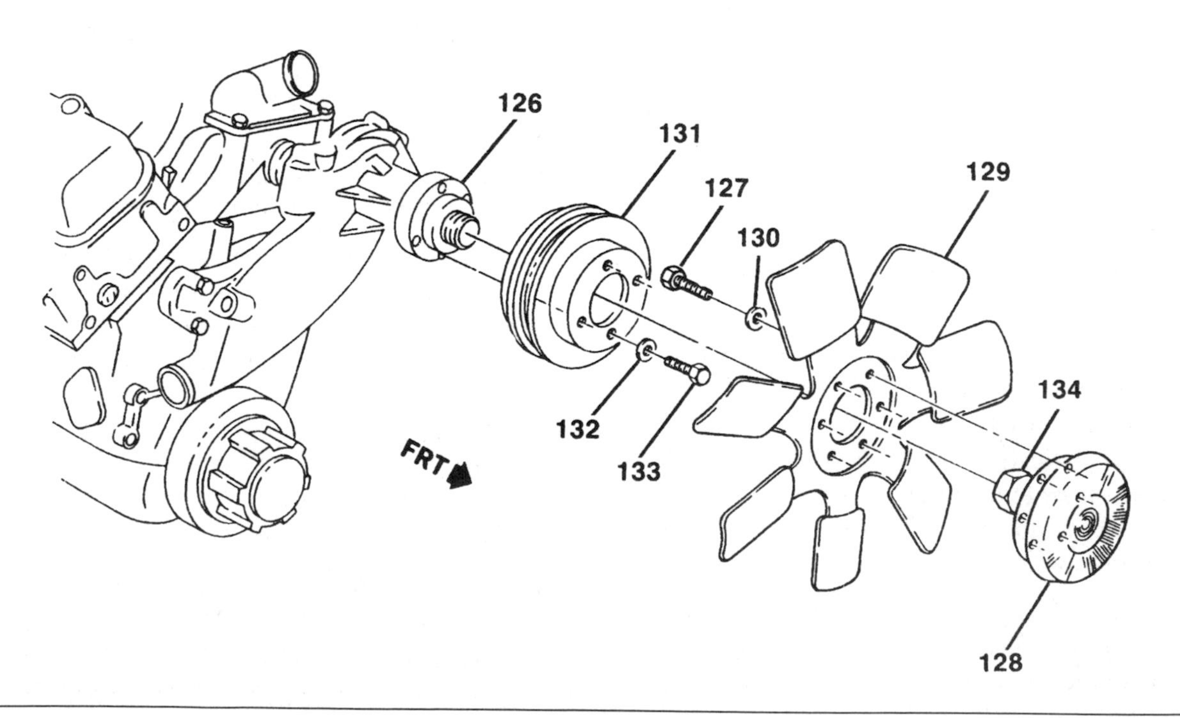

Figure 2–31 Cooling fan and related components. *(Courtesy of General Motors Corporation, Service Technology Group)*

138. On an air-operated fan clutch system:
   A. air pressure is supplied to the fan clutch to engage the clutch.
   B. the air pressure supplied to the fan clutch is controlled by an air regulator valve.
   C. if a defect occurs in the air control circuit, the fan clutch is disengaged continually.
   D. current flow through the fan clutch solenoid is controlled by a thermostatic switch and an electronic timer.

Hint

Fan assemblies must be inspected. Clean the fan and related parts with clean fuel oil and dry them with compressed air. Shielded bearings must not be washed because dirt may be washed in and the cleaning fluid cannot be entirely removed from the bearing. Examine the bearings for any indications of corrosion or pitting. Hold the inner race or cone so it does not turn and revolve the outer race or cup slowly by hand. If rough spots are found, replace the bearings. Check the fan blades for cracks. Replace the fan if the blades are cracked or bent. Straightening may weaken the blades, particularly in the hub area.

Some heavy-duty trucks have an air-operated fan clutch. An air-operated solenoid controls the air supply to the clutch. When the solenoid is not energized electrically, air pressure is shut off from the fan clutch and the clutch is engaged. To disengage the fan clutch the solenoid is energized electrically, and air pressure is supplied to the fan clutch. This action disengages the clutch. Current flow through the solenoid is controlled by a thermostatic switch in the cooling system and an electronic timer.

Task 14

Inspect, repair/replace radiator shutter assembly and controls.

139. A thermostatic piston-controlled radiator shutter relies on all of the following to regulate shutter operation EXCEPT:
   A. linkage adjustments.
   B. ambient air temperature.
   C. radiator temperature.
   D. air pressure.

140. Air shutters operated by a thermostatic piston close automatically but must be manually opened. What component of the system should you inspect?
   A. Shutter drive piston
   B. Shutter assembly springs
   C. Engine thermostat
   D. Shutter thermostatic switch

Hint

Radiator shutters are used to maintain coolant-operating temperature in cold climates, and to speed up warmup of the diesel engine. By partially blocking off airflow through the radiator shutters, excess cooling of the radiator can be prevented. There are two basic ways to control shutters used on trucks. One method is a thermostatic piston that is mounted on the shutter assembly and reacts to outside temperature and radiator heat. Inspection of this type of shutter is usually limited to checking the linkage and shutters for freedom of movement, and testing the thermostatic piston for proper operation.

Another method of opening and closing shutters is to use a thermostat regulated by engine temperature to supply compressed air from the engine compressor. The regulated air pressure is used by a piston on the shutter assembly to counteract spring tension that holds the shutter open. Maintenance of this system will include inspection of all air lines for leaks and blockages, and the shutter piston and thermostat. Check the shutter linkages for binding and wear.

Air Induction and Exhaust Systems Diagnosis and Repair

Task 1

Inspect, service/replace air induction piping, air cleaner, and element; check air restriction.

141. *Technician A* says the oil in the bottom chamber of an oil bath air cleaner is to lubricate air intake components. *Technician B* says this type of air filter has a reduced efficiency at low rpm. Who is right?
   A. A only
   B. B only
   C. Both A and B
   D. Neither A nor B

142. All of the following are steps in an air inlet restriction test EXCEPT:
   A. connect a manometer to the air intake.
   B. check the manometer with the air cleaner and duct removed.
   C. check the manometer reading with the air filter installed.

D. place a piece of cardboard over the air intake to calibrate the manometer.

Hint

Some air cleaners have a design that causes air to circulate around the filter element. This cyclonic action causes airborne particles, like sand and dust, to be slung against the sides of the canister, where they fall and collect in the bottom chamber. This reduces the clogging of the filter element and extends its life. The filter element is impregnated with a resin to trap most of the remaining particles.

The oil bath air filter uses a specially designed oil-filled pan to trap large particles. As intake air is forced through the oil, the heavier particles are slowed and fall to the bottom in the oil. A filter chamber is filled with porous material to catch and return airborne oil and particles to the oil bath. Because the filter relies on the velocity of air flowing through the oil, efficiency is reduced at lower engine rpm.

Excessive restriction of the air inlet affects the flow of air to the cylinders and results in poor combustion and lack of power. Consequently, the restriction must be kept as low as possible considering the size and capacity of the air cleaner. An obstruction in the air inlet system, and dirty or damaged air cleaners results in a high blower inlet restriction. Check the air inlet restriction with a water manometer connected to a removable fitting in the air inlet ducting on nonturbocharged engines or the compressor inlet on turbocharged engines. With the engine running at the specified speed, measure the air inlet restriction with the air cleaner removed and with the air cleaner installed and the manometer connected to the intake

Task 2

Inspect, repair/replace turbocharger, wastegate, boost sensor, overboost protection valve, and piping system.

143. While discussing a turbocharged diesel engine, *Technician A* says that an exhaust leak between the engine and the turbocharger affects the performance of the engine. *Technician B* says that leaks in the intake pipe can cause damage to the intake side of the turbo. Who is right?
    A. A only
    B. B only
    C. Both A and B
    D. Neither A nor B

144. *Technician A* says that on a turbocharged diesel engine, the turbocharger bearings are lubricated by engine oil. *Technician B* says that turboshaft bearing failure may be caused by contaminated engine oil. Who is right?
    A. A only
    B. B only
    C. Both A and B
    D. Neither A nor B

Hint

A turbocharger contains exhaust gas driven rotor blades in the exhaust side of the turbocharger. A shaft is connected from the exhaust driven rotor to the compressor blades in the intake side of the compressor. Because the rotor and compressor are connected to a common shaft and built to close tolerances, the assembly can achieve speeds between 60,000 to 100,000 rpm. The resulting increase in air density from this air induction above atmospheric pressure increases engine power. At high engine rpm a wastegate in the turbocharger is used to bypass a portion of the exhaust gas pressure, thereby preventing overboosting of the cylinders. On some engines the wastegate is operated by manifold pressure supplied to a diaphragm. On other engines the wastegate is electronically controlled. The output air is less dense because the extremely hot turbocharger superheats the air. To reduce the effects of superheating an aftercooler is installed between the turbocharger and the cylinders. The aftercooler is an air-to-air heat exchanger. Some engine manufacturers use an air-to-water heat exchanger called an intercooler. The aftercooler or intercooler reduces air temperature and increases air density.

Task 3

Inspect, repair, rebuild/replace engine-driven blowers.

145. While inspecting a rotor-type blower, the Technician notices an oil film on the blowers rotors. Which of the following is LEAST likely to cause this?
    A. Blower bearing seal leaks
    B. Worn rotor bearing
    C. Restriction in the oil feed line
    D. Worn end plate gaskets

146. The required clearance between the rotor lobes on an engine-driven blower is obtained by which of the following?
    A. Adding or removing shims
    B. Changing the end cover thrust plate
    C. Using over- or undersized gears
    D. Using over- or undersized bearings

Hint

Oil from the engine sump is supplied to the bearings of the rotors for lubrication and cooling. Seals are installed on the rotor shafts to prevent vacuum from drawing oil into the blower and blower air pressure from forcing oil from the bearing back into the engine.

Proper air box pressure is required to maintain sufficient air for combustion and scavenging of the burned gases. Low air box pressure is caused by a high air inlet restriction, damaged blower rotors, an air leak in the air box or a clogged blower air inlet screen. Lack of power, black, or gray exhaust smoke are indications of low air box pressure. High air box pressure can be caused by partially plugged cylinder liner ports. Check the air box pressure with a manometer connected to an air box drain tube. Check the readings obtained at various speeds with the engine operating at the specified rpm.

Task 4

Inspect, repair/replace intake manifold, gaskets, sensors, and connections.

147. *Technician A* says the air intake restriction system may contain pressure sensor and an electronic module. *Technician B* says air intake restriction does not affect engine power. Who is right?
    A. A only
    B. B only
    C. Both A and B
    D. Neither A nor B

148. All of these statements about air intake systems are true EXCEPT:
    A. restrictions cause a decrease in vacuum in the air intake system.
    B. scored cylinders and piston skirts may be caused by air intake leaks.
    C. some air intake restriction indicators contain a vacuum tube.
    D. a restricted air intake reduces the volume of air entering the engine.

Hint

Intake piping, tubing, and hoses are very durable yet very flexible to allow for movement of the engine. Solid piping in the air inlet system is usually made of aluminum or ridged plastic to reduce weight. Connections between the solid sections of piping and major components are made using rubber sections clamped at both ends. When these connections become loose, unfiltered air is allowed to enter the intake system. Airborne abrasive or corrosive elements can then cause damage to critical components. On the other hand, as the rubber or plastic components age or become exposed to extreme heat they may collapse and cause restrictions in the airflow. To provide a visual warning to the operator of problems in the air intake system, most manufacturers include an air intake vacuum indicator. This could be as simple as a vacuum tube attached to the intake piping that runs to an instrument panel indicator, or an electronic sensor that is monitored by an electronic module. If the vacuum increases, check for a restriction such as a collapsed pipe or clogged filter. A drop in vacuum indicates a possible leak, clogged vacuum indicator tube, or faulty sensor.

Task 5

Inspect, test, clean, repair/replace intercooler (aftercooler) and charge air cooler assembly.

149. Reduced power and high coolant temperature are noted on an engine with an intercooler. *Technician A* says you need to bleed the air out of the intercooler. *Technician B* says the air tubes in the intercooler may be clogged. Who is right?
    A. A only
    B. B only
    C. Both A and B
    D. Neither A nor B

150. On a diesel engine with an intercooler, the Technician notices coolant in the intake side of the turbocharger. What should the Technician check first?
    A. The intercooler
    B. The water pump
    C. The intake manifold
    D. The head gasket

Hint

The aftercooler resembles a radiator except the fin tubes appear to be larger and the inlet and outlet connections are larger to allow a large volume of airflow. As the hot pressurized air passes through the small fin tubes, heat is transferred from the pressurized air to the tubes, then to the fins, and finally to the outside airs passing between the fins.

Aftercoolers can be inspected visually while still in the vehicle. Wash fins with low-pressure water and straighten as necessary. Inspect for missing or loose fins, dented or kinked tubes, corrosion, or holes in tubes. Leaks can be detected on turbocharged engines by running the engine, spraying the aftercooler tubes with soap and water, and watching for bubbles.

An intercooler cools the pressurized air using liquid coolant in the opposite way the radiator works. Heated air is passed through small tubes inside a chamber through which coolant is circulated. Heat from the air is transferred to the tubes and on to the coolant, which is cooled by the radiator. A leaking intercooler causes traces of water in the oil or the engine may produce a white smoke when running. The turbocharger may have signs of liquid leakage from its compressor outlet or corrosion on the fins. This would also result in unexplained loss of coolant in the radiator. Often the internal leaking of the intercooler can be misdiagnosed as a cracked head or blown head gasket. One common problem with intercoolers is air locks. Usually intercoolers are located at the top of the engine. When maintenance is performed on the engine resulting in refilling, the coolant air must be bled from the intercooler coolant chamber. Failure to do so could reduce the effectiveness of the cooler and cause overheating due to steam pressure buildup.

Task 6

Inspect, repair/replace exhaust manifold, piping, mufflers, exhaust backpressure regulator, catalytic converter, and mounting hardware.

151. Excess exhaust backpressure may result in all of the following EXCEPT:
   A. lower engine power.
   B. higher exhaust temperature.
   C. increased intake vacuum.
   D. poor combustion.

152. While discussing exhaust systems, *Technician A* says carbon dioxide in the exhaust causes nausea. *Technician B* says operating a truck for a prolonged period with a restricted exhaust system reduces engine life. Who is right?
   A. A only
   B. B only
   C. Both A and B
   D. Neither A nor B

Hint

Since carbon monoxide (CO) in the exhaust is a poisonous gas that causes nausea and even death, exhaust leaks must be eliminated. The exhaust system must be checked for leaks and potential leaks, and exhaust system components must be replaced as required.

A slight pressure in the exhaust system is normal. However, excessive exhaust backpressure seriously affects engine operation. It may cause an increase in the air box pressure with a resultant loss of blower efficiency. This means less air for scavenging, which results in poor combustion and higher temperatures. Causes of high exhaust back-pressure are an improper muffler, an exhaust pipe that is too long or too small in diameter, an excessive number of sharp bends in the exhaust system, or obstructions such as excessive carbon formation or foreign matter in the exhaust system. Check the exhaust backpressure, measured in inches of mercury, with a manometer. Connect the manometer to the exhaust manifold on nonturbocharged engines by removing a pipe plug that is provided for that purpose. If no opening is provided, drill an 11/32 in. (8.73 mm) hole in the exhaust manifold companion flange and tap the hole to accommodate a pipe plug. On turbocharged engines, check the exhaust backpressure in the exhaust piping 6 to 12 in. (15.2 cm to 30 cm) from the turbine outlet. The tapped hole must be in a comparatively straight pipe area for an accurate measurement.

Some engines are equipped with catalytic converters to reduce exhaust emissions. If the converter is working properly the outlet is about 100°F (38°C) hotter compared to the inlet.

Task 7

Inspect, repair/replace preheater/inlet air heater, or glow plug system and controls.

153. All of the following are used to control glow plugs in modern diesel engines EXCEPT:
   A. engine oil temperature.
   B. electronic module.
   C. powertrain control module (PCM).
   D. exhaust gas recirculation (EGR) valve.

154. When the glow plug indicator light illuminates continuously:
   A. the glow plug system is operational.
   B. the glow plug system has experienced a short circuit.
   C. the glow plug in one cylinder is burned out.
   D. the outside temperature requires the driver to turn on the glow plug system.

Hint

Modern diesel engines use glow plugs that are controlled by the PCM. A glow plug relay is used to energize the glow plugs for assisting cold engine startup. Engine oil temperature, battery voltage, and barometric (BARO) pressure are used by the PCM to calculate glow plug on time. On time normally varies between 10 and 120 seconds. With colder oil temperatures and lower barometric pressures, the plugs are on longer. If battery voltage is abnormally high, the duty cycle is shortened to extend glow plug life. The glow plug relay only cycles on and off repeatedly when there is a system high-voltage condition greater than 16 volts. An open in the glow plug relay cir-

cuit renders the glow plugs inoperative. A short circuit may result in a glow plug always on condition. The glow plug WAIT TO START indicator light is located on the instrument panel. When the light goes off, the engine is ready to be started. The light comes on every time a key on reset occurs. On time normally varies between 1 and 10 seconds. The WAIT TO START light on time is independent of the glow plug relay on time, because the glow plugs may stay on to improve performance until the engine reaches a specific temperature.

Task 8

Inspect, repair/replace ether/starting fluid system and controls.

155. With the capsule-type fluid starting system shown in Figure 2–32, which of the following is LEAST likely?
    A. Capsules are not refillable.
    B. Check valves allow the plunger to develop fluid pressure for injection.
    C. Pump plunger is used to operate check valves.
    D. A solenoid valve is used to release starting fluid.

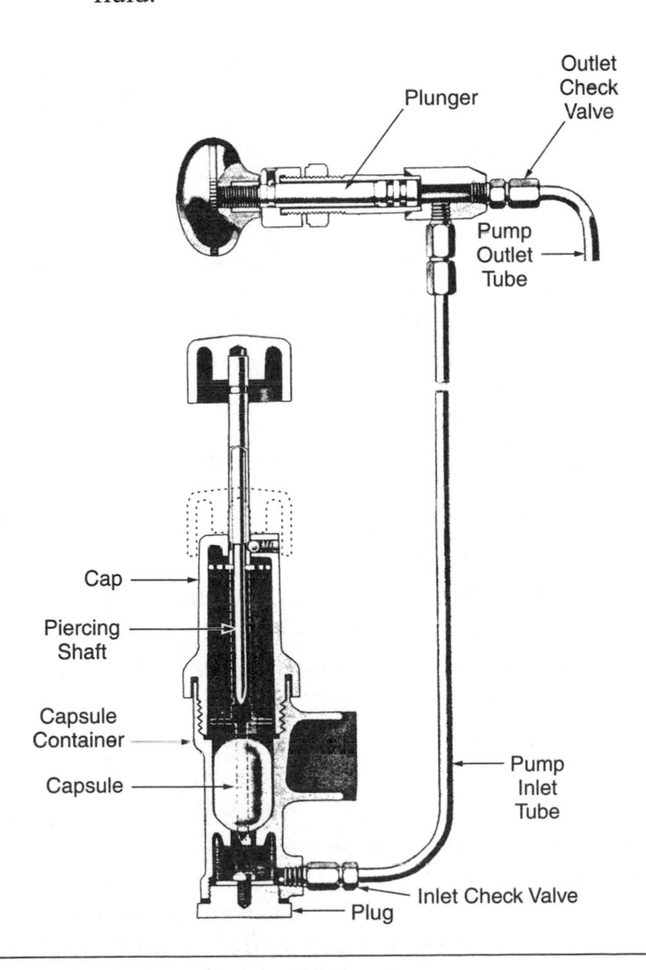

Figure 2–32 Capsule-type fluid starting system.

156. All of the following are parts in a starting aid pump system EXCEPT:
    A. the pump body.
    B. the inlet check valve.
    C. the outlet check valve.
    D. the fan pulley.

Hint

The principal parts of the starting aid pump are the body, plunger, and the spring-loaded ball-type inlet and outlet check valves. The pump body is threaded externally at one end for mounting purposes. One end of the plunger is threaded into the operating knob. Two seal rings of oil-resistant material are located in grooves at the other end of the plunger. The inlet check valve, which opens on the suction stroke of the plunger and seats under pressure, is located in the side opening of the pump body. An arrow indicating the direction of flow is also stamped on each check valve. The fluid starting aid is designed to inject a highly volatile fluid into the air intake system to assist ignition of the fuel at low ambient temperatures.

Task 9

Inspect, repair/replace emergency air induction shut-off system.

157. Which of the following is LEAST likely to have the air shutdown valve mounted to it?
    A. Blower
    B. Inlet side of the turbocharger
    C. Intake manifold
    D. Exhaust manifold

158. *Technician A* says constant use of the air shutdown valve is not harmful to the blower. *Technician B* says the air shutdown valve on a blower is used as a normal shutdown device. Who is right?
    A. A only
    B. B only
    C. Both A and B
    D. Neither A nor B

Hint

The shutdown valve is used to release a door inside the blower air box. This valve completely restricts the air intake. When the air valve is closed, the pressure in the air box side of the rotors increases sharply, and the air pressure to the cylinders drops. Without air for combustion the engine stops. When the engine stops, the vacuum pressure in the air box returns to normal. Oil from the engine sump is supplied to the bearings of the rotor bearings for lubrication and cooling. Seals are installed on the rotor shafts to prevent vacuum from drawing oil into the blower and blower air pressure from forcing oil from the bearing back into the engine.

Continual use of the air box shutdown can cause the seals to rupture and oil to leak into the blower. Excess oil in the blower may cause blue exhaust smoke or excessive engine rpm.

Fuel System Diagnosis and Repair

Mechanical Components

Task 1

Inspect, repair/replace fuel tanks, vents, cap(s), mounts, valves, screens, supply, crossover, and return lines and fittings.

159. *Technician A* says the presence of water in a fuel tank may be tested with an aluminum welding rod coated with water detection paste. *Technician B* says the fuel tank sending unit contains a Hall effect switch. Who is right?
 A. A only
 B. B only
 C. Both A and B
 D. Neither A nor B

160. In diesel fuel systems the:
 A. primary filter is on the charge side of the transfer pump.
 B. the secondary filter is on the suction side of the transfer pump.
 C. fuel tank pickup pipes rest against the bottom of the fuel tank.
 D. a plugged fuel tank vent may cause the engine to stop running.

Hint

In diesel fuel systems, a clear distinction exists between the suction side and the charge side of the fuel transfer pump. A primary filter is most often located on the suction side of the transfer pump while the secondary filter is located on its charge side. However, there are some fuel systems where all movement of fuel through the fuel subsystem is under suction. When such a fuel system uses multiple filters, the terms "primary" and "secondary" are not used.

Excess fuel that is not injected through the injector into the cylinder is returned through the return fuel system to the fuel tank. This return fuel flow helps to cool injectors and other injection system components. Fuel tank vents should be routinely inspected for restrictions and should be protected from ice buildup. A plugged fuel tank vent will rapidly shut down an engine, creating a suction side inlet restriction that the transfer pump is not able to overcome.

To check for the presence of water in fuel tanks, insert a clean aluminum welding rod lightly coated with water detection paste through the fill neck until it bottoms in the base of the tank. Withdraw the rod and examine the water detection paste for a change in color. This test will give some idea of the quantity of water in the tank by indicating the height of color change on the probe. When only the tip of the probe changes color, the slight amount of water will not necessarily present any problems.

The fuel tank sending unit consists of a float and arm connected to a variable resistor. This variable resistor controls current flow to the dash gauge proportional to the fuel level in the tank. Fuel sending unit problems can be diagnosed with an ohmmeter by moving the float arm through its arc and observing readings.

Fuel pickup tubes are positioned so that they draw on fuel slightly above the base of the tank and thereby avoid picking up water and sediment. Pickup tubes are quite often welded into the tank.

Task 2

Inspect, clean, test, repair/replace fuel transfer (lift) pump, pump drives, screens, fuel/water separators/indicators, filters, heaters, and associated mounting hardware.

161. The single-acting plunger-type fuel supply pump can be equipped with all of the following EXCEPT a:
 A. sediment bowl.
 B. sediment strainer.
 C. hand primer.
 D. pop-off valve.

162. An engine cranks but fails to start and the transfer pump fails to build pressure, and the fuel tank is full. The cause of this problem could be:
 A. seized injectors.
 B. an air leak in the primary filter housing.
 C. a restricted secondary filter.
 D. a plugged fuel line between the transfer pump and the injectors.

Hint

Fuel charging or transfer pumps are positive displacement pumps driven directly or indirectly by the engine. On most truck and bus fuel systems transfer pumps are plunger or gear types. A hand primer pump may be located on the fuel transfer pump body or on a filter mounting pad. The function of a hand primer pump is to prime the fuel system whenever prime is lost.

The function of a fuel filter is to entrap fine sediment in the diesel fuel. Some secondary filters will not allow water to pass through the filtering media. Fuel filters may be spin-on disposable-type or cartridge-type.

The transfer pump usually charges secondary filters. Testing charging pressure downstream from the transfer pump is normally performed with an accurate pressure gauge plumbed in series between the transfer pump and injection pump.

Most water separators have a clear sump through which the presence of water may be observed. All water separators are equipped with a drain valve; the purpose of this valve is to siphon water from the sump. Water should be routinely removed through the drain valve. The filter elements used in combination water separator/primary filter units should be replaced during routine maintenance. Whenever a water separator is fully drained, it should be primed before attempting to start the engine.

Many trucks are equipped with fuel heaters. The fuel heater may be combined with the water separator. A fuel heater contains an electric heating element that uses battery current to heat fuel in the subsystem. Electric element fuel heaters may be thermostatically managed so that fuel is only heated as much as required and not to a point that compromises some of its lubricating properties.

Task 3

Check fuel system for air; determine needed repairs.

163. Air bubbles in the fuel return line may be caused by:
    A. a leak at the secondary filter.
    B. a leak in the fuel line between the primary filter and the transfer pump.
    C. a leak at a unit injector inlet line.
    D. a leak in the fuel line from the secondary filter to the injectors.

164. *Technician A* says check for air in the fuel system by checking for bubbles in the fuel return line with the engine shut down. *Technician B* says if there are bubbles in the fuel line from the tank to the primary filter, the line from the tank to the primary filter has a leak. Who is right?
    A. A only
    B. B only
    C. Both A and B
    D. Neither A nor B

Hint

Check for air in the fuel system by installing a clear plastic hose in the return line and look for consistent large bubbles. To troubleshoot the source of air leaks, a diagnostic sight glass can be used. It consists of a clear section of tubing with hydraulic hose couplers at either end and it is fitted in series with the fuel flow. However, the process of uncoupling the fuel hoses always admits some air into the fuel system, so the engine should be run for a while before reading the sight glass.

Task 4

Prime and bleed fuel system; check, repair/replace primer pump.

165. All of the following can be used to bleed air out of the fuel system EXCEPT:
    A. by cranking the engine and bleeding the injector lines.
    B. by pressurizing the fuel system with the primer pump and bleeding the injector lines.
    C. by replacing the fuel filter.
    D. by pressurizing the fuel tank with a low-volume air line and bleeding the injector lines.

166. *Technician A* says when bleeding air from the fuel system it is necessary to pressurize the system with the hand primer. *Technician B* says bleeding air from the fuel system should be done any time fuel filters are changed. Who is right?
    A. A only
    B. B only
    C. Both A and B
    D. Neither A nor B

Hint

A fuel system must be sealed to operate properly. Any introduction of air may cause air locks, which prevent fuel from reaching the injectors. To bleed air from the fuel system, manufacturers install primer pumps to pressurize the system. Once pressurized, opening the end of the lines forces trapped air out of the line. Some injectors incorporate a return line that allows excess fuel to return to the tank. This type of system is self-priming, but some manufacturers strongly recommend against priming this way. The excess fuel is used to cool and lubricate fuel system components. Most manufacturers will install a hand primer to make priming systems easier.

When a vehicle runs out of fuel and the fuel system requires priming, remove the filters and fill them with filtered fuel. Follow these steps to prime the fuel system:

1. *Locate a bleed point in the system, and open the bleed valve. On an in-line injection pump system, this will be at the exit of the charging gallery.*

2. *Next, if the system is equipped with a hand primer pump, actuate it until air bubbles cease to exit from the open valve. If the system is not equipped with a hand primer pump, fit one upstream from the secondary filter and actuate it until air bubbles cease to exit from the open valve.*
3. *Close the bleed valve. Crank engine for 30-second segments with at least 2-minute intervals between crankings until it starts. This procedure allows for starter motor cool-down. In most diesel fuel systems, the high-pressure circuits will self-prime once the subsystem is primed. Some engine manufacturers recommend priming the fuel system by supplying air pressure to the fuel tank. When using this procedure air pressure above 55 psi (8 kPa) may cause fuel tank damage.*

Task 5

Inspect, adjust, repair/replace throttle linkage, and controls.

167. While discussing Cummins PT throttle linkage adjustments, *Technician A* says the rear and front throttle screws should be adjusted after the idle speed adjustment. *Technician B* says the rear throttle screw should be adjusted before the forward throttle screw. Who is right?
   A. A only
   B. B only
   C. Both A and B
   D. Neither A nor B

168. *Technician A* says if the engine lugs down and does not respond properly to load changes, the throttle linkage may require adjusting. *Technician B* says engine valve and unit injector adjustments should be completed after throttle linkage adjustments. Who is right?
   A. A only
   B. B only
   C. Both A and B
   D. Neither A nor B

Hint

Before making any critical adjustments, ensure proper fuel pressure and flow is available. Check fuel filter restriction, sticking injector and throttle linkages, and low fuel pressure from worn pumps. Engine valve and unit injector adjustments should be performed before the throttle linkage adjustments. Improper setting of the throttle can be detected when checking the engine response to load changes. The need for a throttle linkage adjustment is indicated when the throttle is not responding to changes in demand. This leads to a rich air-fuel ratio, a sudden loss of power, and engine lugging. Throttle linkage adjustments have become extremely critical on today's engines. The travel of the throttle controls the engine over its complete operating range. All throttle linkage adjustments must be performed according to the engine manufacturer's specifications and procedures. On a Cummins PT pump these recommended adjustments are rear throttle screw, front throttle screw, and idle speed adjustment.

Task 6

Perform on-engine inspections, tests, adjustments, and time, or replace and time distributor-type injection pump.

169. The timing adjustment on a distributor-type fuel injection pump is accomplished by:
   A. using a timing pin.
   B. using the spill method.
   C. adjusting the rack.
   D. rotating the pump.

170. To perform the fuel injector pump lever (FIPL) sensor adjustments on a distributor injector pump:
   A. the key must be on and the engine running.
   B. the key must be on and the engine off.
   C. the key must be off and the engine off.
   D. the key must be off and the diagnostic tool set to engine run.

171. All of these statements about the distributor-type computer-controlled injection pump in Figure 2–33 are true EXCEPT:
   A. a cable is connected from the accelerator pedal to the injection pump.
   B. the PCM sends control signals to the fuel solenoid driver module.
   C. the fuel solenoid driver module controls the fuel injection solenoid in the injection pump.
   D. the PCM receives input signals from the accelerator pedal position (APP) module.

Hint

The distributor-type pump operates much like an electrical distributor because it distributes fuel to each cylinder when it is ready to fire. As the internal rotor is rotated by the pump drive, ports in the rotor line up ports in the pump head. These ports are connected through high-pressure lines to the injectors. When the ports in the rotor and head are aligned, fuel flows to the injector, and the fuel pressure opens the injector. Further rotor rotation aligns the rotor ports with the next injector line in the firing order and the next injector opens. Fuel is pres-

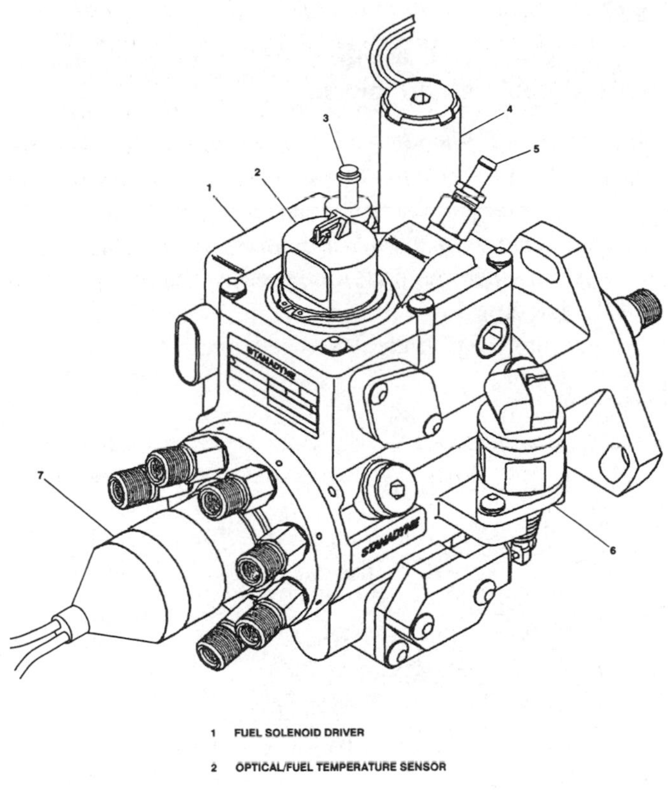

Figure 2–33 Distributor-type computer-controlled injection pump. *(Courtesy of General Motors Corporation, Service Technology Group)*

surized in the injector pump by pumping plungers in the rotor, and fuel is metered by a metering valve. The metering valve position is determined by the throttle position and governor operation. An internal shutdown and/or cold advance solenoid are located under the pump cover. Pump timing is accomplished by loosening the pump mounting bolts and rotating the pump. Never attempt to rotate the pump with the engine running. This action may cause severe pump and injection line damage. Diesel timing lights are available with a pickup that is attached to the number 1 injection line. The pickup senses the high-pressure pulses in the injection line to trigger the timing light. Digital timing meters are also available with a probe that is inserted in a hole in the timing indicator.

On later model distributor pumps the throttle is no longer connected to the injection pump. On these systems the throttle is connected to an accelerator pedal position (APP) module that sends voltage signals in relation to throttle opening to the powertrain control module (PCM). The APP module contains three variable resistors. The PCM sends control signals to the fuel solenoid driver module mounted on the injection pump. The fuel solenoid driver controls the fuel injection solenoid in the injection pump to supply the precise amount of fuel required by the engine.

Task 7

Perform on-engine inspections, tests, adjustments, and time, or replace and time in-line type injection pump and drives.

172. *Technician A* says that when installing an in-line injector pump with an electronic governor, it must be timed with the engine camshaft to ensure proper fuel injection timing. *Technician B* says an internal camshaft in the pump determines the injection firing order. Who is right?
    A. A only
    B. B only
    C. Both A and B
    D. Neither A nor B

173. In a fuel injection system with an in-line pump, *Technician A* says the in-line pump pressurizes, meters, times, and delivers fuel to injectors. *Technician B* says the injector lines between the pump and injectors may have unequal lengths. Who is right?
    A. A only
    B. B only
    C. Both A and B
    D. Neither A nor B

Hint

The in-line injection pump performs the function of metering, pressurizing, timing, and delivering of fuel for injection. The pump is normally gear driven from the engine camshaft. This, in turn, causes an internal camshaft to operate plungers in much the same manner as the engine camshaft operates valves. The fuel pressure developed by the plunger unseats a valve in the top of the barrel, and fuel is delivered to the injector.

An in-line injection pump is driven through one complete rotation (360°) in two engine revolutions (720°). The gear-driven pump drive plate drives the injection pump camshaft. The camshaft is supported by main bearings and rotates within the injection pump cambox. The cambox is the lower portion of the injection pump, which houses the camshaft, tappets, and the integral oil sump. Timing of the pump camshaft to the engine camshaft is critical.

A spill timing procedure is recommended on some engines. This procedure adjusts injection timing by setting the crankshaft position in relation to the injection pump position when the pump spill port closes on number 1 cylinder. Most engines are timed on number 1 cylinder, but a few engines are timed on number 6 cylinder.

Using a timing pin is probably the simplest method of injection pump timing, and least likely to present problems. The engine is located to a specific position by inserting a timing bolt usually in the cam gear but sometimes in the flywheel. Similarly, the injection pump is pinned by a timing tool to a specific location. The injection pump is always removed and installed with the timing tools in position. It goes without saying that the timing tools must be removed before attempting to start the engine.

Task 8

Perform on-engine inspections, tests, and adjustments, or replace PT-type injection pump, drives, and injectors.

174. To perform a maximum delivery check on a Cummins PT pump, *Technician A* says that the torque screw and high idle screws must be backed out about 5 turns. *Technician B* says the pump must rotate through 1,000 strokes to collect the proper amount of fuel. Who is right?
    A. A only
    B. B only
    C. Both A and B
    D. Neither A nor B

175. Of the following parts, which will LEAST likely be found on a pressure-timed (PT) system?
    A. Gear pump
    B. Throttle linkage
    C. Pulsation damper
    D. Transfer pump

Hint

The original pressure time (PT) pump uses a regulator to control the maximum fuel manifold pressure. Later models incorporated a governor control and are referred to as PTG. The most recent improvement (PTG AFC) uses a flow no flow by-pass valve. This improved version air-fuel ratio control (AFC) uses manifold vacuum to control fuel flow during periods of low intake manifold pressure. Regardless of the type of PT pump used, the system has a gear pump to transfer fuel from the tank and a pulsation damper to absorb pulsation caused by operation of the gear pump. The injection pump supplies pressurized fuel to the injectors that are operated mechanically by the camshaft.

Task 9

Perform on-engine inspections, tests, and adjustments, or replace unit injectors.

176. *Technician A* says in a PT injector system when the fuel pressure against the angular surface of the injector needle exceeds the spring tension, the injector fires. *Technician B* says in a unit injector system when the cam lobe presses the plunger of the injector down, it pressurizes fuel and injects fuel into the cylinder. Who is right?
    A. A only
    B. B only
    C. Both A and B
    D. Neither A nor B

177. Which component of a unit injector system is LEAST likely to be the cause of an improper injector spray pattern?
    A. Plugged nozzle tip holes
    B. A stuck injection nozzle
    C. Improper timing of injectors
    D. Excess injector-to-follower clearance

Hint

In a mechanical unit injector fueled engine, each engine cylinder has its own unit injector. Each mechanical unit injector (MUI) has fuel delivered to it at charging pressure between 30 and 70 psi (200 kPa and 470 kPa), dependent on engine speed.

Charging pressure is generated by a positive displacement, gear pump, driven by the engine. In most truck and bus applications, engine output is controlled by a mechanical governor that regulates fuel quantity by controlling the unit injector fuel racks.

When the camshaft lobe, pushrod, and rocker arm force the injector plunger downward, high-pressure fuel is injected through the injector spray tip into the cylinder. The governor and rack rotate the injector plungers in relation to the spill port in the injector housing to control the amount of fuel injected, and thus supply the proper amount of fuel for all engine speeds and loads.

Before timing a unit injector, all valve adjustments must be completed. The governor speed control lever must be in the idle position, and the stop lever must be in the stop position. On a two-cycle engine the exhaust valves must be fully depressed in the cylinder where the injector adjustment is being performed. Some unit injectors are timed by inserting the specified timing gauge into the hole in top of the injector, and the push rod length is adjusted to provide the proper injector timing.

Task 10

Inspect, test, adjust, and repair/replace fuel injectors.

178. While discussing the nozzle shown in Figure 2–34, *Technician A* says screw R is the closing pressure adjusting screw. *Technician B* says screw S is the maximum fuel screw. Who is right?
    A. A only
    B. B only
    C. Both A and B
    D. Neither A nor B

179. When discussing the injector service in Figure 2–35, *Technician A* says the Technician is cleaning the injector spray orifices. *Technician B* says the Technician is measuring the plunger-to-tip clearance. Who is right?
    A. A only
    B. B only
    C. Both A and B
    D. Neither A nor B

Hint

Injection nozzles are simple hydraulic valves operated by fuel pressure. Fuel flow generated by the injection pump enters the nozzle holder at the fuel inlet and proceeds down the feed channel and into the nozzle sac chamber. When the pressure of the fuel against the annular area of the needle valve exceeds the preset pressure of the pressure spring, the needle valve is raised from its seat. Then a metered amount of fuel is injected through the orifices into the combustion chamber. When any diesel fuel system lines are disconnected, the lines and fittings should be capped to keep dirt out of the system.

Some engines have mechanical unit injectors (MUI). These injectors are opened mechanically by the engine camshaft, pushrod, and rocker arm. MUI injectors can be disassembled cleaned, tested, and adjusted. Some later model engines have hydraulically actuated, electronically controlled unit injectors. These injectors have an electric solenoid mounted on top of the injector. When the powertrain control module (PCM) inputs indicate that a piston is in the proper position for combustion to begin, the PCM energizes the injector solenoid in that cylinder. When the solenoid is energized, engine oil pres-

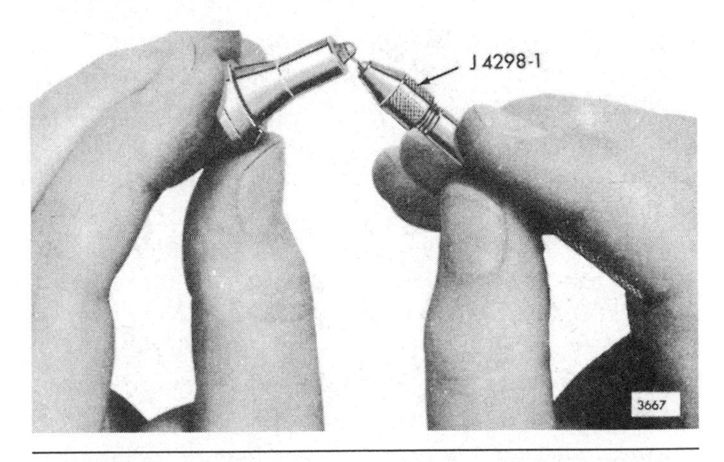

Figure 2–35 Unit injector service. *(Courtesy Detroit Diesel Corporation)*

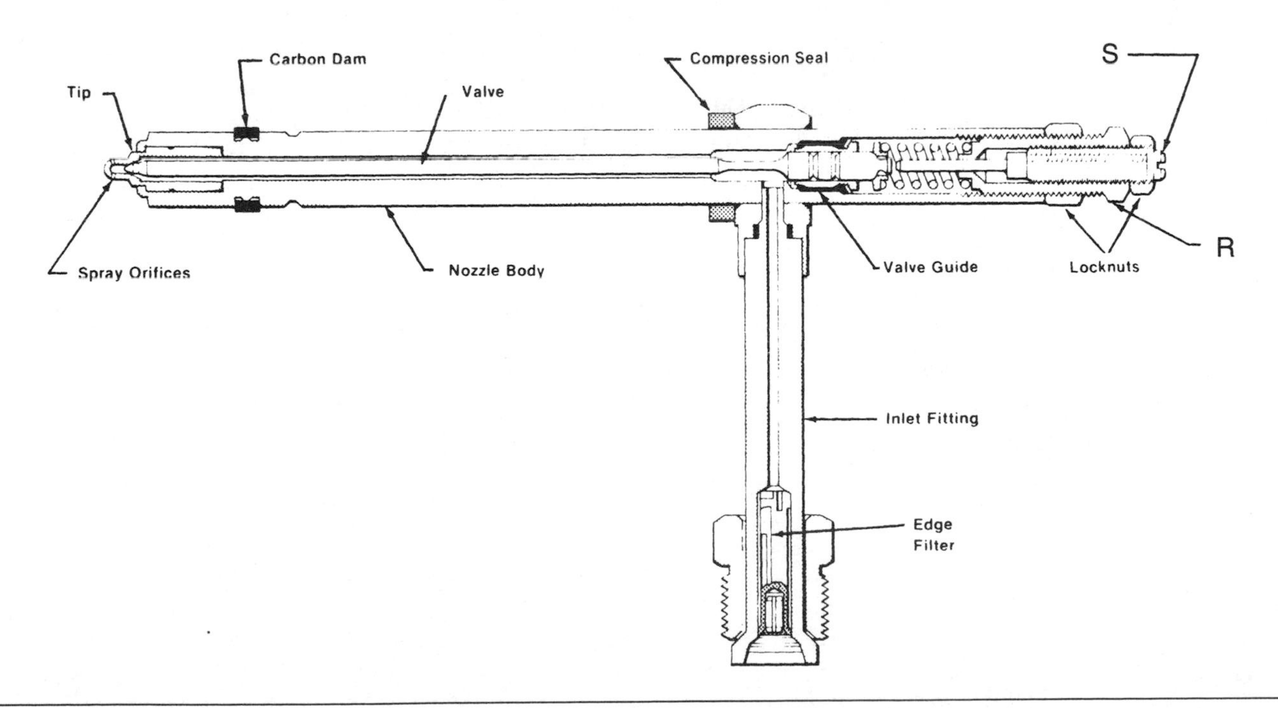

Figure 2–34 Pencil-type nozzle. *(Courtesy of Stanadyne/Hartford Division)*

sure is supplied to the injector piston and forces the plunger downward to pressurize the fuel. The pressurized fuel is then discharged through the injector orifices into the combustion chamber.

The term "popping pressure" is also used to describe the normal opening pressure (NOP) of a nozzle. The NOP value is always one of the first performance specifications to be evaluated when testing injector nozzles on a bench test fixture (pop tester).The actual NOP value is defined by the mechanical spring tension of the injector spring. The two main adjustments on nozzles are opening pressure and lift. The injector tip is placed in a clear plastic container and the pop tester pump is operated to open the nozzle. Observing the nozzle spray pattern when the injector opens indicates the condition of the spray orifices. The spray pattern should be equal from all the orifices. MUI injectors should be tested for fuel output. Special test fixtures and adapters are available to test MUI injectors.

Task 11

Inspect, adjust, repair/replace smoke limiters (air-fuel ratio controls).

180. All of these statements about air-fuel ratio control (AFC) in Figure 2–36 are true EXCEPT:
   A. intake manifold pressure is supplied to the diaphragm in the AFC unit.
   B. the diaphragm in the AFC unit is connected directly or indirectly to the fuel control mechanism.
   C. the AFC device adapts full-load fuel delivery to charge-air pressure.
   D. the AFC device is used on naturally aspirated engines.

181. While discussing air-fuel ratio control (AFC) devices, *Technician A* says if there are puffs of black smoke in the exhaust during transmission shifting, the AFC device may require adjusting.

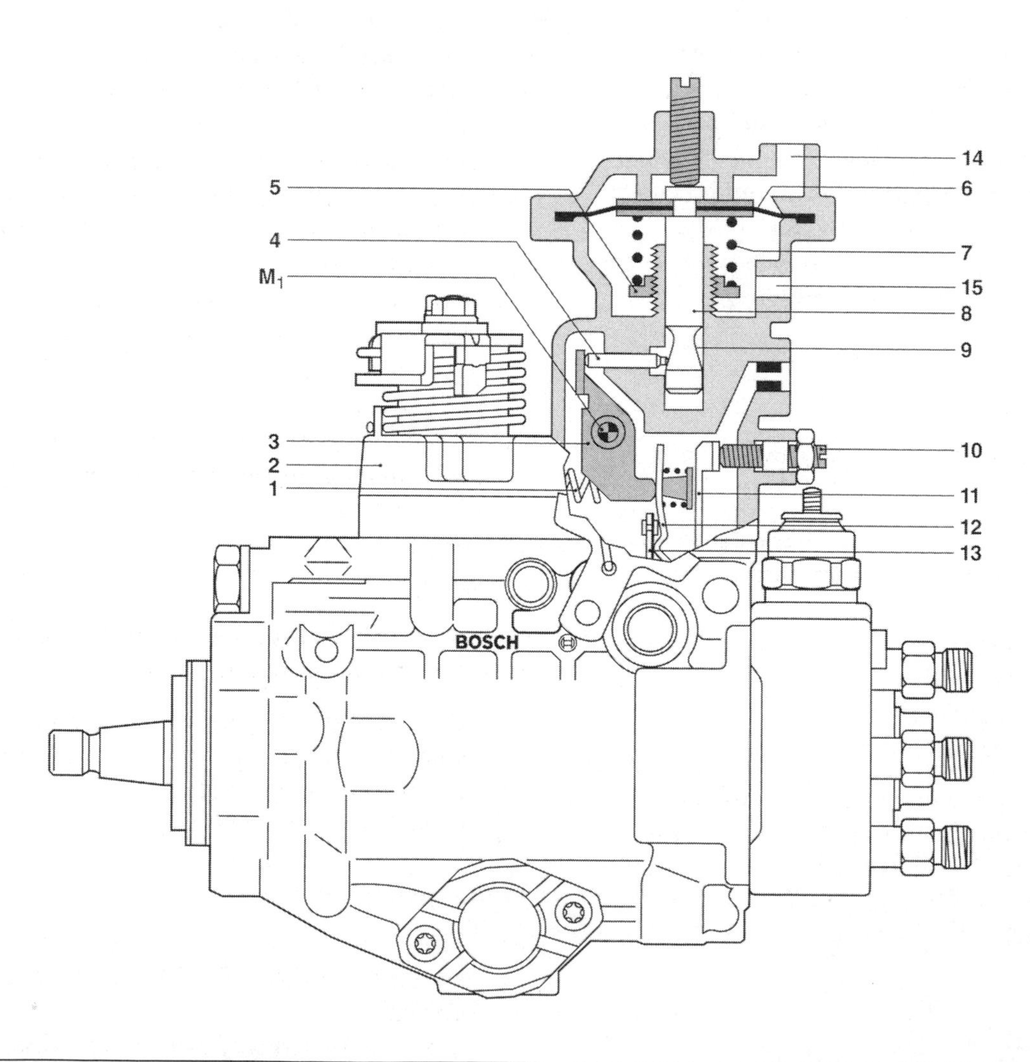

Figure 2–36 Injection pump with air-fuel ratio control. *(Reprinted with permission from Robert Bosch Corporation)*

Technician B says at low engine speeds and boost pressure the AFC diaphragm is forced downward by boost pressure. Who is right?
A. A only
B. B only
C. Both A and B
D. Neither A nor B

Hint

An air-fuel ratio control (AFC) device is used on turbocharged diesel engines to measure manifold boost and limit fueling until the boost pressure achieves a predetermined value.

When an AFC device is used on an in-line, port-helix metering pump, it is usually a mechanism consisting of a manifold, spring, and control rod. The AFC device is fitted with a port and a steel line connects it directly to the engine intake manifold. In this way, boost pressure is delivered to the diaphragm in the AFC device. A diaphragm linkage is connected to the fuel control rack. A spring loads the diaphragm toward the closed position, which limits fueling by preventing the rack from moving into the full fuel position. When manifold boost acting on the diaphragm is sufficient to overcome the spring pressure, it acts on the linkage permitting the rack full travel and thus maximum fueling. Such systems are easily and commonly tampered with by operators in the mistaken belief that aneroid systems reduce engine power. The emission of black smoke from the exhaust stack at each gear shift point is an indication that someone has tampered with the AFC mechanism.

Task 12

Inspect, reinstall/replace high-pressure injection lines, fittings, and seals.

182. After replacing an injector line, what will the Technician have to do first before starting the engine?
A. Readjust the low idle.
B. Bleed all air out of the line.
C. Readjust the high idle.
D. Replace the transfer pump.

183. The injector pump system is being discussed in Figure 2–37. *Technician A* says the lines shown are high-pressure fuel lines. *Technician B* says all fuel connections must be capped when they are disconnected. Who is right?
A. A only
B. B only
C. Both A and B
D. Neither A nor B

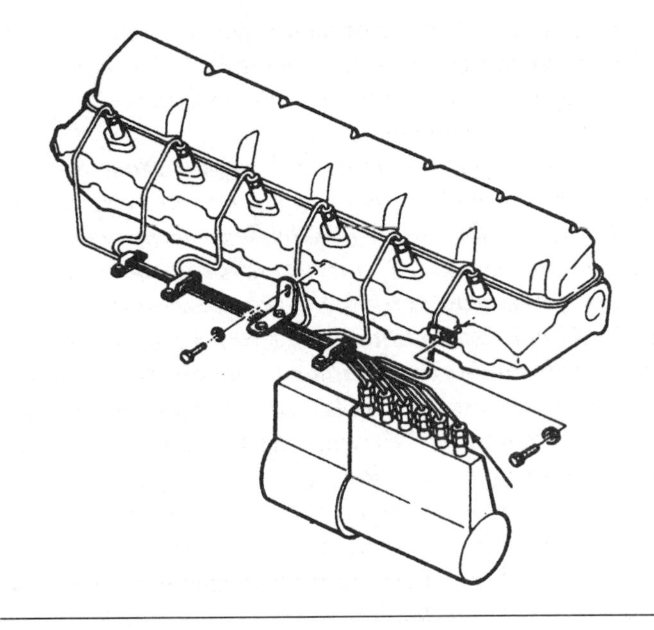

Figure 2–37 Fuel injection system.

Hint

Each high-pressure line is different and cannot be interchanged. Make sure that the clamps holding the lines together are tight. High-pressure fuel lines must be installed as a set. It is much easier than installing them individually. Always cap disconnected fuel lines and fittings. Be sure the brackets holding the high-pressure fuel lines to the intake manifold are properly tightened. Do not remove the caps from the fuel lines or components until they are to be connected. This will help prevent the entry of dirt into the system. After replacing high-pressure fuel lines, it may be necessary to bleed air from the lines by loosening the injector connector one-half to one turn and cranking the engine until solid fuel, free from bubbles, sprays from the connection.

Task 13

Inspect, repair, reinstall/replace low-pressure fuel lines, cooling plate, fittings, and seals.

184. All of the following should be inspected when removing low-pressure fuel lines EXCEPT:
A. any kinks in the lines.
B. connectors on the lines.
C. cracks in the lines.
D. power steering pump.

185. When low-pressure fuel lines are removed they are corroded internally. The MOST likely cause of this problem is:
A. a partially clogged primary fuel filter.
B. water in the fuel.

C. improper grade of diesel fuel.
D. continual low speed operation.

Hint

Before removing any fuel lines, clean the exterior of the fuel lines with fuel oil or solvent to prevent the entry of dirt into the fuel system when the fuel lines are removed. Blow dry the lines with compressed air. Disconnect the battery ground cables from all batteries before removing the fuel lines. Inspect the rubber spacers at the ends of the lines for distortion. Replace all fuel line seals, and make sure the lines have fully seated into their correct fittings before tightening retaining nuts to specifications. Secure lines to prevent damage and kinking. Some modern diesel engines use newer automotive design quick connectors. These connectors snap and clip together providing a good seal when properly connected. To disconnect and reconnect the lines, a special insert tool must be used. Forcing lines apart can cause damage to the connections.

Task 14

Inspect, test, adjust, repair/replace engine governor systems.

186. When the governor is unstable, what is the resulting engine operation?
A. Underrun
B. High idle
C. Hunting
D. Low idle

187. Which of the following is LEAST likely to be a symptom of a minimum-maximum speed governor problem?
A. Low engine power
B. High idle overrun
C. Low idle underrun
D. A rough idle operation

Hint

The mechanical governor is an integral part of the pump or fuel rack that controls the injector. It is either belt or gear driven from the engine. Mechanical governors contain pivoted flyweights that move outward in relation to engine speed. A linkage is connected from the flyweights to the rack so flyweight position can control rack position. On heavy-duty truck applications, the throttle linkage is also connected to the governor linkage. Trucks often have a minimum-maximum speed governor. The governor ensures that the engine does not stall in the idle position, and also governs the maximum engine speed. Between idle and maximum speed, the driver controls the torque by the accelerator pedal, which is connected to the pump linkage. Variable speed governors also control engine speed in the medium speed range. Variable speed governors are also torque sensitive. If more torque is required because the engine load is increased, the governor moves the rack to inject more fuel.

Some engines are equipped with hydraulic governors that use fuel pressure to control the governor. This pressure comes from a constant displacement pump built into the pump housing and driven by the injection pump camshaft. The fuel pressure from the constant displacement pump is supplied to a servo piston connected to the rack to provide governor action. Engine speed and load are computer-controlled on some later model engines. Various input signals inform the powertrain control module (PCM) regarding engine-operating conditions. The PCM controls the amount of fuel injected by the unit injectors or distributor-type pump.

The word "hunting" refers to an unstable governor causing eratic engine operation. This is generally caused by the governor overcorrecting for load. This usually happens when a no-load condition exists, or the governor oil is low. Although some hydraulic governors incorporate a droop function, which effectively causes the governor to become sluggish, thus reducing overcompensation during load changes.

Task 15

Inspect, test, adjust, repair/replace engine fuel shutdown devices and controls.

188. In the internal shutdown solenoid in Figure 2–38, *Technician A* says when an ohmmeter is connected across the solenoid winding and the meter reading is less than specified, the winding is open. *Technician B* says when the ignition switch is turned off the solenoid positions the metering valve in the full fuel position. Who is right?
A. A only
B. B only
C. Both A and B
D. Neither A nor B

189. In the pneumatic fuel shutoff device in Figure 2–39:
A. vacuum is supplied to item 1 when the ignition switch is turned on.
B. item 5 is the governor rod.
C. item 2 is a hand primer.
D. a leak in item 1 may cause failure of the engine to shut off.

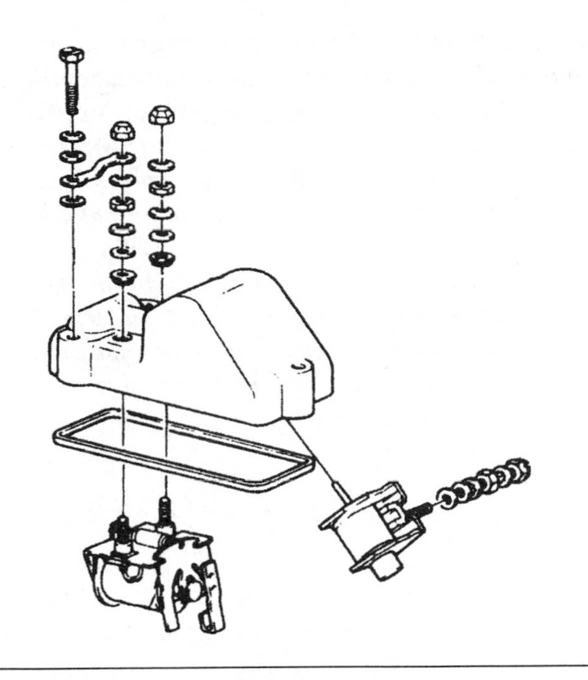

Figure 2–38 Fuel shutdown and vent solenoids.

Hint

Engine fuel shutdown devices may be operated electrically, mechanically, or pneumatically. In a distributor-type injection pump, the fuel shutdown solenoid pushes the metering valve into the no-fuel position when the ignition switch is shut off. In an in-line injection pump, the fuel shutdown solenoid pushes the rack into the no-fuel position. Some fuel shutoff solenoids have a pull-in winding and a hold-in winding. Both windings are energized to move the solenoid away from the no-fuel position. Once the engine is started, only the hold-in winding is energized. In a mechanically operated system, a lever pushes the rack into the no-fuel position. If the engine is equipped with a vacuum pump, a pneumatic fuel shutdown device may be used. When the switch is shut off, vacuum is supplied to the shutoff diaphragm, and a linkage connected to this diaphragm pulls the rack into the no-fuel position.

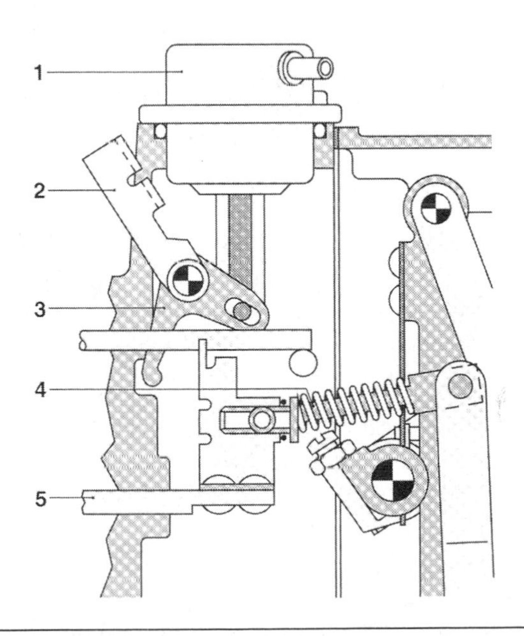

Figure 2–39 Pneumatic shutdown device. *(Reprinted with permission from Robert Bosch Corporation)*

Task 16

Inspect, test, adjust, repair/replace safety shutdown devices, circuits, and sensors.

190. On an engine with an electric fuel shutdown valve and an in-line fuel injection pump, which of the following is not part of this system?
    A. Oil pressure gauge
    B. Key switch
    C. Electric solenoid
    D. Rack

191. An engine with an electric fuel shutdown solenoid and an in-line injection pump starts and immediately stops. The engine cranks normally, and the specified fuel pressure is available at the injection pump. *Technician A* says the hold-in coil in the shutdown solenoid may be open. *Technician B* says there may be higher than normal resistance in the wire connected to the shutdown solenoid windings. Who is right?
    A. A only
    B. B only
    C. Both A and B
    D. Neither A nor B

Hint

A typical fuel shutoff solenoid adjustment procedure on an in-line injection pump follows:

1. *With the key in the off position, pull the ball adjuster off the rack control lever. Manually bottom the solenoid plunger and check to ensure manufacturer's specifications are met for travel before the lever bottoms out. If the lever bottoms before the solenoid, use an adjuster to set the proper clearance. Attach the ball adjuster to the rack control lever.*
2. *Turn the stop screw in, so that the rack control lever will not contact the external stop when the key is in the off position. Turn the key to the off position. The rack control should go all the way to the internal pump stop.*

3. *Adjust the external stop screw to the point where it touches the rack control lever. Then, turn the screw 1/2 to 1 turn beyond this point, so that travel of the rack control is stopped completely by the external stop.*

Electronic Components

Task 1

Check and record engine electronic diagnostic codes and trip/operational data; clear codes; determine needed repairs.

192. *Technician A* says a scan tool is needed to clear a DTC from a vehicle's PCM. *Technician B* says most generic scan tools can access DTCs on any OBD II vehicle. Who is right?
   A. A only
   B. B only
   C. Both A and B
   D. Neither A nor B

193. All of the following are acronyms associated with retrieving faults from a vehicle EXCEPT:
   A. Malfunction indicator lamp (MIL).
   B. Diagnostic trouble code (DTC).
   C. On-board diagnostic ll (OBD II).
   D. Single bit diagnostic subroutine (SBDS).

Hint

Modern engines are controlled and monitored by electronic modules. Most of these modules are capable of storing diagnostic trouble codes (DTCs). To retrieve these codes, the Technician connects a scan tool to a diagnostic connector. This connector is a standard 6-pin connector approved by the Society of Automotive Engineers (SAE) and the American Trucking Associations (ATA). Scan tools may be generic such as the Prolink 9000, or manufacturer specific such as the Cummins Compulink or Echeck. The proper module must be installed in the scan tool for the truck and system being diagnosed. Some vehicles will display a limited readout of DTCs by blinking a malfunction indicator light (MIL) in the instrument panel in a two-digit code called a flash code. Once the problems have been analyzed and corrected, the DTCs need to be cleared. Scan tools can be used to clear these codes. Another approach is to disconnect the battery voltage from the PCM.

The acronym message identifier (MID) is used to describe a major vehicle electronic system, usually with independent processing capability. The acronym parameter identifier (PID) is used to code components within an electronic subsystem. The acronym subsystem identifier (SID) is used to identify the major subsystems of an electronic circuit. Failure mode identifiers (FMIs) are indicated whenever an active or historic code is read using ProLink scan tool or a personal computer (PC). Active DTCs represent faults that are present all the time, and history DTCs are caused by intermittent faults that are not present at the time of diagnosis. A history DTC was caused by an intermittent fault at some time in the past, but the DTC was stored in the PCM memory.

Task 2

Inspect, adjust, repair/replace electronic throttle and PTO control devices, circuits, and sensors.

194. The throttle position sensor is being adjusted on a fuel injection pump with electronic controls. *Technician A* says that on some applications, a gauge block is placed between the gauge boss and the maximum throttle travel screw. *Technician B* says the scan tool will signal when the setting is too low or too high. Who is right?
   A. A only
   B. B only
   C. Both A and B
   D. Neither A nor B

195. When diagnosing a throttle position sensor:
   A. the reference voltage supplied to the TPS should be 5V with the ignition switch off.
   B. the TPS signal voltage increases as the injection pump lever is moved to the maximum speed position.
   C. the TPS contains a magnet and a Hall effect switch.
   D. an improper TPS adjustment may cause lack of power during acceleration.

Hint

The throttle position sensor (TPS) is a variable resistor that supplies the power control module (PCM) with throttle position information so the PCM can regulate engine performance. A 5V reference voltage is sent from the PCM to the TPS, and the TPS varies this voltage in relation to injection pump lever movement. Many TPS sensors supply about 5V at idle and a very low voltage when the pump linkage is moved to the maximum speed position.

To adjust this sensor, it is necessary to connect a scan tool to the vehicle and clear all faults. Disconnect the linkages to the throttle and turn on the key, but do not start the engine during the diagnostic routine. On some applications a gauge block must be installed between the gauge boss and the maximum travel screw. The scan tool will sound a beeping indication of the position of the sensor. If the setting is too high, the beeps will be fast; if it is too low, the beeps will be slow. An ideal setting will have a steady tone. To make these adjustments, it may be necessary to remove the screws holding the TPS sensor to the mounting bracket. Do not try to adjust the maximum throttle travel screw.

Task 3

Perform on-engine inspections, tests, and adjustments on distributor-type injection pump electronic controls.

196. While discussing electronic distributor-type injection pumps, *Technician A* says in some pumps the PCM controls the control collar position to supply the amount of fuel required by the engine. *Technician B* says the PCM does not receive any feedback signal regarding control collar position. Who is right?
   A. A only
   B. B only
   C. Both A and B
   D. Neither A nor B

197. When diagnosing an electronic distributor-type injection pump:
   A. a defective engine speed sensor may cause improper maximum engine rpm.
   B. a defective coolant temperature sensor has no effect on engine starting.
   C. a defective airflow sensor has no effect on fuel delivery.
   D. the needle motion sensor in each injector contains a variable resistor.

Hint

In some nonelectronic distributor-type injection pumps, rollers on the pump drive force the cam plate and plunger forward to compress the fuel in front of the plunger. The compressed fuel is delivered through the discharge ports to the proper injector. The amount of fuel injected is determined by control collar position in relation to the transverse cutoff bore. When the edge of the control collar uncovers the transverse cutoff bore, fuel pressure drops quickly and injection stops (Figure 2–40). The control collar is linked to the throttle linkage and governor.

In an electronically controlled distributor-type injection pump, the control collar position is controlled by a solenoid actuator that is controlled by the PCM (Figure 2–41). The control collar performs the same function as it did in a nonelectronic pump. The PCM receives input signals from the accelerator sensor, atmospheric pressure sensor, vehicle speed sensor, engine speed sensor, airflow sensor, control collar position sensor, water, air and fuel temperature sensor, and needle motion sensor in the injection nozzle. When the PCM receives this input information, it operates the solenoid actuator and the control collar to provide the precise amount of fuel required by the engine under all speeds and loads.

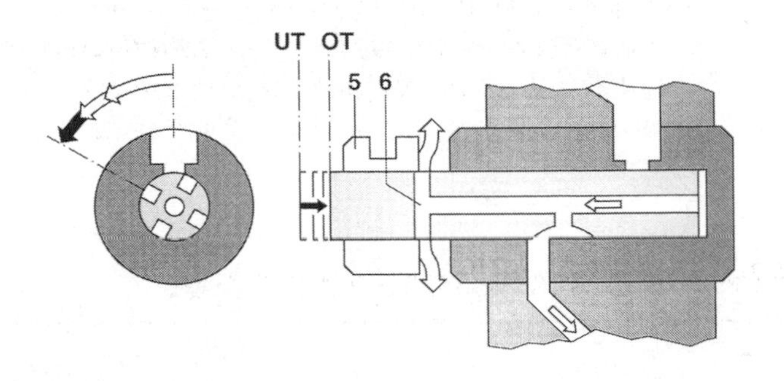

Figure 2–40 Injection pump plunger and control collar. *(Reprinted with permission from Robert Bosch Corporation)*

Task 4

Perform on-engine inspections, tests, and adjustments on in-line type injection pump electronic controls.

198. While discussing electronic in-line injection pumps, *Technician A* says the PCM controls a fuel solenoid in each injector. *Technician B* says the PCM controls all the injectors simultaneously. Who is right?
   A. A only
   B. B only
   C. Both A and B
   D. Neither A nor B

199. When checking the timing on an electronic in-line injection pump:
   A. the fixed timing sensor probe is connected to the timing sensor opening in the flywheel housing.
   B. begin the static timing procedure with number 1 piston at TDC on the compression stroke.
   C. rotate the engine with a barring tool until both lights on the fixed timing sensor are off.
   D. if the timing is not within specifications check the position of the timing indicator.

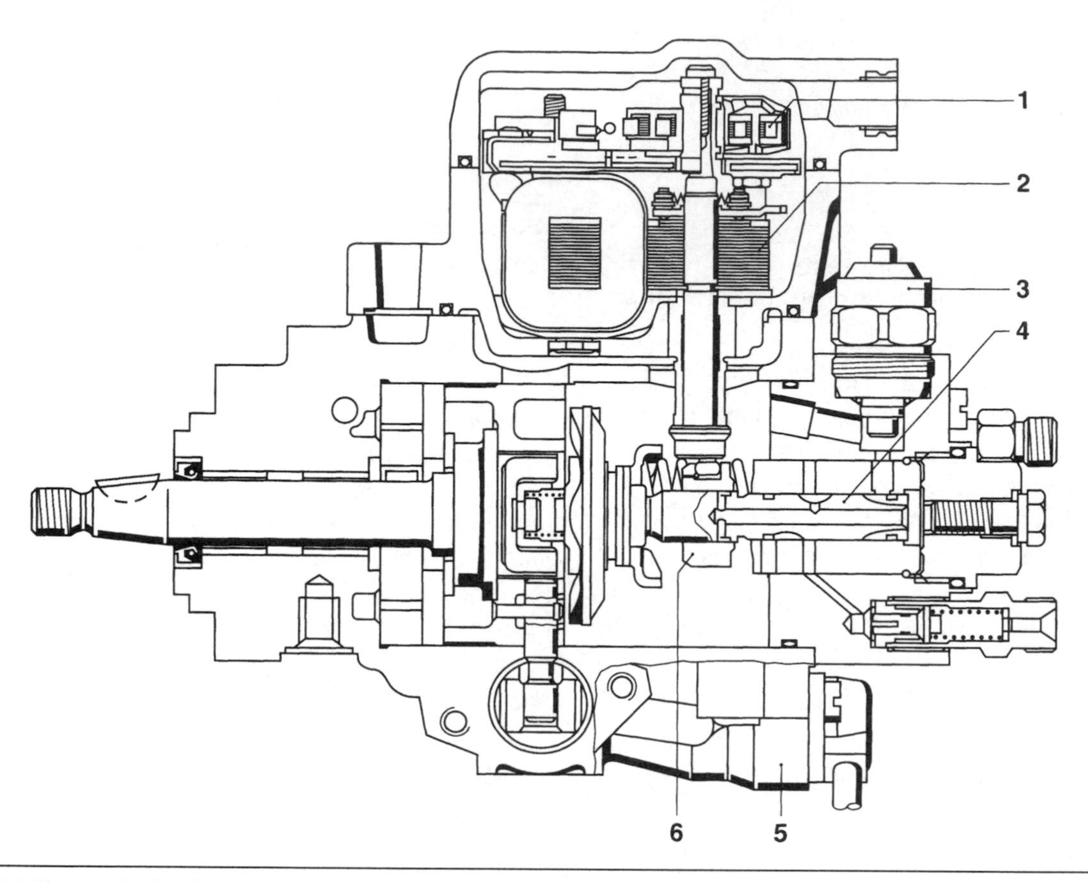

Figure 2–41 Electronic distributor-type injection pump. *(Reprinted with permission from Robert Bosch Corporation)*

Hint

Electronically controlled in-line injection pump systems have a linear motion solenoid connected to the rack in the injection pump (Figure 2–42). The PCM controls this solenoid to provide the precise amount of fuel required by the engine. The PCM calculates the amount of fuel required by the engine from the input signals. These signals include switches for brakes, exhaust brake, and clutch, accelerator pedal sensor, vehicle speed sensor, rack travel sensor, substitute speed signal, pump speed sensor, coolant, air, and fuel temperature sensors, and charge air pressure sensor.

The following is a static timing procedure for an electronically controlled injection pump. A fixed timing sensor probe is installed in the timing event marker (TEM) port in the injection pump. Ground the fixed timing tool ground clamp to a good ground on the engine. The procedure outlined here will be based on that used for a rear-located timing indicator. Rotate the engine using the barring tool to position number 1 cylinder about 45 degrees BTDC. The fixed timing sensor tool is designed to determined the exact location of port closure in the injection pump by reading the timing notch located on the reluctor wheel in the rack actuator housing. There are two lights on the fixed timing sensor tool marked A and B. Each will illuminate for approximately 10 crank angle degrees on either side of the notch off the reluctor wheel. If the starting point is 45 degrees BTDC on the cylinder being timed, and the engine is rotated in correct direction of rotation (CW), the A light will illuminate at 10 degrees before port closure. The objective is to stop at the precise moment that both A and B lights illuminate. Never back the engine up to locate pump port closure due to the engine timing gear train backlash variables.

Next, check the engine calibration scale and timing marker. The reading should be within a 1/2 degree crank angle of specification with both lights on the fixed timing sensor tool illuminated. If the reading is within the

Figure 1: Fuel-injection system with electronically controlled in-line fuel-injection pump
1 Fuel tank, 2 Supply pump, 3 Fuel filter, 4 In-line fuel-injection pump, 5 Electrical shutoff device (ELAB), 6 Fuel-temperature sensor, 7 Rack-travel sensor, 8 Actuator with linear-motion solenoid, 9 Pump-speed sensor, 10 Injector 11 Coolant-temperature sensor, 12 Accelerator-pedal sensor, 13 Switches for brakes, exhaust brake, clutch, 14 Operator panel, 15 Warning lamp and diagnosis connection, 16 Tachograph or vehicle-speed sensor, 17 ECU, 18 Air-temperature sensor, 19 Charge-air-pressure sensor, 20 Exhaust-gas turbocharger, 21 Battery, 22 Glow-plug and starter switch.

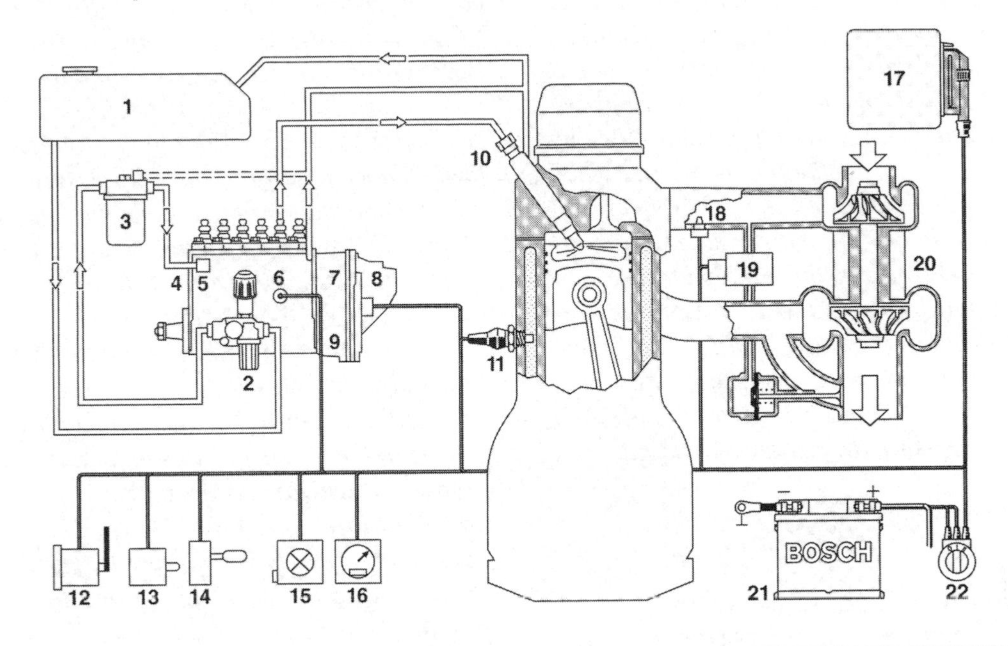

Figure 2–42 Electronically controlled in-line injection pump. *(Reprinted with permission from Robert Bosch Corporation)*

required specification, remove the barring and timing tools and reinstall the TEM sensor. If the reading is out of specification, verify that the timing marker is correctly positioned, that is, correctly reading TDC when the engine is at true TDC. NEVER assume that the pump is incorrectly timed to an engine without first checking the timing marker. To correct static timing, proceed as follows:

Position the engine at the correct static timing location ensuring that it barred CW to this position. Remove the injection pump drive gear access cover. Loosen the pump drive gear capscrew and fit a hub rotation tool onto two of the capscrews. This permits the timing gear hub to be rotated independently of the timing gear. Using the hub rotation tool, turn the timing gear hub counterclockwise until the capscrew locks against the slotted holes in the timing gear. At this point, neither of the timing lights should be illuminated.

Carefully rotate the gear hub CW until both the A and B lights are illuminated. Torque all four timing gear capscrew to specification. During the torquing procedure, it is normal for one of the lights to extinguish. Lubricate the timing gear access cover and torque to specification. Remove the barring and timing tools and reinstall the TEM sensor using the appropriate Mack procedure.

Task 5

Perform on-engine inspections, tests, and adjustments on PT-type injection pump electronic controls.

200. When discussing electronic unit injectors, *Technician A* says this injector is cam actuated. *Technician B* says a rack is used with electronic unit injectors. Who is right?
    A. A only
    B. B only
    C. Both A and B
    D. Neither A nor B

201. When servicing and diagnosing electronic unit injectors:
    A. the amount of fuel injected is determined by the helix in the plunger.
    B. the PCM operates a linear motion solenoid in each injector.
    C. each injector is opened by solenoid force.
    D. the injector solenoid determines the fuel metering and the injection timing functions.

Hint

Cummins introduced Cummins electronic engine control (CELECT) to their full authority electronic management system in 1991 on L10 and N14 engines, and later on their M11 engine when it was introduced in 1994. CELECT Plus electronics were released into the marketplace in 1996 and are used to manage current M11 Plus and N14 Plus engines.

CELECT Plus is a full-authority, computerized engine management system that uses cam-actuated, electronically controlled CELECT injectors. In a mechanical unit injector, the amount of fuel injected is determined by the position of the ports and helices, which are controlled by the rack position. The portion of the plunger stroke when both ports are closed determines the amount of fuel injected. In an electronic unit injector, a solenoid-operated control valve determines the fuel metering and injection timing functions, and the rack is no longer required. When the solenoid-controlled valve is closed, the fuel is pressurized and injected as the plunger is forced down by the cam. When the solenoid controlled valve is opened, fuel pressure drops quickly and injection stops.

Task 6

Perform on-engine inspections, tests, and adjustments on hydraulic electronic unit injector (common rail or rail pressure control) system electronic controls.

202. In a hydraulic electronic unit injector:
    A. the oil pressure is controlled by the oil pump that normally supplies oil to the engine components.
    B. the PCM operates an oil pressure regulator to control oil pressure supplied to the injectors.
    C. the pressure exerted on the fuel by the plunger is four times higher than the oil pressure supplied to the injector.
    D. when the PCM de-energizes the injector solenoid, injection stops slowly.

203. While discussing hydraulic electronic unit injectors, *Technician A* says if the injector oil pressure pump fails, the engine runs in a limp-in mode. *Technician B* says if the injector solenoid is energized, the plunger is forced downward. Who is right?
    A. A only
    B. B only
    C. Both A and B
    D. Neither A nor B

Hint

A typical electronic unit injector system has many improvements over older mechanical ones. In the hydraulic electronic unit injector (HEUI) system, injectors are powered by lubricating oil, which is pressurized by a special pump in the engine valley. The pump output pressure ranges from 450 to 3,000 psi (3,102 to 20,685 kPa). This oil pressure is controlled by the powertrain control module (PCM) with a spill valve called the injector pressure regulator. The high-pressure lubricating oil is delivered to oil rails in the cylinder heads. An injection control pressure sensor mounted on one of the oil rails sends a 0.5V to 5.0V analog voltage signal to the PCM for feedback control of the oil pressure. The piston-actuated second stage of the tandem fuel pump supplies 40 to 70 psi (276 to 486 kPa) fuel pressure to the rear of each cylinder head, where it flows through a fuel rail machined in each cylinder head.

When the PCM inputs indicate that fuel should be injected into the cylinder, the PCM energizes the solenoid in the injector. This action supplies oil pressure to the top of the intensifier piston, and this piston forces the plunger downward to pressurize the fuel and open the nozzle to inject the fuel (Figure 2–43). The top of the intensifier piston is seven times larger than the bottom of the plunger, and this design allows the injector to multiply the pressure under the plunger so it is seven times higher than the oil pressure supplied to the intensifier piston. For example, if the oil pressure on the intensifier piston is 3,000 psi (20,685 kPa), the pressure on the fuel under the plunger is 21,000 psi (144,795 kPa). When the PCM de-energizes the solenoid, the oil pressure on the intensifier piston decreases quickly, and the injector nozzle closes.

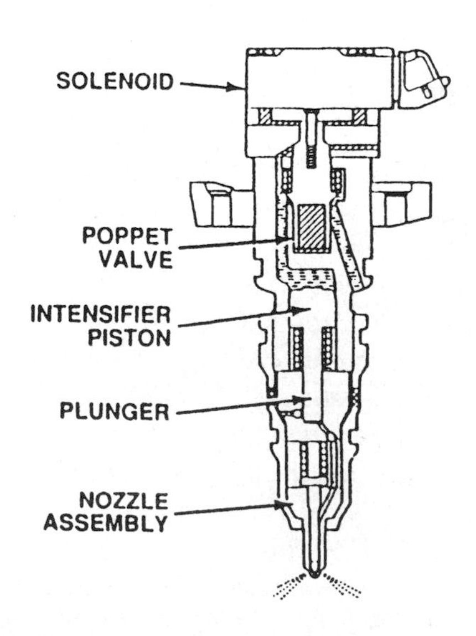

Figure 2–43 Hydraulic electronic unit injector. *(Courtesy of Ford Motor Company)*

Task 7

Perform on-engine inspections, tests, and adjustments on unit injector electronic controls.

204. While discussing unit injector electronic controls, *Technician A* says the electronic distributor unit (EDU) is a stand-alone unit that operates the injectors. *Technician B* says the injectors operate on a pulse width principle. Who is right?
   A. A only
   B. B only
   C. Both A and B
   D. Neither A nor B

205. When diagnosing and servicing electronic unit injectors:
   A. if the injector solenoid is de-energized fuel, injection can occur.
   B. the end of injection is determined by a mechanically operated spill port.
   C. the PCM sends a pulse width voltage signal to the injectors.
   D. if the poppet control valve in the injector opens, fuel injection begins.

Hint

Some electronic unit injector systems have an electronic distributor unit (EDU) that controls regulated current pulses to the injector solenoids at the command of the PCM (Figure 2–44). The PCM informs the EDU regarding engine fuel requirements, and the EDU supplies regulated current pulses to the injectors to open the injectors and supply the precise amount of fuel required by the engine. These regulated current pulses may be called pulse width, which is the on time of the injector. When injector pulse width is increased, more fuel is injected.

On some engines an injector cutout test may be performed by the PCM with the proper scan tool connected to the system. The PCM operates the engine at 1,000 rpm, and the injector pulse width is displayed on the scan tool. As each functioning EUI was cut out in sequence, the average PW would have to increase to maintain the test rpm. However, when the defective EUI was cut out, there would be no change in the average PW, because this cylinder was not contributing to the engine rpm before the injector was cutout.

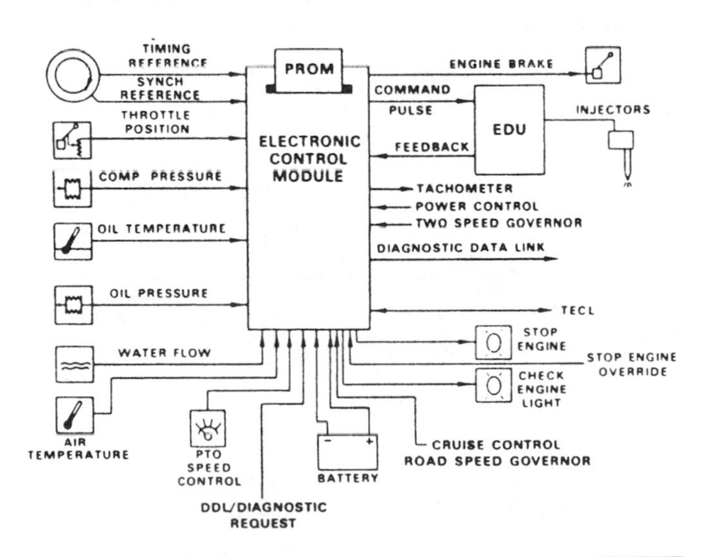

Figure 2–44 Electronic unit injector system. *(Reprinted with permission from SAE Paper 850542 c 19 Society of Automotive Engineers, Inc.)*

Task 8

Inspect, adjust, repair/replace electronic air-fuel ratio controls, circuits, and sensors.

206. While discussing the accelerator pedal position (APP) module in Figure 2–45, *Technician A* says three-pedal position sensors in the APP module should provide the same voltage signal as the accelerator pedal is depressed. *Technician B* says if a fault occurs in the APP module, the MIL light is illuminated in the instrument panel. Who is right?

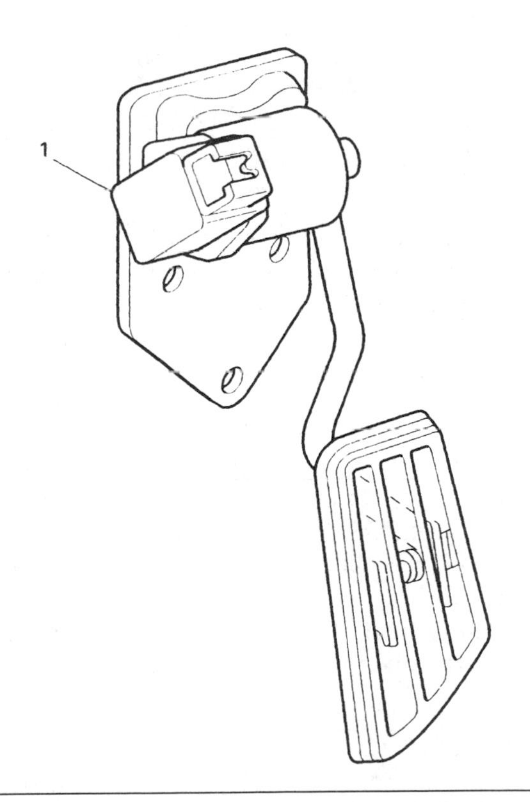

Figure 2–45 Accelerator pedal position module. *(Courtesy of General Motors Corporation, Service Technology Group)*

A. A only
B. B only
C. Both A and B
D. Neither A nor B

207. When diagnosing and servicing electronic diesel injection systems:
A. the APP 2 variable resistor voltage signal should be 4.5V at idle.
B. if one variable resistor is defective in the APP module, engine power is reduced.
C. if one variable resistor is defective in the APP module, the service throttle soon light is on.
D. if all three variable resistors are defective in the APP module, the engine speed is limited to 1,000 rpm.

Hint

On engines that have no direct linkage between the accelerator pedal and the injection pump, the accelerator pedal is connected to an accelerator pedal position (APP) module. This module contains three variable resistors actuated by accelerator pedal movement. Each of these variable resistors send different voltage signals to the PCM as the accelerator pedal is depressed. The APP 1 variable resistor voltage goes from 0.5V at idle to a higher voltage as the accelerator pedal is depressed. The APP 2 variable resistor voltage goes from 4.5V at idle to 1.5V with the accelerator pedal fully depressed. The APP 3 variable resistor voltage goes from 4.0V at idle to 2.0V with the accelerator pedal fully depressed. If one variable resistor in the APP module is defective, the engine will run normally, but a DTC is set in the PCM memory. When two variable resistors in the APP module are defective, a DTC is set in the PCM memory, and the service throttle soon light is illuminated. Under this condition decreased engine performance also occurs. If all three variable resistors in the APP module are defective, a DTC is set in the PCM memory, the service throttle soon light is on, and the engine only runs at idle speed. A scan tool connected to the diagnostic connector may be used to test the APP module. The scan tool also displays data from all the other inputs and outputs.

Task 9

Inspect, test, adjust, repair/replace engine governor electronic controls, circuits, and sensors.

208. While discussing electronic governor action, *Technician A* says the governor functions are programmed into the PCM software. *Technician B* says a rack position sensor sends feedback information to the PCM. Who is right?
A. A only
B. B only
C. Both A and B
D. Neither A nor B

209. A diesel fuel injection system with an electronic governor has all these advantages EXCEPT it:
A. provides improved air-fuel ratio control.
B. reduces exhaust emissions.
C. can be reprogrammed with customer data to tailor the engine for varying engine and chassis applications.
D. reduces diesel knock intensity.

Hint

When the port-helix metering injection pump is managed by a computer, engine governing depends on how the PCM is programmed. The governor housing attached to the rear of the injection pump is replaced by a PCM-controlled linear solenoid, output devices, and sensors. This linear solenoid sets rack position on PCM command. A rack position sensor reports the exact rack position to the PCM.

Additionally, the rack actuator housing may house other sensors to report rotational speed, engine position, and timing data to the PCM. Where a linear solenoid is fitted to the rack in an in-line, port-helix metering injection pump, the engine governing functions are undertaken by the software programmed into the PCM.

Electronic governing offers infinitely more control over engine fueling, and it allows OEMs to meet statutory emission requirements and achieve better fuel economy. This type of system can be easily programmed/reprogrammed with customer data to tailor the engine for varying engine and chassis applications.

Task 10

Inspect, test, adjust, repair/replace engine electronic fuel shutdown devices, circuits, and sensors.

210. In an engine shutdown system with a nonelectronic injection pump:
A. engine shutdown occurs immediately if a low coolant level is sensed.
B. a separate module shuts off the voltage to the fuel shutoff relay winding if the coolant overheats.
C. engine shutdown occurs if the engine speed is above the maximum governed rpm.
D. if the engine shutdown system causes engine shutdown, the engine will not restart until the defect is corrected.

211. A diesel engine nonelectronic in-line injection pump and a shutdown system stops running (Figure 2–46). There are no warning lights illuminated in the instrument panel, and fuel pressure is normal at the injection pump. *Technician A* says the fuel shutoff relay winding may be grounded after the winding. *Technician B* says there may be an open circuit between the warning module and the fuel shutoff relay. Who is right?
   A. A only
   B. B only
   C. Both A and B
   D. Neither A nor B

Hint

Most diesel engines have some type of shutdown system if a defect occurs that may cause engine damage. On some engines this system provides engine shutdown if the oil pressure is low, the coolant temperature is high, or the coolant level is low. When any of these conditions are present a visual warning is provided for the driver. If the condition is present for 30 to 40 seconds, engine shutdown occurs. Key-off is required to restart the engine,

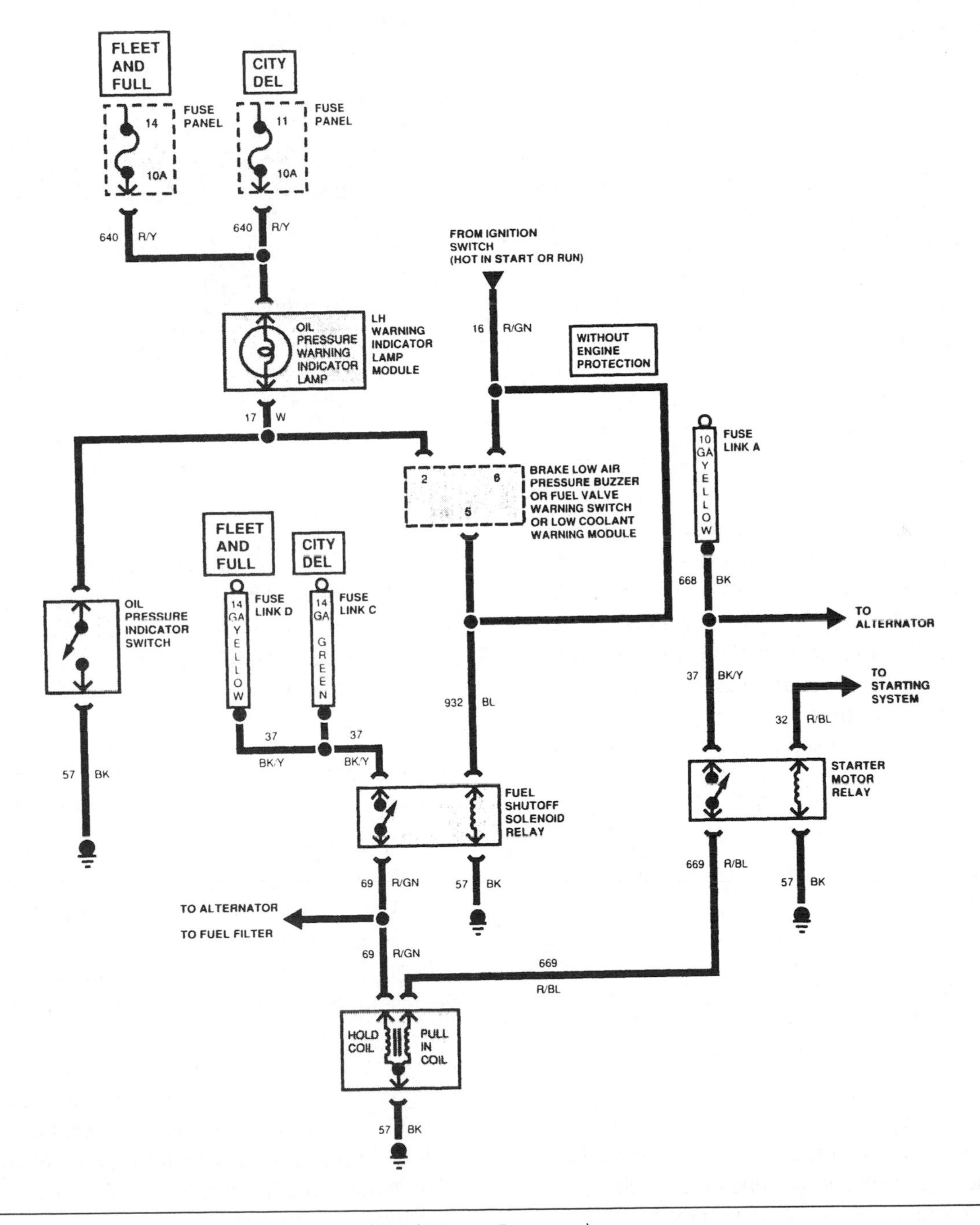

Figure 2–46 Engine shutdown system. *(Courtesy of Ford Motor Company)*

but if the shutdown trigger persists, the engine will be shut down again after 30 seconds.

In some shutdown systems the PCM is directly responsible for engine shutdown when the PCM senses one of the above conditions. The PCM can shut the engine down by moving the linear rack control solenoid to the no-fuel position. On other systems without a PCM-controlled injection pump, a separate module senses the defective conditions and de-energizes the fuel shutoff relay winding. This action opens the fuel shutoff relay contacts, which cut off voltage to the pull-in coil in the fuel shutdown solenoid. Under this condition the fuel shutdown solenoid pushes the rack into the no-fuel position.

Task 11

Inspect, test, adjust, repair/replace electronic safety shut-down devices, circuits, and sensors.

212. While discussing electronic shutdown systems, *Technician A* says the level of engine protection can be programmed into some PCMs. *Technician B* says the engine shutdown function is the first level of engine protection. Who is right?
   A. A only
   B. B only
   C. Both A and B
   D. Neither A nor B

213. On a Detroit diesel electronic control (DDEC) shutdown system:
   A. the shut engine off (SEO) light is illuminated when any electronic fault is sensed.
   B. the malfunction indicator light (MIL) is turned on by a MIL relay.
   C. a stop engine override switch allows the driver to temporarily override the engine shutdown function.
   D. monitored parameters include the turbocharger boost pressure.

Hint

Engine protection strategy can be programmed to three levels of protection: a driver alert, engine speed and power ramp down, and a complete shutdown of the engine. The level of protection is selected by the vehicle owner and programmed to the Detroit diesel electronic control (DDEC) PCM as a customer option. As a customer option, the failure strategy can be changed at any time by a Technician equipped with the access password and a scan tool with the DDEC software.

DDEC III monitors a range of vehicle and chassis functions and is programmed with engine protection strategy that may derate or shut down the engine when a running condition could result in catastrophic failure. The operator is alerted to system problems by illuminating the malfunction indicator light (MIL) and the stop engine light (SEL). The MIL is illuminated when a system fault is logged to alert the operator that DDEC III fault codes have been generated. The SEL is illuminated when the PCM detects a problem that could result in a more serious failure. DDEC III may also be programmed to shut down or ramp down to idle speed after such a problem has been logged. A shutdown override switch known as a stop engine override (SEO) switch permits a temporary override of the shutdown command.

Parameters monitored include low coolant level, high coolant temperature, low oil pressure, high oil temperature, auxiliary digital input.

Task 12

Inspect and test power and ground circuits and connections; determine needed repairs.

214. *Technician A* says to use an ohmmeter to check the resistance between battery negative terminal and ground while cranking the engine. *Technician B* says to check the voltage drop between battery positive terminal and the starter main terminal stud while the engine is cranking. Who is right?
   A. A only
   B. B only
   C. Both A and B
   D. Neither A nor B

215. When using a voltmeter to perform a voltage drop test in a circuit, the leads should be connected in what way?
   A. To the battery terminals
   B. From the positive battery terminal to ground
   C. In series with the circuit being tested
   D. In parallel with the circuit being tested

Hint

With modern electronic modules and low-level signals, proper grounding between modules has become extremely important. Even a slight resistance in a computer ground can cause a fault to occur. Test grounds for voltage drop with power applied using a digital voltmeter with the leads connected in parallel to the circuit being tested. Do not use a standard voltmeter because its internal resistance can affect the internal circuit of the module. Once a defective ground has been detected, clean the ground connection and treat it with conductive grease. Fasten all leads securely and connect a suitable wire between common grounds. Retest all circuits to ensure proper operation.

Task 13

Inspect and replace electrical connector terminals, seals, and locks.

216. When changing weather pack connectors, the Technician should be careful to:
 A. use the insertion/removal tool.
 B. bend the pins to fit.
 C. replace with standard connectors.
 D. ground all wiring before changing.

217. *Technician A* says the EMC connectors should never be probed through the weather pack seals. *Technician B* says a test kit should be used to service connectors. Who is right?
 A. A only
 B. B only
 C. Both A and B
 D. Neither A nor B

Hint

The PCM harness electrically connects the PCM to other modules in the vehicle engine and passenger compartments. When signal wires are spliced into a harness, use wire with high-temperature insulation only. With the low current and voltage levels found in the system, it is important that the best possible bond at all wire splices be made by soldering the splices. Molded connectors require complete replacement of the connector. This means splicing a new connector assembly into the harness. A connector test adapter kit is used to probe terminals during diagnosis. Some connectors, such as the coolant sensor, use terminals called microconnectors. They are changed by inserting a fine pick tool through the front of the connector to bend a locking tab on the terminal. To remove, bend the tab toward the terminal, pull the terminal and wire out of the connector. To install, insert the tool into a new connector and bend the tab away from the terminal. A weather pack connector can be identified by a rubber seal at the rear of the connector. This connector protects against moisture and dirt, which could create oxidation and deposits on the terminals. This protection is important because of the very low voltage and current levels found in the electronic system. To repair a weather pack terminal, use a special tool to remove the pin and sleeve terminals. If removal is attempted with an ordinary pick there is a good chance that the terminal will be bent or deformed. Unlike standard blade-type terminals, these terminals cannot be straightened once they are bent.

Task 14

Connect computer programming equipment to vehicle/engine; access and change customer parameters; determine needed repairs.

218. *Technician A* says that the customer option codes can be changed in the electronic systems with a scan tool. *Technician B* says in some cases changing these codes could result in poor engine performance. Who is right?
 A. A only
 B. B only
 C. Both A and B
 D. Neither A nor B

219. In electronic diagnostics, the acronym PID stands for:
 A. parameter identifier.
 B. part identification number.
 C. positive induction device.
 D. partial induction delivery.

Hint

Modern automotive electronic systems have stored information pertaining to customer options. These are used to change the operational characteristics of the control modules. To access and reprogram this information, connect a scan tool to the diagnostic connector usually located under the dash on the driver's side. This will provide a link with all programmable modules either directly or over a network protocol line. These links must be automatically compatible to generic scan tools by standards established by all automotive manufacturers and international governments. Each manufacturer will provide necessary codes to make required changes to achieve proper operation of the vehicle and customer loaded requirements. A Technician using a generic scan tool must use manufacturers' listed codes.

The PCM reset allows the scan tool to command the PCM to clear all emission-related diagnostic information. When resetting the PCM, a code will be stored in the PCM until all the OBD II system monitors or components have been tested to satisfy a drive cycle without any other faults occurring. The following events occur when PCM reset is performed: clears the DTCs, clears the scan tool freeze frame data, and resets status of the OBD II system monitors.

To reset the random access memory (RAM), disconnect the battery ground cable for a minimum of five minutes. Resetting the RAM will clear learned values the PCM has stored for adaptive systems such as idle and fuel trim. After RAM has been reset, the vehicle may exhibit certain driveability concerns. It will be necessary to drive the vehicle to allow the PCM to relearn values for optimum driveability and performance.

Starting System Diagnosis and Repair

Task 1

Perform battery state-of-charge test; determine needed service.

220. While discussing hydrometers in sealed batteries, *Technician A* says if the hydrometer is yellow the battery should be tested to determine the battery condition. *Technician B* says if the hydrometer is green it should only be fast charged. Who is right?
    A. A only
    B. B only
    C. Both A and B
    D. Neither A nor B

221. When testing and servicing batteries:
    A. a battery with 1.225 specific gravity on all cells is fully charged.
    B. if the battery temperature reaches 150°F (65°C) the battery will be damaged.
    C. oxygen gas only is emitted from the battery while charging.
    D. a battery with an open circuit voltage of 12V is fully charged.

Hint

Battery state of charge may be tested with a hydrometer if the battery has removable filler caps. A fully charged battery has a specific gravity of 1.265. In a fully charged battery the maximum variation in specific gravity readings is 0.050 specific gravity points. A voltmeter connected across the battery terminals indicates the battery state of charge. A fully charged battery should have 12.6V with no load on the battery. This may be called an open circuit voltage test. Sealed batteries have a hydrometer in the top of the battery. If the hydrometer is green the battery is sufficiently charged for testing. A yellow hydrometer indicates the battery should be charged before testing. Battery temperature should not exceed 120°F (49°C) during the charging process. A battery gives off oxygen and hydrogen gas while charging. Because hydrogen is explosive, sources of ignition must be kept away from the battery.

Task 2

Perform battery capacity (load, high-rate discharge) test; determine needed service.

222. When performing a load test on a fully charged battery, the Technician finds that the battery voltage drops below manufacturer's specifications. The Technician should do which one of the following?
    A. Recharge the battery and return it to service.
    B. Recharge the battery and retest it.
    C. Replace the battery.
    D. Replace the voltage regulator.

223. While performing a battery load test with the battery fully charged at 70°F (21°C):
    A. the battery is satisfactory if the voltage remains above 10V.
    B. the battery should be discharged at one-half the ampere hour rating.
    C. the load should be applied to the battery for 15 seconds.
    D. the test voltage is read after the load is turned off.

Hint

Before performing a battery load test, the battery should be fully charged, and the battery temperature should be above 40°F (4.4°C). The battery discharge rate for a capacity test is usually one-half of the cold cranking rating. The battery is discharged at the proper rate for 15 seconds to eliminate the surface charge. Then, after about 1 to 2 minutes, the test is repeated. The battery voltage must remain above 9.6V with the battery temperature at 70°F (21°C) or above. If the battery is fully charged and the voltage at the end of the load test is less than specified, the battery should be replaced.

Task 3

Charge battery using slow or fast charge method as appropriate.

224. *Technician A* says if the battery temperature exceeds 125°F (52°C) the charging rate should be reduced. *Technician B* says the battery is fully charged when the specific gravity on all cells is 1.230. Who is right?
    A. A only
    B. B only
    C. Both A and B
    D. Neither A nor B

225. When testing battery specific gravity after the battery has been fully charged, five cells have 1.265 specific gravity and one cell has 1.200 specific gravity. *Technician A* says the battery should be replaced. *Technician B* says the separators between the plates may be damaged in the cell with 1.200 specific gravity. Who is right?
    A. A only
    B. B only
    C. Both A and B
    D. Neither A nor B

Hint

Battery charging consists of a charge current in amperes for a period in hours. Thus, a 25-ampere charging rate for 2 hours would be a 50-ampere hour charge to the battery. Charge rates between 3 and 50 amperes are satisfactory as long as spewing of the electrolyte does not occur and the battery temperature does not exceed 125°F (52°C). If spewing occurs or the temperature exceeds 125 degrees F (52°C), the charging rate must be reduced or temporarily halted to permit cooling. Most fast chargers are capable of a 50-ampere charging rate on a 12V battery. A control on the charger allows the Technician to adjust the charging rate. Slow chargers usually provide a charging rate up to 10 amperes. Most shop slow chargers are capable of charging several batteries connected in series. A control on the charger allows the Technician to adjust the charger voltage depending on the number of batteries being charged.

The battery is sufficiently charged when all cells are above 1.265 specific gravity or the green dot in the built-in hydrometer is visible. The maximum variation in the cell specific gravity readings is 0.050 point.

Task 4

Start a vehicle using jumper cables, a booster battery, or auxiliary power supply.

226. On a vehicle with the two-battery 12-volt system, the battery's connection is which one of the following?
    A. Series circuit
    B. Parallel circuit
    C. DC circuit
    D. AC circuit

227. When using jumper cables to start a vehicle with a dead battery, always do which of these items?
    A. Connect positive to negative to create a series circuit.
    B. Connect positive to positive and negative to negative to create a parallel circuit.
    C. Connect the negative cable first and the positive last.
    D. Make sure both vehicles are electronically grounded together before making any connections.

Hint

If the vehicle will not start due to a discharged battery, it can often be started by using energy from another battery. This procedure is called "jump-starting." If the vehicle has a 12-volt starting system and a negative ground electrical system, make sure the other vehicle also has a 12-volt starting system and that it is a negative ground system. Diesel engine vehicles have more than one battery because of the higher torque required to start a diesel engine. This procedure can be used to start a single battery vehicle from any of the batteries of the diesel vehicle. However, at low temperatures it may not be possible to start a diesel engine from a single battery in another vehicle.

Position the vehicle with the good battery so that the booster cables will reach but never let the vehicles touch. Also, be sure the booster cables do not have loose or missing insulation. Follow this jump-starting procedure:

1. *Turn off the ignition (engine control switch) and all lamps and accessories except the hazard flasher or any lamps needed for the work area.*
2. *Apply the parking brake firmly. Shift the automatic transmission to park or manual transmission to neutral.*
3. *Make sure the cable clamps do not touch any other metal parts. Clamp one end of the positive booster cable to the positive terminal on one battery, and the other end to the positive terminal on the other battery. Never connect positive to negative.*
4. *Clamp one end of the negative booster cable to the negative terminal of the good battery. Connect the other end of the negative booster cable to the frame rail, chassis, or to any solid, stationary, unpainted metal object on the engine at least 18 inches (450 mm) from the discharged battery. Make sure the cables are not on or near pulleys, fans, or other parts that will move when the engine is started.*
5. *Start the engine of the vehicle with the good (charged) battery and run the engine at a moderate speed for several minutes. Then, start the engine of the vehicle that has the discharged battery.*

6. *Remove the booster cables by reversing the above installation sequence exactly. While removing each clamp, take care it does not touch any another metal while the other end remains attached.*

Task 5

Inspect, clean, repair/replace battery, battery cables, and clamps.

228. A battery has a low electrolyte level. *Technician A* says the charging circuit voltage may be too high. *Technician B* says there may be excessive resistance in the wire from the alternator battery terminal to the positive battery terminal. Who is right?
    A. A only
    B. B only
    C. Both A and B
    D. Neither A nor B

229. When servicing battery terminals:
    A. when removing battery cables, remove the positive cable first.
    B. removing the battery cables erases the PCM adaptive memory.
    C. install the negative cable first.
    D. a battery cable may be removed with the engine running.

Hint

If battery voltage is disconnected from a computer, the adaptive memory in the computer is erased. In a powertrain control module (PCM), disconnecting power may cause erratic engine operation or erratic transmission shifting when the engine is restarted. After the vehicle is driven for about 20 miles, the computer relearns the system, and normal operation is restored. Radio station presets will also be erased. A 12V-power supply from a dry cell battery can be connected through the cigarette lighter or power point connector to maintain voltage to the electrical system when the battery is disconnected.

A battery can be cleaned with a baking soda and water solution. Always wear hand and eye protection when servicing batteries and their components. If the built-in hydrometer indicates light yellow or clear, the electrolyte level is low, and the battery should be replaced. The low electrolyte level may be caused by a high-voltage regulator setting that causes overcharging. When disconnecting battery cables, always disconnect the negative battery cable first. The order is reversed when installing the battery and the cables. Never remove a battery cable with the engine running. This action will severely damage electronic components.

It is always wise to spray the cable clamps with a protective coating to prevent corrosion. A little grease or petroleum jelly will also prevent corrosion. Also available are protective pads that go under the clamp and around the terminal to prevent corrosion.

When replacing a battery, it is very important to check the charging system. A faulty alternator or regulator may cause a discharged or overcharged battery condition, resulting in a damaged battery.

Task 6

Inspect, test, and reinstall/replace starter relays, safety switch(s), and solenoids.

230. When installing a new starter relay, all of the following steps are required EXCEPT:
    A. install the positive lead to the relay.
    B. install the negative lead to the relay.
    C. install the control wires on the relay.
    D. install the ground on the battery.

231. While discussing starter relays, *Technician A* says the starter relay opens and closes the circuit between the battery and the starting motor armature. *Technician B* says if the starter relay contacts stick closed, the starting motor continues to run after the engine starts. Who is right?
    A. A only
    B. B only
    C. Both A and B
    D. Neither A nor B

Hint

Some starter relays are an integral part of the starter and are replaced as a unit. Other starter relays are attached to the starter or mounted as a separate unit or to the fender well. The starter relay opens and closes the circuit between the battery and the starter solenoid windings. Removal of the starter relay entails removing the two positive leads and the control wires. The starter relay winding is grounded to the relay case. Therefore, a relay ground wire is not required. Any time maintenance is to be performed on the starter, remove the negative cable from the battery to prevent accidental grounding of the wiring.

Task 7

Perform starter current draw test; determine needed repairs.

232. *Technician A* says that when performing a starter current draw test, do not crank the engine for more than 30 seconds at a time. *Technician B* says you

can perform this test with a battery charger connected to the battery. Who is right?

A. A only
B. B only
C. Both A and B
D. Neither A nor B

233. While performing a starter current draw test and analyzing the results:

A. a high current draw may be caused by excessive resistance in the starting motor cables.
B. a starter current draw test results are accurate if the vehicle batteries are partly charged.
C. low current draw may be caused by worn bushings in the starting motor.
D. during the test disconnect the fuel shutdown solenoid.

Hint

The starting circuit carries the high current flow within the system and supplies power for the actual engine cranking. Starting circuit components are the battery, battery cables, magnetic switch or solenoid, and the starter motor. The cranking current test measures the amount of current, in amperes, that the starter circuit draws to crank the engine. The battery should be fully charged before performing the starter current draw test. Disconnect the fuel shutdown solenoid to prevent the engine from starting during the test. High starter current draw, low cranking speed, and low cranking voltage usually indicate a faulty starter motor. Low current draw, low cranking speed, and high cranking voltage usually indicate excessive resistance in the starting circuit. Open circuits, failed switches, springs, or relays cause a no-current condition and an inoperative starting motor.

Task 8

Perform starter circuit voltage drop tests; determine needed repairs.

234. Which tool is most likely to be used to determine the starter circuit voltage drop test?

A. Starter shunt resistor
B. Voltmeter
C. Ohmmeter
D. Milliammeter

235. *Technician A* says when measuring voltage drop there should be no current flow in the circuit being tested. *Technician B* says as the resistance in a circuit decreases the voltage drop increases. Who is right?

A. A only
B. B only
C. Both A and B
D. Neither A nor B

Hint

To check the resistance in the ground circuit, use a voltmeter to check voltage drops. Measure the voltage drop across each component in the starter circuit to check the resistance in that part of the circuit. Read this voltage drop while the starter motor is operating. For example, connect the voltmeter leads between the positive battery terminal and the positive cable on the starter solenoid, and crank the engine to measure the voltage drop across the positive battery cable. A ground circuit voltage drop test is performed by connecting a voltmeter between the ground side of the battery and the starter housing. Then, the engine is cranked and the voltage read. Voltage should not exceed 100 millivolts. When selecting new cables, be sure they are at least as large as the ones being replaced.

Excessive resistance caused by poor terminal connections result in an abnormal voltage drop in the starter cable. Low voltage at the starter prevents normal starter operation and causes hard starting.

Task 9

Remove and replace starter.

236. When installing starter, which is LEAST likely to be done?

A. Connect control wiring to solenoid.
B. Align starter motor with flywheel.
C. Install starter mounting bolts.
D. Replace starter bushings.

237. When removing the starter from a vehicle, what should the Technician remove first?

A. The starter bolts
B. The starter solenoid
C. The negative battery cable
D. The inspection cover

Hint

Disconnect the negative cable of the battery to prevent accidental grounding while removing and replacing the starting motor. Disconnect the positive lead and ground cable from the starter motor. Remove the leaf shield, if installed. Remove the control wires from the solenoid. Remove the starter mounting bolts and starter motor. When installing the starting motor always be sure the original shim thickness is installed between the starting motor mounting flange and the flywheel housing.

Engine Brakes

Task 1

Inspect, test, and adjust engine brakes.

238. Refer to Figure 2–47. All of these statements about a compression brake are true EXCEPT:
    A. the driver's foot must be off the accelerator pedal to activate the brake.
    B. the brake will not operate if the driver has the clutch pedal depressed.
    C. the slave cylinder pushes downward on the exhaust valve crosshead.
    D. the master cylinder supplies higher pressure to the slave cylinder when the pushrod begins to move downward.

239. To eliminate pulsation in the air intake system caused by exhaust brakes, install:
    A. an air intake suppressor.
    B. a wastegate.
    C. an oil bath air filter.
    D. a variable camshaft.

Hint

When the compression brake is energized, it causes the engine to perform like a power absorbing air compressor. The brake opens the exhaust valves before the compression stroke is complete and combustion does not occur in the cylinder. The compressed air is released into the engine exhaust system and energy is not returned to the engine through the power stroke.

When the solenoid valve is energized, engine-lubricating oil flows under pressure through the control valve. Then the oil flows to both master piston and slave piston. The oil pressure causes the master piston to move down against the adjusting screw of the injector rocker lever. The pushrod moves the adjusting screw end of the rocker lever up during the injection cycle. This causes the adjusting screw to push against the master piston. The movement of the master piston causes high oil pressure in the oil passage from the master piston to the slave piston. The ball check valve in the control valve holds the high pressure in the oil flow from the master piston to the slave piston.

The high pressure in the oil applied to the slave piston moves this piston down against the crosshead and opens the exhaust valves. The exhaust valves open as the

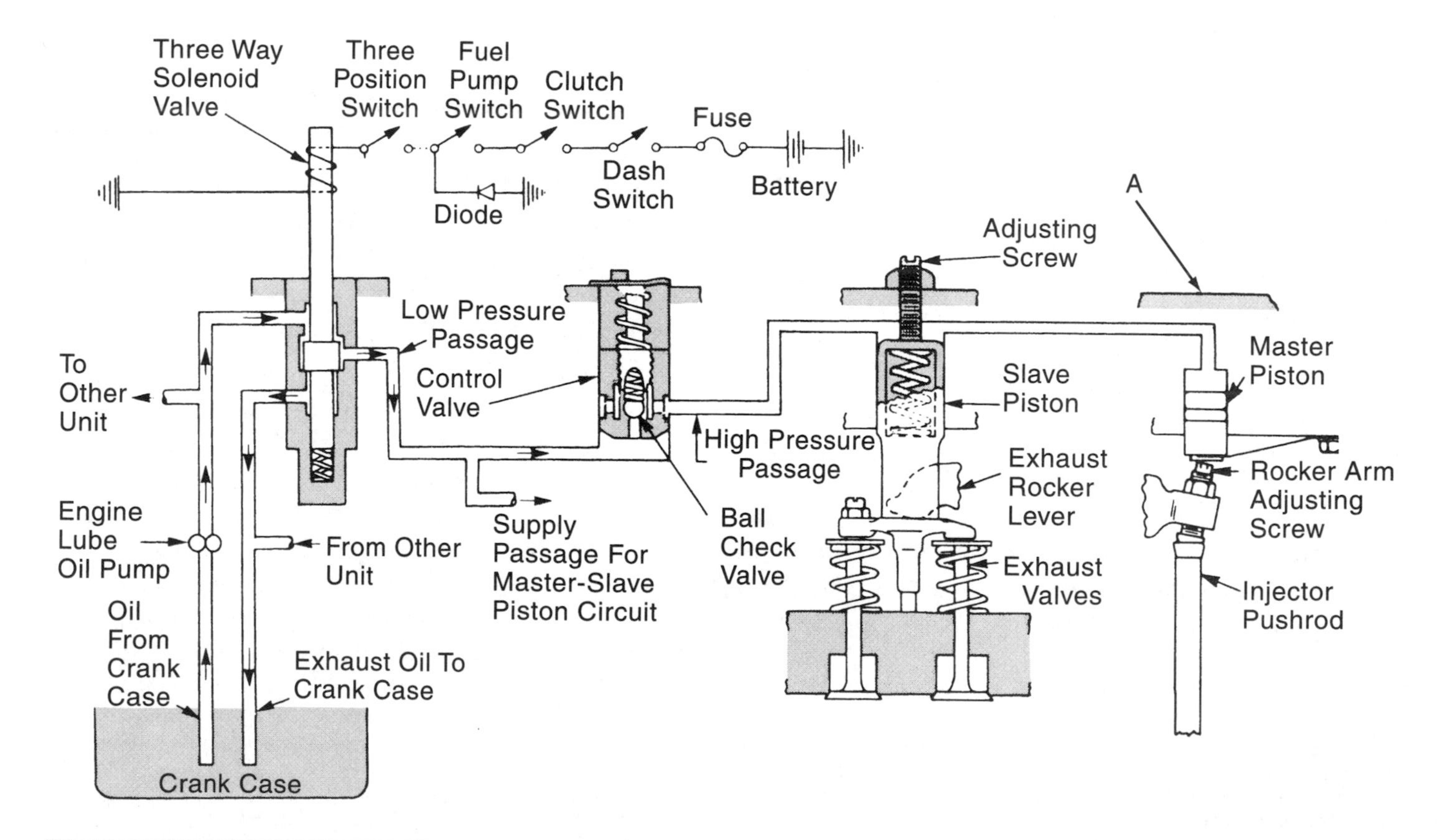

Figure 2–47 Engine compression brake system.

piston moves to near the top of the compression stroke. The compression braking cycle is completed as the compressed air is released from the cylinder.

Task 2

Inspect, test, adjust, repair/replace engine brake control circuits, switches, and solenoids.

240. As shown in Figure 2–47, the diode in the circuit is used to:
  A. apply the ground to the battery.
  B. rectify the signal applied to the battery.
  C. shunt the stored potential when the coil is de-energized.
  D. provide a radio frequency interference shunt.

241. As shown in Figure 2–47, the device at position A must:
  A. activate the injector below it.
  B. sense the pushrod upward motion to open exhaust valves.
  C. sense the pushrod downward motion to open exhaust valve.
  D. push the rocker arm below it to open the intake valves.

Hint

Engine brakes are designed to complement the main vehicle brake system, not replace it. An engine compression brake's ability to absorb power is low compared to the brake capacity of the vehicle. However, engine brakes can greatly extend the life of the vehicle foundation brakes.

Most engine brakes are controlled by an electric circuit that requires a series of normally closed switches. These switches include the control switch, a clutch switch, and an accelerator or governor switch.

The operating principle of an internal engine compression brake is to make the piston perform the work of compressing the cylinder air charge and then negate or cancel the power stroke by releasing the cylinder charge by opening the exhaust valves near TDC on the compression stroke. This changes the engine's role from that of a power-producing pump to that of a power-absorbing pump.

Task 3

Inspect, repair/replace engine brake housing, valves, seals, screens, lines, and fittings.

242. *Technician A* says that a leaking ball check valve in the control valve does not affect compression brake operation. *Technician B* says that the compression brake opens the exhaust valves before the compression stroke is complete. Who is right?
  A. A only
  B. B only
  C. Both A and B
  D. Neither A nor B

243. Which of the following is LEAST likely to be found in a compression brake system?
  A. Master piston
  B. Slave piston
  C. Solenoid valve
  D. Brake shoes

Hint

The driver activates an electric dash switch to operate the compression brake. Voltage is supplied through a fuse to the dash switch. A clutch switch is in series with the dash switch, and the clutch switch opens the circuit and prevents compression brake operation if the driver has the clutch pedal depressed. If the clutch pedal is released, the clutch switch is closed. A fuel switch in series with the dash switch prevents compression brake operation if the driver has the accelerator pedal depressed.

Answers To Questions

1. Answer A is wrong because this would reduce the scavenging and fresh air needed for combustion, thereby reducing the air-to-fuel mixture and power output.

   Answer B is right because high-density air supplied to the cylinder is what is required and therefore the least likely cause of a power loss.

   Answer C is wrong because restrictions in the airflow through the tubes will reduce the compression heat, thereby reducing the combustion temperature and power output. Restricted fins cause inefficient cooling, leading to hotter but less dense air supply to the cylinder. This less dense air reduces the compression heat, reducing the efficiency of the combustion and power output.

   Answer D is wrong because this would reduce the supply of necessary fuel at a power rating where maximum fuel supply is critical to injection and cooling of the injectors. There is a high probability that injectors will start failing, as well as fuel starvation and loss of power.

2. Answer A is wrong because a plugged air filter will restrict airflow to the engine and cause a power loss.

   Answer B is wrong because a fuel leak will cause a power loss.

   Answer C is wrong because plugged fuel filter will cause a power loss.

   Answer D is right because low coolant level will NOT cause a power loss.

3. **Answer A is right** because coolant may form a pool of green or blue liquid below the component that is leaking.

   Answer B is wrong because not all leaks are external. Sometimes, like a defective head gasket, the engine may consume fluids or expel them out as part of the exhaust. Monitoring of fluid levels and engine performance may be the only way to detect most internal leaks.

   Answer C is wrong because *Technician A* is right and *Technician B* is wrong.

   Answer D is wrong because *Technician A* is right and *Technician B* is wrong.

4. Answer A is wrong because small leaks can easily be detected, because antifreeze does not evaporate and generally leaves a red or green colored buildup or stain nearby, but *Technician B* is also right.

   Answer B is wrong because a defective thermostat is most frequently caused by repeated overheating of the engine, but *Technician A* is also right.

   Answer C is right because both *Technician A* and *Technician B* are right.

   Answer D is wrong because both *Technician A* and *Technician B* are right.

5. **Answer A is right** because a connector found to be distorted from heat but when checked with an ohmmeter the electrical connection is still good can be reused.

   Answer B is wrong because electrical connectors that have been heated may appear to provide good electrical connection when checked using direct current (DC) meters. Fractures in the wire may adversely affect the operation of data flow at frequencies of greater than 10 kHz.

   Answer C is wrong because *Technician A* is right and *Technician B* is wrong.

   Answer D is wrong because *Technician A* is right and *Technician B* is wrong.

6. **Answer A is right** because a microconnector is shown.

   Answer B is wrong because the figure shows a microconnector.

   Answer C is wrong because the figure shows a microconnector.

   Answer D is wrong because the figure shows a microconnector.

7. **Answer A is right** because a fan belt is the least likely cause of an engine knock and does track with engine speed.

   Answer B is wrong because camshaft noise increases with engine rpm.

   Answer C is wrong because crankshaft noise increases with engine rpm.

   Answer D is wrong because rocker arm noise increases with engine rpm.

8. Answer A is wrong because worn camshaft bearings do not cause a thumping noise when the engine is first started.

   Answer B is wrong because worn rocker arms do not cause a thumping noise when the engine is first started.

   Answer C is right because loose main bearings may cause a thumping noise when the engine is first started.

   Answer D is wrong because excessive crankshaft end play does not cause a thumping noise when the engine is first started.

9. Answer A is wrong because excess unburned fuel that has not reached combustion temperature will produce white smoke.

   Answer B is right because faulty or improperly installed rings may result in the engine producing a white smoke.

   Answer C is wrong because this will cause burning of oil or water that causes blue smoke and white smoke.

   Answer D is wrong because this condition will cause low operating temperature and failing of the injectors that produce black smoke.

10. **Answer A is right** because overfueling can cause very black smoke to come out of the exhaust pipe. Fuel burned with insufficient air causes black smoke.

    Answer B is wrong because the coolant level has no effect on an overfueling problem.

    Answer C is wrong because *Technician A* is right and *Technician B* is wrong.

    Answer D is wrong because *Technician A* is right and *Technician B* is wrong.

11. Answer A is wrong because a plugged fuel line will cut off the fuel supply to the pump.

    Answer B is wrong because a low idle will cause the pump to run slower causing lower fuel pressure.

    Answer C is wrong because an injector that is sticking open will cause the pressure from the pump to be lower.

    Answer D is right because low coolant level is the least likely cause of low fuel pressure.

12. Answer A is wrong because high-pressure fuel lines should be bled of any air that might be trapped after replacement, but *Technician B* is also right.

    Answer B is wrong because you can check for air in the fuel system by installing a clear plastic hose and looking for bubbles, but *Technician A* is also right.

    Answer C is right because both *Technician A* and *Technician B* are right.

    Answer D is wrong because both *Technician A* and *Technician B* are right.

13. Answer A is wrong because if the air restriction is too excessive, the filter or intake tubes need to be inspected, but *Technician B* is also right.

    Answer B is wrong because the engine speed should be the same for both readings, but *Technician A* is also right.

    Answer C is right because both *Technician A* and *Technician B* are right.

    Answer D is wrong because both *Technician A* and *Technician B* are right.

14. Answer A is wrong because manufacturers allow for movement in the intake system components by using durable rubber couplings, which are clamped at both ends, but *Technician B* is also right.

    Answer B is wrong because if an air intake pipe, tube, or hose is loose or missing a clamp, an inspection should be made of the turbocharger compressor vanes, but *Technician A* is also right.

    Answer C is right because both *Technician A* and *Technician B* are right.

    Answer D is wrong because both *Technician A* and *Technician B* are right.

15. **Answer A is right** because you use a manometer to check pressure.

    Answer B is wrong because failures may not occur at an idle, all operating speeds need to be checked.

    Answer C is wrong because *Technician A* is right and *Technician B* is wrong.

    Answer D is wrong because *Technician A* is right and *Technician B* is wrong.

16. Answer A is wrong because low air box pressure will NOT cause an air inlet restriction.

    Answer B is wrong because you can check air box pressure with a water manometer.

    Answer C is wrong because both *Technician A* and *Technician B* are wrong.

    Answer D is right because both *Technician A* and *Technician B* are wrong.

17. Answer A is wrong because to check exhaust backpressure, you connect a manometer to the exhaust manifold and check against the manufacturer's specifications, but *Technician B* is also right.

    Answer B is wrong because if a plug has not been provided for this test, drill and tap the exhaust manifold pipe, but *Technician A* is also right.

    Answer C is right because both *Technician A* and *Technician B* are right.

    Answer D is wrong because both *Technician A* and *Technician B* are right.

18. Answer A is wrong because the wrong muffler will cause excessive backpressure.

    Answer B is wrong because excessive carbon will cause excessive backpressure.

    Answer C is wrong because a collapsed exhaust pipe will cause excessive backpressure.

Answer D is right because low oil pressure is least likely to cause high exhaust backpressure.

19. Answer A is wrong because problems may occur at a high-rated speed at load.

    Answer B is right because worn piston rings may cause excessive crankcase pressure.

    Answer C is wrong because *Technician A* is wrong and *Technician B* is right.

    Answer D is wrong because *Technician A* is wrong and *Technician B* is right.

20. Answer A is wrong because a leaking oil pan gasket does not cause excessive crankcase pressure.

    Answer B is wrong because a collapsed piston skirt may cause excessive noise, but as long as the rings are sealing properly this problem does not cause excessive crankcase pressure.

    Answer C is right because a restricted crankcase ventilator causes excessive crankcase pressure.

    Answer D is wrong because a burned exhaust valve causes cylinder missing but it does not cause excessive crankcase pressure.

21. **Answer A is right** because hard starting may be caused by air in the fuel system.

    Answer B is wrong because low cylinder compression may cause hard engine starting.

    Answer C is wrong because *Technician A* is right and *Technician B* is wrong.

    Answer D is wrong because *Technician A* is right and *Technician B* is wrong.

22. Answer A is wrong because a faulty injector may cause the engine to run rough, but it should still start.

    Answer B is wrong because a defective starter cutout relay would cause the starter to stay engaged after the engine has started.

    Answer C is right because if the shutdown solenoid fails, no fuel will flow to the metering valve and pumping plungers, resulting in a no-start.

    Answer D is wrong because a leaky intake hose will cause possible future damage to turbocharger and valves, but the engine should start.

23. Answer A is wrong because suction leaks in fuel supply system will cause air to enter the fuel system, resulting in low power.

    Answer B is wrong because incorrect throttle travel will prevent the fuel rack from achieving correct fuel position for full load.

    Answer C is wrong because a partially restricted fuel filter restricts fuel flow to injectors, resulting in low power.

    Answer D is right because a leak in the injector return line does not cause low fueling, and this is the least likely cause of low power.

24. **Answer A is right** because it is not necessary to replace the blower when the emergency shutdown device is serviced.

    Answer B is wrong because these must be flat to make a good seal.

    Answer C is wrong because these must be flat to make a good seal.

    Answer D is wrong because you must replace blower screens.

25. **Answer A is right** because a vibration occurrence rate that is equal to half the engine rpm might be caused by a bent cam, because it turns at half the engine speed.

    Answer B is wrong because this will create a vibration with the same occurrence rate as the engine speed.

    Answer C is wrong because *Technician A* is right and *Technician B* is wrong.

    Answer D is wrong because *Technician A* is right and *Technician B* is wrong.

26. Answer A is wrong because the figure shows a vibration damper.

    Answer B is right because the figure shows a vibration damper.

    Answer C is wrong because the figure shows a vibration damper.

    Answer D is wrong because the figure shows a vibration damper.

27. Answer A is wrong because the pressure raises the boiling point of the coolant and if the boiling point is lower, the engine may overheat, but *Technician B* is also right.

    Answer B is wrong because the cap is designed to release pressure above its rating, but *Technician A* is also right.

Answer C is right because both *Technician A* and *Technician B* are right.

Answer D is wrong because both *Technician A* and *Technician B* are right.

28. Answer A is wrong because defective caps must be replaced.

    Answer B is wrong because radiator tubes cannot stand high pressure.

    Answer C is wrong because hot hoses are very flexible and will collapse when a vacuum is applied to the system.

    Answer D is right because you cannot adjust or replace the spring in the cap.

29. Answer A is wrong because canister oil filters should be prefilled to protect the bearings, but *Technician B* is also right.

    Answer B is a wrong because you need to remove the filter housing mounted by-pass valve and inspect itπ when replacing the filter, but *Technician A* is also right.

    Answer C is right because both *Technician A* and *Technician B* are right.

    Answer D is wrong because both *Technician A* and *Technician B* are right.

30. Answer A is wrong because the indication is coolant in the oil, and a cracked block or cylinder head causes this condition.

    Answer B is wrong because the indication is coolant in the oil, and a blown head gasket causes this condition.

    Answer C is wrong because the indication is water in the oil, and a leaky oil cooler causes this condition.

    Answer D is right because a worn oil pump is the least likely cause of this condition, because there is no connection between the cooling system and the oil pump.

31. Answer A is wrong because any generic scan tool must be able to access on board diagnostic and necessary user programmable selections.

    Answer B is wrong because any generic scan tool must be capable of connecting to the data link connector.

    Answer C is wrong because both *Technician A* and *Technician B* are wrong.

    Answer D is right because both *Technician A* and *Technician B* are wrong.

32. Answer A is wrong because DTC stands for diagnostic trouble codes.

    Answer B is right because DTC stands for diagnostic trouble codes.

    Answer C is wrong because DTC stands for diagnostic trouble codes.

    Answer D is wrong because DTC stands for diagnostic trouble codes.

33. **Answer A is right** because this is an easy check to make and the only thing wrong may be improperly torqued head bolts.

    Answer B is wrong because for best results, crack detectors should be used only after the head has been removed and cleaned.

    Answer C is wrong because the head bolts should have been checked before removing the head.

    Answer D is wrong because a check of the head bolt torque should be accomplished before removing the head to remove deposits or perform a pressure test.

34. Answer A is wrong because the exhaust manifold mating surface on the cylinder head should be checked for warpage with a straightedge and a feeler gauge.

    Answer B is right because there is a limit to the amount of metal that can be machined from the cylinder head to block mating surface without adversely affecting the compression ratio.

    Answer C is wrong because *Technician A* is wrong and *Technician B* is right.

    Answer D is wrong because *Technician A* is wrong and *Technician B* is right.

35. Answer A is wrong because before installation, head bolts must be cleaned and inspected for erosion or pitting, but *Technician B* is also right.

    Answer B is wrong because the cylinder needs to be checked for foreign objects before installing the head, but *Technician A* is also right.

    Answer C is right because both *Technician A* and *Technician B* are right.

    Answer D is wrong because both *Technician A* and *Technician B* are right.

36. **Answer A is right** because oil in a holddown bolt hole of a head can throw off the torque setting of that bolt.

    Answer B is wrong because the bolt should be replaced to prevent warping of the head from possible uneven head bolt torque.

    Answer C is wrong because *Technician A* is right and *Technician B* is wrong.

    Answer D is wrong because *Technician A* is right and *Technician B* is wrong.

37. Answer A is wrong because the cylinder head warpage should be measured transversely across the cylinder head at each end and between each cylinder, but *Technician B* is also right.

    Answer B is wrong because a blown head gasket between two cylinders may indicate a transversely warped cylinder head, but *Technician A* is also right.

    Answer C is right because both *Technician A* and *Technician B* are right.

    Answer D is wrong because both *Technician A* and *Technician B* are right.

38. Answer A is wrong because a straightedge and a feeler gauge are used to measure cylinder head warpage.

    Answer B is wrong because the cylinder head mating surface should be measured longitudinally and transversely for warpage.

    Answer C is wrong because both *Technician A* and *Technician B* are wrong.

    Answer D is right because both *Technician A* and *Technician B* are wrong.

39. Answer A is wrong because when testing a cylinder head for potential coolant leakage, the coolant needs to be heated to operating temperature, but *Technician B* is also right.

    Answer B is wrong because the coolant needs to be under pressure, but *Technician A* is also right.

    Answer C is right because both *Technician A* and *Technician B* are right.

    Answer D is wrong because both *Technician A* and *Technician B* are right.

40. Answer A is wrong because discarded injectors are used to seal the ports at the injector sleeves.

    Answer B is wrong because compressed air provides the medium for indicating leaks, because air will detect smaller leaks that liquids can escape through.

    Answer C is wrong because threaded plugs are used to plug the return-line or crossover port at the other end of the head.

    Answer D is right because injection pumps are not used with unit injectors that have injector tubes.

41. Answer A is wrong because you replace springs that are fractured or broken, out of square, or do not meet the specifications, but *Technician B* is also right.

    Answer B is wrong because companion springs under the same valve bridge as defective valves must also be replaced, but *Technician A* is also right.

    Answer C is right because both *Technician A* and *Technician B* are right.

    Answer D is wrong because both *Technician A* and *Technician B* are right.

42. **Answer A is right** because this is not a test that is performed by static measurement. It is a dynamic measurement.

    Answer B is wrong because an uneven free height check of valve springs will reveal a weak spring preventing possible burnt valves.

    Answer C is wrong because this test is performed to detect valve springs that might fail under load. If not performed the result could be a dropped valve.

    Answer D is wrong because this inspection is performed to detect springs with slopped tops. If springs are not square, valves and guides may wear faster or stems may bind in guides.

43. **Answer A is right** because mechanical rotators on a valve spring will help eliminate carbon deposits on the valve seat.

    Answer B is wrong because they cannot prevent oil leakage from the valve guide.

    Answer C is wrong because *Technician A* is right and *Technician B* is wrong.

    Answer D is wrong because *Technician A* is right and *Technician B* is wrong.

44. Answer A is wrong because missing or defective rotators may cause premature valve failure, but this problem is not likely to cause a valve to be dropped on top of a piston.

    Answer B is right because if the keepers are not seated properly the valve could drop in the time specified.

    Answer C is wrong because *Technician A* is wrong and *Technician B* is right.

    Answer D is wrong because *Technician A* is wrong and *Technician B* is right.

45. **Answer A is right** because when the manufacturer does not supply specifications, press the new guide in head to the same position the old guide was in.

    Answer B is wrong because guide bores that are severely scored should be reamed out and an oversized guide installed.

    Answer C is wrong because *Technician A* is right and *Technician B* is wrong.

    Answer D is wrong because *Technician A* is right and *Technician B* is wrong.

46. Answer A is wrong because the inside diameter can be checked with a snap gauge.

    Answer B is right because the inside diameter cannot be checked with a feeler gauge.

    Answer C is wrong because the inside diameter can be checked with a ball gauge.

    Answer D is wrong because the inside diameter can be checked with a dial indicator.

47. Answer A is wrong because the end of the valve stem should be ground flat.

    Answer B is wrong because any time the valve is to be reused, it must be inspected and reground to ensure proper sealing of the valves.

    Answer C is wrong because both *Technician A* and *Technician B* are wrong.

    Answer D is right because both *Technician A* and *Technician B* are wrong.

48. Answer A is wrong because the figure shows a stretched valve.

    Answer B is wrong because the figure shows a stretched valve.

    Answer C is wrong because the figure shows a stretched valve.

    Answer D is right because the figure shows a stretched valve.

49. Answer A is wrong because when the ball of the ball peen hammer is lightly tapped on the seat and head adjacent to the seat the sounds are compared. Similar ringing sound indicate a tight fitting seat, a dull sound or movement indicates a loose fitting seat.

    Answer B is right because sharpening a chisel to remove defective seats is not an approved procedure and therefore the least likely one to use.

    Answer C is wrong because after grinding the seats, ensure the opening is round using a concentricity indicator. If the seat is not round it will leak.

    Answer D is wrong because the valve seat grinding stone needs to be dressed for the specific valve seat angle that is being used.

50. Answer A is wrong because the cracks allow compression to leak and wear the valve seats.

    Answer B is wrong because oblonged seats allow a compression leak and cause poor performance.

    Answer C is wrong because discoloration of the back of the valve head will indicate valve seat leakage.

    Answer D is right because worn piston rings will NOT cause a valve seat leak.

51. **Answer A is right** because a straightedge and feeler gauge is the right combination.

    Answer B is wrong because the dial indicator and step block may provide a possibility of obtaining the measurement but the caliper and ruler are unnecessary.

    Answer C is wrong because a micrometer is used for taking outside measurements and the point gauge is used to check orifices on the injector. Neither will provide the necessary type of measurement.

    Answer D is wrong because these tools will not provide measurements to the tolerances required for a valve head height check.

52. **Answer A is right** because before installing the valves into the cylinder head you should first check valve-to-seat contact.

    Answer B is wrong because the rocker arms are installed after the head is installed.

Answer C is wrong because the lift on the camshaft does not change when replacing valves.

Answer D is wrong because the head gasket installation is a separate process from valve installation.

53. Answer A is wrong because injector tip height must be checked to confirm proper clearance, but *Technician B* is also right.

    Answer B is wrong because you perform a cylinder head leak test, but *Technician A* is also right.

    Answer C is right because both *Technician A* and *Technician B* are right.

    Answer D is wrong because both *Technician A* and *Technician B* are right.

54. Answer A is wrong because it is necessary to refer to the manufacturer's procedures for removing and installing sleeves because not all sleeves are the same and special tools may be required.

    Answer B is wrong because this test is performed after the sleeves are installed to ensure the seal is good.

    Answer C is wrong because this check is required to confirm that the injector position is correct. An incorrect height can cause improper cylinder operation by causing the injector spray pattern to strike the cylinder wall instead of the piston.

    Answer D is right because the replacement of an injector nozzle, while possible, is the least likely needed repair.

55. Answer A is wrong because the prechamber is driven into the cylinder head with a plastic hammer and light pressure.

    Answer B is wrong because there is no adjustment of the prechamber.

    Answer C is right because prechambers are fitted flush to the cylinder head or plus 0.002 in. (0.050 mm) above it.

    Answer D is wrong because 400 grit sandpaper is not used to grind the prechamber to fit.

56. Answer A is wrong because a precombustion chamber protruding too far out of the cylinder head may damage the piston.

    Answer B is wrong because a precombustion chamber protruding too far out of the cylinder head may damage the intake valve.

    Answer C is right because a precombustion chamber protruding too far out of the cylinder head does not damage the block.

    Answer D is wrong because a precombustion chamber protruding too far out of the cylinder head may damage the exhaust valve.

57. Answer A is wrong because crosshead guide pins should be at right angles to the heads' milled surface, but *Technician B* is also right.

    Answer B is wrong because the Technician should check the crosshead guide pin diameter with a micrometer and compare to manufacturer's specifications, but *Technician A* is also right.

    Answer C is right because both *Technician A* and *Technician B* are right.

    Answer D is wrong because both *Technician A* and *Technician B* are right.

58. Answer A is wrong because four valve heads on a four-stroke engine use two intake valves and two exhaust valves in each cylinder.

    Answer B is right because a four-stroke diesel engine with valve crossheads will use a four-valve head.

    Answer C is wrong because *Technician A* is wrong and *Technician B* is right.

    Answer D is wrong because *Technician A* is wrong and *Technician B* is right.

59. **Answer A is right** because you should first check the bolt holes for foreign matter and cross threading.

    Answer B is wrong because the water pump has no effect on head installation.

    Answer C is wrong because the front crank seal is not disturbed when working on a head.

    Answer D is wrong because the rear crank seal has no effect on this repair.

60. **Answer A is right** because cylinder heads are always torqued in sequence according to specifications set by the manufacturer.

    Answer B is wrong because bent pushrods must always be replaced.

    Answer C is wrong because *Technician A* is right and *Technician B* is wrong.

    Answer D is wrong because *Technician A* is right and *Technician B* is wrong.

61. Answer A is wrong because worn rocker arm bushings must be replaced.

    Answer B is wrong because a worn rocker arm shaft must be replaced.

    Answer C is wrong because scored and pitted rocker arms may cause damage after hours of operation and should be replaced.

    Answer D is right because you do not tighten down the rocker arm tappet screws.

62. Answer A is wrong because a sticking valve may cause a bent pushrod, but *Technician B* is also right.

    Answer B is wrong because pushrods or tubes must be checked for proper fit to cam followers, but *Technician A* is also right.

    Answer C is right because both *Technician A* and *Technician B* are right.

    Answer D is wrong because both *Technician A* and *Technician B* are right.

63. Answer A is wrong because lifter bottoms should be convex.

    Answer B is right because if a lifter foot is pitted, the camshaft mating lobe should be inspected.

    Answer C is wrong because *Technician A* is wrong and *Technician B* is right.

    Answer D is wrong because *Technician A* is wrong and *Technician B* is right.

64. **Answer A is right** because if the cam follower is worn, the injector will be mistimed and have an insufficient stroke, resulting in low fueling and a misfirc.

    Answer B is wrong because a restricted fuel filter will cause fuel starvation and low power for the entire engine, not just one cylinder.

    Answer C is wrong because high fuel pressure will not affect injector operation in a rocker arm activated unit injector system like Detroit Diesel.

    Answer D is wrong because an injector tube (sleeve) failure will not affect a misfire.

65. **Answer A is right** because the valve adjustment is being performed on an exhaust valve with a spring-loaded bridge.

    Answer B is wrong because the valve adjustment is being performed on an exhaust valve with a spring-loaded bridge.

    Answer C is wrong because the valve adjustment is being performed on an exhaust valve with a spring-loaded bridge.

    Answer D is wrong because the valve adjustment is being performed on an exhaust valve with a spring-loaded bridge.

66. **Answer A is right** because the valve bridge is being adjusted.

    Answer B is wrong because this adjustment must be performed before the valve clearance is adjusted.

    Answer C is wrong because *Technician A* is right and *Technician B* is wrong.

    Answer D is wrong because *Technician A* is right and *Technician B* is wrong.

67. Answer A is wrong because a dial indicator is most commonly used to check backlash in the timing gear train.

    Answer B is also wrong because a feeler gauge can be used to measure backlash.

    Answer C is right because the least used tool for this process in the depth gauge.

    Answer D is wrong because a ruler can be used to check bashlash.

68. Answer A is wrong because the figure shows checking the preload on an idler gear that transfers the drive from the crankshaft gear to the camshaft gear.

    Answer B is right because the figure shows checking the preload on an idler gear that transfers the drive from the crankshaft gear to the camshaft gear.

    Answer C is wrong because *Technician A* is wrong and *Technician B* is right.

    Answer D is wrong because *Technician A* is wrong and *Technician B* is right.

69. Answer A is wrong because it says before installing die cast rocker covers on a cylinder head, ensure the silicone gasket is secure in the cover groove to ensure a good seal, but *Technician B* is also right.

    Answer B is wrong because it says to always use new gaskets when installing valve covers, but *Technician A* is also right.

Answer C is right because both *Technician A* and *Technician B* are right.

Answer D is wrong because both *Technician A* and *Technician B* are right.

70. **Answer A is right** because the head rail is not part of the valve cover assembly.

Answer B is wrong because the breather can be mounted on top of the valve cover.

Answer C is wrong because the filter cap is usually part of the valve cover.

Answer D is wrong because the strainer is mounted under the breather.

71. Answer A is wrong because the block should always be checked for cracks first before any assembly.

Answer B is right because the block should always be checked for cracks first before any assembly.

Answer C is wrong because the sleeves should be removed before cleaning the block.

Answer D is wrong because the block should always be checked for cracks first before any assembly.

72. Answer A is wrong because an intercooler failure could cause the egress of water (coolant) into one of the cylinders, but *Technician B* is also right.

Answer B is wrong because a hairline crack might not have been noticed in the previous repair without a dye check and could cause coolant in a cylinder, but *Technician A* is also right.

Answer C is right because both *Technician A* and *Technician B* are right.

Answer D is wrong because both *Technician A* and *Technician B* are right.

73. Answer A is wrong because when pressure testing an engine block, the block need only be submerged in water for about 20 to 30 minutes and the water checked for bubbles, but *Technician B* is also right.

Answer B is wrong because it says cracked cylinder blocks should be replaced, but *Technician A* is also right.

Answer C is right because both *Technician A* and *Technician B* are right.

Answer D is wrong because both *Technician A* and *Technician B* are right.

74. Answer A is wrong because to properly test an engine block for cracks, it should be submerged in a tank of water heated to between 180 and 200°F (82–93°C).

Answer B is right because an engine block can be checked for water leaks by pressurizing it for about two hours without loss of pressure.

Answer C is wrong because *Technician A* is wrong and *Technician B* is right.

Answer D is wrong because *Technician A* is wrong and *Technician B* is right.

75. Answer A is wrong because welding may cause warping of the head surface of the block and weaken the metal between cylinders.

Answer B is right because the block needs to be sent out to have the counterbore resleeved.

Answer C is wrong because *Technician A* is wrong and *Technician B* is right.

Answer D is wrong because *Technician A* is wrong and *Technician B* is right.

76. Answer A is wrong because with a wet sleeve type cylinder, the cylinder only seals at the top and bottom.

Answer B is right because in this type of liner the coolant can pass around the cylinder.

Answer C is wrong because *Technician A* is wrong and *Technician B* is right.

Answer B is wrong because *Technician A* is wrong and *Technician B* is right.

77. **Answer A is right** because when replacing a wet-type sleeve you should always replace the seals on the sleeve.

Answer B is wrong because to remove the sleeve you use a sleeve puller, not tap it out with a hammer.

Answer C is wrong because *Technician A* is right and *Technician B* is wrong.

Answer D is wrong because *Technician A* is right and *Technician B* is wrong.

78. Answer A is wrong because the inside micrometer is used for measuring inside measurement, and comes with precision extensions that can be added to span great diameters such as cylinder sleeves.

Answer B is wrong because a cylinder gauge is typically used to check the cylinder sleeve for out of roundness.

Answer C is right because a dial indicator is not typically used to measure sleeve diameter. It is used to check liner height.

Answer D is wrong because a snap gauge or bore gauge is used to check the cylinder sleeve (liner) for out of roundness.

79. **Answer A is right** because after honing, the liner should have a 60-degree crossover angle pattern with a 15 to 20-microinch crosshatch.

Answer B is wrong because the liner surface should be washed with soapy water.

Answer C is wrong because *Technician A* is right and *Technician B* is wrong.

Answer D is wrong because *Technician A* is right and *Technician B* is wrong.

80. Answer A is wrong because after wet-type sleeves have been removed, light rust and scale must be removed from the packing ring and counterbore lip.

Answer B is right because 100 to 120 grit emery paper should be used to remove rust and scale from the packing ring and counterbore lip, and these areas should be inspected for erosion and pitting.

Answer C is wrong because *Technician A* is wrong and *Technician B* is right.

Answer D is wrong because *Technician A* is wrong and *Technician B* is right.

81. Answer A is wrong because an aluminum block should be heated by submerging it in a tank of hot water for 20 minutes before the insert and liner are installed.

Answer B is wrong because the block bore and counterbore must be clean before insert and liner installation.

Answer C is right because hand pressure must be used to install the liner in the block.

Answer D is wrong because the liner must be pushed downward until the liner flange contacts the insert.

82. Answer A is wrong because the liner flange must be the specified distance below the top of the block surface.

Answer B is wrong because the maximum variation in cylinder liner heights is 0.002 in. (0.050 mm).

Answer C is wrong because both *Technician A* and *Technician B* are wrong.

Answer D is right because both *Technician A* and *Technician B* are wrong.

83. Answer A is wrong because the Technician is measuring the cam journal.

Answer B is right because the Technician is measuring the cam journal

Answer C is wrong because the Technician is measuring the cam journal.

Answer D is wrong because cam duration has no connection to the figure.

84. **Answer A is right** because the bearing clearance is the difference between the camshaft journal diameter and the camshaft bearing diameter.

Answer B is wrong because the bearing must be installed so the oil holes in the block and bearing are aligned.

Answer C is wrong because the bearing must be flush with the front block surface.

Answer D is wrong because the bearing bore must be in satisfactory condition before bearing installation.

85. Answer A is wrong because if the camshaft lobes are pitted, camshaft replacement is necessary.

Answer B is wrong because if the camshaft has edge deterioration for 20 degrees on either side of the cam lobe tip, camshaft replacement is required.

Answer C is right because if the transfer pump cam lobe measures 1.63 in. (41.4 mm) the camshaft is satisfactory.

Answer D is wrong because if one intake camshaft lobe measures 1.95 in. (49.53 mm), the camshaft must be replaced.

86. **Answer A is right** because the gear is not properly seated.

Answer B is wrong because there is no need to replace the camshaft, but the gear should be pressed onto the camshaft until it contacts the camshaft shoulder.

Answer C is wrong because *Technician A* is right and *Technician B* is wrong.

Answer D is wrong because *Technician A* is right and *Technician B* is wrong.

87. Answer A is wrong because the cause may be found by carefully inspecting the vibration damper, but *Technician B* is also right.

    Answer B is wrong because the flywheel or torque converter needs to be closely inspected as a possible cause, but *Technician A* is also right.

    Answer C is right because both *Technician A* and *Technician B* are right.

    Answer D is wrong because both *Technician A* and *Technician B* are right.

88. Answer A is wrong because the measurements at A and B indicate vertical taper.

    Answer B is wrong because the measurements at C and D indicate horizontal taper.

    Answer C is right because the measurements at A and C indicate out-of-round.

    Answer D is wrong because the measurements at A and D are not valid.

89. **Answer A is right** because the crankshaft journals should be machined to fit the bearings.

    Answer B is wrong because the main caps should not always be line bored.

    Answer C is wrong because *Technician A* is right and *Technician B* is wrong.

    Answer D is wrong because *Technician A* is right and *Technician B* is wrong.

90. **Answer A is right** because the numbered side of the cap goes toward the blower side of the engine.

    Answer B is wrong because this bearing cap must be installed on the main bearing from which it was removed.

    Answer C is wrong because *Technician A* is right and *Technician B* is wrong.

    Answer D is wrong because *Technician A* is right and *Technician B* is wrong.

91. **Answer A is right** because main bearing shells should receive a light coat of oil before final installation.

    Answer B is wrong because only one journal should be checked at a time unless others are readily available. Under no circumstance turn the crank while plastigage is installed.

    Answer C is wrong because *Technician A* is right and *Technician B* is wrong.

    Answer D is wrong because *Technician A* is right and *Technician B* is wrong.

92. **Answer A is right** because the crankshaft end play is decreased by installing thicker thrust washers on each side of the rear main bearing cap.

    Answer B is wrong because the crankshaft end play is decreased by installing thicker thrust washers on each side of the rear main bearing cap.

    Answer C is wrong because the crankshaft end play is decreased by installing thicker thrust washers on each side of the rear main bearing cap.

    Answer D is wrong because the crankshaft end play is decreased by installing thicker thrust washers on each side of the rear main bearing cap.

93. **Answer A is right** because all gears in the timing geartrain must be inspected for tooth wear, including the crankshaft gear.

    Answer B is wrong because this condition is not normal and the gears must be replaced, repair should not be considered.

    Answer C is wrong because *Technician A* is right and *Technician B* is wrong.

    Answer D is wrong because *Technician A* is right and *Technician B* is wrong.

94. **Answer A is right** because the engine manufacturer used this key to offset the timing and it must be replaced with the same type.

    Answer B is wrong because this is an offset key used by the manufacturer to offset the timing of the cam. The shift in timing may have been necessary for proper operation of some special option added to the standard engine or a special purpose application.

    Answer C is wrong because *Technician A* is right and *Technician B* is wrong.

    Answer D is wrong because *Technician A* is right and *Technician B* is wrong.

95. Answer A is wrong because pistons should be laid out in the order in which they came out, but *Technician B* is also right.

    Answer B is wrong because installing an improper injector nozzle could cause scoring and cracking of pistons skirts, but *Technician A* is also right.

Answer C is right because both *Technician A* and *Technician B* are right.

Answer D is wrong because both *Technician A* and *Technician B* are right.

96. **Answer A is right** because excessive lubrication levels in the crankcase is the least likely cause.

    Answer B is wrong because this causes heat buildup and metal-to-metal contact, which causes scoring and cracks.

    Answer C is wrong because this causes uneven pressures and excessive heat, causing the piston to become fatigued and fracture.

    Answer D is wrong because this causes the piston to become fatigued and fracture.

97. **Answer A is right** because if a 6 lb. (2.72 kg) pull is required to remove a 0.004 in. (0.101 mm) feeler gauge, the piston-to-cylinder clearance is 0.005 in. (0.127 mm).

    Answer B is wrong because the piston should NOT be placed in hot water before the measurement.

    Answer C is wrong because *Technician A* is right and *Technician B* is wrong.

    Answer D is wrong because *Technician A* is right and *Technician B* is wrong.

98. Answer A is wrong because piston scoring may be caused by less than specified piston-to-cylinder clearance.

    Answer B is wrong because excessive piston-to-cylinder clearance may cause piston slapping.

    Answer C is wrong because the piston-to-cylinder clearance on cam ground pistons must be measured at 90 degrees to the piston pin.

    Answer D is right because a scored piston does not cause lower than specified oil pressure.

99. Answer A is wrong because if a compression ring gap is less than specified, the ends of the ring may be filed or stoned.

    Answer B is right because the piston ring should be positioned part way down in the normal ring travel in the cylinder.

    Answer C is wrong because the ends of the piston ring must remain square if they are filed or stoned.

    Answer D is wrong because a piston should be used to push the ring down in the cylinder.

100. Answer A is wrong because the top piston ring may be called a fire ring.

     Answer B is wrong because fire rings are keystone or rectangular shaped.

     Answer C is wrong because both *Technician A* and *Technician B* are wrong.

     Answer D is right because both *Technician A* and *Technician B* are wrong.

101. **Answer A is right** because a piston skirt crack can be caused by improperly installed connecting rods with the connecting rod and bottom of the cylinder line damaged.

     Answer B is wrong because cylinder liner-to-piston clearance will not cause skirt cracking.

     Answer C is wrong because *Technician A* is right and *Technician B* is wrong.

     Answer D is wrong because *Technician A* is right and *Technician B* is wrong.

102. Answer A is wrong because wear on the edges of a connecting rod bearing indicates connecting rod misalignment.

     Answer B is right because wear on the edges of a connecting rod bearing indicates connecting rod misalignment.

     Answer C is wrong because wear on the edges of a connecting rod bearing indicates connecting rod misalignment.

     Answer D is wrong because wear on the edges of a connecting rod bearing indicates connecting rod misalignment.

103. Answer A is wrong because the markings on the connecting rod and bearing cap may be matched on the same side of the connecting rod.

     Answer B is wrong because the markings on the connecting rod and bearing cap may face the specified side of the engine.

     Answer C is right because on two-cycle diesel engines, the piston and connecting rod assembly should be installed in the liner before installation in the engine.

     Answer D is wrong because the piston ring gaps should be staggered around the piston before it is installed in the cylinder.

104. Answer A is wrong because the top bushing must be removed before the spray nozzle.

Answer B is wrong because when installing the nozzle, the holes on the nozzle must be in the position specified by the engine manufacturer.

Answer C is right because a press, sleeve, and remover are required to remove the spray nozzle.

Answer D is wrong because the spray nozzle should be installed bottomed in the counterbore.

105. Answer A is wrong because the vibration damper is not serviceable.

Answer B is right because most engine manufacturers recommend damper replacement during a major engine overhaul.

Answer C is wrong because *Technician A* is wrong and *Technician B* is right.

Answer D is wrong because *Technician A* is wrong and *Technician B* is right.

106. **Answer A is right** because a puller must be used to remove the damper, and this statement is not true.

Answer B is wrong because if the viscous fluid is leaking from the damper, damper replacement is necessary.

Answer C is wrong because dents in the damper housing may render the damper ineffective.

Answer D is wrong because if the damper is bulged or split, the fluid in the damper has exploded.

107. Answer A is wrong because the dial indicator stem must be positioned so it does not drop into a capscrew hole, but *Technician B* is also right.

Answer B is wrong because a flywheel housing face runout measurement of 0.005 in. (0.127 mm) is acceptable, but *Technician A* is also right.

Answer C is right because both *Technician A* and *Technician B* are right.

Answer D is wrong because both *Technician A* and *Technician B* are right.

108. Answer A is wrong because the measurement in the figure is flywheel housing bore alignment.

Answer B is wrong because the measurement in the figure is flywheel housing bore alignment.

Answer C is right because the measurement in the figure is flywheel housing bore alignment.

Answer D is wrong because the measurement in the figure is flywheel housing bore alignment.

109. **Answer A is right** because the flywheel-to-crankshaft mating surfaces should be checked for metal burrs.

Answer B is wrong because this condition does not cause the transmission to jump out of gear.

Answer C is wrong because *Technician A* is right and *Technician B* is wrong.

Answer D is wrong because *Technician A* is right and *Technician B* is wrong.

110. **Answer A is right** because on a pot-type flywheel, the flywheel bore runout may be measured with a dial indicator.

Answer B is wrong because on a pot-type flywheel if the flywheel face is resurfaced, an equal amount of metal must be machined from the pot face.

Answer C is wrong because if 0.250 in. (6.35 mm) is machined from the flywheel face the clutch and pressure plate are moved toward the engine and improper clutch operation results.

Answer D is wrong because on a pot-type flywheel if the bore runout is excessive, the flywheel and crankshaft mating surfaces, and mounting bolts should be checked.

111. Answer A is wrong because the starter ring gear is a shrink fit on the flywheel assembly.

Answer B is wrong because the ring gear must be installed with the beveled side of the gear teeth facing toward the starting motor.

Answer C is wrong because both *Technician A* and *Technician B* are wrong.

Answer D is right because both *Technician A* and *Technician B* are wrong.

112. Answer A is wrong because the flywheel mounting bolts would not be loose enough to allow the starter drive gear to miss the flywheel ring gear.

Answer B is right because the ring gear may be loose and turning on the flywheel.

Answer C is wrong because *Technician A* is wrong and *Technician B* is right.

Answer D is wrong because *Technician A* is wrong and *Technician B* is right.

113. Answer A is wrong because low voltage will cause false readings.

Answer B is wrong because shorted or open wires will cause false readings.

Answer C is right because low coolant level has little effect on oil pressure.

Answer D is wrong because a defective sending unit will cause false readings.

114. Answer A is wrong because the passage would have to be clogged to create this problem, not open.

Answer B is wrong because incorrect oils do not cause this condition.

Answer C is right because depending on the design, an open in the wiring to the sensor causes a zero pressure reading.

Answer D is wrong because the wrong sensor causes an inaccurate reading, but not the lack of any reading.

115. **Answer A is right** because inner rotor diameter is not a normal oil pump component measurement.

Answer B is wrong because clearance between the rotors is a normal oil pump component measurement.

Answer C is wrong because inner and outer rotor thickness is a normal oil pump component measurement.

Answer D is wrong because outer rotor to housing clearance is a normal oil pump component measurement.

116. Answer A is wrong because the oil pump measurement in the figure is rotor end play.

Answer B is wrong because the oil pump measurement in the figure is rotor end play.

Answer C is wrong because the oil pump measurement in the figure is rotor end play.

Answer D is right because the oil pump measurement in the figure is rotor end play.

117. Answer A is wrong because a visual inspection for cracks is a must.

Answer B is wrong because a visual inspection of the gasket's surface will ensure a tight seal.

Answer C is right because magna fluxing of housing assembly and metal parts to detect small cracks is typically not performed.

Answer D is wrong because a visual inspection of all passageways is necessary to ensure oil flows.

118. Answer A is wrong because there are two different types of filters used, the cartridge and the spin-on type, but *Technician B* is also right.

Answer B is wrong because they also use a partial type and a full-flow type, but *Technician A* is also right.

Answer C is right because both *Technician A* and *Technician B* are right.

Answer D is wrong because both *Technician A* and *Technician B* are right.

119. Answer A is wrong because low oil pressure has no effect on cooler operation.

Answer B is wrong because very high oil pressure would not cause a cooler failure.

Answer C is right because cooler failures result in mixing oil and water.

Answer D is wrong because bubbles in the cooling system with the engine running does not indicate a leaking oil cooler.

120. Answer A is wrong because the figure shows an engine oil cooler with oil-to-engine coolant cooling.

Answer B is wrong because component G is the by-pass valve.

Answer C is wrong because both *Technician A* and *Technician B* are wrong.

Answer D is right because both *Technician A* and *Technician B* are wrong.

121. **Answer A is right** because a restricted turbocharger oil drain line causes excessive oil accumulation in the turbo and this oil may be forced past the seals into the exhaust.

Answer B is wrong because a pitted turbine wheel does not cause oil to enter the exhaust.

Answer C is wrong because *Technician A* is right and *Technician B* is wrong.

Answer D is wrong because *Technician A* is right and *Technician B* is wrong.

122. Answer A is wrong because excessive turbocharger shaft end play is not the most likely cause of repeated turbocharger bearing failure.

Answer B is right because contaminated engine oil is the most likely cause of repeated turbocharger bearing failure.

Answer C is wrong because a leak in the air intake system may cause damage to the compressor wheel, but this problem is not the most likely cause of repeated turbocharger bearing failure.

Answer D is wrong because a leak in the exhaust system before the turbo decreases turbocharger efficiency, but this is not the most likely cause of repeated turbocharger bearing failure.

123. Answer A is wrong because alignment between pulleys must not exceed 1/16 inch (1.59 mm) for each 12 inches (30.5 cm) of distance between pulley centers, but *Technician B* is also right.

Answer B is wrong because drive belts must not touch the bottom of grooves and protrude more than 3/32 inch (2.38 mm) above the top of the groove, but *Technician A* is also right.

Answer C is right because both *Technician A* and *Technician B* are right.

Answer D is wrong because both *Technician A* and *Technician B* are right.

124. Answer A is wrong because all identical belts should be changed as a pair.

Answer B is wrong because starting the belt on the edge of the pulley and cranking the engine will cause damage to belts and/or personal injury.

Answer C is wrong because both *Technician A* and *Technician B* are wrong.

Answer D is right because both *Technician A* and *Technician B* are wrong.

125. Answer A is wrong because you place heat sensor tape on the radiator upper tank.

Answer B is wrong because you place the thermometer in the radiator after the engine has reached normal operating temperature.

Answer C is wrong because you point the heat gun at the upper radiator hose.

Answer D is right because voltmeters are not used to check mechanical temperature gauges.

126. Answer A is wrong because the water temperature sender does not contain a thermostatic switch, but it does contain a thermistor.

Answer B is right because the water temperature sender increases in resistance value as the coolant temperature decreases.

Answer C is wrong because the water temperature sender decreases in resistance value as the coolant temperature increases.

Answer D is wrong because the water temperature sender decreases in voltage drop as the coolant temperature increases.

127. **Answer A is right** because a stuck thermostat could cause this problem.

Answer B is wrong because an overfilled radiator could NOT cause this problem.

Answer C is wrong because *Technician A* is right and *Technician B* is wrong.

Answer D is wrong because *Technician A* is right and *Technician B* is wrong.

128. **Answer A is right** because failure to install one of the thermostats (item 5) causes engine overheating.

Answer B is wrong because the thermostats must be installed only in one direction.

Answer C is wrong because *Technician A* is right and *Technician B* is wrong.

Answer D is wrong because *Technician A* is right and *Technician B* is wrong.

129. **Answer A is right** because when the cooling system pressure is decreased, the coolant boiling point decreases.

Answer B is wrong because if more antifreeze is added to the coolant, the boiling point is increased.

Answer C is wrong because a good quality ethylene glycol antifreeze contains antirust and corrosion inhibitors.

Answer D is wrong because equipment is available to recover, recycle, and reuse coolant solutions.

130. Answer A is wrong because when filling the cooling system, the air bleeder valves should be open.

Answer B is right because if the coolant solution contains over 67 percent antifreeze, heat transfer from the cylinders to the coolant is adversely affected.

Answer C is wrong because *Technician A* is wrong and *Technician B* is right.

Answer D is wrong because *Technician A* is wrong and *Technician B* is right.

131. Answer A is wrong because SCA stands for supplemental coolant additive, which is an corrosion inhibitor additive. Adding more antifreeze does not change SCA reading.

Answer B is wrong because continuing to run the truck until the next PMI will not change the SCA reading.

Answer C is right because one needs to drain the entire coolant system and add the proper SCA mixture to achieve the correct percentage of SCA.

Answer D is wrong because this will not change SCA readings.

132. Answer A is wrong because a cooling system filled with water causes excessive corrosion in the system.

Answer B is wrong because a 50/50 water and antifreeze solution has a higher boiling point compared to water.

Answer C is wrong because both *Technician A* and *Technician B* are wrong.

Answer D is right because both *Technician A* and *Technician B* are wrong.

133. Answer A is wrong because a collapsed upper radiator hose may be caused by a defective pressure release valve in the radiator cap, but *Technician B* is also right.

Answer B is wrong because a collapsed upper radiator hose may be caused by a plugged hose from the radiator filler neck to the coolant recovery reservoir, but *Technician A* is also right.

Answer C is right because both *Technician A* and *Technician B* are right.

Answer D is wrong because both *Technician A* and *Technician B* are right.

134. Answer A is wrong because a defective water pump bearing may cause a growling noise with the engine idling, but *Technician B* is also right.

Answer B is wrong because the water pump bearing may be ruined by coolant leaking past the water pump seal, but *Technician A* is also right.

Answer C is right because both *Technician A* and *Technician B* are right.

Answer D is wrong because both *Technician A* and *Technician B* are right.

135. Answer A is wrong because the pressure in the radiator is too great, the pressure cap will release the excess pressure and fluid into an expansion tank, but *Technician B* is also right.

Answer B is wrong because most radiator pressure caps have a release pressure of 15 psi or lower, but *Technician A* is also right.

Answer C is right because both *Technician A* and *Technician B* are right.

Answer D is wrong because both *Technician A* and *Technician B* are right.

136. Answer A is wrong because backflushing the radiator in the vehicle has limited success and can even clog passages or cause a rupture.

Answer B is wrong because the system will bleed off trapped air and is the preferred method when refilling cooling system.

Answer C is wrong because both *Technician A* and *Technician B* are wrong.

Answer D is right because both *Technician A* and *Technician B* are wrong.

137. Answer A is wrong because the figure shows a thermostatic fan hub.

Answer B is wrong because the figure shows a thermostatic fan hub.

Answer C is right because the figure shows a thermostatic fan hub.

Answer D is wrong because the figure shows a thermostatic fan hub.

138. Answer A is wrong because air pressure is supplied to the fan clutch to disengage the clutch.

Answer B is wrong because the air pressure supplied to the fan clutch is controlled by a thermostatic switch and an electronic timer.

Answer C is wrong because if a defect occurs in the air control circuit, the fan clutch is engaged continually.

Answer D is right because current flow through the fan clutch solenoid is controlled by a thermostatic switch and an electronic timer.

139. Answer A is wrong because the linkage adjustments are used to calibrate closure.

Answer B is wrong because the ambient temperature affects thermostatic piston movement.

Answer C is wrong because the radiator temperature affects the thermostatic piston movement.

Answer D is right because air pressure is not used to control shutters operated by a thermostatic piston.

140. Answer A is wrong because the shutter drive piston is used to close the shutters.

Answer B is right because the shutter assembly springs open the shutters.

Answer C is wrong because the engine thermostat does not directly operate the shutter.

Answer D is wrong because a thermostatic switch is not used on a thermostatic piston-type shutter system.

141. Answer A is wrong because the oil in an oil bath air cleaner is not for lubrication of the intake system components.

Answer B is right because this type of air filter has a reduced efficiency at low rpm.

Answer C is wrong because *Technician A* is wrong and *Technician B* is right.

Answer D is wrong because *Technician A* is wrong and *Technician B* is right.

142. Answer A is wrong because you do use a manometer to check the amount of vacuum or pressure.

Answer B is wrong because you do remove the air filter to calibrate the manometer.

Answer C is wrong because you do test with the air filter installed for restriction.

Answer D is right because placing a piece of cardboard over the air intake to calibrate the manometer is not part of an intake system restriction test.

143. Answer A is wrong because an exhaust leak will affect the performance of the engine, but *Technician B* is also right.

Answer B is wrong because leaks in the intake pipe can cause damage to the intake side of the turbo, but *Technician A* is also right.

Answer C is right because both *Technician A* and *Technician B* are right.

Answer D is wrong because both *Technician A* and *Technician B* are right.

144. Answer A is wrong because the turbocharger bearings are lubricated by engine oil, but *Technician B* is also right.

Answer B is wrong because turboshaft bearing failure may be caused by contaminated engine oil, but *Technician A* is also right.

Answer C is right because both *Technician A* and *Technician B* are right.

Answer D is wrong because both *Technician A* and *Technician B* are right.

145. Answer A is wrong because if the blower bearings leak, it could cause this condition.

Answer B is wrong because the rotor bearings are fed by engine oil.

Answer C is right because a restriction in the oil feed line is the least likely cause of film on the rotor.

Answer D is wrong because a worn end plate gasket could cause this condition.

146. **Answer A is right** because adding or removing shims moves the rotor lobs closer or farther apart, adjusting the clearance.

Answer B is wrong because changing the end cover will not adjust the rotor clearance.

Answer C is wrong because using over- or under-sized gears will not affect rotor clearance and these parts do not exist.

Answer D is wrong because changing bearings is not the appropriate process to adjust rotor clearance.

147. **Answer A is right** because the air intake restriction system may contain a pressure sensor and an electronic module.

Answer B is wrong because air intake restriction reduces engine power.

Answer C is wrong because *Technician A* is right and *Technician B* is wrong.

Answer D is wrong because *Technician A* is right and *Technician B* is wrong.

148. **Answer A is right** because restrictions cause an increase in vacuum in the air intake system.

Answer B is wrong because scored cylinders and piston skirts may be caused by air intake leaks.

Answer C is wrong because some air intake restriction indicators contain a vacuum tube.

Answer D is wrong because a restricted air intake reduces the volume of air entering the engine.

149. Answer A is wrong because with reduced power you may need to bleed the air out of the intercooler, but Tachnician B is also right.

Answer B is wrong because when you have reduced power and high coolant temperature are noted on an engine with an intercooler, the air tubes in the cooler may be clogged, but *Technician A* is also right.

Answer C is right because both *Technician A* and *Technician B* are right.

Answer D is wrong because both *Technician A* and *Technician B* are right.

150. **Answer A is right** because the Technician should first check the intercooler when coolant is present in the intake side of the turbocharger.

Answer B is wrong because the water pump is not directly connected to the turbocharger.

Answer C is wrong because the intake manifold gasket is connected after the turbocharger.

Answer D is wrong because the turbocharger is not directly connected to the head.

151. Answer A is wrong because excessive exhaust backpressure reduces airflow through the engine and prevents good scavenging and reduces power out.

Answer B is wrong because of ineffective scavenging, the cylinder temperature will increase.

Answer C is right because excessive exhaust backpressure reduces airflow through the engine and decreases intake vacuum.

Answer D is wrong because poor combustion will result from excess exhaust backpressure.

152. Answer A is wrong because carbon dioxide in the exhaust does not cause nausea. Carbon monoxide has this effect.

Answer B is right because operating a truck for a prolonged period with a restricted exhaust system reduces engine life. This condition causes high exhaust temperature and reduced exhaust valve life.

Answer C is wrong because *Technician A* is wrong and *Technician B* is right.

Answer D is wrong because *Technician A* is wrong and *Technician B* is right.

153. Answer A is wrong because oil temperature is used to determine if the engine is warm or not.

Answer B is wrong because an electronic module may be used to control the glow plugs.

Answer C is wrong because the PCM uses sensor inputs to vary the operation of glow plugs and indicator light.

Answer D is right because the exhaust gas recirculation (EGR) valve has no function in the glow plug system, so it is the exception. EGR actually lowers combustion temperature.

154. Answer A is wrong because a glow plug light that is illuminated continually indicates the glow plug circuit is not operating properly.

Answer B is right because most glow plug systems will have the glow plug light on when the system has experienced a short circuit.

Answer C is wrong because one burned-out glow plug does not illuminate the glow plug light.

Answer D is wrong because the PCM or engine temperature sensor controls the glow plug system.

155. Answer A is wrong because capsules are punctured when first installed.

Answer B is wrong because inlet check valve only opens during a vacuum condition and the outlet check valve only opens under a pressure condition.

Answer C is wrong because the plunger is used to draw fluid through one check valve and force it back out through the other check valve.

Answer D is right because a solenoid valve is not used to release starting fluid.

156. Answer A is wrong because the pump is part of the system.

Answer B is wrong because the inlet check valve is part of the system.

Answer C is wrong because the outlet check valve is part of the system.

Answer D is right because the fan pulley is not part of the system.

157. Answer A is wrong because the air shutdown valve can be mounted to the blower housing.

Answer B is wrong because the air shutdown valve can be mounted to the turbocharger inlet side.

Answer C is wrong because the air shutdown valve can be mounted to the intake manifold.

Answer D is right because the air shutdown valve is least likely to be mounted to the exhaust manifold.

158. Answer A is wrong because constant use of the air shutdown valve may damage the blower seals.

Answer B is wrong because the air shutdown valve on a blower is used as an emergency shutdown device.

Answer C is wrong because both *Technician A* and *Technician B* are wrong.

Answer D is right because both *Technician A* and *Technician B* are wrong.

159. **Answer A is right** because the presence of water in a fuel tank may be tested with an aluminum welding rod coated with water detection paste.

Answer B is wrong because the fuel tank sending unit contains a variable resistor.

Answer C is wrong because *Technician A* is right and *Technician B* is wrong.

Answer D is wrong because *Technician A* is right and *Technician B* is wrong.

160. Answer A is wrong because the primary filter is on the suction side of the transfer pump.

Answer B is wrong because the secondary filter is on the charge side of the transfer pump.

Answer C is wrong because the fuel tank pickup pipe is positioned a short distance from the bottom of the fuel tank.

Answer D is right because a plugged fuel tank vent may cause the engine to stop running.

161. Answer A is wrong because this type of pump has a glass or metal bowl to catch sediments.

Answer B is wrong because this type of pump has a plastic or metal strainer to block sediments before entering the pump and fuel lines.

Answer C is wrong because this type of pump has a hand-operated plunger-type pump.

Answer D is right because this type of pump does not have a pop-off valve.

162. Answer A is wrong because seized injectors would not cause failure of the transfer pump to build pressure.

Answer B is right because an air leak in the primary filter housing could cause the transfer pump not to build pressure.

Answer C is wrong because a restricted secondary filter would not cause failure of the transfer pump to build pressure.

Answer D is wrong because a plugged fuel line between the transfer pump and the injectors would not cause failure of the transfer pump to build pressure.

163. Answer A is wrong because the secondary filter contains transfer pump pressure, and a leak at this location causes fuel to leak out of the system.

Answer B is right because the fuel line between the primary filter and the transfer pump contains vacuum from this pump, and a leak at this point causes air to enter the system.

Answer C is wrong because a unit injector inlet line is subjected to transfer pump pressure, and a leak at this point causes fuel to leak out of the system.

Answer D is wrong because the fuel line from the secondary filter to the injectors is subjected to transfer pump pressure and a leak at this point causes fuel to leak out of the system.

164. Answer A is wrong because air in the fuel system may be checked by watching for bubbles in the fuel return line with the engine running.

Answer B is right because if there are bubbles in the fuel line from the tank to the primary filter with the engine running, the line from the tank to the primary filter has a leak.

Answer C is wrong because *Technician A* is wrong and *Technician B* is right.

Answer D is wrong because *Technician A* is wrong and *Technician B* is right.

165. Answer A is wrong because the injector lines should be bled of all air.

Answer B is wrong because the primer pump can be used to bleed the injector lines.

Answer C is right because replacing fuel filter is not necessary in the bleeding process.

Answer D is wrong because pressurizing the tank with a small amount of air pressure is acceptable.

166. Answer A is wrong because when bleeding air from the fuel system it is necessary to pressurize the system with the hand primer, but *Technician B* is also right.

Answer B is wrong because bleeding air from the fuel system should be done any time fuel filters are changed, but *Technician A* is also right.

Answer C is right because both *Technician A* and *Technician B* are right.

Answer D is wrong because both *Technician A* and *Technician B* are right.

167. Answer A is wrong because the rear and front throttle screws should be adjusted before the idle speed adjustment.

Answer B is right because the rear throttle screw should be adjusted before the forward throttle screw.

Answer C is wrong because *Technician A* is wrong and *Technician B* is right.

Answer D is wrong because *Technician A* is wrong and *Technician B* is right.

168. **Answer A is right** because if the engine lugs down and does not respond properly to load changes, the throttle linkage may require adjusting.

Answer B is wrong because engine valve and unit injector adjustments should be completed before throttle linkage adjustments.

Answer C is wrong because *Technician A* is right and *Technician B* is wrong.

Answer D is wrong because *Technician A* is right and *Technician B* is wrong.

169. Answer A is wrong because you time in-line pumps with a timing pin, not distributor pumps.

Answer B is wrong because the spill method is used on in-line pumps.

Answer C is wrong because the rack adjustment has no effect on timing in a distributor system.

Answer D is right because a distributor pump is timed by rotating the pump.

170. Answer A is wrong because the key must be on and the engine must be off.

Answer B is right because the key must be on and the engine off.

Answer C is wrong because the key must be on to power the diagnostic.

Answer D is wrong because the key must be on and the engine off.

171. **Answer A is right** because there is no mechanical connection between the accelerator pedal and the injection pump.

Answer B is wrong because the PCM sends control signals to the fuel solenoid driver module.

Answer C is wrong because the fuel solenoid driver module controls the fuel injection solenoid in the injection pump.

Answer D is wrong because the PCM receives input signals from the accelerator pedal position (APP) module.

172. Answer A is wrong because when installing an in-line injector pump with an electronic governor, it must be timed with the engine camshaft to ensure proper injection timing, but *Technician B* is also right.

Answer B is wrong because an internal camshaft in the pump determines the firing order, but *Technician A* is also right.

Answer C is right because both *Technician A* and *Technician B* are right.

Answer D is wrong because both *Technician A* and *Technician B* are right.

173. **Answer A is right** because the in-line pump pressurizes the meter's timing and delivers fuel to injectors.

Answer B is wrong because the injector lines between the pump and injectors must have equal lengths or injection timing is adversely affected.

Answer C is wrong because *Technician A* is right and *Technician B* is wrong.

Answer D is wrong because *Technician A* is right and *Technician B* is wrong.

174. Answer A is wrong because on a PT system you back out the torque screw and high idle screws must be backed out about 5 turns, but *Technician B* is also right.

Answer B is wrong because the pump must rotate to produce 1,000 strokes to collect the proper amount of fuel, but *Technician A* is also right.

Answer C is right because both *Technician A* and *Technician B* are right.

Answer D is wrong because both *Technician A* and *Technician B* are right.

175. Answer A is wrong because the gear pump is necessary to draw fuel from the tank.

Answer B is wrong because the throttle linkage provides a link from the pump to the accelerator to determine the mean operating speed.

Answer C is wrong because the pulsation damper is necessary to correct for pressure spikes caused by the gear pump as it draws fuel from the tank.

Answer D is right because Cummins PT systems do not use transfer or lift pumps.

176. Answer A is wrong because in a PT injector system when that pressure of fuel against the angular surface of the injector needle exceeds the spring tension, the injector fires, but *Technician B* is also right.

Answer B is wrong because in a unit injector system when the cam lobe presses the plunger of the injector down, it pressurizes fuel and injects fuel into the cylinder, but *Technician A* is also right.

Answer C is right because both *Technician A* and *Technician B* are right.

Answer D is wrong because both *Technician A* and *Technician B* are right.

177. Answer A is wrong because plugged nozzles will cause improper spray patterns.

Answer B is wrong because a stuck injection nozzle will cause injectors to dribble.

Answer C is right because improper timing of injectors will not affect spray pattern and is the least likely cause of this condition.

Answer D is wrong because excessive injector-to-follower clearance will prevent adequate fuel pressurization for injection.

178. Answer A is wrong because screw R is the opening pressure adjusting screw.

Answer B is wrong because screw S is the lift adjusting screw.

Answer C is wrong because both *Technician A* and *Technician B* are wrong.

Answer D is right because both *Technician A* and *Technician B* are wrong.

179. **Answer A is right** because the Technician is cleaning the injector spray orifices.

Answer B is wrong because the Technician is cleaning the injector spray orifices.

Answer C is wrong because *Technician A* is right and *Technician B* is wrong.

Answer D is wrong because *Technician A* is right and *Technician B* is wrong.

180. Answer A is wrong because intake manifold pressure is supplied to the diaphragm in the AFC unit.

Answer B is wrong because the diaphragm in the AFC unit is connected to directly or indirectly to the fuel control mechanism.

Answer C is wrong because the AFC device adapts full-load fuel delivery to charge-air pressure.

Answer D is right because the AFC device is used on turbocharged engines.

181. **Answer A is right** because if there are puffs of black smoke in the exhaust during transmission shifting, the AFC device may require adjusting.

Answer B is wrong because at low engine speeds and boost pressure, the boost pressure is not high enough to force the diaphragm downward.

Answer C is wrong because *Technician A* is right and *Technician B* is wrong.

Answer D is wrong because *Technician A* is right and *Technician B* is wrong.

182. Answer A is wrong because you bleed the injector line first.

Answer B is right because you bleed the injector line first.

Answer C is wrong because you bleed the injector line first.

Answer D is wrong because you bleed the injector line first.

183. Answer A is wrong because the lines shown are high-pressure fuel lines, but *Technician B* is also right.

Answer B is wrong because all fuel connections must be capped when they are disconnected, but *Technician A* is also right.

Answer C is right because both *Technician A* and *Technician B* are right.

Answer D is wrong because both *Technician A* and *Technician B* are right.

184. Answer A is wrong because any kinks in the lines should be checked.

Answer B is wrong because corroded fuel connectors should be checked.

Answer C is wrong because you should check the lines for cracks.

Answer D is right because there is no need to check the power steering pump when inspecting low-pressure fuel lines.

185. Answer A is wrong because the fuel line corrosion is most likely caused by water in the fuel.

Answer B is right because the fuel line corrosion is most likely caused by water in the fuel.

Answer C is wrong because the fuel line corrosion is most likely caused by water in the fuel.

Answer D is wrong because the fuel line corrosion is most likely caused by water in the fuel.

186. Answer A is wrong because underrun is not an effect of governor instability.

Answer B is wrong because high idle is the effect of a high idle setting.

Answer C is right because hunting is the result of an unstable governor.

Answer D is wrong because low idle is the effect of a low idle setting.

187. **Answer A is right** because the governor affects a low power complaint the least.

Answer B is wrong because the high idle overrun can be affected by governor operation.

Answer C is wrong because low idle underrun can be affected by governor operation.

Answer D is wrong because a rough idle operation may be caused by the governor.

188. Answer A is wrong because when an ohmmeter is connected across the solenoid winding and the meter reading is less than specified, the winding is shorted.

Answer B is wrong because when the ignition switch is turned off, the solenoid positions the metering valve in the no-fuel position.

Answer C is wrong because both *Technician A* and *Technician B* are wrong.

Answer D is right because both *Technician A* and *Technician B* are wrong.

189. Answer A is wrong because vacuum is supplied to item 1 when the ignition switch is shut off.

Answer B is wrong because item 5 is the rack.

Answer C is wrong because item 2 is a manual shutoff lever.

Answer D is right because a leak in item 1 may cause failure of the engine to shut off.

190. **Answer A is right** because on a vehicle with an electric fuel shutoff valve, the oil pressure gauge is not part of the system.

Answer B is wrong because the key switch is part of the system.

Answer C is wrong because the solenoid is part of the system.

Answer D is wrong because the rack is part of the system.

191. **Answer A is right** because if the hold-in coil in the shutdown solenoid is open, the pull-in coil has enough magnetism to pull the solenoid into the start position, but when this winding is shut off as the engine starts, the hold-in coil has no magnetism to hold the solenoid in the run position.

Answer B is wrong because if higher than normal resistance in the wire connected to the shutdown solenoid windings still provides enough current flow to energize the pull-in winding and start the engine momentarily, this resistance is not the cause of the problem.

Answer C is wrong because *Technician A* is right and *Technician B* is wrong.

Answer D is wrong because *Technician A* is right and *Technician B* is wrong.

192. Answer A is wrong because DTCs may be cleared from most PCMs by disconnecting battery voltage from the PCM.

Answer B is right because most generic scan tools can access any OBD II vehicle DTCs.

Answer C is wrong because *Technician A* is wrong and *Technician B* is right.

Answer D is wrong because *Technician A* is wrong and *Technician B* is right.

193. Answer A is wrong because this acronym refers to the OBD II codes used to turn on instrument panel indicator lights.

Answer B is wrong because this acronym refers to the code assigned and stored for a specific vehicle fault.

Answer C is wrong because this acronym refers to the standards established for all new vehicles. These standards establish requirements for the detection and storage of faults while the vehicle is operating.

Answer D is right because is not a term used in truck electronics.

194. Answer A is wrong because when the throttle position sensor is adjusted on a fuel injection pump with electronic controls, on some applications a gauge block is placed between the gauge boss and the maximum throttle travel screw, but *Technician B* is also right.

Answer B is wrong because the scan tool will signal when the setting is too low or too high, but *Technician A* is also right.

Answer C is right because both *Technician A* and *Technician B* are right.

Answer D is wrong because both *Technician A* and *Technician B* are right.

195. Answer A is wrong because the reference voltage supplied to the TPS should be 5V with the ignition switch on.

Answer B is wrong because on many truck applications the TPS signal voltage decreases as the injection pump lever is moved to the maximum speed position.

Answer C is wrong because the TPS contains a variable resistor.

Answer D is right because an improper TPS adjustment may cause lack of power during acceleration.

196. **Answer A is right** because in some pumps, the PCM controls the control collar position to supply the amount of fuel required by the engine.

Answer B is wrong because the PCM does receive feedback signals regarding control collar position from the control collar position sensor.

Answer C is wrong because *Technician A* is right and *Technician B* is wrong.

Answer D is wrong because *Technician A* is right and *Technician B* is wrong.

197. **Answer A is right** because a defective engine speed sensor may cause improper maximum engine rpm.

Answer B is wrong because a defective coolant temperature sensor may cause hard engine starting.

Answer C is wrong because a defective airflow sensor may cause an improper air-fuel ratio.

Answer D is wrong because the needle motion sensor in each injector contains a winding and a pressure pin.

198. Answer A is wrong because the PCM controls a linear motion solenoid connected to the pump rack.

Answer B is wrong because the PCM controls a linear motion solenoid connected to the pump rack.

Answer C is wrong because both *Technician A* and *Technician B* are wrong.

Answer D is right because both *Technician A* and *Technician B* are wrong.

199. Answer A is wrong because the fixed timing sensor probe is connected to the timing event marker port in the pump.

Answer B is wrong because the static timing procedure begins with number 1 piston at 45 degrees BTDC on the compression stroke.

Answer C is wrong because the engine should be rotated with a barring tool until both lights on the fixed timing sensor are on.

Answer D is right because if the timing is not within specifications, the position of the timing indicator should be checked.

200. **Answer A is right** because the electronic unit injector is cam actuated.

Answer B is wrong because a rack is not used with electronic unit injectors.

Answer C is wrong because *Technician A* is right and *Technician B* is wrong.

Answer D is wrong because *Technician A* is right and *Technician B* is wrong.

201. Answer A is wrong because the amount of fuel injected is not determined by the helix in the plunger.

Answer B is wrong because the PCM operates a vertical motion solenoid in each injector.

Answer C is wrong because each injector is opened by cam action.

Answer D is right because the injector solenoid determines the fuel metering and the injection timing functions.

202. Answer A is wrong because the oil pressure for the injectors is not supplied from the oil pump that normally supplies oil to the engine components.

Answer B is right because the PCM operates an oil pressure regulator to control oil pressure supplied to the injectors.

Answer C is wrong because the pressure exerted on the fuel by the plunger is seven times higher than the oil pressure supplied to the injector.

Answer D is wrong because when the PCM de-energizes the injector solenoid, injection stops quickly.

203. Answer A is wrong because if the injector oil pressure pump fails, the engine will not run.

Answer B is right because if the injector solenoid is energized, the plunger is forced downward.

Answer C is wrong because *Technician A* is wrong and *Technician B* is right.

Answer D is wrong because *Technician A* is wrong and *Technician B* is right.

204. Answer A is wrong because the electronic distributor unit (EDU) operates on commands from the PCM.

Answer B is right because the injectors operate on a pulse width principle.

Answer C is wrong because *Technician A* is wrong and *Technician B* is right.

Answer D is wrong because *Technician A* is wrong and *Technician B* is right.

205. **Answer A is right** because when the injector solenoid is deenergized fuel injection can occur.

Answer B is wrong because the end of injection is determined by solenoid opening.

Answer C is wrong because the EDU sends a pulse width voltage signal to the injectors.

Answer D is wrong because if the poppet control valve in the injector opens, fuel injection stops.

206. Answer A is wrong because the three-pedal position sensors in the APP module should provide different voltage signals as the accelerator pedal is depressed.

Answer B is wrong because if a fault occurs in the APP module, a separate service throttle soon light is illuminated in the instrument panel.

Answer C is wrong because both *Technician A* and *Technician B* are wrong.

Answer D is right because both *Technician A* and *Technician B* are wrong.

207. **Answer A is right** because the APP 2 variable resistor voltage signal should be 4.5V at idle.

Answer B is wrong because if one variable resistor is defective in the APP module, engine power is normal.

Answer C is wrong because if one variable resistor is defective in the APP module, the service throttle soon light is not on.

Answer D is wrong because if all three variable resistors are defective in the APP module, the engine speed is limited to idle.

208. Answer A is wrong because the governor functions are programmed into the PCM software, but *Technician B* is also right.

Answer B is wrong because a rack position sensor sends feedback information to the PCM, but *Technician A* is also right.

Answer C is right because both *Technician A* and *Technician B* are right.

Answer D is wrong because both *Technician A* and *Technician B* are right.

209. Answer A is wrong because the electronic governor system provides improved air-fuel ratio control.

Answer B is wrong because the electronic governor system reduces exhaust emissions.

Answer C is wrong because the electronic governor system can be reprogrammed with customer data to tailor the engine for varying engine and chassis applications.

Answer D is right because the electronic governor system does not reduce diesel knock intensity.

210. Answer A is wrong because engine shutdown occurs 30 to 40 seconds after a defect is sensed.

Answer B is right because a separate module shuts off the voltage to the fuel shutoff relay winding if the coolant overheats.

Answer C is wrong because engine shutdown does not occur if the engine speed is above the maximum governed rpm.

Answer D is wrong because if the engine shutdown system causes engine shutdown, the engine will restart after a key-off, but it is shut down again in 30 to 40 seconds if the defect is still present.

211. Answer A is wrong because the fuel shutoff relay winding is normally grounded after the winding.

Answer B is right because there may be an open circuit between the warning module and the fuel shutoff relay.

Answer C is wrong because *Technician A* is wrong and *Technician B* is right.

Answer D is wrong because *Technician A* is wrong and *Technician B* is right.

212. **Answer A is right** because the level of engine protection can be programmed into some PCMs.

Answer B is wrong because a visual warning to the driver is the first level of engine protection.

Answer C is wrong because *Technician A* is right and *Technician B* is wrong.

Answer D is wrong because *Technician A* is right and *Technician B* is wrong.

213. Answer A is wrong because when the shut engine off (SEO) light is illuminated, a serious fault is sensed.

Answer B is wrong because the malfunction indicator light (MIL) is turned on by the PCM.

Answer C is right because a stop engine override switch allows the driver to temporarily override the engine shutdown function.

Answer D is wrong because monitored parameters do not include the turbocharger boost pressure.

214. Answer A is wrong because this will overload the meter fuse and could cause damage to the ohmmeter. Use a voltmeter.

Answer B is right because checking the voltage drop between the battery positive terminal and the starter terminal stud while the starter is cranking is a voltage drop check on cable connection resistance.

Answer C is wrong because *Technician A* is wrong and *Technician B* is right.

Answer D is wrong because *Technician A* is wrong and *Technician B* is right.

215. Answer A is wrong because a Technician should only connect a voltmeter in parallel with or across the circuit tested.

Answer B is wrong because positive terminal-to-ground only tests that circuit.

Answer C is wrong because a Technician should only connect a voltmeter in parallel to the circuit being tested.

Answer D is right because a Technician should only connect a voltmeter in parallel with or across the circuit tested.

216. **Answer A is right** because the use of improper tools can bend pins.

Answer B is wrong because pins cannot be straightened; the connector must be replaced again.

Answer C is wrong because improper or standard connectors may produce higher resistance.

Answer D is wrong because grounding some circuits will cause damage to the PCM.

217. Answer A is wrong because the EMC connectors should never be probed through the weather pack seals, but *Technician B* is also right.

Answer B is wrong because a test kit should be used to service connectors, but *Technician A* is also right.

Answer C is right because both *Technician A* and *Technician B* are right.

Answer D is wrong because both *Technician A* and *Technician B* are right.

218. Answer A is wrong because the customer option codes can be changed in the electronic systems with a scan tool, but *Technician B* is also right.

Answer B is wrong because in some cases changing these codes could result in poor engine performance, but *Technician A* is also right.

Answer C is right because both *Technician A* and *Technician B* are right.

Answer D is wrong because both *Technician A* and *Technician B* are right.

219. **Answer A is right** because PID stands for parameter identifier.

Answer B is wrong because PID stands for parameter identifier.

Answer C is wrong because PID stands for parameter identifier.

Answer D is wrong because PID stands for parameter identifier.

220. Answer A is wrong because if the hydrometer is yellow the electrolyte is low, and the battery should be replaced.

Answer B is wrong because if the hydrometer is green, it may be fast or slow charged.

Answer C is wrong because both *Technician A* and *Technician B* are wrong.

Answer D is right because both *Technician A* and *Technician B* are wrong.

221. Answer A is wrong because a battery with 1.265 specific gravity on all cells is fully charged.

Answer B is right because if the battery temperature reaches 150°F (65°C) the battery will be damaged.

Answer C is wrong because oxygen and hydrogen gas are emitted from the battery while charging.

Answer D is wrong because a battery with an open circuit voltage of 12.6V is fully charged.

222. Answer A is wrong because a battery that shows readings below specifications should be replaced.

Answer B is wrong because a battery that shows readings below specifications should be replaced.

Answer C is right because the battery should be replaced.

Answer D is wrong because voltage regulation has no effect on the battery during a load test.

223. Answer A is wrong because the battery is satisfactory if the voltage remains above 9.6V.

Answer B is wrong because the battery should be discharged at one half the cold cranking rating.

Answer C is right because the load should be applied to the battery for 15 seconds.

Answer D is wrong because the test voltage is read with the load still applied at the end of the test.

224. **Answer A is right** because if the battery temperature exceeds 125°F (52°C) the charging rate should be reduced.

Answer B is wrong because the battery is fully charged when the specific gravity on all cells is 1.265.

Answer C is wrong because *Technician A* is right and *Technician B* is wrong.

Answer D is wrong because *Technician A* is right and *Technician B* is wrong.

225. Answer A is wrong; the battery should be replaced because there is 0.065 specific gravity points difference in the cell readings, but *Technician B* is also right.

Answer B is wrong because the separators between the plates may be damaged in the cell with 1.200 specific gravity, but *Technician A* is also right.

Answer C is right because both *Technician A* and *Technician B* are right.

Answer D is wrong because both *Technician A* and *Technician B* are right.

226. Answer A is wrong because the batteries are connected in parallel to maintain a 12-volt system.

Answer B is right because the batteries are connected in parallel.

Answer C is wrong because the batteries are connected in parallel to maintain a 12-volt system.

Answer D is wrong because the batteries are connected in parallel to maintain a 12-volt system.

227. Answer A is wrong because this will cause a short circuit and will damage both batteries.

Answer B is right because a parallel circuit allows the maximum current flow.

Answer C is wrong because this connection can cause gassing, and if the batteries are gassing they may explode.

Answer D is wrong because if vehicles are making an electrical ground when the positive cable is attached, it might arc and cause battery explosion.

228. **Answer A is right** because the charging circuit voltage may be too high.

Answer B is wrong because excessive resistance in the wire from the alternator battery terminal to the positive battery terminal reduces the charging rate and causes an undercharged battery.

Answer C is wrong because *Technician A* is right and *Technician B* is wrong.

Answer D is wrong because *Technician A* is right and *Technician B* is wrong.

229. Answer A is wrong because when removing battery cables, remove the negative cable first.

Answer B is right because removing the battery cables erases the PCM adaptive memory.

Answer C is wrong because the positive cable should be installed first.

Answer D is wrong because a battery cable must never be removed with the engine running.

230. Answer A is wrong because the positive must be supplied to the relay.

Answer B is right because installing negative lead to relay is the exception.

Answer C is wrong because these must be connected to operate the relay.

Answer D is wrong because this must be connected to provide a ground connection to all electrical circuit back to the battery.

231. Answer A is wrong because the starter relay opens and closes the circuit between the battery and the solenoid windings.

Answer B is right because if the starter relay contacts stick closed, the starting motor continues to run after the engine.

Answer C is wrong because *Technician A* is wrong and *Technician B* is right.

Answer D is wrong because *Technician A* is wrong and *Technician B* is right.

232. **Answer A is right** because when performing a started load test, you should not crank the engine for no more than 30 seconds at a time.

Answer B is wrong because this test should not be performed with a battery charger connected to the battery.

Answer C is wrong because *Technician A* is right and *Technician B* is wrong.

Answer D is wrong because *Technician A* is right and *Technician B* is wrong.

233. Answer A is wrong because excessive resistance in the starting motor cables causes slow cranking speed with low current draw.

Answer B is wrong because starter current draw test results are not accurate if the vehicle batteries are partly charged.

Answer C is wrong because high current draw may be caused by worn bushings in the starting motor.

Answer D is right because during the test, the fuel shutdown solenoid should be disconnected to prevent the engine from starting.

234. Answer A is wrong because a voltmeter is used to measure voltage drop in the starting circuit.

Answer B is right because a voltmeter is used to measure voltage drop in the starting circuit.

Answer C is wrong because a voltmeter is used to measure voltage drop in the starting circuit.

Answer D is wrong because a voltmeter is used to measure voltage drop in the starting circuit.

235. Answer A is wrong because when measuring voltage drop, there should be a normal current flow in the circuit being tested.

Answer B is wrong because when the resistance in a circuit decreases, the voltage drop also decreases.

Answer C is wrong because both *Technician A* and *Technician B* are wrong.

Answer D is right because both *Technician A* and *Technician B* are wrong.

236. Answer A is wrong because this is a required task when installing a starter.

Answer B is wrong because this is a required task when installing a starter.

Answer C is wrong because this is a required task when installing a starter.

Answer D is right because starter bushings are replaced when overhauling a starter.

237. Answer A is wrong because the negative cable is removed first to prevent arcing.

Answer B is wrong because the negative cable is removed first to prevent arcing.

Answer C is right because the negative cable is removed first.

Answer D is wrong because the negative cable is removed first to prevent arcing.

238. Answer A is wrong because the driver's foot must be off the accelerator pedal to activate the brake.

Answer B is wrong because the brake will not operate if the driver has the clutch pedal depressed.

Answer C is wrong because the slave cylinder pushes downward on the exhaust valve crosshead.

Answer D is right because the master cylinder supplies higher pressure to the slave cylinder when the pushrod begins to move upward.

239. **Answer A is right** because an air intake suppressor can eliminate pulsation.

Answer B is wrong because wastegates are used to prevent overboost.

Answer C is wrong because an oil bath air cleaner will not stop the pulsation.

Answer D is wrong because variable cam timing will not stop the pulsation.

240. Answer A is wrong because when battery potential is applied, the diode is reverse biased.

Answer B is wrong because the applied voltage in this circuit is direct current.

Answer C is right because the diode is used to shunt the stored potential when the coil is de-energized.

Answer D is wrong because this is a diode, not an RFI filter.

241. Answer A is wrong because the pushrod is below the master piston component A.

Answer B is right because the master piston at position A senses the pushrod's upward motion to open exhaust valves.

Answer C is wrong because the slave piston senses the upward motion to open exhaust valves.

Answer D is wrong because the master piston does not operate the rocker arm, the rocker arm moves the plunger in the master piston.

242. Answer A is wrong because a leaking check valve allows the high pressure from the master piston to bleed back into the system, and the compression brake does not operate on that cylinder.

Answer B is right because the compression brake opens the exhaust valves before the compression stroke is complete.

Answer C is wrong because *Technician A* is wrong and *Technician B* is right.

Answer D is wrong because *Technician A* is wrong and *Technician B* is right.

243. Answer A is wrong because the master piston is part of the system.

Answer B is wrong because the slave piston is part of the system.

Answer C is wrong because the solenoid valve is part of the system.

Answer D is right because brake shoes are not part of a compression brake.

3 Drivetrain

Pretest

The purpose of this pretest is to determine the amount of review you may require before taking the ASE Medium/Heavy Truck Drivetrain Test. If you answer all the pretest questions correctly, complete the questions and study the information in this chapter to prepare for this test.

If two or more of your answers to the pretest questions are incorrect, complete a study of all the chapters in *Today's Technician Medium/Heavy-Duty Truck Drivetrain Systems* Classroom and Shop Manuals published by Delmar Publishers, plus a study of the questions and information in this chapter.

The pretest answers are located at the end of the pretest; these answers also are in the answer sheets supplied with this book.

1. A single-disc, push-type clutch does not disengage properly when the clutch pedal is fully depressed. *Technician A* says the clutch pedal free play may be insufficient. *Technician B* says the clutch disc splines may be sticking on the transmission input shaft. Who is right?
   A. A only
   B. B only
   C. Both A and B
   D. Neither A nor B

2. A clutch release bearing on a medium-duty pickup and delivery truck became noisy 20,000 miles (32,000 km) after it was replaced. *Technician A* says the driver may have been riding the clutch pedal too much with his foot. *Technician B* says the clutch pedal free play may have been improperly adjusted. Who is right?
   A. A only
   B. B only
   C. Both A and B
   D. Neither A nor B

3. A burnt pressure plate may be caused by all of the following EXCEPT:
   A. oil on the clutch disc.
   B. not enough clutch pedal free play.
   C. binding linkage.
   D. a damaged pilot bearing.

4. To install a pull-type clutch, a Technician will need to do all the following EXCEPT:
   A. align the clutch disc.
   B. adjust the self-adjusting release bearing.
   C. resurface the limited torque clutch brake.
   D. lubricate the pilot bearing.

5. A vehicle with the type of clutch as shown in Figure 3–1 is brought into the shop because it goes through ceramic friction discs too frequently. What should the Technician do?
   A. Instruct the driver not to rest his foot on the clutch pedal.
   B. Switch to a fiber-type disc.
   C. Replace the pressure plate.
   D. Replace the release bearing.

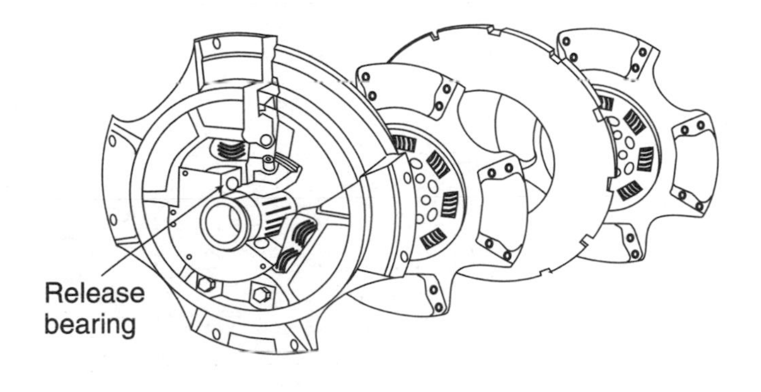

Figure 3–1 Multiple-disc clutch.

6. The transmission shift cover detents on the shift bar housing shown in Figure 3–2 are being inspected and serviced. *Technician A* says to check for worn or oblong detent recesses. *Technician B* says to place the specified lubricant in the detent spring channels. Who is right?
   A. A only
   B. B only
   C. Both A and B
   D. Neither A nor B

7. The LEAST likely damage to check for on an input shaft is:
   A. pilot bearing shaft cracking.
   B. gear teeth damage.
   C. input spline damage.
   D. cracking or other fatigue wear to the input shaft spline.

8. Which of the following damage is LEAST likely to happen in a two-gear drive combination?
   A. Tooth chipping
   B. Bottoming
   C. Climbing
   D. Spalling

9. *Technician A* says that pilot bore runout should be measured when a vehicle makes a grinding noise that cannot be identified in any other drivetrain member. *Technician B* says that pilot bore runout should be measured only when the pilot bearing is found to be in poor shape. Who is right?
   A. A only
   B. B only
   C. Both A and B
   D. Neither A nor B

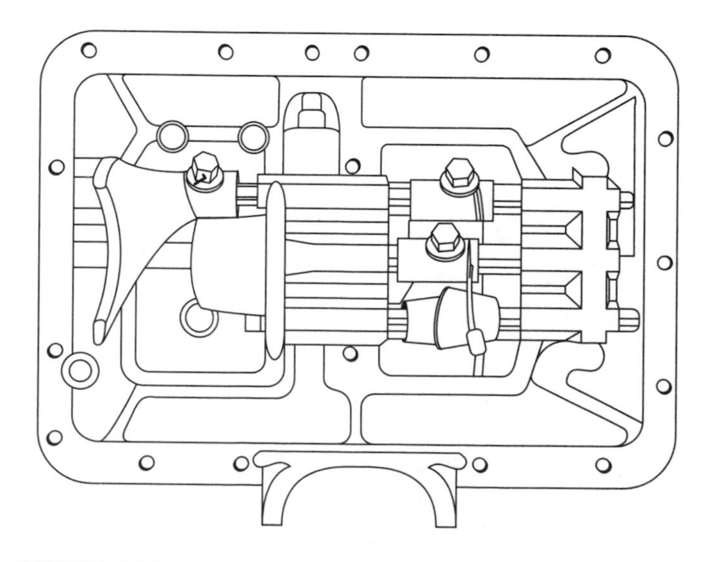

Figure 3–2 Transmission shift rail assembly.

10. A Technician notices overheated oil coating the seals of the transmission. *Technician A* says that you must replace all the seals in the transmission. *Technician B* says that changing to a higher grade of transmission oil is all that is necessary. Who is right?
    A. A only
    B. B only
    C. Both A and B
    D. Neither A nor B

11. A truck with a power divider has power applied to the tandem forward rear drive axle while no power is applied to the rearward rear drive axle. The MOST likely cause of the malfunction is:
    A. broken teeth of the forward drive axle ring gear.
    B. broken teeth of the rear drive axle ring gear.
    C. stripped output shaft splines.
    D. damaged interaxle differential.

12. The axle range and interaxle differential lockout system shown in Figure 3–3 will not shift from high range to low range. The MOST likely cause is:
    A. a faulty air compressor.
    B. an air leak at the axle shift unit.
    C. a defective quick release valve.
    D. a plugged air filter.

13. A truck has a high-speed vibration. *Technician A* says the best way to determine whether the vibration is from the wheels or the drive shaft is to disengage the clutch while the vehicle is exhibiting the vibration. If the vibration goes away, the vibration is in the wheels. *Technician B* says the best way is to lightly apply the brakes while maintaining the vehicle's speed. If the vibration goes away with the brakes lightly applied, the problem is in the drive shaft. Who is right?
    A. A only
    B. B only
    C. Both A and B
    D. Neither A nor B

14. A Technician is replacing the U-joints in a drive shaft. After removal the technician realized that he forgot to scribe a line onto the yokes for proper phasing. The likely method of testing for proper phasing of the yokes would be to:
    A. drive the truck on level terrain at a steady speed below 40 mph (64 km) and check for driveline vibration.
    B. drive the truck on level terrain at a steady speed above 40 mph (64 km) and check for driveline vibration.

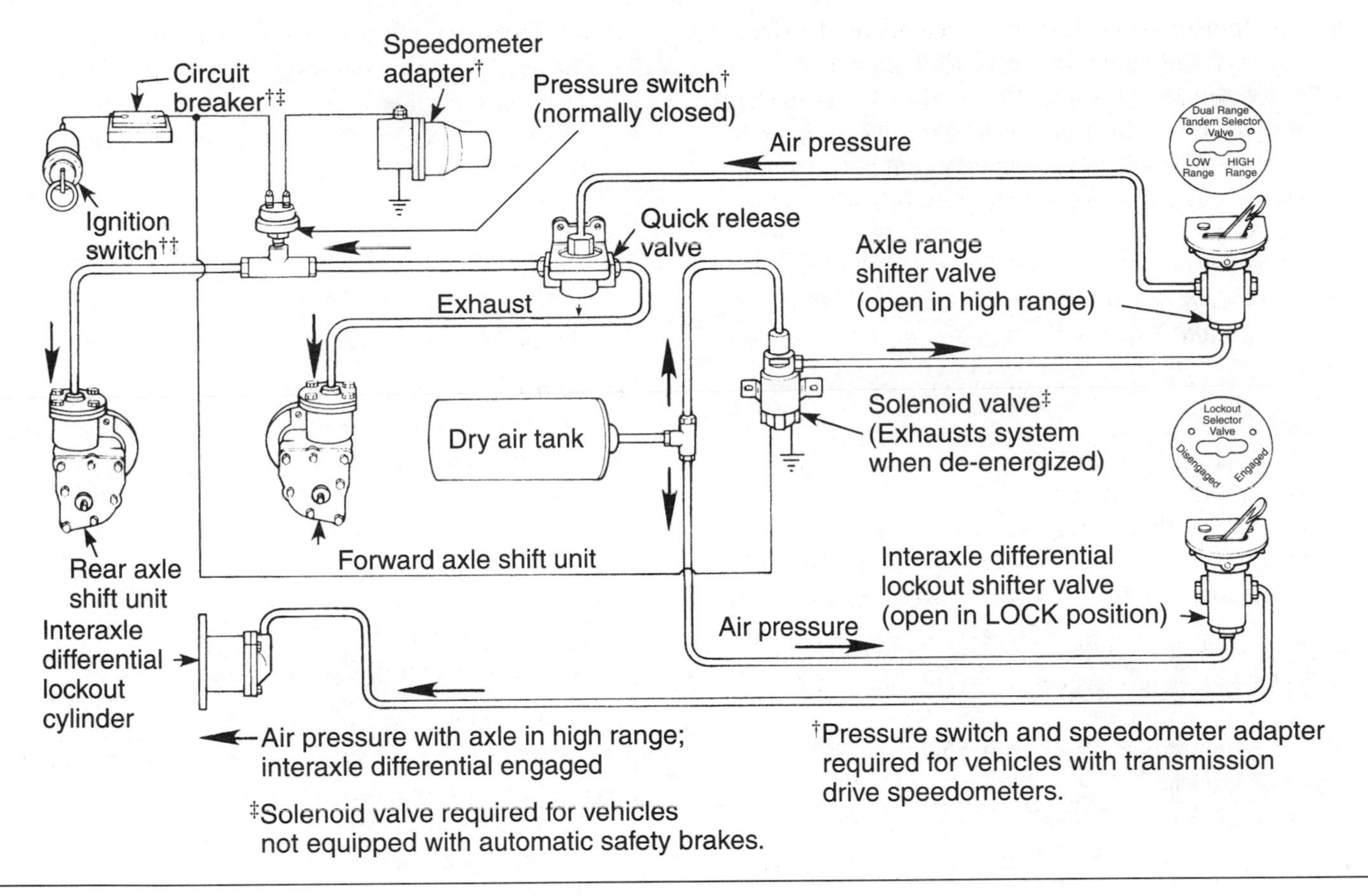

Figure 3–3 Axle range and interaxle differential lockout system.

C. place the truck under load by driving up an incline and check for driveline vibration.
D. depress the clutch pedal with the truck traveling at 35 mph (56 km) and allow the vehicle to coast while checking for vibration.

Answers to Pretest

1. B, 2. C, 3. D, 4. C, 5. A, 6. A, 7. A, 8. B, 9. B, 10. B, 11. C, 12. C, 13. C, 14. B

Clutch Diagnosis and Repair

Task 1

Diagnose clutch noises, binding, slippage, pulsation, vibration, grabbing, and chatter problems; determine cause of failure and needed repairs.

1. Clutch slippage may be caused by:
   A. a worn or rough clutch release bearing.
   B. excessive input shaft end play.
   C. a leaking rear main seal.
   D. a weak or broken torsional springs.
2. A vehicle has a burnt friction disc in the clutch. *Technician A* says it could be caused by too much clutch pedal free play. *Technician B* says binding linkage may have caused it. Who is right?
   A. A only
   B. B only
   C. Both A and B
   D. Neither A nor B
3. All of the following may cause premature clutch disc failure EXCEPT:
   A. oil contamination of the disc.
   B. worn torsion springs.
   C. worn U-joints.
   D. a worn clutch linkage.
4. The LEAST likely cause for clutch slippage is:
   A. a sticking release bearing.
   B. oil contamination on the clutch disc.
   C. a worn pilot bearing.
   D. a worn clutch linkage.

Hint

The most frequent cause of clutch failure is excess heat. The heat generated between the flywheel, driven discs, intermediate plate, and pressure plate may be intense enough to cause the metal to melt and the friction material to be destroyed. Heat or wear is practically nonexistent when the clutch is fully engaged. However, considerable heat can be generated at clutch engagement when the clutch is picking up the load. An improperly adjusted or slipping clutch rapidly generates sufficient heat to self-destruct. Causes of clutch slippage include improper adjustment of an external linkage, a worn or damaged pressure plate, worn clutch disc, and grease or oil contamination of the clutch disc.

Task 2

Inspect, adjust, repair, or replace clutch linkage, pedal height, cables, levers, brackets, bushings, pivots, springs, and clutch safety switch (includes push-type and pull-type assemblies).

5. The transmission grinds when shifting into reverse. The LEAST likely cause would be:
   A. air in the hydraulic clutch control system.
   B. not enough clutch pedal free play.
   C. a noisy pilot bearing.
   D. a leaking or weak air servo cylinder.

6. When adjusting a clutch linkage as shown in Figure 3–4, *Technician A* says the pedal free travel should be about 1.5 to 2 inches (38.1 to 50.8 mm). *Technician B* says insufficient clutch pedal free travel may cause hard transmission shifting. Who is right?
   A. A only
   B. B only
   C. Both A and B
   D. Neither A nor B

Hint

"Riding" the clutch pedal is another name for operating the vehicle with the clutch partially engaged. This is very destructive to the clutch, as it permits slippage and generates excessive heat. Riding the clutch also puts constant thrust load on the release bearing, which can thin out the lubricant and cause excessive wear on the pads. Release bearing failures are often the result of this type of driving practice. The best way to determine if clutch disc failure is due to driver error or mechanical failure is to speak with the driver of the vehicle.

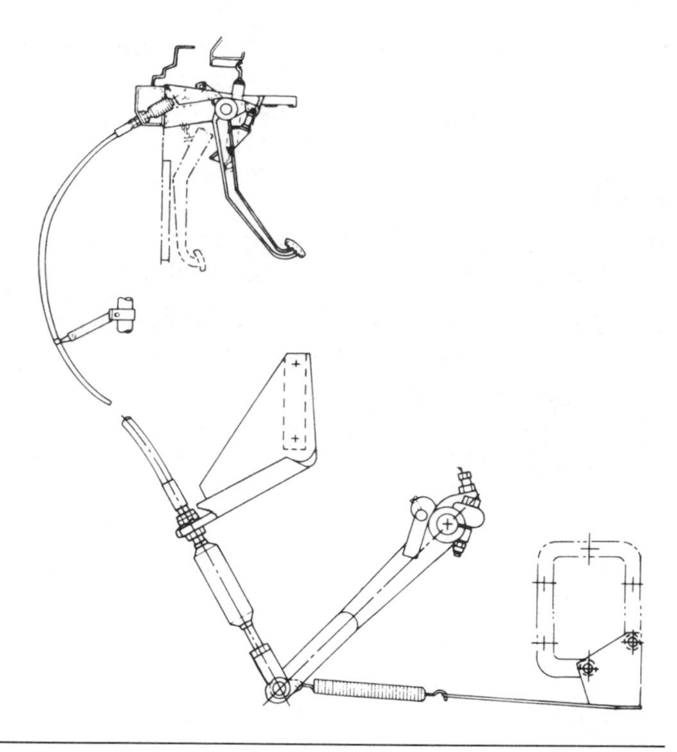

Figure 3–4 Clutch cable and linkage. *(Courtesy of Dana Corp.)*

Task 3

Inspect, adjust, repair, or replace hydraulic clutch slave and master cylinders, lines, and hoses; bleed system.

7. A driver complains that with the clutch pedal pressed all the way to the floor, the transmission will not disengage from the engine. The Technician has checked the fluid in the clutch master cylinder reservoir and found it to be above the MIN mark. The LEAST likely cause would be:
   A. poorly adjusted linkage.
   B. poor adjusting of the hydraulic slave cylinder.
   C. a frozen pilot bearing.
   D. a worn clutch disc.

8. On a vehicle with a hydraulic clutch, the following components are all in the system EXCEPT:
   A. the master cylinder.
   B. metal and flexible tubes.
   C. a slave cylinder.
   D. a clutch cable.

Hint

A typical clutch is controlled and operated by hydraulic fluid pressure and assisted by an air servo cylinder. More specifically, it consists of a master cylinder, hydraulic fluid reservoir, and an air-assisted servo cylinder. These components are all connected using metal and flexible tubes. When the clutch pedal is depressed, the plunger forces the piston in the master cylinder to move forward, causing the hydraulic fluid to act upon the air servo cylinder, which in turn activates the release fork.

Task 4

Inspect, adjust, or replace release (throw-out) bearing, sleeve, bushing, springs, levers, shafts, and seals.

9. When installing a pull-type clutch release bearing as shown in Figure 3–5, the LEAST likely inspection would be:
   A. to inspect the quality and amount of lubrication.
   B. to measure for proper free travel.
   C. to inspect for inner race spalling or scoring.
   D. to measure for clutch brake surface runout.

10. With the engine running, the transmission in neutral, and the clutch pedal partially or fully depressed a growling noise is heard. This noise disappears when the clutch pedal is released. The cause of this noise could be:
    A. a worn release bearing sleeve and bushing.
    B. a defective bearing on the transmission input shaft.
    C. a defective clutch release bearing.
    D. a worn, loose clutch release fork.

Hint

Both push-type and pull-type clutches are disengaged through the movement of a release bearing. The release bearing is a unit within the clutch consisting of bearings that mount on the transmission input shaft sleeve but do not rotate with it. A fork attached to the clutch pedal linkage controls the movement of the release bearing. As the release bearing moves, it forces the pressure plate away from the clutch disc.

Manually adjusted clutches have an adjusting ring that permits the clutch to be manually adjusted to compensate for wearing of the friction linings. The ring is positioned behind the pressure plate and is threaded into the clutch cover. A lock strap or lock plate secures the ring so that it cannot move. The levers are seated in the ring. When the lock strap is removed, the adjusting ring is rotated in the cover so that it moves toward the engine.

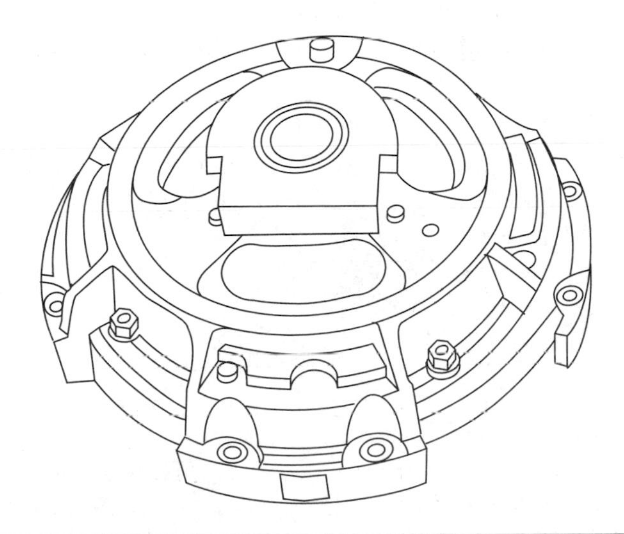

Figure 3–5 Pressure plate and release bearing. *(Courtesy of Rockwell Automotive)*

Task 5

Inspect, adjust, or replace single-disc clutch pressure plate assembly.

11. A burnt pressure plate may be caused by all of the following EXCEPT:
    A. oil on the clutch disc.
    B. not enough clutch pedal free play.
    C. binding linkage.
    D. a damaged pilot bearing.

12. What is the Technician measuring in Figure 3–6?
    A. Pressure plate runout
    B. Thickness of the pressure plate
    C. Thickness of the flywheel
    D. Flywheel runout

Hint

When the clutch is applied, the pressure plate springs jam the clutch disc friction surfaces between the pressure plate friction surface and the flywheel friction surface. As the clutch pedal is depressed to release the clutch, the cable and linkage force the clutch release bearing against the pressure plate levers or diaphragm. Because these levers are pivoted, the release bearing action moves the pressure plate friction surface away from the clutch disc friction surface to release the clutch. A worn pressure plate friction surface or weak pressure plate springs cause clutch slipping.

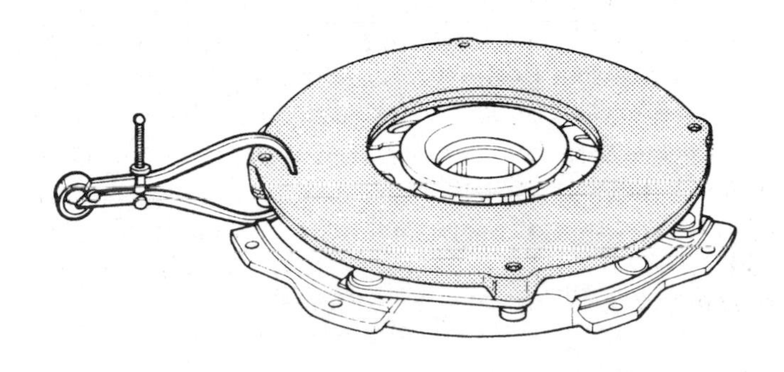

Figure 3–6 Clutch system component measurement. *(Courtesy of Rockwell Automotive)*

Task 6

Inspect and replace single-disc clutch disc.

13. When discussing single-disc clutch service, *Technician A* says a depth gauge may be used to measure the lining depth from the rivet heads to the clutch facing surface. *Technician B* says a single-disc clutch may be installed in either direction between the pressure plate and the flywheel. Who is right?
    A. A only
    B. B only
    C. Both A and B
    D. Neither A nor B

14. A single-disc clutch has a very harsh application each time the clutch pedal is released. The cause of this problem could be:
    A. a scored surface on the flywheel.
    B. broken and weak torsional springs.
    C. a scored surface on the pressure plate.
    D. worn clutch facings.

Hint

The disc linings are critical to clutch life and performance because they directly receive the full torque of the engine each time the clutch is engaged. There are basically two types of linings: organic and ceramic. Both types of linings should be inspected for thickness, hot spots, cracking, and uneven wear. The organic type should additionally be checked for oil or grease contamination.

Task 7

Inspect, adjust, measure, align, or replace two-plate clutch assembly (includes intermediate plate and drive pins).

15. A truck has a broken intermediate plate. *Technician A* says a broken intermediate plate can be caused by improper driver technique. *Technician B* says that a truck pulling loads that are too heavy can cause a broken intermediate plate. Who is right?
    A. A only
    B. B only
    C. Both A and B
    D. Neither A nor B

16. An intermediate plate shows cracks in the surface on only one side. All of the following could be causes EXCEPT:
    A. a release bearing that is not moving freely.
    B. an improperly manufactured friction disc.
    C. improper driver clutching technique.
    D. an improperly adjusted intermediate plate.

17. While discussing the clutch system in Figure 3–7, *Technician A* says this system is used for high-torque applications. *Technician B* says this clutch system uses an intermediate plate in the system. Who is right?
    A. A only
    B. B only
    C. Both A and B
    D. Neither A nor B

Hint

If the clutch has two driven discs, an intermediate plate or center plate separates the two clutch friction discs. This plate is machined smooth on both sides because it is pressed between two friction surfaces. An intermediate plate increases the torque capacity of the clutch by increasing the friction area, allowing more area for the transfer of torque.

Task 8

Inspect, adjust, or replace clutch brake assembly; inspect or replace input shaft splines.

18. To install a new pull-type clutch, a Technician will need to do all the following EXCEPT:
    A. align the clutch disc.
    B. adjust the self-adjusting release bearing.

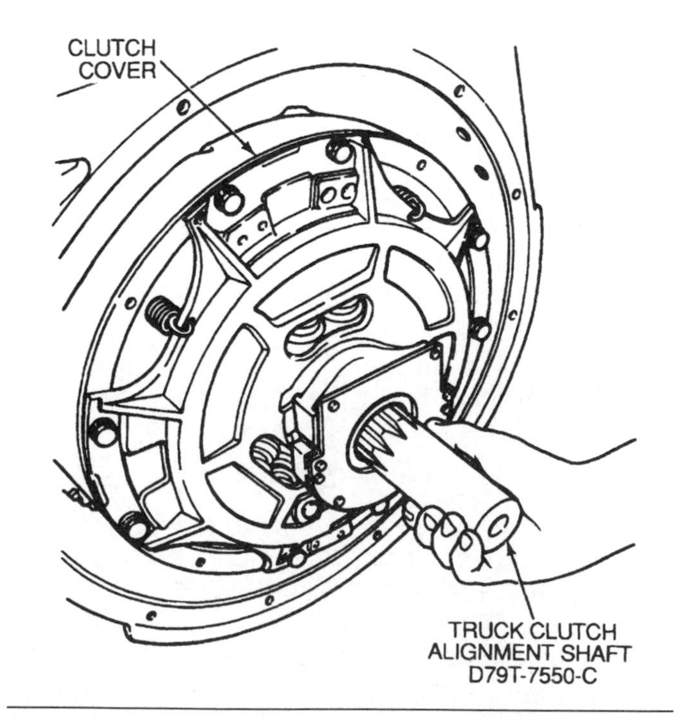

Figure 3–7 Truck clutch assembly. *(Courtesy of Ford Motor Company)*

C. resurface the limited torque clutch brake.
D. lubricate the pilot bearing.

19. *Technician A* says that a clutch brake needs to be inspected for wear and fatigue in the same manner as the pressure plate. *Technician B* says that as long as the component is in place, it will function properly. Who is right?
A. A only
B. B only
C. Both A and B
D. Neither A nor B

Hint

The clutch brake is a circular disc with a friction surface that is mounted on the transmission input spline shaft between the release bearing and the transmission. Its purpose is to slow or stop the transmission input shaft from rotating in order to allow gears to be engaged without clashing and to keep transmission gear damage to a minimum. Clutch brakes are used only on vehicles with nonsynchronized transmissions.

Task 9

Inspect, adjust, or replace self-adjusting clutch mechanisms.

20. A self-adjusting clutch is found to be out of adjustment. *Technician A* says the adjuster ring may be defective. *Technician B* says the clutch pedal linkage may be binding. Who is right?
A. A only
B. B only
C. Both A and B
D. Neither A nor B

21. *Technician A* says the wear compensator is a replaceable part. *Technician B* says the wear compensator keeps the free travel in the clutch pedal within specifications. Who is right?
A. A only
B. B only
C. Both A and B
D. Neither A nor B

Hint

The wear compensator is a replaceable component that automatically adjusts for facing wear each time the clutch is actuated. Once facing wear exceeds a predetermined amount, the wear compensator allows the adjusting ring to be advanced toward the engine, keeping the pressure plate to clutch disc clearance within proper operating specification. This also keeps the pedal free play adjustment within specification.

Task 10

Inspect pilot bearing; replace as necessary.

22. A damaged pilot bearing may cause a rattling or growling noise when:
A. the engine is idling and the clutch pedal is fully depressed and clutch released.
B. the vehicle is decelerating in high gear with the clutch pedal released and clutch engaged.
C. the vehicle is accelerating in low gear with the clutch pedal released and clutch engaged.
D. the engine is idling, the transmission is in neutral, and the clutch pedal is released and clutch engaged.

23. While discussing the removal of the pilot bearing, *Technician A* says a slide hammer can be used to remove the bearing. *Technician B* says you can also use a drill motor to remove the pilot bearing. Who is right?
A. A only
B. B only
C. Both A and B
D. Neither A nor B

Hint

Every time the clutch assembly is serviced or the engine is removed, the pilot bearing in the flywheel should be removed and replaced. Use an internal puller or a slide hammer to remove the pilot bearing. The best way to ensure proper installation of the transmission input shaft into the pilot bearing is to use wheel bearing grease on the rollers of the pilot bearing. This will keep the rollers in place long enough to install the transmission input shaft into place.

Task 11

Inspect flywheel mounting area on crankshaft and check crankshaft end play; determine needed repairs.

24. In Figure 3–8 the Technician is measuring:
A. crankshaft end play.
B. flywheel runout.
C. flywheel housing bore runout.
D. runout on the flywheel housing outer surface.

25. The most common damage that occurs on the flywheel mounting surface is:
A. cracking of the boltholes.
B. elongating of the boltholes.
C. warping of the mounting surface.
D. heat checking and pitting.

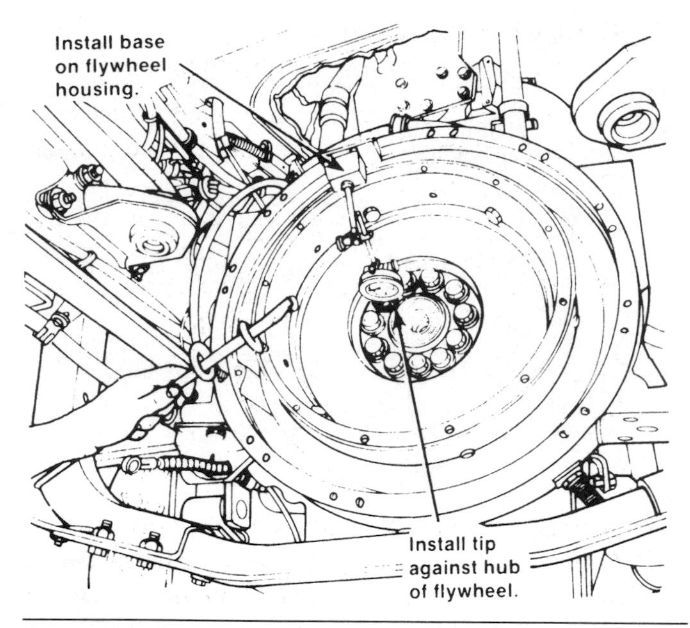

Figure 3–8 Dial indicator measurement. *(Courtesy of Rockwell Automotive)*

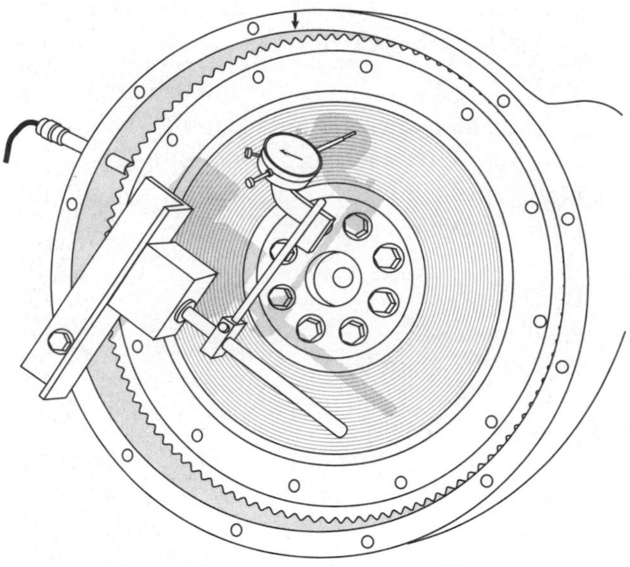

Figure 3–9 Flywheel measurement. *(Courtesy of Dana Corp.)*

Hint

Inspect the surface of the flywheel for wear or damage. Make sure the flywheel is not cracked. Heat marks are a normal operating condition that can be removed with an emery cloth. Some wear or damage can be removed by grinding a new surface on the flywheel. If wear or damage on the surface of the flywheel cannot be removed, the flywheel must be replaced.

Task 12

Inspect, measure, service, or replace flywheel and starter ring gear.

26. When changing a flywheel ring gear, what steps should be taken?
    A. Cool the ring gear in a freezer overnight.
    B. Heat the ring gear in an oven to 400°F (204°C).
    C. Cool the ring gear and heat the flywheel.
    D. Heat the ring gear and cool the flywheel.

27. While performing the measurement in Figure 3–9, *Technician A* says the crankshaft should be pulled rearward and then rotated. *Technician B* says machining an excessive amount of metal from the flywheel friction surface may adversely affect clutch operation. Who is right?
    A. A only
    B. B only
    C. Both A and B
    D. Neither A nor B

Hint

Inspect the teeth of the ring gear on the outer surface of the flywheel. If the teeth are worn or damaged, replace the ring gear or the flywheel and inspect the starter drive teeth. If there is any damage evident on the starter drive gear, the starter or starter drive must be replaced.

Task 13

Measure flywheel face runout and pilot bore runout; determine needed repairs.

28. When performing the measurement in Figure 3–10, the maximum runout on the dial indicator is:
    A. 0.003 in. (0.073 mm).
    B. 0.005 in. (0.127 mm).
    C. 0.008 in. (0.203 mm).
    D. 0.010 in. (0.254 mm).

29. The bore runout measure in Figure 3–10 is excessive. *Technician A* says the flywheel must be replaced. *Technician B* says the bore in the flywheel may be built up and machined to the specified bore size. Who is right?
    A. A only
    B. B only
    C. Both A and B
    D. Neither A nor B

30. *Technician A* says that pilot bore runout should be measured when a vehicle makes a grinding noise that cannot be identified in any other drivetrain member. *Technician B* says that pilot bore runout

Figure 3–10 Clutch system measurement. *(Courtesy of Dana Corp.)*

should be measured only when the pilot bearing is found to be in poor shape. Who is right?

A. A only
B. B only
C. Both A and B
D. Neither A nor B

Hint

The acceptable runout on the outer surface of the flywheel is a specified amount multiplied by the diameter of the flywheel in inches. For example, maximum permissible runout may be listed as 0.0005 inch (0.0127 mm) per inch of flywheel diameter with the total indicated difference between the high and low points being 0.007 inch (0.178 mm) or less for a 14-inch (356 mm) clutch (0.0005 [0.0127 mm] x 14 [356 mm] = 0.007 [0.178 mm]). For a 15.5-inch (394 mm) clutch the total allowable variation in flywheel surface is 0.008 inch (0.203 mm) or less. The maximum pilot bore runout in the crankshaft is usually 0.005 inch (0.127 mm). Always check service manual specifications for exact tolerances.

Task 14

Inspect engine block, flywheel housing(s), and transmission housing mating surface(s); determine needed repairs.

31. Flywheel housing face misalignment may cause:
    A. a growling noise with the clutch pedal depressed.
    B. the transmission to jump out of gear.
    C. clutch chatter and grabbing.
    D. wear on the clutch release bearing.

32. A Technician measures and finds the flywheel housing bore face runout to be out of specification. The likely cause is:
    A. overtightening of the transmission, causing undue pressure on the housing face.
    B. extreme overheating of the clutch, causing warpage in the flywheel housing.
    C. an improper torque sequence by the previous Technician.
    D. a manufacturing imperfection.

Hint

The mating surfaces of the transmission clutch housing and the engine flywheel housing should be inspected for signs of wear or damage. Any appreciable wear on either housing will cause misalignment. Most wear is found on the lower half of these surfaces, with the most common wear occurring between the 3 o'clock and 8 o'clock positions. If any signs of wear are evident, the housing must be replaced.

Task 15

Measure flywheel housing bore runout and face runout; repair as necessary.

33. *Technician A* says that a worn clutch housing can cause misalignment of the mating surfaces of the clutch assembly. *Technician B* says that if the clutch housing is worn it can be repaired and reused. Who is right?
    A. A only
    B. B only
    C. Both A and B
    D. Neither A nor B

34. When measuring flywheel housing runout, what should the Technician do first?
    A. Attach a dial indicator to the center of the flywheel.
    B. Attach a dial indicator to the crankshaft.
    C. Attach a dial indicator to the input shaft.
    D. Attach a dial indicator to the clutch housing.

Hint

When measuring flywheel housing face or bore runout, first attach a dial indicator to the center of the flywheel. Zero the needle on whichever surface is being measured. Turn the flywheel and take special note of the readings, using soapstone, or another similar marker to indicate high and low points. As with other runout measurements, subtract the low measurement from the high measure-

ment to get the runout dimension. For both runout dimensions there should be no more than approximately 0.010 inch (0.254 mm) of runout. If runout value is more than specified, service as necessary.

Transmission Diagnosis and Repair

Task 1

Diagnose transmission component failure cause(s), both before and during disassembly procedures; determine needed repairs.

35. A broken transmission mount can cause any of the following EXCEPT:
    A. a thudding noise each time the clutch pedal is released during a shift.
    B. vibration at highway speeds.
    C. vibration at speeds below 30 mph (48 km/h).
    D. a growling noise at speeds below 50 mph (80 km/h).

36. The damage to the gear in Figure 3–11 may have been caused by:
    A. improper handling outside of the transmission.
    B. the transmission "walking" between gears.
    C. worn sleeve-type bearings.
    D. normal wear.

37. A Technician is dismantling a transmission counter shaft and notices that a bearing outer race is marred as shown in Figure 3–12, but the bearing race bore is not scored. What could cause this type of bearing race damage?
    A. Dirty transmission fluid
    B. Normal vibration of the transmission
    C. The bearing has been "spun."
    D. Manufacturing defects

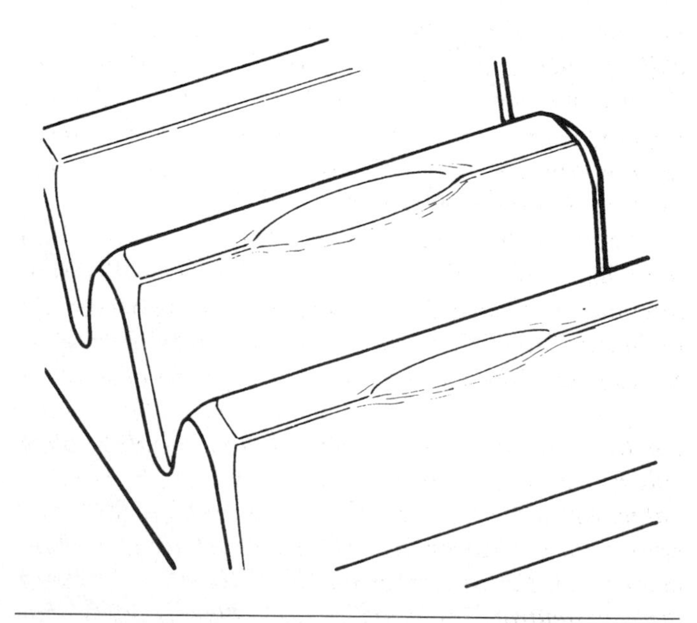

Figure 3–11 Gear wear.

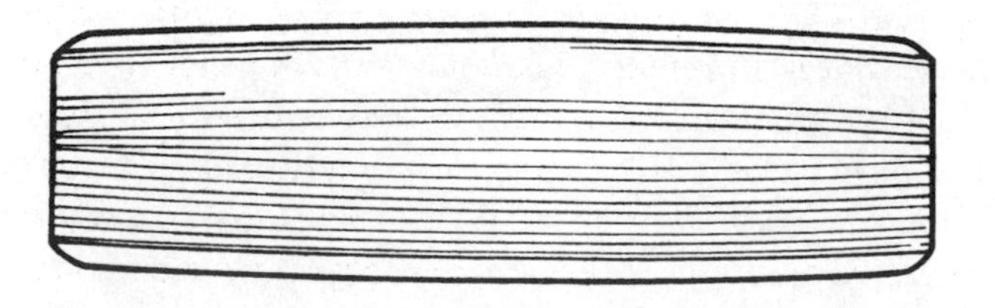

Figure 3–12 Bearing race defect.

Hint

Overheating of transmission gears and bearings may be caused by low lubricant level, an improper lubricant, or overloading the truck. Chipped and damaged transmission gears may be caused by improper shifting of the transmission, or improper clutch release during shifts. Improper transmission shifting by the driver may also result in excessive wear on synchronizer assemblies.

Task 2

Diagnose transmission noise, shifting, lockup, jumping out of gear, overheating, and vibration problems; determine needed repairs.

38. A transmission jumps out of gear while traveling down the road. The following are all possible causes EXCEPT:
    A. worn bearings that allow excessive shaft end play.
    B. worn detents on the shift rails.
    C. worn gear teeth on synchronizers.
    D. excessive crankshaft end play.

39. Which of the following is LEAST likely to cause noise in a manual transmission?
    A. A broken detent spring
    B. A worn or pitted input bearing
    C. A worn or pitted output bearing
    D. A worn counter shaft bearing.

Hint

The truck should be road tested to determine if the driver's complaint of noise is actually in the transmission. Also, Technicians should try to locate and eliminate noise by means other than transmission removal or overhaul. If the noise does seem to be in the transmission, try to break it down into classifications. If possible, determine what position the gearshift lever is in when the noise occurs. If the noise is evident in only one gear position, the cause of the noise is generally traceable to the gears in operation. Jumping out of gear is usually caused by excessive end play on gears or synchronizer assemblies. This problem may also be caused by weak or broken detent springs and worn detents on the shifter rails.

Task 3

Inspect, adjust, repair, or replace transmission remote shift linkages, cables, brackets, bushings, pivots, and levers.

40. *Technician A* says the best way to test for a binding or stuck shift linkage is to shift between gear positions with the truck standing still. If there is any resistance while shifting into gear, the shift linkage is binding. *Technician B* says you have to disconnect the linkage at the transmission and check the linkage inside the transmission separately from checking the linkage outside the transmission. Who is right?
    A. A only
    B. B only
    C. Both A and B
    D. Neither A nor B

41. *Technician A* says that if the detents in the shifting tower are not aligned, it could cause clutch wear. *Technician B* says that on a cab-over-engine (COE) tractor, a remote shift lever is connected to the shift bar housing on top of the transmission. Who is right?
    A. A only
    B. B only
    C. Both A and B
    D. Neither A nor B

Hint

Manual adjustment of the manual gear range selector valve linkage is important. The shift tower detents must correspond exactly to those in the transmission. Failure to obtain proper detent in DRIVE, NEUTRAL, or REVERSE gears can adversely affect the supply of transmission oil at the forward or fourth (reverse) clutch. The resulting low-apply pressure can cause clutch slippage and decreased transmission life.

Task 4

Inspect, test operate, adjust, repair, or replace air shift controls, lines, hoses, valves, regulators, filters, and cylinder assemblies.

42. A truck equipped with a pneumatic high/low shift system shown in Figure 3–13 will not shift into high range. What may be the cause?
    A. A dirty or plugged air filter
    B. A blown fuse
    C. A worn synchronizer in the auxiliary portion of the transmission
    D. Worn gear teeth

43. The range shift system shown in Figure 3–13 only allows range shifts to occur when the transmission is:
    A. operating in any forward gear.
    B. operating in any forward gear during engine deceleration.
    C. in neutral or passing through neutral.
    D. in a forward gear and the engine speed is above 1,000 rpm.

Hint

In an air shift system the air filter prevents dirt and moisture from entering the system. The filter is combined with a pressure regulator that maintains system pressure from 57 to 62 psi (393 to 427 kPa). The range valve controls air pressure applied to the slave valve. The slave valve directs air pressure to the appropriate port on the range shift cylinder to supply the shift selected by the driver. If the driver selects high range, the air pressure from the range valve supplies air pressure to the appropriate side of the slave valve, and this valve moves so it supplies air pressure to the high range side of the piston in the range shift cylinder. This air pressure on the range shift cylinder piston moves the piston and yoke bar forward to the high range position. Under this condition air pressure on the low range side of the range shift cylinder piston is exhausted through the slave valve.

When the driver selects low range, air pressure from the range valve moves the slave valve so this valve supplies air pressure to the low range port on the range shift cylinder. This action supplies air pressure to the low range side of the range shift cylinder piston and moves this piston rearward to provide low range. Under this condition air pressure is exhausted from the high range side of the range shift cylinder piston through the slave valve. An air valve shaft extending from the shift bar housing to the slave valve only allows slave valve movement when the transmission is in neutral or passing through neutral. Therefore, range shifts can only occur under this condition.

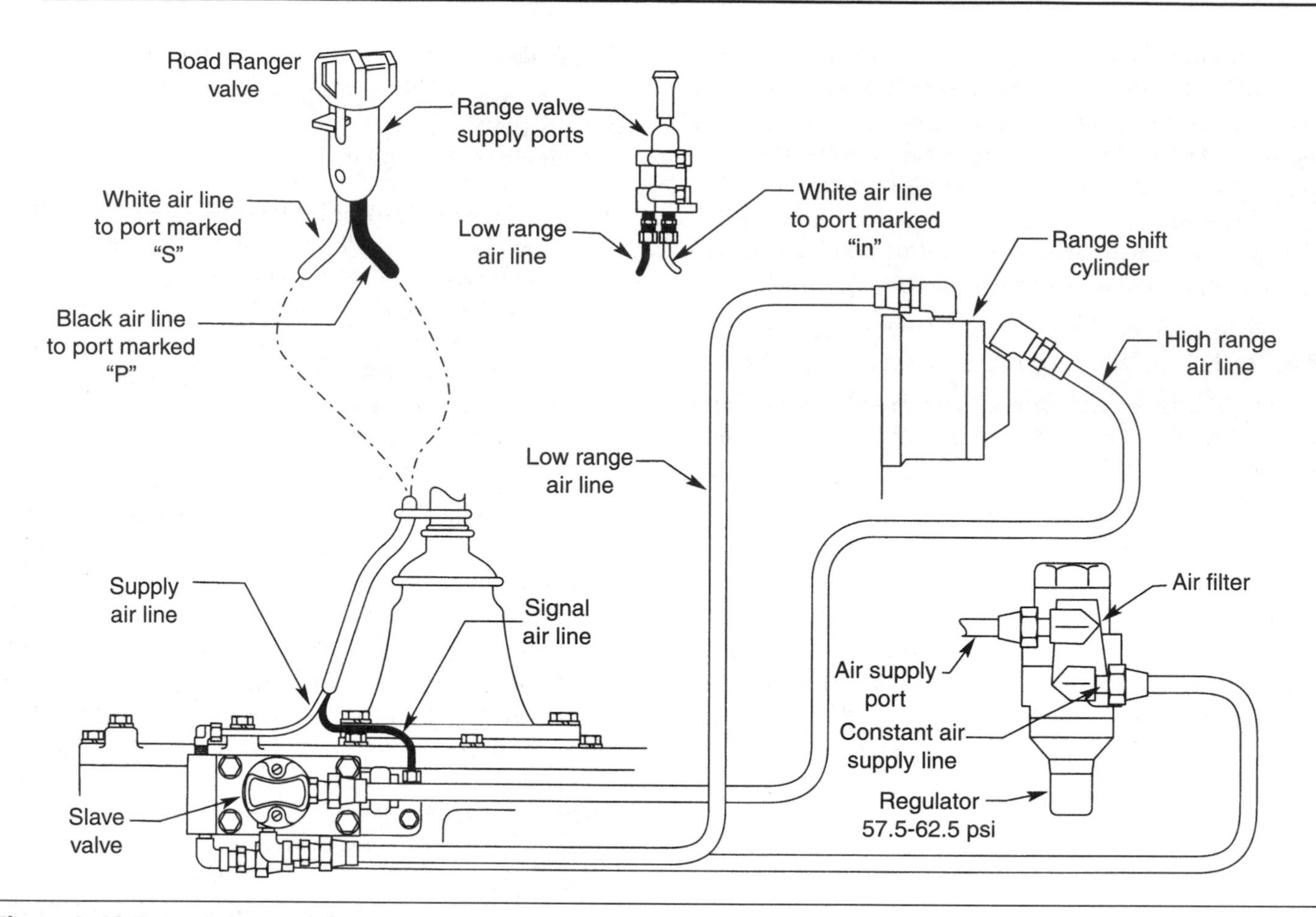

Figure 3–13 Transmission air shift system.

Task 5

Inspect or replace transmission mounts, insulators, and mounting bolts.

44. To thoroughly inspect a transmission mount, a Technician should:
    A. remove the mount and put opposing tension on the two mounting plates while inspecting for cracking or other signs of damage.
    B. remove the mount and place the mount into a vise while inspecting for cracking or other signs of damage.
    C. visually inspect the mount while still in the vehicle.
    D. replace the mount if it is suspect.

45. *Technician A* says that transmission mounts are used to absorb torque from the engine. *Technician B* says that transmission mounts absorb drivetrain vibration. Who is right?
    A. A only
    B. B only
    C. Both A and B
    D. Neither A nor B

Hint

Transmission mounts and insulators play an important role in keeping drivetrain vibration from transferring to the chassis of the vehicle. If the vibration were allowed to transmit to the chassis of the vehicle, the life of the vehicle would be reduced. Driving comfort is another reason why insulators are used in transmissions. The most important reason is the ability of the mounts to absorb shock and torque. If the transmission was mounted directly to a stiff and rigid frame, the entire torque associated with hauling heavy loads would need to be absorbed by the transmission and its internal components, causing increased damage and a much shorter service life. Broken transmission mounts are not readily identifiable by any specific symptoms. They should be visually inspected any time a Technician is working near them.

Task 6

Inspect for leakage and replace transmission cover plates, gaskets, seals, and cap bolts as necessary; inspect seal surfaces.

46. When replacing an output shaft seal in a manual transmission, a Technician should always check the following EXCEPT:
    A. excessive output shaft radial movement.
    B. excessive wear on transmission gears.
    C. proper transmission venting.
    D. wear on the sealing surface contacting the seal lip.

47. While inspecting a transmission for leaks, the Technician notices that the gaskets appear to be blown out of their mating sealing surfaces. What should the Technician check first?
    A. The transmission breather
    B. The shifter cover
    C. The type of lubricant
    D. The rear seal

Hint

In diagnosing and correcting fluid leaks, finding the exact cause of the leakage can be difficult because evidence of the leakage may occur in an area other than the source of the leakage. Always check the transmission breather filters when repairing any transmission leaks. Under normal operating conditions the transmission can build up pressure inside the case if the vents or breathers are not functioning correctly. This pressure can cause fluid to leak past seals that do not need replacement or seals that are otherwise in good working order.

Task 7

Check transmission fluid level, type, and condition; determine needed service.

48. What is the most common cause of bearing failure in a transmission?
    A. Extended high torque operating conditions
    B. Dirt and contaminants in the lubricant
    C. Operating the tractor in high temperature situations
    D. Poor quality of lubricant

49. *Technician A* says overfilling a manual transmission could result in overheating of the transmission. *Technician B* says overfilling a transmission could cause excessive aeration of the transmission fluid. Who is right?
    A. A only
    B. B only
    C. Both A and B
    D. Neither A nor B

50. When checking transmission fluid level on a manual transmission, what is the proper procedure?
    A. Follow the guidelines stamped on the transmission dipstick.
    B. Check for proper oil level by using your finger to feel for oil through the filler plug hole.
    C. Make sure the oil level is even with the bottom of the filler plug hole.
    D. Remove the specified rear housing retaining bolt, and check for proper fluid level at the bottom of this bolt hole.

Hint

Most manufacturers suggest a specific grade and type of transmission oil, heavy-duty engine oil, or straight mineral oil, depending on the ambient air temperature during operation. Do not use mild EP gear oil or multipurpose gear oil when operating temperatures are above 230°F (110°C). Many of these gear oils break down above 230°F (110°C) and coat seals, bearings, and gear with deposits that might cause premature failures. If these deposits are observed (especially on seal areas where they can cause oil leakage), change to heavy-duty engine oil or mineral gear oil to ensure maximum component life.

Always follow the manufacturer's exact hydraulic fluid specifications. For example, several transmission manufacturers recommend DEXRON, DEXRON II, and type C-3 (ATD approved SAE 10W or SAE 30) oils for their automatic transmissions. Type C-3 fluids are the only fluids usually approved for use in off-highway applications. Type C-3 SAE 30 is specified for all applications where the ambient temperature is consistently above 86°F (30°C). Some, but not all, DEXRON II fluids also qualify as type C-3 fluids. If type C-3 fluids must be used, be sure all materials used in tubes, hoses, external filters, and seals are C-3 compatible.

Task 8

Inspect, adjust, or replace transmission shift lever, cover, rails, forks, levers, bushings, sleeves, detents, interlocks, springs, and lock bolts/safety wires.

51. In Figure 3–14 *Technician A* says component H is mounted between the shift rail openings in the transmission. *Technician B* says the shift tower in Figure 3–14 is from a tractor with the engine compartment in front of the cab. Who is right?
    A. A only
    B. B only

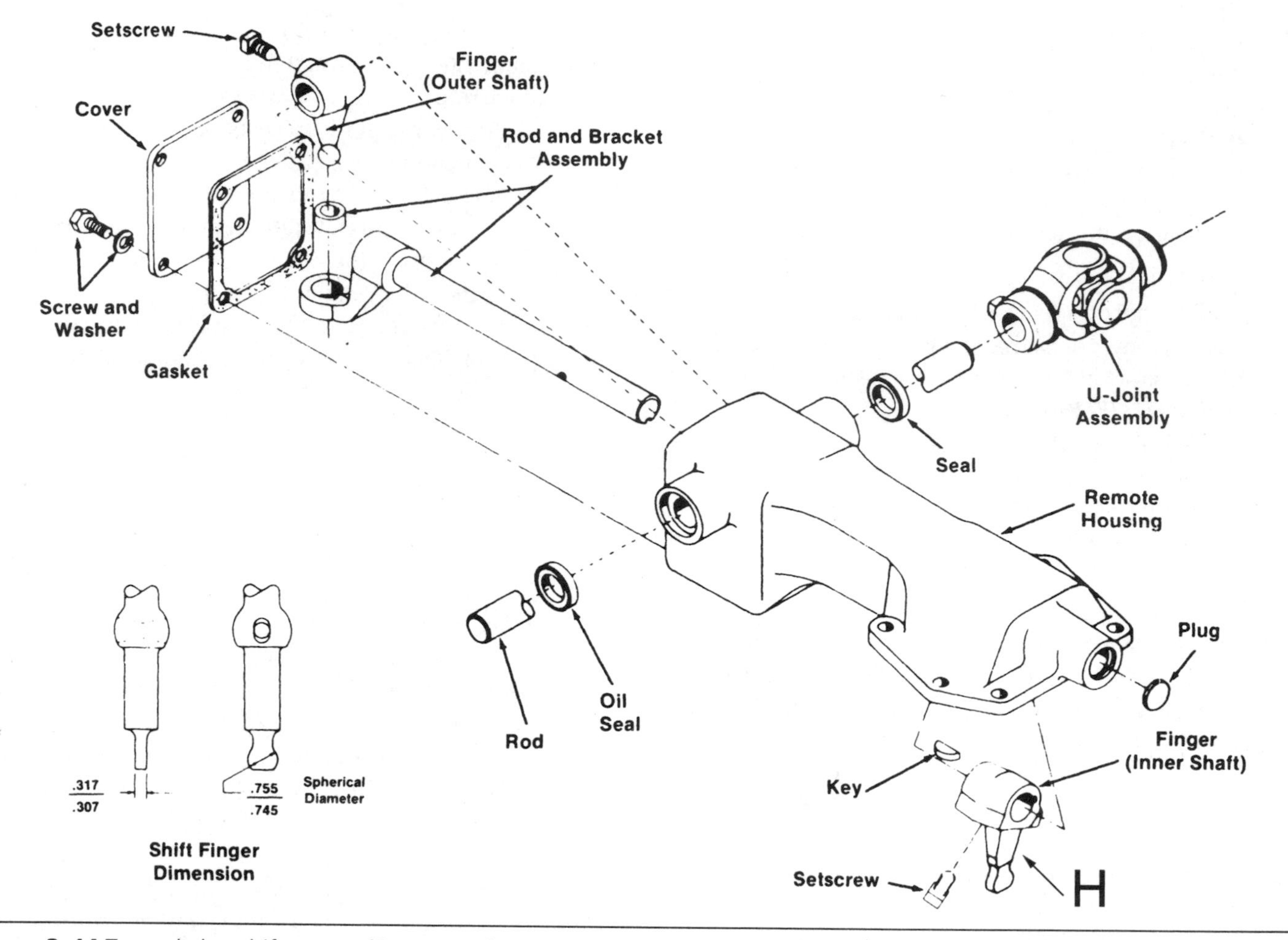

Figure 3–14 Transmission shift tower. *(Courtesy of Dana Corp.)*

C. Both A and B
D. Neither A nor B

52. The LEAST likely reason for a single counter shaft transmission to jump out of fifth gear would be:
    A. damaged friction rings on the synchronizer blocker rings.
    B. a broken detent spring.
    C. a worn shift fork.
    D. worn blocker ring teeth and worn dog teeth on the fifth speed gear.

Hint

When a sliding clutch is moved to engage with a main shaft gear, the mating teeth must be parallel. Tapered or worn clutching teeth try to "walk" apart as the gears rotate, causing the sliding clutch and gear to slip out of engagement. Slipout generally occurs when pulling with full power or decelerating with the load pushing. Different from slipout, jumpout occurs when a fully engaged gear and sliding clutch are forced out of engagement. It generally occurs when a force sufficient to overcome the detent spring pressure is applied to the yoke bar, moving the sliding clutch to a neutral position. Keep in mind that the whipping action of extra long or heavy shift levers can cause the transmission to jump out of gear.

Task 9

Inspect or replace input shaft, gear, spacers, bearings, retainers, and slingers.

53. The LEAST likely damage to check for on an input shaft is:
    A. pilot bearing shaft cracking.
    B. gear teeth damage.
    C. input spline damage.
    D. cracking or other fatigue wear to the input shaft spline.

54. A growling noise occurs in a ten-speed manual transmission with the engine running, the transmission in neutral, and the clutch pedal released. The noise disappears when the clutch pedal is fully depressed. *Technician A* says the

bearing that supports the input shaft in the transmission housing may be worn. *Technician B* says the needle bearings that support the front of the output shaft in the rear of the input shaft may be scored. Who is right?

A. A only
B. B only
C. Both A and B
D. Neither A nor B

Hint

The input shaft can be affected by many different parts of a drivetrain. On a vehicle with a manual transmission, the input shaft fits into the pilot bearing and is splined into the clutch disc or discs. This vital part of the drivetrain should be inspected for any abnormal wear of the splines, as well as the gear portion of the input shaft. This is one of the few parts of the drivetrain that carries the entire torque load of the vehicle.

Task 10

Inspect main shaft, gears, sliding clutches, washers, spacers, bushings, bearings, auxiliary drive assemblies, retainers, and keys; determine needed repairs.

55. The dial indicator in Figure 3–15 is installed to measure:
    A. counter shaft end play.
    B. input shaft end play.
    C. output shaft runout.
    D. output shaft end play.

56. In Figure 3–15 the end play is more than specified. *Technician A* says to remove one of the shims between the rear bearing end cover and the transmission case. *Technician B* says to install a thicker shim between the rear of the input shaft and the front of the output shaft. Who is right?
    A. A only
    B. B only
    C. Both A and B
    D. Neither A nor B

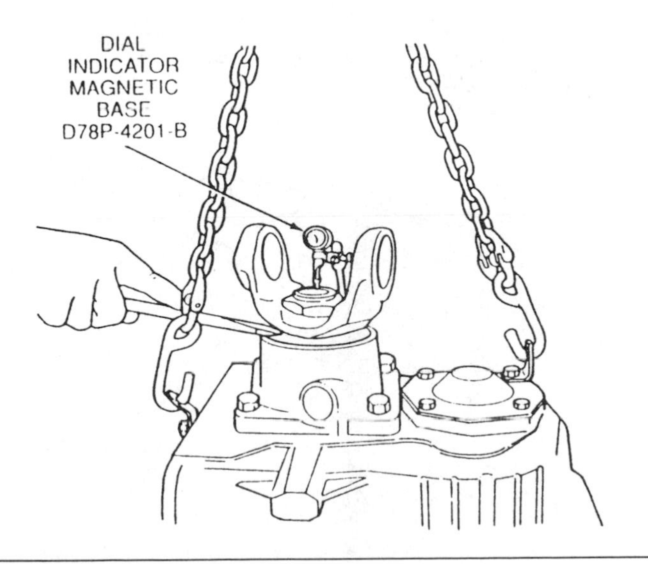

Figure 3–15 End play measurement. *(Courtesy of Ford Motor Company)*

Hint

Two types of gear damage are bottoming and climbing. Bottoming occurs when the teeth of one gear touch the lowest point between the teeth of a mating gear. Bottoming does not occur in a two-gear drive combination but can occur in multiple-gear drive combinations. A simple two-gear drive combination always tends to force the two gears apart; therefore, bottoming cannot occur in this arrangement. Climbing is caused by excessive wear in gears, bearings, and shafts. It occurs when the gears move sufficiently apart to cause the apex (or point) of teeth on one gear to climb over the apex of the teeth on another gear with which it is meshed. This results in a loss of drive until other teeth are engaged, and causes rapid destruction of the gears.

Task 11

Inspect counter shafts, gears, bearings, retainers, and keys; adjust bearing preload/clearance, and time multiple counter shaft gears; determine needed repairs.

57. When timing a three-counter shaft transmission as shown in the Figure 3–16, what should the Technician do?
    A. Align the timing marks provided by the manufacturer.

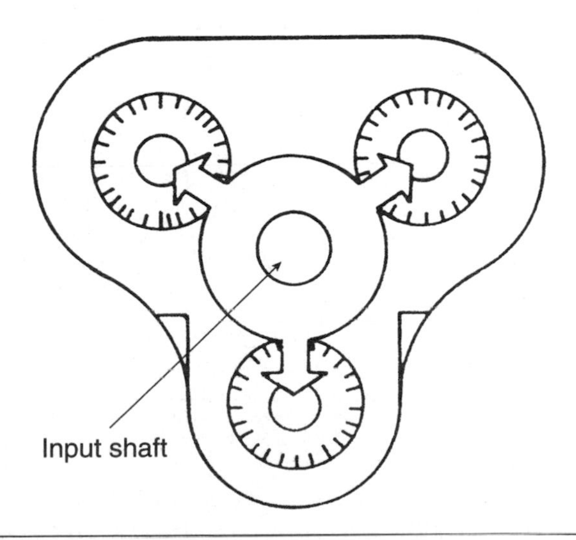

Figure 3–16 Three-counter shaft transmission.

B. Mark the gears before disassembly, then align those marks during assembly.
C. Align the keyway so that all counter shaft keyway align with the main shaft.
D. Align the timing tooth of each counter shaft with the corresponding timing mark on the main shaft.

58. The end play in Figure 3–17 is less than specified. To correct this problem it is necessary to:
A. install a thicker shim between the rear counter shaft bearing retainer and the rear counter shaft outer bearing race.
B. install a thinner shim between the front counter shaft bearing retainer and the front counter shaft outer bearing race.
C. install a thinner spacer between the rear of the counter shaft and the transmission case.
D. install a thicker snap ring on the front of the counter shaft.

Hint

All twin counter shaft transmissions are "timed" at assembly. It is important that the manufacturer's timing procedures are followed when reassembling the transmission. Timing ensures that the counter shaft gears contact the mating main shaft gears at the same time, allowing main shaft gears to center on the main shaft and equally divide the load. Timing is a simple procedure of marking the appropriate teeth of a gearset before removal (while still in the transmission). In the front section, it is necessary to time only the drive gearset. Depending on the model, the low range, deep reduction, or splitter gearset is timed in the auxiliary section.

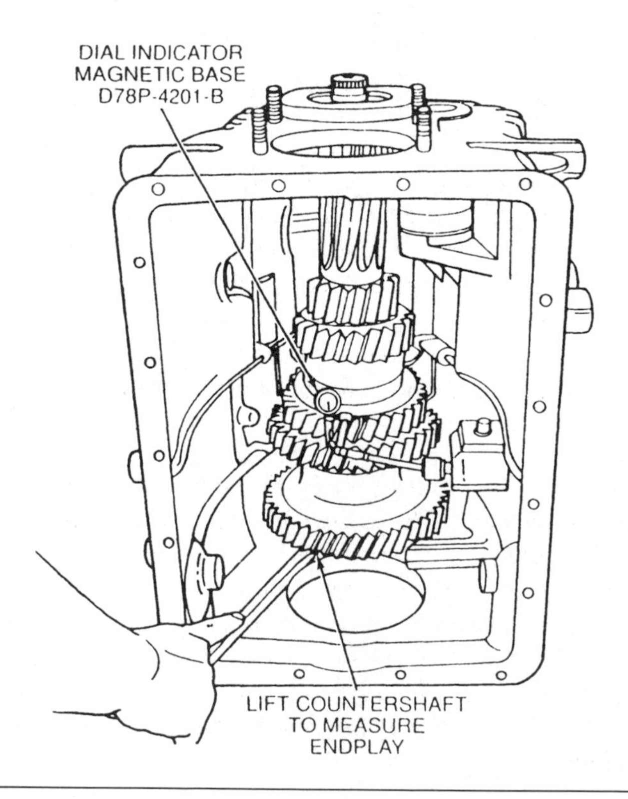

Figure 3–17 Measuring transmission end play. *(Courtesy of Ford Motor Company)*

Task 12

Inspect output shafts, gears, washers, spacers, bearings, retainers, and keys; determine needed repairs.

59. Component R being installed on the transmission output shaft in Figure 3–18 is a:
A. bearing end play spacer.
B. speedometer rotor.
C. bearing retainer.
D. output shaft yoke spacer.

60. The output shaft shown in Figure 3–18 is bent. *Technician A* says this condition may cause driveline vibration. *Technician B* says this condition may cause premature wear on the output shaft support bearing in the transmission case. Who is right?
A. A only
B. B only
C. Both A and B
D. Neither A nor B

Hint

Leakage in transmission rear seals is perhaps the most common problem in truck transmissions. The problem is more than a nuisance because if not repaired, a leaking seal can lead to catastrophic transmission failure. On the other hand, it can be very time-consuming and expensive

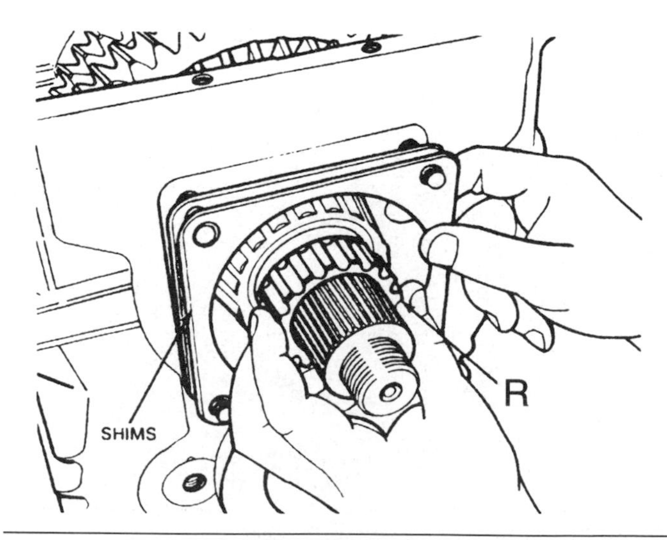

Figure 3–18 Transmission output shaft and related components. *(Courtesy of Ford Motor Company)*

to replace a rear seal system only to find the seal system was not causing the leak. More often, rear main leaks are due to worn or damaged output shaft bearings or thrust washers.

Task 13

Inspect reverse idler shafts, gears, bushings, bearings, thrust washers, and retainers; check reverse idler gear end play (where applicable); determine needed repairs.

61. In the manual transmission shown in Figure 3–19, the reverse idler gear is:
    A. item 14.
    B. item 15.
    C. item 20.
    D. item 21.

62. With the manual transmission gears positioned as illustrated in Figure 3–19, the transmission is in:
    A. reverse gear.
    B. low gear.
    C. second gear.
    D. third gear.

63. The manual transmission in Figure 3–19 is noisy only in reverse gear. *Technician A* says the teeth on the low reverse sliding gear may be worn and chipped. *Technician B* says the reverse idler gear teeth may be worn and chipped. Who is right?
    A. A only
    B. B only
    C. Both A and B
    D. Neither A nor B

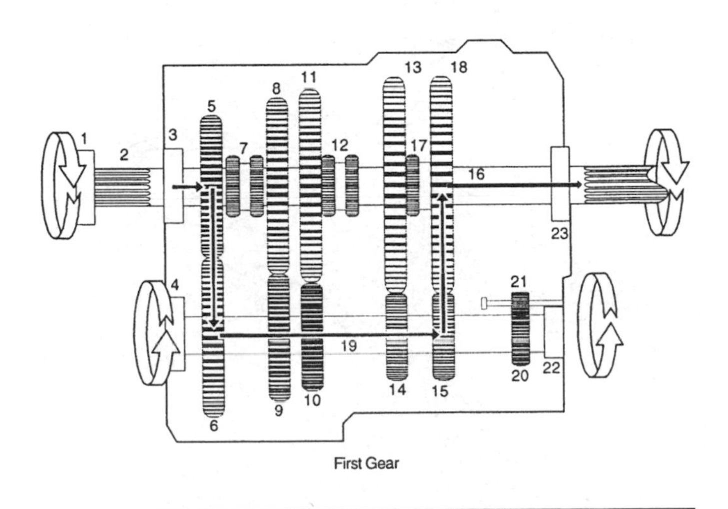

Figure 3–19 Manual truck transmission.

Hint
More than 90 percent of all ball bearing failures are caused by dirt, which is always abrasive. Dirt can enter the bearings during assembly of the units or be carried into the bearing by the lubricant while in service. Dirt can enter through the seals through the breather or from dirty containers used for the addition or change of lubricant. One of the few ways to measure for bearing failure while the transmission is still assembled is to spin the shaft or check for any vertical movement, particularly toward the ends of the shaft.

Task 14

Inspect synchronizer hub, sleeve, keys (inserts), springs, blocking rings, synchronizer plates, blocker pins, and sliding clutches; determine needed repairs.

64. A vehicle that uses cone-type synchronizers has a worn blocker ring and worn dog teeth on the matching gear. *Technician A* says that you must replace both the gear and the blocker ring to restore proper operation. *Technician B* says that because the blocker ring slows down the gear before the shift, you only need to replace the blocker ring. Who is right?
    A. A only
    B. B only
    C. Both A and B
    D. Neither A nor B

65. While inspecting synchronizer assemblies:
    A. the dog teeth on the blocker rings should be flat with smooth surfaces.
    B. the threads on the cone area of the blocker rings should be sharp and not dulled.
    C. the clearance is not important between the blocker rings and the matching gear's dog teeth.
    D. the sleeve should fit snugly on the hub and offer a certain amount of resistance to movement.

Hint
Check the synchronizer for burrs, uneven and excessive wear at contact surfaces, and metal particles. Check the blocker pins for excessive wear or looseness. Check the synchronizer contact surfaces for excessive wear. If the vehicle is equipped with cone-type synchronizers, check to see that the blocker ring is within tolerance by twisting the ring onto the matching gear cone. If the blocker ring "locks" itself onto the gear surface, the ring is still useable.

Task 15

Inspect the transmission case, including mating surfaces, bores, bushings, pins, studs, magnetic plugs, and vents; determine needed repairs.

66. The best way to clean a transmission case breather is to:
    A. replace the breather.
    B. soak the breather in gasoline.
    C. use solvent and blow-dry with compressed air.
    D. use a rag to wipe the orifice clean.

67. In Figure 3–20, item 16 is:
    A. the transmission breather.
    B. a transmission cover capscrew.
    C. a detent ball and spring plug.
    D. an air valve shaft retainer plug.

Hint

Two of the more simple items on a transmission that often get overlooked during servicing are the transmission case and breather(s). The transmission case must be checked for any signs of fatigue. Cracking is a symptom that is usually accompanied by fluid leakage. Plugged breathers are also associated with fluid leakage but not at the location of the breather. When a breather becomes plugged, fluid is often forced past seals in the transmission. If a Technician jumps to a conclusion when he or she notices fluid leakage at a seal, he or she may mistakenly replace the seal and think the problem is solved. A thorough job requires the Technician to check the transmission breathers during any transmission diagnosis.

Task 16

Inspect, service, and replace transmission lubrication system components (i.e., pumps, troughs, collectors, slingers, coolers, filters, and lines).

68. To remove an automatic transmission oil pump, a Technician must:
    A. remove the transmission, then the torque converter, then the oil pump.
    B. remove the transmission pan and filter, then remove the oil pump.
    C. remove the transmission pan and filter, then remove the main control valve body, then the oil pump.
    D. remove the transmission, then remove the torque converter, then remove the bell housing, then remove the oil pump.

69. An automatic truck transmission has been overhauled because of burned clutch plates. *Technician A* says the external transmission oil cooler should be flushed with an agitator-type cleaner. *Technician B* says the torque converter must be flushed using an agitator-type cleaner. Who is right?
    A. A only
    B. B only
    C. Both A and B
    D. Neither A nor B

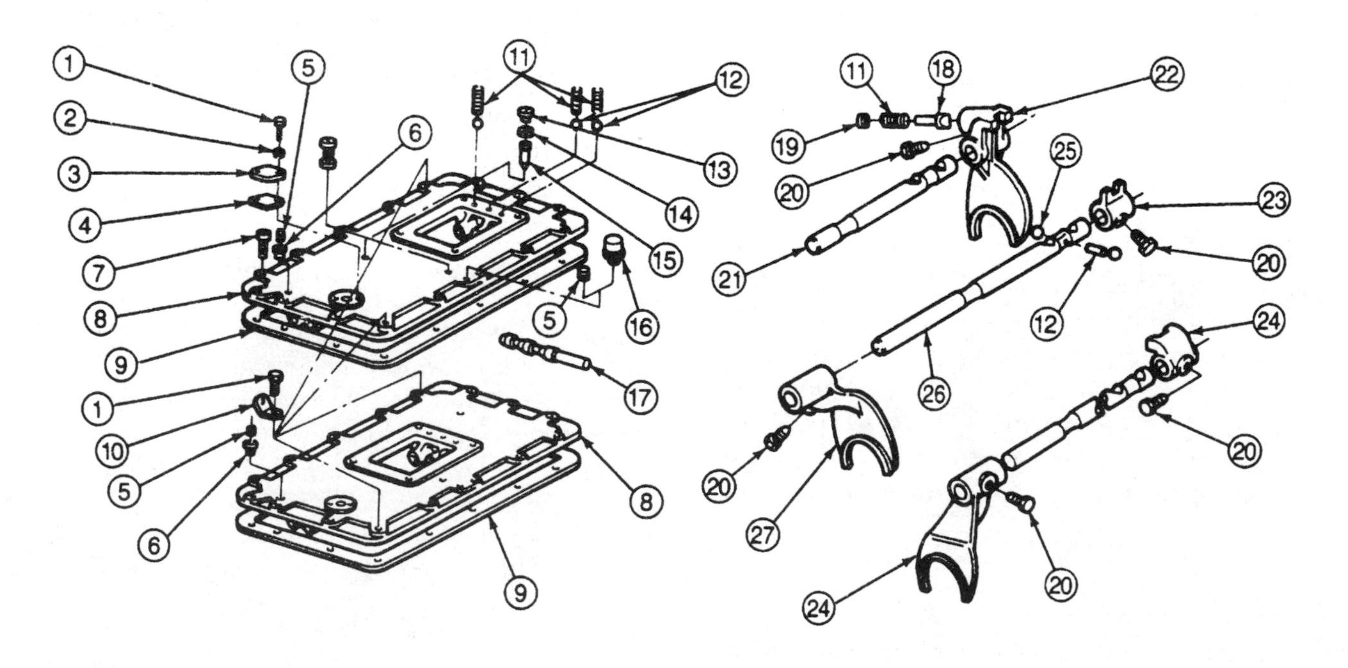

Figure 3–20 Manual transmission cover. *(Courtesy of Ford Motor Company)*

Hint

In an automatic transmission, the converter-cooler-lubrication circuit originates at the main pressure regulator valve. Converter-in oil flows to the torque converter. Oil must flow through the converter continuously to keep it filled and to carry off the heat generated by the converter. The converter pressure regulator valve controls converter-in pressure by bypassing excessive oil to the sump. Converter-out oil, leaving the torque converter, flows to an external cooler (supplied by the vehicle or engine manufacturer). A flow of air or water over or through the cooler removes the heat from the transmission oil. A thorough rebuild of a transmission includes a flushing of the transmission cooling system.

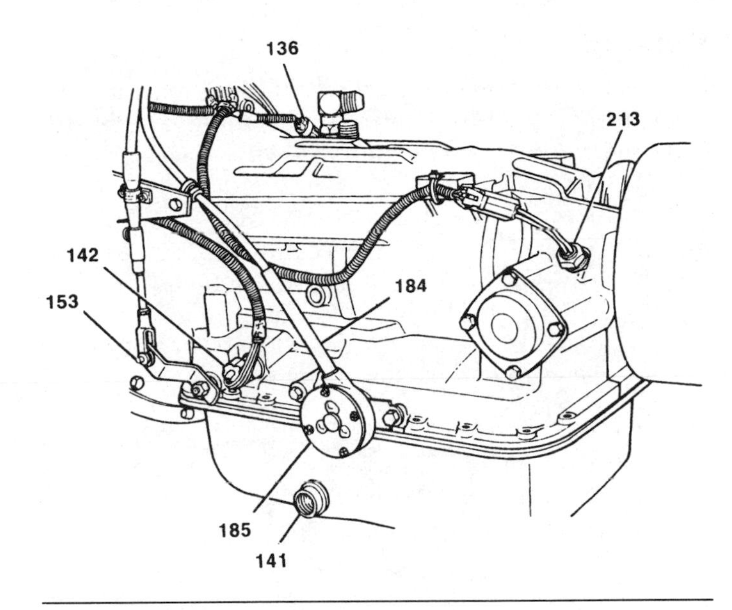

Figure 3–21 External automatic transmission components. *(Courtesy of General Motors Corporation, Service Technology Group)*

Task 17

Inspect, test, replace, or adjust speedometer components (mechanical and electronic).

70. A truck has a nonfunctional electronic speedometer while the odometer operates correctly. The LEAST likely cause of the problem would be:
    A. a defective vehicle speed sensor.
    B. loose wiring at the instrument cluster.
    C. a broken speedometer gauge.
    D. an open circuit in the wiring behind the instrument cluster.

71. In Figure 3–21, item 213:
    A. is a transmission temperature sensor.
    B. sends a digital voltage signal to the speedometer.
    C. sends an analog voltage signal to the vehicle speed sensor (VSS) calibrator module.
    D. sends pulse width modulated (PWM) voltage signal to the transmission temperature gauge.

Hint

The best way to determine whether or not a problem in any system is due to electrical or mechanical failure is to gather information about the system. Considerations like whether the power supply to that system is shared with another system, or whether grounding to the component is shared, are both quick ways to help pinpoint the source of the fault. Mechanical failure in a system is almost never intermittent and usually can be investigated by a simple visual inspection.

Task 18

Inspect, adjust, service, repair, or replace power take-off assemblies, controls, and PTO shafts.

72. When discussing power take-off (PTO) systems, all of these statements are true EXCEPT:
    A. some PTO shafts are driven from the transmission.
    B. some PTO shafts are used to drive a hydraulic hoist pump.
    C. some PTO shafts are driven from the transfer case.
    D. most PTO systems are designed for continual operation.

73. While discussing the PTO in Figure 3–22, *Technician A* says this type of PTO is mounted on the transfer case. *Technician B* says this type of PTO is shifted electrically. Who is right?
    A. A only
    B. B only
    C. Both A and B
    D. Neither A nor B

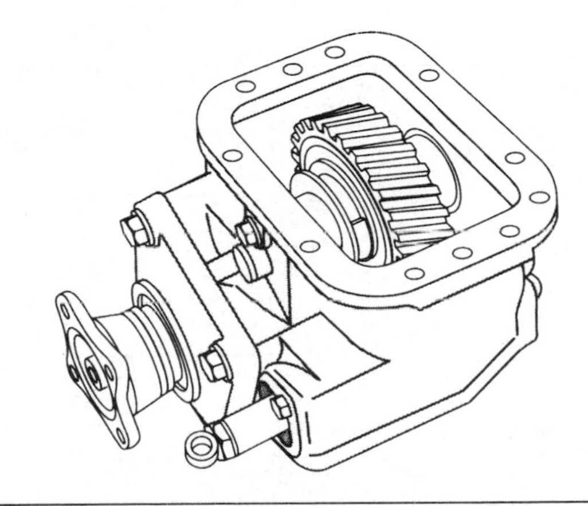

Figure 3–22 Power take-off (PTO).

Hint

PTO drive shafts are very similar in design to the vehicle's powertrain drive shaft. Universal joints and slip joints allow the working angles between the PTO and the driven accessory to change due to movements in the powertrain from torque reactions and chassis deflections. Because most PTO drive shaft applications are for strictly intermittent service, a precisely balanced shaft is rarely used.

Task 19

Inspect and test function of back-up light, neutral start, and warning device circuit switches.

74. Refer to Figure 3–23. When the transmission is placed in reverse with the ignition switch on or the engine running, the right hand back-up light is illuminated dimly, but the left hand back-up light is on normally. *Technician A* says there may be high resistance in circuit 150 from junction S401

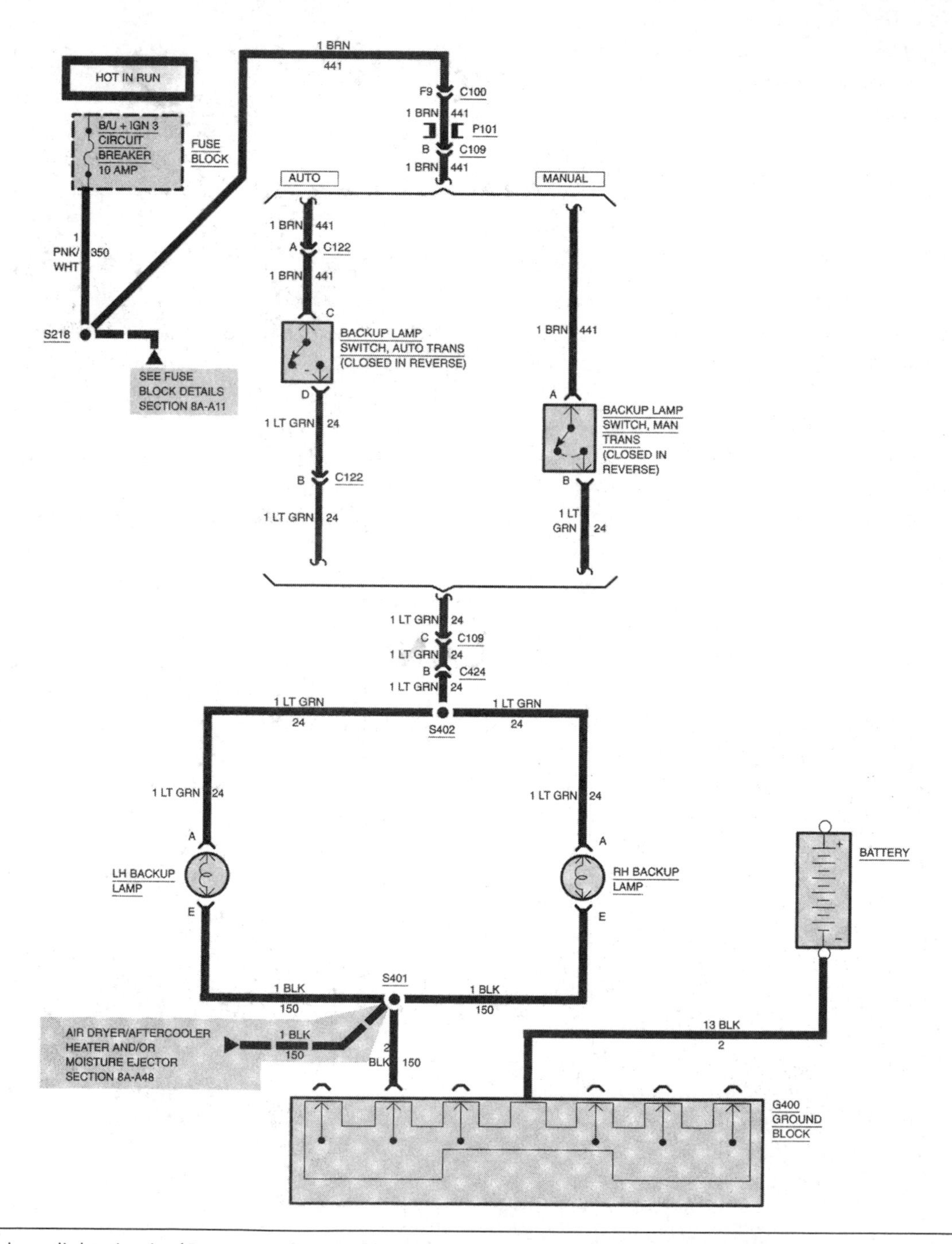

Figure 3–23 Back-up light circuit. *(Courtesy of General Motors Corporation, Service Technology Group)*

to the ground block. *Technician B* says there may be high resistance in circuit 24 from the back-up light switch to junction S402. Who is right?

A. A only
B. B only
C. Both A and B
D. Neither A nor B

75. Refer to Figure 3–23. When the ignition switch is turned on, the back-up lights are illuminated continually in any gear. The cause of this problem may be:
    A. a continually open back-up light switch.
    B. circuit 24 between the back-up light switch and junction S402 is shorted to the battery positive.
    C. circuit 150 from the left hand back-up light to junction S401 is shorted to ground.
    D. circuit 24 between junction S402 and the right hand back-up light is shorted to ground.

Hint

The neutral safety switch or starting safety switch prevents a vehicle from being started in gear. These safety switches may be located in the transmission for both automatic and manual transmissions, or on the clutch pedal or its associated linkage. Placing the transmission in either park or neutral closes the switch so current can flow through the starting relay. The back-up lamp switch is another switch that is associated with transmission linkage. The switch can be diagnosed like any other two-position switched circuit.

Task 20

Inspect and test transmission temperature gauge circuit for accuracy; determine needed repairs.

76. Refer to Figure 3–24. The transmission temperature gauge reads cold under all driving conditions, and 12 volts are supplied through circuit 341 to the gauge with the ignition switch on. The cause of this problem could be:
    A. an open circuit in circuit 585 from the transmission temperature indicator light to junction S206.
    B. an open circuit in the 10-ampere circuit breaker.
    C. an open circuit in circuit 341 from the 10-ampere circuit breaker to the transmission temperature gauge.
    D. an open circuit from junction S206 to the transmission temperature gauge sender.

77. When diagnosing the transmission temperature gauge circuit in Figure 3–24, *Technician A* says the transmission temperature gauge sender contains a thermistor. *Technician B* says the resistance of the transmission temperature gauge sender should decrease as the temperature of the transmission fluid decreases. Who is right?
    A. A only
    B. B only
    C. Both A and B
    D. Neither A nor B

Hint

The most reliable way to test a transmission temperature sensor for accuracy is to obtain a temperature-to-resistance chart from the manufacturer. This allows the Tech-

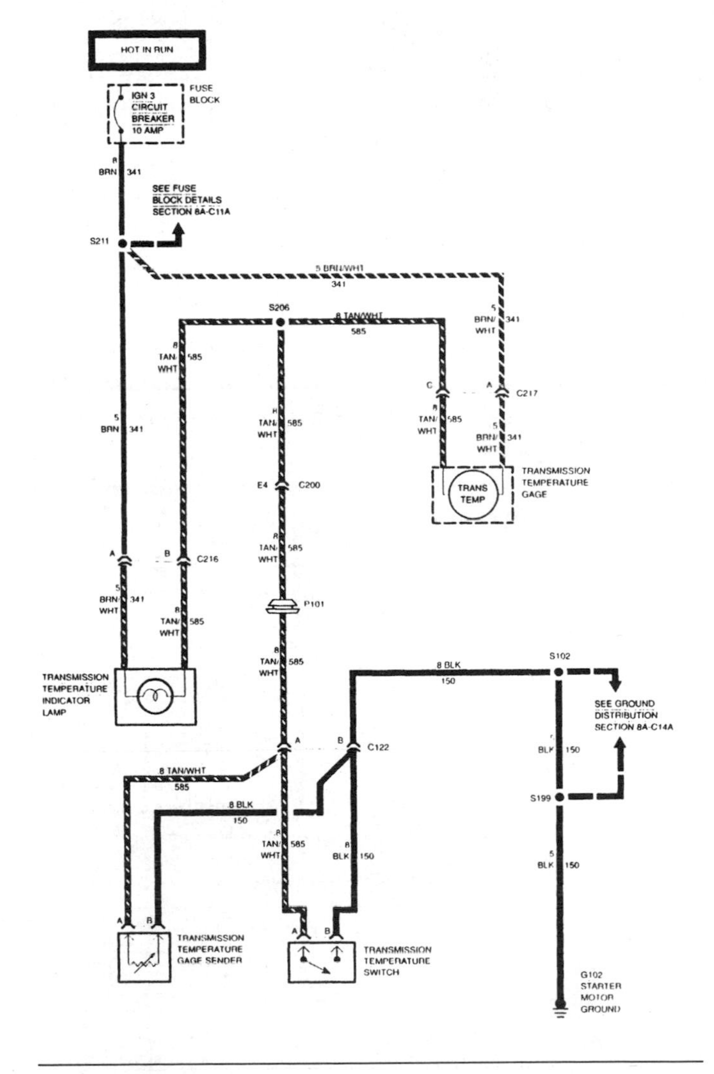

Figure 3–24 Transmission temperature gauge circuit. *(Courtesy of General Motors Corporation, Service Technology Group)*

nician to determine if the temperature sensor is sending a signal appropriate for the temperature it is encountering. Another way of diagnosing the temperature indicating circuit is to place a variable resistance in the place of the sensor. Varying a known resistance and checking the temperature display for proper operation is also a good way of helping pinpoint the source of the problem.

Task 21

Inspect, adjust, repair, or replace transfer case assemblies.

78. A conventional three-shaft drop box designed transfer case shows signs of extreme heat damage to the transfer case gears. The most LIKELY cause for this is:
    A. poor quality bearings.
    B. poor quality lubricant.
    C. inferior quality input gears.
    D. continuous overloading of the drivetrain.

79. A transfer case with straight air will not shift from the low gear ratio up to 1:1. The likely cause is:
    A. a broken shift shaft spring.
    B. a faulty air compressor.
    C. a damaged air line.
    D. a broken shift shaft.

Hint

A transfer case is simply an additional gearbox located between the main transmission and the rear axle. The transfer case may be equipped with an optional parking brake and a speedometer drive gear that can be installed on the idler assembly. Most transfer cases that use the counter shaft design in their gearing is of the constant mesh helical cut type. Most counter shafts are mounted on ball or roller bearings. All rotating and contact components of the transfer case are lubricated by oil from gear throw-off during operation. However, some units are provided with an auxiliary oil pump, externally mounted to the transfer case. To diagnose components in the transfer case, use the same logic as drive axle or transmission gearing.

Task 22

Inspect, adjust, repair, or replace transfer case controls.

80. While discussing the transfer case in Figure 3–25, *Technician A* says that if the transfer case is shifted into low range, the annulus gear on the planetary gearset is locked to the case. *Technician B* says low range gear reduction is obtained by the sun gear driving the planetary carrier. Who is right?
    A. A only
    B. B only
    C. Both A and B
    D. Neither A nor B

81. Refer to Figure 3–25. *Technician A* says the transfer case can be shifted into low range with the vehicle moving forward at 10 mph (16 km/h). *Technician B* says operating the transfer case in 4WD while driving long distances on a paved road causes excessive tire wear. Who is right?
    A. A only
    B. B only
    C. Both A and B
    D. Neither A nor B

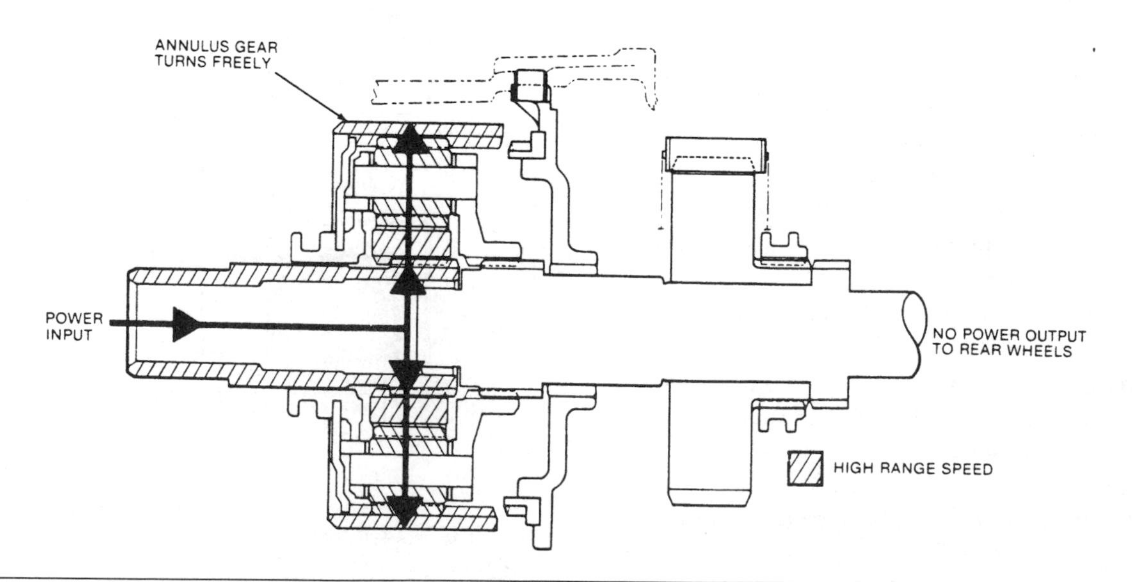

Figure 3–25 Transfer case gears. *(Courtesy of Ford Motor Company)*

Hint

Most complaints regarding transfer case shifting are with remote-type linkages used in cab-over-engine vehicles. Before checking for hard shifting, the remote linkages should be inspected. Linkage problems stem from worn connections or bushings, binding, or improper adjustment, lack of lubrication on the joints, or an obstruction that restricts free movement. A transfer case may be shifted by air pressure, mechanical linkage, electronically, or by engine vacuum if the truck has a gasoline engine.

Drive Shaft and Universal Joint Diagnosis and Repair

Task 1

Diagnose drive shaft and universal joint noise, vibration, and runout problems; determine cause of failure and needed repairs.

82. If a vehicle has an out-of-balance drive shaft, when would vibration likely be noticeable?
    A. Between 500 to 1,200 rpm under load
    B. Between 1,200 to 2,000 rpm under load
    C. Under load below 30 mph (48 km/h)
    D. Above 50 mph (80km/h) with no load

83. When the clutch pedal is released to start moving the vehicle, a single clanging noise is heard under the truck. The MOST likely cause of this noise is:
    A. a worn drive shaft center support bearing.
    B. a worn universal joint.
    C. a loose differential flange connected to the driveshaft.
    D. excessive end play on the transmission output shaft.

84. A fast cycling squeaking noise is heard under a truck at low speeds. *Technician A* says one of the U-joints may be worn and dry. *Technician B* says the splines on the drive shaft slip joint may be worn and dry. Who is right?
    A. A only
    B. B only
    C. Both A and B
    D. Neither A nor B

Hint

Oftentimes, vibration is too quickly attributed to the drive shaft. Before condemning the drive shaft as the cause of vibration, the vehicle should be thoroughly road tested to isolate the vibration cause. To assist in finding the source, ask the operator to determine what, where, and when the vibration is encountered. It is very helpful to keep in mind some of the causes of driveline vibration. U-joints are the most common source if the vibration is coming from the drive shaft, while drive shaft balancing is the next most common. Pay special attention to phasing when removing or installing a drive shaft.

Task 2

Inspect, service, or replace drive shaft, slip joints, yokes, drive flanges, universal joints, and retaining hardware; properly phase yokes.

85. A Technician is replacing the U-joints in a drive shaft. After removal, the Technician realized that he or she forgot to scribe a line onto the yokes for proper phasing. The likely method of testing for proper phasing of the yokes would be to:
    A. drive the truck on level terrain at a steady speed below 40 mph (64 km/h) and check for driveline vibration.
    B. drive the truck on level terrain at a steady speed above 40 mph (64 km/h) and check for driveline vibration.
    C. place the truck under load by driving up an incline and check for driveline vibration.
    D. depress the clutch pedal with the truck traveling at 35 mph (56 km/h) and allow the vehicle to coast while checking for vibration.

86. *Technician A* says that lubricating slip splines requires special lithium-based grease. *Technician B* says good quality U-joint grease can also be used on slip splines. Who is right?
    A. A only
    B. B only
    C. Both A and B
    D. Neither A nor B

Hint

The following are descriptions of common visually evident damage to U-joints:

- *Cracking shows up as stress lines due to metal fatigue.*
- *Galling occurs when metal is cropped off or displaced due to friction between surfaces, most commonly found on trunnion ends.*
- *Spalling occurs when chips, scales, or flakes of metal break off due to fatigue rather than wear. Spalling is usually found on splines and U-joint bearings.*

- *Pitting appears as small pits or craters in metal surfaces due to corrosion. Pitting can lead to surface wear and eventual failure.*
- *Brinelling is evidenced by grooves worn in the bearing race surface, often caused by improper installation of the U-joints. Do not confuse the polishing of a surface where no structural damage occurs with actual brinelling.*

Task 3

Inspect, repair, and replace drive shaft center support bearings and mounts.

87. What should a Technician do when replacing a support bearing assembly?
    A. Lubricate the bearing.
    B. Apply lubricant to the outer bearing race to help press the bearing into place.
    C. Fill the entire cavity around the bearing with grease.
    D. Pay close attention to proper orientation of the support bearing assembly.

88. When replacing drive shaft center support bearings, a Technician should always:
    A. look for any shims during removal of the old bearing.
    B. replace the center support assembly mounting bolts.
    C. measure driveline angles once the new component is installed.
    D. use hand tools; air tools could twist the mounting cage and shorten bearing life.

Hint

Center bearings are lubricated by the manufacturer and are not serviceable. However, when replacing a support bearing assembly, be sure to fill the entire cavity around the bearing with waterproof grease to shield the bearing from water and contaminants. Grease must fill the cavity to the extreme edge of the slinger surrounding the bearing. Use only waterproof lubricants. Consult a grease supplier or the bearing manufacturer for recommendations. Also pay attention when removing the existing center support bearing. Any shims used to adjust the driveline angle must be reinstalled when installing a replacement bearing.

Task 4

Measure and adjust driveline angles (vehicle loaded and unloaded).

89. When measuring driveline angles, *Technician A* says that the driveline angle measurement is from the center of the rear axles to the parallel center of the chassis. *Technician B* says that driveline angle measurement is given from the front of the vehicle to the rear. Who is right?
    A. A only
    B. B only
    C. Both A and B
    D. Neither A nor B

90. When measuring driveline angles as illustrated in Figure 3–26, the angle at point E should not exceed.
    A. 12 degrees.
    B. 15 degrees.
    C. 18 degrees.
    D. 21 degrees.

Hint

With the vehicle on a level surface, tire pressures equalized, and the output yoke on the transmission in a vertical position, use either a magnetic base protractor or an electronic inclinometer to measure drive shaft angle. Always measure drive shaft angle from front to rear. Always take driveline angle measurements with the vehicle loaded and unloaded.

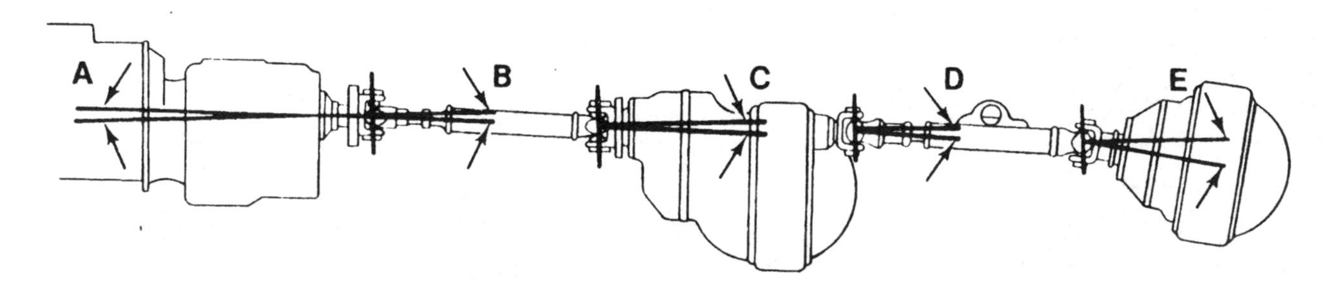

Figure 3–26 Driveline angles. *(Courtesy of General Motors Corporation, Service Technology Group)*

Drive Axle Diagnosis and Repair

Task 1

Diagnose rear axle drive unit noise and overheating problems; determine cause of failure and needed repairs.

91. A Technician has an overheating problem with both axles of a tandem drive axle configuration. *Technician A* says the tractor may be continually overloaded. *Technician B* says some of the interaxle differential components may be damaged. Who is right?
    A. A only
    B. B only
    C. Both A and B
    D. Neither A nor B

92. A two-speed differential makes a whining noise only while driving straight ahead in high range at 55 to 65 mph (88 to 105 km/h). During deceleration and at lower speeds the noise disappears. The fluid level and type of lubricant are correct. The MOST likely cause of this problem is:
    A. ring gear and pinion worn or improperly adjusted.
    B. worn differential side gears.
    C. worn planetary gearset.
    D. worn pinion bearings.

Hint

Because of the similar nature of transmissions, drive axles, and transfer cases, determining exactly where a noise is coming from and which component is causing the noise may be very difficult. A set of guidelines may work very well for certain combinations of components, while not working at all for others. Sometimes it may be beneficial to keep a list describing sounds that the truck exhibited and the component that caused it. This list could prove to be a good way to assist a Technician's memory when the vehicle is currently in for repair and the cause is not evident.

Task 2

Check for, and repair, fluid leaks; inspect or replace rear axle drive housing cover plates, gaskets, vents, magnetic plugs, and seals.

93. Fluid leaks are evident between the axle housing and the differential carrier in Figure 3–27. The differential contains the proper type and lubricant level. The MOST likely cause is:
    A. normal deterioration of the gasket.
    B. repeated overloading of the drivetrain.
    C. plugged axle housing breather vent.
    D. normal loss of viscosity due to aging lubricant.

94. On a tractor with a tandem rear axle, the right hand inner hub seal in the forward axle has been replaced three times because it was leaking grease. There are no other leaks in the forward differential, and the wheel bearings are adjusted to specifications each time the seal is replaced. *Technician A* says the seal bore in the hub should be checked for scratches, nicks, and burrs. *Technician B* says the hub end play may be excessive. Who is right?
    A. A only
    B. B only
    C. Both A and B
    D. Neither A nor B

Hint

A thorough Technician can usually determine the operating condition of the drive axle differential by the fluid. Pay special attention to the condition of the fluid during the scheduled fluid changes. Most drive axles are equipped with a magnetic plug that is designed to attract any metal particles suspended in the gear oil. A nominal amount of "glitter" is normal because of the high torque environment of the drive axle; however, too much "glitter" indicates a problem that requires further investigation. Carefully investigate any source of leaks found on the axle differential. Replacing the seal if the axle differ-

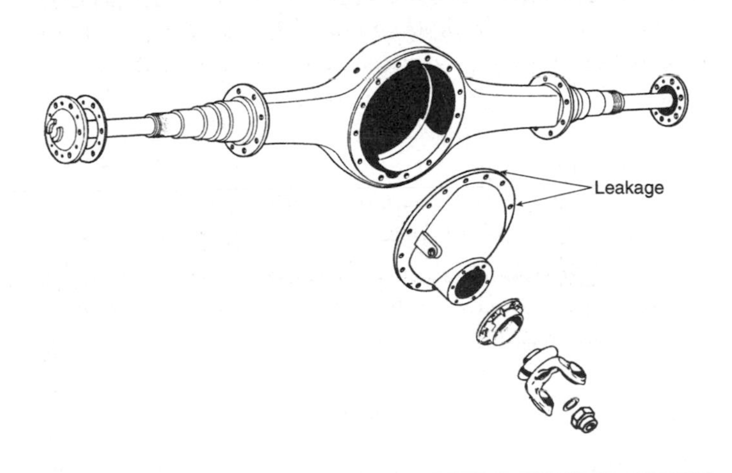

Figure 3–27 Rear axle fluid leaks. *(Courtesy of Dana Corporation)*

ential breather is plugged does not cure leaking seals. A plugged breather results in high pressure, which can result in leakage past seals and gaskets.

Task 3

Check rear axle drive unit fluid level and condition; determine needed service, and add proper type of lubricant.

95. During a routine drive axle oil change, a Technician notices a few metal particles on the magnetic plug of the drive axle. What should the Technician do?
    A. Inform the customer that further investigation is needed.
    B. Inform the customer of the condition and tell him or her to monitor the amount of particles.
    C. Do not follow up with the customer, as some metal particles are normal.
    D. Begin to disassemble the drive axle to find the cause.

96. While discussing tandem rear axle lubrication, *Technician A* says the proper fluid level is when you can feel the lubricant with your finger in the differential filler plug hole. *Technician B* says in some rear axle hubs the proper lubricant must be installed through hub filler plugs for wheel bearing lubrication. Who is right?
    A. A only
    B. B only
    C. Both A and B
    D. Neither A nor B

Hint

Drain and flush the factory-fill axle lubricant of a new or reconditioned axle after the first 1,000 miles (621 km) and never later than 3,000 miles (1,864 km). This is necessary to remove fine particles of wear material generated during break-in that would cause accelerated wear on gears and bearings if not removed. Draining the lubricant while the unit is still warm ensures that any contaminants are still suspended in the lubricant. Flush the axle with clean axle lubricant of the same viscosity as used in service. Do not flush axles with solvents such as kerosene. Avoid mixing lubricants of a different viscosity or oils made by different manufacturers.

Task 4

Remove and replace differential carrier assembly; lubricate as required.

97. While removing the differential carrier from the axle housing, where should the Technician place the jack stands?
    A. Under the spring seats
    B. Under the most outer surface of the axle housing
    C. As close as possible to the axle housing cover
    D. Under the brake chamber mounting bracket

98. The likely time an axle housing or differential cover mating surface may be damaged is during:
    A. axle assembly.
    B. axle disassembly.
    C. gasket surface preparation.
    D. continuous high torque driving conditions.

99. When installing a new differential carrier into an existing axle housing, a Technician should check:
    A. for differential carrier mounting flange squareness.
    B. the axle housing mating flange for nicks, scratches, or burrs.
    C. the differential carrier mating flange for nicks, scratches, or burrs.
    D. both mating surfaces for a good, clean, and tight fit.

Hint

When removing the differential carrier, leave two of the mounting bolts in the axle housing to hold the differential in place until a hydraulic jack can be placed under the differential carrier. The extreme weight of the differential carrier must be safely strapped onto the hydraulic jack for removal of the carrier. When the carrier is ready for installation, inspect the mounting surfaces for nicks, scratches, or gouges.

Task 5

Inspect and replace differential case assembly, including spider gears, cross shaft, side gears, thrust washers, case halves, and bearings.

100. A Technician notices excessive end play in the differential side gears. How should the Technician fix the problem?
    A. Split the differential case and replace the side gear thrust washers.
    B. Split the differential case and replace the side gears.
    C. Split the differential case, replace the differential pinion gears and thrust washers.
    D. Loosen the differential side gear retaining caps and install new side gear thrust washers.

101. A two-speed single axle in a medium-duty truck makes a clunking noise when turning a corner at low speeds. *Technician A* says the drive pinion and ring gear teeth may be damaged. *Technician B* says the differential side gears and pinion gears may have damaged teeth. Who is right?
   A. A only
   B. B only
   C. Both A and B
   D. Neither A nor B

Hint

There are many types of damage that can occur in a differential assembly, including shock failure. This damage occurs when the gear teeth or pinion are stressed beyond the strength of the material from which they are machined. The failure may be immediate from a sudden shock or it may be a progressive failure after the initial shock cracks the teeth or pinion. As in any other situation, early detection of damage can prevent additional damage.

Task 6

Inspect and replace components of locking differential assembly.

102. A truck with a driver-controlled main differential lock will not lock. The following are all possible causes EXCEPT:
   A. a broken shift fork.
   B. a sticking shift fork.
   C. a damaged air solenoid.
   D. a broken disengagement spring.

103. When discussing locking differentials, *Technician A* says that when the differential lock is engaged one of the axle shafts is locked to the differential case. *Technician B* says the differential lock should be engaged to prevent tire wear during highway driving. Who is right?
   A. A only
   B. B only
   C. Both A and B
   D. Neither A nor B

Hint

The typical single reduction carrier with differential lock has the same type of gears and bearings as the standard-type carriers. However, an air-actuated shift assembly that is mounted on the carrier and operated from the truck's cab operates the differential lock. By actuating an air plunger or electric switch, usually mounted on the instrument panel, the driver or operator can lock the differential to achieve positive traction and control of the truck on poor or slippery road or highway conditions.

Task 7

Inspect differential carrier case and caps, side bearing bores, and pilot (spigot, pocket) bearing bore; determine needed service.

104. While rebuilding a differential that has 200,000 miles of service on it, a Technician notices faint, equally spaced grooves on the bearing caps. *Technician A* says the marks are from the original machining process and the caps do not need to be replaced. *Technician B* says the bearing caps must be marked and reinstalled in their original position. Who is right?
   A. A only
   B. B only
   C. Both A and B
   D. Neither A nor B

105. The Technician in Figure 3–28 is measuring:
   A. case warpage caused by excessive heat.
   B. squareness to ensure that the bearing caps did not get switched during assembly.
   C. differential housing twist commonly associated with damage caused by excessive overloading or shock load damage.
   D. differential bearing cap preload.

106. The following are all reasons for replacement of the spigot bearing EXCEPT:
   A. rough action.
   B. binding action.
   C. spalled outer race.
   D. excessive end play.

Hint

Differential bearing caps should be marked before removal to ensure that they are reassembled onto the proper leg. Put the differential bearing outer race and adjusting nut in place before installing the bearing cap onto the leg of the differential. Placing the bearing race and adjusting nut in position before the bearing cap ensures proper alignment of the bearing cap. This also makes the process of setting the differential bearing preload easier because the adjusting nut will move more freely inside of the leg and bearing cap.

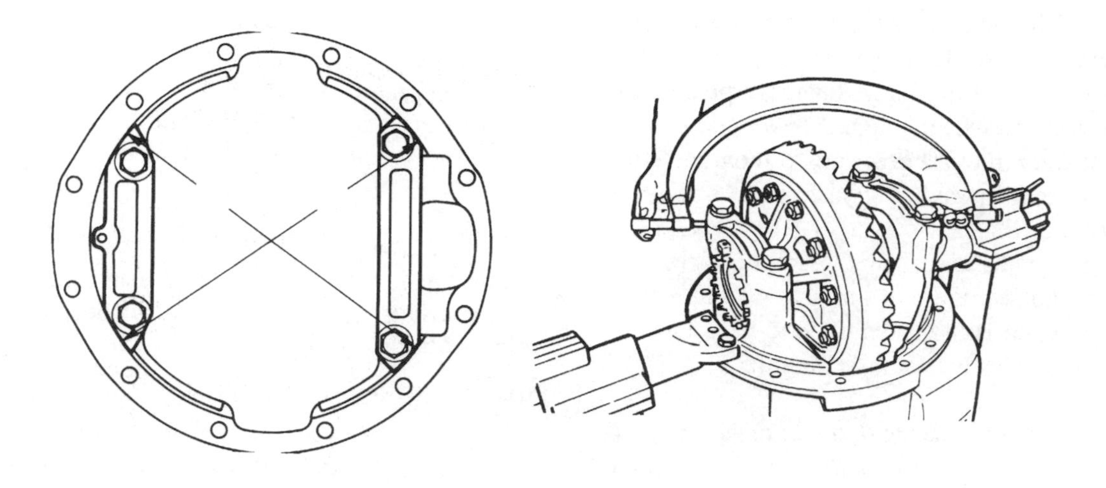

Figure 3–28 Differential measurement. *(Courtesy of Rockwell Automotive)*

Task 8

Measure ring gear runout; determine needed repairs.

107. A Technician measuring ring gear runout obtains a reading of 0.010 inch (0.254 mm). The Technician will likely:
    A. replace the ring and pinion because of a warped condition.
    B. continue with the installation; the measurement is within specification.
    C. remove the ring gear from the mounting flange, rotate the ring gear 90 degrees, and reinstall on the flange.
    D. reset the differential bearing preload and measure again.

108. *Technician A* says the differential side bearing preload must be properly adjusted before performing the measurement in Figure 3–29. *Technician B* says worn drive pinion gear teeth will affect the measurement in Figure 3–29. Who is right?
    A. A only
    B. B only
    C. Both A and B
    D. D. Neither A nor B

Hint

Runout of the ring gear is measured with a dial indicator attached to the axle housing. If the runout of the ring gear exceeds the specification, remove the differential and ring gear assembly from the carrier. Check the differential parts, including the carrier, for the problem that caused the runout of the gear to exceed specifications. Repair or replace defective parts. After the parts are repaired or replaced, install the differential and ring gear into the carrier and repeat the runout measurement.

Task 9

Inspect and replace ring and drive pinion gears, spacers, sleeves, bearing cages, and bearings.

109. A differential shows spalling on the teeth of the ring gear, while the gear teeth of the pinion have no signs of wear. What should the Technician do?
    A. Replace the ring gear and pinion.
    B. Replace the ring gear.
    C. Properly rinse out the differential housing and switch to a higher viscosity of lubricant.
    D. Correctly adjust the ring gear and pinion backlash before any further damage takes place.

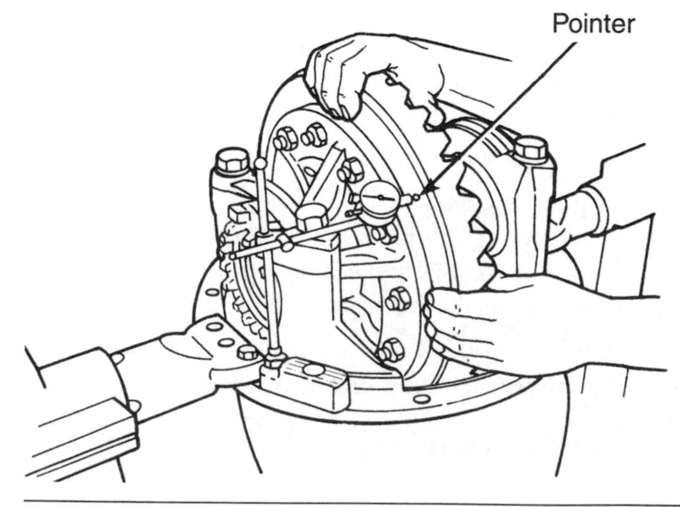

Figure 3–29 Ring gear measurement. *(Courtesy of Rockwell Automotive)*

110. How would a Technician replace a ring gear once the rivets are removed?
    A. Press out the old one, heat the new ring gear in water, and assemble.
    B. Simply allow the old ring gear to separate from the differential case and install the new one.
    C. Pry the old ring gear off the differential and install the replacement ring gear with a press.
    D. Lightly hammer the old ring gear off the differential case and use a torch to heat the differential case before installing the new ring gear.

Hint

Correct tooth contact between the pinion and the ring gear cannot be overemphasized because improper tooth contact can lead to early failure of the axle and noisy operation. Used gearing usually does not display the square, even contact pattern found in new gearsets. The gear normally has a pocket at the toe-end of the gear tooth that falls into a contact line along the root of the tooth. The more a gear is used, the more the line becomes the dominant characteristic of the pattern. If a ring gear and pinion is to be reused, measure the tooth contact pattern and backlash before disassembling the differential.

Task 10

Measure and adjust drive pinion bearing preload.

111. The pinion preload turning torque is more than specified on the pinion shaft assembly is Figure 3–30. To correct this problem it is necessary to:
    A. install a thicker spacer between the two pinion bearing inner races.
    B. install a thicker spacer between the rear bearing and the pinion gear.
    C. install a thinner spacer between the rear bearing and the pinion gear.
    D. install a thinner spacer between the two pinion bearing inner races.

112. *Technician A* says drive pinion depth should be set once you properly preload the pinion bearing cage. *Technician B* says that setting the drive pinion depth requires adjustment of the ring gear. Who is right?
    A. A only
    B. B only
    C. Both A and B
    D. Neither A nor B

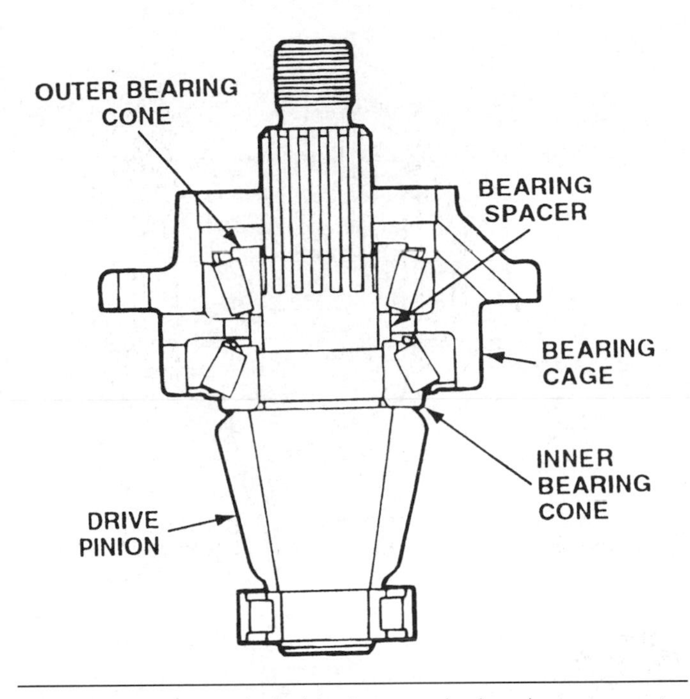

Figure 3–30 Pinion shaft, bearings, and related components. *(Courtesy of Rockwell Automotive)*

Hint

There are a few different methods of measuring pinion bearing preload. One method is to use a spring scale to measure the rolling resistance of the assembly while the pinion bearings are under pressure from a hydraulic press. It is always best to refer to applicable manufacturer's specific information whenever testing for pinion bearing preload. A second method of measuring pinion bearing preload is to lubricate, assemble, and install the pinion bearing components. Be sure the pinion shaft nut is tightened to the specified torque. The pinion bearing preload is determined by measuring the pinion bearing turning torque with a torque wrench and socket attached to the pinion shaft nut. A shim adjustment is provided to increase or decrease pinion bearing preload.

Task 11

Adjust drive pinion depth.

113. While adjusting the pinion depth on the pinion assembly in Figure 3–31, *Technician A* says that if a thicker shim is installed between the differential carrier and the pinion bearing cage, the pinion gear is moved inward toward the ring gear. *Technician B* says to increase the pinion depth, install a thicker shim between the rear pinion bearing and the pinion gear. Who is right?
    A. A only
    B. B only

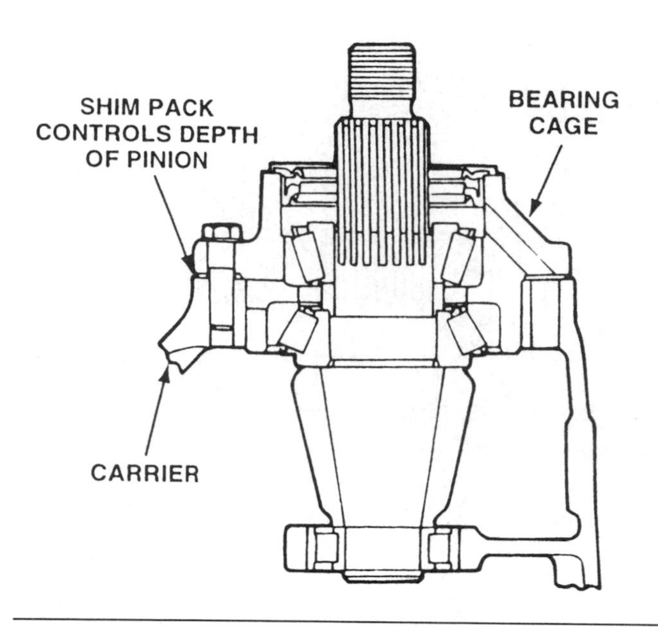

Figure 3–31 Pinion gear, bearings, and bearing cage. *(Courtesy of Rockwell Automotive)*

C. Both A and B
D. Neither A nor B

114. The marking on the pinion gear in Figure 3–32 is used to calculate:
 A. pinion bearing preload.
 B. ring gear backlash.
 C. pinion gear depth.
 D. pinion gear turning torque.

Hint

A correct tooth pattern is clear of the toe, and centers evenly along the face width between the top land and the root. If necessary, adjust the contact pattern by moving the ring gear and drive pinion. Ring gear position controls contact pattern along the face width of the gear tooth. Pinion position is determined by the size of the pinion bearing cage shim pack, and controls contact on the depth of the gear tooth.

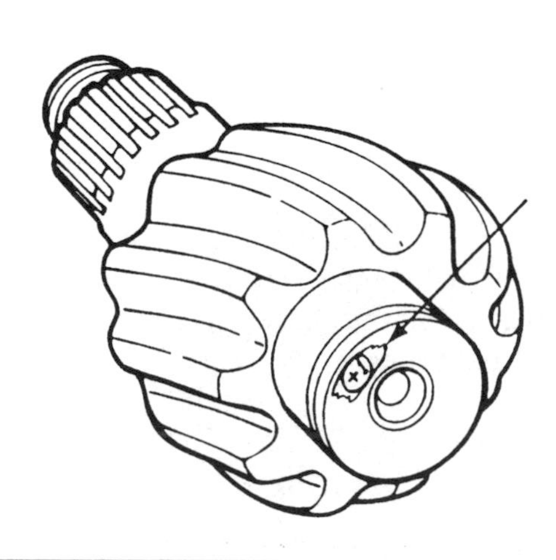

Figure 3–32 Pinion gear marking. *(Courtesy of Rockwell Automotive)*

Task 12

Measure and adjust side bearing preload and ring and pinion backlash.

115. When performing the adjustment in Figure 3–33:
 A. rotate the differential adjusting rings until there is zero end play on the ring gear, and then tighten each ring one notch.
 B. rotate the differential adjusting rings until there is 0.010 in. (0.254 mm) end-play on the ring gear.
 C. rotate the differential adjusting rings until the ring gear turning torque is 15 ft.-lb. (20 Nm).
 D. rotate the differential adjusting rings until tooth pattern on the ring gear teeth is correct.

116. Insufficient differential side bearing preload may cause all of these problems EXCEPT:
 A. rapid differential side bearing wear.
 B. excessive ring gear tooth wear.
 C. excessive pinion bearing wear.
 D. excessive pinion gear tooth wear.

117. When performing the adjustment in Figure 3–34, the dial indicator reading is more than specified. To correct this problem it is necessary to:
 A. tighten adjuster ring A and loosen the opposite adjuster ring an equal amount.
 B. tighten both adjuster rings one or two notches.

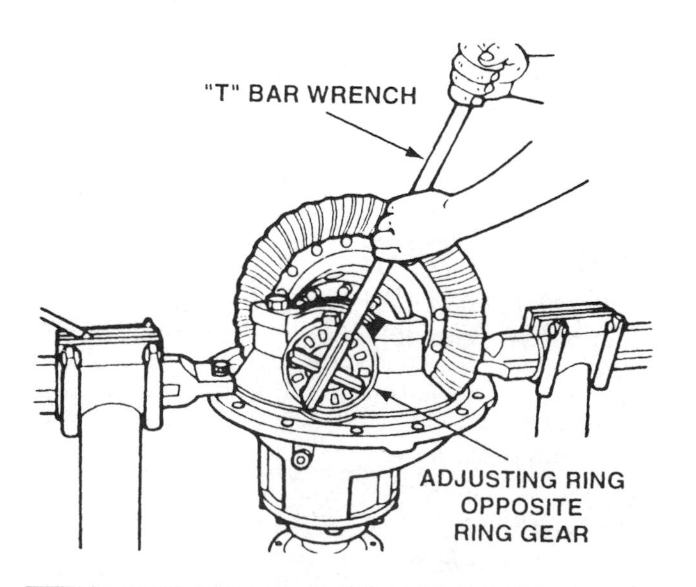

Figure 3–33 Differential adjustment. *(Courtesy of Rockwell Automotive)*

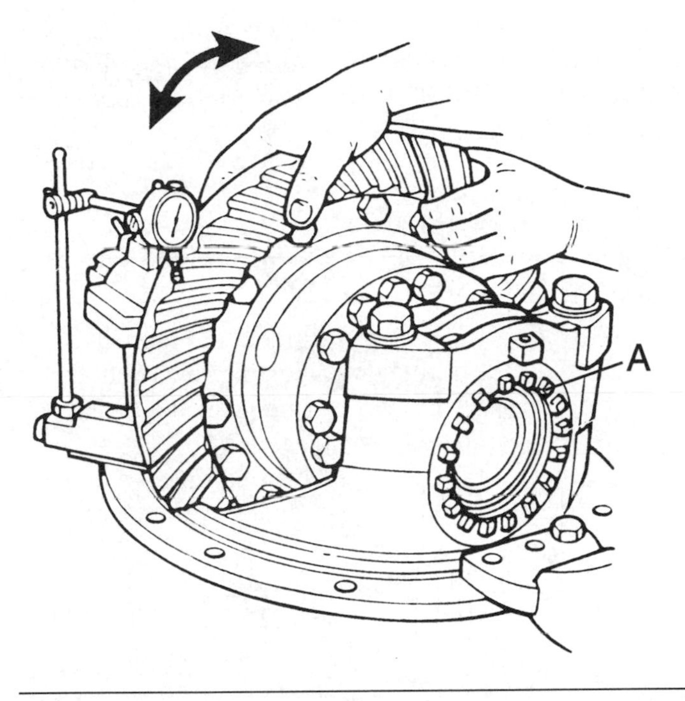

Figure 3–34 Ring gear measurement. *(Courtesy of Rockwell Automotive)*

C. retorque the ring gear bolts and side bearing cap bolts.
D. loosen adjuster ring A and tighten the opposite adjuster ring an equal amount.

Hint

There are a few different ways to measure and adjust differential bearing preload. Most of the methods use the same logic, although they go about it in different fashions. The idea is to place a dial indicator on the back face of the ring gear and adjust the differential bearing adjusting rings until there is no runout measured on the ring gear. When there is no runout measured on the ring gear, tighten the adjusting rings one or two notches further. This adds the proper amount of pressure to "load" the bearings.

Task 13

Check ring and pinion gear set tooth contact pattern; interpret pattern, and adjust to manufacturer's specifications.

118. A Technician is checking the tooth contact pattern in Figure 3–35. *Technician A* says that the ring and pinion needs to be readjusted because the contact pattern is too close to the tooth root. *Technician B* says that the ring and pinion needs to be readjusted because the contact pattern is too close to the tooth toe. Who is right?

Figure 3–35 Differential tooth contact pattern number 1. *(Courtesy of Rockwell Automotive)*

A. A only
B. B only
C. Both A and B
D. Neither A nor B

119. To correct the improper ring gear tooth contact pattern shown in Figure 3–36 *Technician A* says it is necessary to install a thinner shim between the pinion bearing cage and the differential housing. *Technician B* says it is necessary to readjust the side bearing preload. Who is right?
A. A only
B. B only
C. Both A and B
D. Neither A nor B

Hint

With the axle differential assembled, use a marking compound to paint at least twelve teeth of the ring gear. After rolling the ring gear, examine the marks left by the pinion gear contacting the ring gear teeth. A correct pattern is one that comes close to, but does not touch, the ends of the gear. As much of the pinion gear as possible should contact the ring gear tooth face without contacting or going over any edges of the ring gear.

Figure 3–36 Ring gear tooth contact pattern number 2. *(Courtesy of Rockwell Automotive)*

Task 14

Inspect, adjust, or repair ring gear thrust block/bolt.

120. When discussing the ring gear thrust bolt, *Technician A* says the thrust bolt maintains proper contact between the ring gear and pinion gear teeth. *Technician B* says the thrust bolt prevents excessive movement fore-and-aft movement of the pinion gear. Who is right?
  A. A only
  B. B only
  C. Both A and B
  D. Neither A nor B

121. When adjusting a ring gear thrust bolt, it is necessary to:
  A. turn the thrust bolt in until it contacts the ring gear, back it off 2 turns, and tighten the locknut.
  B. turn the thrust bolt in until it contacts the differential case and tighten the locknut.
  C. turn the thrust bolt in until it contacts the ring gear, back it off one-half turn, and tighten the locknut.
  D. turn the thrust bolt in until the lever contacts the rear of the pinion bearing, back it off one-half turn, and tighten the locknut.

Hint

A thrust screw is incorporated into some axle differentials to allow the axle to withstand more torque. The thrust screw is designed to be a "stop" if the torque demanded from the ring and pinion ever causes the ring gear to deflect. If any deflection occurs, the thrust screw prevents the ring gear from deflecting to a point of tooth slip, which would cause extensive damage to the drive axle. To adjust the thrust screw, simply thread the screw into the axle housing until it rests on the ring gear. Then, back the thrust screw off one-half turn and tighten the jam nut. The thrust screw is not designed to be in constant contact with the ring gear, serving only as a backup measure for extreme torque conditions.

Task 15

Inspect, adjust, repair, or replace planetary gear-type two-speed axle assembly, including case, idler pinion, pins, thrust washers, sliding clutch gear, shift fork, pivot, seals, cover, and springs.

122. If a two-speed rear axle with a planetary gear set shown in Figure 3–37 is operating in low range, additional gear reduction is obtained when:

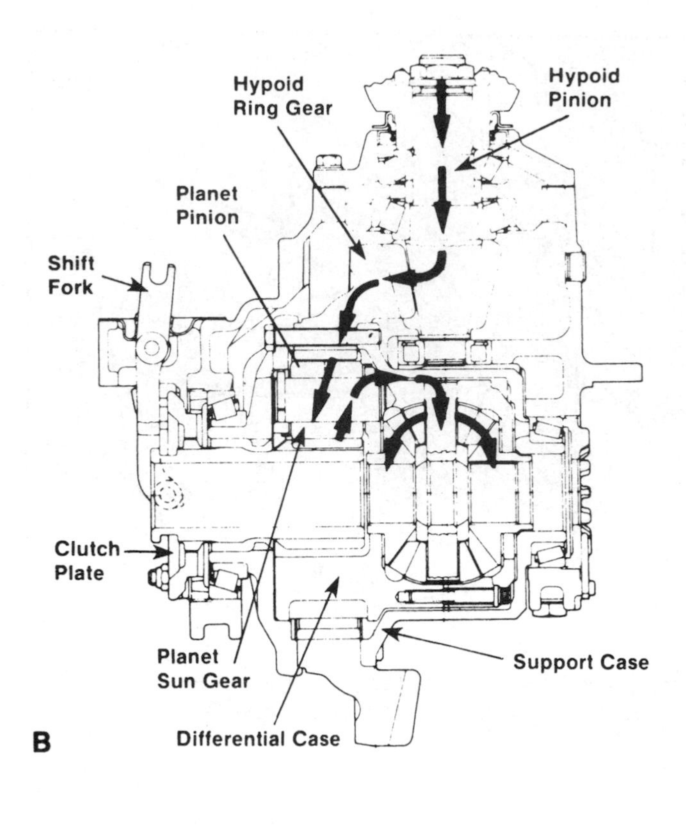

Figure 3–37 Two-speed rear axle. *(Courtesy of Dana Corporation)*

  A. the annulus gear attached to the ring is driving the planetary pinions attached to the differential case, and the sun gear is locked.
  B. the sun gear attached to the differential case is driving the planetary pinions attached to the ring gear, and the annulus gear is locked.
  C. the planetary pinions attached to the ring gear are driving the sun gear attached to the differential case, and the annulus gear is locked.
  D. the planetary pinions attached to the ring gear are locked, and the annulus gear is driving the sun gear attached to the differential case.

123. The rear axle being serviced in Figure 3–38 is a:
  A. single-speed type.
  B. double reduction type.
  C. two-speed type.
  D. double reduction two-speed type.

Hint

A two-speed axle is similar to the double reduction axle except that a two-speed axle can be operated in either the high or low range. Both are similar in design in that

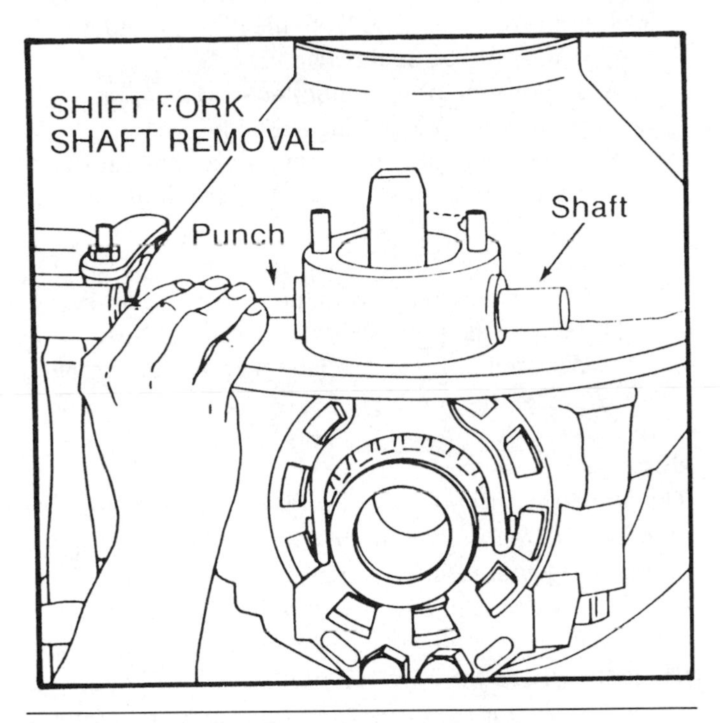

Figure 3–38 Rear axle assembly. *(Courtesy of Dana Corporation)*

they use a planetary gearset as the second reduction gearset. In a two-speed rear axle the annulus gear is part of the ring gear, and the planetary pinions and carrier are attached to the differential case. The sun gear can be locked or unlocked by the shift mechanism. In low range the sun gear is locked and the annulus gear drives the planetary pinions and carrier. Because the annulus gear is smaller than the planetary pinions and carrier, a reduction is provided. An air or electric shift unit is mounted on the carrier and a driver-operated switch in the cab operates this unit. The air shift unit prevents the driver from downshifting to a lower gear before the transmission is at a safe speed, to prevent any possible damage from improper downshifting. Damage can still be incurred to the shift mechanism during upshifting if the driver is not paying attention to the conditions in which the vehicle is being operated.

Task 16

Inspect, repair, or replace air, electric, and vacuum two-speed axle shift control switches, speedometer adapters, motors, axle shift units, wires, connectors, hoses, and diaphragms.

124. While discussing the two-speed rear axle shift system controlled by air pressure shown in Figure 3–39, *Technician A* says the speedometer adapter and pressure switch prevent axle shifting above a

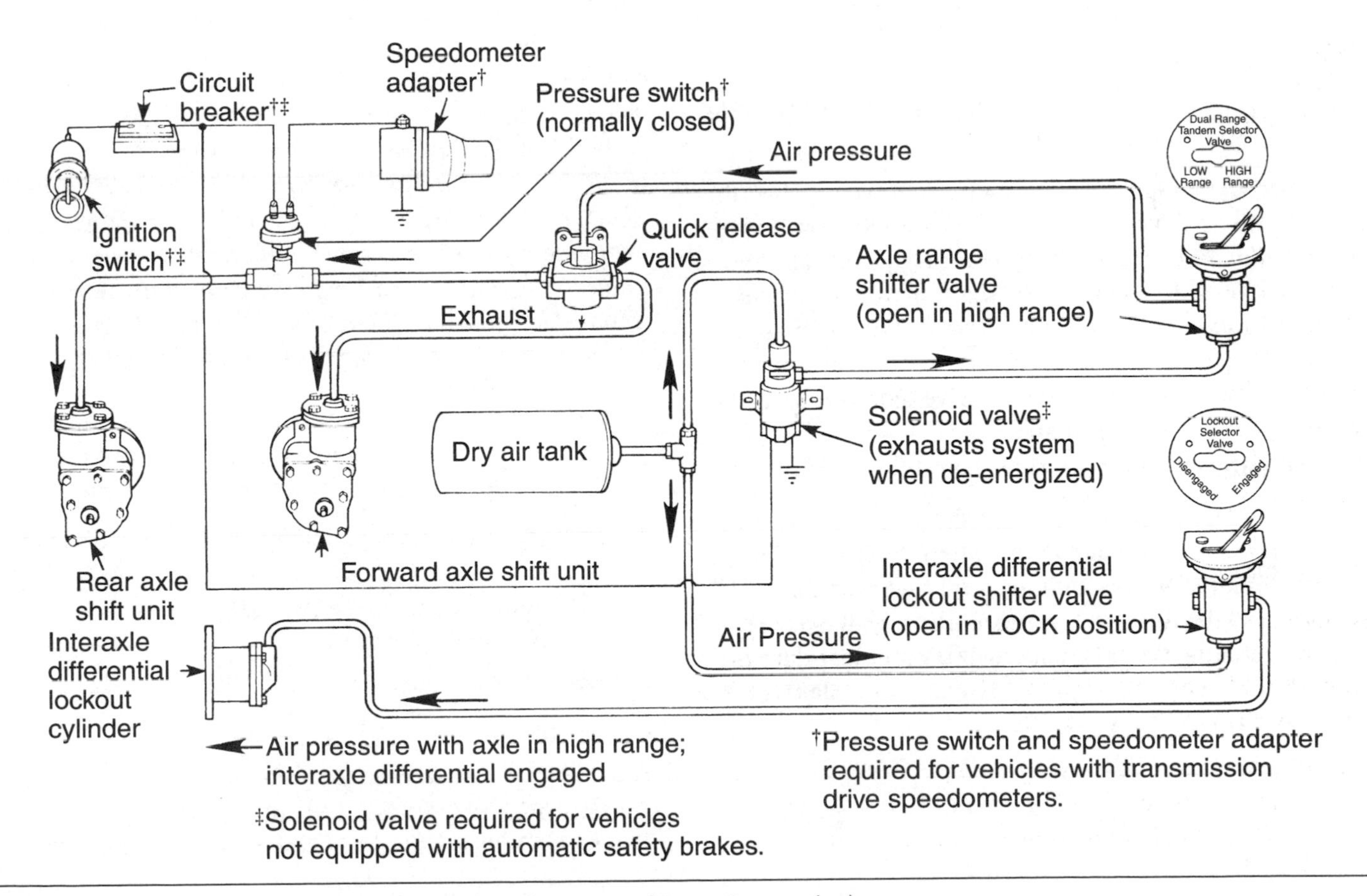

Figure 3–39 Rear axle air-operated shift system. *(Courtesy of Dana Corporation)*

predetermined speed. *Technician B* says air pressure is supplied to the front and rear axle shift units when the axles are in high range. Who is right?

A. A only
B. B only
C. Both A and B
D. Neither A nor B

125. Refer to Figure 3–39. When shifting from high to low range, both axles shift too slowly. The cause of this problem could be:

A. a restricted exhaust passage in the quick release valve.
B. an air leak in the air line from the range shifter valve to the quick release valve.
C. low air pressure in the air supply tank.
D. a restricted air line between the solenoid valve and the range shifter valve.

Hint

Although some vehicles are equipped with electrical shift units, most axles are equipped with pneumatic shift systems. There are two air-activated shift systems predominantly used to select the range of a dual range tandem axle or to engage a differential lockout. Usually the air shift unit is not serviceable. If it is found defective, it should be replaced.

Task 17

Inspect, repair, or replace power divider (interaxle differential) assembly.

126. A truck with a power divider has power applied to the tandem forward rear drive axle while no power is applied to the rearward rear drive axle. The MOST likely cause of the malfunction is:

A. broken teeth on the forward drive axle ring gear.
B. broken teeth on the rear drive axle ring gear.
C. stripped output shaft splines.
D. damaged interaxle differential.

127. A power divider differential shows extremely high temperature damage to the interaxle differential. *Technician A* says a plugged oil line probably caused the damage. *Technician B* says the driver not locking the two tandem axles together during slippery conditions probably caused the damage. Who is right?

A. A only
B. B only
C. Both A and B
D. Neither A nor B

128. To remove a side gear from the power divider in Figure 3–40, the Technician must:

A. remove the power divider cover and all applicable gears as an assembly.
B. disconnect the airline; remove the output shaft yoke, power divider cover, and all applicable gears as an assembly.
C. remove the differential carrier and separate the differential gears from the power divider gears.
D. remove the power divider cover and begin disassembling and separating the gears of the power divider.

Hint

The interaxle differential is an integral part of the forward rear axle differential carrier. Components of the interaxle differential are the same as in a regular differential: a spider (or cross), differential pinion gears, a case, washers, and the side gears. The interaxle differential assembly containing the pinion gears and case is splined to the input shaft. Therefore, engine torque is transmitted from the input shaft to the case and pinion gears. Engine torque is transmitted from the pinion gears to the front side gear and a helical gear attached to this gear. Because this helical gear is meshed with another helical gear that is splined to the front differential pinion shaft, engine torque is transmitted through the helical gears to the front differential pinion gear, ring gear, and axles in the front differential. Engine torque is also transmitted from the pinion gears and case in the interaxle differential to the rear side gear that is splined to the output shaft that is in turn connected to the interaxle propeller shaft. Therefore, engine torque is transmitted through the rear side gear to the output shaft and interaxle propeller shaft to the pinion shaft and ring gear in the rear differential. The interaxle differential divides the engine torque between the front and rear differentials. If one rear wheel

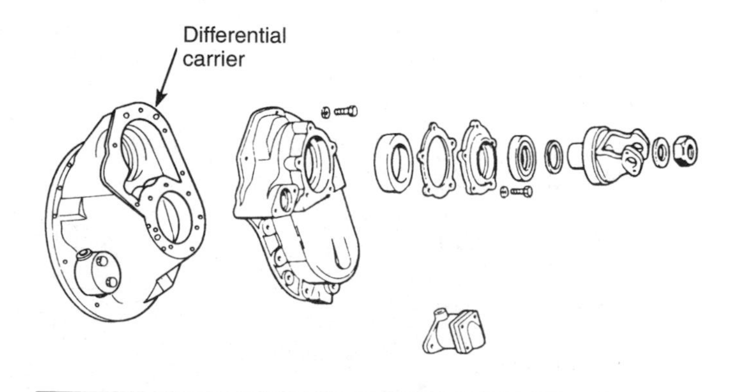

Figure 3–40 Differential carrier and power divider cover.

is on a slippery road surface, the interaxle differential allows that wheel to turn at a different speed than the other wheels.

Task 18

Inspect, adjust, repair, or replace air-operated power divider (interaxle differential) lockout assembly, including diaphragms, seals, springs, yokes, pins, lines, hoses, fittings, and controls.

129. A truck driver complains that his or her tractor will not shift out of interaxle differential lock. Which of these could be the cause?
   A. Shift shaft spring
   B. A broken shift shaft
   C. An open or damaged air line
   D. A stripped differential clutch collar

130. While discussing the operation of the interaxle differential and two-speed rear axle shift system in Figure 3–41, *Technician A* says the axle range shifter valve is locked in position when the interaxle differential lockout is engaged. *Technician B* says the differential lockout should be engaged to prevent excessive wheel slip when driving on slippery road surfaces. Who is right?
   A. A only
   B. B only
   C. Both A and B
   D. Neither A nor B

Hint

A air-operated interaxle differential locking mechanism contains a shifter fork and a sliding lockout clutch that are splined to the input shaft. When the lockout lever in the cab is operated, air pressure is supplied to a piston in the shift assembly. This air pressure moves the piston, shifter fork, and sliding lockout clutch. Notches on the lockout clutch are now engaged with matching notches on the front side gear in the interaxle differential. Under this condition, the input shaft is locked to the side gear and the interaxle differential is locked and must turn as a unit. In the locked mode all four rear wheels must turn together at the same speed, and wheel slip on slippery road surfaces is eliminated.

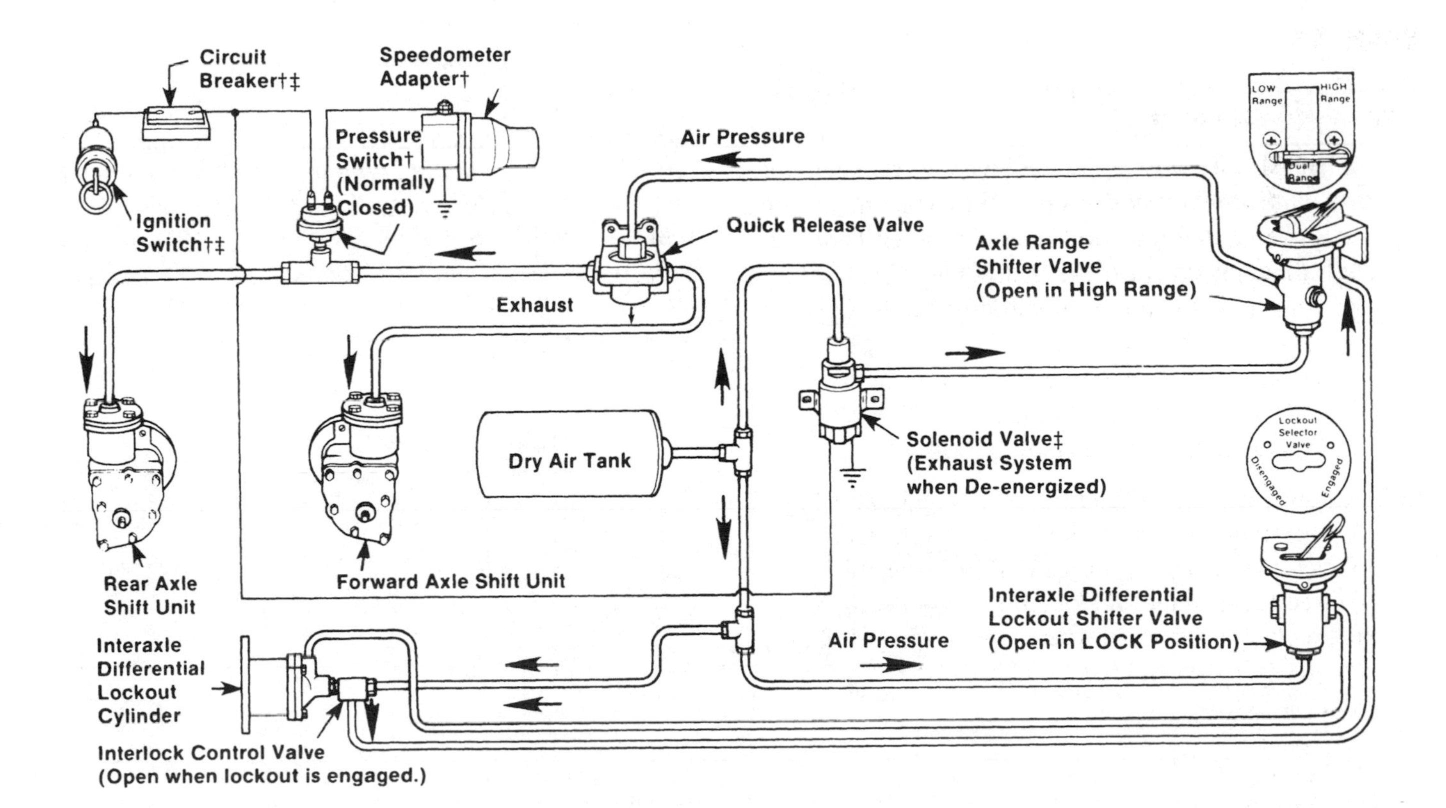

Figure 3–41 Interaxle lockout and axle range shift systems. *(Courtesy of Dana Corporation)*

Task 19

Inspect and measure drive axle housing mating surfaces and alignment; determine needed repairs.

131. The MOST likely result of metal burrs and gouges on the axle housing and differential carrier mating surfaces is:
    A. worn ring and pinion gear because of misalignment.
    B. excessive wear on pinion bearings.
    C. lubricant leaks.
    D. ring and pinion gear noise.

132. A recently overhauled rear axle assembly has a lubricant leak between the differential carrier and the axle housing. *Technician A* says a piece of the old gasket may have been left on the differential carrier mating surface. *Technician B* says the differential carrier mating surface may be warped from continual high torque driving conditions. Who is right?
    A. A only
    B. B only
    C. Both A and B
    D. Neither A nor B

Hint

The differential carrier assembly can be steam cleaned while mounted in the housing as long as all openings are tightly plugged. Once removed from its housing, do not steam clean the differential carrier or any axle components. Steam cleaning at this time could allow water to be trapped in cored passages, leading to rust, lubricant contamination, and premature component wear. Once the axle housing and differential are properly cleaned, inspect for any signs of cracking and check the mating surfaces for notches, visible steps, or grooves. Most damage done to the differential and axle housing mating surfaces is caused by poor assembly practices. Most of this type of damage can be repaired by filing or slightly grinding the surface smooth and using an appropriate sealant.

Task 20

Inspect, repair, or replace drive axle lubrication system components (i.e., pump, troughs, collectors, slingers, tubes, and filters).

133. What should a Technician do when servicing a differential oil pump?
    A. Replace all internal hoses or lines.
    B. Pack the pump full of lithium-based grease to ensure priming of the system after installation.
    C. Replace all external hoses.
    D. Check the pump for smooth operation and blow forced air through all passages.

134. The rear axle oil pump in Figure 3–42 is driven by the rear axle:
    A. input shaft.
    B. interaxle differential output shaft.
    C. pinion shaft on the front differential.
    D. lockout sliding clutch.

Hint

Drive axle lubrication system components are, in most cases, a splash feed type system. In such systems the lubrication level is at a point where vital components are lubricated by the simple action of rotating parts coming into contact with the lubricant and then throwing off that lubricant as they are spun around. Additionally, some axles incorporate a pump and hoses or tubes to disperse

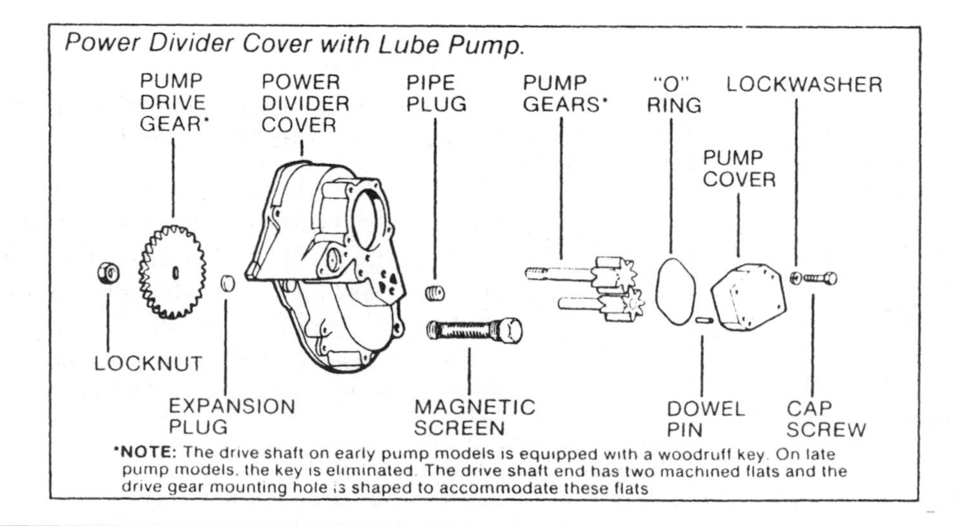

Figure 3–42 Rear axle oil pump. *(Courtesy of Dana Corporation)*

the lubricant to critical parts in the axle differential. Always remember that passages need to be kept clear. Usually forcing air through the passages during rebuilding is sufficient. Some rear axles have a spin-on lubricant filter.

Task 21

Inspect and replace drive axle shafts.

135. The following are all reasons to replace an axle shaft EXCEPT:
   A. minute surface cracks in the axle shaft.
   B. a bent axle shaft.
   C. pitting of the axle shaft.
   D. twisting of the axle shaft.

136. When installing a rear drive axle, *Technician A* says a gasket is not required between the axle shaft flange and the hub mating surface. *Technician B* says the tapered dowels must be installed in the stud openings before the washers and axle stud nuts are installed. Who is right?
   A. A only
   B. B only
   C. Both A and B
   D. Neither A nor B

Hint

For longer life, the surface of axle shafts are case hardened for wear resistance. A lower hardness, ductile core is retained for toughness. Fatigue failures can occur in either or both of these areas. Failures can be classified into three types that are noticeable during a close visual inspection: surface, torsional (or twisting), and bending. Overloading the truck beyond the rated capacity or abusive operation of the truck over rough terrain generally causes fatigue failures.

Task 22

Remove and replace wheel assembly; check drive axle wheel/hub seal and axle flange gasket for leaks; determine needed repairs.

137. A truck has an extremely hot wheel hub after a test drive. The LEAST likely cause would be:
   A. axle shaft damage.
   B. axle shaft bearing damage.
   C. air line damage to the brakes.
   D. poor quality lubricant.

138. A rear hub seal has repeated failures, allowing grease to contaminate the brake shoes. The seal wear sleeve has been replaced each time a new seal is installed. *Technician A* says to check the fit of the outer seal housing in the hub bore. *Technician B* says to decrease the hub end play when the bearings are adjusted. Who is right?
   A. A only
   B. B only
   C. Both A and B
   D. Neither A nor B

Hint

There are slight differences in bearing and seal service between grease and oil lubricated systems and front and rear drive axles. On a spoke wheel the brake drum is mounted on the inboard side of the wheel and is held in place with nuts. Servicing inboard brake drums on spoke wheels involves removing the single or dual wheel and drum as a single assembly. This involves removing the hub nut and disturbing hub components, so bearing and seal service are required. On disc wheels, the brake drum usually is mounted on the outboard side of the disc hub. The drum fits over the wheel studs and is secured between the wheel and hub. This means the wheel and drum can be dismounted without disturbing the hub nut. Outboard drums can be serviced without servicing the bearings and seals.

The wheel bearings are important components of the rear axle housing. The seals that protect the bearings can fail and allow oil to leak thus causing damage to other components. Under certain circumstances, the dirt or road water may enter the seal and cause more extensive damage. If possible seal damage is suspected, the bearing(s) should be removed, cleaned, and repacked. Always use a new seal whenever the housing or the bearings are removed. Be sure to check the housing vent and refill with the proper type and amount of lubricant after service.

Task 23

Diagnose drive axle wheel bearing noises and damage; determine needed repairs.

139. While discussing front and rear wheel bearing diagnosis, *Technician A* says the growling noise produced by a defective front wheel bearing is most noticeable while driving straight ahead. *Technician B* says the growling noise produced by a defective rear wheel bearing is most noticeable during acceleration. Who is right?
   A. A only
   B. B only
   C. Both A and B
   D. Neither A nor B

140. While discussing rear wheel bearing service, *Technician A* says bearing cups in an aluminum hub may be removed by heating the hub with an oxyacetylene torch. *Technician B* says if the rear wheel hub has a two-piece seal, the wear sleeve should be installed so it is even with the spindle shoulder. Who is right?
    A. A only
    B. B only
    C. Both A and B
    D. Neither A nor B

Hint

Problems associated with different models of axles and types of gearing can be specific to one model only. However, in most cases one problem area generally can be caused by the same malfunction. The Technician must bear in mind that universal joints, transmissions, tires, and drivelines can create noises that are often blamed on the drive axles. Experience and a keen ear are two of the most valuable tools in diagnosing driveline noises. Keep a record of past repair information to help diagnose future problems that are not obvious.

Task 24

Clean, inspect, relubricate, and replace wheel bearings; replace seals and wear rings; adjust drive axle wheel bearings.

141. While discussing truck rear wheel bearing adjustment, *Technician A* says the rear wheel bearings should be preloaded on rear axles over a certain axle weight rating (AWR). *Technician B* says the rear wheel lockwashers may be tang type or dowel type. Who is right?
    A. A only
    B. B only
    C. Both A and B
    D. Neither A nor B

142. While servicing rear axle wheel bearings and seals, *Technician A* says the lips on the rear hub seal must face toward the center of the hub. *Technician B* says after installing a new bearing cup, a 0.015 in. (0.0381 mm) feeler gauge should fit between the cup and the hub. Who is right?
    A. A only
    B. B only
    C. Both A and B
    D. Neither A nor B

Hint

Under normal operating conditions, axle wheel bearings are protected by lubricant carried into the wheel ends by the motion of axle shafts and gearing. Lubricant becomes trapped in the cavities of the wheel end and remains there, ensuring that lubricant is instantly available when the vehicle is in motion. In cases where wheel equipment is being installed, either new or after maintenance activity, these cavities are empty. Bearings must be manually supplied with adequate lubrication or they will be severely damaged before the normal motion of gearing and axle shafts can force lubricant to the hub ends of the housing. Improper wheel bearing end play can result in looseness in the bearings or steering problems. When properly adjusted, the hub and wheel should rotate freely with the specified end play.

Task 25

Inspect and test for accuracy drive axle temperature gauge sensor; determine needed repairs.

143. A drive axle temperature gauge is not accurate. *Technician A* says to remove the instrument panel gauge and test for proper movement. *Technician B* says to disconnect the drive axle temperature sensor and substitute with a variable resistance to check for proper movement of the needle. Who is right?
    A. A only
    B. B only
    C. Both A and B
    D. Neither A nor B

144. In the rear axle temperature gauge circuit in Figure 3–43 the gauge indicates cold axle temperature at all times. *Technician A* says the wire between the gauge and the sender may have an open circuit. *Technician B* says the sender may have lower than specified resistance. Who is right?
    A. A only
    B. B only
    C. Both A and B
    D. Neither A nor B

Hint

The most reliable way to test a drive axle temperature sensor for accuracy is to obtain a temperature-to-resistance chart from the manufacturer. This allows the Technician to determine if the temperature sensor is sending a signal appropriate for the temperature it is encountering. Another way of diagnosing the temperature-indicating circuit is to place a variable resistance in the place of the sensor. Varying a known resistance and checking the tem-

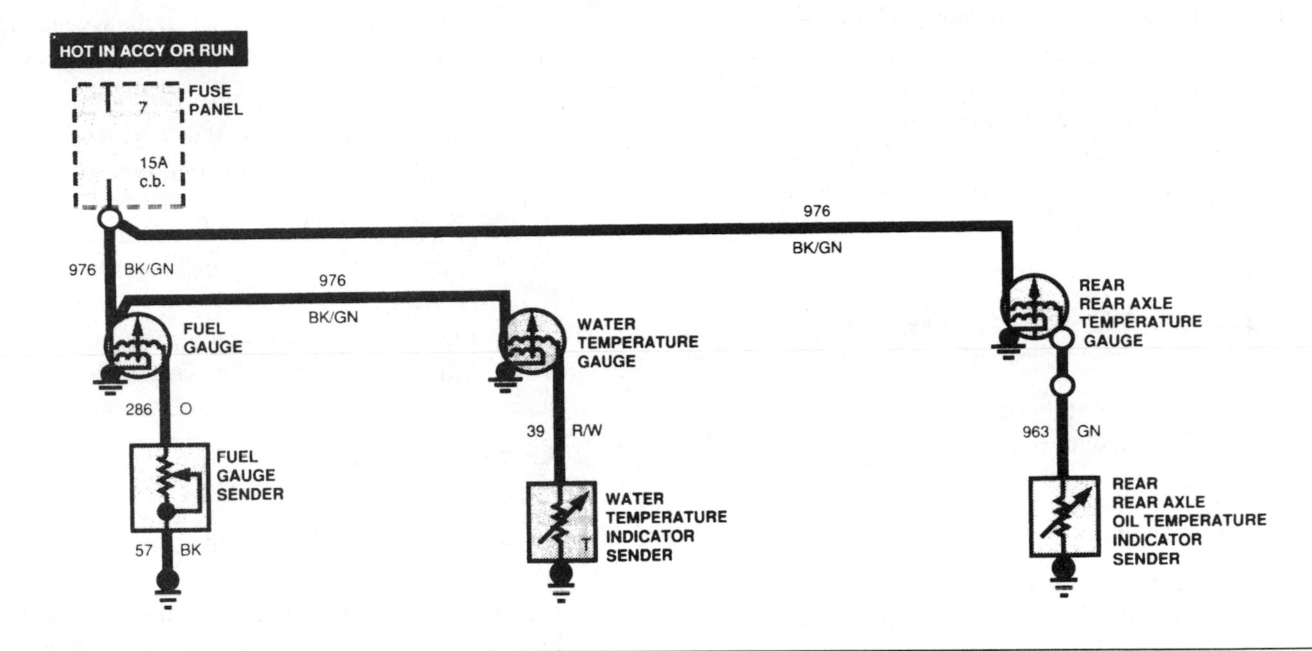

Figure 3–43 Rear axle temperature gauge circuit. *(Courtesy of Ford Motor Company)*

perature display for proper operation is also a good way of helping pinpoint the source of the problem.

Task 26

Check, adjust, and replace wheel speed sensor(s).

145. While discussing wheel speed sensors, *Technician A* says a broken wheel speed sensor wire can be fixed with a crimp splice or equivalent. *Technician B* says an ohmmeter may be used to test the wheel speed sensor signal while rotating the wheel. Who is right?
 A. A only
 B. B only
 C. Both A and B
 D. Neither A nor B

146. While diagnosing wheel speed sensors, *Technician A* says that if one of the teeth is missing on the toothed ring, this ring must be replaced. *Technician B* says mud and dirt on the toothed ring can affect the wheel speed sensor signal. Who is right?
 A. A only
 B. B only
 C. Both A and B
 D. Neither A nor B

Hint

Wheel speed sensors are very simple looking but rather sophisticated components. Wheel speed sensor produces an alternating voltage signal that is sent to the control module. The wiring needs to be of the absolute best integrity to be a reliable conductor for the AC signal to accurately reach the module. A break in the wire at the wheel can actually cause total module failure by drawing water up to the connector pins on the module.

Task 27

Clean, inspect, relubricate, and replace wheel end-locking hubs.

147. A vehicle with wheel end-locking hubs will not switch into high. *Technician A* says to replace the shift collar return spring. *Technician B* says to clean out and repack the hub. Who is right?
 A. A only
 B. B only
 C. Both A and B
 D. Neither A nor B

148. A Technician finds dirt when removing the wheel end-locking hub. What should the Technician do?
 A. Replace the entire hub.
 B. Replace the bearings and repack the hub.
 C. Repack the bearings.
 D. Clean, inspect, and repack the bearings, and replace the seal.

Hint

Some 4WD systems on trucks and utility vehicles have front wheel drive hubs that can be disengaged when the vehicle is operating in 2WD. When these front wheel hubs are unlocked in 2WD, the front drivetrain components stop turning to reduce wear on these components. The front wheel hubs must be locked in 4WD. Some front hubs lock automatically, but other hub designs require the driver to manually turn a knob or lever at the center of each front hub to lock the hubs.

Answers to Questions

1. Answer A is wrong because this would cause a growling noise with the clutch pedal depressed.

   Answer B is wrong because this would not cause clutch slipping.

   Answer C is right because a leaking rear main seal will wet the friction surface of the clutch and cause slippage.

   Answer D is wrong because this would cause chatter upon engagement.

2. Answer A is wrong because too much clutch pedal free play would affect disengagement, not engagement.

   Answer B is right because binding linkage can cause slippage and a burnt friction disc.

   Answer C is wrong because Technician A is wrong and Technician B is right.

   Answer D is wrong because Technician A is wrong and Technician B is right.

3. Answer A is wrong because oil contamination may cause disc failure.

   Answer B is wrong because worn torsion springs may cause disc failure.

   Answer C is right because worn U-joints will not cause premature clutch failure.

   Answer D is wrong because worn clutch linkage may cause disc failure.

4. Answer A is wrong because a sticking release bearing causes slippage.

   Answer B is wrong because oil contamination causes slippage.

   Answer C is right because a worn pilot bearing is the LEAST likely cause of clutch slippage.

   Answer D is wrong because a worn linkage causes slippage.

5. Answer A is wrong because air in the hydraulic clutch control system could affect shifting into reverse.

   Answer B is wrong because not enough clutch pedal free play could cause grinding in reverse.

   Answer C is right because a noisy pilot bearing would not affect clutch release and shifting into reverse.

   Answer D is wrong because a leaking or weak air servo cylinder is a likely cause of grinding when shifting into reverse.

6. **Answer A is right** because the pedal free travel should be about 1.5 to 2 inches (38.1 to 50.8 mm).

   Answer B is wrong because insufficient clutch pedal free travel does not cause hard transmission shifting. This problem may be caused by excessive clutch pedal free travel.

   Answer C is wrong because Technician A is right and Technician B is wrong.

   Answer D is wrong because Technician A is right and Technician B is wrong.

7. Answer A is wrong because an improperly adjusted linkage may cause the clutch to remain engaged.

   Answer B is wrong because improper adjustment of the hydraulic slave cylinder may cause the clutch to remain engaged.

   Answer C is wrong because a frozen pilot bearing may cause the clutch to remain engaged.

   Answer D is right because worn clutch disc is the LEAST likely cause of a clutch remaining engaged.

8. Answer A is wrong because the master cylinder is part of the system.

   Answer B is wrong because the tubes are part of the system.

   Answer C is wrong because a slave cylinder is part of the system.

   Answer D is right because a clutch cable is not used in a hydraulic clutch system.

9. **Answer A is right** because you do not need to inspect the quality and amount of lubrication. Just lube the release bearing unless it is a sealed unit.

   Answer B is wrong because measuring free travel is a necessary step during a clutch release bearing replacement.

Answer C is wrong because inspection for spalling or scoring is a necessary part of a clutch release bearing replacement.

Answer D is wrong because clutch brake runout measurement is a necessary step during a clutch release bearing replacement.

10. Answer A is wrong because a worn release bearing sleeve and bushing does not cause a growling noise with the clutch pedal depressed.

    Answer B is wrong because a defective bearing on the transmission input shaft causes a growling noise only when the clutch pedal is released and the clutch is applied.

    Answer C is right because a defective release bearing causes a growling noise when the clutch pedal is depressed.

    Answer D is wrong because a worn, loose release fork does not cause a growling noise with the clutch pedal depressed.

11. Answer A is wrong because oil on the clutch disc will cause clutch slippage, causing high temperatures.

    Answer B is wrong because this will not allow the clutch to fully engage, causing high temperatures.

    Answer C is wrong because this may not allow proper clutch engagement, resulting in a burned pressure plate.

    Answer D is right because a damaged pilot bearing will not cause clutch slippage or a burned pressure plate.

12. Answer A is wrong because the Technician is checking the thickness of the pressure plate.

    Answer B is right because the Technician is checking the pressure plate thickness.

    Answer C is wrong because the Technician is checking the thickness of the pressure plate.

    Answer D is wrong because the Technician is checking the thickness of the pressure plate.

13. **Answer A is right** because a depth gauge may be used to measure the depth of the clutch facing from the rivet heads to the facing surface.

    Answer B is wrong because the direction of a single disc clutch is not reversible between the pressure plate and the flywheel. The side with the torsion spring protrusion must face toward the pressure plate.

    Answer C is wrong because Technician A is right and Technician B is wrong.

    Answer D is wrong because Technician A is right and Technician B is wrong.

14. Answer A is wrong because a scored flywheel may cause clutch slipping and rapid facing wear but it does not cause harsh application.

    Answer B is right because broken or weak torsion springs cause harsh clutch application.

    Answer C is wrong because a scored pressure plate may cause clutch slipping and rapid facing wear, but it does not cause harsh application.

    Answer D is wrong because worn clutch facings may cause clutch slipping, but this problem does not cause harsh application.

15. Answer A is wrong because a broken intermediate plate can be caused by poor driver technique, but Technician B is also right.

    Answer B is wrong because a truck pulling loads that are too heavy can cause a broken intermediate plate, but Technician A is also right.

    Answer C is right because both Technician A and Technician B are right.

    Answer D is wrong because both Technician A and Technician B are right.

16. **Answer A is right** because a release bearing that is not moving freely will not cause cracks.

    Answer B is wrong because an improperly manufactured friction disc could cause cracking on the intermediate plate.

    Answer C is wrong because improper clutching technique would affect both surfaces of the intermediate plate.

    Answer D is wrong because this would affect clutch pedal feel, and have no effect on the intermediate plate.

17. Answer A is wrong because a twin disc clutch as shown in Figure 3–7 is used for high-torque applications, but Technician B is also right.

    Answer B is wrong because the twin disc clutch system uses an intermediate plate in the system, but Technician A is also right.

    Answer C is right because both Technician A and Technician B are right.

Answer D is wrong because both Technician A and Technician B are right.

18. Answer A is wrong because aligning the disc is necessary to install a pull-type clutch.

    Answer B is wrong because adjusting the release bearing is necessary to install a pull-type clutch.

    Answer C is right because you would not resurface the limited torque clutch brake unless there is surface damage or unevenness.

    Answer D is wrong because lubricating the pilot bearing is necessary to install a pull-type clutch.

19. **Answer A is right** because a clutch brake needs to be inspected for wear and fatigue in the same manner as the pressure plate.

    Answer B is wrong because the friction face needs to be inspected.

    Answer C is wrong because Technician A is right and Technician B is wrong.

    Answer D is wrong because Technician A is right and Technician B is wrong.

20. **Answer A is right** because a defective adjuster ring may cause improper clutch adjustment.

    Answer B is wrong because a binding linkage may cause improper clutch release or application, but it should not affect clutch adjustment.

    Answer C is wrong because Technician A is right and Technician B is wrong.

    Answer D is wrong because Technician A is right and Technician B is wrong.

21. Answer A is wrong because the wear compensator is a replaceable part, but Technician B is also right.

    Answer B is wrong because the wear compensator will keep the free travel in the clutch pedal within specifications, but Technician A is also right.

    Answer C is right because both Technician A and Technician B are right.

    Answer D is wrong because both Technician A and Technician B are right.

22. **Answer A is right** because the engine is idling and the clutch pedal is fully depressed and the clutch is released. At this point the transmission input shaft is turning inside the pilot bearing, and the clutch is not keeping the input shaft centered.

    Answer B is wrong because with the pedal released and the clutch engaged, the clutch will keep the input shaft centered, and the pilot bearing, input shaft, and clutch turn as an assembly. Therefore, the input shaft is no longer rotating in the pilot bearing.

    Answer C is wrong because with the pedal released and the clutch engaged, the clutch will keep the input shaft centered, and the pilot bearing, input shaft, and clutch turn as an assembly. Therefore, the input shaft is no longer rotating in the pilot bearing.

    Answer D is wrong because with the pedal released and the clutch engaged, the clutch will keep the input shaft centered, and the pilot bearing, input shaft, and clutch turn as an assembly. Therefore, the input shaft is no longer rotating in the pilot bearing.

23. **Answer A is right** because a slide hammer can be used to remove the pilot bearing.

    Answer B is wrong because a pilot bearing is removed with a slide hammer or a puller.

    Answer C is wrong because Technician A is right and Technician B is wrong.

    Answer D is wrong because Technician A is right and Technician B is wrong.

24. **Answer A is right** because crankshaft end play is being measured in Figure 3–8.

    Answer B is wrong because crankshaft end play is being measured in Figure 3–8.

    Answer C is wrong because crankshaft end play is being measured in Figure 3–8.

    Answer D is wrong because crankshaft end play is being measured in Figure 3–8.

25. **Answer A is right** because cracking of the boltholes is the MOST likely damage.

    Answer B is wrong because although this damage occurs, it is not as common as cracking.

    Answer C is wrong because warping of the flywheel mounting surface is not a common problem.

    Answer D is wrong because heat checking occurs on the flywheel face, not the mounting surface.

26. Answer A is wrong because you do not cool the ring gear in a freezer overnight.

Answer B is wrong because heating the ring gear is only part of the correct procedure.

Answer C is wrong because heating the flywheel and cooling the ring gear is opposite of the correct procedure.

Answer D is right because the ring gear should be heated to expand it, and the flywheel should be cooled to shrink it before ring gear installation on the flywheel.

27. Answer A is wrong because the flywheel should be pushed forward and then rotated.

    Answer B is right because removing an excessive amount of metal from the flywheel friction surface may adversely affect clutch operation because this process moves the clutch plate toward the flywheel and away from the pressure plate. Many truck manufacturers and machinists agree that 0.040 in. (1.01 mm) is the maximum amount of metal that should be machined from a flywheel.

    Answer C is wrong because Technician A is wrong and Technician B is right.

    Answer D is wrong because Technician A is wrong and Technician B is right.

28. Answer A is wrong because the maximum pilot bore runout is 0.005 in. (0.127 mm).

    Answer B is right because the maximum pilot bore runout is 0.005 in. (0.127 mm).

    Answer C is wrong because the maximum pilot bore runout is 0.005 in. (0.127 mm).

    Answer D is wrong because the maximum pilot bore runout is 0.005 in. (0.127 mm).

29. Answer A is wrong because the pilot bearing bore is in the crankshaft, and the crankshaft must be repaired or replaced if this bore runout is excessive.

    Answer B is wrong because the pilot bearing bore is in the crankshaft, and the crankshaft must be repaired or replaced if this bore runout is excessive.

    Answer C is wrong because both Technician A and Technician B are wrong.

    Answer D is right because both Technician A and Technician B are wrong.

30. Answer A is wrong because any grinding noise should be diagnosed by the location of the noise and the driving conditions when it occurs.

    Answer B is right because pilot bore runout should be measured only when the pilot bearing is found to be worn.

    Answer C is wrong because Technician A is wrong and Technician B is right.

    Answer D is wrong because Technician A is wrong and Technician B is right.

31. Answer A is wrong because clutch housing face misalignment does not cause a growling noise with the clutch pedal depressed.

    Answer B is wrong because clutch housing face misalignment does not cause the transmission to jump out of gear.

    Answer C is right because clutch housing face misalignment causes clutch chatter and grabbing.

    Answer D is wrong because clutch housing face misalignment does not cause wear on the clutch release bearing.

32. **Answer A is right** because overtightening of the transmission can cause undue pressure on the housing face.

    Answer B is wrong because an overheated clutch disc and pressure plate would indicate an overheating a clutch.

    Answer C is wrong because improper torque sequence would have no effect on the clutch housing.

    Answer D is wrong because overtightening creates heat inside the transmission, not in the housing.

33. **Answer A is right** because a worn clutch housing can cause misalignment of the mating surfaces of the clutch assembly.

    Answer B is wrong because the clutch housing must be replaced if it is worn.

    Answer C is wrong because Technician A is right and Technician B is wrong.

    Answer D is wrong because Technician A right and Technician B is wrong.

34. **Answer A is right** because you attach a dial indicator to the center of the flywheel because you are using the flywheel to measure against the housing.

    Answer B is wrong because the dial indicator must be attached to the flywheel.

Answer C is wrong because the dial indicator must be attached to the flywheel.

Answer D is wrong because the dial indicator must be attached to the flywheel.

35. Answer A is wrong because a broken transmission mount may cause a thudding noise each time the clutch pedal is released during a shift.

Answer B is wrong because a broken transmission mount can cause high-speed vibrations.

Answer C is wrong because a broken transmission mount can cause moderate speed vibrations.

Answer D is right because a broken transmission mount does not cause a growling noise below 50 mph (80 km/h).

36. **Answer A is right** because improper handling outside of the transmission caused the gear damage. Normal operation could not cause this type of damage.

Answer B is wrong because a transmission "walking" between gears causes a different kind of gear tooth damage.

Answer C is wrong because worn bearings would cause a different kind of gear tooth damage.

Answer D is wrong because normal wear has a different pattern.

37. Answer A is wrong because a dirty transmission would not leave any marks here.

Answer B is right because normal transmission vibrations will cause fretting on the bearing cup.

Answer C is wrong because a spun bearing would cause severe scoring of the bearing bore.

Answer D is wrong because this is not a sign of manufacturing defects.

38. Answer A is wrong because bearings can cause the transmission to jump out of gear.

Answer B is wrong because detents can cause the transmission to jump out of gear.

Answer C is wrong because worn gear teeth on the synchronizers may cause the transmission to jump out of gear.

Answer D is right excessive crankshaft end play may affect clutch pedal free play and clutch operation, but this problem does not cause the transmission to jump out of gear.

39. **Answer A is right** because a broken detent spring may cause the transmission to jump out of gear, but this problem does not cause noise in the transmission.

Answer B is wrong because a worn input bearing may cause a noise in the transmission.

Answer C is wrong because a defective output bearing may cause a noise in the transmission.

Answer D is wrong because a worn counter shaft bearing may cause a noise in the transmission.

40. Answer A is wrong because this is not a reliable method of testing shift linkage.

Answer B is right because you have to disconnect the linkage at the transmission and check the linkage inside the transmission separately from checking the linkage outside the transmission.

Answer C is wrong because Technician A is wrong and Technician B is right.

Answer D is wrong because Technician A is wrong and Technician B is right.

41. Answer A is wrong because worn detents in the shifting tower do not cause clutch wear.

Answer B is right because on a cab-over-engine (COE) tractor, a remote shift lever is connected to the shift bar housing on top of the transmission.

Answer C is wrong because Technician A is wrong and Technician B is right.

Answer D is wrong because Technician A is wrong and Technician B is right.

42. **Answer A is right** because a dirty or plugged air filter will cause this transmission, not shift into high range.

Answer B is wrong because the figure shown is not an electric shift unit.

Answer C is wrong because synchronizers are not part of the high/low range shift system.

Answer D is wrong because wear on the gear teeth will affect shifting quality, not render the system unable to shift.

43. Answer A is wrong because transmission range shifts can only occur when the transmission is in neutral or passing through neutral.

Answer B is wrong because transmission range shifts can only occur when the transmission is in neutral or passing through neutral.

Answer C is right because transmission range shifts can only occur when the transmission is in neutral or passing through neutral.

Answer D is wrong because transmission range shifts can only occur when the transmission is in neutral or passing through neutral.

44. Answer A is wrong because removal of the mount is not necessary to thoroughly inspect it.

    Answer B is wrong because removal of the mount is not necessary to thoroughly inspect it.

    Answer C is right because you visually inspect the mount while still in the vehicle.

    Answer D is wrong because replacing the mount if it is only suspected is not an accepted practice.

45. Answer A is wrong because transmission mounts absorb torque from the engine, but Technician B is also right.

    Answer B is wrong because transmission mounts absorb drivetrain vibration, but Technician A is also right.

    Answer C is right because both Technician A and Technician B are right.

    Answer D is wrong because both Technician A and Technician B are right.

46. Answer A is wrong because it is important to check the output shaft radial movement because this may have caused the seal failure.

    Answer B is right because it is not necessary to check gear wear when replacing an output shaft seal.

    Answer C is wrong because it is important to check for proper transmission venting, because a restricted vent may have caused the seal failure.

    Answer D is wrong because it is important to check sealing surface wear on the surface that contacts the seal lip, because a scored sealing surface may have caused the seal failure.

47. **Answer A is right** because when the breather is plugged, internal pressure can build inside the transmission and force the gaskets out of place.

    Answer B is wrong because the quickest and easiest check is to inspect the breather, and this is the most likely cause of the problem.

    Answer C is wrong because the quickest and easiest check is to inspect the breather, and this is the most likely cause of the problem.

    Answer D is wrong because the quickest and easiest check is to inspect the breather, and this is the most likely cause of the problem.

48. Answer A is wrong because extended high torque operating conditions may be a cause of bearing failure, but not the leading cause.

    Answer B is right because dirt and contaminants in the lubricant are the most common cause of transmission bearing failure.

    Answer C is wrong because operating the tractor in high temperature situations may be a cause of bearing failure, but not the leading cause.

    Answer D is wrong because poor quality of lubricant may be a cause of bearing failure, but not the leading cause.

49. Answer A is wrong because overfilling a manual transmission could result in overheating of the transmission, but Technician B is also right.

    Answer B is wrong because overfilling a transmission could cause excessive aeration of the transmission fluid, but Technician A is also right.

    Answer C is right because both Technician A and Technician B are right.

    Answer D is wrong because both Technician A and Technician B are right.

50. Answer A is wrong because the proper procedure is to be sure the lubricant is level with the bottom of the filler plug hole.

    Answer B is wrong because the proper procedure is to be sure the lubricant is level with the bottom of the filler plug hole.

    Answer C is right because the proper procedure is to be sure the lubricant is level with the bottom of the filler plug hole.

    Answer D is wrong because the proper procedure is to be sure the lubricant is level with the bottom of the filler plug hole.

51. **Answer A is right** because item H is mounted between the shifter rail openings in the transmission.

    Answer B is wrong because the shift tower in Figure 3–14 is from a cab-over-engine truck.

    Answer C is wrong because Technician A is right and Technician B is wrong.

    Answer D is wrong because Technician A is right and Technician B is wrong.

52. **Answer A is right** because damaged friction rings on the synchronizer blocking rings is the LEAST likely cause of the transmission jumping out of gear. This problem causes hard shifting.

    Answer B is wrong because a broken detent spring may cause a transmission to jump out of gear.

    Answer C is wrong because a worn shift fork may cause a transmission to jump out of gear.

    Answer D is wrong because worn teeth on the synchronizer blocker ring and worn dog teeth on the fifth speed gear may cause the transmission to jump out of fifth gear.

53. **Answer A is right** because pilot bearing shaft cracking is the LEAST likely damage to check on an input shaft.

    Answer B is wrong because checking for gear teeth damage is a valid check.

    Answer C is wrong because checking for input spline damage is a valid check.

    Answer D is wrong because checking for cracking or other fatigue wear to the input shaft spline is a valid check.

54. Answer A is wrong because a worn input shaft bearing may cause a growling noise in neutral with clutch pedal released, but Technician B is also right.

    Answer B is wrong because worn needle bearings that support the output shaft in the rear of the input shaft may cause a growling noise in neutral with the clutch pedal released, but Technician A is also right.

    Answer C is right because both Technician A and Technician B are right.

    Answer D is wrong because both Technician A and Technician B are right.

55. Answer A is wrong because the dial indicator is installed to measure the output shaft end play.

    Answer B is wrong because the dial indicator is installed to measure the output shaft end play.

    Answer C is wrong because the dial indicator is installed to measure the output shaft end play.

    Answer D is right because the dial indicator is installed to measure the output shaft end play.

56. **Answer A is right** because a shim must be removed between the end cover and the transmission case to reduce the end play.

    Answer B is wrong because there is no shim adjustment between the front of the output shaft and the rear of the input shaft.

    Answer C is wrong because Technician A is wrong and Technician B is right.

    Answer D is wrong because Technician A is wrong and Technician B is right.

57. Answer A is wrong because it is not standard for manufacturers to provide timing marks on the gears.

    Answer B is right because you mark the gears before disassembly, then align those marks during assembly.

    Answer C is wrong because this is not an appropriate way to time the gears.

    Answer D is wrong because a specific tooth is not identified as a timing tooth, nor is there a mark on the main shaft.

58. **Answer A is right** because a thinner shim must be installed between the rear counter shaft bearing retainer and the rear counter shaft outer bearing race to increase counter shaft end play.

    Answer B is wrong because a thinner shim must be installed between the rear counter shaft bearing retainer and the rear counter shaft outer bearing race to increase counter shaft end play.

    Answer C is wrong because a thinner shim must be installed between the rear counter shaft bearing retainer and the rear counter shaft outer bearing race to increase counter shaft end play.

    Answer D is wrong because a thinner shim must be installed between the rear counter shaft bearing retainer and the rear counter shaft outer bearing race to increase counter shaft end play.

59. Answer A is wrong because the component being installed on the transmission output shaft is the speedometer rotor.

    Answer B is right because the component being installed on the transmission output shaft is the speedometer rotor.

    Answer C is wrong because the component being installed on the transmission output shaft is the speedometer rotor.

    Answer D is wrong because the component being installed on the transmission output shaft is the speedometer rotor.

60. Answer A is wrong because a bent output shaft may cause driveline vibration, but Technician B is also right.

    Answer B is wrong because a bent output shaft may cause premature wear on the output shaft bearing, but Technician A is also right.

    Answer C is right because both Technician A and Technician B are right.

    Answer D is wrong because both Technician A and Technician B are right.

61. Answer A is wrong because the reverse idler gear in Figure 3–19 is item 21.

    Answer B is wrong because the reverse idler gear in Figure 3–19 is item 21.

    Answer C is wrong because the reverse idler gear in Figure 3–19 is item 21.

    Answer D is right because the reverse idler gear in Figure 3–19 is item 21.

62. Answer A is wrong because the transmission gears in Figure 3–20 are positioned to provide low gear.

    Answer B is right because the transmission gears in Figure 3–20 are positioned to provide low gear.

    Answer C is wrong because the transmission gears in Figure 3–20 are positioned to provide low gear.

    Answer D is wrong because the transmission gears in Figure 3–20 are positioned to provide low gear.

63. Answer A is wrong because chipped, worn teeth on the low reverse sliding gear would result in noise when the transmission is in low or reverse gear.

    Answer B is right because chipped, worn teeth on the reverse idler gear would cause noise only in reverse, because this is the only time the reverse idler gear is driving another gear.

    Answer C is wrong because Technician A is wrong and Technician B is right.

    Answer D is wrong because Technician A is wrong and Technician B is right.

64. **Answer A is right** because you must replace both the gear and the blocker ring to restore proper operation.

    Answer B is wrong because worn dog teeth on the gear may cause hard shifting or jumping out of gear.

    Answer C is wrong because Technician A is right and Technician B is wrong.

    Answer D is wrong because Technician A is right and Technician B is wrong.

65. Answer A is wrong because the dog teeth on the blocker rings should be pointed with smooth surfaces.

    Answer B is right because the threads on the blocker rings should be sharp and not dulled.

    Answer C is wrong because the clearance is important between the blocker ring and the matching gear dog teeth.

    Answer D is wrong because the sleeve should slide freely on the hub splines.

66. Answer A is wrong because replacing the breather is not cleaning the breather.

    Answer B is wrong because soaking the breather in gasoline is not a recommended procedure.

    Answer C is right because the best way to clean a transmission breather is to use solvent and then blow-dry with compressed air.

    Answer D is wrong because wiping the orifice will not clean any dirt trapped inside the breather.

67. **Answer A is right** because item 16 in Figure 3–20 is the transmission breather.

    Answer B is wrong because item 16 in Figure 3–20 is the transmission breather.

    Answer C is wrong because item 16 in Figure 3–20 is the transmission breather.

    Answer D is wrong because item 16 in Figure 3–20 is the transmission breather.

68. **Answer A is right** because you remove the transmission, then the torque converter, then the oil pump.

    Answer B is wrong because the transmission oil pump is not accessible through the oil pan.

    Answer C is wrong because the transmission oil pump is not accessible through the oil pan.

    Answer D is wrong because the transmission oil pump is not accessible through the oil pan.

69. Answer A is wrong because the external oil cooler must be flushed with an agitator-type cleaner, but Technician B is also right.

Answer B is wrong because the torque converter must be flushed with an agitator-type cleaner, but Technician A is also right.

Answer C is right because both Technician A and Technician B are right.

Answer D is wrong because both Technician A and Technician B are right.

70. **Answer A is right** because a broken vehicle speed sensor is the LEAST likely cause when the odometer still functions.

    Answer B is wrong because loose wiring is a possible cause.

    Answer C is wrong because a broken speedometer gauge is a possible cause.

    Answer D is wrong because an open circuit is a possible cause.

71. Answer A is wrong because item 213 in Figure 3–21 is a vehicle speed sensor that sends an analog voltage signal to the VSS calibrator module.

    Answer B is wrong because item 213 in Figure 3–21 is a vehicle speed sensor that sends an analog voltage signal to the VSS calibrator module.

    Answer C is right because item 213 in Figure 3–21 is a vehicle speed sensor that sends an analog voltage signal to the VSS calibrator module.

    Answer D is wrong because item 213 in Figure 3–21 is a vehicle speed sensor that sends an analog voltage signal to the VSS calibrator module.

72. Answer A is wrong because the PTO may be driven from the transmission.

    Answer B is wrong because the PTO may drive a hydraulic hoist pump.

    Answer C is wrong because the PTO may be driven from the transfer case.

    Answer D is right because the PTO is designed for intermittent use.

73. Answer A is wrong because the type of transfer case in Figure 3–22 is mounted on the transmission.

    Answer B is wrong because this type of transfer case is shifted with a mechanical linkage.

    Answer C is wrong because both Technician A and Technician B are wrong.

    Answer D is right because both Technician A and Technician B are wrong.

74. Answer A is wrong because high resistance in circuit 150 from junction S401 to the ground block affects both back-up lights.

    Answer B is wrong because high resistance in circuit 24 from the back-up light switch to junction S402 affects both back-up lights.

    Answer C is wrong because both Technician A and Technician B are wrong.

    Answer D is right because both Technician A and Technician B are wrong.

75. Answer A is wrong because a continually open back-up light switch causes inoperative back-up lights.

    Answer B is right because if circuit 24 from the back-up light switch to junction S402 is shorted to battery positive, the back-up lights are on continually because this condition supplies 12 volts to the lights and the other side of the lights is grounded.

    Answer C is wrong because if circuit 150 from the left hand back-up light to junction S401 is grounded, it has no effect because this side of the lights is normally grounded.

    Answer D is wrong because if circuit 24 from junction S402 to the right hand back-up light is grounded, the resistance of the back-up light bulb is eliminated from the circuit and excessive current blows the fuse.

76. Answer A is wrong because an open circuit in circuit 585 from the transmission temperature indicator light to junction S206 only affects the operation of the indicator light and this defect does not affect the operation of the gauge.

    Answer B is wrong because an open 10-ampere breaker results in 0V at the gauge and the question informs us that 12V are available at the gauge.

    Answer C is wrong because an open circuit in circuit 341 from the circuit breaker to the gauge results in 0V at the gauge and the question informs us that 12V are available at the gauge.

    Answer D is right because an open circuit from junction S206 to the gauge sender results in zero current flow from the gauge through the sender to ground, and this provides a continually low gauge reading.

77. **Answer A is right** because the transmission temperature gauge sender contains a thermistor.

Answer B is wrong because the resistance of the transmission temperature gauge sender increases as the transmission fluid temperature decreases.

Answer C is wrong because Technician A is right and Technician B is wrong.

Answer D is wrong because Technician A is right and Technician B is wrong.

78. Answer A is wrong because poor quality bearings would be evidenced by damage to the bearings themselves.

Answer B is right because poor quality lubricant is the MOST likely cause.

Answer C is wrong because inferior quality parts would be evidenced by damage other than overheating.

Answer D is wrong because overloading of the drivetrain would be evident in many areas of the drivetrain, not just the transfer case gears.

79. Answer A is wrong because a broken spring will allow the unit to shift into high but not to stay in high.

Answer B is wrong because a faulty air compressor will affect more systems than just the transfer case shifting.

Answer C is wrong because a damaged air line will most likely have the same effect as a faulty air compressor.

Answer D is right because a broken shift shaft is the MOST likely cause.

80. Answer A is wrong because the annulus gear is locked to the case in low range, but Technician B is also right.

Answer B is wrong because low range gear reduction is obtained when the sun gear attached to the input shaft drives the planetary carrier attached to the output shaft, but Technician A is also right.

Answer C is right because both Technician A and Technician B are right.

Answer D is wrong because both Technician A and Technician B are right.

81. Answer A is wrong because the vehicle cannot be shifted into four wheel drive low range with the vehicle moving forward at 10 mph (16 km/h).

Answer B is right because operating the vehicle in 4WD for long distances on a paved road causes excessive tire wear.

Answer C is wrong because Technician A is wrong and Technician B is right.

Answer D is wrong because Technician A is wrong and Technician B is right.

82. Answer A is wrong because usually drive shaft vibration will not be evident in the lower speed range.

Answer B is wrong because drive shaft vibration will not be evident in the moderate speed range.

Answer C is wrong because drive shaft vibration will not be evident in the moderate speed range.

Answer D is right because when the truck is above 50 mph (80 km/h) with no load, the drive shaft is spinning nearest to its maximum speed range.

83. Answer A is wrong because a worn center support bearing causes a growling noise, but not a clanging noise when the clutch is released.

Answer B is right because a worn U-joint may cause a single clanging noise when the clutch is released.

Answer C is wrong because a loose differential flange causes differential gear noise when accelerating and decelerating.

Answer D is wrong because excessive end play on the transmission output shaft does not cause a single clanging noise when the clutch pedal is released.

84. **Answer A is right** because a worn, dry U-joint may cause a fast cycling squeaking noise at low speeds.

Answer B is wrong because worn, dry slip joint splines may cause a clunking noise, but this problem does not cause a fast cycling squeaking noise.

Answer C is wrong because Technician A is right and Technician B is wrong.

Answer D is wrong because Technician A is right and Technician B is wrong.

85. Answer A is wrong because driveline phasing problems are usually not evident at low speeds.

Answer B is right because you drive the truck on level terrain at a steady speed above 40 mph (64 km/h) and check for driveline vibration.

Answer C is wrong because driveline phasing problems are usually not dependent on load conditions.

Answer D is wrong because driveline phasing problems are usually not evident at low speeds.

86. Answer A is wrong because a special lithium-based grease will work for slip splines but is not necessary.

Answer B is right because good quality U-joint grease can also be used on slip splines.

Answer C is wrong because Technician A is wrong and Technician B is right.

Answer D is wrong because Technician A is wrong and Technician B is right.

87. Answer A is wrong because support bearings are sealed.

Answer B is wrong because there is usually no pressing necessary for support bearings.

Answer C is right because you fill the entire cavity around the bearing with grease.

Answer D is wrong because orientation of the support bearing is not critical.

88. **Answer A is right** because when you disassemble, you do look for any shims during removal of the old bearing to maintain the original position.

Answer B is wrong because it is not necessary to replace the center support assembly mounting bolts.

Answer C is wrong because driveline angle adjustment should not be necessary if the procedure is done correctly.

Answer D is wrong because installation with hand or air tools should not have any effect on quality.

89. Answer A is wrong because this is not a correct statement.

Answer B is right because driveline angle measurement is given from the front of the vehicle to the rear.

Answer C is wrong because Technician A is wrong and Technician B is right.

Answer D is wrong because Technician A is wrong and Technician B is right.

90. **Answer A is right** because the angle at point E should not exceed 12 degrees.

Answer B is wrong because the angle at point E should not exceed 12 degrees.

Answer C is wrong because the angle at point E should not exceed 12 degrees.

Answer D is wrong because the angle at point E should not exceed 12 degrees.

91. **Answer A is right** because continual overloading may cause overheating of both axles.

Answer B is wrong because the interaxle differential is in the forward axle, and damaged interaxle differential components only cause overheating of this axle.

Answer C is wrong because Technician A is right and Technician B is wrong.

Answer D is wrong because Technician A is right and Technician B is wrong.

92. **Answer A is right** because a worn or improperly adjusted ring and pinion may cause a whining noise only at 55 to 65 mph (88 to 105 km/h).

Answer B is wrong because worn differential side gears usually cause a noise when turning a corner.

Answer C is wrong because in high range the drive is not going through the planetary gearset.

Answer D is wrong because worn pinion bearings cause a growling noise at speeds other than 55 to 65 mph (88 to 105 km/h).

93. Answer A is wrong because there is very little or no gasket deterioration, assuming all the fasteners are properly torqued.

Answer B is wrong because overloading of the drivetrain may cause overheating and excessive wear, but this problem does not cause fluid leakage.

Answer C is right because plugged axle housing breather vent is a very likely cause to this condition.

Answer D is wrong because differential fluid viscosity loss is usually not a problem.

94. **Answer A is right** because scratches, nicks, or burrs on the seal bore in the hub may be causing the lubricant leak.

Answer B is wrong because the question states that the wheel bearings were adjusted to specifications, which provides the proper hub end play.

Answer C is wrong because Technician A is right and Technician B is wrong.

Answer D is wrong because Technician A is right and Technician B is wrong.

95. Answer A is wrong because a few particles indicate normal wear, not a problem needing immediate resolution.

Answer B is right because you do inform the customer of the condition and tell them to monitor the amount of particles.

Answer C is wrong because it is always best to inform the customer of any condition on their vehicle that may require extra attention.

Answer D is wrong because a few particles indicate normal wear, not a problem needing immediate resolution.

96. Answer A is wrong because the differential lubricant must be level with the bottom of the filler plug hole.

Answer B is right because in some rear axle hubs, the proper lubricant for the wheel bearings must be installed through hub filler plug holes.

Answer C is wrong because Technician A is wrong and Technician B is right.

Answer D is wrong because Technician A is wrong and Technician B is right.

97. **Answer A is right** because you place the jacks under the spring seats.

Answer B is wrong because under the axle housing is not an appropriate place to put the jack stands.

Answer C is wrong because under the axle housing is not an appropriate place to put the jack stands.

Answer D is wrong because under the brake chamber is not an appropriate place to put the jack stands.

98. Answer A is wrong because during assembly, the differential carrier is securely placed onto the hydraulic jack.

Answer B is right because during disassembly is the MOST likely time for damaging the mating surfaces of the axle housing or differential carrier.

Answer C is wrong because usually Technicians do not use tools that can cause damage to the mating surfaces during this step.

Answer D is wrong because if assembled correctly, the axle will not become damaged in this area as a result of high torque conditions.

99. Answer A is wrong because this check is only necessary if a problem is indicated.

Answer B is wrong because although this is one of the checks, it is not the only check that should be made.

Answer C is wrong because although this is one of the checks, it is not the only check that should be made.

Answer D is right because you should check both mating surfaces for a good, clean, and tight fit.

100. **Answer A is right** because worn thrust washers typically cause excessive end play in the side gears.

Answer B is wrong because the gears do not need to be replaced unless the gear teeth are worn or otherwise damaged.

Answer C is wrong because there is no need to replace the differential pinion gears and thrust washers unless they are worn or damaged.

Answer D is wrong because it is necessary to split the case for access to the side gears.

101. Answer A is wrong because worn teeth on the drive pinion and ring gear cause a noise when driving straight ahead.

Answer B is right because worn teeth on the differential side gears and pinion gears may cause a clunking noise when turning a corner.

Answer C is wrong because Technician A is wrong and Technician B is right.

Answer D is wrong because Technician A is wrong and Technician B is right.

102. Answer A is wrong because a broken shift fork can cause the main differential lock to not operate.

Answer B is wrong because a sticking shift fork can cause the main differential lock to not operate.

Answer C is wrong because a damaged air solenoid may not allow the main differential to lock.

Answer D is right because a broken disengagement spring will not cause a disengagement, therefore the exception.

103. **Answer A is right** because when the differential lock is engaged, one of the axles is locked to the differential case.

Answer B is wrong because the differential lock is used to increase traction on slippery road surfaces or during off-road hauling.

Answer C is wrong because Technician A is right and Technician B is wrong.

Answer D is wrong because Technician A is right and Technician B is wrong.

104. Answer A is wrong because the marks on the bearing caps are from the original machining process and the caps do not need to be replaced, but Technician B is also right.

Answer B is wrong because the bearing caps must be marked and installed in their original positions, but Technician A is also right.

Answer C is right because both Technician A and Technician B are right.

Answer D is wrong because both Technician A and Technician B are right.

105. Answer A is wrong because the Technician is not measuring for case warpage.

Answer B is wrong because the Technician is not measuring for bearing cap squareness.

Answer C is wrong because the Technician is not measuring for differential housing twist.

Answer D is right because the Technician is measuring bearing cap preload.

106. Answer A is wrong because rough action is a valid reason to replace the spigot bearing.

Answer B is wrong because binding action is a valid reason to replace the spigot bearing.

Answer C is wrong because a spalled outer race is a valid reason to replace the spigot bearing.

Answer D is right because end play is not measured on a spigot bearing.

107. Answer A is wrong because this reading does not prove that the ring gear is warped.

Answer B is wrong because for the reading to be within specification it should be less than 0.008 in. (0.203 mm).

Answer C is wrong because this is not an accepted method of fixing an out-of-range runout reading for the ring gear.

Answer D is right because the Technician will reset the differential bearing preload and measure again.

108. **Answer A is right** because the differential side bearing preload must be properly adjusted before measuring the ring gear runout in Figure 3–29.

Answer B is wrong because worn drive pinion gear teeth do not affect the ring gear runout.

Answer C is wrong because Technician A is right and Technician B is wrong.

Answer D is wrong because Technician A is right and Technician B is wrong.

109. **Answer A is right** because the Technician should replace the ring gear and pinion for a tooth spalling condition.

Answer B is wrong because the ring gear should always be replaced as a set.

Answer C is wrong because changing lubrication will not fix the problem of tooth spalling.

Answer D is wrong because this does not fix the problem of tooth spalling.

110. **Answer A is right** because you do press out the old one, heat the new ring gear in oil or water, and reassemble.

Answer B is wrong because the ring gear is a friction fit component; therefore, force will be necessary.

Answer C is wrong because pressing a new ring gear onto the differential case may damage the ring gear or case.

Answer D is wrong because it is not acceptable to use a hammer to remove the old ring gear, and heating the differential case will make it impossible to successfully mount the new ring gear.

111. **Answer A is right** because it is necessary to install a thicker shim between the two pinion bearing inner races to reduce preload.

Answer B is wrong because it is necessary to install a thicker shim between the two pinion bearing inner races to reduce preload.

Answer C is wrong because it is necessary to install a thicker shim between the two pinion bearing inner races to reduce preload.

Answer D is wrong because it is necessary to install a thicker shim between the two pinion bearing inner races to reduce preload.

112. **Answer A is right** because drive pinion depth should be set once you properly preload the pinion bearing.

Answer B is wrong because setting the drive pinion depth does not require adjustment of the ring gear.

Answer C is wrong because Technician A is right and Technician B is wrong.

Answer D is wrong because Technician A is right and Technician B is wrong.

113. Answer A is wrong because installing a thicker shim between the differential carrier and the pinion bearing cage moves the pinion gear outward away from the ring gear.

Answer B is wrong because a shim is not installed between the rear pinion bearing and the pinion gear.

Answer C is wrong because both Technician A and Technician B are wrong.

Answer D is right because both Technician A and Technician B are wrong.

114. Answer A is wrong because the marking on the pinion gear is used to calculate pinion gear depth.

Answer B is wrong because the marking on the pinion gear is used to calculate pinion gear depth.

Answer C is right because the marking on the pinion gear is used to calculate pinion gear depth.

Answer D is wrong because the marking on the pinion gear is used to calculate pinion gear depth.

115. **Answer A is right** because the differential adjusting rings should be rotated until there is zero end play on the ring gear, and then tighten each ring one notch.

Answer B is wrong because the differential adjusting rings should be rotated until there is zero end play on the ring gear, and then tighten each ring one notch.

Answer C is wrong because the differential adjusting rings should be rotated until there is zero end play on the ring gear, and then tighten each ring one notch.

Answer D is wrong because the differential adjusting rings should be rotated until there is zero end play on the ring gear, and then tighten each ring one notch.

116. Answer A is wrong because insufficient differential side bearing preload causes rapid side bearing wear.

Answer B is wrong because insufficient differential side bearing preload causes excessive ring gear tooth wear.

Answer C is right because insufficient differential side bearing preload does not cause excessive pinion bearing wear.

Answer D is wrong because insufficient differential side bearing preload causes excessive pinion gear tooth wear.

117. Answer A is wrong because if the ring gear backlash is more than specified, it is necessary to loosen adjuster ring A and tighten the opposite adjuster ring an equal amount.

Answer B is wrong because if the ring gear backlash is more than specified, it is necessary to loosen adjuster ring A and tighten the opposite adjuster ring an equal amount.

Answer C is wrong because if the ring gear backlash is more than specified, it is necessary to loosen adjuster ring A and tighten the opposite adjuster ring an equal amount.

Answer D is right because if the ring gear backlash is more than specified, it is necessary to loosen adjuster ring A and tighten the opposite adjuster ring an equal amount.

118. Answer A is wrong because the tooth contact pattern is correct; thus no further adjustment is needed.

Answer B is wrong because the tooth contact pattern is correct; thus no further adjustment is needed.

Answer C is wrong because both Technician A and Technician B are wrong.

Answer D is right because both Technician A and Technician B are wrong.

119. **Answer A is right** because to lower the tooth contact pattern on the ring gear teeth, it is necessary to move the pinion gear rearward by installing a thinner shim between the pinion gear cage and the differential housing.

Answer B is wrong because the tooth contact pattern is lowered by moving the pinion gear toward the ring gear. The side bearing preload and ring gear backlash are adjusted before checking the tooth contact pattern.

Answer C is wrong because Technician A is right and Technician B is wrong.

Answer D is wrong because Technician A is right and Technician B is wrong.

120. Answer A is wrong because the thrust bolt and block prevents excessive ring gear deflection under extremely heavy load conditions.

Answer B is wrong because the thrust bolt and block prevents excessive ring gear deflection under extremely heavy load conditions.

Answer C is wrong because both Technician A and Technician B are wrong.

Answer D is right because both Technician A and Technician B are wrong.

121. Answer A is wrong because to properly adjust the thrust bolt, it is necessary to turn the thrust bolt in until it contacts the ring gear, back it off one-half turn, and tighten the locknut.

Answer B is wrong because to properly adjust the thrust bolt, it is necessary to turn the thrust bolt in until it contacts the ring gear, back it off one-half turn, and tighten the locknut.

Answer C is right because to properly adjust the thrust bolt, it is necessary to turn the thrust bolt in until it contacts the ring gear, back it off one-half turn, and tighten the locknut.

Answer D is wrong because to properly adjust the thrust bolt, it is necessary to turn the thrust bolt in until it contacts the ring gear, back it off one-half turn, and tighten the locknut.

122. **Answer A is right** because in low range, the annulus gear attached to the ring gear is driving the planetary pinions attached to the differential case, and the sun gear is locked.

Answer B is wrong because in low range, the annulus gear attached to the ring is driving the planetary pinions attached to the differential case, and the sun gear is locked.

Answer C is wrong because in low range, the annulus gear attached to the ring is driving the planetary pinions attached to the differential case, and the sun gear is locked.

Answer D is wrong because in low range, the annulus gear attached to the ring is driving the planetary pinions attached to the differential case, and the sun gear is locked.

123. Answer A is wrong because the differential shown in Figure 3–38 has a shift fork, and therefore it is a two-speed rear axle.

Answer B is wrong because the differential shown in Figure 3–38 has a shift fork, and therefore it is a two-speed rear axle.

Answer C is right because the differential shown in Figure 3–38 has a shift fork, and therefore it is a two-speed rear axle.

Answer D is wrong because the differential shown in Figure 3–38 has a shift fork, and therefore it is a two-speed rear axle.

124. Answer A is wrong because the speedometer adapter and pressure switch compensate speedometer readings in low range if the tractor has a transmission-driven speedometer.

Answer B is right because air pressure is supplied to the front and rear shift units when the axles are in high range.

Answer C is wrong because Technician A is wrong and Technician B is right.

Answer D is wrong because Technician A is wrong and Technician B is right.

125. **Answer A is right** because air must be exhausted from the shift units through the quick release valve to shift the axles into low range. Therefore, a restricted exhaust passage in the quick release valve causes slow or delayed shifting.

Answer B is wrong because shifting from high to low range requires exhausting of the air pressure from the shift units through the quick release valve.

Answer C is wrong because shifting from high to low range requires exhausting of the air pressure from the shift units through the quick release valve.

Answer D is wrong because shifting from high to low range requires exhausting of the air pressure from the shift units through the quick release valve.

126. Answer A is wrong because broken teeth on the forward drive axle ring gear can cause the front axle to be nonpowered.

Answer B is wrong because broken ring gear teeth on the rear differential may cause no rear axle drive, but this is not the MOST likely cause of this problem.

Answer C is right because the MOST likely cause are stripped output shaft splines.

Answer D is wrong because a damaged interaxle differential can render both axles powerless.

127. Answer A is wrong because a plugged oil line may cause heat damage to the interaxle differential, but Technician B is also right.

Answer B is wrong because the driver not locking the two tandem axles together during slippery conditions may cause heat damage to the interaxle differential, but Technician A is also right.

Answer C is right because both Technician A and Technician B are right.

Answer D is wrong because both Technician A and Technician B are right.

128. Answer A is wrong because the air lines would need to be taken off.

Answer B is right because you do disconnect the air line, remove the output shaft yoke, power divider cover, and all applicable gears as an assembly.

Answer C is wrong because it is not necessary to remove the entire differential carrier.

Answer D is wrong because the air lines would need to be taken off, as well as the input yoke.

129. **Answer A is right** because a broken shift spring will not cause the transmission not to shift out of interlock.

Answer B is wrong because a broken shift shaft would not allow the unit to shift into differential lock.

Answer C is wrong because an air line problem would not allow the unit to shift into differential lock.

Answer D is wrong because a stripped clutch collar would not allow the unit to shift into differential lock.

130. Answer A is wrong because the shifter valve is locked in position when the interaxle lockout is engaged, but Technician B is also right.

Answer B is wrong because the interaxle lockout should be engaged to prevent excessive wheel slip on slippery road surfaces, but Technician A is also right.

Answer C is right because both Technician A and Technician B are right.

Answer D is wrong because both Technician A and Technician B are right.

131. Answer A is wrong because the MOST likely cause of metal burrs and gouges on the axle housing and differential carrier mating surfaces is lubricant leaks.

Answer B is wrong because the MOST likely cause of metal burrs and gouges on the axle housing and differential carrier mating surfaces is lubricant leaks.

Answer C is right because the MOST likely cause of metal burrs and gouges on the axle housing and differential carrier mating surfaces is lubricant leaks.

Answer D is wrong because the MOST likely cause of metal burrs and gouges on the axle housing and differential carrier mating surfaces is lubricant leaks.

132. **Answer A is right** because a piece of the old gasket left on the differential carrier mating surface could be the cause of the lubricant leak.

Answer B is wrong because continual high torque driving conditions may damage rear axle splines, gears, and bearings, but it is very unlikely that this condition would warp the differential carrier.

Answer C is wrong because Technician A is right and Technician B is wrong.

Answer D is wrong because Technician A is right and Technician B is wrong.

133. Answer A is wrong because replacing the hoses is not necessary.

Answer B is wrong because the pump does not need to be packed because it is submerged.

Answer C is wrong because replacing the hoses is not necessary.

Answer D is right because you do check the pump for smooth operation and blow forced air through all passages.

134. **Answer A is right** because the oil pump is driven from the rear axle input shaft.

Answer B is wrong because the oil pump is driven from the rear axle input shaft.

Answer C is wrong because the oil pump is driven from the rear axle input shaft.

Answer D is wrong because the oil pump is driven from the rear axle input shaft.

135. Answer A is wrong because cracks are a valid reason to replace the axle shaft.

Answer B is wrong because a bent axle shaft is a valid reason to replace the axle shaft.

Answer C is right because pitting of the axle shaft is not a valid reason to replace the axle shaft, therefore the exception.

Answer D is wrong because a twisted axle shaft is a valid reason to replace the axle shaft.

136. Answer A is wrong because a gasket is required between the axle shaft flange and the hub mating surface.

Answer B is right because the tapered dowels must be installed in the axle stud openings followed by the washers and axle shaft retaining nuts.

Answer C is wrong because Technician A is wrong and Technician B is right.

Answer D is wrong because Technician A is wrong and Technician B is right.

137. **Answer A is right** because axle shaft damage is the least likely cause of a hot hub, because it is most probably the result of the heat, not the cause.

Answer B is wrong because bearing damage may cause an overheated axle.

Answer C is wrong because air line damage may cause a dragging brake and an overheated axle.

Answer D is wrong because poor lubricant quality may cause an overheated axle.

138. **Answer A is right** because an out-of-round hub seal bore may cause a lubricant leak between the seal housing and this bore.

Answer B is wrong because the hub end play must be maintained at specifications to provide normal wheel bearing life.

Answer C is wrong because Technician A is right and Technician B is wrong.

Answer D is wrong because Technician A is right and Technician B is wrong.

139. Answer A is wrong because the noise produced by a defective front wheel bearing is most noticeable while turning a corner.

Answer B is wrong because the noise produced by a defective rear wheel bearing is most noticeable while driving at low speeds.

Answer C is wrong because both Technician A and Technician B are wrong.

Answer D is right because both Technician A and Technician B are wrong.

140. Answer A is wrong because aluminum hubs must never be heated with an oxyacetylene torch to remove the bearing cups.

Answer B is right because the wear sleeve should be installed so it is even with the shoulder on the spindle.

Answer C is wrong because Technician A is wrong and Technician B is right.

Answer D is wrong because Technician A is wrong and Technician B is right.

141. Answer A is wrong because rear wheel bearings are never preloaded.

Answer B is right because rear wheel lockwashers may be tang type or dowel type.

Answer C is wrong because Technician A is wrong and Technician B is right.

Answer D is wrong because Technician A is wrong and Technician B is right.

142. **Answer A is right** because the seal lip must face toward the center of the hub.

Answer B is wrong because a 0.015 in. (0.0381 mm) feeler gauge should not fit between the bearing cup and the hub.

Answer C is wrong because Technician A is right and Technician B is wrong.

Answer D is wrong because Technician A is right and Technician B is wrong.

143. Answer A is wrong because removing the instrument panel gauge and testing it is not the place to begin diagnosing. The problem is MOST likely in the sensor or wiring.

 Answer B is right because disconnecting the sensor and substituting a variable resistance in place of this sensor is the proper diagnostic procedure.

 Answer C is wrong because Technician A is wrong and Technician B is right.

 Answer D is wrong because Technician A is wrong and Technician B is right.

144. **Answer A is right** because an open circuit in the wire from the gauge to the sender causes zero current flow through the hot coil in the gauge and the sender. Under this condition all the current flows through the cold coil in the gauge to ground, and the magnetism of this coil pulls the gauge pointer to the cold position.

 Answer B is wrong because less resistance than specified in the sender increases the current flow through the hot coil and sender, and this makes the gauge read higher than normal.

 Answer C is wrong because Technician A is right and Technician B is wrong.

 Answer D is wrong because Technician A is right and Technician B is wrong.

145. Answer A is wrong because the wheel speed sensor must be replaced if the lead wires are damaged.

 Answer B is wrong because the wheel speed sensor produces an AC voltage signal, and this signal cannot be tested with an ohmmeter while rotating the wheel.

 Answer C is wrong because both Technician A and Technician B are wrong.

 Answer D is right because both Technician A and Technician B are wrong.

146. **Answer A is right** because if there is one tooth missing on the toothed ring, the ring must be replaced.

 Answer B is wrong because the electromagnetic operation of the wheel speed sensor is not affected by mud or dirt in the toothed ring.

 Answer C is wrong because Technician A is right and Technician B is wrong.

 Answer D is wrong because Technician A is right and Technician B is wrong.

147. Answer A is wrong because you do replace the shift collar return spring when wheel end-locking hubs will not switch into high, but Technician B is also right.

 Answer B is wrong because you can clean out and repack the hub to correct this condition, but Technician A is also right.

 Answer C is right because both Technician A and Technician B are right.

 Answer D is wrong because both Technician A and Technician B are right.

148. Answer A is wrong because it is not necessary to replace the entire hub.

 Answer B is wrong because the seal that most likely allowed the dirt to enter needs to be replaced as well.

 Answer C is wrong because the seal that most likely allowed the dirt to enter needs to be replaced as well.

 Answer D is right because you clean, inspect, and repack the bearings, and replace the seal.

4 Brakes

Pretest

The purpose of this pretest is to determine the amount of review you may require before taking the ASE Medium/Heavy Truck Brakes Test. If you answer all the pretest questions correctly, complete the questions and study the information in this chapter to prepare for this test.

If two or more of your answers to the pretest questions are incorrect, complete a study of all the chapters in *Today's Technician Medium/Heavy-Duty Truck Brake Systems* Classroom and Shop Manuals published by Delmar Publishers, plus a study of the questions and information in this chapter.

The pretest answers are located at the end of the pretest; these answers also are in the answer sheets supplied with this book.

1. While discussing compressor operation, *Technician A* says the unloader pistons are operated by unloader pressure from the governor. *Technician B* says that in the unloaded cycle the compressor is pumping air to the supply reservoir. Who is right?
   A. A only
   B. B only
   C. Both A and B
   D. Neither A nor B

2. While discussing compressors and governors, *Technician A* says the adjuster screw in the end of the governor is rotated to set the governor cut-in pressure. *Technician B* says an off-vehicle source of air pressure may be connected to the inflator valve air connection. Who is right?
   A. A only
   B. B only
   C. Both A and B
   D. Neither A nor B

3. A truck equipped with air brakes has parking brake release problems, but the brakes operate normally during a brake application. *Technician A* says the cause could be a defective spring brake chamber. *Technician B* says the camshaft bushings may be worn. Who is right?
   A. A only
   B. B only
   C. Both A and B
   D. Neither A nor B

4. A safety valve in the supply reservoir:
   A. opens at 150 psi (1,034 kPa).
   B. is operated by governor unloader pressure.
   C. is operated by one of the unloader pistons.
   D. allows water to escape from the air brake system.

5. An automatic drain valve:
   A. is opened when the reservoir pressure reaches the governor cut-out pressure.
   B. is mounted in a threaded fitting near the top of the supply reservoir.
   C. is opened by decreasing supply reservoir pressure when air is used out of the reservoir.
   D. may contain a 12V heater to prevent water from freezing in the valve.

6. All of these statements about pressure protection valves are true EXCEPT:
   A. when the pressure in the air-operated devices equals the pressure in the air brake system, the pressure protection valve remains open.
   B. air is supplied through the pressure protection valve to the air-operated devices on the truck.
   C. if a leak occurs in the driver's air seat, the pressure protection valve maintains airflow to the seat.
   D. if a leak occurs in the two-speed rear axle control, the pressure protection valve closes.

7. When discussing desiccant-type air dryers, *Technician A* says the purge valve in the dryer opens when the compressor enters the unloaded cycle. *Technician B* says the electric heater in the air dryer is operated by a module and relay. Who is right?
   A. A only
   B. B only
   C. Both A and B
   D. Neither A nor B
8. During normal operation of the brake application valve on an air brake system:
   A. if the primary section in the brake application valve is in a balanced condition, modulated air pressure is supplied to the rear brakes.
   B. if the secondary section in the brake application valve is in a balanced condition, the secondary exhaust valve is open.
   C. if the primary section in the brake application valve is in a balanced condition, the primary inlet valve is open.
   D. when the driver depresses the brake pedal fully during a panic stop, the primary inlet valve is open and the secondary inlet valve is closed.
9. A separate quick release valve in an air brake system:
   A. is usually connected in the rear brake system.
   B. has an open exhaust port when the brakes are applied.
   C. releases air faster than the brake application valve.
   D. has an open inlet port when the brakes are released.
10. The MOST likely result of a complete secondary circuit failure in a truck air brake system would be:
   A. full wheel lockup.
   B. complete brake failure.
   C. visible and audible driver alert.
   D. loss of trailer brakes.
11. The function of the control valve in Figure 4–1 is to:
   A. provide tractor front wheel braking.
   B. enable smooth bobtail stops.
   C. apply trailer parking brakes.
   D. apply trailer service brakes.
12. Refer to the air brake chamber in Figure 4-2. *Technician A* says that this assembly should never be disassembled without first caging the spring. *Technician B* says that this type of brake chamber

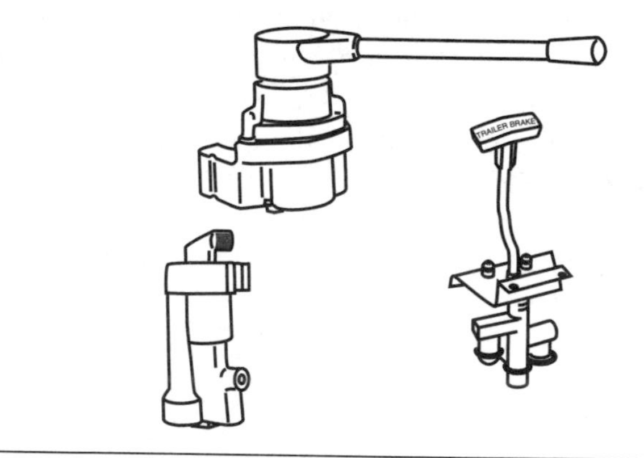

Figure 4–1 Air brake system control valve. *(Courtesy of Haldex Midland Services.)*

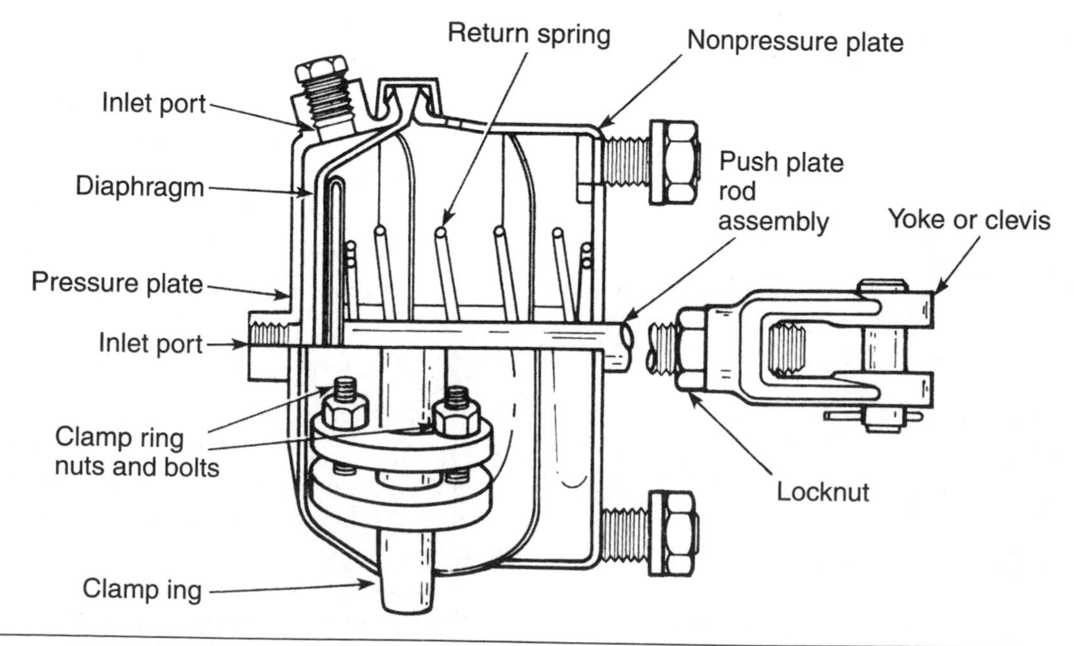

Figure 4–2 Air brake chamber.

is used only for parking brake applications. Who is right?
A. A only
B. B only
C. Both A and B
D. Neither A nor B

13. When discussing parking brake control valves, *Technician A* says the air supply line is connected from the park control valve to the double check valve between the front axle and rear axle reservoirs. *Technician B* says when the park control valve knob is pushed inward, air is vented in the line from the park control valve to the tractor parking brake relay valve. Who is right?
A. A only
B. B only
C. Both A and B
D. Neither A nor B

14. When an antilock brake system (ABS) is operating normally:
A. the ABS system provides 5 percent wheel slip without wheel lockup in the antilock mode.
B. the ABS system enters the antilock mode each time the brakes are applied.
C. if there are no electrical defects in the ABS system, the ABS warning light should go out after the engine is started and the vehicle reaches 6 mph (10 km/h).
D. each wheel speed sensor in the ABS system produces a DC voltage signal.

Answers to Pretest

1. A, 2. B, 3. A, 4. A, 5. D, 6. C, 7. A, 8. A, 9. C, 10. C, 11. D, 12. D, 13. A, 14. C

Air Supply and Service Systems

Task 1

Diagnose poor stopping, premature wear, brake noise, air leaks, pulling, grabbing, or dragging problems caused by supply and service system malfunctions; determine needed repairs.

1. Which of the following causes are correct when diagnosing air brake systems?
A. A chassis vibration during braking is caused by excessive radial tire runout.
B. Brake grab on one wheel is caused by an improper governor cut-out adjustment.
C. Brake drag is caused by glazed brake linings.
D. An improper governor cut-in adjustment may cause the safety valve to open.

2. An air brake system has slower than specified air pressure build-up time. The cause of this problem could be:
A. a smaller than specified supply reservoir.
B. an air leak in the compressor air intake filter.
C. carbon buildup in the compressor discharge hose.
D. an excessive amount of water in the supply reservoir.

3. All of these statements about air brake supply systems are true EXCEPT:
A. the supply reservoir tends to collect more moisture than the front or rear axle reservoirs.
B. water must be drained daily from the reservoirs if an air dryer is used in the air brake system.
C. a small amount of oil in the supply reservoir is a normal condition.
D. some air dryers and automatic moisture ejectors contain electric heaters o prevent moisture from freezing in these components.

4. A truck with an air brake system has reduced braking force on one rear wheel. Which of the following is the LEAST likely cause?
A. Restricted airflow or low air pressure to the service brake chamber
B. Improper adjustment of slack adjuster and brake chamber pushrod
C. Grease on shoe linings/faces
D. A ruptured brake chamber diaphragm

5. All of the following could cause excessive leakage in an air brake system with the service brakes applied EXCEPT:
A. a leaking brake chamber diaphragm.
B. a leaking hose, tube, or fittings.
C. defective compressor gaskets.
D. a bad relay valve.

Hint

The potential energy of an air brake system is compressed air. The action of the driver's foot on a brake pedal contributes nothing to developing this potential energy which is developed by an engine driven air compressor.

A tractor trailer air brake system can be divided into three systems: the supply system, the control system, and the foundation assembly. The supply system is responsi-

ble for supplying compressed air at the correct pressure to the control system. The control system manages the service and parking brake functions of the vehicle. The foundation brakes affect the braking managed by the control system. A failure within any of these three systems can cause anything from a minor malfunction to a complete brake failure.

Truck brakes must be balanced. A balanced brake system can be defined as one in which the braking pressure reaches each actuator at the same moment and at the same pressure level. Factors that affect brake balance are application and release times. To meet balanced performance requirements, vehicle manufacturers match all the brake system valves and components, including the hose size and fitting geometry used in the system. Air application and release performance is dependent both on the size and volume of chambers and the distance the air must travel.

Inspection for leaks in an air brake system is an important part of any brake system inspection. Every component and the entire plumbing system of the air brake system should be checked.

Air-related causes of dragging brakes on a truck air brake system can include a leaking hold-off diaphragm, low hold-off pressure in the spring brake section of the brake chamber, spring brake control valve problems, low system pressure causing partial application of the spring brakes, and sticking service application and relay valves.

The majority of brake applications on a typical highway involves application pressures of 20 psi (138 kPa) or less. The air brake system must be capable of full pressure stops even though these stops are rare. A truck air brake system must be able to accommodate four to six full reserve stops with the brakes properly adjusted.

Brakes that are mechanically out of adjustment require a greater volume of application air.

Unlike hydraulic fluid, air is compressible. The larger the volume of air, the higher its compressibility. Brake timing lags are greater in air brake systems than in hydraulic brake systems. When a brake control signal has to travel from the application valve in the tractor to an actuator valve in the trailer, the lag time depends on the distance the signal has to travel. Tractor trailer brake systems compensate for this by pneumatically balancing the relay valves in the system by crack pressure: the valves farthest from the application valve are designed with the lowest crack pressures, with those located on the tractor itself having slightly higher crack pressure values. The objective is to have simultaneous application of the brakes in each wheel assembly on the rig at both low- and high-application pressures. Current Federal Motor Vehicle Safety Standard (FMVSS) 121 brake application timing requirements are:

Trailer:

Application: 0.39 seconds Release: 0.65 seconds

Truck:

Application: 0.45 seconds Release: 0.55 seconds

Pneumatic imbalance may cause uneven braking in a truck air brake system. Causes of pneumatic imbalance may be defective valves, or plumbing irregularities, such as restricted lines, the fitting of a 90-degree fitting in place of a 45-degree fitting, or using brake hose with smaller or larger than specified inside diameters.

To perform a full pressure balance and pressure build-up (timing) test, a Technician needs a pair of test hoses, and a duplex gauge. For multiple combinations, two additional hoses are needed per trailer. Ideal balance timing in a combination vehicle is defined as each axle receiving identical air pressure simultaneously on a brake application.

Task 2

Check air system build-up time; determine needed repairs.

6. What is the minimum performance (psi/sec) a compressor should have?
   A. 85–100 psi in 25 seconds or less
   B. 60–80 psi in 20 seconds or less
   C. 90–110 psi in 30 seconds or less
   D. 50–110 psi in 25 seconds or less

7. During a brake inspection, a Technician tests the air supply system and finds that the buildup is slow. *Technician A* says the compressor outlet hose may be restricted. *Technician B* says there may be a leak in one of the brake chambers. Who is right?
   A. A only
   B. B only
   C. Both A and B
   D. Neither A nor B

Hint

System air buildup times are defined by federal legislation, specifically FMVSS 121. This legislation defines the required build-up times and values. A common check performed by enforcement agencies requires that the supply system on a vehicle be capable of raising air system pressure from 85 to 100 psi (586 to 689 kPa) in 25 seconds or less.

Failure to achieve this build-up time indicates a worn compressor, defective compressor unloader assembly, supply system leakage, a defective governor, or a restricted compressor outlet hose.

Task 3

Drain air reservoir tanks; check for oil, water, and foreign material; determine needed repairs.

8. *Technician A* says that daily draining of air reservoirs that are not equipped with automatic drain valves is highly recommended. *Technician B* says that you should check all automatic drain valves and moisture removing devices periodically for proper operation. Who is right?
   A. A only
   B. B only
   C. Both A and B
   D. Neither A nor B

9. An automatic reservoir drain valve:
   A. opens when the reservoir pressure reaches the governor pressure and is unloaded.
   B. is mounted in a threaded fitting near the top of the reservoir.
   C. opens by decreasing supply reservoir pressure.
   D. drains water if the sump pressure in the valve is 2 psi (13.7 kPa) higher than reservoir pressure.

Hint

Air reservoirs store and provide air for the truck air brake system. Servicing of the reservoirs consists of inspection, draining the tanks, and performing leakage tests. Air brake systems must have at least three air reservoirs. They may, and usually do, have more. Compressed air from the compressor is delivered to the supply reservoir. The supply reservoir is also known as a wet tank. The supply reservoir supplies air to the primary and secondary reservoirs, which supply the brake system with air.

A wet tank or supply tank is so named for the moisture that forms in such a tank when hot, compressed, moisture laden air cools and condenses on the tank inside walls.

Reservoirs may be equipped with either automatic or manual drain valves. Daily draining of manual drain valves is recommended to keep the system free of contaminants and moisture. Automatic reservoir drain valves should be checked for proper operation, also on a daily basis.

Supply reservoirs are equipped with a safety valve also known as a pop-off valve. This valve is designed to open at 150 psi (1,034 kPa). This protects the system in the event of a governor failure.

Air reservoirs on trucks are pressure vessels. They are hydrostatically tested after manufacture. The inside wall is treated with a corrosion protection coating. Truck air reservoirs should never be repair welded, firstly because they are pressure vessels and secondly because the internal coating is destroyed. Leaking tanks should be replaced.

Evidence of oil in a wet tank often indicates the compressor is pumping oil through the system. Oil can damage valving throughout the system, so the source of oil contamination must be determined immediately and repaired.

Task 4

Inspect, adjust, align, or replace compressor drive belts and pulleys.

10. If you notice belt contact on the bottom of the pulley when adjusting the tension of the compressor belt, you should:
    A. decrease tension to original specification level and recheck contact surface for excessive wear.
    B. measure the level of deflection at a section of belt that is longest between pulleys.
    C. replace the belt.
    D. ignore it because some applications require this additional contacting area.

11. While discussing compressor drive belts, *Technician A* says a slipping drive belt may cause slow compressor build-up time. *Technician B* says some compressors have an adjustable idler pulley that may be moved to adjust belt tension. Who is right?
    A. A only
    B. B only
    C. Both A and B
    D. Neither A nor B

Hint

Some compressors, especially those on smaller engines or in those cases where an air brake system has been retrofit, are driven by the engine using a belt and pulley. Belt sets, pulleys, and idlers should be routinely inspected for indications of wear and axial runout. Belt tension should be set to specifications using a belt tension gauge. Cracks and nicks in a drive belt require that it be replaced.

Task 5

Inspect, time, or replace compressor drive gear and coupling.

12. You are replacing an air compressor drive on a truck with a two-stroke diesel engine. *Technician A* says it is necessary to retime the engine when this replacement process is complete. *Technician B*

says you inspect the gear for worn or chipped teeth. Who is right?
A. A only
B. B only
C. Both A and B
D. Neither A nor B

13. A driver complains that the compressor on his truck cuts in after every brake application. *Technician A* says the compressor drive may be worn, creating a lag time. *Technician B* says the air reservoirs may be loaded up with moisture. Who is right?
A. A only
B. B only
C. Both A and B
D. Neither A nor B

Hint

Most compressors have two cylinder pumps and they are balanced units. These compressors do not have to be timed to the engine on installation. Ensure that the oil feed tube is correctly aligned in the compressor crankshaft and that the gear teeth are not damaged.

Some single cylinder compressors are not self-balanced units. These compressors must be timed to the engine on installation. Ensure that the manufacturer's service literature is consulted when installing compressors that must be timed to the engine.

Task 6

Inspect, repair, or replace air compressor, air cleaner, oil supply, water lines, hoses, and fittings.

14. All of the following are part of an air compressor EXCEPT:
A. piston rings.
B. crankshaft.
C. discharge valve.
D. roller bearing.

15. What is the LEAST likely cause of a noisy air compressor?
A. A loose drive pulley
B. Restrictions in cylinder head or discharge line
C. A defective head gasket
D. Inadequate lubrication of the unit

Hint

A compressor is an air pump. The basic air compressor operates much like an internal combustion engine. It consists of a crankcase, cylinder block, and cylinder head. The crankshaft is supported in the crankcase by main bearings. Connecting rods are connected to the throws on the crankshaft at their big end. The piston assemblies are connected to the small end of the connecting rod by means of a wrist pin. As the crankshaft rotates, the pistons reciprocate in the cylinder bores in the compressor block. Piston rings seal the piston in the cylinder bore and control an oil film on the cylinder wall.

The cylinder head assembly is equipped with discharge valves and discharge ports. It also has reed-type inlet valves. On the piston downstroke, the inlet valve is unseated by lower pressure in the cylinder, and a charge of air is induced into the cylinder. When the piston passes through the bottom of its travel and begins the upward stroke, the inlet valve is seated by the cylinder pressure. As the piston continues to be driven upward, pressure in the cylinder rises and when it exceeds the pressure in the discharge port, the discharge valve opens and compressed air is forced out of the cylinder into the discharge line.

An unloading mechanism in the cylinder head is controlled by a governor. The governor uses an air signal to cycle the compressor through the loaded (pumping) and unloaded (not pumping) cycles. The air signal from the governor acts on the unloader assembly, which holds the inlet valves open throughout the cycle. When this occurs, the compressor is unable to compress air as the cylinder cannot be sealed. Therefore, the compressor is unloaded.

The compressor is pressure lubricated by the engine lubrication system. Pressurized oil is usually delivered to the compressor crankshaft by means of a tube connected to the engine lubrication system. The crankshaft main bearings, crankshaft, and piston assemblies require lubrication. Compressors are liquid cooled by the engine cooling system through inlet and outlet lines located in the cylinder head. When the air is compressed, the air temperature increases. Compressors require cooling to keep air discharge temperatures at 300°F (148°C) or less.

When troubleshooting a noisy compressor, a Technician must determine the source of the noise. Squeaking drive belts or pulleys, failed bearings, and lubrication related problems are often the cause of a noisy compressor.

An air compressor is the source of compressed air to all the vehicle pneumatic components, including windshield wipers, steering assist units, suspension systems, and air starters.

Task 7

Inspect, test, adjust, or replace system pressure controls (governor/relief valve), unloader assembly valves, filters, lines, hoses, and fittings.

16. A governor cut-out test is being performed. *Technician A* says the cut-in pressure should be 50 psi (345 kPa) less than the cut-out pressure. *Technician B* says both the cut-in and cut-out pressures are adjustable. Who is right?
    A. A only
    B. B only
    C. Both A and B
    D. Neither A nor B

17. A tractor with the compressor in Figure 4–3 has a problem with its air pressure rising above the specified governor cut-out pressure. *Technician A* says the check valve on the governor could be stuck closed. *Technician B* says there could be too much clearance at the compressor unloading valves. Who is right?
    A. A only
    B. B only
    C. Both A and B
    D. Neither A nor B

18. When testing and adjusting the governor cut-out and cut-in pressures, which of these statements apply?
    A. The cut-out pressure may be increased by rotating the adjusting screw counterclockwise.
    B. Turn the adjusting screw clockwise to increase cut-in pressure.
    C. Turn the adjustment screw 1/4 turn to change cut-in or cut-out pressure 7 psi (48.27 kPa).

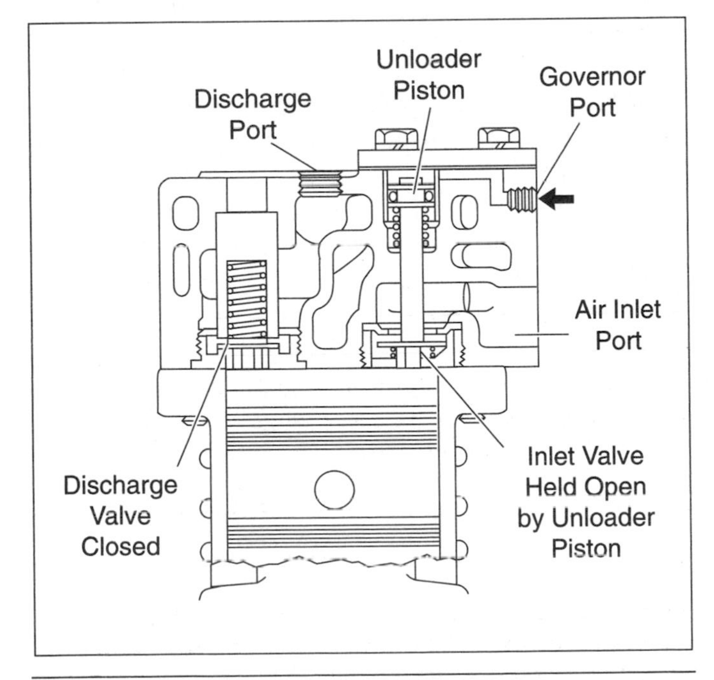

Figure 4–3 Compressor valves and unloader piston. *(Courtesy of Allied Signal Truck Brake Systems Co.)*

    D. A restricted air line from the supply reservoir to the governor decreases the cut-out pressure.

Hint

Many different air valves are used to control, regulate, or condition the air in a brake system. These valves guide the direction of flow, control the amount of pressure, and remove contaminants from the air system(s) in which they serve.

The governor manages system pressure. It monitors pressure in the supply tank by means of a line directly to it. This pressure acts on a diaphragm and spring within the governor. The spring tension is adjustable. The governor manages compressor loaded and unloaded cycles. Loaded cycle is the compressor effective cycle, that is, the compressor is pumping air. The unloaded cycle occurs when the compressor is driven by the engine but it is not actually compressing air. The compressor is in the loaded cycle until it receives an air signal from the governor to put it into unloaded cycle. This signal acts on the unloader assembly in the compressor cylinder head. The function of the unloader assembly is to hold the inlet valves open. In the unloaded cycle the compressor pulls a charge of air through the inlet ports on piston downstroke and pushes it out through the same inlet port on the upstroke.

The governor simply controls whether the compressor is in the loaded or unloaded operating mode. The governor determines the system pressure. Air brake system pressure in most trucks is set at values between 110 and 130 psi (758 and 896 kPa), with 120 psi (827 kPa) being typical. System pressure is known as cut-out pressure, the pressure at which the governor outputs the unloader signal to the compressor. The unloader signal is maintained until pressure in the supply tank drops to the cut-in value. Cut-in pressure is required by FMVSS 121 to be no more than 25 psi (172 kPa) less that the cut-out value. The difference on most systems ranges between 20 and 25 psi (137 and 172 kPa).

Governor operation can be easily checked. One method is to drop the air pressure in the supply tank to below 60 psi (413 kPa) and with the vehicle's engine running, allow the pressure to increase. A master gauge should be used to record the cut-out pressure value. This should be exactly at the specification value. If not, remove the dust boot at the top of the governor, release the locknut, and turn the adjusting screw either CW or CCW to either lower or raise the cut-out pressure.

If the unloader signal is not delivered to the compressor unloader assembly, high system pressures will result. If the safety valve on the supply tank opens, this is usually an indication of governor or compressor unloader malfunction.

The only adjustment on an air governor is the cut-out pressure value. If the difference between governor cut-out and cut-in is out of specification, the governor must be replaced.

Task 8

Inspect, repair, or replace air system lines, hoses, fittings, and couplings.

19. *Technician A* uses double flared tubing when replacing brake lines. *Technician B* uses ISO tubing when replacing brake lines. Who is right?
    A. A only
    B. B only
    C. Both A and B
    D. Neither A nor B

20. *Technician A* says a 45-degree elbow in an air brake line may be replaced with a 90-degree elbow without affecting brake operation. *Technician B* says that if a trailer service system has an air fitting that is smaller than the original line, the air timing/balance is affected. Who is right?
    A. A only
    B. B only
    C. Both A and B
    D. Neither A nor B

Hint

Brake hose must be replaced with hose that conforms to the same standard as the original hose or brake performance may be compromised. Sizing of lines on both the tractor and the trailer will affect both application and release timing of the brakes. Every fitting used in air brake system plumbing affects the fluid dynamics, and care should be taken to always replace fittings with those that are the same as the original fittings. Replacing a 45-degree elbow with a 90-degree elbow is equivalent to adding an additional 7 feet of brake hose and this will affect pneumatic timing.

Brake hose should be securely clamped away from moving components when installed. When reusing dry-seal fittings, the nipples and seats should be inspected. Department of Transportation (DOT) approved hose should be used when replacing defective hose. Sometimes brake hose can fail internally and form a rubber flap that acts as a check in the line, permitting air to flow toward a valve, but trapping it there.

Task 9

Inspect, test, clean, or replace air tank relief (pop-off) valves, check valves, draincocks, spitter valves, heaters, wiring, and connectors.

21. Component E in Figure 4–4 is a/an:
    A. automatic drain valve.
    B. manual drain valve.
    C. one-way check valve.
    D. safety valve.

22. When diagnosing an air brake system, the technician discovers that the air pressure in the supply reservoir is 170 psi (1,172 kPa). *Technician A* says the governor is not working properly. *Technician B* says the safety valve is malfunctioning. Who is right?
    A. A only
    B. B only
    C. Both A and B
    D. Neither A nor B

23. While discussing the one-way check valves in Figure 4–5, *Technician A* says these check valves protect the air supply in the primary or secondary reservoir if an air leak occurs in the supply reservoir. *Technician B* says these check valves prevent excessive air pressure from reaching the primary and secondary reservoirs. Who is correct?
    A. A only
    B. B only
    C. Both A and B
    D. Neither A nor B

Hint

A system safety valve is usually located in the supply tank. It is designed to open at a pressure value of 150 psi (1,034 kPa). It is non-adjustable and consists of a ball seat and spring. The safety valve function is to relieve system air if the pressure builds to a dangerously high level, such as when the governor is defective.

The supply tank supplies air to the primary and secondary reservoirs of the brake system. The primary and secondary reservoirs are pressure protected by means of a one-way check valve. One-way check valve operation can be verified by removing the supply and checking

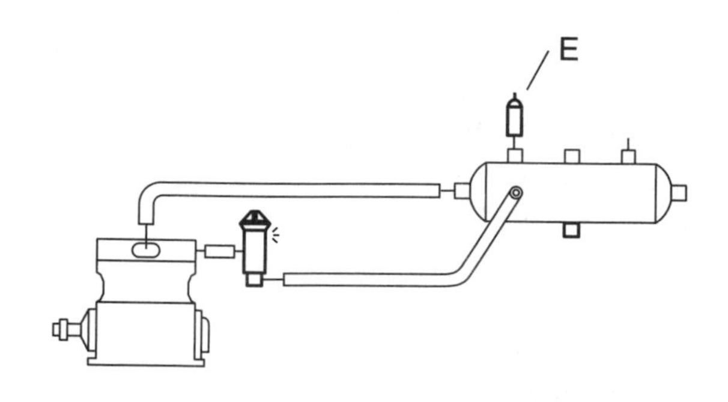

Figure 4–4 Air brake supply system. *(Courtesy of Allied Signal Truck Brake Systems Co.)*

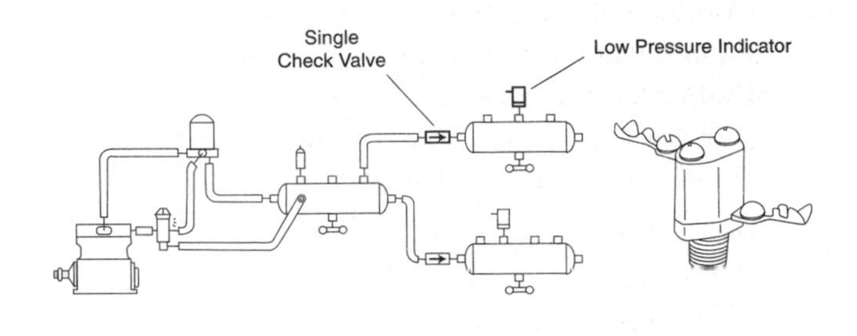

Figure 4–5 Air brake supply, primary, and secondary reservoirs. *(Courtesy of Allied Signal Truck Brake Systems Co.)*

back leakage. One-way check valves are used variously throughout an air brake system to pressure protect and isolate portions of the system.

Automatic drain valves can become plugged with sludge (oil and water residues) and may require periodic cleaning. Should oil and excessive water be evident at the drain valves, check the air dryer and/or compressor.

Task 10

Inspect, test, clean, repair, and replace air dryer systems, filters, valves, heaters, wiring, and connectors.

24. Component C in Figure 4–6 is the:
    A. unloader valve.
    B. purge valve.
    C. solenoid valve.
    D. desiccant holder.

25. While discussing desiccant-type air dryers, *Technician A* says the purge valve opens when the governor enters the loaded cycle. *Technician B* says the one-way check valve in the air dryer discharge port prevents airflow from the supply reservoir into the dryer during the purge cycle. Who is right?
    A. A only
    B. B only
    C. Both A and B
    D. Neither A nor B

26. While discussing condensation-type air dryers, *Technician A* says this type of air dryer has a desiccant bed. *Technician B* says this type of air dryer has a purge valve operated by governor pressure. Who is right?
    A. A only
    B. B only
    C. Both A and B
    D. Neither A nor B

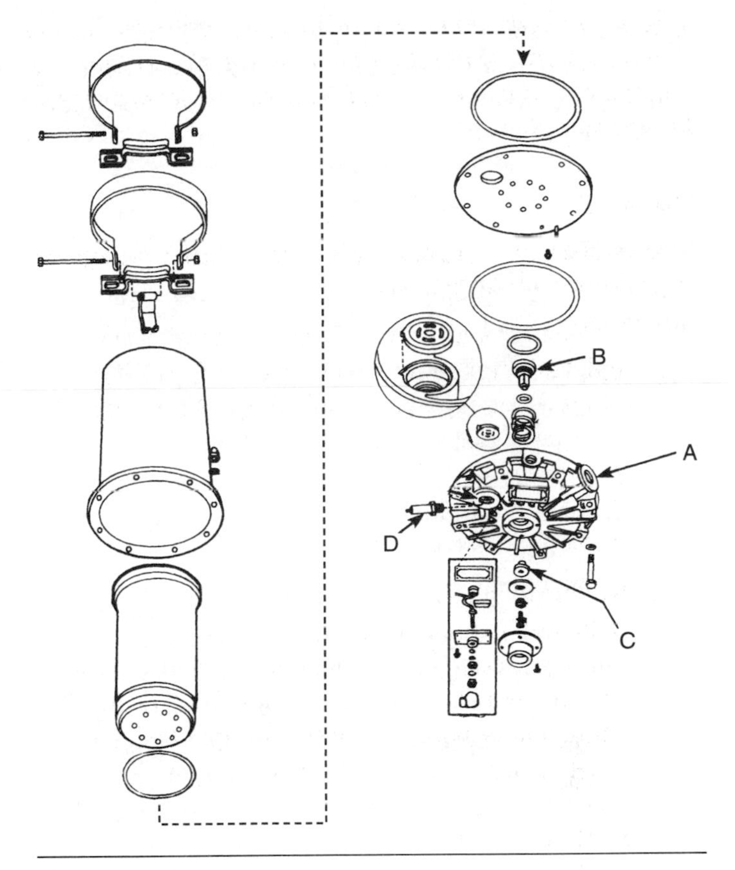

Figure 4–6 Air dryer.

Hint

Moisture in an air system can be very damaging. Ambient moisture is a problem any time the relative humidity is high, and so it is both a summer and a winter problem. The airborne moisture condenses in the reservoirs as the compressed air cools. Most current systems use air dryers to remove moisture from the compressed air before it gets to the supply tank. These use two principles to remove moisture from the compressed air. The first type is the desiccant type. During the charge cycle, hot compressed air passes through a desiccant pack in which moisture adheres. The dry air exits to the discharge port and the supply reservoir. When the air dryer purge valve receives a governor cut-out signal, the purge valve opens. Under this condition the purge valve exhausts moisture to the atmosphere. The second type of air dryer is the condensation-type. This type of air dryer attempts to cool the compressed air to the point that the moisture is condensed. Once the moisture is condensed, it can be separated and dumped to the atmosphere through the purge valve. Some air dryers use a combination of both principles.

Many air dryers use a heater to prevent icing in cold weather. This heater is thermostat regulated. In cold weather, the thermostat controls the heater cycles to maintain a temperature exceeding 45°F (7.2°C).

Oil can destroy the desiccant pack in an air dryer. When an air compressor fails and pumps its lubricating oil through the system, the desiccant pack becomes contaminated and requires replacement.

Task 11

Inspect, test, adjust, or replace brake application (foot) valve, fittings, and mounts.

27. All of the following steps in removing a brake application valve are correct EXCEPT:
    A. the truck should be on a level surface.
    B. mark or label the brake lines.
    C. mark the valve body in relation to the mounting plate.
    D. maintain brake system pressure.

28. The brake application valve in Figure 4–7 is being repaired. *Technician A* says that the brake valve has two pistons: a reaction or rear modulating piston, and a primary or front modulating piston. *Technician B* says that a reaction spring between the two pistons of a brake valve is responsible for the basic operation of the valve. Who is right?
    A. A only
    B. B only
    C. Both A and B
    D. Neither A nor B

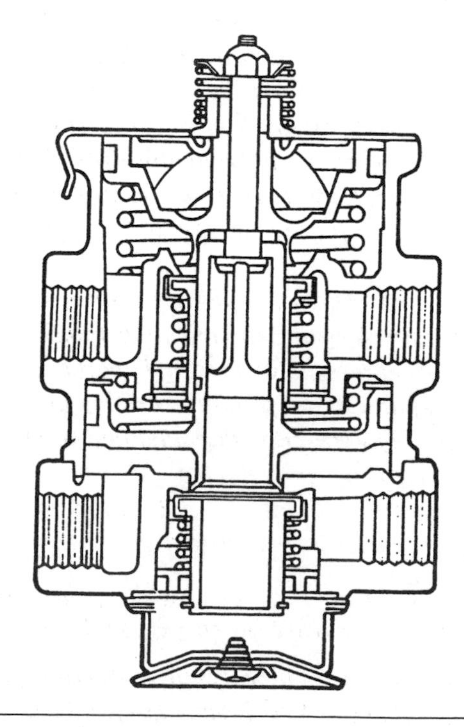

Figure 4–7 Brake application valve.

29. When diagnosing the brake application valve, *Technician A* says that the air pressure at the brake application valve delivery port must be proportional to treadle movement. *Technician B* says that with the brakes applied, a 1 inch (2.54 cm) bubble in 3 seconds at the brake application valve exhaust port indicates excessive leakage. Who is right?
    A. A only
    B. B only
    C. Both A and B
    D. Neither A nor B

Hint

The service brake application valve in a truck brake system is a floor-mounted foot valve, known as a treadle valve. The treadle valve is actually two valves in one. The upper portion of the valve is the primary section and the lower portion is the secondary section. Each section has a dedicated feed and its own exhaust port. The upper or primary section of the treadle valve is supplied directly from the primary reservoir. The lower secondary section of the valve is supplied by the secondary reservoir.

The treadle valve is actuated mechanically, that is, by foot pressure from the driver. When the driver's foot acts on the treadle valve, the primary piston is forced downward. This movement first closes the primary exhaust port and then modulates air proportional to piston travel to:

1. *actuate whatever brakes are plumbed into the primary circuit: this is usually, but not exclusively, the drive axles on a typical tandem drive tractor unit.*
2. *actuate the relay or secondary piston, located below the primary section in a dual circuit application valve.*
3. *act against the mechanical pressure (foot pressure) to provide brake feel.*

The secondary section of the dual brake application valve is actuated pneumatically by primary system air. This section operates similarly to a relay valve in that a signal pressure value (the air from the primary section) is used to displace a relay piston that modulates secondary system air to whatever brakes/valves are located in the secondary system. In a typical tractor air brake system, this would normally be the front axle brakes. Like a relay valve, the secondary section is designed to modulate air pressure to the secondary system, identical to the signal pressure.

Each portion of the valve has its own exhaust port. Primary and secondary system air never come into contact with each other in the treadle valve.

In an emergency application of the treadle valve, both the primary and secondary inlet valves are held open and full reservoir pressure is applied to each of the two systems.

In the event of a total primary system failure, no air is available to the primary section of the treadle valve. If this condition occurs, the low-pressure warning alerts are on, and when the treadle valve is depressed, foot pressure drives the primary piston downward until it mechanically contacts the relay piston to actuate the secondary system. Under this condition there is zero brake "feel" and greater pedal travel.

In the event of a total secondary system failure, the primary section of the treadle valve would function normally. However, the low air pressure alert is on, and the vehicle would have to be brought to a halt using only primary system air pressure.

Task 12

Inspect, test, or replace two-way (double) check valves and anticompounding valves.

30. When testing single and double check valves:
    A. with air pressure applied to the inlet side of a single check valve, leakage must not exceed specifications at the outlet side.
    B. with air pressure applied to one of the inlet ports on a double check valve, the air pressure should increase slowly at the outlet port.
    C. with air pressure applied to one inlet port in a double check valve, the test gauge at the opposite inlet port should indicate air pressure.
    D. when air pressure is released at one inlet port on an double check valve, the air pressure should decrease slowly at the outlet port.

31. While discussing single and double check valves, *Technician A* says single check valves are connected between the primary and secondary reservoirs. *Technician B* says a double check valve allows air pressure to flow from the lowest of two pressure sources. Who is correct?
    A. A only
    B. B only
    C. Both A and B
    D. Neither A nor B

32. What is anticompounding?
    A. Reversing gladhand positions on the trailer and tractor lines
    B. Running the tractor in a bobtail position
    C. Rapidly releasing the air in the spring brake chamber to allow the brakes to be applied
    D. Preventing simultaneous application of the service and emergency side of the spring brake chambers

Hint

Two-way (double) check valves play an important role as a safeguard in dual circuit air brake systems. The typical two-way check valve is a T with two inlets and a single outlet. It outputs the higher of the two source pressures to the outlet port and checks the lower value source. This valve will shuttle in the event of a change in the source pressure value. In other words it will always prioritize the higher source pressure. Two-way check valves provide a means of providing the primary system with secondary system air and vice versa in the event of a system failure. In the event that both source pressures are equal, as would be the case in a properly functioning dual air brake system, the valve will prioritize the first source to act on it.

Compounding occurs when a foundation brake is subjected to both mechanical and pneumatic force. A spring brake chamber in park mode has air exhausted from the parking brake chamber. This enables the mechanical force of the spring in the chamber to act on the slack adjuster and apply the brakes. It takes an air pressure of approximately 60 psi (413 kPa) to cage the spring brake. Therefore, the spring brake chamber is capable of applying approximately the equivalent of that amount of force to the foundation brakes when no air is acting on the hold-off diaphragm. In this parked condition, if a driver made a full service application of the brakes, the mechanical spring force on the foundation brakes is compounded by a further 120 psi (827 kPa) acting on the service diaphragm. The result would be a total application pressure 50 percent greater than the specified maximum. To prevent this from happening, anticompounding valves are used. These valves prevent simultaneous application of the service and parking brakes.

Anticompounding valve operation can be easily verified with a pair of air pressure gauges fitted to the service and hold-off lines and the vehicle parked.

Failure of an anticompounding valve can result in twisted S-cam shafts, spline damage, and slack adjuster damage.

Task 13

Inspect, test, repair, or replace stop and parking brake light circuit switches, wiring, and connectors.

33. The stoplights are inoperative in the stop light circuit in Figure 4–8. *Technician A* says the 10

Amp Hyd Brk circuit breaker may have an open circuit. *Technician B* says there may be an open circuit at terminal A on the stoplight switch. Who is right?

A. A only
B. B only
C. Both A and B
D. Neither A nor B

34. Refer to Figure 4–8. When the brakes are applied or when the right turn signal is on, the right rear brake and signal light is dim. The right rear taillight is also dim when the taillights are on. The cause of this problem could be:

A. excessive resistance in the wire from the right rear signal light to junction S401.
B. excessive resistance in the wire from the signal light switch to connection C211 between the stoplight switch and the signal light switch.
C. excessive resistance in the wire from the stoplight switch to connection C211.
D. excessive resistance in the stoplight switch contacts connected to terminals A and B.

Hint

Functioning brake lights are required in all highway vehicles. Trucks with air brakes require two brake light switches, either of which can illuminate the vehicle brake light circuit.

The service stoplight switch is a normally open, air-actuated switch, plumbed into the service brake circuit. The switch is closed by a small application pressure acting on it.

The parking brake light switch works oppositely. This switch is designed to close the electrical circuit when no air is acting on it, and open when air is charged to the hold-off chambers.

The operation of both switches can be verified with a DVOM. The voltmeter should be used when testing either switch in circuit, and the ohmmeter may be used to test the switch when it is disconnected from the circuit.

Task 14

Inspect, test, repair, or replace hand brake (trailer) control valve, lines, hoses, fittings, and mountings.

35. When testing the operation of the trailer control valve, a Technician should:

A. move the handle to the fully applied position and record the air pressure.
B. drain the air system to 0 psi, then allow pressure to build up again.
C. record the gauge reading of the air pressure first, then drain the system.
D. listen for air leakage around the handle as the handle is moved to the applied position.

36. When diagnosing a trailer control valve with a pressure gauge connected to the valve delivery port, all of these statements are true EXCEPT:

A. the air pressure at the delivery port should be proportional to valve handle movement.
B. when the valve handle is fully applied, full reservoir pressure should be indicated on the gauge.
C. with the valve handle fully applied, a 1 in. (2.54 cm) bubble in 7 seconds at the valve exhaust port indicates excessive leakage.
D. when the valve handle is released, the pressure on the gauge should drop quickly to zero.

Hint

The trailer control valve is used to actuate the trailer service brakes on a trailer independently of the tractor's service brakes. This valve is also known by the terms ***trolley valve****,* ***broker brake****, and* ***spike.*** *The source of air supplied to the trailer control valve is usually the secondary circuit. This system air pressure is modulated proportionally with control valve travel.*

Two-way check valves are used in the application circuit of the trailer brakes. The two-way check valve is located downstream from both the trailer control valve and the treadle valve. In a typical system, the trailer application valve uses secondary system air, and the treadle valve trailer service signal uses primary system air to activate the trailer service brakes. If a driver was using the trailer application valve to stop the vehicle and during this process an emergency occurred that required a panic application of the treadle valve, the two-way check valve would permit the air source for the trailer control valve to change from the secondary system to the higher source pressure value from the treadle valve.

Trailer control valve performance can be verified with an accurate air pressure gauge. The trailer is supplied with air by means of air hoses. Trailer service and supply air line connections are connected to a tractor protection valve. The air pressure is transferred to the trailer by couplers known as gladhands. Gladhands enable easy coupling of the service and supply air lines between the tractor and trailer.

Gladhand seals are usually manufactured out of rubber. They are retained in an annular groove in the gladhand and are easily replaced when they fail.

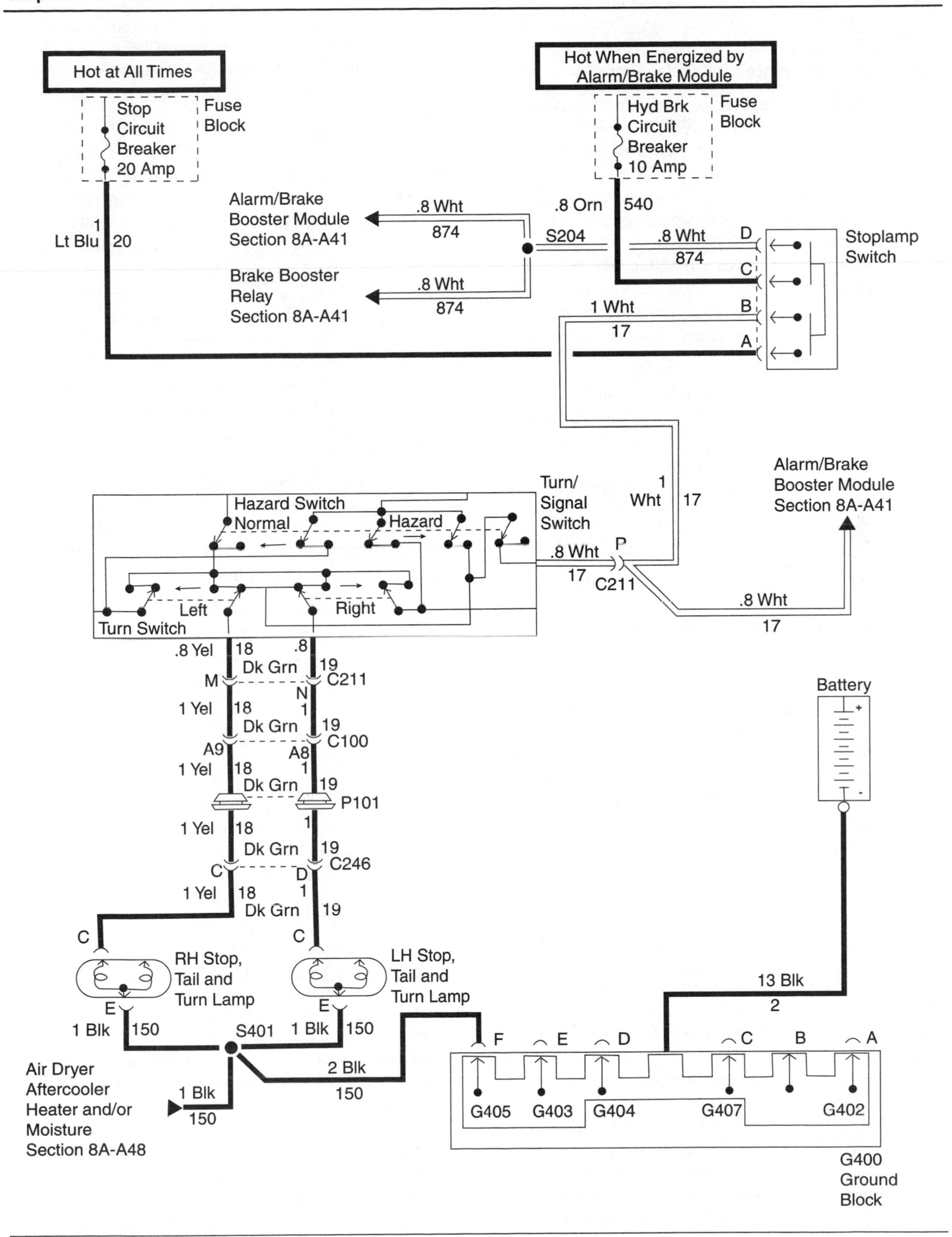

Figure 4–8 Stop light circuit. *(Courtesy of General Motors Corporation, Service Technology Group)*

Task 15

Inspect, test, or replace brake relay valve.

37. When diagnosing a service brake relay valve, all of the following apply EXCEPT:
    A. inlet valve leakage is tested with the service brakes released.
    B. apply a soap solution to the area around the inlet and exhaust valve retaining ring to check exhaust valve leakage.
    C. exhaust valve leakage is tested with the brakes applied.
    D. the control port on the service brake relay valve is connected to the supply port on the brake application valve.

38. On a tractor service brake relay valve:
    A. air lines are connected from the service brake relay valve delivery ports to the rear axle service brake chambers.
    B. when the brakes are released, the inlet valve is open in the service brake relay valve.
    C. when the brakes are released, the exhaust valve is closed in the service brake relay valve.
    D. the service brake relay valve is in a balanced position when the air pressure in the rear brake chambers equals reservoir pressure.

Hint

Relay valves permit a remote air signal (from the treadle or trailer valve) to effect service braking with an air supply close to the brake chambers. The relay valve is controlled by a signal that is plumbed to its service port. System air pressure is available at the relay valve supply port from a nearby air reservoir. When the relay valve receives an air signal, this pressure acts on the relay piston. Relay piston movement modulates the air available at its supply port, to actuate the service brake chambers connected to its delivery ports. One relay valve can typically manage two to four service brake chambers. The valve is usually designed so that the signal value is identical to the output value modulated to the delivery ports. When the signal pressure is dropped or relieved, the relay valve exhausts the service supply air returning from the service chambers, and a retraction spring returns the relay valve piston to its neutral position.

A sticking relay valve piston causes supply air to be modulated to the service chambers when there is no signal pressure. This condition causes dragging brakes or locked brakes, depending on the location of the relay piston. Moisture in the air system may freeze and prevent the relay retraction spring from returning the relay piston.

Relay valve operation can be verified using a pair of gauges that monitor signal pressure and delivery pressure.

Task 16

Inspect, test, and replace quick release valves.

39. All of these statements about limiting quick release valve diagnosis are true EXCEPT:
    A. with the control valve in the dry road position, the limiting quick release valve reduces air pressure to the front brakes.
    B. with the control valve in the slippery road position and the service brakes applied, leakage at the control valve exhaust port should not exceed a 1 inch (2.54 cm) bubble in 3 seconds.
    C. With the control valve in the slippery road position and the service brakes applied, leakage at the limiting quick release valve exhaust port should not exceed a 1 inch (2.54 cm) bubble in 3 seconds.
    D. with the control valve in the slippery road position and the service brakes applied, it limits front brake pressure to 50 percent of the application valve pressure.

40. Technicians are discussing quick-release valves. *Technician A* says that a bad tractor protection valve will cause wheel lockup during a brake application. *Technician B* says a defective quick-release valve can cause slow front pressure buildup and poor stopping ability. Who is right?
    A. A only
    B. B only
    C. Both A and B
    D. Neither A nor B

41. Where is a quick-release valve mounted?
    A. Close to the brake chambers
    B. In the cab, close to the trailer protection valve
    C. In the middle of a combination vehicle's axles for easier line routing
    D. On the outside backwall of the tractor next to the gladhands

Hint

Quick-release valves are mounted close to the brake chambers or components they serve. They are used throughout the air brake system and other vehicle air systems to speed release times.

The typical quick release valve has a single inlet port and two outlet ports. When air is charged to the inlet port, the valve acts like a T fitting and simply divides the air to the two outlets. However, when the air pressure supplied to the inlet is exhausted at the application or control valve, the air supplied from the quick release to the brake chambers is exhausted at the quick-release valve.

A quick-release valve may also be used on the service gladhand supplying the trailer(s). This feature greatly reduces the release lag that can occur in service braking, by exhausting the service signal at a much higher speed.

Because of its simplicity, a quick-release valve seldom fails. They are vulnerable to external damage because of their location, and contaminated air that may cause exhaust port leakage.

Task 17

Inspect, test, and replace front and rear axle limiting (proportioning) valves.

42. A heavy truck brake system is being inspected. *Technician A* says that the proportioning valve should be inspected every time the brakes are serviced. *Technician B* uses gauges ahead of and behind the proportioning valve to test its function. Who is right?
    A. A only
    B. B only
    C. Both A and B
    D. Neither A nor B

43. With 50 psi (344.75 kPa) supplied from the brake application valve to the limiting valve inlet port, the pressure read on a test gauge at the limiting valve outlet port should be:
    A. 40 psi (275.8 kPa).
    B. 15 psi (103.42 kPa).
    C. 25 psi (172.37 kPa).
    D. 30 psi (206.85 kPa).

Hint

Many trucks are fitted with automatic front wheel proportioning valves also known as ratio valves. A ratio valve manages air pressure to the front brake chambers by proportioning the application pressure delivered to it. Typically, the ratio valve will proportion as follows:

Application pressure	*Pressure to front brakes*
0–10 psi	*0*
10–40 psi	*50 percent application pressure*
40–60 psi	*50 pecent graduating to 100 percent*
60 psi and higher	*100 pecent application pressure*

The bobtail proportioning relay valve is integral with the tractor service brake relay valve. This valve functions as a relay valve under normal operation. When the tractor is driven without the trailer, the bobtail proportioning relay valve bleeds some of the air pressure supplied from the brake application valve to the modulating piston in the bobtail proportioning relay valve. This action reduces air pressure supplied to the rear axle brake chambers to prevent rear wheel lockup.

Task 18

Inspect, test, and replace tractor protection valve.

44. When diagnosing a pressure protection valve, *Technician A* says that when the air pressure is reduced to zero on the delivery side of the pressure protection valve, the air pressure should decrease to zero on the supply side of this valve. *Technician B* says a 2-inch bubble in 5 seconds is acceptable around the cap on the pressure protection valve. Who is right?
    A. A only
    B. B only
    C. Both A and B
    D. Neither A nor B

45. While discussing tractor protection valves, *Technician A* says that when the trailer air supply valve is pulled outward, the air pressure is supplied to the tractor protection valve. *Technician B* says that if a trailer breakaway occurs, service air pressure from the application valve can no longer pass through the tractor protection valve. Who is right?
    A. A only
    B. B only
    C. Both A and B
    D. Neither A nor B

46. The purpose of the tractor protection valve in Figure 4–9 is to:
    A. ensure against air leakage at the gladhands.
    B. provide a means of preserving air pressure in sufficient amount to stop the tractor in the event of a trailer breakaway.
    C. provide a means of holding off the spring brake in case of a spring brake chamber air loss.
    D. sense that the tractor is running without a trailer and automatically reduces the amount of air pressure that can be applied to the tractor's drive axle(s).

Hint

The tractor protection valve protects the tractor air supply under a trailer breakaway condition or severe air

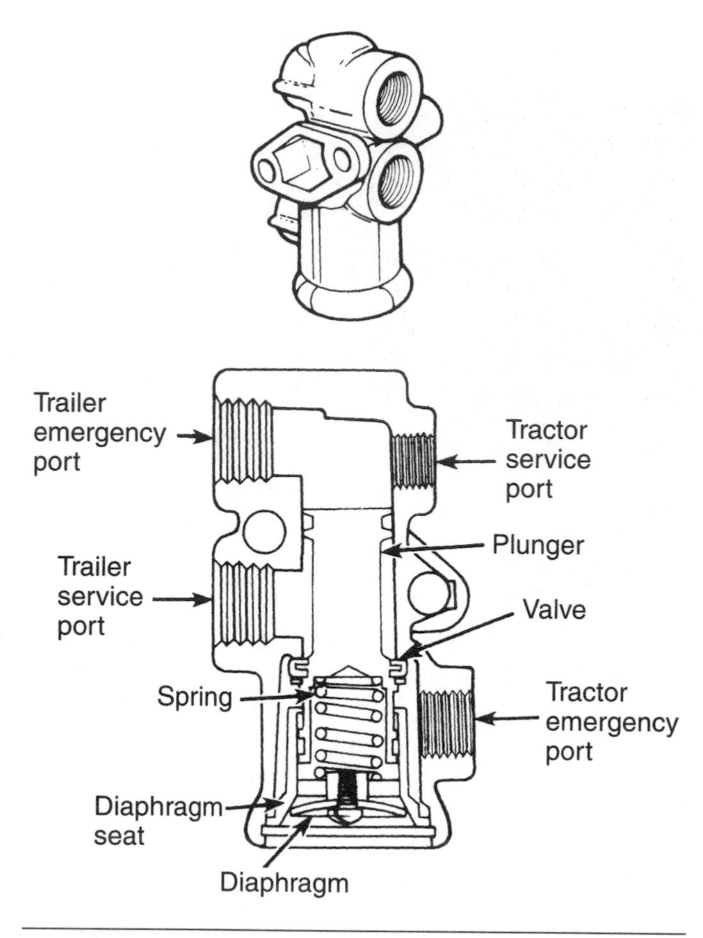

Figure 4–9 Tractor protection valve. *(Courtesy of Allied Signal Truck Brake Systems Co.)*

leakage. Two air lines connect the tractor air brake system with the trailer air brake system. The trailer supply line supplies the trailer with air for braking and any other pneumatic systems such as an air suspension system. The trailer service line is the service brake signal line. The tractor service and supply air lines are connected to the tractor protection valve. When the trailer air supply dash valve is pulled outward, air pressure is vented in the trailer supply line and tractor protection valve. In this condition, when the driver makes a service brake application, no air will exit the tractor protection valve service signal line to the trailer. When the trailer air supply valve is pushed inward, air pressure opens the tractor protection valve and supplies air pressure to the trailer air systems. In this condition, when the driver makes a service brake application, the service brake signal air is transmitted through the tractor protection valve to apply the trailer service brakes.

It should be noted that the trailer air supply controls the trailer park brakes. Whenever this air supply is interrupted, whether intentionally, such as when the trailer is being parked, or unintentionally, such as in a breakaway, the spring brakes will apply regardless of how much pressure is in the trailer air reservoirs.

The tractor protection valve is designed to isolate the tractor air system from that of the trailer at a predetermined value that ranges from 20 to 45 psi depending on the system. It should be noted that at these pressures, the spring brakes on both the tractor and trailer would be partially applied.

Task 19

Inspect, test, and replace emergency (spring) brake control valve(s).

47. Technicians are diagnosing the tractor spring brake control valve. *Technician A* says that if the rear axle reservoir is drained to zero air pressure and the brake application valve is applied, the tractor air brake pressure should be at 120 psi. *Technician B* says that when the tractor parking brakes are applied, the air pressure at the tractor spring brake valve delivery port should drop quickly to zero. Who is right?
    A. A only
    B. B only
    C. Both A and B
    D. Neither A nor B

48. While discussing the air brake system valve in Figure 4–10, *Technician A* says that if the air pressure in the #1 (rear axle) reservoir drops below 55 psi (379 kPa), this valve allows partial application of the spring brakes when the service brakes are applied. *Technician B* says that if piston B in this valve is moved upward, air pressure is supplied to the spring brake chambers to release the spring brakes. Who is right?
    A. A only
    B. B only
    C. Both A and B
    D. Neither A nor B

Hint

The spring brake valve supplies air to the hold-off chambers to release the spring brakes and enable the vehicle to move. During normal operation, it limits the hold-off pressure to the spring brake chambers to a value around 90 psi (620 kPa). This action speeds application times in the event of an emergency and permits a consistent hold-off pressure because the system pressure fluctuates between cut-in and cut-out pressure values.

Some systems use an inversion valve. In the event of a primary air system failure, the inversion valve allows for a modulated application of the emergency air brake

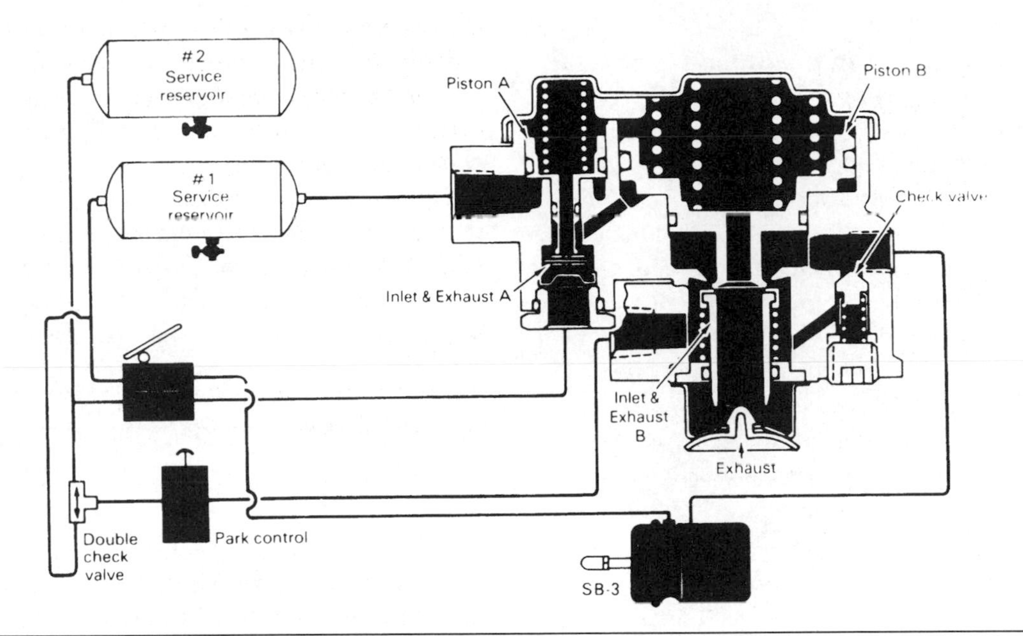

Figure 4–10 Air brake system valve. *(Courtesy of Allied Signal Truck Brake Systems Co.)*

chambers by relieving air from the hold-off chambers in direct proportion to the application pressure at the treadle valve.

Spring brake control valves are often incorporated in multifunction valves that contain parking, service, and inversion functions.

A. A only
B. B only
C. Both A and B
D. Neither A nor B

Task 20

Inspect, test, or replace low-pressure warning devices.

49. On a tractor with low air pressure warning system as shown in Figure 4–11, the red low pressure warning light on a vehicle does not go out after startup and the buzzer continues buzzing. *Technician A* says the problem may be caused by a loss of air pressure in one section of the dual system. *Technician B* says the wire from one of the low-pressure switches to the in-line diode may be grounded. Who is right?
    A. A only
    B. B only
    C. Both A and B
    D. Neither A nor B

50. Refer to Figure 4–11. *Technician A* says that an inoperative warning light and buzzer below 60 psi (414 kPa) is caused by a defective bulb. *Technician B* says that an inaccurate dash pressure gauge can be the cause of a warning light on above 60 psi (414 kPa) supply pressure indicated on the dash gauge. Who is right?

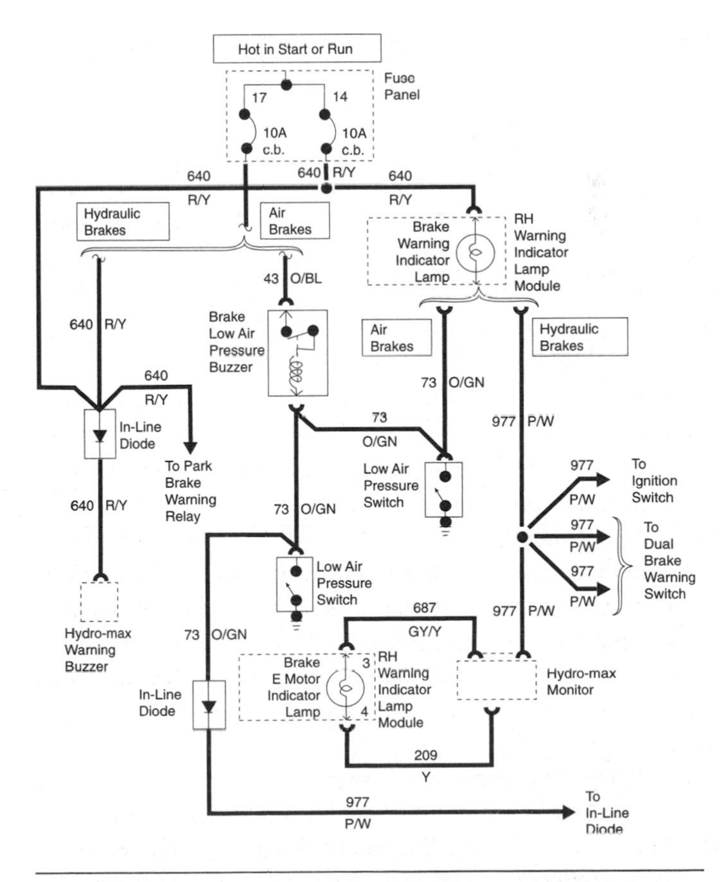

Figure 4–11 Low air pressure warning system. *(Courtesy of Ford Motor Company)*

Hint

FMVSS 121 requires that a driver must receive a visible alert when the system pressure drops below 60 psi (414 kPa). In most cases this is accompanied by an audible alert, usually a buzzer. A low air pressure warning device is fitted to both the primary and secondary systems. This is a simple electrical switch that can be plumbed anywhere into a system that requires monitoring. The switch is electrically closed whenever the air pressure being monitored is below 60 psi (414 kPa). When the air pressure value exceeds 60 psi (414 kPa), the switch opens.

Verifying the operation of a low air pressure warning switch can be done by pumping the service application valve until the system pressure drops to the trigger value. Both the primary and secondary circuit air pressure must be monitored by a dash-mounted gauge. The required visible warning is usually a dash warning light.

Task 21

Inspect, test, repair, and replace air pressure gauges, lines, hoses, and fittings.

51. A trailer service brake relay valve was replaced because it did not release air pressure quickly after a brake application, and the trailer brakes dragged for a short time after the brakes were released. The tractor and trailer brakes applied normally. After the relay valve replacement the driver complains that the tractor brakes lock up before the trailer brakes during a hard brake application. *Technician A* says there may be a restriction in the trailer service brake line. *Technician B* says the replacement trailer relay valve may have the wrong crack pressure. Who is right?
    A. A only
    B. B only
    C. Both A and B
    D. Neither A nor B

52. Why is it very difficult to have the same actuation pressure on all axles of a combination vehicle?
    A. Because there are more valves between the foot valve and the trailer axles than the tractor axles, thus increasing restrictions and lowering those respective pressures.
    B. Because the increased length of the lines to the trailer cause a corresponding pressure drop or decrease.
    C. The same actuation pressure is impossible to achieve due to the fact that both the tractor and trailer air systems are two separate and distinct systems.
    D. Because the treadle valve operates the tractor air system, while the trailer supply valve operates the trailer air system.

Hint

Air pressure gauge operation can be verified by using a master gauge, a good quality, liquid-filled gauge that uses a Bourdon principle of operation. When troubleshooting vehicle air pressure management problems, the vehicle gauges should not be relied on.

Mechanical/Foundation

Task 1

Diagnose poor stopping, premature wear, brake noise, pulling, grabbing, or dragging complaints caused by foundation brake components, slack adjuster, and brake chamber problems; determine needed repairs.

53. All of the following could cause sluggish service brake release EXCEPT:
    A. larger than specified brake chamber diameter.
    B. weak chamber return spring.
    C. broken or weak brake return springs.
    D. obstructed brake chamber air passage.

54. An ideal air braking system can be defined as:
    A. one in which braking forces are proportional on all axles.
    B. one in which the braking pressure reaches each actuator simultaneously and at the same pressure level.
    C. one in which each axle receives braking force in a sequential manner, starting with the closest axle to the governor.
    D. one in which each wheel receives a graduated braking force over a preset range of the most probable braking conditions.

55. A driver complains about poor stopping on a tractor air brake system. The system air pressure is within specifications and none of the wheels lock up. *Technician A* says there may be excessive air leakage at the exhaust valve in the service brake relay valve. *Technician B* says different types of brake linings may have been installed on some of the tandem axle wheels. Who is right?
    A. A only
    B. B only
    C. Both A and B
    D. Neither A nor B

Hint

The S-cam, wedge, or disc brake mechanism, linings or pads, and related parts such as brake chamber(s), slack adjuster(s), and parking brake components generally make up what is called the foundation brake assembly.

All vehicle braking requires that kinetic energy (the energy of motion) be converted to heat energy by means of friction, and the heat energy must then be dissipated to atmosphere. The foundation brake assembly consists of those brake components that are responsible for providing the retarding effort required to stop a vehicle.

In a typical S-cam, shoe/drum assembly, slack adjusters connect the brake actuation chambers with the foundation brake assembly. Slack adjusters are levers so the length of the slack adjuster arm is critical. The greater the distance between the centerline of the S-cam and the point at which the brake chamber clevis connects with the slack adjuster, the greater its leverage. Slack adjusters also convert the linear force produced by the brake chamber into torque (twisting force). Slack adjusters are simple components whose parts include a worm gear, a pushrod, an actuator, and an actuator piston. Current slack adjusters are required to be automatic adjustment type.

The geometry of the relationship between the brake chamber and slack adjuster requires that the angle between the slack adjuster arm and the chamber pushrod be 90 degrees when the brake is in the fully applied position. In any position other than 90 degrees, the slack adjuster has less mechanical advantage. As the brake shoe friction surfaces wear, this 90-degree angle must be maintained be periodic adjustment of manual slack adjusters or functioning automatic slack adjusters.

Both parking and service brake performance relate directly to brake adjustment. Foundation brake problems can cause unbalanced braking, grabbing, not releasing, or failure to apply.

Some causes of foundation brake problems are use of improper replacement parts, cargo weight overloads, lining contamination, poorly maintained or installed brakes, and broken or malfunctioning brake components, brakes out of adjustment, overheating.

Task 2

Inspect, test, adjust, repair, or replace service brake chambers, diaphragm, clamp, spring, pushrod, clevis, and mounting brackets.

56. *Technician A* says that a slack adjuster must be adjusted so the angle between the slack adjuster arm and the chamber pushrod is 90 degrees when the brakes are fully applied. *Technician B* says that the chamber pushrod should be adjusted with the shortest possible stroke, without dragging the brakes. Who is right?
    A. A only
    B. B only
    C. Both A and B
    D. Neither A nor B

57. The component marked "B" in Figure 4–12 is the:
    A. S-cam.
    B. power spring.
    C. check valve.
    D. pressure plate.

58. In Figure 4–13 the clevis is being reinstalled onto its pushrod. *Technician A* says that the clevis should be correctly positioned on the pushrod in order to facilitate proper brake adjustment.

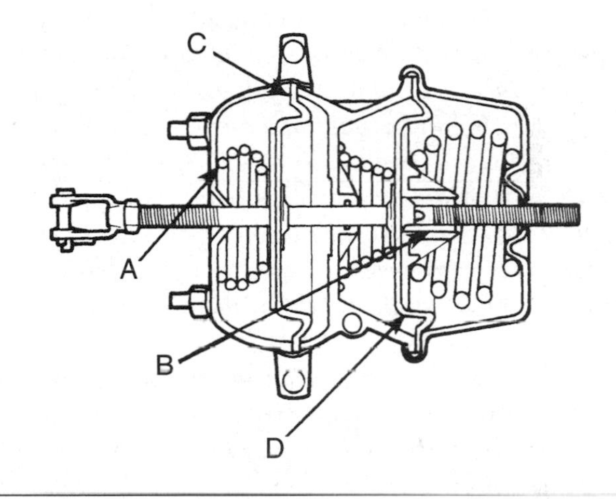

Figure 4–12 Brake chamber components.

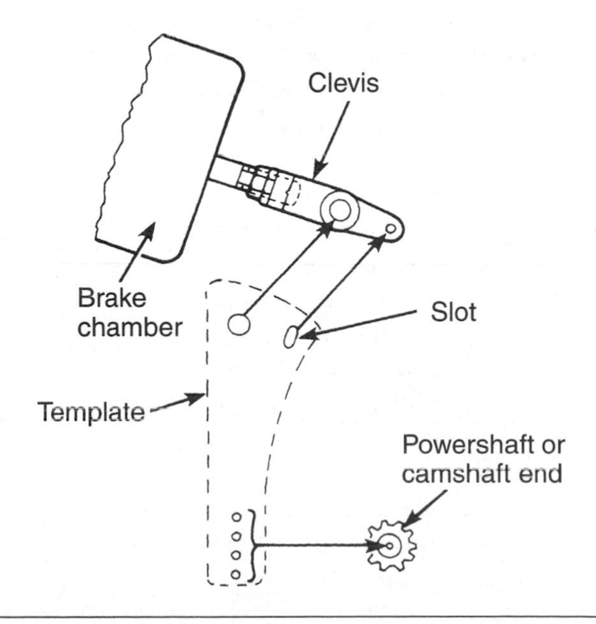

Figure 4–13 Single brake chamber and clevis. *(Courtesy of Rockwell International Corporation)*

Technician B says that the clevis on a brake chamber pushrod is keyed. Who is right?
A. A only
B. B only
C. Both A and B
D. Neither A nor B

Hint

To service a spring brake chamber, it will be necessary to "cage" or release the compression spring before any service may be performed. A spring brake is normally manually released by tightening a nut on a supplied release tool. After the spring brake chamber is caged, air can be released from the chamber, thus facilitating service on the spring brake.

If the clevis is not positioned properly onto its pushrod, the brake will not completely release. The clevis is threaded onto the brake chamber pushrod. Most manufacturers furnish slack adjuster installation templates with units, thus facilitating correct installation.

A .060 inch gap between the clevis and the collar is considered the maximum allowable before replacement. The automatic slack adjuster automatically adjusts the clearance between the brake linings and the brake drum or rotor. The slack adjuster controls the clearance by sensing the length of the stroke of the air brake chamber pushrod.

Problems concerning foundation brake geometry are one of the most common causes of sluggish service brake operation. Problems such as an obstruction in a brake chamber, or poor alignment of the brake linkage often produce apply and release rate performance problems.

When removing the spring chamber from the adapter, it is necessary to slide it sideways while holding the adapter. The diaphragm can be removed from the spring chamber by rotating it counterclockwise from the chamber. The clamps or any other part of the adapter or chambers should not be struck with a hammer or heavy object. Finally, the air pressure should be exhausted if it was used to aid in caging the spring brake. Rotochambers were designed to replace the conventional-type brake chamber. They operate similarly, and are identified by types. A type 30 Rotochambers would have an effective diaphragm area of 30 inches.

Spring brake chamber assemblies should not be repaired but replaced as an assembly, after they have been manually caged. When the park control valve is pulled outward, the air in the rear portion of the spring chamber is exhausted thus allowing the compression spring to apply the parking brake.

The pressure plate on a spring brake chamber is located between the adapter return spring and the compression spring. The pressure plate is what holds off the compression spring when the spring brake chamber is charged, thus releasing the spring (parking) brake. When air is released from the service brake air chambers, the pushrod return spring, in combination with the brake shoe return spring, returns the diaphragm, push plate and pushrod assembly, slack adjuster, and brake cam (or wedge) to their released positions, thereby releasing the brakes.

Task 3

Inspect, adjust, repair, or replace manual and automatic slack adjusters.

59. The slack adjuster in Figure 4–14 should be adjusted so that the brake chamber pushrod is:
A. at an 85-degree angle in relation to the slack adjuster arm when fully applied.
B. parallel to the slack adjuster arm when the brakes are fully applied.
C. at a 95-degree angle in relation to the slack adjuster arm when fully applied.
D. at a 90-degree angle in relation to the slack adjuster arm when fully applied.

60. The pushrod stroke is being measured. *Technician A* says that the applied stroke uses an 80 psi brake application. *Technician B* says the brakes are adjusted (if needed) to achieve the proper free stroke. Who is right?
A. A only
B. B only
C. Both A and B
D. Neither A nor B

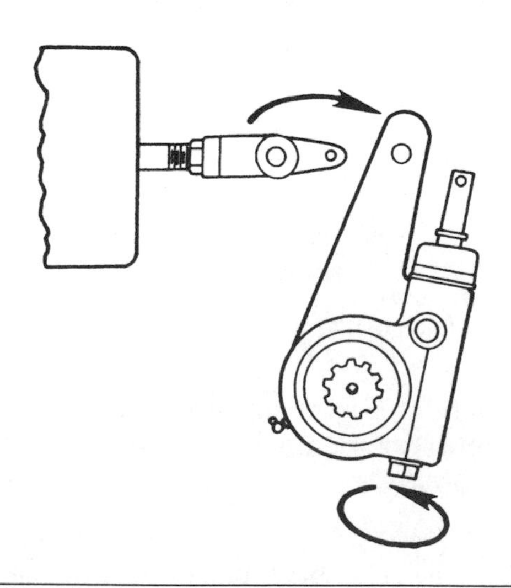

Figure 4–14 Brake chamber and slack adjuster. *(Courtesy of Rockwell International Automotive)*

61. A slack adjuster is being adjusted. *Technician A* says excessive wear limits between the clevis and the collar will cause the stroke to be too long. *Technician B* says improper brake adjustment may cause premature wheel lockup during a brake application. Who is right?
    A. A only
    B. B only
    C. Both A and B
    D. Neither A nor B

Hint

The slack adjuster is critical in maintaining the required free stroke and adjustment angle. Automatic slack adjusters are currently required on air brake systems, but their operation must be verified routinely.

Slack adjusters connect the brake chambers with the foundation assemblies on each wheel. They are connected to the pushrod of the brake chamber by means of a clevis yoke that is threaded onto the pushrod. Slack adjusters are spline mounted to the S-cams and positioned by shims and an external snap ring.

The objective of brake adjustment with a manual or automatic slack adjuster is to maintain a specified drum-to-lining clearance and a specified amount of free stroke. Free stroke is the amount of slack adjuster stroke that occurs before the linings contact the drum. The first step in determining the free stroke is to measure the distance from the brake chamber face to the center of the large clevis pin. The second step is to insert a prybar into the clevis, and pull on the prybar to apply the brake until the linings contact the drum. While holding the clevis and slack adjuster in this position, measure the distance again from the brake chamber face to the center of the large clevis pin. The free stroke is the difference in the distance from the brake chamber face to the center of the large clevis pin with the brakes released and the brakes applied by hand so the linings contact the drum. Rotate the adjusting nut on the slack adjuster to obtain the specified free stroke. A locking pawl must be released or removed before rotating this adjusting nut. The applied stroke must also be measured. The applied stroke is the difference in the distance from the brake chamber face to the center of the large clevis pin with the brake released, and the brake applied with 80 to 90 psi (552 to 620 kPa) air pressure. The adjusting nut on the slack adjuster is rotated to adjust the applied stroke if it is not within specifications. When the brakes are applied at 80 to 90 psi (552 to 620 kPa), there should be a 90-degree angle between the brake chamber pushrod and the slack adjuster arm. Automatic slack adjusters may require periodic adjustment. When a wheel-up adjustment is performed, the free stroke should be as short as possible without any brake drag.

Slack adjusters are lubricated by grease or automatic lubing systems. The seals in slack adjuster are always installed with the lip angle facing outward. When grease is pumped into the slack adjuster, grease exits easily past the seal lip when the internal lubrication circuit has been charged.

When replacing a failed slack adjuster, the distance between the axis of the S-cam bore and the clevis pin bore must be maintained. A difference of one-half inch can greatly alter the brake torque and unbalance the brakes.

Task 4

Inspect, or replace cams, rollers, shafts, bushing, seals, spacers, retainers, brake spiders, shields, anchor pins, and springs.

62. The brake linings are worn and there is excessive clearance between the linings and the brake drum on a cam-type foundation brake. *Technician A* says this condition may cause the cam to roll over during a brake application. *Technician B* says this condition may cause loss of steering control. Who is right?
    A. A only
    B. B only
    C. Both A and B
    D. Neither A nor B

63. When assembling a camshaft all of the following apply EXCEPT:
    A. both camshaft seal lips must face toward the slack adjuster.
    B. use the proper driver when installing camshaft seals.
    C. lubricate the cam head with petroleum jelly.
    D. measure camshaft radial movement.

Hint

When performing foundation brake service, all of the hardware mounted on the axle spider should be inspected and replaced if required. Radial and axial play on the S-camshaft must be within specifications. Rollers and cams should be inspected for flat spots and it is good practice to replace any spring steel components, such as retraction springs and snap rings. Spider fastener integrity should be checked at each brake job, and spiders should be inspected for cracks.

S-cam bushing seals should be installed so that both seal lips face toward the slack adjuster. This seal position prevents grease from being forced past the seal next to the S-cam into the foundation assembly. S-cams should never be lubricated.

Task 5

Inspect, adjust, repair, or replace wedge-type brake housing, plungers, and wedge assembly.

64. While discussing wedge brake diagnosis and service, *Technician A* says installing a new wedge with a different angle does not affect brake operation. *Technician B* says no lubricant should be used on the wedge or wedge ramps. Who is correct?
    A. A only
    B. B only
    C. Both A and B
    D. Neither A nor B

65. When servicing wedge brakes, all of these apply EXCEPT:
    A. minor scoring may be removed from the actuator bores with fine emery cloth.
    B. if the threaded actuator opening is not chamfered, install the collet nut so the flat side faces away from the actuator threaded opening.
    C. the clearance between each wedge brake shoe lining and the drum should be 0.020 to 0.040 in. (.51 to 1.02 mm).
    D. if the threaded actuator opening is chamfered, the taper on the collet nut faces the taper on the actuator opening.

Hint

Drum brakes can be classified as either cam actuated or wedge actuated. Wedge brakes can be actuated on two planes and these have higher theoretical braking efficiencies than cam-actuated brakes. They are used today in certain applications, often as front brakes on a highway tractor.

In a wedge-actuated brake the brake chamber pushrod drives a wedge between a pair of plungers that force the shoes outward against the drums. Many wedge-type braking systems are self-adjusting. However, these adjusting mechanisms are vulnerable to seizure when equipment is not regularly used.

A specified lining-to-drum clearance must be maintained, and this is checked with a thickness gauge inserted at either end of each shoe. The drum-to-lining clearance should be consistent in all four measuring locations.

Task 6

Inspect, repair, or replace wedge brake spiders, manual adjuster plungers, and automatic adjuster plungers.

66. *Technician A* says many wedge brakes are self-adjusting. *Technician B* says wedge brakes are equipped with automatic slack adjusters. Who is right?
    A. A only
    B. B only
    C. Both A and B
    D. Neither A nor B

67. In a wedge brake foundation assembly:
    A. the wedge angle determines the amount of brake shoe movement in relation to the brake chamber diaphragm and pushrod movement.
    B. all adjusting plungers contain automatic adjusting mechanisms.
    C. the tapered wedge surfaces contact the straight surfaces on the bottom of the anchor and adjusting pistons.
    D. the brake chamber diaphragm pushrod contacts the wedge return spring.

Hint

Single-actuated wedge brakes use a single plunger housing mounted on the spider assembly. This type of brake has shoes with fixed anchors mounted to the spider.

Double-actuated wedge brakes have a pair of floating shoes actuated by two sets of plunger housings and brake chambers. The shoes float and are retained to the spider by clips and to the plunger housings by retraction springs that link each shoe. In a wedge brake the brake chamber pushrod pushes directly on the wedge, and the inner end of this wedge is positioned between two plungers that contact the brake shoes. Inward brake chamber pushrod and wedge movement forces the brake shoes outward against the drum. Because the brake chamber pushrod pushes directly on the wedge, the slack adjuster is not required.

Wedge brakes may have manual or automatic adjusting mechanisms. A manual adjusting mechanism located on each plunger assembly has a star wheel that rotates a threaded plunger from the housing either inward or outward. The automatic adjuster consists of a spring-loaded pawl that rides against helical grooves on the exterior of the plunger actuator. As the brake friction linings wear, pawl ridges ratchet to a new position on the plunger actuator thus limiting retraction travel and defining free play.

When the plunger assemblies seize, the automatic adjusters cease to operate, and the plunger assembly must be serviced. Plunger assemblies must be lubricated with high-temperature lubricant on reassembly.

Task 7

Inspect, clean, rebuild/replace, and adjust air disc brake caliper assemblies.

68. On an air disc brake in Figure 4–15 short outboard lining life is caused by:
    A. a piston seized in its bore.
    B. a damaged rotor.
    C. a damaged or defective brake hose/line.
    D. a caliper seized on the slide pins.

69. When inspecting a 4-wheel disc hydraulic brake system, *Technician A* says seized caliper pistons may cause brake drag. *Technician B* says that rotors machined beyond the maximum diameter may cause brake pedal fade when you first apply the brakes. Who is right?
    A. A only
    B. B only
    C. Both A and B
    D. Neither A nor B

70. Some manufacturers use two different free stroke settings for their air disc brakes. Which of the following is one of the settings?
    A. Full-clevis free stroke
    B. Half-clevis free stroke
    C. Full free stroke
    D. Initial free stroke

Hint

Air-actuated heavy-duty truck disc brakes have some similarities to automobile, hydraulically actuated, disc brakes. Air disc brakes uses a brake chamber and pushrod to apply braking torque to a powershaft. The powershaft has an external helical gear that acts on an actuator nut to create the clamping force required by the caliper to effect retarding effort on a rotor that turns with the wheel assembly. In other words, axial force applied to the powershaft is converted to clamping force by the caliper assembly.

Within the caliper assembly, pistons transmit the clamping force to brake pads, which provide retarding effort by converting the kinetic energy (energy of motion) to friction and then heat.

Short outboard friction pad service life is usually an indication of caliper assembly seizure on the caliper slide pins.

The rotors used on truck air disc brakes are usually vented to aid in the dissipation of brake heat.

Task 8

Inspect and replace brake shoes, linings, or pads; determine needed repair.

71. When replacing brake shoes on a tandem tractor axle, the primary shoe for the left front tandem (with the cam in front of the axle) is positioned at what point in Figure 4–16?
    A. To the left.
    B. To the right.
    C. On the top.
    D. On the bottom.

72. When replacing the pads on both brakes of a single axle or all four brakes of a tandem axle you should:
    A. replace all at the same time.
    B. replace only one side at a time.
    C. replace only the worn pads, while measuring the rest.
    D. resurface all discs that have had their pad(s) replaced.

73. A truck is brought in with combination lining sets. *Technician A* says that combination linings are linings on the same drum that are of different sizes. *Technician B* says differences in the method of attachment (i.e., riveted, bolted) is a definition of combination linings. Who is right?
    A. A only
    B. B only

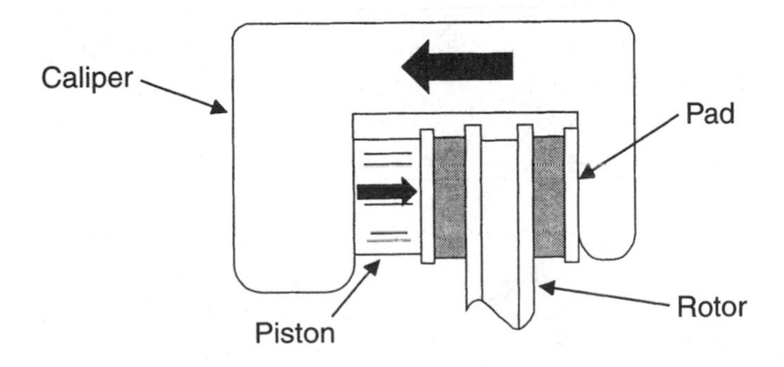

Figure 4–15 Brake caliper and pads.

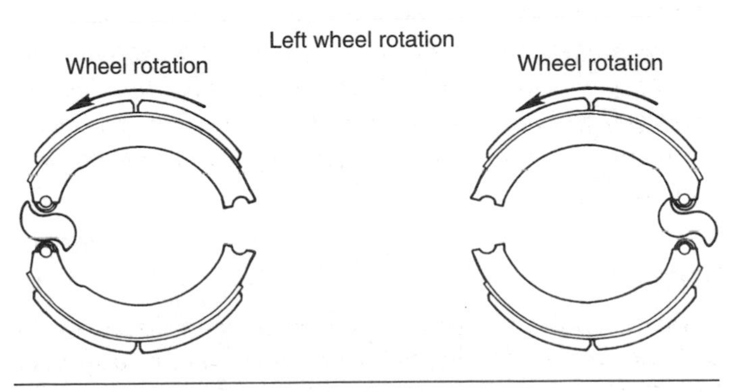

Figure 4–16 Brake shoes and wheel rotation. *(Courtesy of Rockwell International Automotive)*

C. Both A and B
D. Neither A nor B

74. *Technician A* says that on brakes with vented discs, the inboard pad could be thicker than the outboard pad. *Technician B* says that on brakes with solid discs, the inboard and outboard pads could be the same thickness. Who is right?
A. A only
B. B only
C. Both A and B
D. Neither A nor B

Hint

Truck brake shoes are in most cases fixed anchor assemblies mounted to the axle spider and actuated by an S-camshaft.

Brake shoes for a heavy-duty truck can have their linings mounted to the shoe by bonding, riveting, or by fasteners. In most current applications, the shoes are remanufactured and replaced as an assembly, that is, with the new friction facing already installed. When reusing shoes, they must be checked for arc deformities usually caused by prolonged operation with out of adjustment brakes. The lining blocks are tapered and seldom require machine arcing.

The brake shoes must be fitted with the correct linings. Lining requirements can change with different vehicles and different applications. When reconditioning shoes with bolted linings, if the original fasteners are reused, the lockwashers should at least be replaced. Riveted linings should be riveted in the manufacturer's recommended sequence. The friction rating of linings is coded by letter codes stamped on the edge of the lining. Combination sets of shoe linings are occasionally used, and these linings use different friction ratings on the primary and secondary shoes. When combination friction lining sets are used, care should be taken to install the lining blocks in the correct locations on the brake shoes.

It is good practice to replace the brake linings on all four wheels of a tandem drive axle truck on a PM schedule. When the linings on a single wheel are damaged, such as in the event of hub seal failure, the linings of both wheel assemblies on the axle should be replaced to maintain brake balance.

The friction pads in air disc brake assemblies should also be changed in paired sets, that is, both wheels on an axle. Vented disc brake rotors may have thicker inner pads than outer pads, whereas solid disc brake rotors usually have equal thickness inner and outer pads. Heat transfer is generally more uniform in the solid disc assemblies than vented ones, requiring thicker inboard pads.

Task 9

Inspect, reface, or replace brake drums or rotors.

75. *Technician A* says oversize drums may be machined as long as the drum manufacturer's recommendation for machining dimensions are followed. *Technician B* says that turning an oversize drum does not sacrifice its strength. Who is right?
A. A only
B. B only
C. Both A and B
D. Neither A nor B

76. An air disc brake system is being inspected. *Technician A* says that thickness variation is being checked in Figure 4–17. *Technician B says that* thickness variation should be measured at twelve locations around the rotor. Who is right?
A. A only
B. B only
C. Both A and B
D. Neither A nor B

Hint

If brake drums are to be reused they should be inspected for size specification, heat discoloration, scoring, glazing, threading, concaving convexing, bellmouthing, and heat checking. Used truck brake drums can seldom be successfully machined due to heat tempering.

A drum micrometer is used to check brake drum diameter. The maximum legal service limit for 14-inch diameter brake drums is 0.120 inches. The machining limit is 0.060 inch and the discard limit is 0.090 inch.

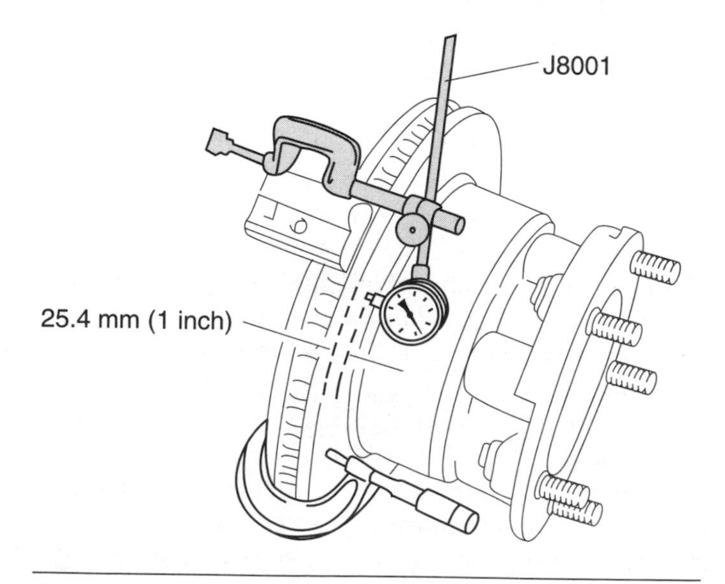

Figure 4–17 Brake rotor measurement. *(Courtesy of General Motors Corporation, Service Technology Group)*

Brake rotors must be visually inspected for heat checking, scoring, and cracks. Rotors must be measured for thickness with a micrometer. A dial indicator is used to check rotor runout and parallelism. In highway applications, rotors tend to outlast drums and are often reused after a brake job. They must be turned within legal service specifications using a heavy-duty rotor lathe.

Parking Brakes

Task 1

Inspect and test parking (spring) brake chamber, diaphragm, and seals; replace parking (spring) brake chamber.

77. What should you not do when replacing the diaphragm on a spring brake chamber?
    A. Slide the chamber sideways off the adapter.
    B. Rotate the chamber to break the diaphragm loose.
    C. Loosen the clamps by lightly striking them with a soft hammer.
    D. Exhaust the air pressure.

78. When servicing a piggyback spring brake assembly, *Technician A* says no attempt should be made to repair/replace any part of the piggyback assembly, but it should be completely replaced as a unit. *Technician B* says that the spring chamber should be disarmed (manually released) before discarding. Who is right?
    A. A only
    B. B only
    C. Both A and B
    D. Neither A nor B

Hint

FMVSS 121 requires that all trucks be equipped with a mechanical parking and emergency brake system. The parking brake system is a generally separate system from the service brake system, with its own lines, chambers, and valves. The mechanical braking force is obtained by springs located in spring brake chambers. These chambers use air acting on a hold-off diaphragm to release the spring brake.

The spring in a spring brake chamber requires approximately 60 psi (414 kPa) acting on its hold-off chamber to fully release it. It is therefore capable of applying the equivalent of a 60 psi (414 kPa) air brake application when there is no air acting on the hold-off chamber.

Spring brake chambers are dual chamber assemblies that combine a service application chamber in the front section and a spring chamber in their rear section. Their size rating, such as 30/30, 24/24, 30/24, refer to the surface area of the service and hold-off diaphragms.

The caging port at the rear of the spring diaphragm is sealed with a plastic seal. If this seal is not in place, the internal components can rapidly corrode. Spring brake chamber assemblies fail when either the service or hold-off diaphragms puncture or when the spring breaks. Spring breakage is often indicated by an air leak at the hold-off diaphragm that occurs when the fractured spring punctures this diaphragm. It is essential that spring brakes be handled with caution even when it is known that the main spring has fractured.

Task 2

Inspect, test, or replace parking (spring) brake check valves, lines, hoses, and fittings.

79. *Technician A* says that the park and emergency braking system is a separate and distinct air system, completely isolated from the regular service air system. *Technician B* says that spring brake chambers use air pressure in the opposite way from the service brake chambers usage. Who is right?
    A. A only
    B. B only
    C. Both A and B
    D. Neither A nor B

80. When testing a parking brake system on a truck equipped with air brakes, *Technician A* says that when the primary or secondary system fails, federal law requires that the parking brakes must release at least once. *Technician B* says that in order for the truck to be mobile, air must be released from the spring brake chambers. Who is right?
    A. A only
    B. B only
    C. Both A and B
    D. Neither A nor B

Hint

The supply of air to the parking brake system is pressure protected so that in the event of a complete primary or secondary circuit failure, there is sufficient air for a complete apply/release sequence.

Restrictions in the lines supplying the hold-off chambers can slow both application and release times.

Spring brake valve operation can be verified with an air pressure gauge. Lines supplying the hold-off chambers on each wheel should be of equal length. Release-timing is unbalanced if this is not the case.

Task 3

Inspect, test, or replace parking (spring) brake application and release valve.

81. In Figure 4–18, what does the trailer spring brake valve do?
    A. Releases air in the reservoir to apply the service brake
    B. Allows the buildup of air pressure to allow operation of spring brake system
    C. Redirects air to the spring brake chambers for initial application
    D. Releases the trailer parking brakes with supply air pressure above 85 psi (586 kPa)

82. When the park control valve knob in Figure 4–19 is pulled outward:
    A. the spring brake releases.
    B. air begins to build up in the piggyback.
    C. it will not reopen until the treadle valve is depressed and opened.
    D. air in the rear portion of the spring brake chamber is exhausted.

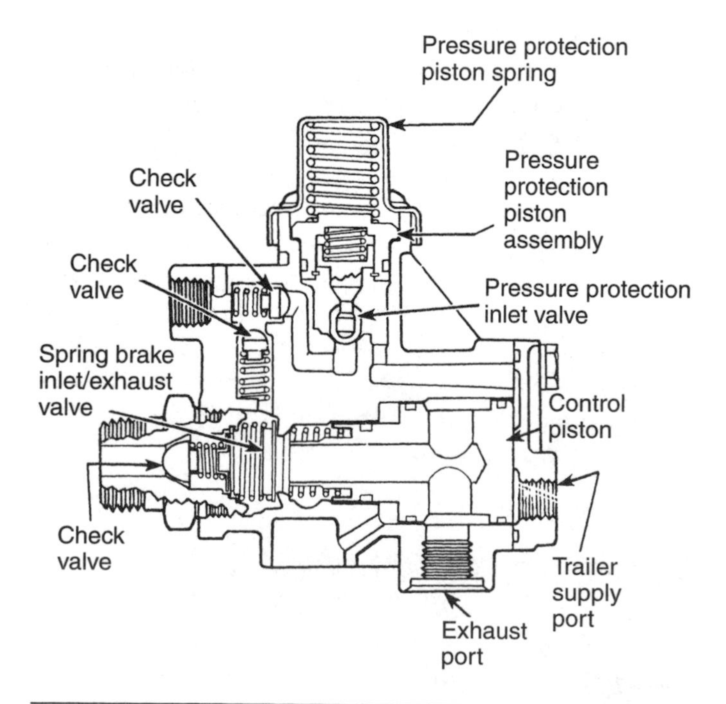

Figure 4–18 Trailer air brake system valve. *(Courtesy of Navistar International Transportation Corp.)*

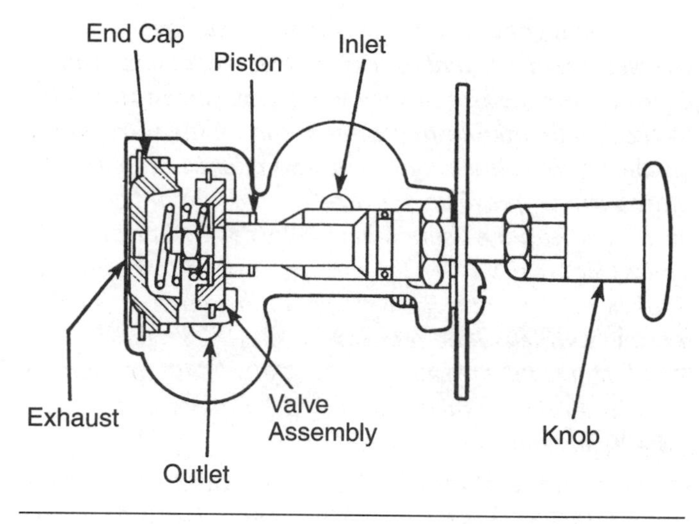

Figure 4–19 Park control valve. *(Courtesy of Navistar International Corp.)*

83. Technicians are diagnosing the tractor spring brake relay valve. *Technician A* says an air leak in a spring brake diaphragm may cause brake dragging. *Technician B* says that when the tractor parking brakes are applied, the air pressure at the tractor spring brake relay valve supply port should drop quickly to zero. Who is right?
    A. A only
    B. B only
    C. Both A and B
    D. Neither A nor B

84. While discussing a tractor park valve, *Technician A* says the tractor park valve has a red, diamond-shaped knob that is pulled out to apply the trailer parking brakes. *Technician B* says the tractor park valve may be used to apply the trailer parking brakes before uncoupling the trailer from the tractor. Who is right?
    A. A only
    B. B only
    C. Both A and B
    D. Neither A nor B

Hint

Spring brake valves are often incorporated in multifunction valves. It is critically important that when valves malfunction, they be replaced by matching the part numbers remembering that two valves that appear identical may have widely different performance characteristics.

The parking brake system of a tractor-trailer combination manages hold-off pressure delivered to the spring brake chambers. Two and three dash valve systems are used. FMVSS 121 requires that the shape and color of each valve be as follows:

Red octagonal knob: the trailer air supply valve. The system spring brake park control valve must have already been actuated before actuating this valve. This valve is responsible for supplying all system air to the trailer(s), and it must be pushed inward to release the trailer spring brakes.

Yellow diamond/square knob: the system spring brake (park) control valve. This valve masters the parking brakes on a tractor trailer combination. It is pushed inward to release the parking brakes. When pulled outward, it cuts the air supply to the trailer thus applying its park/emergency brakes, and puts the tractor into park/emergency mode.

Blue round knob: isolates the tractor park brake function from that of the trailer. This valve allows the driver to apply the tractor parking brakes while charging the trailer brakes with air pressure.

Task 4

Manually release and reset parking (spring) brakes.

85. A truck is in for parking brake service. *Technician A* says the spring brake chambers should be "caged" before disconnecting any air line or hose. *Technician B* recommends that draining the service air system be one of the first operations in servicing truck parking brakes. Who is right?
    A. A only
    B. B only
    C. Both A and B
    D. Neither A nor B

86. While servicing the combined service and spring brake chamber in Figure 4–20, *Technician A* says the spring brake chamber is in the uncaged position. *Technician B* says after the spring in the spring brake chamber is caged, the clamp may be removed from the spring brake chamber. Who is right?
    A. A only
    B. B only
    C. Both A and B
    D. Neither A nor B

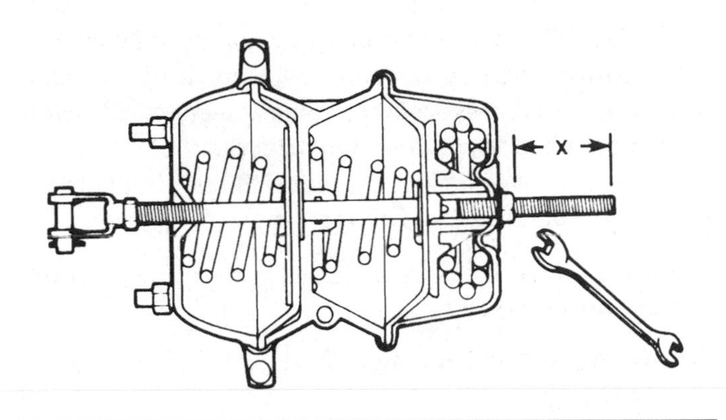

Figure 4–20 Dual service brake and parking brake chamber. *(Courtesy of Holland Group Inc., Holland Anchorlok Division.)*

Hint

Spring brake chambers are usually equipped with a cage bolt. The cage bolt is threaded and has a pair of lugs that engage to an internal cage plate. When a nut is turned down the cage bolt, the cage plate is pulled inward compressing the main spring of the spring brake chamber. This releases the parking brake. If the vehicle on-board air supply is available, a spring brake may be caged by chocking the wheels and releasing the parking brakes (supplying air to the hold-off chambers) and then installing the cage bolt to the cage plate with the main spring already compressed. It is important to ensure that the lugs in the cage bolt are properly engaged in the cage plate. Cage plates are manufactured out of aluminum alloy, and those in older spring brake assemblies were susceptible to corrosion. Caged spring brake assemblies should always be handled with a great amount of care. Spring brake chambers should always be fully caged before removal and installation. In the past, failed hold-off diaphragms were routinely replaced. Currently when a spring brake hold-off diaphragms fails, a piggy-back assembly (spring brake assembly minus the service chamber), or the entire spring brake chamber and service brake chamber assembly is replaced. The spring brake chamber band clamps should never be removed with the main spring in uncaged condition. Many current spring brake assemblies have chamber clamps that cannot be unbolted.

It is illegal to discard spring brakes without first disarming them. Disarming a spring brake means releasing the main spring. This is achieved by placing the entire spring brake assembly in a disarmament chamber with the main spring uncaged, torch cutting the spring brake chamber clamps and separating the chamber. The entire assembly should remain in the disarmament chamber until the main spring can be observed to be free.

Antilock Brake System (ABS)

Task 1

Observe antilock brake system (ABS) warning light operation; determine if further diagnosis is needed.

87. When discussing the ABS warning light in the instrument panel, *Technician A* says that if the ABS warning light is on longer than 4 seconds after the engine is started, there is a defect in the ABS. *Technician B* says that if the ABS warning light is on at vehicle speeds above 4 mph (6.43 km/h) there is a defect in the ABS. Who is right?
    A. A only
    B. B only
    C. Both A and B
    D. Neither A nor B

88. While diagnosing an ABS with an ABS check switch and a light in the instrument panel:
    A. when the ABS check switch and the ignition switch are turned on, the ABS light flashes blink codes.
    B. the first set of flashes in the blink code sequence indicates whether the system is ABS only or ABS and ATC.
    C. when there are no electrical defects in the ABS, a 000 blink code is received.
    D. a blink code indicates whether an electrical defect is in a wheel speed sensor or the connecting wires.

Hint

All current ABS are electronically controlled. A typical ABS adapts a standard air or hydraulic system for electronic control. All current systems are designed to default to conventional non-ABS operation in the event of an electrical or electronic malfunction. ABSs are now mandatory on new tractors and trailers.

An ABS consists of a means of monitoring the speed of each wheel, an electronic control unit (ECU) to manage the system, and a modulator assembly to modulate the service application pressures to each wheel.

ABS brakes are especially effective at managing split coefficient braking conditions. Split coefficient braking occurs when road surface conditions (icy, wet, gravel, dry, etc.), differ from one wheel to another.

In some ABSs a fail relay coil is installed between the vehicle power source and the modulator valve assembly and is designed to illuminate a warning light in the cab in the event of a system failure. When this occurs the system fault must be diagnosed.

Task 2

Diagnose antilock brake system (ABS) electronic control(s) and components using self-diagnosis and/or recommended test equipment; determine needed repairs.

89. While diagnosing a tractor ABS with a scan tool:
    A. the wheel speed sensor readings may be displayed on the scan tool with the tires contacting the shop floor.
    B. if NO DATA RECEIVED is displayed on the scan tool, there are no fault codes in the ECU.
    C. the scan tool should be connected to the ABS diagnostic connector with the ignition switch on.
    D. the scan tool may be used to test the data link wires from the ECU to the powertrain control module (PCM).

90. *Technician A* says that in some ABSs the ECU uses a series of LEDs in the diagnostic window to indicate defects in the system. *Technician B* says that the number of LEDs in the diagnostic window vary depending on the system.
    A. A only
    B. B only
    C. Both A and B
    D. Neither A nor B

Hint

Although ABSs are available in 2, 4, or 6 channel designs, the 4 channel system is commonly used today in tandem drive axle trucks. The four channel system has a sensor on each of the steer-axle wheels, and on two of the four rear wheels, both on the same axle. The sensors input wheel speed data to the ABS module. The ABS module outputs an electrical signal to the four modulator valves. Two modulator valves control each wheel on the steer axle. The other two modulator valves control service braking on each side of the tandem drive axles.

The ABS wheel sensors use a rotating toothed wheel attached to the hub and a fixed pickup to send a small AC electrical signal to the ABS module. This signal voltage *value increases proportionally with wheel speed. In this way, a lock-up condition can be sensed by the ABS module. The ABS module controls the system modulator valves. The modulator valves control the service application pressures to each of the front wheels and to each side of the tandem drive assembly. When a lock-up condition is sensed by the ABS module, the air to the service chamber(s) controlled by the modulator can be momentarily relieved or pulsed. In current air brake systems, the pulsing speed of each modulator peaks at about 4 times per second.*

A 6 channel system on a tandem drive axle tractor monitors and modulates each of the 6 wheels on the vehicle. Two channel systems monitor each side of one of the drive axles and modulate the service brakes on each side of the drive axle assembly by having each modulator valve actuate a pair of service chambers. Most trailers use a 2 channel system.

All electronic ABSs are equipped with self-diagnostics. These can be accessed by flash or blink codes using the dash warning light or a scan tool connected either at the diagnostic connector in the cab or directly to the ABS module. All truck ABSs use J1939 communication protocols, which means that one manufacturer's system can be read by any other.

Task 3

Diagnose poor stopping and wheel lockup caused by failure of the antilock brake system (ABS); determine needed repairs.

91. While discussing ABSs, *Technician A* says that in the antilock mode, ABS provides about 5 percent tire slip without wheel lockup. *Technician B* says ABS enters the antilock mode each time the brakes are applied. Who is right?
    A. A only
    B. B only
    C. Both A and B
    D. Neither A nor B

92. While discussing tractor and trailer ABSs, *Technician A* says the same ECU operates the ABSs on the tractor and the trailer. *Technician B* says that if replacement tires are a different size from the original tires, the wheel speed sensor signals are affected. Who is right?
    A. A only
    B. B only
    C. Both A and B
    D. Neither A nor B

Hint

When the ABS malfunctions, the brake system reverts to functioning as any non-ABS managed brake system. When diagnosing a poor stopping and wheel lockup complaint, the Technician should determine whether the problem is an electrical problem by scanning the ABS module. It should always be remembered that an air ABS is still essentially an air brake system, and the modulator valves while capable of being controlled electronically, are fundamentally relay valves.

Task 4

Inspect, test, and service antilock brake system (ABS) air, electrical, and mechanical components.

93. An ABS on a truck is being inspected. *Technician A* says that there is generally one modulator per axle, which means that each axle brakes as one unit. *Technician B* says that a truck's ABS is designed to prevent wheel lockup even when the tractor wheels are on different road surface conditions. Who is right?
    A. A only
    B. B only
    C. Both A and B
    D. Neither A nor B

94. All these statements about ABS are true EXCEPT:
    A. the gap is not adjustable between the wheel speed sensor and the toothed ring.
    B. the average gap is 0.100 in. (25.4 mm) between the wheel speed sensor and the toothed ring.
    C. in the antilock mode, the ECU operates the solenoids in the modulator to provide about 20 percent tire slip without wheel lockup.
    D. in an individual modulator valve, the exhaust solenoid is normally open and the inlet solenoid is normally closed.

Hint

All current ABSs manage subcomponent failures with safety first failure strategy. Minor system malfunctions result in the illumination of the antilock warning lamp on the dash and the setting of a fault code in the ECU. When the ECU detects a fault in the ABS, in most cases the ECU shuts down the antilock function and conventional non-ABS braking is maintained.

Manufacturers use sequential troubleshooting charts to troubleshoot ABS circuits and their components. Perform each step in the sequence exactly as prescribed. The most common tools used to troubleshoot ABSs are scan tools and digital volt ohmmeters (DVOMs).

Task 5

Service, test, and adjust antilock brake system (ABS) speed sensors following manufacturer's recommended procedures.

95. A wheel sensor from an ABS is being inspected. *Technician A* says that as the toothed wheel rotates, the sensor generates a simple DC signal. *Technician B* says that the frequency of the wheel speed sensor signal decreases as the wheel speed increases. Who is right?
    A. A only
    B. B only
    C. Both A and B
    D. Neither A nor B

96. When an AC voltmeter is connected to the wheel speed sensor terminals and the wheel is rotated, the wheel speed sensor signal is erratic on the voltmeter. The cause of this problem could be:
    A. an open circuit in the wheel speed sensor winding.
    B. a damaged toothed ring that triggers the wheel speed sensor.
    C. a grounded circuit in the wheel speed sensor lead wires.
    D. excessive wheel speed sensor gap.

Hint

Wheel speed sensors use a toothed wheel and pickup principle to produce an AC signal to the ECU. Wheel sensors will only function properly if the air gap between the rotating toothed wheel and the stationary pickup is properly set. Wheel sensor performance can also be affected by road dirt contamination and impact damage. A wheel sensor malfunction produces a fault code in the ECU.

Hydraulic Brakes Diagnosis and Repair

Hydraulic System

Task 1

Diagnose poor stopping, pulling, premature wear, noise, or dragging complaints caused by hydraulic system problems; determine needed repairs.

97. What of the following conditions can cause brake pedal fade?
    A. A seized brake caliper piston
    B. Brake drum machined beyond its limit
    C. Low fluid level in the master cylinder
    D. Air in the hydraulic system

98. When inspecting a 4-wheel disc hydraulic brake system, *Technician A* says seized caliper pistons may cause brake drag. *Technician B* says that rotors machined to less than the minimum thickness may cause brake pedal fade when you first apply the brakes. Who is right?
    A. A only
    B. B only
    C. Both A and B
    D. Neither A nor B

99. On a medium-duty truck equipped with front disc and rear drum brakes pedal, pulsations occur when the brakes are applied. The cause of this problem could be:
    A. sticking wheel cylinder pistons on one of the rear wheels.
    B. leaking caliper piston seals on one of the front calipers.
    C. an out-of-round brake drum.
    D. a rotor with less than the minimum thickness.

Hint

The potential energy of a hydraulic brake system is mechanical force created by the action of a driver's foot acting on a brake pedal, usually proportionally assisted by pedal geometry leverage and a power assist system.

In any hydraulic system, it can be assumed that the brake fluid is not compressible. If force is mechanically applied to a liquid in a closed system, it will be transmitted equally by the liquid to all parts of the system. Force applied by a master cylinder is transmitted equally throughout the hydraulic system though it may be modulated by valves in parts of that system.

The hydraulic brake system consists of a master cylinder, proportioning valves, metering valves, pressure differential valve, and wheel cylinders. All truck hydraulic brake systems are dual systems, meaning that in the event of a failure in one of the systems, the other will back it up to effect at least one stop. As in truck air brake systems, the systems are defined as the primary and secondary circuits.

A failure of any moving part within the hydraulic system may cause pressure to become entrapped in a portion of the system and this may cause dragging brakes or slow release times.

Task 2

Pressure test hydraulic system and inspect for leaks.

100. When road testing a truck with hydraulic brakes, the brake pedal fades during a brake application. *Technician A* says there may be a fluid leak in the brake system. *Technician B* says one of the flexible brake hoses may be bulging. Who is right?
    A. A only
    B. B only
    C. Both A and B
    D. Neither A nor B

101. When discussing dual master cylinders, *Technician A* says if there is no brake fluid in the primary section, a complete brake system failure occurs. *Technician B* says force is applied to the primary

piston from the brake booster pushrod. Who is right?
A. A only
B. B only
C. Both A and B
D. Neither A nor B

Hint

Pressure values within the hydraulic system may be tested with pressure gauges. The hydraulic system can be pressurized simply by starting the vehicle engine and applying the brakes by foot pressure.

External leaks may be verified by cleaning the externally visible portions of the system and applying the brakes. Internal leaks are more difficult to locate. Internal leakage within a master cylinder can be verified by using gauges plumbed to each portion of the hydraulic system.

Task 3

Check and adjust brake pedal pushrod length.

102. While road testing a truck after a master cylinder overhaul, the truck brakes drag. *Technician A* says that the compensation ports may be blocked. *Technician B* says that excessive pedal free play could be the cause. Who is right?
A. A only
B. B only
C. Both A and B
D. Neither A nor B

103. All of these problems may cause a low, spongy brake pedal EXCEPT:
A. improper pushrod adjustment and excessive brake pedal free play.
B. air in the hydraulic brake system.
C. brake fluid contaminated with moisture.
D. low brake fluid level.

Hint

Most truck hydraulic brake systems use the brake pedal assembly to provide added leverage to the mechanical force provided by the driver's foot pressure. Brake pedal pushrod adjustment should always be made according to manufacturer's specifications.

Task 4

Inspect, test, or replace master cylinder.

104. A truck with a hydraulic brake system is being overhauled. *Technician A* uses a vacuum gauge to test master cylinder output. *Technician B* says that all the wheel cylinders must be replaced when an overhaul is being performed. Who is right?
A. A only
B. B only
C. Both A and B
D. Neither A nor B

105. As shown in the Figure 4–21, after a brake application, what component could prevent the master cylinder piston from returning the brake pedal to its original unapplied position?
A. A check valve
B. The metering valve
C. A sliding rubber seal
D. A return spring

106. *Technician A* says master cylinder parts may be washed in an oil-based solvent. *Technician B* says an aluminum master cylinder bore may be honed if it is scored. Who is right?
A. A only
B. B only
C. Both A and B
D. Neither A nor B

Hint

The master cylinder converts the mechanical force applied to it by the driver's foot pressure and the brake booster system into hydraulic pressure to actuate the primary and secondary systems of the brake system. The master cylinder has dual reservoirs for the primary and secondary systems, and a cylinder housing, compensating ports, return springs, and primary and secondary pistons. The primary piston is actuated mechanically. Pressure developed in the primary portion of the master cylinder charges the primary system that can actuate either the front or rear brakes. When the master cylinder is operating normally, the secondary piston is actuated

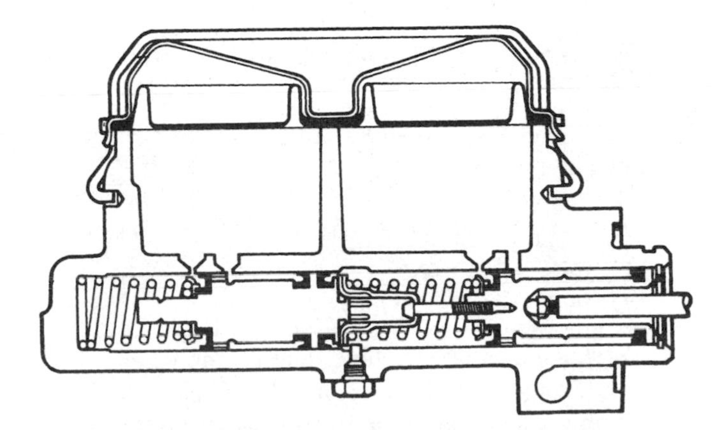

Cross-section of a typical dual master cylinder

Figure 4–21 Master cylinder components.

hydraulically, by whatever pressure value is developed in the primary portion of the cylinder.

When the mechanical force applied to the primary piston is relieved, return springs acting on both the primary and secondary pistons return them to their original positions permitting the fluid applied to each system to return to the reservoirs. Both sections of the master cylinder are supplied with brake fluid by fill and compensating ports. Each piston is sealed in its bore by rubber seals. The primary and secondary sections are not interconnected.

When a failure occurs in either system, the pressure differential light in the dash illuminates the first time the brakes are applied following the failure. If the failure occurs in the primary system, the primary piston is forced through its travel without generating any fluid pressure until it contacts the secondary piston. The secondary piston is then mechanically actuated by the primary piston. If the failure occurs in the secondary system, the primary system functions normally but the actuation of the secondary piston results in no pressure delivered to the secondary system. In either case, the vehicle should be brought to an immediate standstill and not operated until a repair has been undertaken.

When testing a master cylinder in a hydraulic brake system, a liquid-filled hydraulic test gauge should be used. Deteriorated fluid, deteriorated seals, mixture of incompatible fluids may cause sludge and particulate that can plug fill and compensating ports, resulting in slow application times, slow release times and brake failure.

Master cylinders in truck hydraulic brake systems are routinely reconditioned by replacing seals and springs and honing the master cylinder bore. When cleaning master cylinders before reassembly, only isopropyl alcohol should be used to clean components; use of solvents can swell the seals and leave behind corrosive residues.

Task 5

Inspect, test, or replace brake lines, flexible hoses, and fittings.

107. The inside of a new hydraulic brake line can be cleaned with:
    A. soap and water.
    B. isopropyl alcohol.
    C. mineral spirits.
    D. hydraulic brake fluid.

108. *Technician A* uses double flare tubing when replacing brake lines. *Technician B* uses ISO tubing when replacing brake lines. Who is right?
    A. A only
    B. B only
    C. Both A and B
    D. Neither A nor B

Hint

Steel tubing should be checked for wear, dents, kinks, and corrosion. Preflared and preformed tubing helps reduce custom cuts such as bending and flaring of new tubing. Two types of flaring styles and seats are used: ISO and double flare. ISO uses an outward flare, whereas a double flare creates a double wall at the nipple seat for greater strength.

When cleaning brake tubing, only isopropyl alcohol should be used due to its noncorrosive characteristics, ability to evaporate rapidly, and residue-free drying. Residues remain when using other cleaning agents such as soap and water, mineral spirits, and hydraulic brake fluid. When bending tubing, a tube bender should be used to prevent the line from collapsing.

Task 6

Inspect, test, and replace metering (hold-off), load sensing/proportioning, proportioning, and combination valves.

109. When discussing metering valves, *Technician A* says the metering valve delays fluid pressure to the front brakes during a hard brake application. *Technician B* says a metering valve prevents front wheel lockup during a hard brake application. Who is right?
    A. A only
    B. B only
    C. Both A and B
    D. Neither A nor B

110. All of these statements about combination valves are true EXCEPT:
    A. a combination valve may contain a quick take-up valve.
    B. a combination valve may contain a metering valve.
    C. a combination valve may contain a proportioning valve.
    D. a combination valve may contain a pressure differential switch.

Hint

The metering valve is used on vehicles equipped with front disc and rear drum brakes. It is required to achieve brake timing balance during light brake applications by withholding the delivery of application pressure to the front disc brakes until pressure exceeds a predetermined value in the system responsible for actuating the rear brakes. This lag or delay is required so that hydraulic pressure builds sufficiently in the rear brake hydraulic system to overcome the tension of the rear brake shoe return springs and the free travel of the shoes. The objective is to enable simultaneous application of both front and rear brakes. For this reason, the metering valve is sometimes known as a hold-off valve.

When a pressure bleeder is used to bleed any system equipped with a metering valve, the manufacturer's instructions as to how to open the valve must be observed. When manually bleeding a brake system, application of the brake pedal develops sufficient pressure to overcome the metering valve opening pressure.

A proportioning valve is also used on systems combining front disc and rear drums. The proportioning valve is installed in the system supplying the rear brakes. Its function is to reduce the application pressure to the rear wheel cylinders and prevent rear wheel lockup.

The proportioning valve operation should be verified at each brake inspection or if the vehicle has a rear wheel lock-up condition. To check valve operation, hydraulic gauges should be installed ahead and behind the valve, or alternatively to each of the two systems. A load or height-sensing valve is used on some systems to sense vehicle load transfer affect during a brake application. The valve proportions front and rear braking correlating it to weight transfer during braking. The valve is located on the vehicle frame cross member, and it is activated through a linkage system connected to the rear axle housing.

Task 7

Inspect, test, repair, or replace brake pressure differential valve and warning light circuit switch, bulbs, wiring, and connectors.

111. While discussing pressure differential valves, *Technician A* says the pressure differential valve grounds the brake warning light circuit. *Technician B* says that if the primary and secondary sections of the master cylinder have equal pressure, the pressure differential valve electrical contact is closed. Who is right?
    A. A only
    B. B only
    C. Both A and B
    D. Neither A nor B

112. A Technician has just replaced a wheel cylinder. *Technician A* says that the pressure differential valve will require manual resetting. *Technician B* says that after manually resetting the pressure differential valve, you should replace the proportioning valve. Who is right?
    A. A only
    B. B only
    C. Both A and B
    D. Neither A nor B

Hint

The pressure differential valve consists of a cylinder through which primary and secondary hydraulic pressures act on either side of a spool. When the pressure in both the primary and secondary systems is equal, the spool floats in a neutral position. Should a pressure imbalance occur in either the primary or secondary system, the spool shuttles to one side of the cylinder. This spool movement grounds an electrical signal and illuminates a dash warning light.

The pressure differential valve will normally recenter automatically upon the first application of the brakes after repairs are completed. Some pressure differential valves will require manual resetting.

The pressure differential valve should be inspected whenever the brakes are serviced. This inspection should include the electrical warning light circuit.

Task 8

Inspect, clean, and rebuild or replace wheel cylinders.

113. A wheel cylinder:
    A. does not need any maintenance.
    B. needs to be inspected at a PM.
    C. needs to be replaced periodically.
    D. can be constructed of aluminum.

114. When servicing adjuster wheel cylinders:
    A. adjuster wheel cylinders provide automatic brake shoe adjustment during a brake application.
    B. a foundation brake assembly on one wheel may have two adjuster wheel cylinders.
    C. brake fluid is forced into the area between the cylinder pistons when the brakes are applied.
    D. the manual override wheel is operated by the parking brake system.

Hint

A wheel cylinder is the actuator of a hydraulic brake system. It is supplied with hydraulic brake fluid from the master cylinder, and converts hydraulic pressure to mechanical force at the foundation brake assembly. Wheel cylinders are usually constructed of cast iron for higher durability and lower manufacturing costs.

Most wheel cylinders are double acting, and house two pistons within a cylinder bore. The pistons are sealed in the cylinder by rubber seals. When the pistons are subjected to hydraulic pressure they are forced outward to actuate brake shoes, forcing them against the drum. Manual and automatic adjusting mechanisms are integral in the wheel cylinder assembly.

Wheel cylinders are commonly reconditioned. The cylinder bores can be honed and most of the internal components replaced if required. Wheel cylinders are vulnerable to contaminants in the hydraulic fluid. When the cylinder bores become scored, the result is fluid leakage from the cylinder past the seals. Seals may fail if exposed to chemical contaminants.

Each wheel cylinder is equipped with a bleed port. A bleeder screw in this port is loosened to purge air from the hydraulic circuit in the bleeding process.

Task 9

Inspect, clean, and rebuild or replace disc brake caliper assemblies.

115. If the floating caliper on a hydraulic brake truck does not slide freely, which of these conditions will result?
   A. Brake pedal fade.
   B. Brake pedal pulsations.
   C. Reduced braking force.
   D. The brakes will grab.

116. When servicing disc brake assemblies, which of these applies to the pads?
   A. Thickness is indicated by a code on the lining edge.
   B. All semimetallic linings are edge stamped FF.
   C. All asbestos-type linings are edge stamped EE.
   D. Linings should use the same edge brand as OEM.

117. When inspecting the brake pads on a truck equipped with hydraulic disc brakes, *Technician A* says you must remove the wheel in most cases to measure pad thickness. *Technician B* says that if the brakes were applied shortly before measuring the lining-to-rotor clearance, this clearance will be greater than specified. Who is right?
   A. A only
   B. B only
   C. Both A and B
   D. Neither A nor B

Hint

Fixed and sliding or floating calipers are used in hydraulic brake systems. In a fixed-type caliper, the rotor sits in between two or four pistons, and the clamping action of opposed pistons acting on the rotor is responsible for retarding the rotor. In a floating or sliding caliper, one or more pistons sit on one side of the rotor, and the caliper housing is designed to slide when the piston is subjected to hydraulic pressure. This action effects the clamping action of the rotor. The friction surfaces that contact the rotor are brake pads. Disc brakes are not self-energizing and generally require greater application pressures. They operate at higher mechanical efficiencies than drum brakes.

Disc brake calipers are reconditioned in the same manner as wheel cylinders. Manufacturer's tolerance specifications must be observed during reconditioning.

Task 10

Inspect/test brake fluid; bleed and/or flush system.

118. Never use brake fluid out of a container:
   A. that is made out of glass.
   B. that has been tightly stored.
   C. that has been used to store any other liquid.
   D. that has had brake fluid in it for more than one year.

119. When bleeding a hydraulic brake system with a metering valve, *Technician A* applies a device to hold the metering valve open when he or she uses a pressure bleeder to bleed the system. *Technician B* bleeds the brakes manually with no problems. Who is right?
   A. A only
   B. B only
   C. Both A and B
   D. Neither A nor B

120. Which of the following procedure is correct when pressure bleeding a hydraulic brake system?
   A. You bleed the right front caliper or wheel cylinder first.
   B. The pressure bleeder is pressurized to 20 to 25 psi (137.9 to 172.37 kPa).
   C. If equipped, the proportioning valve must be closed.
   D. You bleed the left rear caliper or wheel cylinder last.

Hint

It is good practice to replace the system brake fluid at each major brake overhaul, especially if the failure has been caused by seal failure or fluid contamination. Using the manufacturer's specified brake fluid is important to ensure proper operation of the system. The approved heavy-duty brake fluid should retain the proper consistency at all operating temperatures. It will not damage rubber cups and helps to protect the metal parts of the brake system against corrosion.

Water, mineral spirits, and gasoline should never be used to flush a hydraulic brake system due to incompatibility with other materials and corrosive effects. Alcohol or compatible brake fluid should always used to flush hydraulic braking systems, and these fluids should not be reused after the flushing is complete.

A container storing brake fluid must always be tightly sealed when not in use to prevent moisture from being absorbed into the brake fluid. Mineral oil, alcohol, antifreeze, cleaning solvents, and water, even in very small quantities, will contaminate most brake fluids. Brake fluid has a shelf life of one year. This shelf life diminishes if the storage container is not completely full.

Mixing of different brake fluid may cause coagulation and result in hydraulic failure. It is important to ensure that fluids are compatible when topping up system reservoirs.

Brake systems may be bled using a pressure bleeder or manually. The first method is preferred as the operation can be performed by one person. In all cases, system air is purged from the valves and wheel cylinders in a sequence defined by the manufacturer.

Task 11

Inspect, test, repair, or replace hydraulically controlled parking brake components and systems.

121. In the hydraulic parking brake system in Figure 4–22:
   A. the fluid pressure is supplied from the control valve delivery port to the unitized valve control port.
   B. the unitized valve supplies fluid pressure to the parking brake chambers in the apply mode.
   C. the parking brake chambers are pressurized when the dash control knob is pulled outward.
   D. the pump supplies pressure directly to the control valve in the release mode.

122. While discussing hydraulic parking brake service, *Technician A* says the parking brake chambers should be caged in the applied position. *Technician B* says the parking brake chambers should be caged with an impact wrench. Who is right?
   A. A only
   B. B only
   C. Both A and B
   D. Neither A nor B

Hint

Some truck hydraulic brake systems are equipped with a hydraulic parking brake system. These systems use a dedicated hydraulic system and have a parking brake cylinder located in the foundation assembly on the opposite side of the shoe to the wheel cylinder. The parking brake is actuated by a cab located foot or hand lever. Hydraulic pressure is supplied from a pump to the parking brake chambers to release the parking brakes. When the parking brakes are applied, the hydraulic pressure is released from the parking brake chambers. This action allows the springs in the parking brake chambers to apply the parking brakes.

Mechanical System

Task 1

Diagnose poor stopping, brake noise, premature wear, pulling, grabbing, dragging, or pedal feel complaints caused by drum and disc brake mechanical assembly problems; determine needed repairs.

123. During hydraulic brake system diagnosis:
   A. a chassis vibration during a brake application may be caused by excessive radial runout on a tire.
   B. brake grab on one wheel may be caused by improper brake pedal pushrod adjustment.
   C. brake drag may be caused by glazed brake linings.
   D. a low, firm pedal may be caused by air in the hydraulic brake system.

124. A truck equipped with hydraulic power brakes has excessive pedal effort and poor stopping ability. The hydraulic brake booster is tested and proven to be operating normally, and the pedal height is normal and firm. This problem may be caused by all of the following defects EXCEPT:
   A. wheel caliper pistons seized.
   B. glazed brake linings.
   C. air in the hydraulic brake system.
   D. a binding brake pedal.

125. A truck with front disc and rear drum hydraulic brakes pulls to the right when the brakes are applied. The cause of this problem could be:
   A. a seized left front caliper piston.
   B. a broken brake shoe return spring in the left front brake assembly.
   C. loose wheel lug nuts.
   D. a cupped right front tire.

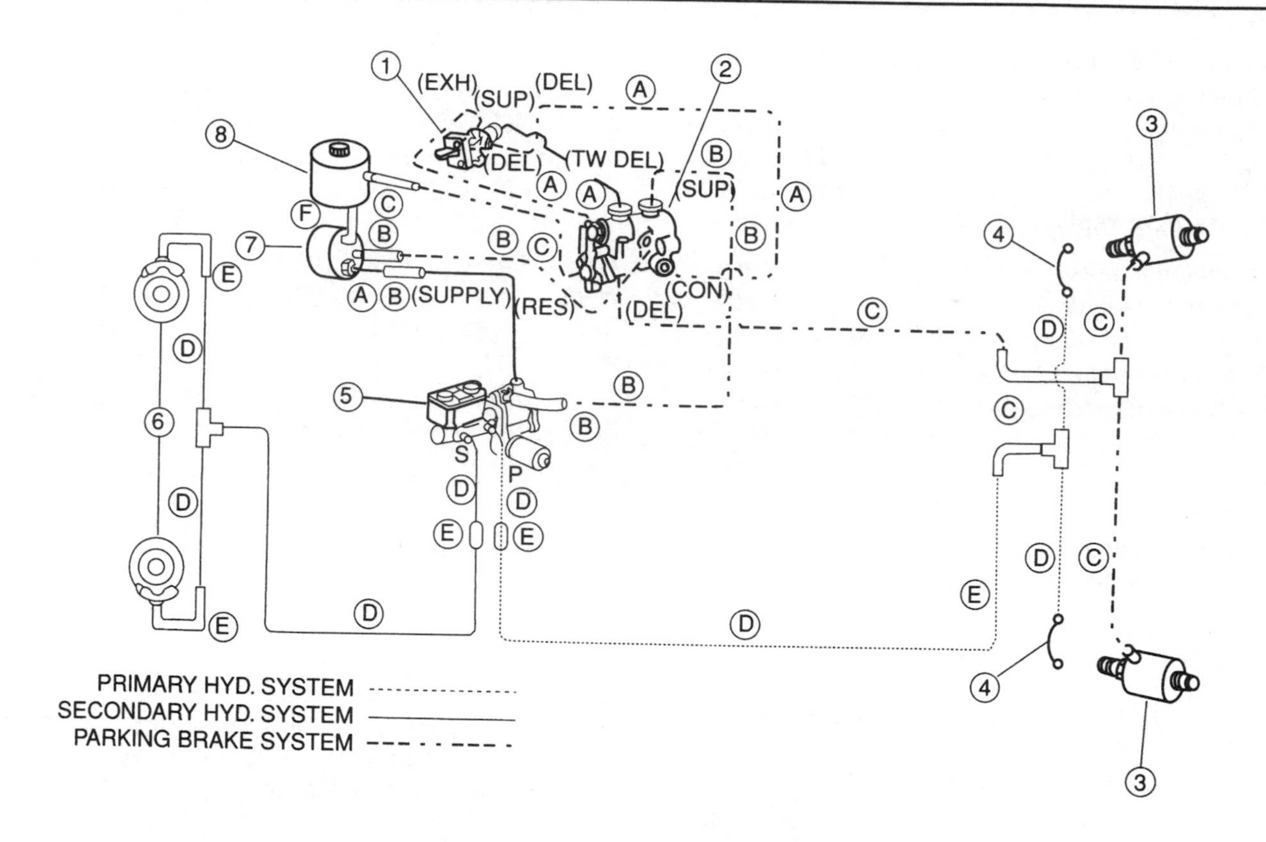

Item	Part Number	Description
1	2K806	Parking Brake Control Valve (TW-11)
2	2A449	Parking Brake Unitized Vave (HR-1)
3	–	Spring-Set Parking Brake Chamber (Part of 2225)
4	1126	Drum Brake Assembly
5	2B195	Hydro-Max Booster and Mini-Master Cylinder

Item	Part Number	Description
6	–	Disc Brake (part of 2B 120)
7	2N211	Parking Brake Pump
8	2N148	Parking Brake Reservoir
A	–	3/8-Inch Hose
B	–	1/2-Inch Hose
C	–	5/8-Inch Hose
D	–	1/4-Inch Hose
E	–	1/8-Inch Hose
F	–	3/4-Inch Hose

Figure 4–22 Hydraulic parking brake system. *(Courtesy of Ford Motor Company)*

Hint

Hydraulic foundation brakes may use servo and non-servo principles. Servo action occurs when the action of one shoe is guided by the movement of the other, permitting both shoes to act as a single unit. Self-energization occurs when the shoes are driven into the drum and rotate slightly with the drum to create more application force.

Non-servo action is also a term generally used for certain drum/shoe type brakes. It is defined as shoes working independently of each other to stop the vehicle. These shoes are anchored separately. When actuated by the wheel cylinder, each shoe pivots on the anchor and is forced against the drum.

Disc brakes require more force to achieve the same braking effort as servo drum brakes. However, they have superior mechanical efficiency and are often used on the front axle brakes of straight trucks. Front wheel brakes perform a higher percentage of braking on straight trucks due to load transfer and suspension effects.

Grabbing and pulling can be caused by broken hardware components, especially return springs, drum failures, malfunctions in the adjusting mechanism, and parking brake related problems.

A primary cause of a pulsating pedal condition is warped disc brake rotors. Rotors warp as a result of overheating. This can be caused by vehicle overloading or machining rotors too thin at overhaul. Thickness vari-

ations on disc brake rotors can also cause pedal pulsation and loss of braking power.

Task 2

Inspect, reface, or replace brake drums or rotors.

126. *Technician A* says that when performing a brake overhaul on a hydraulic rear drum brake system, you replace the return springs. *Technician B* says the maximum allowable out-of-round specification on a brake drum should be 0.025 inch. Who is right?
   A. A only
   B. B only
   C. Both A and B
   D. Neither A nor B

127. Extreme rotor thickness variation may result in:
   A. low brake output.
   B. pedal pulsation.
   C. loss of brake fluid.
   D. loss of directional control.

Hint

Brake drums may be reused if they are within the manufacturer's specifications. The critical specifications are the maximum wear limit, machine limit, and maximum permissible diameter. Drums are measured with a drum gauge and should be checked for out-of-round, bell-mouth, convex, concave, and taper conditions. Drums should be inspected for heat checks and cracks before machining.

Disc brake rotors may be reused if they are within the manufacturer's specifications. They should be measured for thickness with a micrometer and checked for parallelism and runout with a dial indicator. If it is within machine limits, the rotor may be turned on a rotor lathe.

Task 3

Inspect, adjust, or replace drum brake shoes/linings, mounting hardware, adjuster mechanisms, and backing plates.

128. When adjusting the hydraulic drum brakes in Figure 4–23, the self-adjusting mechanism should keep the lining-to-drum clearance to:
   A. 0.010–0.025 in. (0.254 0.635 mm).
   B. 0.008–0.016 in. (0.203–0.406 mm).
   C. 0.005–0.010 in. (0.127–0.254 mm).
   D. 0.004–0.012 in. (0.101–0.304 mm).

129. *Technician A* says that a servo-type brake uses different primary and secondary shoe linings.

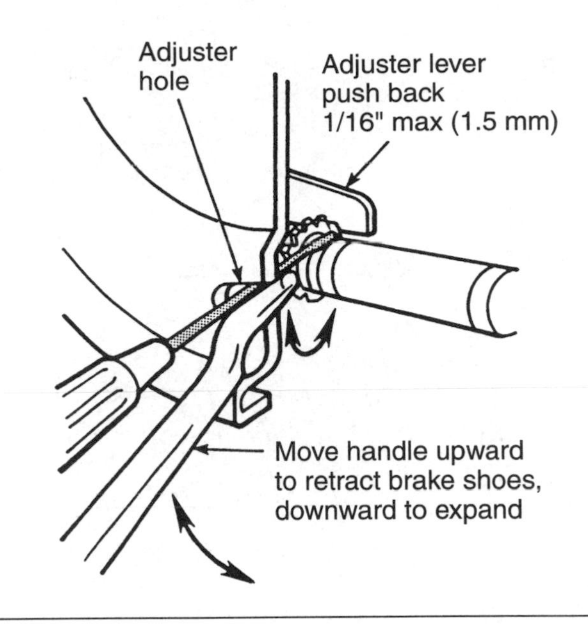

Figure 4–23 Drum brake adjustment.

Technician B says that the primary shoe in a servo-type brake system has a weaker return spring. Who is right?
   A. A only
   B. B only
   C. Both A and B
   D. Neither A nor B

Hint

When a brake job is performed, the brake shoes, return springs, and fastening hardware are replaced. The friction edge codes should be observed when replacing brake shoes. Most brake shoes today use bonded friction blocks, but riveted and bolted types are still in existence. Ensure that primary and secondary shoes are installed in their correct locations.

Task 4

Inspect, service, or replace disc brake pads, hardware, and mounts.

130. When disc brakes are released, an acceptable lining-to-rotor clearance would be:
   A. 0.003 in. (0.076 mm).
   B. 0.010 in. (0.254 mm).
   C. 0.012 in. (0.304 mm).
   D. 0.015 in. (0.381 mm).

131. In Figure 4–24, the measurement being performed on the disc foundation brake assembly is:
   A. lining-to-rotor clearance.
   B. caliper gap.

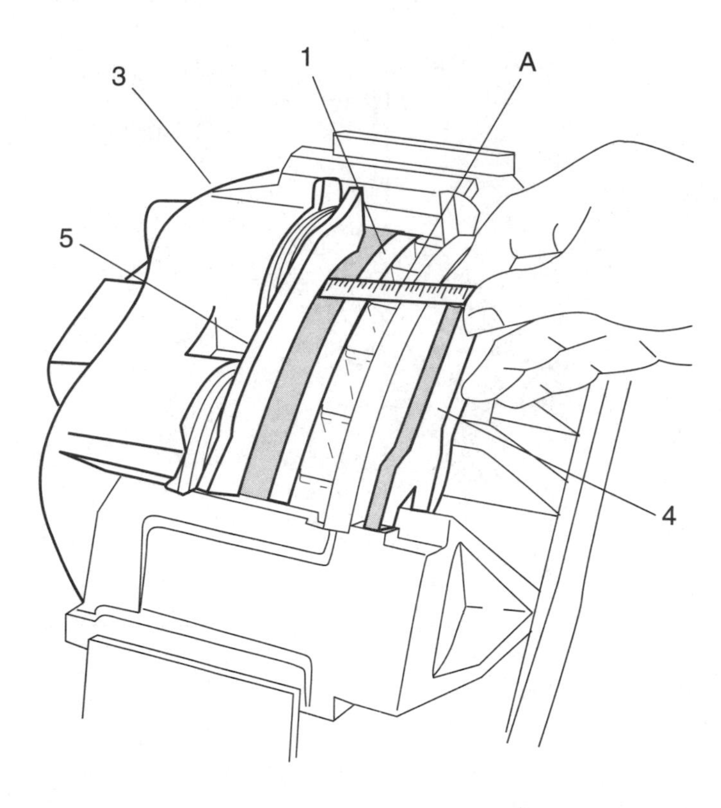

A. Ruler
1. Rotor
3. Caliper
4. Lining, Outer
5. Lining, Inner

Figure 4–24 Disc brake measurement. *(Courtesy of General Motors Corporation, Service Technology Group)*

C. lining thickness.
D. V-way wear.

Hint

Most disc brake assemblies have wear indicators that produce a noise when the wear limit is exceeded. Servicing disc brake pads usually involves removing the caliper assembly, ensuring that float pins are not seized, and installing a new pair of brake pads. Whenever the brake pads are replaced, the brake rotors should be both measured and visually inspected.

Task 5

Inspect, adjust, repair, or replace driveline parking brake drums, rotors, bands, shoes, mounting hardware, and adjusters.

132. The diameter on a driveline parking brake drum is greater at the edges of the friction surface than in the center. Which of these faults describes this description?
    A. Bell-mouth drum
    B. Concave brake drum
    C. Drum out-of-round
    D. Convex drum

133. In the parking brake assembly in Figure 4–25, the component that moves the brake shoes outward against the drum during a brake application is:
    A. item 45.
    B. item 42.
    C. item 41.
    D. item 28.

Hint

Cable-actuated, driveline parking brakes are used in some hydraulic truck brake systems, especially in air-over-hydraulic brake systems. The unit is a band and rotor assembly mounted at the rear of the transmission. When this parking brake is engaged, the rear wheels are locked stationary by means of the drive shaft.

Task 6

Inspect, adjust, repair, or replace driveline parking brake application system pedal, cables, linkage, levers, pivots, and springs.

134. While discussing the parking brake adjustments on a mechanical propeller shaft parking brake, *Technician A* says the brake lining wear adjustment must be completed before the linkage adjustment. *Technician B* says the lever adjustment must be completed before the linkage adjustment. Who is right?
    A. A only
    B. B only
    C. Both A and B
    D. Neither A nor B

135. The parking brake adjustment being performed in Figure 4–26 is the:
    A. linkage adjustment.
    B. lining wear adjustment.
    C. lever adjustment.
    D. camshaft adjustment.

Hint

Some driveline parking brakes have three adjustments. The brake lining wear adjustment is performed first, followed by the linkage and lever adjustments. All highway vehicles are legally required to have functioning parking brakes.

26. Lining, Parking Brake
27. Spring
28. Camshaft
29. Pin, Anchor
30. Plate, Support
31. Bracket, Camshaft
32. Washer
33. Nut
34. Washer
35. Bolt
36. Spring
37. Lever
38. Washer
39. Nut
40. Spring
41. Nut, Pivot
42. Screw, Adjusting
43. Socket
44. Link
45. Eccentric, Adjusting
46. Washer, Curved Spring
47. Washer
48. Nut

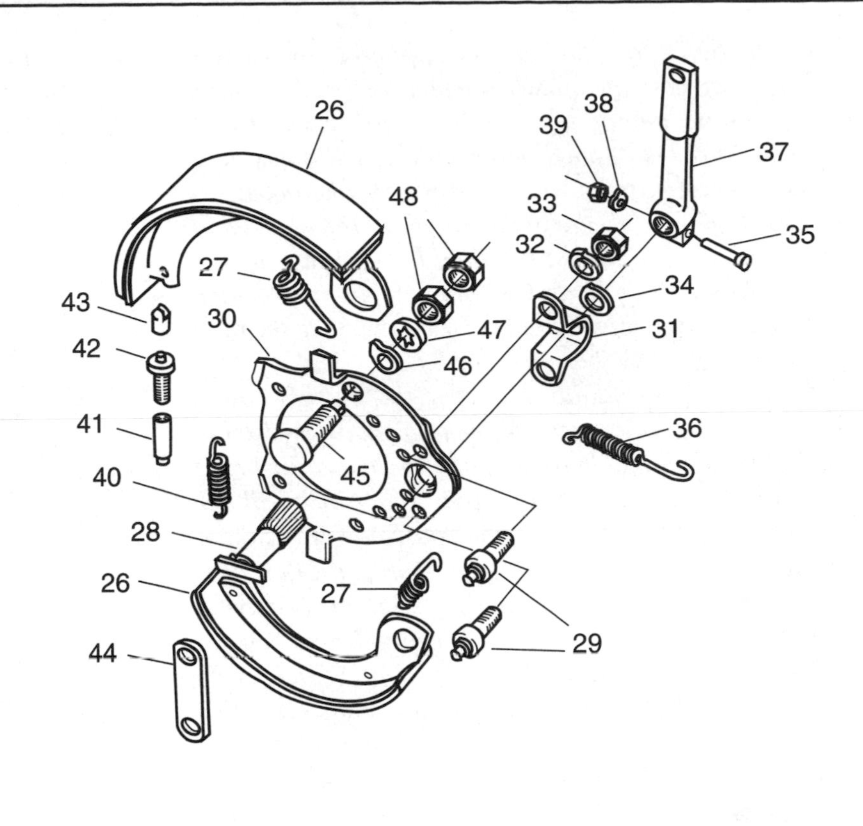

Figure 4–25 Manual propeller shaft parking brake assembly. *(Courtesy of General Motors Corporation, Service Technology Group)*

Power Assist Units and Miscellaneous

Task 1

Diagnose poor stopping complaints caused by power brake booster problems; determine needed repairs.

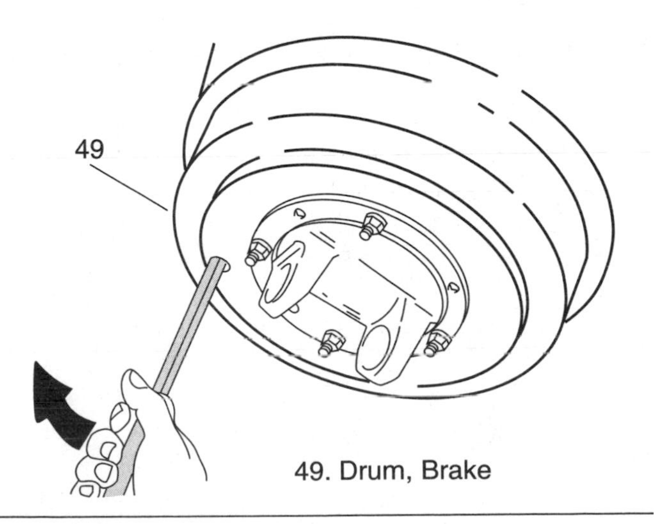

Figure 4–26 Parking brake adjustment. *(Courtesy of General Motors Corporation, Service Technology Group)*

136. A truck with a hydrovac brake booster requires excessive brake pedal effort. *Technician A* says the one-way check valve may be restricted in the vacuum hose to the hydrovac unit. *Technician B* says a sticking power piston in the hydrovac unit could be the cause. Who is right?
    A. A only
    B. B only
    C. Both A and B
    D. Neither A nor B

137. While discussing normal operation of a hydraulic brake system with a hydraulic brake booster, *Technician A* says that during a brake application, the pressure for brake power assist is supplied by the hydraulic booster. *Technician B* says that during a brake application, the electohydraulic pump should not be running. Who is right?
    A. A only
    B. B only
    C. Both A and B
    D. Neither A nor B

Hint

In truck hydraulic brake systems, the use of a vacuum or hydraulically assisted brake booster is required to reduce the foot effort that must be applied to the master cylinder to actuate the brakes. Vacuum-assisted boosters are not found on today's medium-duty trucks using hydraulic brakes. Some older trucks with gasoline engines were equipped with Hydrovac brake boosters. The Hydrovac units were mounted under the truck, and these units were supplied with fluid pressure from the master cylinder. The Hydrovac unit uses engine vacuum to operate a large diameter piston and increase the fluid pressure supplied by the master cylinder.

A hydraulic brake booster uses the truck's power steering pump or a dedicated pump to supply hydraulic pressure to the brake booster unit. The hydraulic brake booster has an open center valve, reaction feedback mechanism, a large diameter boost or power piston, a reserve electrically powered pump, an integral flow control switch, and a steering gear operating in series with the hydraulic brake booster. A hydraulic brake booster system may be called a hydroboost system. Vehicles with manual steering gear must use hydraulic boosters with a dedicated hydraulic pump.

Power brake boosters operate by assisting brake pedal effort with hydraulic pressure in proportion to pedal travel. Malfunctioning hydroboost units require greatly increased brake pedal effort. The brake system cannot be effectively operated in this condition.

Task 2

Inspect, test, repair, or replace power brake booster, hoses, and control valves.

138. If the reserve electrical motor for the hydraulic power brake system failed, the result would be:
  A. there is no hydraulic pressure for the brake booster without the engine running.
  B. there is no reserve electrical power in case of an electrical failure in the booster.
  C. fluid circulation in the steering gear will fail.
  D. braking power will be reduced with the engine running.

139. In the hydraulic brake booster electrical circuit in Figure 4–27:
  A. the electrical system voltage is supplied to the brake booster relay contacts through the ignition switch.
  B. the electrical system voltage is supplied to the brake switch at all times.
  C. the brake booster relay winding is connected directly to the battery ground.
  D. the alternator voltage signal informs the alarm/brake booster module if the alternator operation is normal.

Hint

In a hydroboost power brake booster, if flow from the hydraulic pump is interrupted, an electric pump backs up the system. When replacing hydraulic hoses in the boost system, the lines must conform to the SAE J189 standard.

If the fluid level in the hydroboost becomes too low, air may enter the system. If this condition occurs, the air must be bled from the system using the truck manufac-

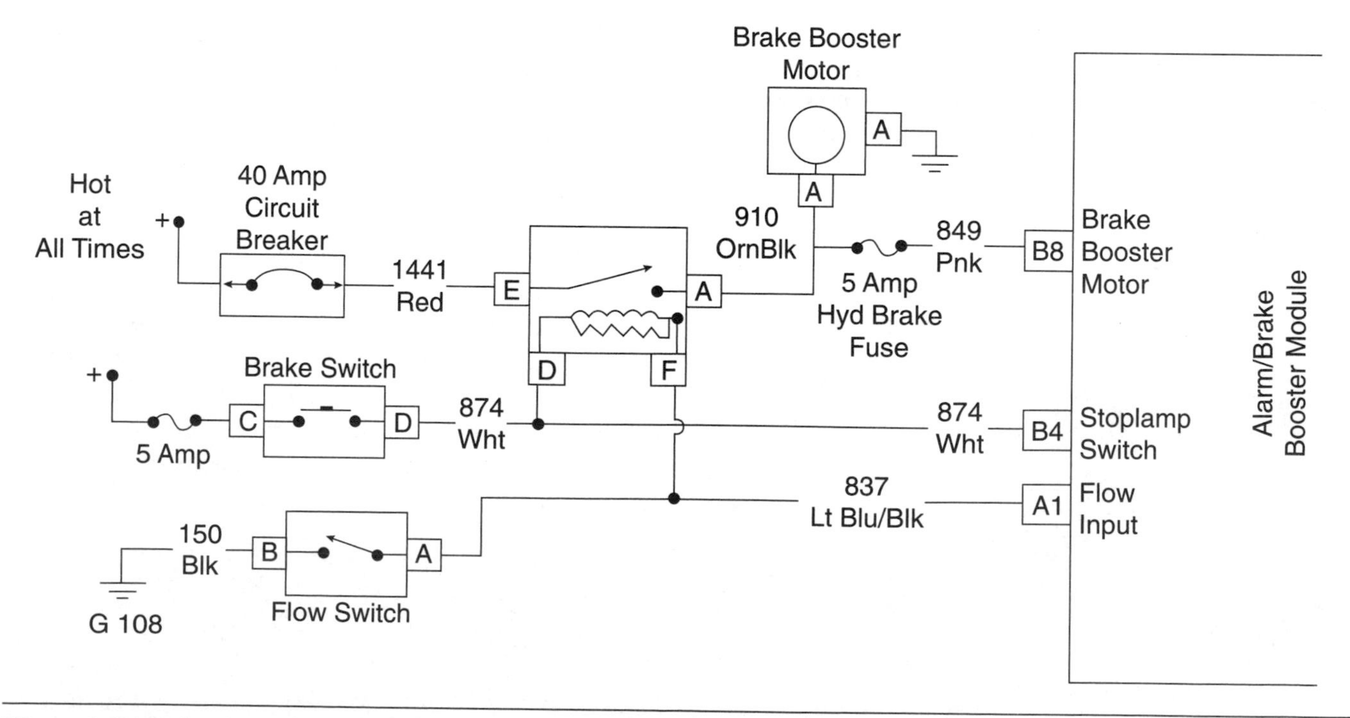

Figure 4–27 Hydraulic brake booster electrical circuit. *(Courtesy of General Motors Corporation, Service Technology Group)*

turer's recommended procedure. The hydroboost brake system reservoir must be filled with the recommended fluid. The refilling procedure may have to be repeated during the bleeding process.

Task 3

Test, adjust, and replace brake stoplight switch, bulbs, wiring, and connectors.

140. In the stoplight circuit on a truck with a hydraulic brake booster as shown in Figure 4–28:
   A. the rear stoplights and signal lights have separate bulb filaments.
   B. during a left turn with the brakes applied, the right rear stoplight remains illuminated.
   C. the stoplight switch is connected parallel to the signal light switch.
   D. when the ignition switch is on and no turn is indicated, there is no continuity through the signal light switch.

141. Refer to Figure 4–28. The right rear signal light and stoplight are inoperative, but all the other lights work properly. The cause of this problem could be:
   A. an open ground circuit from the right rear light to the ground block.
   B. an open wire from the signal light switch to the right rear signal and stoplight.
   C. an open wire from the ground block to the battery ground in the engine compartment.
   D. an open wire from the stoplight switch to the signal light switch.

Hint

Stoplights on hydraulic brake circuits may be actuated electromechanically or electrohydraulically. In either case the switch simply closes the brake light circuit. The brake light circuit may be tested with a DVOM.

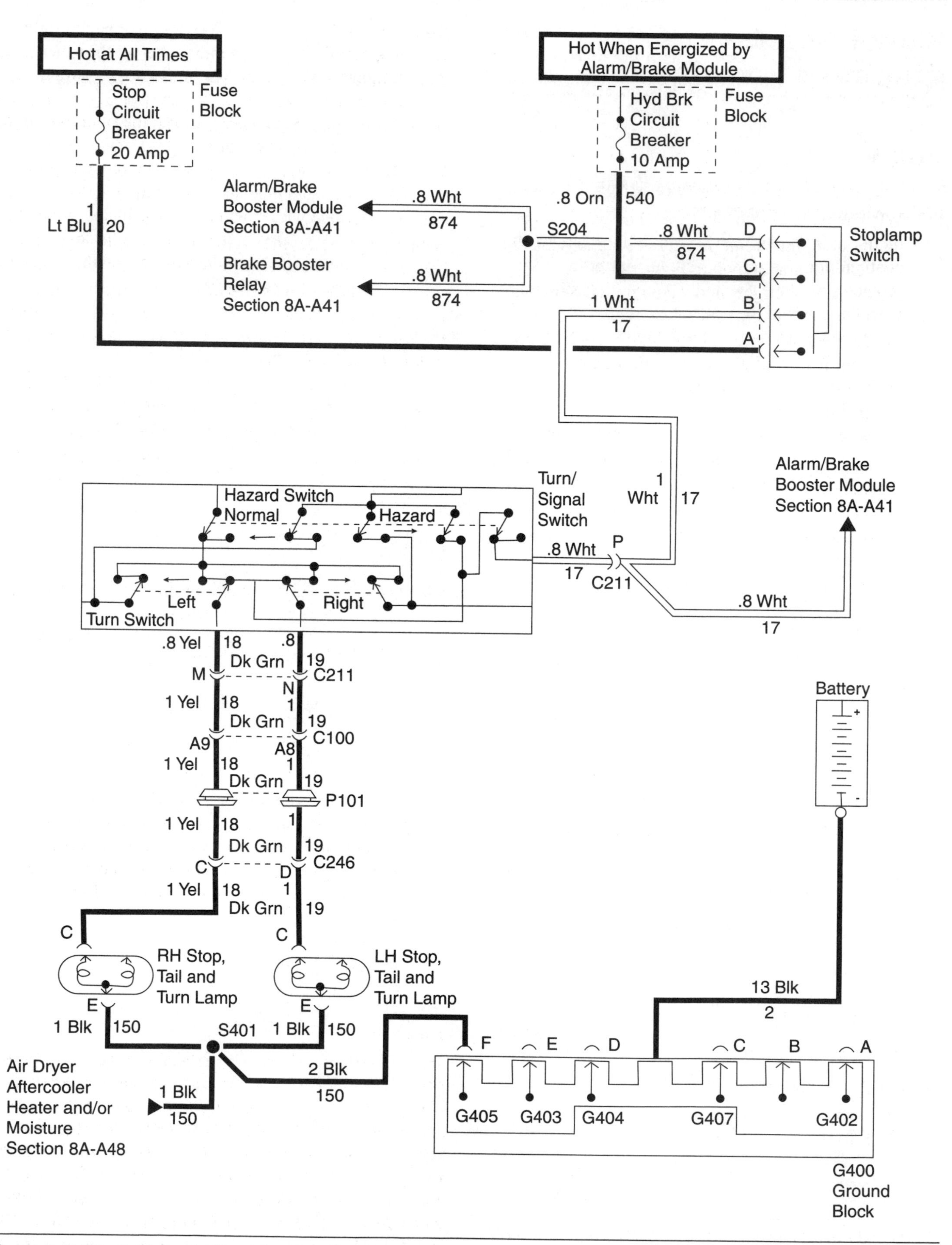

Figure 4–28 Stoplight circuit. *(Courtesy of General Motors Corporation, Service Technology Group)*

Wheel Bearings Diagnosis and Repair

Task 1

Remove and replace axle wheel assembly.

142. When reinstalling the dual wheel assembly on a tractor axle, *Technician A* says to place a light coating lubricant on the spindle, and install a seal protector over the threads. *Technician B* says rear wheel hub assemblies do not use oil filler plugs to fill the wheel cavities. Who is right?
   A. A only
   B. B only
   C. Both A and B
   D. Neither A nor B

143. When removing and replacing truck rim and tire assemblies as shown in Figure 4–29:
   A. the stud nuts should be removed before the rim clamps are loosened on demountable rims.
   B. the rims should be checked for wobble after the stud nuts are tightened to the specified torque.
   C. the rim spacer should be installed on spoke wheels with a rolling action.
   D. on demountable rims, the stud nut being tightened should be at the bottom of the rim.

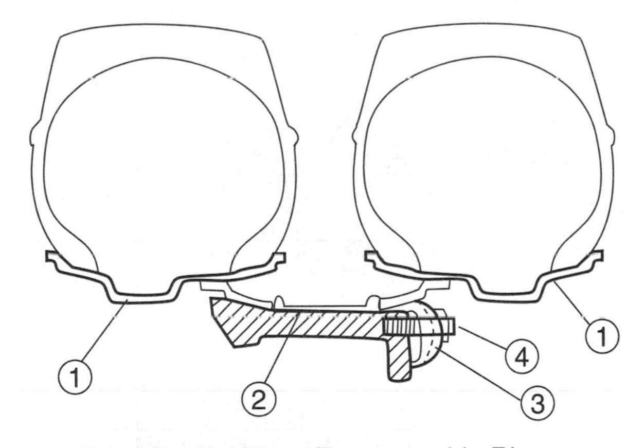

Item	Part Number	Description
1	1030	Rim
2	10444	Spacer
3	1043	Clamp
4	—	Stud (Part of 1014)

Figure 4–29 Dual wheel mounting. *(Courtesy of Ford Motor Company)*

Hint

The wheel assembly should be removed using a wheel dolly. The wheel seals should be removed using a heal bar or drift and a hammer. It is good practice to replace wheel seals each time they are removed from the hub. Unitized seals require replacement when removed, because the rubber sealing surface is damaged on removal. Removing the wheel seal enables the bearing cone to be removed for cleaning and inspection. The bearing cup must also be inspected in the hub. If the bearing assembly has to be replaced, the cups may be driven out of their bores using an appropriately sized bearing driver. Alternatively, a mild steel drift and hammer may be used. The proper driver must be used to install the bearing cups.

Hubs should be prelubed when installing the wheel assembly. On non-driving axles, the bearing and hub assembly must be filled to a prescribed level in the calibrated inspection cover. In some drive axles, the bearing and hub assembly is supplied with lubricant from the differential carrier.

Task 2

Clean, inspect, lubricate, or replace wheel bearings; replace seals and seal wear rings.

144. The bearing damage shown in Figure 4–30, F is called:
   A. frettage.
   B. brinelling.
   C. fatigue spalling.
   D. etching.

145. When servicing a truck's wheel bearings, *Technician A* says to use a high-quality, high-temperature grease when repacking is necessary. *Technician B* says wheel bearings may be switched from one wheel to another. Who is right?
   A. A only
   B. B only
   C. Both A and B
   D. Neither A nor B

Hint

Wheel bearings should be inspected at each brake job and at each PM service that requires wheel removal. The use of tapered roller bearings has become almost universal. A tapered wheel bearing assembly consists of a cone assembly and a cup or race. The cone assembly consists of tapered rollers mounted in a roller cage. The bearing race or cup is interference-fit to the hub. A pair of taper roller bearings support the load in each wheel hub. The cones and cups of taper roller bearing are not inter-

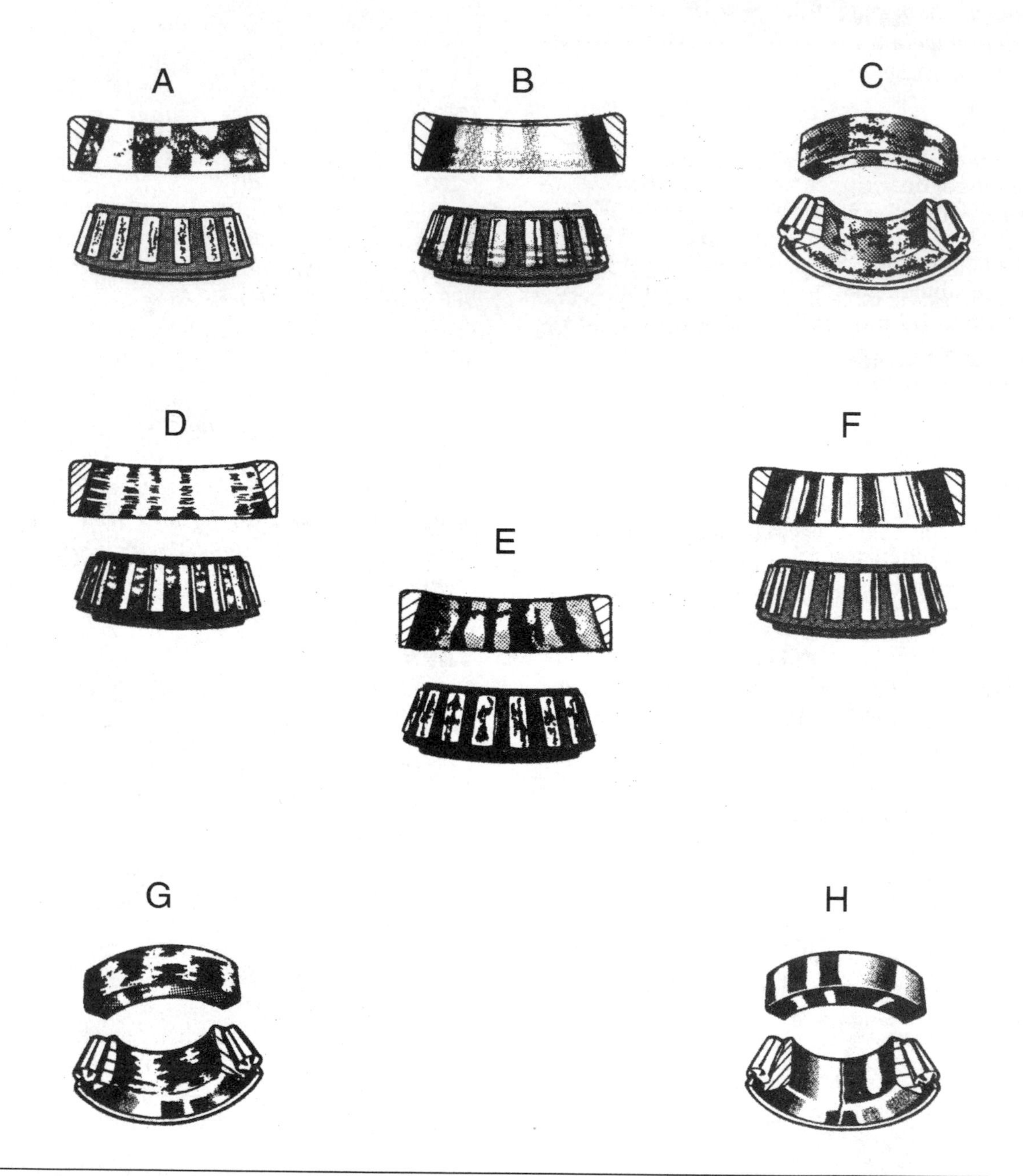

Figure 4–30 Defective bearing conditions. *(Courtesy of General Motors Corporation, Service Technology Group)*

changeable. When damage is evident in either the cup or the cone, both must be replaced.

Bearings should be cleaned with solvent and air dried, ensuring that the cone is not spun by the compressed air. Both the rollers and the cone should be inspected for spalling, galling, scoring, heat discoloration, and any sign of hard surface failure. When reinstalling oil-lubricated bearings, they should be prelubed with the same oil to be used in the differential. When grease-packed bearings are used, the bearing cone must be packed with grease. This procedure may be performed by hand or using a cone packer. Grease-lubricated wheel bearings require the use of a high-temperature grease.

Bearings fail primarily because of dirt contamination. This may be due to the conditions a vehicle is operated under or poor service practice. Lubrication failures caused by an inappropriate lubricant or lack of lubricant also account for a large number of bearing failures. Lack of lubricant results in rapid failure and a bearing possibly welded to the axle.

Misadjusted bearings can fail rapidly, especially when preloaded. The result of high preload on a bearing can be to friction weld it to the axle. Overloaded bearings fail over time rather than rapidly.

Task 3

Adjust axle wheel bearings in accordance with manufacturer's procedures.

146. When adjusting wheel bearings, *Technician A* says torque the adjusting nut to 50 ft.-lb., then back off the nut 1/6 to 1/3 turn and install the lock ring. *Technician B* says backing off the nut will cause the bearing to overheat. Who is right?
   A. A only
   B. B only
   C. Both A and B
   D. Neither A nor B

147. When performing the measurement in Figure 4–31, an acceptable dial indicator reading would be:
   A. 0.012 in. (0.304 mm).
   B. 0.010 in. (0.254 mm).
   C. 0.008 in. (0.203 mm)
   D. 0.003 in. (0.076 mm).

Hint

The proper method for adjusting a tapered roller wheel bearing is that outlined by the Truck Maintenance Council (TMC) division of the American Trucking Association (ATA). All bearing manufacturers in North America have endorsed this method, which supersedes any previous adjustment procedures outlined by them.

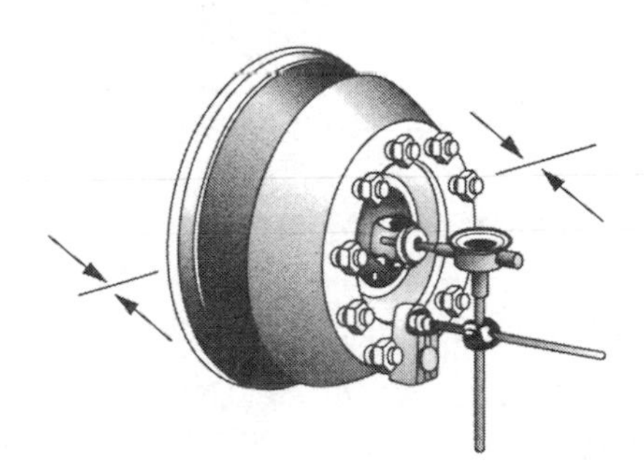

Figure 4–31 Hub measurement. *(Illustration courtesy The Maintenance Council (TMC) of the American Trucking Association. Copyright 1999.)*

The TMC bearing adjustment method requires that the bearing adjusting nut be tightened to 200 ft.-lb. (271 Nm) while rotating the wheel to seat the bearing. Next, the adjusting should be backed off one turn and then tightened to 50 ft.-lb. (68 Nm). The adjusting nut is then backed off between 1/6 and 1/3 of a turn depending on the application. The locking mechanism is then installed on the adjusting nut. Finally, the bearing end play must be measured with a dial indicator. The required specification is between 0.001 and 0.005 inch. If at least 0.001 inch endplay is not present, the adjustment procedure should be repeated.

The bearing adjusting nut is locked in place on the axle spindle by a lock or jam nut that should be torqued to specification after the adjustment procedure. In some axles, a split forged or castellated nut and cotter pin are used to retain the wheel assembly.

Current recommended practice in the truck service industry is defined by the TMC, and the above method was developed to establish a single standard to counter numerous highway wheel-off incidents.

Answers to Questions

1. **Answer A is right** because a chassis vibration during braking can be caused by excessive radial tire runout.

   Answer B is wrong because brake grab is not caused by improper governor cut-out adjustment.

   Answer C is wrong because brake drag is not caused by glazed brake linings. Glazed linings result in reduced stopping ability.

   Answer D is wrong because an improper governor cut-in adjustment does not cause the safety valve to open.

2. Answer A is wrong because a smaller than specified supply reservoir causes a faster air pressure build-up time.

   Answer B is wrong because an air leak in the compressor air intake filter does not cause a longer than specified compressor air pressure build-up time.

   Answer C is right because carbon buildup in the compressor discharge hose causes a longer than specified compressor air pressure build-up time.

   Answer D is wrong because an excessive amount of water in the supply reservoir causes a faster air pressure build-up time.

3. Answer A is wrong because the supply reservoir tends to collect more moisture than the front or rear axle reservoirs.

   Answer B is right because water does not need to be drained daily from the reservoirs if an air dryer is used in the air brake system.

   Answer C is wrong because a small amount of oil in the supply reservoir is a normal condition.

   Answer D is wrong because some air dryers and automatic moisture ejectors contain electric heaters to prevent moisture from freezing in these components.

4 Answer A is wrong because a condition causing a lowering of air pressure will subsequently reduce service brake application force.

   Answer B is wrong because an incorrect or lack of adjustment to a slack adjuster will cause a reduction in overall brake application forces.

   Answer C is right because grease on shoe linings/faces causes grabbing. It is the least likely cause of insufficient braking.

   Answer D is wrong because a ruptured brake chamber diaphragm will severely decrease an air brake system's service application force.

5. Answer A is wrong because a damaged or defective brake chamber diaphragm may cause excessive leakage.

   Answer B is wrong because a damaged hose or fitting may cause leakage when pressurized.

   Answer C is right because defective compressor gaskets will not cause brake system leakage.

   Answer D is wrong because a defective relay valve may cause an excessive air leak.

6. **Answer A is right** because 85 to 100 psi in 25 seconds or less is the acceptable specification.

   Answer B is wrong because 60 to 80 psi in 20 seconds is not the generally accepted standard for compressor performance.

   Answer C is wrong because 90 to 110 psi in 30 seconds is not the generally accepted standard for compressor performance.

   Answer D is wrong because 50 to 110 psi in 25 seconds is not the generally accepted standard for compressor performance.

7. **Answer A is right** because a restricted compressor outlet hose causes slow compressor build-up time.

   Answer B is wrong because a leaking brake chamber diaphragm does not cause slow compressor build-up time, because there is no air pressure supplied to the brake chamber diaphragms until the brakes are applied.

   Answer C is wrong because Technician A is right and Technician B is wrong.

   Answer D is wrong because Technician A is right and Technician B is wrong.

8. Answer A is wrong because daily draining of air reservoirs is highly recommended, but Technician B is also right.

   Answer B is wrong because you should check all automatic drain valves and moisture removing devices periodically for proper operation, but Technician A is also right.

   Answer C is right because both Technician A and Technician B are right.

   Answer D is wrong because both Technician A and Technician B are right.

9. Answer A is wrong because an automatic drain valve opens when the sump pressure in the valve is 2 psi (13.7 kPa) higher than reservoir pressure.

   Answer B is wrong because an automatic drain valve is mounted in a threaded fitting near the bottom of the reservoir.

   Answer C is wrong because an automatic drain valve opens by increasing supply reservoir pressure.

   Answer D is right because an automatic drain valve drains water if the sump pressure in the valve is 2 psi (13.7 kPa) higher than reservoir pressure.

10. Answer A is wrong because decreasing tension to original specification level and recheck contact surface for excessive wear will not cure the condition.

    Answer B is wrong because measuring the level of deflection at a section of belt that is longest between pulleys will not solve the problem.

    Answer C is right because you replace the belt if you notice contact at the bottom of the pulley with the belt.

    Answer D is wrong because if you ignore it, the problem will persist.

11. Answer A is wrong because a slipping compressor drive belt may cause slow compressor build-up time, but Technician B is also right.

    Answer B is wrong because some compressors have an adjustable idler pulley that may be moved to adjust belt tension, but Technician A is also right.

    Answer C is right because both Technician A and Technician B are right.

    Answer D is wrong because both Technician A and Technician B are right.

12. Answer A is wrong because it is NOT necessary to retime the engine when this replacement process is complete.

    Answer B is right because you do inspect the drive gear for worn or chipped teeth.

    Answer C is wrong because Technician A is wrong and Technician B is right.

    Answer D is wrong because Technician A is wrong and Technician B is right.

13. Answer A is wrong because a worn compressor drive may cause noise, but it does not cause the excessive compressor cycling.

    Answer B is right because air brake reservoirs containing excessive amounts of moisture reduce the volume of air available to the air brake system and cause the compressor to cycle frequently.

    Answer C is wrong because Technician A is wrong and Technician B is right.

    Answer D is wrong because Technician A is wrong and Technician B is right.

14. Answer A is wrong because piston rings are part of a brake system's air compressor.

    Answer B is wrong because the crankshaft is a fundamental part of an air compressor.

    Answer C is wrong because the discharge valve is part of an air compressor.

    Answer D is right because air compressors do not use roller bearings.

15. Answer A is wrong because a loose drive pulley may cause excessive noise.

    Answer B is wrong because a restriction in the unit or outlet hose may cause the compressor to make more noise.

    Answer C is right because a defective head gasket is the least likely cause of noise.

    Answer D is wrong because improper lubrication causes excessive compressor noise.

16. Answer A is wrong because the cut-in pressure should be 25 psi (172 kPa) less than the cut-out pressure.

    Answer B is wrong because only the governor cut-out pressure is adjustable.

    Answer C is wrong because both Technician A and Technician B are wrong.

    Answer D is right because both Technician A and Technician B are wrong.

17. Answer A is wrong because governors do not have check valves.

    Answer B is right because too much clearance at the compressor unloading valves could cause higher than normal air pressure.

    Answer C is wrong because Technician A is wrong and Technician B is right.

    Answer D is wrong because Technician A is wrong and Technician B is right.

18. **Answer A is right** because the governor cut-out pressure may be increased by rotating the adjusting screw counterclockwise.

    Answer B is wrong because you cannot adjust cut-in pressure.

    Answer C is wrong because you turn the adjustment screw 1/4 turn to change cut-out pressure 4 psi (27.5 kPa).

    Answer D is wrong because if you restrict the air line from the supply reservoir to the governor, the cut-out pressure increases.

19. Answer A wrong because you do use double flare tubing when replacing brake lines, but Technician B is also right.

    Answer B is wrong because ISO tubing is used when replacing a brake line, but Technician A is also right.

    Answer C is right because both Technician A and Technician B are right.

    Answer D is wrong because both Technician A and Technician B are right.

20. Answer A is wrong because if a 45-degree elbow in an air brake line is replaced with a 90-degree elbow, brake timing is adversely affected.

    Answer B is right because a smaller than specified air line slows the brake application.

    Answer C is wrong because Technician A is wrong and Technician B is right.

    Answer D is wrong because Technician A is wrong and Technician B is right.

21. Answer A is wrong because component E in Figure 4–4 is a safety valve.

    Answer B is wrong because component E in Figure 4–4 is a safety valve.

    Answer C is wrong because component E in Figure 4–4 is a safety valve.

    Answer D is right because component E in Figure 4–4 is a safety valve.

22. Answer A is wrong because the governor cut-out pressure should be 120 to 130 psi (827 to 896 kPa), and if the governor pressure is higher the governor is not operating properly, but Technician B is also right.

    Answer B is wrong because if the safety valve does not vent air pressure above 150 psi (1,034 kPa) this valve is malfunctioning, but Technician A is also right.

    Answer C is right because both Technician A and Technician B are right.

    Answer D is wrong because both Technician A and Technician B are right.

23. **Answer A is right** because these check valves protect the air supply in the primary or secondary reservoir if an air leak occurs in the supply reservoir.

    Answer B is wrong because these check valves protect the air supply in the primary or secondary reservoir if an air leak occurs in the supply reservoir.

    Answer C is wrong because Technician A is right and Technician B is wrong.

    Answer D is wrong because Technician A is right and Technician B is wrong.

24. Answer A is wrong because component C is the purge valve.

    Answer B is right because component C is the purge valve.

    Answer C is wrong because component C is the purge valve.

    Answer D is wrong because component C is the purge valve.

25. Answer A is wrong because the purge valve in the air dryer opens when the governor enters the unloaded cycle.

    Answer B is right because the one-way check valve in the air dryer discharge port prevents airflow from the supply reservoir into the dryer during the purge cycle.

    Answer C is wrong because Technician A is wrong and Technician B is right.

    Answer D is wrong because Technician A is wrong and Technician B is right.

26. Answer A is wrong because a condensation-type air dryer does not have a desiccant bed.

    Answer B is right because a condensation-type air dryer has a purge valve operated by governor pressure.

    Answer C is wrong because Technician A is wrong and Technician B is right.

    Answer D is wrong because Technician A is wrong and Technician B is right.

27. Answer A is wrong because when removing an application valve, the truck should be on a level surface.

    Answer B is wrong because when removing an application valve, you do mark or label the brake lines.

    Answer C is wrong because when removing an application valve, you do mark the valve body in relation to the mounting plate.

    Answer D is right because you do not maintain system air pressure when removing a brake application valve, instead you exhaust all the air pressure in the reservoirs.

28. **Answer A is right** because the application valve shown has two pistons: a reaction or rear modulating piston, and a primary or front modulating piston.

    Answer B is wrong because the reaction spring between the two pistons of a brake valve is NOT responsible for the basic operation of the valve.

Answer C is wrong because Technician A is right and Technician B is wrong.

Answer C is wrong because Technician A is right and Technician B is wrong.

29. **Answer A is right** because the air pressure at the brake application valve delivery port must be proportional to treadle movement.

 Answer B is wrong because with the brakes applied, a 1 inch (2.54 cm) bubble in 3 seconds at the brake application valve exhaust port indicates normal leakage.

 Answer C is wrong because Technician A is right and Technician B is wrong.

 Answer D is wrong because Technician A is right and Technician B is wrong.

30. Answer A is wrong because you apply air pressure applied to the outlet side of a single check valve.

 Answer B is wrong because with air pressure applied to one of the inlet ports on a double check valve, the air pressure should be the same at the outlet port.

 Answer C is right because with air pressure applied to one inlet port in a double check valve, the test gauge at the opposite inlet port should not indicate air pressure.

 Answer D is wrong because when air pressure is released at one inlet port on an double check valve, the air pressure should drop quickly at the outlet port.

31. **Answer A is right** because single check valves are connected between the primary and secondary reservoirs.

 Answer B is wrong because a double check valve allows air pressure to flow from the highest of two pressure sources.

 Answer C is wrong because Technician A is right and Technician B is wrong.

 Answer D is wrong because Technician A is right and Technician B is wrong.

32. Answer A is wrong because reversing the gladhand positions only reverses the switching for those valves.

 Answer B is wrong because a tractor "running bobtail" is a tractor running without a trailer.

 Answer C is wrong because the quick-release valve releases air from a chamber rapidly.

 Answer D is right because preventing simultaneous application of the service and emergency side of the spring brake chambers is anticompounding.

33. Answer A is wrong because the 10 Amp Hyd Brk circuit breaker only supplies voltage through the dual stoplight switch to the alarm/brake booster module.

 Answer B is right because voltage is supplied through terminal A on the dual brake light switch to the stoplights.

 Answer C is wrong because Technician A is wrong and Technician B is right.

 Answer D is wrong because Technician A is wrong and Technician B is right.

34. **Answer A is right** because high resistance between the right rear signal light and junction S401 reduces the current flow in the right rear stoplight or signal light and the right rear taillight.

 Answer B is wrong because high resistance between the signal light switch and connection C211 reduces current flow through both rear stoplights.

 Answer C is wrong because high resistance between stoplight switch and connection C211 reduces current flow through both rear stoplights.

 Answer D is wrong because high resistance at stoplight switch terminals A and B reduces current flow through both rear stoplights.

35. **Answer A is right** because you do move the handle to the fully applied position and record the air pressure.

 Answer B is wrong because the air system does not have to be drained to test the trailer control valve.

 Answer C is wrong because as long as the trailer brakes are fully charged, the initial gauge reading is unnecessary.

 Answer D is wrong because air leakage around the control handle should not occur.

36. Answer A is wrong because the air pressure at the delivery port should be proportional to valve handle movement.

Answer B is wrong because when the valve handle is fully applied, full reservoir pressure should be indicated on the gauge.

Answer C is right because with the valve handle fully applied, a 1 in. (2.54 cm) bubble in less than 3 seconds at the valve exhaust port indicates excessive leakage.

Answer D is wrong because when the valve handle is released, the pressure on the gauge should drop quickly to zero.

37. Answer A is wrong because the inlet valve leakage is tested with the service brakes released.

    Answer B is wrong because you do apply a soap solution to the area around the inlet and exhaust valve retaining ring to check exhaust valve leakage.

    Answer C is wrong because exhaust valve leakage is tested with the brakes applied.

    Answer D is right because the control port on the service brake relay valve is connected to the delivery port on the brake application valve.

38. **Answer A is right** because air lines are connected from the service brake relay valve delivery ports to the rear axle service brake chambers.

    Answer B is wrong because when the brakes are released, the inlet valve is closed, not open in the service brake relay valve.

    Answer C is wrong because when the brakes are released, the exhaust valve is open, not closed in the service brake relay valve.

    Answer D is wrong because the service brake relay valve is in a balanced position when the air pressure in the rear brake chambers equals system pressure.

39. **Answer A is right** because with the control valve in the dry road position, the limiting quick-release valve does not reduce air pressure to the front brakes.

    Answer B is wrong because with the control valve in the slippery road position and the service brakes applied, leakage at the control valve exhaust port should not exceed a 1 in. (2.54 cm) bubble in 3 seconds.

    Answer C is wrong because with the control valve in the slippery road position and the service brakes applied, leakage at the limiting quick-release valve exhaust port should not exceed a 1 in. (2.54 cm) bubble in 3 seconds.

    Answer D is wrong because with the control valve in the slippery road position and the service brakes applied, it limits front brake pressure to 50 percent of the application valve pressure.

40. Answer A is wrong because a bad tractor protection valve will not cause wheel lockup during a brake application.

    Answer B is right because a defective quick-release valve can cause slow front pressure buildup and cause poor stopping ability.

    Answer C is wrong because Technician A is wrong and Technician B is right.

    Answer C is wrong because Technician A is wrong and Technician B is right.

41. **Answer A is right** because quick-release valves are located close to the brake chambers.

    Answer B is wrong because quick-release valves are not mounted in the cab.

    Answer C is wrong because quick-release valves must be mounted close to the chambers they serve.

    Answer D is wrong because quick-release valves must be mounted close to the chambers they serve.

42. Answer A is wrong because the proportioning valve should be inspected every time the brakes are serviced, but Technician B is also right.

    Answer B is wrong because you use gauges ahead of and behind the proportioning valve to test its function, but Technician A is also right.

    Answer C is right because both Technician A and Technician B are right.

    Answer D is wrong because both Technician A and Technician B are right.

43. **Answer A is right** because when you supply 50 psi (344.75 kPa) from the brake application valve to the limiting valve inlet port, the pressure on a gauge at the limiting valve outlet port should be 40 psi (275.8 kPa).

    Answer B is wrong because 15 psi (103.42 kPa) is the wrong value.

    Answer C is wrong because 25 psi (172.37 kPa) is the wrong value.

    Answer D is wrong because 30 psi (206.85 kPa) is the wrong value.

44. Answer A is wrong because when the air pressure is reduced to zero on the delivery side of the pressure protection valve, the air pressure on the supply side should show no further pressure loss.

    Answer B is wrong because the leak rate is a 1-inch bubble in 3 seconds.

    Answer C is wrong because both Technician A and Technician B are wrong.

    Answer D is right because both Technician A and Technician B are wrong.

45. Answer A is wrong because when the trailer air supply valve is pulled outward, the air pressure is not supplied to the tractor protection valve.

    Answer B is right because if a trailer breakaway occurs, service air pressure from the application valve can no longer pass through the tractor protection valve.

    Answer C is wrong because Technician A is wrong and Technician B is right.

    Answer D is wrong because Technician A is wrong and Technician B is right.

46. Answer A is wrong because gladhands seals ensure leakage at the gladhands connections to the trailer.

    Answer B is right because tractor protection valve provides a means of preserving air pressure in sufficient amount to stop the tractor in the event of a trailer breakaway.

    Answer C is wrong because the tractor protection valve does not hold off the spring brake in case of sudden air loss in the spring brake chamber.

    Answer D is wrong because the bobtail proportioning valve senses when the tractor is running trailerless.

47. Answer A is wrong because if the rear axle reservoir is drained to zero air pressure and the brake application valve is applied, the tractor air brake pressure should be at what ever is left in the front reservoir.

    Answer B is right because when the tractor parking brakes are applied, the air pressure at the spring brake control valve delivery port should drop quickly to zero. This zero air pressure allows the springs in the parking brakes to apply the brakes.

    Answer C is wrong because Technician A is wrong and Technician B is right.

    Answer D is wrong because Technician A is wrong and Technician B is right.

48. **Answer A is right** because if the air pressure in the #1 (rear axle) reservoir drops below 55 psi (379 kPa), this valve allows partial application of the spring brakes when the service brakes are applied.

    Answer B is wrong because if piston B in this valve is moved upward, air pressure is exhausted from the spring brake chambers to apply the spring brakes.

    Answer C is wrong because Technician A is right and Technician B is wrong.

    Answer D is wrong because Technician A is right and Technician B is wrong.

49. Answer A is wrong because the system has low-pressure switches in both the primary and secondary reservoirs, and low-pressure in either system operates the warning light and buzzer, but Technician B is also right.

    Answer B is wrong because a grounded wire between one of the low-pressure switches and the in-line diode operates the low pressure warning light and the buzzer, but Technician A is also right.

    Answer C is right because both Technician A and Technician B are right.

    Answer D is wrong because both Technician A and Technician B are right.

50. Answer A is wrong because a defective low-pressure indicator bulb only causes the bulb to be inoperative. The buzzer would still operate normally with a defective bulb.

    Answer B is right because an inaccurate dash pressure gauge can be the cause of a warning light on above 60 psi (414 kPa) supply pressure.

    Answer C is wrong because Technician A is wrong and Technician B is right.

    Answer D is wrong because Technician A is wrong and Technician B is right.

51. Answer A is wrong because the tractor and trailer brakes applied normally before the relay valve replacement, and therefore the trailer service line is not restricted.

Answer B is right because a trailer relay valve with the wrong crack pressure may cause the trailer brakes to apply too slowly compared to the tractor brakes.

Answer C is wrong because Technician A is wrong and Technician B is right.

Answer D is wrong because Technician A is wrong and Technician B is right.

52. **Answer A is right** because there are more valves between the foot valve and the trailer axles than the tractor axles, thus increasing restrictions and lower those respective pressures.

    Answer B is wrong because the increased length of air line increases the time for air delivery.

    Answer C is wrong because the combination vehicle service lines are integrated via relay valves.

    Answer D is wrong because the treadle valve operates both tractor and trailer air systems.

53. **Answer A is right** because a brake chamber with a larger than specified diameter increases braking force, but it does not affect brake release time.

    Answer B is wrong because a weak chamber return spring causes sluggish brake release.

    Answer C is wrong because broken or weak brake return springs cause sluggish brake release.

    Answer D is wrong because an obstruction in a brake chamber air passage causes sluggish brake release.

54. Answer A is wrong because ideal braking forces require proportionality, correct timing, and identical pressure levels.

    Answer B is right because an ideal brake system is one in which the braking pressure reaches each actuator simultaneously and at the same pressure level.

    Answer C is wrong because wheels will lock up receiving pressure first if given in sequential order.

    Answer D is wrong because an ideal air brake system is one in which the braking pressure reaches each actuator simultaneously and at the same pressure level.

55. **Answer A is right** because excessive air leakage at the rear service brake relay valve with the brakes applied reduces air pressure supplied to the rear brakes and causes poor stopping.

    Answer B is wrong because different types of brake linings installed on some of the rear tandem wheels may decrease braking capability, but this condition also causes wheel lockup.

    Answer C is wrong because Technician A is right and Technician B is wrong.

    Answer D is wrong because Technician A is right and Technician B is wrong.

56. Answer A is wrong because a slack adjuster must be adjusted so the angle between it and the chamber pushrod is 90 degrees when the brakes are fully applied, but Technician B is also right.

    Answer B is wrong because the chamber pushrod should be adjusted with the shortest possible stroke without dragging the brakes, but Technician A is also right.

    Answer C is right because both Technician A and Technician B are right.

    Answer D is wrong because both Technician A and Technician B are right.

57. Answer A is wrong because component B is the pressure plate.

    Answer B is wrong because component B is the pressure plate.

    Answer C is wrong because component B is the pressure plate.

    Answer D is right because component B is the pressure plate.

58. **Answer A is right** because the clevis should be properly positioned on the pushrod in order to facilitate proper brake adjustment.

    Answer B is wrong because the clevis is not keyed, but threaded, onto the brake pushrod.

    Answer C is wrong because Technician A is right and Technician B is wrong.

    Answer D is wrong because Technician A is right and Technician B is wrong.

59. Answer A is wrong because the overall braking force is decreased if the angle is less than 90 degrees between the brake chamber pushrod and the slack adjuster arm.

    Answer B is wrong because a brake chamber pushrod cannot be parallel to the slack adjuster to be functional.

Answer C is wrong because overall braking force is decreased if there is more than a 90-degree angle between the brake chamber pushrod and the slack adjuster arm.

Answer D is right because the braking force is at the maximum if there is a 90-degree angle between the brake chamber pushrod and the slack adjuster arm.

60. Answer A is wrong because the applied stroke uses an 80 psi brake application, but Technician B is also right.

    Answer B is wrong because the brakes are adjusted (if needed) to achieve the proper free stroke, but Technician A is also right.

    Answer C is right because both Technician A and Technician B are right.

    Answer D is wrong because both Technician A and Technician B are right.

61. Answer A is wrong because excessive wear limits between the clevis and the collar will cause the stroke to be too long, but Technician B is also right.

    Answer B is wrong because improper brake adjustment may cause premature wheel lockup during a brake application, but Technician A is also right.

    Answer C is right because both Technician A and Technician B are right.

    Answer D is wrong because both Technician A and Technician B are right.

62. **Answer A is right** because excessive clearance between the linings and the brake drum may cause the cam to roll over during a brake application.

    Answer B is wrong because this condition will not cause loss of steering control.

    Answer C is wrong because Technician A is right and Technician B is wrong.

    Answer D is wrong because Technician A is right and Technician B is wrong.

63. Answer A is wrong because both camshaft seal lips must face toward the slack adjuster.

    Answer B is wrong because you do use the proper driver when installing camshaft seals.

    Answer C is right because you never lubricate the cam head.

    Answer D is wrong because you do measure camshaft radial movement.

64. Answer A is wrong because installing a new wedge with a different angle will affect brake operation by changing the distance the brake shoes travel.

    Answer B is wrong because lubricant should be placed on the wedge and wedge ramps.

    Answer C is wrong because both Technician A and Technician B are wrong.

    Answer D is right because both Technician A and Technician B are wrong.

65. Answer A is wrong because minor scoring may be removed from the actuator bores with fine emery cloth.

    Answer B is right because when the threaded actuator opening is not chamfered, the flat side of the collet nut must face toward the threaded actuator opening.

    Answer C is wrong because the clearance between each wedge brake shoe lining and the drum should be .020 to 0.040 in. (.51 to 1.02 mm).

    Answer D is wrong because when the threaded actuator opening is chamfered, the taper on this collet nut faces the taper on the actuator opening.

66. **Answer A is right** because many wedge brakes are self-adjusting.

    Answer B is wrong because wedge brakes do not have slack adjusters.

    Answer C is wrong because Technician A is right and Technician B is wrong.

    Answer D is wrong because Technician A is right and Technician B is wrong.

67. **Answer A is right** because the wedge angle determines the amount of brake shoe movement in relation to the brake chamber diaphragm and pushrod movement.

    Answer B is wrong because adjusting plungers may have manual or automatic adjusting mechanisms.

    Answer C is wrong because the tapered wedge surfaces contact grooves on the bottom of the anchor and adjusting pistons.

    Answer D is wrong because the brake chamber diaphragm pushrod contacts the outer end of the wedge.

68. Answer A is wrong because a caliper piston seized in its bore would usually wear both pads evenly.

    Answer B is wrong because a damaged and/or defective rotor will usually wear both pads evenly.

    Answer C is wrong because a damaged brake hose/line will tend to wear both pads evenly.

    Answer D is right because a caliper seized on the slide pins will not release or react to the braking force and wear the outboard pad.

69. **Answer A is right** because seized caliper pistons may cause brake drag.

    Answer B is wrong because rotors machined beyond the maximum diameter may cause rotor overheating and cracking, but this problem does not cause brake pedal fade after initial brake application.

    Answer C is wrong because Technician A is right and Technician B is wrong.

    Answer D is wrong because Technician A is right and Technician B is wrong.

70. Answer A is wrong because full-clevis free stroke is not a type of adjustment or setting.

    Answer B is wrong because half-clevis free stroke is not a type of adjustment or setting.

    Answer C is wrong because full free stroke is not a type of adjustment or setting.

    Answer D is right because initial free stroke is the correct adjustment term.

71. Answer A is wrong because the primary shoe is the first shoe past the cam in the direction of wheel rotation, and therefore the primary shoe goes on the bottom.

    Answer B is wrong because the primary shoe is the first shoe past the cam in the direction of wheel rotation, and therefore the primary shoe goes on the bottom.

    Answer C is wrong because the primary shoe is the first shoe past the cam in the direction of wheel rotation, and therefore the primary shoe goes on the bottom.

    Answer D is right because the primary shoe is the first shoe past the cam in the direction of wheel rotation, and therefore the primary shoe goes on the bottom.

72. **Answer A is right** because you replace all the pads at the same time.

    Answer B is wrong because brake pads should not be replaced on one side only at any one time.

    Answer C is wrong because brake pads should always be changed in pairs.

    Answer D is wrong because resurfacing of a brake rotor when pads are replaced is at the technician's discretion.

73. Answer A is wrong because different size linings are not criteria for a definition of combination linings.

    Answer B is wrong because differences in attachment methods are not criteria for a definition of combination linings.

    Answer C is wrong because both Technician A and Technician B are wrong.

    Answer D is right because both Technician A and Technician B are wrong.

74. Answer A is wrong because on brakes with vented discs, the inboard pad could be thicker than the outboard pad, but Technician B is also right.

    Answer B is wrong because on brakes with solid discs, the inboard and outboard pads could be the same thickness, but Technician A is also right.

    Answer C is right because both Technician A and Technician B are right.

    Answer D is wrong because both Technician A and Technician B are right.

75. **Answer A is right** because oversize drums may be machined as long as the drum manufacturer's recommendation for machining dimensions are followed.

    Answer B is wrong because turning an oversize drum decreases its overall strength.

    Answer C is wrong because Technician A is right and Technician B is wrong.

    Answer D is wrong because Technician A is right and Technician B is wrong.

76. Answer A is wrong because the dial indicator in Figure 4–17 is measuring rotor runout. A micrometer is used to check thickness variation in brake rotors.

    Answer B is right because thickness variation should be measured at twelve locations around the rotor.

    Answer C is wrong because Technician A is wrong and Technician B is right.

Answer D is wrong because Technician A is wrong and Technician B is right.

77. Answer A is wrong because to facilitate removing the spring chamber off the adapter, slide the spring chamber sideways while holding the adapter.

    Answer B is wrong because you remove the diaphragm by turning it counterclockwise from the chamber.

    Answer C is right because you do not loosen the clamps by lightly striking them with a soft hammer.

    Answer D is wrong because the air pressure should be exhausted if it was used to aid in caging the spring brake.

78. Answer A is wrong because no attempt should be made to repair/replace any part of the piggyback assembly, and it should be completely replaced as a unit, but Technician B is also right.

    Answer B is wrong because the spring chamber should be disarmed (manually released), before discarding, but Technician A is also right.

    Answer C is right because both Technician A and Technician B are right.

    Answer D is wrong because both Technician A and Technician B are right.

79. Answer A is wrong because the park and emergency braking system is a separate and distinct air circuit, completely isolated from the regular service air system, but Technician B is also right.

    Answer B is wrong because spring brake chambers use air pressure in the opposite way from the service brake chambers, but Technician A is also right.

    Answer C is right because both Technician A and Technician B are right.

    Answer D is wrong because both Technician A and Technician B are right.

80. **Answer A is right** because when the primary or secondary system fails, federal law requires that the parking brakes must release at least once.

    Answer B is wrong because the parking brake will apply when air is released from the spring brake chamber.

    Answer C is wrong because Technician A is right and Technician B is wrong.

    Answer D is wrong because Technician A is right and Technician B is wrong.

81. Answer A is wrong because air pressure must be built up in the service brake chamber to apply the service brake.

    Answer B is wrong because when air is released from the spring brake chamber, the spring brakes apply.

    Answer C is wrong because a relay valve redirects air to the spring brake chambers.

    Answer D is right because the trailer spring brake valve releases the trailer parking brakes with supply air pressure above 85 psi (586 kPa).

82. Answer A is wrong because the spring does not release, but applies.

    Answer B is wrong because air begins to exhaust from the piggyback chamber.

    Answer C is wrong because the treadle or application valve is not part of the park control valve circuit.

    Answer D is right because air in the rear portion of the spring brake chamber is exhausted and the springs apply the parking brakes.

83. **Answer A is right** because a leak in a spring brake diaphragm causes brake dragging.

    Answer B is wrong because system air pressure should always be available at the spring brake relay valve supply port.

    Answer C is wrong because Technician A is right and Technician B is wrong.

    Answer D is wrong because Technician A is right and Technician B is wrong.

84. Answer A is wrong because the tractor park valve has a blue, round-shaped knob that is pulled out to apply the tractor parking brakes.

    Answer B is wrong because the tractor park valve cannot be used to apply the trailer parking brakes.

    Answer C is wrong because both Technician A and Technician B are wrong.

    Answer D is right because both Technician A and Technician B are wrong.

85. **Answer A is right** because the spring brake chambers should be "caged" before disconnecting any air line or hose due to safety reasons.

Answer B is wrong because the spring brakes apply when the park valves cause the relay valves to drain all air pressure.

Answer C is wrong because Technician A is right and Technician B is wrong.

Answer D is wrong because Technician A is right and Technician B is wrong.

86. Answer A is wrong because the spring is caged in the spring brake chamber.

Answer B is wrong because the spring brake chamber clamp is crimped and cannot be removed.

Answer C is wrong because both Technician A and Technician B are wrong.

Answer D is right because both Technician A and Technician B are wrong.

87. Answer A is wrong because the ABS warning light is illuminated to indicate an electrical defect in the system when the vehicle speed is above 4 mph (6.43 km/h).

Answer B is right because the ABS warning light is illuminated to indicate an electrical defect in the system when the vehicle speed is above 4 mph (6.43 km/h).

Answer C is wrong because Technician A is wrong and Technician B is right.

Answer D is wrong because Technician A is wrong and Technician B is right.

88. **Answer A is right** because when the ABS check switch and the ignition switch are turned on, the ABS light flashes blink codes.

Answer B is wrong because the first set of flashes indicates the ABS configuration.

Answer C is wrong because when there are no electrical defects in the ABS, a 1,00, 2,00, or 4,00 blink code is provided.

Answer D is wrong because a blink code indicates a defect in a certain area, but not necessarily in a specific component.

89. Answer A is wrong because the tire and wheel has to be rotating to check the wheel speed sensor readings.

Answer B is wrong because if NO DATA RECEIVED is displayed on the scan tool, there is a defective connection between the scan tool and the diagnostic connector, or the ECU is not powered.

Answer C is wrong because the ignition switch must be off when connecting the scan tool to the diagnostic connector.

Answer D is right because the scan tool may be used to test the data link wires between the ECU and the PCM.

90. Answer A is wrong because the ABS ECU uses a series of LEDs in the diagnostic window to indicate defects in the system, but Technician B is also right.

Answer B is wrong because the number of LEDs in the diagnostic window varies depending on the system, but Technician A is also right.

Answer C is right because both Technician A and Technician B are right.

Answer D is wrong because both Technician A and Technician B are right.

91. Answer A is wrong because an ABS provides 20 to 30% tire slip without wheel lockup.

Answer B is wrong because the ABS only enters the antilock mode when one or more wheels are approaching a lockup condition, and this condition is usually during a severe brake application.

Answer C is wrong because both Technician A and Technician B are wrong.

Answer D is right because both Technician A and Technician B are wrong.

92. Answer A is wrong because the tractor and trailer ABSs have separate ECUs.

Answer B is right because if replacement tires are a different size compared to the original tires, the wheel speed sensor signals are affected.

Answer C is wrong because Technician A is wrong and Technician B is right.

Answer D is wrong because Technician A is wrong and Technician B is right.

93. Answer A is wrong because a heavy-duty truck with ABS has one modulator per wheel.

Answer B is right because a truck's ABS is designed to prevent wheel lockup even when the tractor wheels are on different road surface conditions.

Answer C is wrong because Technician A is wrong and Technician B is right.

Answer D is wrong because Technician A is wrong and Technician B is right.

94. Answer A is wrong because the wheel speed sensor is retained in the proper position by the retaining clip.

    Answer B is right because this response is not true. The average wheel speed sensor gap is 0.015 in. (0.038 mm).

    Answer C is wrong because in the antilock mode, the ECU operates the solenoids in the modulator to provide about 20 percent tire slip without wheel lockup.

    Answer D is wrong because in an individual modulator valve, the exhaust solenoid is normally open and the inlet solenoid is normally closed.

95. Answer A is wrong because the wheel speed sensor generates an AC voltage signal.

    Answer B is wrong because the frequency of the wheel speed sensor signal increases as the wheel speed increases.

    Answer C is wrong because both Technician A and Technician B are wrong.

    Answer D is right because both Technician A and Technician B are wrong.

96. Answer A is wrong because an open wheel speed sensor winding would result in no signal from this sensor.

    Answer B is right because a damaged toothed ring may cause an erratic wheel speed sensor signal.

    Answer C is wrong because a grounded wheel speed sensor lead wire would result in no signal from this sensor.

    Answer D is wrong because excessive wheel speed sensor gap causes a weak wheel speed sensor signal.

97. Answer A is wrong because a seized brake caliper piston will not cause brake fade, only pad wear and reduced braking power.

    Answer B is right because a brake drum machined beyond its limit causes the shoes to move farther for contact. Brake drum diameter increases with heat when the drum is heated during a severe or prolonged brake application. When heat causes brake drum expansion, one must depress the brake pedal further to force the shoes against the drum. Industry calls this action "brake pedal fade."

    Answer C is wrong because low fluid level in the master cylinder will cause a low pedal, not fade.

    Answer D is wrong because air in the hydraulic system will cause a spongy pedal, not fade.

98. **Answer A is right** because seized caliper pistons may cause brake drag.

    Answer B is wrong because rotors machined to less than the minimum thickness may cause rotor overheating and rotor damage, but this problem does not cause brake pedal fade.

    Answer C is wrong because Technician A is right and Technician B is wrong.

    Answer D is wrong because Technician A is right and Technician B is wrong.

99. Answer A is wrong because sticking wheel cylinder pistons increase brake pedal effort and reduce braking efficiency, but this problem does not cause pedal pulsations.

    Answer B is wrong because leaking caliper piston seals cause loss of brake fluid, but this problem does not cause pedal pulsations.

    Answer C is right because an out-of-round brake drum may cause pedal pulsations.

    Answer D is wrong because a rotor with less than the minimum thickness causes rotor overheating and damage, but this problem may not cause pedal pulsations unless the rotor has excessive thickness variation.

100. Answer A is wrong because a fluid leak may cause pedal fade, but Technician B is also right.

     Answer B is wrong because a bulge in a flexible brake hose may cause pedal fade, but Technician A is also right.

     Answer C is right because both Technician A and Technician B are right.

     Answer D is wrong because both Technician A and Technician B are right.

101. Answer A is wrong because if there is no fluid in the primary master cylinder section, the secondary section still supplies fluid pressure to some of the wheels and provides some braking action.

     Answer B is right because the power booster pushrod applies force to the primary piston.

     Answer C is wrong because Technician A is wrong and Technician B is right.

Answer D is wrong because Technician A is wrong and Technician B is right.

102. **Answer A is right** because blocked compensation ports will cause brake drag.

Answer B is wrong because excessive pedal free play will not cause brake drag. Insufficient pedal free play will cause brake drag.

Answer C is wrong because Technician A is right and Technician B is wrong.

Answer D is wrong because Technician A is right and Technician B is wrong.

103. **Answer A is right** because excessive brake pedal free play causes a low, firm pedal.

Answer B is wrong because air in the hydraulic system causes a low, spongy brake pedal.

Answer C is wrong because brake fluid contaminated with moisture causes a low, spongy brake pedal.

Answer D is wrong because low brake fluid level causes a low, spongy brake pedal.

104. Answer A is wrong because a hydraulic pressure gauge is used to test master cylinder output.

Answer B is wrong because wheel cylinders are generally rebuilt, not replaced.

Answer C is wrong because both Technician A and Technician B are wrong.

Answer D is right because both Technician A and Technician B are wrong.

105. Answer A is wrong because check valves regulate the return flow of brake fluid to the master cylinder.

Answer B is wrong because the metering valve slows the front application of the disc brakes to allow the rear wheel cylinders to build up equal pressure.

Answer C is wrong because sliding rubber seals in the wheel cylinders contain the brake fluid and prevent leakage.

Answer D is right because the return spring as shown will return the brake pedal to its original position.

106. Answer A is wrong because you never wash master cylinder parts in an oil-based solvent.

Answer B is wrong because you never hone an aluminum master cylinder even if it is scored.

Answer C is wrong because both Technician A and Technician B are wrong.

Answer D is right because both Technician A and Technician B are wrong.

107. Answer A is wrong because soap and water causes corrosion and leaves residue.

Answer B is right because isopropyl alcohol is the correct cleaning agent.

Answer C is wrong because mineral spirits leave residue.

Answer D is wrong because you use hydraulic brake fluid for operation of the system, not for cleaning it.

108. Answer A is wrong because you do use double flare tubing when replacing brake lines, but Technician B is also right.

Answer B is wrong because one does use ISO tubing when replacing a brake line, but Technician A is also right.

Answer C is right because both Technician A and Technician B are right.

Answer D is wrong because both Technician A and Technician B are right.

109. Answer A is wrong because a metering valve delays fluid pressure to the front brakes during a light brake application.

Answer B is wrong because a metering valve does not prevent front wheel lockup during a hard brake application.

Answer C is wrong because both Technician A and Technician B are wrong.

Answer D is right because both Technician A and Technician B are wrong.

110. **Answer A is right** because a combination valve does not contain a quick take-up valve. This valve is mounted in some master cylinders.

Answer B is wrong because a combination valve may contain a metering valve.

Answer C is wrong because a combination valve may contain a proportioning valve.

Answer D is wrong because a combination valve may contain a pressure differential switch.

111. **Answer A is right** because the pressure differential valve grounds the brake warning light circuit.

   Answer B is wrong because if the primary and secondary sections of the master cylinder have equal pressure, the pressure differential valve electrical contact is closed.

   Answer C is wrong because Technician A is right and Technician B is wrong.

   Answer D is wrong because Technician A is right and Technician B is wrong.

112. Answer A is wrong because pressure differential valves automatically recenter themselves upon application of the brakes after repairs have been made.

   Answer B is wrong because you do not replace the proportioning valve after resetting the pressure differential valve.

   Answer C is wrong because both Technician A and Technician B are wrong.

   Answer D is right because both Technician A and Technician B are wrong.

113. Answer A is wrong because a hydraulic wheel cylinder may occasionally need to be rebored.

   Answer B is right because you always inspect wheel cylinders on a PM.

   Answer C is wrong because a wheel cylinder does not need replacing periodically, but needs servicing on occasion.

   Answer D is wrong because wheel cylinders are generally constructed of cast iron.

114. Answer A is wrong because adjuster wheel cylinders provide automatic brake shoe adjustment when the brakes are released.

   Answer B is wrong because some foundation brake assemblies have one adjuster cylinder and a parking brake cylinder.

   Answer C is right because brake fluid is forced into the area between the cylinder pistons when the brakes are applied.

   Answer D is wrong because the manual override wheel is operated by a Technician to adjust the brakes if necessary.

115. Answer A is wrong because a seized floating caliper on a hydraulic brake system does not cause brake pedal fade.

   Answer B is wrong because a seized floating caliper on a hydraulic brake system does not cause brake pedal pulsations.

   Answer C is right because a seized floating caliper on a hydraulic brake system causes reduced braking force.

   Answer D is wrong because a seized floating caliper on a hydraulic brake system does not cause brake grabbing.

116. Answer A is wrong because thickness is not indicated by a code on the lining edge.

   Answer B is wrong because not all semimetallic linings are edge stamped FF. The FF designation represents the cold and hot coefficient of friction.

   Answer C is wrong because asbestos-type linings are no longer used.

   Answer D is right because linings should use the same edge brand as OEM.

117. **Answer A is right** because you must remove the wheel in most cases to measure pad thickness.

   Answer B is wrong because if the brakes were applied shortly before measuring the lining-to-rotor clearance, this clearance will be LESS than specified.

   Answer C is wrong because Technician A is right and Technician B is wrong.

   Answer D is wrong because Technician A is right and Technician B is wrong.

118. Answer A is wrong because glass is not generally used for containing brake fluid.

   Answer B is wrong because all containers must be tightly sealed when storing brake fluid.

   Answer C is right because you never put brake fluid in any container that has been used to store another liquid.

   Answer D is wrong because brake fluid from a tightly sealed container may be used after one year of storage.

119. Answer A is wrong because you do use a tool to hold the metering valve open when using a pressure bleeder to bleed the system, but Technician B is also right.

   Answer B is wrong because you can also bleed this type of system manually, but Technician A is also right.

Answer C is right because both Technician A and Technician B are right.

Answer D is wrong because both Technician A and Technician B are right.

120. Answer A is wrong because first you bleed the right rear wheel cylinder or caliper that is farthest from the master cylinder.

Answer B is right because the pressure bleeder is pressurized to 20 to 25 psi (137.9–172.37 kPa).

Answer C is wrong because the proportioning valve is normally open and does not require any action when bleeding the brakes.

Answer D is wrong because first you bleed the right rear wheel cylinder or caliper that is farthest from the master cylinder.

121. **Answer A is right** because the fluid pressure is supplied from the control valve delivery port to the unitized valve control port.

Answer B is wrong because the unitized valve supplies fluid pressure to the parking brake chambers in the release mode.

Answer C is wrong because the parking brake chambers are pressurized when the dash control knob is pushed inward.

Answer D is wrong because the pump supplies pressure directly to the unitized valve in the release mode.

122. Answer A is wrong because parking brake chambers should be caged in the released position since the spring is compressed in this mode.

Answer B is wrong because caging parking brake chambers with an impact wrench causes internal chamber damage.

Answer C is wrong because both Technician A and Technician B are wrong.

Answer D is right because both Technician A and Technician B are wrong.

123. **Answer A is right** because a chassis vibration during a brake application may be caused by excessive radial runout on a tire.

Answer B is wrong because brake grab on one wheel is not caused by improper brake pedal pushrod adjustment.

Answer C is wrong because brake drag is not caused by glazed brake linings.

Answer D is wrong because a low, firm pedal is not caused by air in the hydraulic brake system. This problem causes a low, spongy pedal.

124. Answer A is wrong because seized wheel caliper pistons may cause excessive pedal effort and poor stopping ability.

Answer B is wrong because glazed brake linings may cause excessive pedal effort and poor stopping ability.

Answer C is right because air in the hydraulic system causes a low, spongy brake pedal, and the question says the pedal height is normal and firm.

Answer D is wrong because a binding brake pedal may cause excessive pedal effort and poor stopping ability.

125. **Answer A is right** because a seized left front caliper piston causes very little braking action on the left front wheel, and the normal braking action on the right front wheel causes the vehicle to pull to the right during a brake application.

Answer B is wrong because a broken brake shoe return spring in the left front brake assembly does not cause pull to the right during a brake application.

Answer C is wrong because loose wheel lug nuts do not cause pull to the right during a brake application.

Answer D is wrong because a cupped right front tire does not cause pull to the right during a brake application.

126. **Answer A is right** because when performing a brake overhaul on a hydraulic rear drum brake system, you replace the return springs.

Answer B is wrong because the maximum allowable out-of-round specification on a brake drum is 0.015 inch.

Answer C is wrong because Technician A is right and Technician B is wrong.

Answer D is wrong because Technician A is right and Technician B is wrong.

127. Answer A is wrong because high thickness variation does not cause reduced braking power.

Answer B is right because high rotor thickness variation causes pedal pulsation.

Answer C is wrong because high thickness

variation will not cause loss of brake fluid.

Answer D is wrong because loss of directional control is not generally evident with high thickness variation.

128. Answer A is wrong because 0.010–0.025 in. (0.254–0.635 mm) is not an acceptable clearance.

Answer B is right because 0.008–0.016 in. (0.203–0.406 mm) is acceptable.

Answer C is wrong because 0.005–0.010 in. (0.127–0.254 mm) is not an acceptable clearance.

Answer D is wrong because 0.004–0.012 in. (0.101–0.304 mm) is not an acceptable clearance.

129. Answer A is wrong because the duo servo brake system uses different primary and secondary shoe linings, but Technician B is also right.

Answer B is wrong because the primary shoe in a servo-type brake system has a weaker return spring, but Technician A is also right.

Answer C is right because both Technician A and Technician B are right.

Answer D is wrong because both Technician A and Technician B are right.

130. **Answer A is right** because the maximum lining-to-rotor clearance is 0.005 in. (0.127 mm), and 0.003 in. (0.076 mm) is an acceptable clearance.

Answer B is wrong because the maximum lining-to-rotor clearance is 0.005 in. (0.127 mm), and 0.010 in. (0.254 mm) is excessive.

Answer C is wrong because the maximum lining-to-rotor clearance is 0.005 in. (0.127 mm), and 0.012 in. (0.304 mm) is excessive.

Answer D is wrong because the maximum lining-to-rotor clearance is 0.005 in. (0.127 mm), and 0.015 in. (0.381 mm) is excessive.

131. Answer A is wrong because the measurement being performed is lining thickness.

Answer B is wrong because the measurement being performed is lining thickness.

Answer C is right because the measurement being performed is lining thickness.

Answer D is wrong because the measurement being performed is lining thickness.

132. Answer A is wrong because a bell-mouth condition results when the drum diameter is greater at the edge of the drum next to the backing plate compared to the edge near the wheel.

Answer B is wrong because a concave brake drum results when the drum diameter is greater in the center compared to both edges.

Answer C is wrong because an out-of-round condition results when there is a variation in readings 180 degrees apart.

Answer D is right because a convex drum results when the diameter on a driveline parking brake drum is greater at the edges of the friction surface than in the center.

133. Answer A is wrong because the camshaft (item 28) moves the brake shoes outward against the drum during a brake application.

Answer B is wrong because the camshaft (item 28) moves the brake shoes outward against the drum during a brake application.

Answer C is wrong because the camshaft (item 28) moves the brake shoes outward against the drum during a brake application.

Answer D is right because the camshaft (item 28) moves the brake shoes outward against the drum during a brake application.

134. **Answer A is right** because the brake lining wear adjustment must be completed before the linkage adjustment.

Answer B is wrong because the lever adjustment must be completed after the linkage adjustment.

Answer C is wrong because Technician A is right and Technician B is wrong.

Answer D is wrong because Technician A is right and Technician B is wrong.

135. Answer A is wrong because the parking brake adjustment in Figure 4–25 is the lining wear adjustment.

Answer B is right because the parking brake adjustment in Figure 4–25 is the lining wear adjustment.

Answer C is wrong because the parking brake adjustment in Figure 4–25 is the lining wear adjustment.

Answer D is wrong because the parking brake adjustment in Figure 4–25 is the lining wear adjustment.

136. Answer A is wrong because a restricted one-way check valve in the vacuum hose to the hydrovac unit causes excessive brake effort, but Technician B is also right.

Answer B is wrong because a sticking power piston in the hydrovac unit could also cause excessive brake effort, but Technician A is also right.

Answer C is right because both Technician A and Technician B are right.

Answer D is wrong because both Technician A and Technician B are right.

137. Answer A is wrong because the pressure for brake power assist is supplied by the power steering pump or a dedicated pump.

Answer B is right because the electrohydraulic pump should not be running during a normal brake application. This pump should only run if pressure from the power steering pump or dedicated pump is not available.

Answer C is wrong because Technician A is wrong and Technician B is right.

Answer D is wrong because Technician A is wrong and Technician B is right.

138. **Answer A is right** because if the electrohydraulic pump fails, there is no hydraulic pressure for the brake booster without the engine running.

Answer B is wrong because if the electrohydraulic pump fails, voltage is still supplied to the brake booster electrical circuit.

Answer C is wrong because fluid pressure will still flow from the power steering pump through the steering gear if the electrohydraulic pump fails.

Answer D is wrong because hydraulic pressure from the power steering pump or dedicated pump will still be available for brake boosting with the engine running.

139. Answer A is wrong because the electrical system voltage is supplied to the brake booster relay contacts directly from the battery positive terminal.

Answer B is right because the electrical system voltage is supplied to the brake switch at all times.

Answer C is wrong because the brake booster relay winding is connected to ground through the flow switch contacts.

Answer D is wrong because the alternator voltage signal informs the alarm/brake booster module if the engine is running.

140. Answer A is wrong because the rear stoplights and signal lights share the same bulb filaments.

Answer B is right because during a left turn with the brakes applied, the right rear stoplight remains illuminated.

Answer C is wrong because the stoplight switch is connected in series with the signal light switch.

Answer D is wrong because when the ignition switch is on and no turn is indicated, there is continuity through the signal light switch.

141. Answer A is wrong because an open ground circuit from the right rear light to the ground block would also cause the taillight to be inoperative.

Answer B is right because an open wire from the signal light switch to the right rear signal and stoplight would cause these lights to be inoperative.

Answer C is wrong because an open wire from the ground block to the battery ground in the engine compartment would cause all the rear lights to be inoperative.

Answer D is wrong because an open wire from the stoplight switch to the signal light switch would cause all the signal lights and stoplights to be inoperative.

142. **Answer A is right** because you put a light coating lubricant on the spindle and place a seal protector over the threads.

Answer B is wrong because rear wheel hub assemblies do use oil filler plugs to fill the wheel cavities.

Answer C is wrong because Technician A is right and Technician B is wrong.

Answer D is wrong because Technician A is right and Technician B is wrong.

143. Answer A is wrong because removing the stud nuts before the rim clamps are loosened on demountable rims may result in personal injury.

Answer B is right because the rims should be checked for wobble after the stud nuts are tightened to the specified torque.

Answer C is wrong because the rim spacer should be installed on spoke wheels by driving it straight into place.

Answer D is wrong because on demountable rims, the stud nut being tightened should be at the top of the rim.

144. Answer A is wrong because the bearing damage shown in Figure 4–30, F is brinelling.

Answer B is right because the bearing damage shown in Figure 4–30, F is brinelling.

Answer C is wrong because the bearing damage shown in Figure 4–30, F is brinelling.

Answer D is wrong because the bearing damage shown in Figure 4–30, F is brinelling.

145. **Answer A is right** because high-quality, high-temperature grease must be used when repacking is necessary.

Answer B is wrong because a wheel bearing is not transferable from one race to another.

Answer C is wrong because Technician A is right and Technician B is wrong.

Answer D is wrong because Technician A is right and Technician B is wrong.

146. **Answer A is right** because you torque the wheel bearing adjusting nut 50 ft.-lb. then back off the nut 1/6 to 1/3 turn and install the lock ring.

Answer B is wrong because not backing off the nut will cause the bearing to overheat. The nut must be backed off to allow for thermal expansion.

Answer C is wrong because Technician A is right and Technician B is wrong.

Answer D is wrong because Technician A is right and Technician B is wrong.

147. Answer A is wrong because an acceptable hub end play is 0.003 in. (0.076 mm).

Answer B is wrong because an acceptable hub end play is 0.003 in. (0.076 mm).

Answer C is wrong because an acceptable hub end play is 0.003 in. (0.076 mm).

Answer D is right because an acceptable hub end play is 0.003 in. (0.076 mm).

5 Steering and Suspension

Pretest

The purpose of this pretest is to determine the amount of review you may require before taking the ASE Medium/Heavy Truck Suspension and Steering Test. If you answer all the pretest questions correctly, complete the questions and study the information in this chapter to prepare for this test.

If two or more of your answers to the pretest questions are incorrect, complete a study of all the chapters in *Today's Technician Medium/Heavy-Duty Truck Steering and Suspension Systems* Classroom and Shop Manuals published by Delmar Publishers, plus a study of the questions and information in this chapter.

The pretest answers are located at the end of the pretest; these answers also are in the answer sheets supplied with this book.

1. A vehicle exhibits a steering shimmy below 30 mph. *Technician A* says the cause could be a loose pitman arm on the steering gear output shaft. *Technician B* says a misadjusted drag link could cause the steering gear to operate off center. Who is right?
   A. A only
   B. B only
   C. Both A and B
   D. Neither A nor B

2. Too much worm gear end play can cause:
   A. lack of lubrication.
   B. steering fluid leak.
   C. loss of motion within the steering gear.
   D. hard steering condition during cold operation.

3. *Technician A* says it is necessary to change the length of the drag link to correct for an off-center steering gear. *Technician B* says that before drag link adjustment you should check ball joint motion in each end of the drag link tie-rods. Who is right?
   A. A only
   B. B only
   C. Both A and B
   D. Neither A nor B

4. When you are adjusting the tension of the power steering belt, you notice contact at the bottom of the pulley with the belt. You should:
   A. decrease tension to the original specification level and recheck the contact surface for excessive wear.
   B. measure the level of deflection at a section of the belt that is longest between pulleys.
   C. replace the belt.
   D. ignore it because some applications require this additional contacting area.

5. A vehicle suddenly veers to the right or left after striking a bump with the front wheels. Which of these is the LEAST likely cause?
   A. Loose idler arm
   B. Damaged relay rod
   C. Worn tie-rod end
   D. Wheel out of balance

6. A broken leaf spring center bolt can cause which of the following conditions?
   A. Wheel imbalance
   B. Leaf spring degradation
   C. Premature bushing wear
   D. Premature tire wear

7. *Technician A* says a broken leaf spring may cause an off-level vehicle attitude. *Technician B* says if a bushing is not relaxed during assembly, a binding condition can leave a vehicle off level. Who is right?
   A. A only
   B. B only
   C. Both A and B
   D. Neither A nor B

8. A bent torque rod can cause:
   A. wheel alignment concerns.
   B. improper vehicle attitude.

C. premature spring shackle wear.
D. unequal axle shaft preload.

9. An improperly adjusted air suspension height control valve can cause:
   A. an offset in vehicle attitude.
   B. a hissing noise during compressor operation.
   C. backpressure in the supply line.
   D. a leak causing intermittent condition.

10. The two types of equalizing (walking) beam suspensions are:
    A. air spring and spring.
    B. leaf spring type and rubber load cushion type.
    C. torsion bar and load leveling.
    D. progressive and auxiliary spring.

11. In Figure 5–1 component X is the:
    A. torque rod.
    B. leaf spring.
    C. composite spring.
    D. equalizing beam.

12. In Figure 5–2 the Technician is using a tram bar to:
    A. measure wheel bearing end play.
    B. compare the left axle position with the right axle position.
    C. adjust rear toe-in.
    D. measure rear wheel runout.

13. *Technician A* says that if the air seat cushion has a leak, the pressure reduction valve closes to protect the air brake system from air loss. *Technician B* says the pressure protection valve protects the air seat cushion, if the air brake system has excessive pressure. Who is right?
    A. A only
    B. B only
    C. Both A and B
    D. Neither A nor B

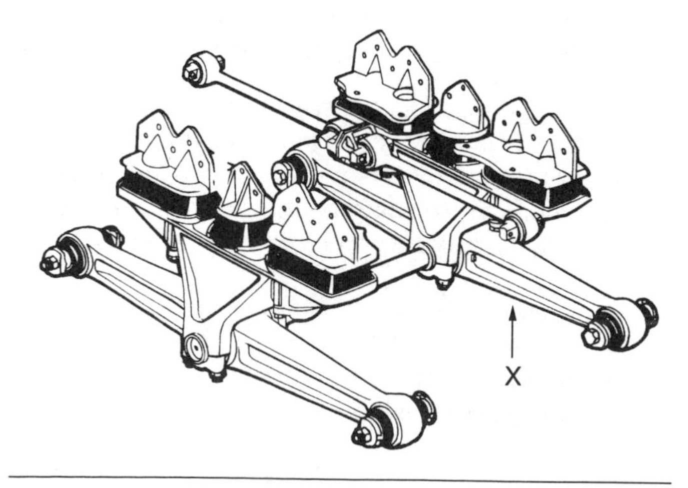

Figure 5–1 Rear suspension system.

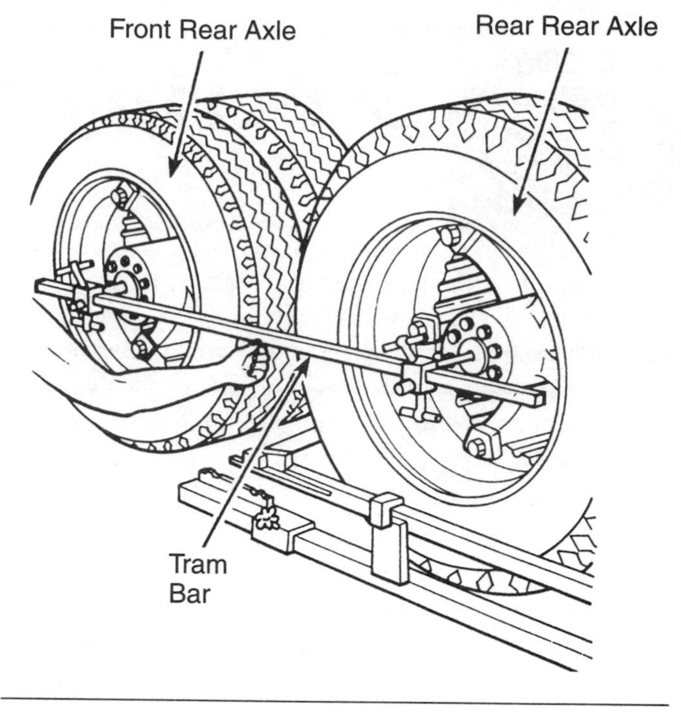

Figure 5–2 Rear axle measurement.

14. The power steering system is being drained and flushed. *Technician A* says you should disconnect the return hose from the gear to the remote reservoir at the remote reservoir to drain the fluid. *Technician B* says that when the fluid begins to discharge from the return hose, shut the engine off. Who is right?
    A. A only
    B. B only
    C. Both A and B
    D. Neither A nor B

Answers to Pretest

1. C, 2. C, 3. C, 4. C, 5. D, 6. D, 7. C, 8. A, 9. A, 10. B, 11. D, 12. B, 13. D, 14. C

Steering System Diagnosis and Repair

Steering Column

Task 1

Diagnose fixed and driver adjustable steering column and shaft noise looseness and binding problems; determine needed repairs.

1. All the following are parts of the steering column EXCEPT:
   A. U-joints.
   B. turn signal switch.
   C. drag link.
   D. boot seal.

2. Which of these will result in excessive play in the mounted steering column assembly on a truck equipped with a tilt steering column?
   A. Faulty anti-lash spring in the centering spheres
   B. Upper bearing race missing
   C. Upper bearing not seating in the race
   D. Loose support lock shoe pin.

3. Two Technicians are discussing steering column connection. *Technician A* says that on some tilt steering columns, the upper steering shaft is connected to the lower steering shaft by a universal joint. *Technician B* says two pivot bolts connect the main housing and the support bracket. Who is right?
   A. A only
   B. B only
   C. Both A and B
   D. Neither A nor B

4. A steering shaft is kept centered in the steering column housing by the:
   A. steering wheel.
   B. upper and lower column bushings.
   C. steering shaft U-joints.
   D. steering column support bracket.

Hint

A worn front steering bushing leaves a gap in the linkage that can cause a loose feeling in the steering wheel when driving straight, noise when hitting a bump in the road because the bushing acts as a cushion, and a pull to the left or right while driving because the wheels will drift to the pitch of the highway.

Task 2

Inspect and replace steering shaft Ujoint(s), slip joints, bearings, bushings, and seals; phase shaft Ujoints.

5. A truck exhibits excessive steering wheel free play. *Technician A* says either a worn steering shaft universal joint or an out-of-adjustment steering gear could cause this free play. *Technician B* says that the cause could be either a bent idler arm, or no lubricant in the steering gear. Who is correct?
   A. A only
   B. B only
   C. Both A and B
   D. Neither A nor B

6. A rattling noise occurs in the steering column when driving on rough roads. *Technician A* says the steering shaft may be bent. *Technician B* says the U-joint in the steering shaft may be worn. Who is right?
   A. A only
   B. B only
   C. Both A and B
   D. Neither A nor B

Hint

All the following are parts of the steering column. Steering systems use U-joints to allow the shafts to mount on an angle. The turn signal switch mounts to the left side of the steering column and the boot seal the steering column where it passes through the body. Loose steering wheel to steering shaft fasteners would cause excessive play and poor steering response.

You should not compress the column steering U-joint with a C-clamp or large pliers when testing a linkage joint for excessive wear. This process will compress the tension spring in the joint and give the impression that there is wear in the joint. Simply push against the joint with the force you can create with your hands. This should be enough to identify excessive wear.

Task 3

Check and adjust cab mounting and ride height.

7. Which of the following statements is correct regarding a cab air suspension?
   A. It has three springs.
   B. It has a leveling valve.
   C. It is used with rubber cab mounts.
   D. It does not have shock absorbers.

8. *Technician A* says air pressure is supplied to an air-suspended seat through a pressure protection valve and a pressure reduction valve. *Technician B* says many cab air suspensions contain two air springs, two shock absorbers, and a leveling valve. Who is right?
   A. A only
   B. B only
   C. Both A and B
   D. Neither A nor B

Hint

Many cab air suspensions contain two air springs, two shock absorbers, and a leveling valve. The leveling valve maintains the proper air pressure in the air springs to provide the correct cab height. You adjust the height control or leveling valve (similar to air suspension systems) to maintain the proper cab height. An air-suspended seat reduces road shock transferred through the chassis, cab, and seat to the driver. Air pressure is supplied to the air-suspended seat through a pressure protection valve and a pressure reduction valve.

Steering Units

Task 1

Diagnose power steering system noise, steering binding, turning effort, looseness, hard steering, overheating, fluid leakage, and fluid aeration problems; determine needed repairs.

9. A truck is experiencing high steering effort. *Technician A* says a sheared shift tube could be the cause. *Technician B* says the column assembly is misaligned. Who is right?
   A. A only
   B. B only
   C. Both A and B
   D. Neither A nor B

10. The power steering on a truck is overheating. *Technician A* says the air passages through the power steering cooler may be restricted. *Technician B* says the power steering return line may be restricted. Who is right?
   A. A only
   B. B only
   C. Both A and B
   D. Neither A nor B

11. *Technician A* says foaming in the remote reservoir may indicate air in the power steering system. *Technician B* says some truck manufacturers recommend checking the power steering fluid level at operating temperature. Who is right?
   A. A only
   B. B only
   C. Both A and B
   D. Neither A nor B

Hint

Installing the steering wheel in the wrong position on the steering shaft causes the steering gear to be off center, causing premature wear to the steering gear. A bent worm gear in the power steering gear can cause a feeling of erratic torque every 360 degrees. Excessive worm gear end play can cause lost motion within the steering gear. Power steering system overheating is caused by the following: underlubricated ball joints, a kink or pinch in the fluid return line, or blocked airflow across the heat exchanger. A dry kingpin pivot bearing will cause a squeak when turning left or right.

A loose sleeve clamp on the drag link adjuster with damaged adjusting threads will cause the steering wheel to become more off center over the course of 500 miles. A worn steering shaft universal joint or an out-of-adjustment steering gear could cause steering wheel free play. Defective kingpins or kingpin bearings will cause a wheel and tire not to return to the straight-ahead position during a front axle and linkage-binding test. Excessive positive caster will cause high steering effort and fast steering wheel return.

Reservoir O-rings, drive shaft seals, high-pressure fittings, and dipstick cap are all possible leak sources. If leaks occur at any of the seal locations, replace the seal. When a leak is present at the high-pressure fitting, first tighten it to the specified torque. If this fails, replace the O-ring at this fitting.

Task 2

Determine recommended type of power steering fluid; check level and condition; determine needed service.

12. All of these could be the source of leaking power steering fluid EXCEPT:
   A. lower sector shaft seal.
   B. submersed style pump-to-reservoir surface.
   C. supply line double-flare fitting.
   D. steering gear input shaft.

13. A routine inspection shows discoloration of power steering fluid. All of the following could cause this EXCEPT:
   A. the wrong type of fluid.
   B. mixed brands of fluid.
   C. water mixed with fluid.
   D. overheated or burned smell condition.

14. Most vehicle manufacturers recommend checking the level of power steering fluid with the system at a working temperature of:
   A. 70°F (21°C).
   B. 90°F (32°C).
   C. 120°F (49°C).
   D. 175°F (79°C).

Hint

Most OEMs recommend the use of power steering fluid or automatic transmission fluid in power steering systems. Low fluid level will cause increased steering effort and erratic steering. It may also cause a growling or cavitation noise in the pump. Foaming in the remote reservoir may indicate air in the power steering system. Most OEM truck manufacturers recommend checking the power steering fluid level at operating or working temperature of 175°F (79°C).

With the engine at 1,000 rpm or less, turn the steering wheel slowly and completely in each direction several times to raise the fluid temperature. Check the reservoir for foaming as a sign of aerated fluid. The fluid level in the reservoir should be at the hot full mark on the dipstick.

Task 3

Flush and refill power steering system; purge air from system.

15. A heavy-duty truck has erratic power steering effort. *Technician A* says that some of the knuckle pin bushings may be dry of lubricant. *Technician B* says there may be air in the power steering system. Who is right?
    A. A only
    B. B only
    C. Both A and B
    D. Neither A nor B

16. A power steering system has been flushed and refilled. *Technician A* says air should be bled from the system by loosening the return hose on the remote reservoir and running the engine. *Technician B* says air should be bled from the system by connecting the power steering analyzer to the power steering system and closing the gate valve for 5 seconds. Who is right?
    A. A only
    B. B only
    C. Both A and B
    D. Neither A nor B

Hint

When you drain and flush the power steering system, disconnect the return hose from the remote reservoir to drain the fluid. When the fluid begins to discharge from the return hose, shut the engine off. After the power steering system has been drained and refilled, the steering wheel should be turned fully in both directions with the engine running to bleed air from the system.

Task 4

Perform power steering system pressure and flow tests; determine needed repairs.

17. A vane-type power steering pump is being tested and the pressure gauge reads low system pressure with the shutoff (load) valve closed. *Technician A* says the pressure relief valve may be frozen open. *Technician B* says the pump vanes may be sticking in their slots. Who is right?
    A. A only
    B. B only
    C. Both A and B
    D. Neither A nor B

18. When using the power steering analyzer in Figure 5–3 to test power steering pump pressure and flow with the engine running, the gate valve should never be closed for more than:
    A. 2 seconds.
    B. 10 seconds.
    C. 15 seconds.
    D. 20 seconds.

19. With the power steering analyzer connected, the gate valve wide open, and the engine idling at the specified rpm, the power steering pressure is higher than specified. The cause of this problem could be:
    A. a restricted power steering return line.
    B. a restricted power steering high-pressure line.
    C. worn rings on the rotary valve in the steering gear.
    D. loose worm shaft bearing adjustment.

Hint

The power steering analyzer should be connected from the power steering pump high-pressure hose to the steer-

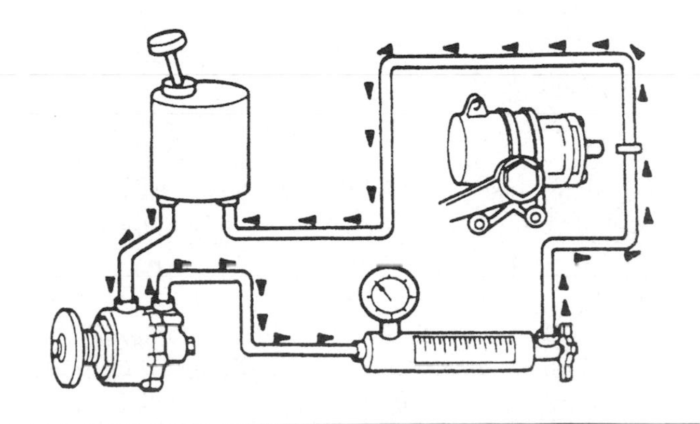

Figure 5–3 Power steering analyzer. *(Courtesy of Sterling Truck Corporation)*

ing gear. The load (gate) valve should be open at the beginning of the power steering pump pressure and flow test. Start the engine, run at idle speed, and observe the pressure and flow on the analyzer. If the power steering pump flow is less than specified at the specified pressure (usually 200 psi [1,379 kPa]), check the high-pressure hose for restrictions. If the pressure is above 200 psi (1,379 kPa) at idle with the load valve open, check the return line for restrictions. If the flow rate is less than 2 gallons per minute (gpm), the pump or the flow control may be defective.

Rotate the gate (load) valve toward a closed position until the pump pressure rises to 700 psi (4,826 kPa). If the flow rate is 1 gpm less than recorded with the gate (load) valve open, the pump or flow control valve are bad.

Task 5

Inspect, service, or replace power steering reservoir including filter, seals, and gaskets.

20. Two Technicians are discussing power steering reservoirs. *Technician A* says that prying on the reservoir may damage or puncture the reservoir. *Technician B* says that a power steering system may have either an integral or a remote fluid reservoir. Who is right?
    A. A only
    B. B only
    C. Both A and B
    D. Neither A nor B

21. All these statements about power steering reservoirs are true EXCEPT:
    A. a remote reservoir is connected in the power steering return line.
    B. a remote reservoir may contain a filter.
    C. the remote reservoir must contain the specified fluid level.
    D. the remote reservoir contains a power steering fluid cooler.

Hint

Prying on the reservoir to adjust the pump drive belt may damage or puncture the reservoir. The power steering system may have either an integral or a remote fluid reservoir. The source of leaking power steering fluid could be a lower sector shaft seal, a submersed style pump-to-reservoir surface, or a supply line double-flare fitting. To remove the remote reservoir, use the following steps:

1. *Stop the engine and remove the cover; using a suction gun remove the fluid from the reservoir.*
2. *Use the appropriate steps to drain fluid out of the hoses and remove the reservoir.*
3. *Remove the spring, filter cap, and filter from the bottom of the reservoir.*
4. *Install a new filter, filter cap, and spring.*
5. *Install the reservoir and torque to specifications.*
6. *Fill and bleed the system.*

Task 6

Inspect, adjust, align, or replace power steering pump belt(s) and pulley(s).

22. All of the following could cause the power steering pump pulley to become misaligned EXCEPT:
    A. an overpressed pulley.
    B. a loose fit from the pulley hub to pump shaft.
    C. a worn or loose pump mounting bracket.
    D. a broken engine mount.

23. Close visual inspection of the power steering belt(s) can reveal all the following EXCEPT:
    A. proper belt tension.
    B. premature wear due to misalignment.
    C. correct orientation of dual belt application.
    D. proper belt seating in the pulley.

24. *Technician A* says elongated mounting holes in the power steering pump bracket may cause a noise while in operation. *Technician B* says worn holes in the power steering pump-mounting bracket could cause premature belt wear. Who is right?
    A. A only
    B. B only
    C. Both A and B
    D. Neither A nor B

Hint

An overpressed pulley, a loose fit from pulley hub to pump shaft, or worn or loose pump-mounting brackets could cause the power steering pump pulley to become misaligned. Close visual inspection of the power steering belt(s) cannot reveal proper belt tension. A belt tension gauge should be used to measure for proper tension. You align the V-belt with the pulley with the sides contacting the pulley, and the lower edge should never touch the bottom of the pulley groove.

Task 7

Inspect, replace as required, power steering pump drive gear and coupling.

25. You are replacing the power steering pump drive on a truck with a two-stroke diesel engine. *Technician A* says it is not necessary to retime the engine when this replacement process is complete. *Technician B* says you inspect the gear for worn or chipped teeth. Who is right?
    A. A only
    B. B only
    C. Both A and B
    D. Neither A nor B

26. All of these tasks are performed when removing and installing a gear-driven power steering pump EXCEPT:
    A. check the pump mounting holes for wear.
    B. remove the hoses from the pump and cap the fittings.
    C. reuse the O-ring if it is in good condition.
    D. bleed the air from the power steering system.

Hint

When you are replacing the power steering pump drive on a truck with a two-stroke diesel engine, it is not necessary to retime the engine when this replacement process is complete. Further, you inspect the driven gear for worn or chipped teeth. When removing and installing a gear-driven power steering pump, check the pump-mounting holes for wear, remove the hoses from the pump and cap the fittings, replace the O-ring, and bleed the air from the power steering system.

Task 8

Inspect, adjust, or replace power steering pump, mountings, and brackets.

27. A belt-driven power steering pump wears out the drive belt in 10,000 miles (16,000 km), but power steering operation is normal. The cause of this problem could be:
    A. a restriction in the high-pressure power steering hose.
    B. a restricted power steering filter.
    C. worn power steering pump mounting boltholes.
    D. worn rings on the steering gear rotary valve.

28. To properly inspect a power steering gear before removal, you should perform all of these EXCEPT:
    A. rotate the input shaft and visually determine if it is true.
    B. clean and inspect for evidence of fluid leakage.
    C. check all mounting fasteners.
    D. adjust the truck toe.

29. When inspecting and adjusting a power steering pump drive belt:
    A. place the belt tension gauge close to the power steering pump pulley.
    B. use a prybar placed against the top of the reservoir to tighten the belt.
    C. loosen the power steering pump-mounting bolts when tightening the belt.
    D. there is no need to replace the belt if the bolt is at the end of the adjusting slot in the pump bracket.

Hint

As the belt rotates the rotor and vanes inside the cam ring, centrifugal force causes the vanes to slide out of the rotor slots. The vanes follow the elliptical surface of the cam ring. When the area between the vanes expands, a low-pressure area occurs between the vanes and fluid flows from the reservoir through the two suction ports into the space between the vanes. As the vanes approach the higher portion of the cam at the two outlet ports, the cam ring pushes the blades inward and the area between the vanes becomes smaller. This action pressurizes the fluid and the fluid is forced out the two discharge ports.

The flow control valve is positioned so some of the fluid discharged from the pump is directed through this valve back to the reservoir. Spring tension and fluid pressure determine the flow control valve position. With low fluid flow, the pressure is high in the outlet fitting venturi. The higher pressure is applied to the spring side of the flow control valve and helps the spring keep this valve nearly closed. With high fluid flow, the pressure in the venturi is low. When the speed increases, the fluid flow increases and creates a decrease in pressure at the venturi. This pressure decrease is sensed at the spring side of the venturi, which allows the pump discharge pressure to force the flow control valve partially open.

You must disassemble the power steering pump to determine the extent of the damage due to overheating. A damaged power steering pump can be identified by score marks in the pump drive gear. A loose steering gear mount could be binding the steering column and cause the power steering to become increasingly harder to turn. Elongated mounting holes in the power steering pump bracket may cause a noise while in operation. Worn holes in the power steering pump-mounting bracket could cause premature belt wear.

Task 9

Inspect and replace power steering system cooler, lines, hoses, and fittings.

30. *Technician A* says you should check the power steering cooler air passages and fins for restrictions

during any PM process. *Technician B* says one should check all power steering hoses for leaks, cracks, dents, and sharp bends. Who is right?

A. A only
B. B only
C. Both A and B
D. Neither A nor B

31. Restricted power steering pump cooler fins may cause:

A. premature power steering pump wear.
B. reduced fluid flow through the power steering system.
C. reduced power steering fluid temperature.
D. deterioration of the power steering filter.

Hint

Power steering fluid flows through the steering gear and then it flows through a hose to the cooler or coolers. You should check the power steering cooler air passages and fins for restrictions during any preventive maintenance (PM) process. You should check all power steering hoses for leaks, cracks, dents, and sharp bends. Further, you torque all pressure fittings to specifications before replacing the O-ring.

Task 10

Inspect, adjust, or replace linkage-assist type power steering cylinder or gear (dual system).

32. The linkage-assist type power steering in Figure 5–4 binds when turning corners in either direction, but with short steering corrections, wheel recovery is normal. Which of these is the LEAST likely cause?

A. Worn kingpins
B. Improper sector lash adjustment
C. A bent worm gear
D. Worn-out tie-rod end assembly

33. The linkage-type power steering system in Figure 5–4 requires more steering effort during a right turn compared to a left turn. The cause of this problem could be:

A. worn, dry knuckle pin bushings.
B. worn, dry drag link ball studs.
C. a defective or misadjusted integral valve.
D. low power steering pump pressure.

Hint

Worn kingpins, improper steering gear mesh prelude, or a bent worm gear will cause a linkage-assist type power steering to bind when turning corners, but with short steering corrections, wheel recovery is normal. Low lubricant level in the steering gear can cause the manual steering gear assembly of a linkage-assist type power steering to become noisy when turning the steering wheel.

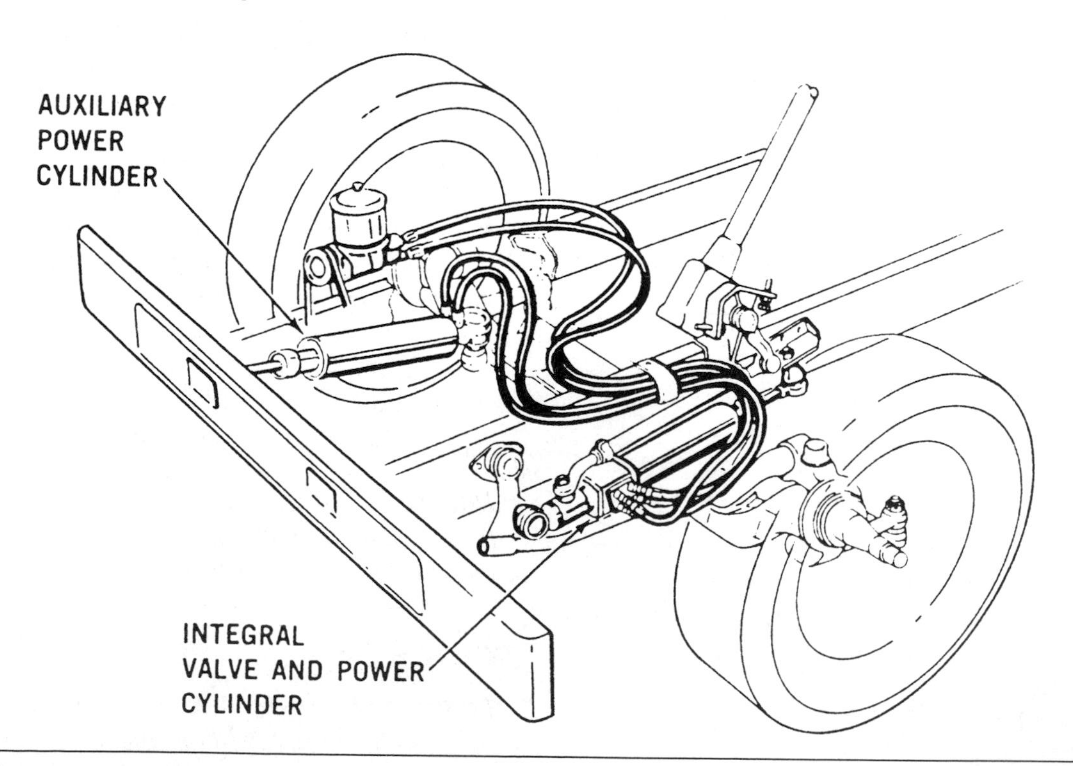

Figure 5–4 Linkage-type power steering system. *(Reprinted with permission from SAE publication SP 374 c 19 The Truck Steering System from Hand Wheel to Road Wheel, Society of Automotive Engineers)*

Task 11

Inspect, adjust, repair, or replace integral-type power steering gear.

34. Improper worm gear bearing preload is determined to be the cause of a binding steering condition in an integral power steering gear. *Technician A* says you should center the sector shaft to measure the correct bearing preload. *Technician B* says both the input and output shafts should be disconnected to measure correct preload torque. Who is right?
    A. A only
    B. B only
    C. Both A and B
    D. Neither A nor B

35. In order to identify the true center of a steering gear pitman shaft, you must:
    A. disassemble the steering gear assembly and mark with paint dot.
    B. drain all lubricant from the gear housing for access.
    C. determine backlash areas on either side of worm gear.
    D. rotate input shaft until full mesh occurs between the worm and sector gears.

36. A driver complains that the power steering has become increasingly harder to turn over the last 1,000 miles. *Technician A* says you may need to replace the system filter because of clogging. *Technician B* says the problem may be a loose steering gear mount that could be binding the steering column. Who is right?
    A. A only
    B. B only
    C. Both A and B
    D. Neither A nor B

37. A power steering system is overheating. The cause of this problem could be:
    A. a worn steering shaft flex joint.
    B. blocked airflow across the power steering cooler.
    C. loose tie-rod ends.
    D. loose sector lash adjustment.

38. The driver will complain about which of the following conditions when the power steering gear poppet valves are misadjusted?
    A. Steering wheel kick
    B. Reduced wheelcut
    C. Directional pull
    D. Nonrecovery

Hint

Both the input and output shafts should be disconnected to measure steering gear preload torque. This process prevents binding in the steering column or linkage from affecting the worn bearing preload and sector lash adjustments. To center the steering gear, turn the worm shaft from stop to stop and count the number of turns. Starting at one end of worm shaft travel, turn the worm shaft back half the number of turns from stop to stop. This action centers the steering gear in preparation for the sector lash adjustment.

When the front wheels are turned in a power steering gear, torsion bar deflection moves the spool valve inside the rotary valve. This valve movement directs power steering fluid to the appropriate side of the piston to provide steering assist. In a TRW/Ross power steering gear, if the steering wheel is turned in either direction so that the steering linkage is 1/3 of a turn from the stops, a poppet valve in the piston opens. The power steering fluid pressure on the piston is then released past the poppet valve. This action stops the power steering assist. Misadjusted poppet valves cause reduced wheel cut.

Steering Linkage

Task 1

Inspect and replace pitman arm.

39. The pitman arm connects the:
    A. steering gear to the drag link.
    B. steering gear to the left upper steering arm.
    C. steering arm to the tie-rod.
    D. drag link to the tie-rod.

40. When replacing a steering or pitman arm, you should perform all of the following EXCEPT:
    A. road test when repairs are completed.
    B. replace both outer tie-rod ends.
    C. check and correct for changes in wheel alignment.
    D. lube the replacement part after installation.

Hint

A loose pitman arm on the steering gear output shaft could cause a steering shimmy below 30 mph. When you replace a steering or pitman arm, you should perform all of the following: a road test when repairs are completed, check and correct for changes in wheel alignment, and lube the replacement part after installation. The pitman arm can be directly responsible for directional stability. A damaged pitman arm can be directly responsible for the steering wheel being off center.

Task 2

Inspect, adjust, or replace drag link (relay rod) and tie-rod ends (ball and adjustable socket type).

41. To correctly center an off-center steering wheel it is necessary to:
    A. measure the toe-out-on-turns and replace the bent steering arm.
    B. adjust the length of the tie-rod between the two steering arms.
    C. remove the steering wheel and reposition it at the input shaft, then retorque to specification.
    D. adjust the length of the drag link then reset toe-in.

42. A driver notices over the course of 500 miles that the steering wheel is increasingly becoming more off center. *Technician A* says a loose sleeve clamp on the drag link adjuster may have damaged adjusting threads. *Technician B* says a damaged steering component from a collision may be the cause. Who is right?
    A. A only
    B. B only
    C. Both A and B
    D. Neither A nor B

Hint

A loose sleeve clamp on the drag link adjuster may allow enough play in the sleeve to damage the adjusting threads on the steering arm, enough to change the steering wheel position. Tie-rod clamp interference with the I-beam on a full turn can cause a noise when turning over bumps. A misadjusted drag link could cause the steering gear to operate off-center and cause a steering wheel shimmy. To center an off-center steering wheel, you adjust the length of the drag link then reset toe-in. You need to thread the tie-rod end in beyond the split. Before you adjust the drag link, check ball joint motion in each end of the drag link tie-rods. A worn relay rod socket or an out-of-adjustment tie-rod may cause a wandering condition while driving.

Task 3

Inspect and replace idler arm, bearings, and bushings.

43. When performing the measurement in Figure 5–5:
    A. a horizontal load of 40 pounds should be applied to the idler arm.
    B. a horizontal load of 75 pounds should be applied to the idler arm.
    C. a vertical load of 25 pounds should be applied to the idler arm.

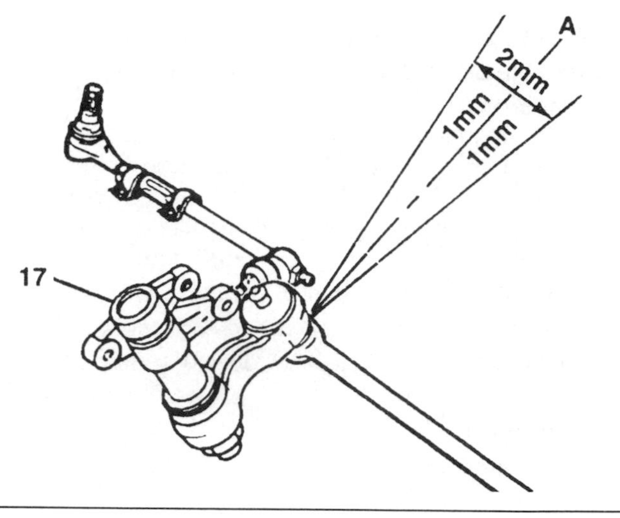

Figure 5–5 Idler arm measurement. *(Courtesy of General Motors Corporation, Service Technology Group)*

    D. a vertical load of 60 pounds should be applied to the idler arm.

44. Worn idler arm bushings and excessive idler arm movement results in:
    A. sudden veering in one direction when the front wheel strikes a bump.
    B. a squeaking noise when turning the front wheels in either direction.
    C. excessive steering effort in either direction.
    D. improper steering wheel return after a turn.

Hint

You never grease an idler arm or bushing before inspection because the grease may hide joint wear. Hard steering, a squawking noise when the wheels turn, and poor steering wheel return are all the results of bad idler arm bushings. A worn front steering bushing can cause all of the following: loose feeling in steering wheel when driving straight, noise when hitting a bump in the road, or a pull to the left while driving.

Task 4

Inspect and replace steering arm and levers and linkage pivot joints.

45. *Technician A* says the tie-rod clamps must be positioned so the slot in the clamp is aligned with the slot in the tie-rod. *Technician B* says the length of the tie-rod determines the front wheel toe. Who is right?
    A. A only
    B. B only
    C. Both A and B
    D. Neither A nor B

46. A damaged or bent steering arm in the front knuckle can cause which of the following conditions?
    A. A change in the steering wheel position
    B. A change in total toe measurement
    C. A change in individual toe measurement
    D. A pull condition when driving

47. When installing the tie-rod end in Figure 5–6, *Technician A* says position A is the correct position for the tie-rod end. *Technician B* says that you need to thread the tie-rod end in beyond the split in position B. Who is right?
    A. A only
    B. B only
    C. Both A and B
    D. Neither A nor B

Hint

A damaged or bent steering arm can cause a change in the steering wheel position. The drag link is the adjustment point to center the steering wheel. Replacing both the tie-rod and knuckle then resetting the toe-in is an effective repair when a tie-rod end is loose in the taper of the steering knuckle. To test a steering linkage joint for excessive wear, you should simply push against the joint with force that you can create with your hands, which should be enough to identify excessive wear.

Task 5

Inspect or replace clamps and retainers; position as needed.

48. When installing the tie-rod end in Figure 5–7:
    A. the front wheels should be turned fully to the right or left.

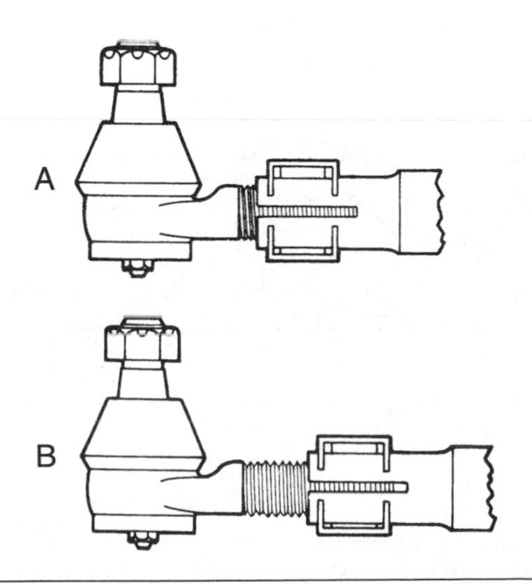

Figure 5–6 Installing tie-rod end.

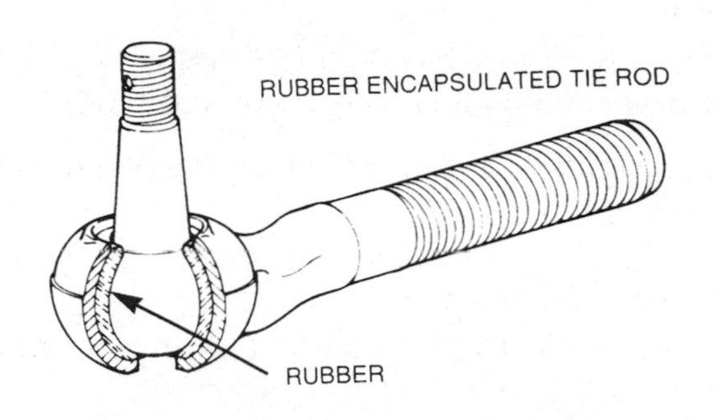

Figure 5–7 Tie rod end. *(Courtesy of Moog Automotive. Inc.)*

    B. the front wheels should be straight ahead.
    C. the front wheel should be turned part way to the right or left.
    D. thc front whcel should be raised off the shop floor.

49. In Figure 5–8, the drag link is connected to:
    A. item 9.
    B. item 21.
    C. item 29.
    D. item 30.

Hint

A tie-rod clamp that interferes with the I-beam on a full turn will cause a noise when turning over bumps. The length of the tie-rod determines the front wheel toe. Tie-rod clamps are installed with the clamping bolt opposite the split in the tie-rod cross tube. The Technician must completely insert the threaded portion of both tie-rod ends into the cross-tube split.

Task 6

Check and adjust steering linkage or wheel stops.

50. When wheel stops are missing, damaged, or out of specified adjustment range, they can cause:
    A. wheel imbalance.
    B. excessive turning radius.
    C. steering wheel to be off center.
    D. steering wheel binding on turns.

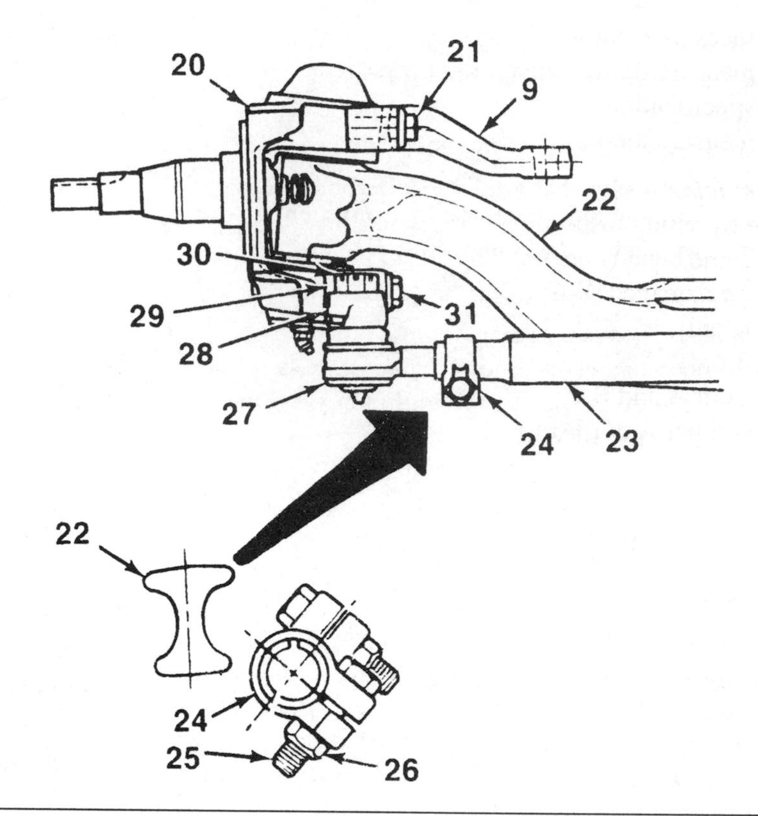

9. Steering Arm
20. Knuckle
21. Steering Arm Bolt
22. Axle
23. Tie Rod Tube
24. Tie Rod Clamp
25. Tie Rod Clamp Bolt
26. Tie Rod Clamp Nut
27. Tie Rod Ball Joint
28. Tie Rod Arm
29. Ball Joint Nut
30. Cotter Pin
31. Tie Rod Arm Bolt

Figure 5–8 Steering linkage components. *(Courtesy of General Motors Corporation, Service Technology Group)*

51. In Figure 5–9, component B is turned outward too much. This will result in:
    A. damage to component B.
    B. insufficient turning radius.
    C. a bent steering arm.
    D. improper front wheel toe.

Hint

When the wheel stops are missing, damaged, or out of specified adjustment range, they can cause excessive turning radius. You adjust the wheel stop bolts to correct turning radius.

Suspension System Diagnosis and Repair

Task 1

Inspect and replace front axle beam, control arms, and mounting hardware.

52. *Technician A* says that the lockbolt and draw key must be removed before the kingpin will disassemble from the knuckle. *Technician B* says after reassembly of the knuckle, you need to grease the kingpin. Who is right?
    A. A only
    B. B only
    C. Both A and B
    D. Neither A nor B

53. You suspect damage to a front axle from a collision. To determine if axle damage exists, you should:
    A. lift both front wheels and rotate the tires to measure for radial runout.

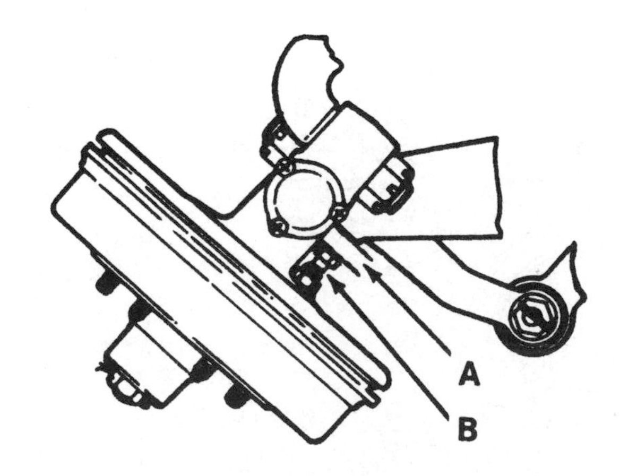

Figure 5–9 Steering adjustment. *(Courtesy of General Motors Corporation, Service Technology Group)*

B. measure front wheel setback.
C. measure tie-rod length and compare to specifications.
D. measure toe-in.

54. *Technician A* says that you can determine a bent axle by visual inspection. *Technician B* says that if the same bend is on both ends of the axle, it must not be damaged. Who is right?
A. A only
B. B only
C. Both A and B
D. Neither A nor B

Hint

An I-beam suspension uses a solid I-beam that is connected or sprung to the vehicle frame by multileaf springs through U-bolts. At the ends of this I-beam are kingpins, which connect the I-beam to the steering knuckle and act as pivots to turn the front wheels. An axle may be bent and not detectable to the eye. Measure front wheel setback to check for a bent axle, and use a machinist's protractor to check the axle for twists. Technicians measure the front wheel setback on both wheels and compare because a difference in measurement could be a bent axle.

Task 2

Inspect, service, adjust, or replace kingpin, steering knuckle bushings, locks, bearings, seals, and covers.

55. In the Figure 5–10 which component must you remove to access the kingpin?
A. Knuckle cap
B. Spindle
C. Ackerman arm
D. Axle

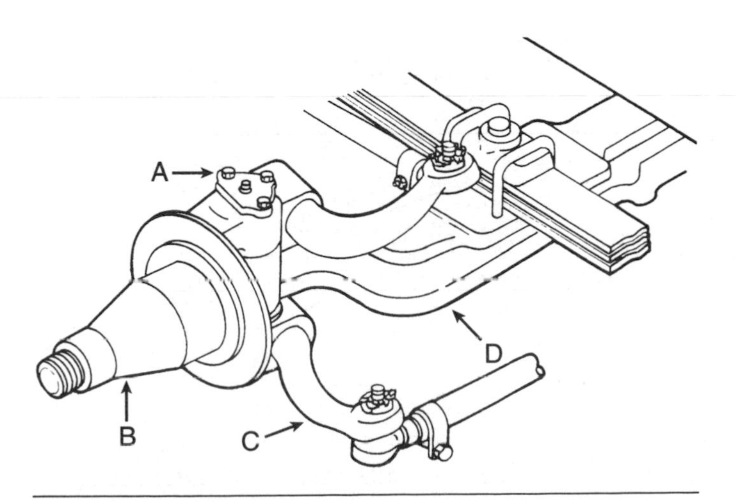

Figure 5–10 Steering knuckle assembly.

56. While replacing the steering knuckle bushings, a Technician finds there is interference during the kingpin installation. All of the following could be the cause EXCEPT:
A. right side kingpin switched with the left side kingpin.
B. burred bushing during installation.
C. improperly aligned bushing.
D. omission of reaming operation.

Hint

Some OEM manufacturers refer to the kingpin as a knuckle pin. The dial indicator will measure vertical play in the steering knuckle when the Technician pries the knuckle up and down. Remove the upper and lower knuckle caps to access the kingpins. Before pressing out the kingpin, remove the lockbolt and drive out the draw key. The draw key keeps the kingpin aligned. Kingpin bushings are the same on either side. A loose kingpin or kingpin bearing could cause a shimmy with slight vibrations. Grease the new kingpin after installation. When greasing a front axle kingpin, grease will leak out through the pivot bearing, which shows a thorough distribution of grease.

Task 3

Inspect and replace shock absorbers, bushings, brackets, and mounts; adjust shock absorbers where applicable.

57. Interference of a shock absorber outer tube to the reservoir can cause:
A. premature bushing wear.
B. scraping noise with travel.
C. main shaft seal failure.
D. excessive fatigue in mount assembly.

58. After a stretch of rough road is driven, the steering wheel continues to shake for a few seconds. Which of these could be the cause?
A. Leaking front shock absorbers
B. Low tire pressure
C. Rusted rear shock absorber
D. Missing jounce bumper

Hint

When a vehicle hits a bump, the wheel and suspension move upward in relation to the chassis. This causes jounce and rebound. Shock absorbers dampen or control spring action from jounce and rebound, reduce body sway, and improve directional stability and driver comfort. Worn-out shock absorbers will allow the front end to bounce, causing the steering wheel to shake for a few seconds. A shock absorber piston shield often gets bent or damaged and will scrape against the shock absorber when the suspension moves up and down. Shock absorbers stop spring oscillation, and they are still needed on air spring suspensions. It is advisable to replace shocks absorbers in pairs. Inspect shock absorbers with one end disconnected for resistance, bent pistons, bushing failure, and bent piston sleeves.

Task 4

Inspect, repair, or replace leaf springs, center bolts, clips, eye bolts and bushings, shackles, slippers, insulators, brackets, and mounts.

59. A truck has a history of breaking the center spring boltholes. *Technician A* says that loose spring shackles could be the cause. *Technician B* says that loose U-bolts could be the cause. Who is right?
    A. A only
    B. B only
    C. Both A and B
    D. Neither A nor B

60. *Technician A* says leaf spring shackles are necessary and allow free lateral movement of the spring. *Technician B* says you adjust leaf spring shackles to compensate for off-center static loads. Who is right?
    A. A only
    B. B only
    C. Both A and B
    D. Neither A nor B

61. A Technician hears a rattling noise on the right side of a medium truck when driving over bumps. Which of these could be the cause?
    A. An overtorqued shock absorber mount
    B. Underinflated right rear tire
    C. Broken mount fastener for a leaf spring
    D. Cracked air brake supply line

Hint

OEM manufacturers design both the leaf spring and rubber cushion types to lower the center of gravity of the axle load by placing the beam below the axle centerline. Loose spring shackles will not break the spring center bolts. Spring center bolts breakage is usually caused by loose spring U-bolts. A broken leaf spring center bolt will not cause wheel imbalance, spring degradation, or premature bushing wear. A broken leaf spring center bolt can cause rear axle shift leading to premature toelike tire wear.

Task 5

Inspect, adjust, or replace torque arms, bushings, and mounts.

62. On rear spring suspensions, *Technician A* says that some designs substitute torque rods for rear shock absorbers. *Technician B* says four to six torque rods are used on spring suspensions. Who is right?
    A. A only
    B. B only
    C. Both A and B
    D. Neither A nor B

63. All of the following apply to an axle alignment adjustment on a spring suspension with torque rods EXCEPT:
    A. adjustment is made through a lower adjustable torque rod.
    B. shims are used between the torque rod front and spring hanger bracket.
    C. adjustment is made with an eccentric bushing at the torque leaf.
    D. adjustment is made through the upper adjustable torque rods.

64. Refer to Figure 5–11. The distance between the front and rear tandem axle centers is 0.75 in. (19.05 mm) more on the left side compared to the right side. The cause of this problem could be:

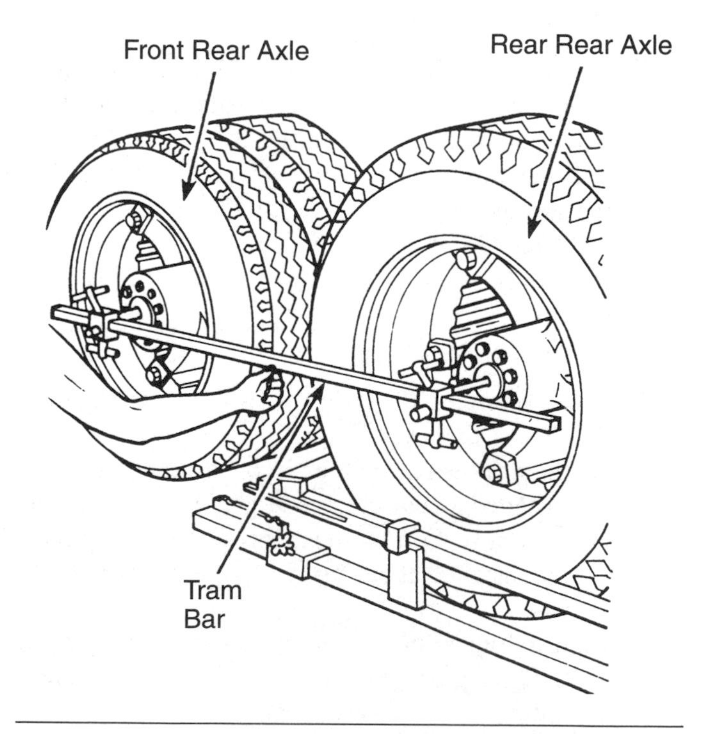

Figure 5–11 Tandem rear axle alignment. *(Courtesy of Navistar International Corp., designer and manufacturer of brand diesel engines)*

A. a bent rear axle shaft.
B. worn rear wheel bearings.
C. a bent lower torque rod.
D. improperly mounted wheel.

Hint

You can twist torsion bars at the crank assembly to increase or decrease pressure. Torsion bars are the same length but not interchangeable left to right. As long as a torsion bar is not cracked or broken, it can be twisted at the crank assembly to level the vehicle. You never heat or bend torsion bars or torque arms. Premature suspension bushing wear will cause noisy operation, directional instability, and excessive wear to adjacent suspension parts. A broken leaf spring can cause an off-level vehicle attitude. A bushing not relaxed during assembly can cause a binding condition, leaving a vehicle off level.

Task 6

Inspect, adjust, or replace axle aligning devices such as radius rods, track bars, stabilizer bars, and related bushings, mounts, shims, and cams.

65. All of the following could correct axle alignment EXCEPT:
   A. radius rod.
   B. axle shaft.
   C. track bar.
   D. adjustment eccentric.

66. *Technician A* says adjustment of the rear position on some spring designs requires the rotation of the lower torque rods. *Technician B* says you adjust some rear axles by rotating eccentric bushings in the spring hangers. Who is right?
   A. A only
   B. B only
   C. Both A and B
   D. Neither A nor B

67. When servicing the torque rods shown in Figure 5–12:
   A. the flat surface on the bar pins should be installed at a 90-degree angle to the torque rod.
   B. the flat surface on the front and rear bar pins are installed at the same angle in the torque rod.
   C. installing another shim between the bar pin and the spring hanger bracket moves the rear axle rearward.
   D. the torque rods absorb braking and acceleration torque.

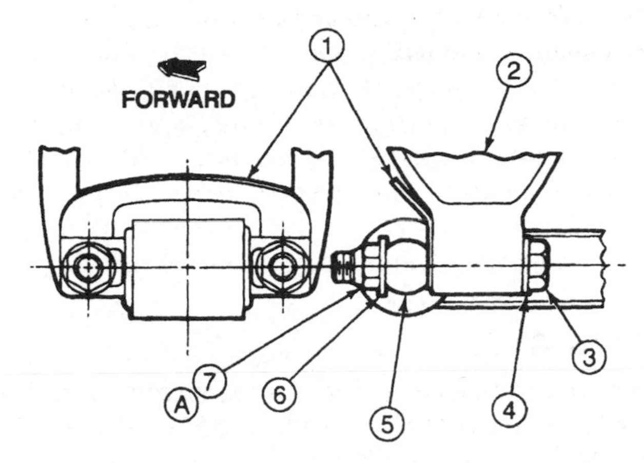

Item	Part Number	Description
1	4A002	1/16-Inch Shim (1/4-Inch Total Max.)
2	5785	Spring Hanger Bracket
3	58731	Bolt
4	44881	Washer
5	4962	Torque Rod Bar Pin
6	44881	Washer
7	34991	Locknut
A	—	Tighten to 204-277 N·m (150-205 Lb-Ft)

Figure 5–12 Rear suspension torque rods. *(Courtesy of Sterling Truck Corporation)*

Hint

Some tandem axle suspensions have four multileaf springs and four torque rods. Between the front and rear springs on each side of the tractor, the springs ride on an equalizer pivot mounted on a sleeve in the equalizer bracket. Torque rods may be substituted for rear shock absorbers. Four to six torque rods on spring suspensions are also used for suspension alignments. Rear axle alignment adjustment on a spring suspension with torque rods can be made through lower adjustable torque rods. Some rear suspensions use shims between the torque rod front and spring hanger bracket to align the rear axle. Other rear suspensions use an eccentric bushing at the torque leaf for alignment adjustment. The torque rods provide absorption of braking force and acceleration torque. Bent torque rods can cause the rear axle wheel alignment to be out of adjustment.

Task 7

Inspect or replace walking beams, center (cross) tube, bushings, mounts, load pads, brackets, caps, and mounting hardware.

68. *Technician A* says you can service the walking (equalizing) beam center and end bushings on the vehicle. *Technician B* says you may need to slightly heat the beam housing to remove the end bushing. Who is correct?
    A. A only
    B. B only
    C. Both A and B
    D. Neither A nor B

69. *Technician A* says there are two types of equalizing beam suspensions: leaf spring and rubber cushion. *Technician B* says the equalizing beam suspension lowers the center of gravity of the axle load. Who is correct?
    A. A only
    B. B only
    C. Both A and B
    D. Neither A nor B

70. In Figure 5–13 the bushings identified as number 15 are severely worn. The MOST LIKELY result of this problem is:
    A. broken spring leafs.
    B. directional instability.
    C. broken spring center bolts.
    D. shock absorber fluid leak.

Hint

The equalizing beam is below the equalizing bracket between the tandem springs. Equalizing beam suspension lowers the center of gravity of the axle load. The two types of equalizing beam suspensions are leaf spring type and rubber load cushion type. Some tandem rear axle suspension systems have equalizer beams on each side of the suspension. Bushings in the equalizer beams are attached to the rear axle housings and a cross tube is mounted between the two equalizer beams. You can service the walking (equalizing) beam center and end bushings on the vehicle using a special tool made by Owatonna. Multileaf springs are mounted on saddles above the equalizer beams.

Some rear tandem axle suspensions (Volvo-GM) have an inverted parabolic tapered leaf spring mounted on a cradle that pivots in a saddle bracket. Rubber springs are mounted between the ends of the springs and saddles on the rear axle housings. Four upper torque rods are mounted between the rear axle housings and the tractor frame, and four lower torque rods are positioned between the rear axle housing and the saddle brackets.

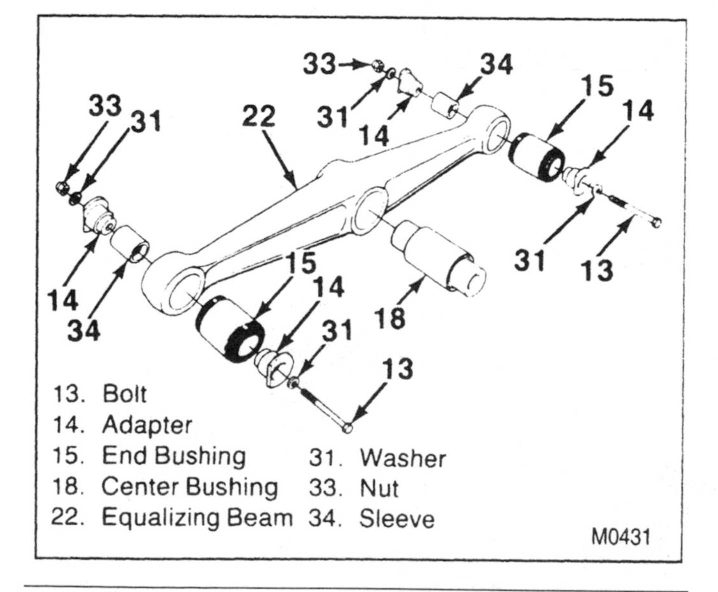

Figure 5–13 Rear suspension walking (equalizing) beam. *(Courtesy of General Motors Corporation, Service Technology Group)*

Task 8

Inspect, test, adjust, repair, or replace air suspension pressure regulator and height control valves, lines, hoses, and fittings.

71. A truck with air spring suspension sits low when you hook up the trailer and sits high without the trailer. *Technician A* says a leaking air spring could be the problem. *Technician B* says an improperly adjusted height control valve could be the problem. Who is right?
    A. A only
    B. B only
    C. Both A and B
    D. Neither A nor B

Hint

Air spring suspensions provide a smooth shock- and vibration-free ride and automatically adjusts riding height. A height control (leveling) valve controls the air supply to the air springs to maintain a constant ride height. These systems may give single- or dual-height control valves. The control and relay valves provide a means of deflating the air springs when uncoupling a trailer from a tractor. Blown-out air springs will not cause a too high ride condition. If the air springs are blown, the suspension rides low. Misadjusted leveling (height control) valves in the rear air suspension system may cause high or low riding height.

Task 9

Inspect, test, repair, or replace air spring assemblies (bags), mounting plates, springs, suspension arms, and bushings.

72. *Technician A* says that air spring suspensions eliminate the need for shock absorbers. *Technician B* says that in most rear axle air suspension systems, the air spring seat is mounted on the cross channel or transverse beam. Who is right?
    A. A only
    B. B only
    C. Both A and B
    D. Neither A nor B

73. A truck that is equipped with the air suspension system in Figure 5–14 has one side rising after unloading, but suspension operation is normal under load. This problem could be caused by:
    A. a plugged exhaust valve.
    B. a loose air spring seat.
    C. a worn main support beam bushing.
    D. a defective pressure protection valve.

74. During road testing, a heavy-duty truck with air suspension exhibits a rougher than normal ride. *Technician A* says that this rough ride could be caused by a slight leak at the governor. *Technician B* says that a defective air suspension relief valve could be the cause. Who is right?
    A. A only
    B. B only
    C. Both A and B
    D. Neither A nor B

Hint

In a rear axle air suspension system, the air springs are mounted between the frame and the outer end of the main support beams, or on the transverse beams. The main support beams are attached to the rear axle. The front of the main support beam is mounted in a frame bracket.

Trailer air suspension systems have a rigid beam on each side of the suspension. The front of this beam is retained in a bracket with a bushing. Some systems have an eccentric pivot bolt in this bracket for axle alignment. These brackets are welded to the trailer axle and retained in the rigid beams by two bushings and retaining bolts. The air springs are mounted between the rear end of the rigid beam and the frame.

Task 10

Measure vehicle ride height; determine needed adjustments or repairs.

75. Which of the following items applies to a vehicle ride height adjustment?

Specifications

MODEL	PART NO.	CLOSING PRESS	OPENING PRESS
STANDARD	905 54 107	65 PSIG min.	75 PSIG
METRIC	905 54 151	5.2 BAR min.	6.1 BAR
METRIC	905 54 174	4.2 BAR min.	5.1 BAR

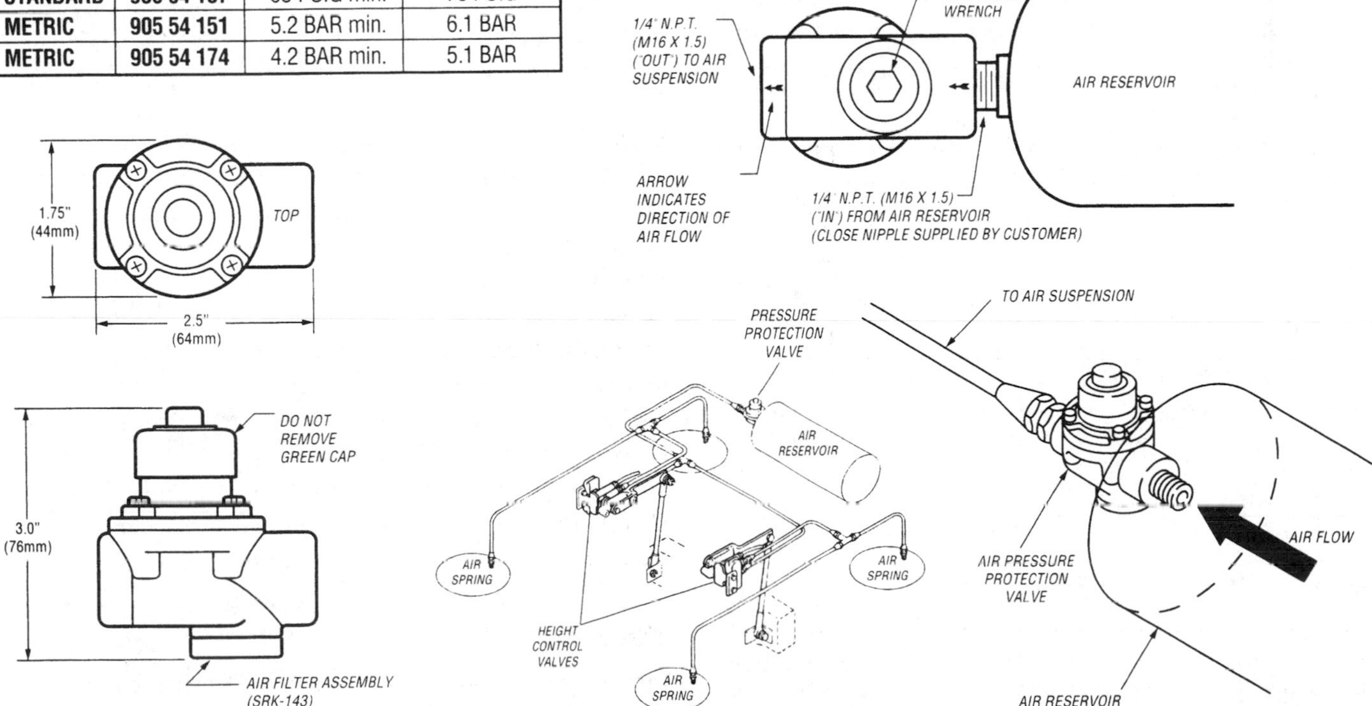

Figure 5–14 Rear axle air suspension system. *(Courtesy of Neway Anchorlok International Inc.)*

A. Truck must be in a laden condition with a full load.
B. Height control valve mounting bolts must be loosened.
C. Use safety stands with the proper weight rating.
D. Move the height control valve lever upward to deflate all air.

76. A height control valve in an air suspension system:
A. is bolted to the transverse beam.
B. has a linkage connected to the truck frame.
C. requires lubrication at regular intervals.
D. may contain an exhaust (dump) valve.

Hint

The air spring suspension has a height control valve that automatically controls the amount of air in the air spring to keep the truck level. Air spring design eliminates interleaf friction and absorbs road shock normally transferred to the frame, cargo, or driver. The height control rod, connected between the control valve and rear axle, maintains riding height under all conditions. An improperly adjusted height control valve can cause the vehicle to sit too low or too high.

Adjustment of the height (leveling) control valve requires the following:

- *The truck must be empty, and the height control valve linkage disconnected.*
- *Move the height control valve upward to raise the vehicle frame until you can place the safety stands set at the specified height under each side of the frame. The Technician must support the truck using safety stands at the specified weight rating.*
- *Move the height control valve lever downward to deflate all air. Hold them in the 45-degree downward position for 10 seconds.*
- *Return the control valve slowly to the center position. Insert the wood locating pins into the adjusting block.*
- *Loosen the 1/4 inch locknut on each height control valve. Connect the linkage and tighten the lock to specifications. Remove the wood locating pins and reverse the other steps.*

Task 11

Diagnose rough ride problems.

77. A driver complains about rough riding on the front suspension of his truck. The cause of this problem could be:
A. excessive positive camber on one or both front wheels.
B. excessive negative caster on the front wheels.
C. fifth wheel too far ahead on the tractor.
D. equal loading of the rear axles.

78. A driver complains about wheel hop and rough ride quality on the rear equalizing beam, leaf spring suspension on his truck. The cause of this problem could be:
A. worn equalizing beam bushings.
B. improper torque on the shock absorber mounting bolts.
C. reduced torque on the spring U-bolts.
D. a bent cross tube.

Hint

Excessive positive caster, leaking or damaged shock absorbers, and loose or worn suspension components can cause rough ride characteristics. Improper wheel balance, worn kingpin bushings, a bent wheel mounting surface, or a shifted belt inside a tire can also cause rough ride characteristics.

Wheel Alignment Diagnosis, Adjustment, and Repair

Task 1

Diagnose vehicle wandering, darting, pulling, shimmy, and hard steering problems; determine needed adjustments and repairs.

79. A vehicle with power steering has a squeak when turning left or right. Which of these could be the cause?
A. A dry kingpin pivot bearing
B. A bent front wheel
C. An overfilled fluid reservoir
D. Worn front shock absorbers

80. Which of the following can cause steering wheel shimmy?
A. Too high a toe-in setting
B. Improper dynamic wheel balance
C. An excessive load in the vehicle
D. Too high a toe-out setting

81. The driver of a truck complains that a truck still has a shimmy after an alignment has been performed. *Technician A* says this could be caused by too much positive caster. *Technician B* says an improper camber setting can be the cause. Who is right?
A. A only
B. B only

C. Both A and B
D. Neither A nor B

Hint

The Technician should road test the truck and inspect the front end before replacing any parts or alignment. The vehicle should also be road tested after the Technician performs the work, to ensure the work completion and performance. The Technician should check the front steering components and wheel balance. Multiple problems can cause abnormal steering conditions. Excessive positive caster causes higher steering effort and fast steering wheel return. Underinflated tires may cause higher steering effort but this will ***not*** *cause the steering wheel to return fast. A loose kingpin or kingpin bearing can cause a front shimmy with slight vibrations. Improper dynamic wheel balance can cause steering wheel shimmy. A shimmy after an alignment has been performed can be caused by excessive positive caster.*

A loose idler arm will cause fluctuations in the steering when striking a bump. Damaged tie-rods or cross tubes can cause veering when striking a bump because of looseness in the ball joint end of the tie-rod. Wheel balance will not cause veering, only a shaking in the steering wheel. An incorrect turning angle will cause excessive noise and tire scuffing wear. Excessive positive caster will cause oversteer and cause the steering wheel to return too quickly. Underinflated tires may cause higher steering effort.

Task 2

Check camber and KPI/SAI (king pin inclination/steering axis inclination); determine needed repairs.

82. If the wheel alignment specifications show a preferred right camber of 1/4 degree positive, and a tolerance of 1/8 degree, all of the following would be an acceptable final measurement EXCEPT:
    A. 1/4 degree positive.
    B. 1/8 degree negative.
    C. 3/8 degree positive.
    D. 1/8 degree positive.

83. In Figure 5–15 the angle between lines 1 and 2 identifies:
    A. toe-out condition.
    B. setback condition.
    C. positive caster condition.
    D. negative camber condition.

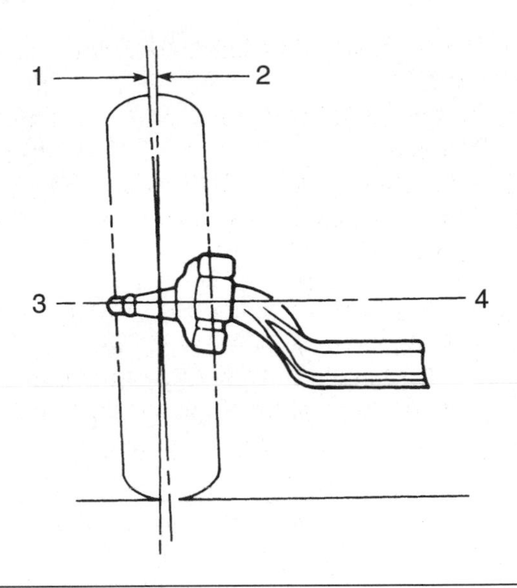

Figure 5–15 Front suspension angle.

84. When diagnosing the I-beam front suspension angles in Figure 5–16:
    A. the kingpin inclination angle (KPI) is adjustable.
    B. the camber angle is adjustable.
    C. the included angle is the sum of the camber and KPI angles.
    D. excessive positive camber wears the inside edge of the tire.

Hint

Camber is the inward or outward tilt of the wheel. Inward is negative movement, outward is positive move-

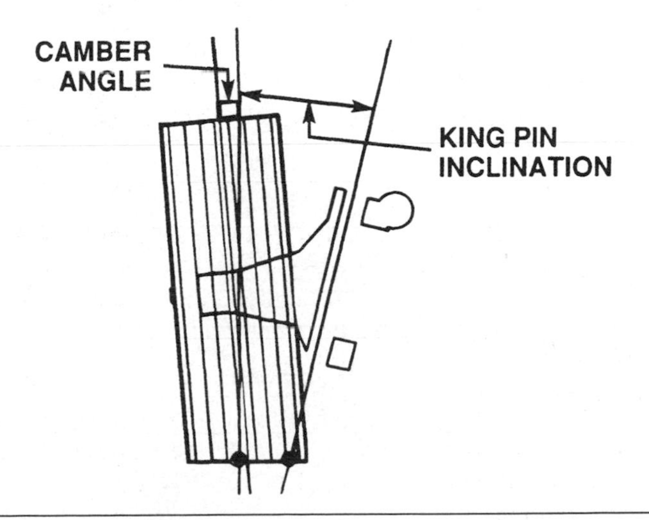

Figure 5–16 Front suspension angles, I-beam axle. (Courtesy of Rockwell Automotive, Inc.)

ment. Kingpin inclination (KPI) is the inward tilt of a line through the center of the kingpins in relation to the true vertical line through the center of the tire and wheel. The included angle is the sum of the KPI angle and the camber angle. KPI helps to return the wheels to the straight-ahead position after a turn. If KPI is increased, steering effort is increased. The distance between the centerline of the tire and the KPI line at the road surface is called kingpin offset. If the I-beam of an axle is designed with some kingpin offset, the front tires tend to roll around the point where the KPI line contacts the road surface when the wheel is turned and the vehicle is stationary. This action reduces static steering effort.

If a front wheel has positive camber, the top of the wheel is tilted outward away from the vertical centerline of the tire, and the vehicle weight is concentrated on the outside edge of the tire. Under this condition, the outside edge has a smaller diameter than the inside edge. Therefore, the outside edge must complete more revolutions to travel the same distance as the inside edge. Because both edges are on the same tire, the outside edge must slip and scuff on the road surface with the resulting wear on the outside edge. Excessive negative camber tilts the wheel inward and concentrates the vehicle weight on the inside edge of the tire tread. Negative camber causes wear on the inside edge of the tire.

Task 3

Check and adjust caster.

85. In Figure 5–17 the shim is installed to adjust:
    A. caster.
    B. camber.
    C. toe-in.
    D. spring sag.

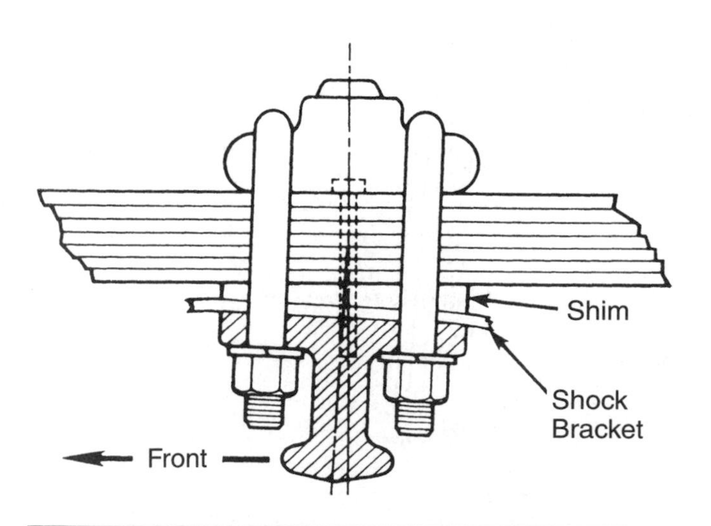

Figure 5–17 Adjustment shim, I-beam axle.

86. When adjusting caster, *Technician A* says stack shims no more than three high. *Technician B* says do not shim one side more than one degree than the other side. Who is right?
    A. A only
    B. B only
    C. Both A and B
    D. Neither A nor B

87. *Technician A* says to make caster more positive, you install a shim between the axle and spring with the high side at the front. *Technician B* says to make the caster more negative, you install a shim between the axle and spring with the low side at the front. Who is right?
    A. A only
    B. B only
    C. Both A and B
    D. Neither A nor B

Hint

Caster is the tilt of a vertical line through the center of the kingpins in relation to the true vertical centerline of the tire and wheel viewed from the side. Caster shims come 1/2 degree increments and you should only use one shim per side. A tolerance of 1/8 degree at 1/4 degree positive would be a reading of 3/8 degree positive or 1/8 degree positive. Positive caster helps to maintain vehicle directional stability. Excessive positive caster increases steering effort and creates a very rapid steering wheel return. Excessive negative caster reduces steering effort and decreases directional stability.

Task 4

Check and adjust toe.

88. In Figure 5–18 distance A is greater than distance B. This is a:
    A. toe-in condition.
    B. camber condition.
    C. toe-out condition.
    D. steering axis inclination.

89. Toe-in settings on most M/HD trucks should be between:
    A. 0 to 1 inch positive.
    B. 1/16 inch to 1/32 inch.
    C. 1/2 to 3/4 inch.
    D. 3/4 to 1 inch.

Hint

Toe-in is when the distance between the rear inside edges of the front tires is greater than the distance between the

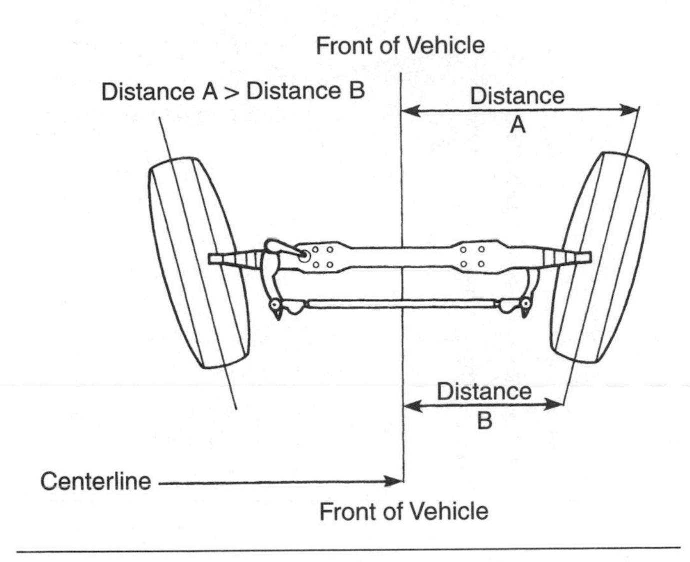

Figure 5–18 Front suspension alignment.

front inside edges of the front tires. Toe-out occurs when the distance between the inside front tires edges exceeds the distance between the inside rear tire edges. A toe-in error of 1/8 inch is the equivalent of dragging the tires crosswise for 11 feet for each mile driven. This crosswise movement causes severe feathered tire tread wear.

Task 5

Check rear axle(s) alignment (thrustline/centerline) and tracking; adjust or determine needed repairs.

90. A term that describes when a vehicle's rear wheels track directly behind the front wheels is:
    A. dog tracking.
    B. proper thrustline.
    C. Ackerman effect.
    D. toe-in.

91. All of the following apply to an axle alignment adjustment on a spring suspension with torque rods EXCEPT:
    A. adjustment is made through a lower adjustable torque rod.
    B. shims are used between the torque rod front and spring hanger bracket.
    C. adjustment is made with an eccentric bushing at the torque leaf.
    D. adjustment is made through the upper adjustable torque rods.

92. An axle alignment is to be performed on a tractor with a walking (equalizing) beam rear suspension. *Technician A* says worn walking (equalizing) beam bushing can cause the rear axles to be out of alignment. *Technician B* says that an incorrect torque leaf eccentric bushing setting can cause an axle alignment problem. Who is right?
    A. A only
    B. B only
    C. Both A and B
    D. Neither A nor B

Hint

When a vehicle's rear wheels track directly behind the front wheels, the rear axle thrustline is positioned at the vehicle's geometric centerline. Worn walking (equalizing) beam bushings can cause the rear axles to be out of alignment. Replacing a bent radius rod could correct axle alignment. An axle shaft replacement will not correct the alignment of the axle assembly. A defective tracking bar could cause an axle alignment problem. Rotating eccentric bushings in the spring hangers does adjust the rear position on some spring designs.

Task 6

Check turning/Ackerman angle (toe-out-on-turns); determine needed repairs.

93. Two Technicians are discussing the subject of front suspension turning radius or toe-out-on-turns (Ackerman angle). *Technician A* says that if the toe-out-on-turns is not within specifications, one of the steering arms may be bent. *Technician B* says improper toe-out-on-turns may cause tire scuffing. Who is right?
    A. A only
    B. B only
    C. Both A and B
    D. Neither A nor B

94. When the front wheel on the outside of a turn is turned 20 degrees, the wheel on the inside of the turn should be turned:
    A. 20 degrees.
    B. 22 degrees.
    C. 25 degrees.
    D. 28 degrees.

Hint

When a vehicle turns a corner, the front and rear wheels must turn around a common center with respect to the turn radius or angle. On a single rear axle, this common center is located at the center of the rear wheels. On most front suspensions, the front wheels pivot independently at different distances from the center of the turn, and therefore the front wheels must turn at different angles. The inside front wheel must turn at a sharper angle compared to the outside wheel. This action is nec-

essary because the inside wheel is actually ahead of the outside wheel.

Wheels and Tires Diagnosis and Repair

Task 1

Diagnose tire wear patterns; determine needed repairs.

95. The rear tires of a vehicle have worn down to the minimum wear indicators after only 8,000 miles of use. Which of the following could be the cause?
    A. Damaged / worn rear axle bearings
    B. Rear caster measurement
    C. Front wheel toe-in out of specification
    D. Bent rear axle housing
96. The tire wear in Figure 5–19 was caused by:
    A. excessive toe-out.
    B. negative camber.
    C. positive caster.
    D. underinflation.
97. The tire wear in Figure 5–20 was caused by:
    A. excessive toe-out.
    B. excessive positive camber.
    C. excessive positive caster.
    D. overinflation.
98. Which of these front suspension angles can cause the greatest tire wear?
    A. Kingpin inclination
    B. Turning radius
    C. Caster
    D. Camber

Hint

Worn shock absorbers allow the tire to bounce causing a cupping mark in the tire. Out-of-specification toe setting causes feathered tire tread wear. Underinflation will wear the edges of the tire. Overinflation will wear the center of the tire. Camber causes wear on one side of the tire. The wear occurs on the inside of the tire with negative camber or the outside of the tire with positive camber.

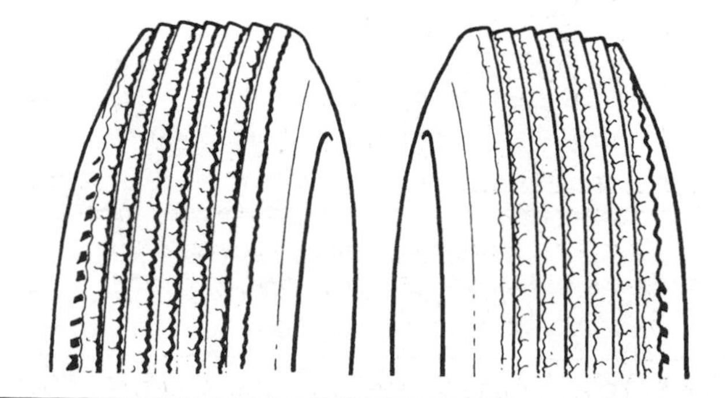

Figure 5–19 Front tire tread feathered wear.

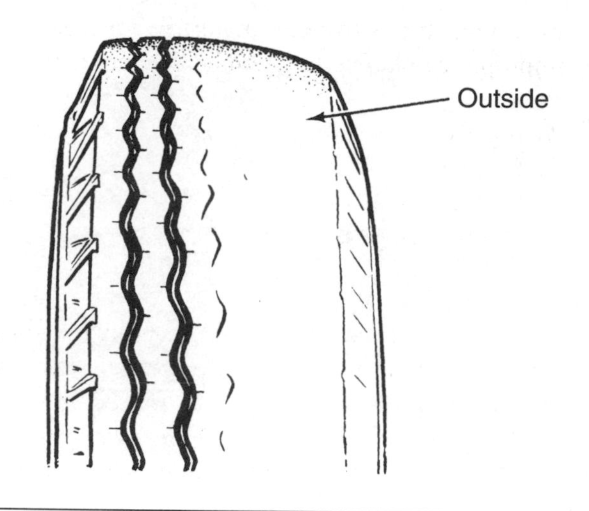

Figure 5–20 Front tire tread wear.

Task 2

Diagnose wheel/tire vibration, shimmy, pounding, hop (tramp) problems; determine needed repairs.

99. While driving a vehicle at 28 mph, a driver notices a vibration in the steering wheel. Which of the following is the LEAST likely cause?
    A. Improper wheel balance
    B. Worn kingpin bushings
    C. Bent wheel mounting surface
    D. Shifted belt inside tire
100. During a tire inspection, a Technician notices cupping marks around a tire. What is the MOST LIKELY cause?
    A. Shock absorbers
    B. Camber
    C. Toe-out-on-turns
    D. Caster
101. Which of the following can cause steering wheel shimmy?
    A. Too high a toe-in setting
    B. Improper dynamic wheel balance
    C. An excessive load in the vehicle
    D. Too high a toe-out setting
102. The driver of a truck complains that a truck still has a shimmy after an alignment has been performed. *Technician A* says this could be caused

by too much positive caster. *Technician B* says an improper camber setting can be the cause. Who is right?
A. A only
B. B only
C. Both A and B
D. Neither A nor B

Hint
While driving a vehicle at 28 mph, a vibration in the steering wheel could be caused by improper wheel balance, bent wheel mounting surface, or shifted belt inside the tire. A separated belt forms a gap under the tread causing a popping sound and the separating causes the wheel to shake.

Task 3

Inspect and replace wheels, spacers, rims, side flanges, locking rings, rim clamps, studs, and nuts.

103. When removing a disc wheel from the truck, *Technician A* says the right side wheel will have right-hand threads and the left side left-hand threads. *Technician B* says wear safety glasses and do not stand in front of a deflating tire. Who is right?
A. A only
B. B only
C. Both A and B
D. Neither A nor B

104. When dismounting a split ring tire, what is the LEAST important thing to check?
A. Excessive rust or corrosion buildup
B. Tire brand
C. Bent flanges
D. Deep tool marks on the rings and gutter area

105. *Technician A* says that dismounting and mounting split side rims is extremely dangerous and should only be performed by trained professionals according to OSHA rules and regulations. *Technician B* says never hammer on split side rim rings. Who is right?
A. A only
B. B only
C. Both A and B
D. Neither A nor B

Hint
A right side wheel has right-hand threads, and a left side wheel is equipped with left-hand threads. Always wear safety glasses and do not stand in front of a deflating tire because dirt particles may be discharged with the air.

When demounting a split ring tire, always check for excessive rust or corrosion buildup, bent flanges, deep tool marks on the rings and gutter area, and damaged and missing rim drive plates. Always check the tire size on the new and old tires, but the brand really does not matter. When adjusting wheel bearings, torque the adjusting nut to 50 ft.-lb. Then back off the nut the amount specified by the truck manufacturer, which is usually 1/6 to 1/3 turn, and then install the lock ring. An excessively tight wheel bearing adjustment causes the wheel bearings to overheat. With the stud nuts slightly loose on demountable rims, strike the wheel clamps with a hammer to loosen these clamps. Different makes of rim components must not be interchanged.

Task 4

Measure wheel and tire radial and lateral runout; determine needed repairs or adjustments.

106. The test procedure with the equipment installed as shown in Figure 5–21 is measuring:
A. radial wheel runout.
B. lateral wheel runout.
C. wheel bearing movement.
D. radial tire runout.

107. A rim on a spoke wheel has excessive lateral runout. The first step to correct this procedure should be:
A. remove the wheel, demount the tire, and straighten the wheel in a hydraulic press.
B. loosen the stud nut at the point of greatest clearance, and tighten the stud nut opposite this nut.

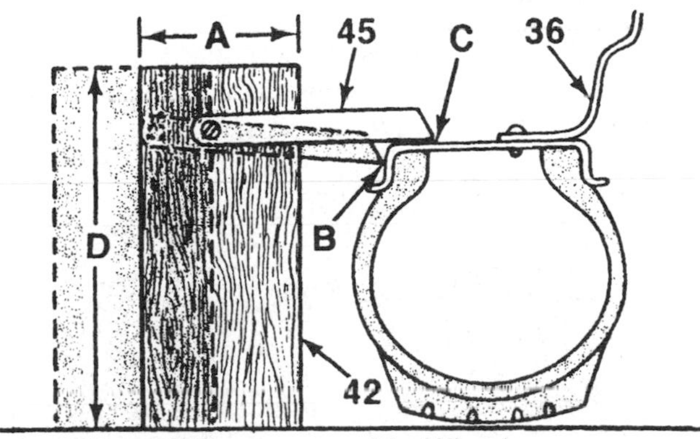

Figure 5–21 Wheel and tire measurement. *(Courtesy of General Motors Corporation, Service Technology Group)*

C. loosen and retorque all the stud nuts to specifications.
D. clean, inspect, lubricate, and adjust the wheel bearings.

Hint

A separated belt in the tire will cause a popping sound and a shake in the steering above 45 mph. Wheel runout should be checked before a wheel balance procedure.

Task 5

Inspect tires; check air pressure.

108. An overinflated tire will cause excessive wear in which of these?
A. Outside edges of the tire
B. Center of the tire
C. Cupping pattern
D. Inside edges of the tire

109. *Technician A* says you cannot mix radial tires with bias tires on the steering axle. *Technician B* says mixing tires on the rear axle is all right. Who is right?
A. A only
B. B only
C. Both A and B
D. Neither A nor B

110. *Technician A* says to inflate mounted tires in a safety cage or using a portable lock ring guard. *Technician B* says to first mount the tire on the truck, then inflate to the proper tire pressure. Who is right?
A. A only
B. B only
C. Both A and B
D. Neither A nor B

111. The driver complains of a popping sound and a shake in the steering above 45 mph. What is the MOST likely cause?
A. Worn shock absorber bushings
B. Out-of-balance tire
C. Incorrect toe-in setting
D. Separated belt in the tire

Hint

Underinflation will wear the edges of the tire. Overinflation will wear the center of the tire. Always inflate mounted tires in a safety cage or using a portable lock ring guard. One never mounts the tire on the truck before inflation because an improperly fitted rim ring could dislodge, causing injury. Dismounting and mounting split side rims is extremely dangerous and only trained professionals should perform this process according to OSHA rules and regulations. You never use a hammer on split side rim rings.

Task 6

Perform static balance of wheel and tire assembly.

112. *Technician A* says the maximum wheel weight per tire should not exceed 18 ounces. *Technician B* says if 16 ounces of weight is required at one spot, use an 8-ounce weight on each side of the rim directly across from each other. Who is right?
A. A only
B. B only
C. Both A and B
D. Neither A nor B

113. During the balancing of a truck tire, a heavy spot is found on the outside edge of the tire. *Technician A* says when using a static balancing process, a wheel weight is installed 180 degrees from this heavy spot on the outside edge. *Technician B* says the Technician does not have to enter any information in an electronic wheel balancer. Who is right?
A. A only
B. B only
C. Both A and B
D. Neither A nor B

Hint

- *Preliminary wheel balancing checks:*
- *Check for objects in tire tread.*
- *Check for objects inside tire on tubeless tires.*
- *Inspect tread and sidewall.*
- *Check inflation pressure.*
- *Measure tire and wheel runout.*
- *Check for mud collected on the inside of the wheel.*

On some types of wheel balancers during the static wheel balance process, the wheel is allowed to rotate by gravity. A heavy spot rotates the tire until this spot is at the ***bottom****. One then adds static balance weights at the* ***top*** *of the wheel, 180-degrees from the heavy spot. When you balance the wheel and tire statically, gravity does not rotate the wheel from the at-rest position. Rotate the tire by hand and check the static balance at 120-degree intervals. A tire that has a vulcanizing repair will need balancing.*

Task 7

Perform dynamic balance of wheel and tire assembly.

114. You are balancing the radial tires on a medium truck. *Technician A* says wheel runout should be measured before the balance procedure. *Technician B* says after the wheels are balanced, mount the tire with the wheel weights 180 degrees from the brake drum weights. Who is right?
A. A only
B. B only
C. Both A and B
D. Neither A nor B

115. A tire has a vulcanizing repair. *Technician A* says that if you place the tire on the rim in the same position as it was removed, the tire will retain dynamic balance. *Technician B* says you need to perform a static balance procedure. Who is right?
A. A only
B. B only
C. Both A and B
D. Neither A nor B

116. *Technician A* says you cannot perform dynamic balance with the tire mounted on the truck. *Technician B* says that dynamically balanced tires should be marked before removal and installed in the same position. Who is right?
A. A only
B. B only
C. Both A and B
D. Neither A nor B

Hint

One can perform dynamic balance with the tire on the truck. Before removing a dynamically balanced tire from the truck, it should be marked and installed in the same position as removed. The electronic-type balancer performs static and dynamic wheel balance calculations simultaneously and indicates the correct weight size and location. When the heavy spot is on the outside edge of the tread, you install the correct size wheel weight at a location 180 degrees from the heavy spot on the outside of the wheel.

Task 8

Measure tire diameter and match tires on tandem axle(s).

117. *Technician A* says you can mix radial tires with bias tires on the steering axle. *Technician B* says mixing tires on the rear axle is prohibited because it is hazardous. Who is right?
A. A only
B. B only

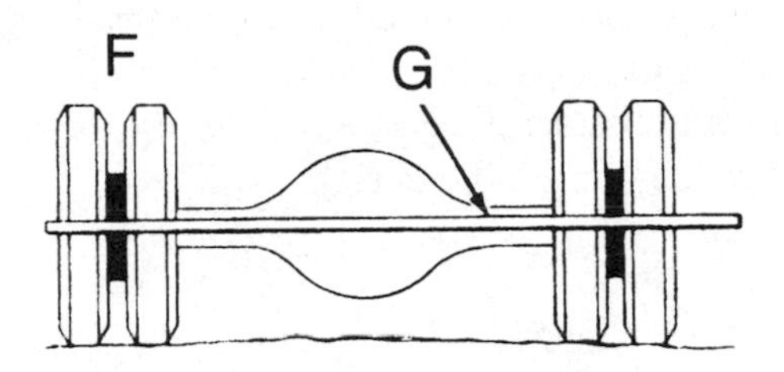

Figure 5–22 Rear axle measurement. *(Courtesy of General Motors Corporation, Service Technology Group)*

C. Both A and B
D. Neither A nor B

118. You are measuring tire diameter on dual wheels for matching purposes. *Technician A* says you match tire sizes on dual wheels to prevent tire tread wear from slippage because of uneven tire surface areas. *Technician B* says dual drive wheels may be measured with a tape measure. Who is right?
A. A only
B. B only
C. Both A and B
D. Neither A nor B

119. The measurement being performed in Figure 5–22 is:
A. tire diameter comparison.
B. rear axle alignment.
C. rear axle toe.
D. rear tire runout.

Hint

Radial tires with bias tires must not be mixed on a steering axle. Mixing tires on the rear axle is dangerous because radial tires expand more than bias tires. Matching tire sizes on dual wheels prevents tire tread wear from slippage from uneven surface areas. Dual drive wheels may be measured with a tape measure, caliper, etc.

Task 9

Remove and replace steering and/or drive axle wheel assembly.

120. The two main designs of truck wheels are:
A. single piece and split side rims.
B. aluminum and hub piloted.
C. disc and drum.
D. hub and spoke.

121. One is reinstalling the dual wheel assembly of a tractor axle. *Technician A* says you put a light

coating lubricant on the spindle and place a seal protector over the threads. *Technician B* says rear wheel hub assemblies use oil filler plugs to fill the wheel cavities. Who is right?

A. A only
B. B only
C. Both A and B
D. Neither A nor B

122. All of these tasks are performed when installing rear wheels and hubs EXCEPT:

A. use a wheel dolly to install the dual wheel assembly.
B. use the OEM recommended lube on the spindle.
C. install a large amount of lubricant in the wheel cavity.
D. install the adjusting nut on the spindle.

Hint

When you reinstall the dual wheel assembly of a tractor axle, you do put a light coating lubricant on the spindle and place a seal protector over the threads. Some OEMs use oil filler plugs in the rear wheel hubs. A Technician should use a wheel dolly to install the dual wheel assembly. Use the OEM recommended lube on the spindle and place only the specified amount of lubricant in the wheel cavity.

Task 10

Clean, inspect, lubricate, or replace wheel bearings; replace seals and wear rings; adjust steering and drive axle wheel bearings.

123. *Technician A* says one repacks rear wheel bearings every time the drum is removed. *Technician B* says you lubricate rear wheel bearings with rear axle fluid. Who is right?

A. A only
B. B only
C. Both A and B
D. Neither A nor B

124. After cleaning wheel bearings with solvent, *Technician A* says to spin the bearings with compressed air and dry them. *Technician B* says that spinning a dry bearing with compressed air could cause the bearing to explode. Who is right?

A. A only
B. B only
C. Both A and B
D. Neither A nor B

125. When servicing and adjusting rear wheel bearings, *Technician A* says the lip on the hub seal must face toward the differential. *Technician B* says to torque the adjusting nut to 50 ft.-lb., then back off the nut the specified amount and install the lock ring and jam nut. Who is right?

A. A only
B. B only
C. Both A and B
D. Neither A nor B

126. After a wheel bearing has been properly adjusted, the hub end play should be:

A. 0.001 to 0.005 in. (0.054 to 0.127 mm).
B. 0.008 to 0.010 in. (0.203 to 0.254 mm).
C. 0.010 to 0.012 in. (0.254 to 0.304 mm).
D. 0.012 to 0.014 in. (0.304 to 0.355 mm).

127. Front and rear wheel bearing diagnosis is being discussed. *Technician A* says the growling noise produced by a defective rear wheel bearing is most noticeable at high speeds. *Technician B* says that a growling noise produced by a defective front wheel bearing is most noticeable while turning a corner. Who is right?

A. A only
B. B only
C. Both A and B
D. Neither A nor B

128. Refer to Figure 5–23. After the rear wheel bearings have been serviced and adjusted, the lubricant level in the differential should be:

A. at the lower edge of the temperature sensor hole.

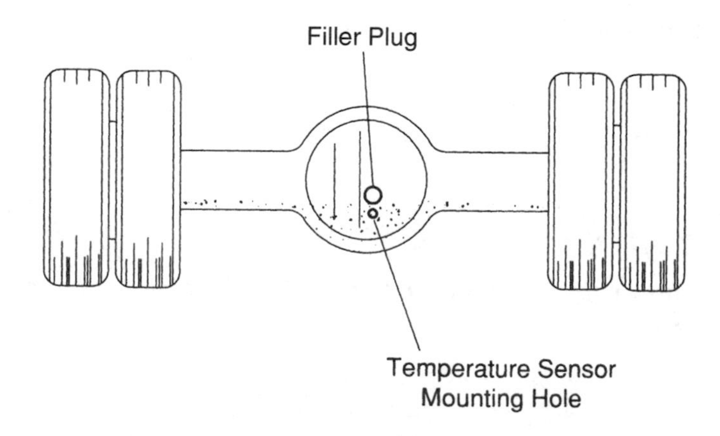

Figure 5–23 Truck rear axle. *(Courtesy of Eaton Corporation)*

B. between the temperature sensor hole and the filler hole.
C. at the lower edge of the filler hole
D. 1/2 in. (12.7 mm) below the edge of the temperature sensor hole.

Hint

Torque the wheel bearing adjusting nut 50 ft.-lb., then back off the nut 1/6 to 1/3 turn depending on the truck manufacturer's specifications and install the lock ring. Defective rear wheel bearings produce growling noises at low speeds. A growling noise produced by a defective front wheel bearing is most noticeable while turning a corner. Rear wheel bearings do not require repacking because rear wheel bearings are lubricated with rear axle fluid. Never spin dry a wheel bearing with compressed air because spinning a bearing with compressed air can cause the bearing to disintegrate.

Frame Service and Repair

Task 1

Inspect, adjust, service, repair, or replace fifth wheel, pivot pins, bushings, locking jaw mechanisms, and mounting bolts.

129. The fifth wheel in Figure 5–24 is a:
   A. semioscillating fifth wheel.
   B. rigid fifth wheel.
   C. compensating fifth wheel.
   D. sliding fifth wheel.

130. The fifth wheel in Figure 5–25 is used on a tractor:
   A. so the tractor can be used for over-the-road hauling and trailer spotting and switching.
   B. that is mainly used for off-road hauling.
   C. that is mainly used for log hauling.
   D. that is used to haul a frameless dump trailer.

Figure 5–24 Fifth wheel. *(Courtesy of Holland Hitch Co.)*

Figure 5–25 Specialized fifth wheel. *(Courtesy of Holland Hitch Co.)*

131. A fifth wheel is being mounted on a tractor. *Technician A* says that if it is located too far forward, hard steering and tire wear can result. *Technician B* says the fifth wheel should be mounted to the frame flanges by welding. Who is right?
   A. A only
   B. B only
   C. Both A and B
   D. Neither A nor B

132. *Technician A* says that the condition of high hitch occurs when the trailer kingpin and bolster plate are positioned too high in the fifth wheel. *Technician B* says this condition is caused by improper bolster plate position during the coupling process. Who is right?
   A. A only
   B. B only
   C. Both A and B
   D. Neither A nor B

Hint

The fifth wheel couples the tractor to the trailer kingpin or pintle. During the tractor-to-trailer coupling process, the fifth wheel should be adjusted so the trailer bolster plate makes initial contact with the fifth wheel at a point 8 inches (20.32 cm) to the rear of the center on the fifth wheel mounting bracket. After the tractor is coupled to the trailer, the driver should get under the tractor and trailer and use a flashlight to visually inspect the position of the kingpin in the fifth wheel jaws.

Task 2

Inspect, adjust, service, repair, or replace sliding fifth wheel, tracks, stops, locking systems, air cylinders, springs, lines, hoses, and controls.

133. When the air slide release fifth wheel in Figure 5–26 is being released:
   A. you raise the landing gear to relieve pressure on the plungers.
   B. if the locking plungers will not release, the air cylinder may be defective.
   C. the tractor should be moving slowly ahead or backward.
   D. the trailer should be unloaded.

134. A sliding fifth wheel position is being discussed regarding steering and frame diagnosis. *Technician A* says that if the fifth wheel is too far ahead, the steering effort is decreased. *Technician B* says that if the fifth wheel is too far rearward, premature wheel lockup may occur during a hard brake application. Who is right?
   A. A only
   B. B only
   C. Both A and B
   D. Neither A nor B

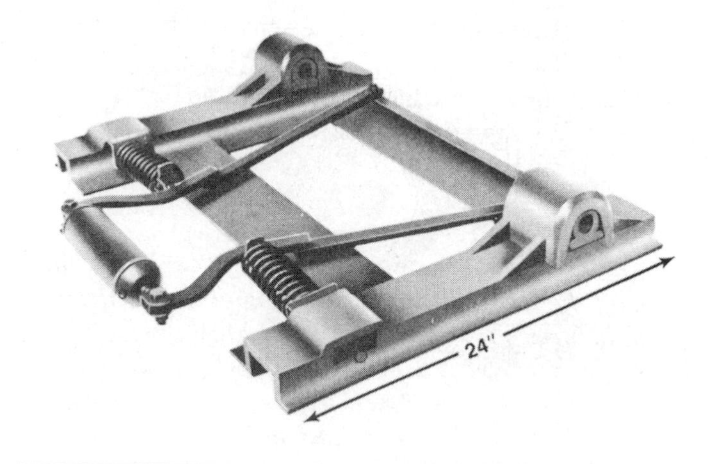

Figure 5–26 Air release sliding fifth wheel. *(Courtesy of Holland Hitch Co.)*

Hint

A sliding fifth wheel is designed to move forward or rearward on its mounting plate. This type of fifth wheel is mounted on tracks and locked in position. The locking mechanism may be released mechanically with a lever, or by air pressure supplied to an air cylinder.

Task 3

Inspect frame and frame members for cracks, breaks, distortion, elongated holes, looseness, and damage; determine needed repairs.

135. Which of these statements is correct when diagnosing truck frame problems?
   A. Tandem axle tractors have the maximum bending moment occur at the bogie centerline.
   B. Single axle trucks with van bodies have the maximum bending moment occur just ahead of the rear axle.
   C. Frame buckle may be caused by too many holes drilled in the frame web.
   D. Frame twist may be caused by uneven loading.

136. Which of the following is the MOST LIKELY cause of the frame condition in Figure 5–27?
   A. Too many holes drilled in the frame
   B. Extreme operating conditions
   C. The wrong bolts used in a repair
   D. A loose fifth wheel mounting

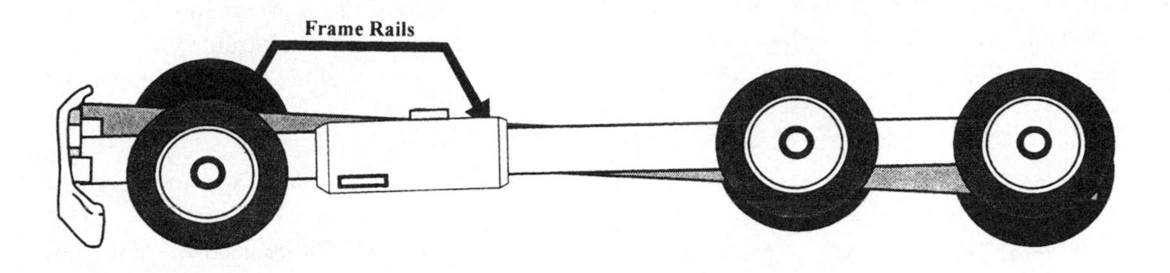

Figure 5–27 Defective frame condition. *(Courtesy of Bee Line Co., Bettendorf, Iowa)*

137. When diagnosing the frame condition in Figure 5–28, *Technician A* says that towing another truck with a chain attached to one corner of the frame could be the cause. *Technician B* says the shifting of the front or rear axles could be the cause. Who is right?
   A. A only
   B. B only
   C. Both A and B
   D. Neither A nor B
138. The frame condition in Figure 5–29 may be caused by all of these conditions EXCEPT:
   A. holes drilled too close together in the frame web.
   B. welding on the frame.
   C. collision damage.
   D. cutting notches in the frame rails.

Hint

Frame sidesway *occurs when one or both frame rails are bent inward or outward. The following are some causes: collision damage, fire damage, and using the truck for other than original design.*

Frame sag *occurs when the frame rails are bent downward in relation to the rail ends. The following are some causes:*

Excessive loads
Uneven weight distribution
Holes drilled in frame flanges
Too many holes drilled in the frame web

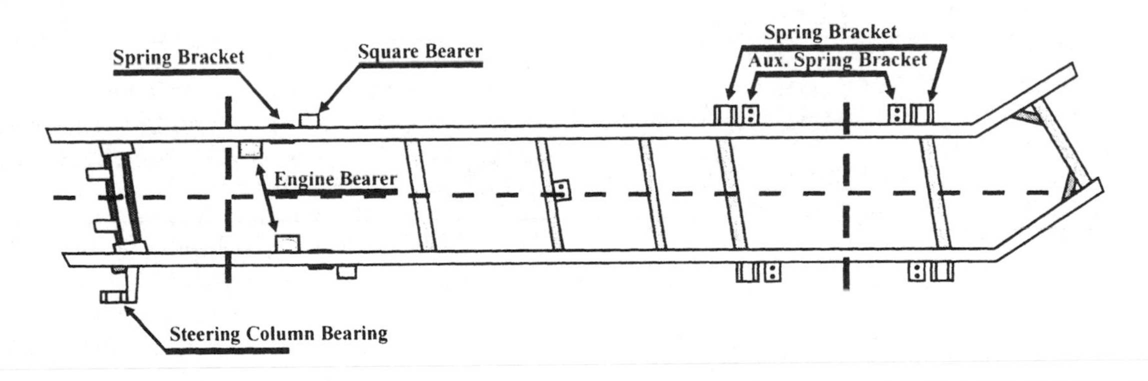

Figure 5–28 Defective frame condition. *(Courtesy of Bee Line Co., Bettendorf, Iowa)*

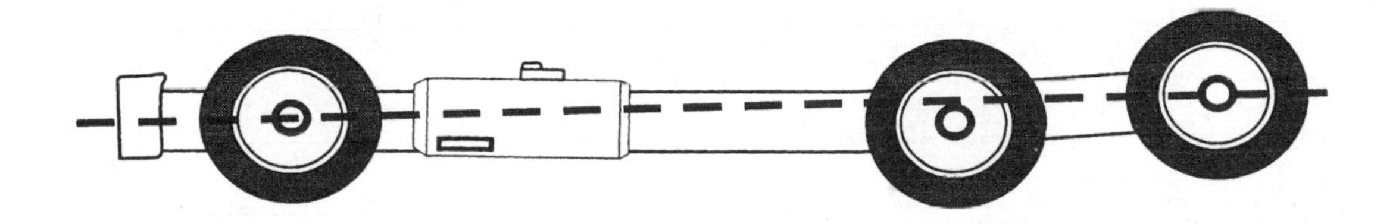

Figure 5–29 Defective frame condition. *(Courtesy of Bee Line Co., Bettendorf, Iowa)*

Holes drilled too close together in the frame web

Welding on the frame

Cutting holes in the frame with a torch

Cutting notches in the frame rails

Fire

Collision damage

Using the truck for other than original design

***Frame buckle** occurs when one or both frame rails are bent upward in relation to the ends of the rails. The following causes can cause buckle: collision, using the truck for other than design intent, and fire.*

***Diamond-shaped** frame occurs when one frame rail is pushed rearward in relation to the opposite frame rail. Vehicle tracking is affected by this condition. The following causes can result in diamond: collision, towing, or being towed with the chain only attached to one side of the truck frame.*

***Frame twist** occurs when the end of one frame rail is bent upward or downward in relation to the opposite frame rail. The following causes can result in a twisted frame: collision damage, rollover, rough terrain operation, and uneven loading.*

Task 4

Repair, lengthen, or shorten frame and frame members by fish (double) plating and welding in accordance with manufacturers' recommended procedures.

139. *Technician A* says fishplate frame reinforcements may only extend below the frame. *Technician B* says channel frame reinforcements are only installed on the outside of the frame. Who is right?
   A. A only
   B. B only
   C. Both A and B
   D. Neither A nor B

140. Which of these statements is correct when working on a truck frame?
   A. An arc welder may be used to install accessories by welding their brackets to the frame.
   B. Deep section frame reinforcement is usually installed near the front of the frame.
   C. Gussets may be used when attaching the reinforcement plate to the frame.
   D. When installing frame reinforcement plates, the original boltholes in the frame should be used.

141. When measuring frame alignment in Figure 5–30, if the frame alignment is satisfactory, the length of any pair of diagonal lines should not vary more than:
   A. 1/4 in. (6.35 mm).
   B. 1/2 in. (12.7 mm).
   C. 5/8 in. (15.8 mm).
   D. 3/4 in. (19.05 mm).

Hint

Some frame reinforcements are a single channel inside the frame rail; this type of reinforcement may be called a two-element frame rail. Other frame reinforcements have a single inside and single outside channel. This type of reinforcement is referred to as a three-element frame rail. Another type of reinforcement is called an L-channel. Some frame reinforcements are a heavy metal plate bolted to the outside of the frame. This type of reinforcement is called a fishplate. Fishplate reinforcements may be installed on the outside of the frame to strengthen it. Fishplates may extend above or below the frame.

Task 5

Inspect, install, or repair frame, hangers, brackets, and cross members in accordance with manufacturers' recommended procedures.

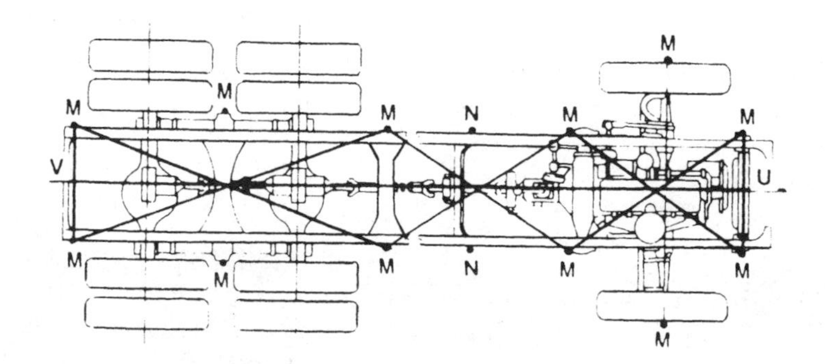

Figure 5–30 Frame measurement. *(Courtesy of Mack Trucks, Inc.)*

142. You are preparing to repair a frame cross member. *Technician A* says to disconnect the truck batteries before welding a frame. *Technician B* says it is permissible to weld across frame flanges. Who is right?
    A. A only
    B. B only
    C. Both A and B
    D. Neither A nor B

143. *Technician A* says you never drill holes in the flanges of a truck frame. *Technician B* says you never drill holes in the web of a truck frame. Who is right?
    A. A only
    B. B only
    C. Both A and B
    D. Neither A nor B

Hint

Do not drill extra holes in a truck frame. Use only frame service procedures recommended in the truck manufacturer's service manual. Do not use an arc welding or cutting torch on a truck frame.

Task 6

Inspect, install, repair, or replace pintle hooks and draw bars.

144. The pintle hook latch is:
    A. electrically operated.
    B. operated by air pressure.
    C. operated by spring tension.
    D. operated manually by a lever.

145. The pintle hook is:
    A. welded to the centerline of the forward trailer.
    B. bolted to the centerline of the rear trailer.
    C. welded to the centerline of the converter dolly.
    D. bolted to the centerline of the forward trailer.

Hint

Always make sure the pintle hook and draw bar are not worn, loose, or bent. The pintle hook must be bolted to the forward trailer and these mounting bolts must be tightened to the specified torque.

Answers to Questions

1. Answer A is wrong because a U-joint is the part on the steering column that allows steering shaft flexibility.

   Answer B is wrong because the turn signal switch is mounted on the left side of the steering column.

   Answer C is right because the drag link is not connected to the steering column; it is connected to the steering gear.

   Answer D is wrong because the boot seal seals the steering column where it goes through the body.

2. Answer A is wrong because excessive play in the mounted steering column will not be caused by a faulty anti-lash spring in the centering spheres.

   Answer B is wrong because excessive play in the mounted steering column assembly will not be caused by an upper bearing race missing.

   Answer C is wrong because excessive play in the mounted steering column assembly will not be caused by an upper bearing not seating in the race.

   Answer D is right because a loose support lock shoe pin will cause excessive play in the mounted steering column assembly on a truck equipped with a tilt steering column.

3. **Answer A is right** because on some tilt steering columns, the upper steering shaft is connected to the lower steering shaft by a universal joint.

   Answer B is wrong because two pivot bolts do not connect the main housing and the support bracket. Pivot bolts are used on tilt designs.

   Answer C is wrong because Technician A is right and Technician B is wrong.

   Answer D is wrong because Technician A is right and Technician B is wrong.

4. Answer A is wrong because you center the steering wheel on the top of the steering column.

   Answer B is right because the bushings mounted at the top and bottom of the steering column keep the steering shaft straight.

   Answer C is wrong because the U-joints allow flexibility in mounting the steering column.

   Answer D is wrong because the support bracket holds the steering column in place.

5. **Answer A is right** because a worn steering shaft universal joint or an out-of-adjustment steering gear could cause excessive steering wheel free play.

   Answer B is wrong because a bent idler arm or no lubricant in the steering gear will not cause excessive free play. A bent idler arm may cause the steering wheel to be off center and lack of lubricant will seize the steering gear.

Answer C is wrong because Technician A is right and Technician B is wrong.

Answer D is wrong because Technician A is right and Technician B is wrong.

6. Answer A is wrong because a bent steering shaft causes a binding condition as the steering wheel is turned.

   Answer B is right because a worn U-joint in the steering shaft may cause a rattling noise on rough road surfaces.

   Answer C is wrong because Technician A is wrong and Technician B is right.

   Answer D is wrong because Technician A is wrong and Technician B is right.

7. Answer A is wrong because a cab air suspension does not use springs.

   Answer B is right because it has a leveling valve.

   Answer C is wrong because a cab air suspension does not use rubber cab mounts.

   Answer D is wrong because a cab air suspension uses shock absorbers.

8. Answer A is wrong because air pressure is supplied to an air-suspended seat through a pressure protection valve and a pressure reduction valve, but Technician B is also right.

   Answer B is wrong because many cab air suspensions contain two air springs, two shock absorbers, and a leveling valve, but Technician A is also right.

   Answer C is right because both Technician A and Technician B are right.

   Answer D is wrong because both Technician A and Technician B are right.

9. Answer A is wrong because a sheared shift tube will not cause a truck to experience high steering effort.

   Answer B is wrong because a misaligned column will not cause a truck to experience high steering effort.

   Answer C is wrong because both Technician A and Technician B are wrong.

   Answer D is right because both Technician A and Technician B are wrong.

10. Answer A is wrong because restricted air passages in the power steering cooler cause overheating of the power steering fluid, but Technician B is also right.

    Answer B is wrong because a restricted power steering return hose causes overheating of the power steering fluid, but Technician A is also right.

    Answer C is right because both Technician A and Technician B are right.

    Answer D is wrong because both Technician A and Technician B are right.

11. Answer A is wrong because foaming in the remote reservoir may indicate air in the power steering system, but Technician B is also right.

    Answer B is wrong because some truck manufacturers recommend checking the power steering fluid level at operating temperature, but Technician A is also right.

    Answer C is right because both Technician A and Technician B are right.

    Answer D is wrong because both Technician A and Technician B are right.

12. Answer A is wrong because the lower sector shaft seal can be the source of leaking power steering fluid.

    Answer B is wrong because the submersed style pump to reservoir surface seal can be the source of leaking power steering fluid.

    Answer C is wrong because the supply line double-flare fitting can be the source of leaking power steering fluid.

    Answer D is right because the steering gear input shaft is in the upper part of the steering gear and is not a part of the power assist components.

13. Answer A is wrong because an incompatible fluid will contaminate and discolor the power steering fluid.

    Answer B is right because mixed brands of fluid are compatible. Dye is added to different brands for production runs and identification. Mixing different brands will not discolor existing fluids.

Answer C is wrong because water in power steering fluid will cause foaming and a milky look to the fluid.

Answer D is wrong because overheated fluid will change the fluid to brown and it will have a burnt smell.

14. Answer A is wrong because most truck manufacturers recommend checking the power steering fluid with the fluid at an operating temperature of 175°F (79.4°C).

    Answer B is wrong because most truck manufacturers recommend checking the power steering fluid with the fluid at an operating temperature of 175°F (79.4°C).

    Answer C is wrong because most truck manufacturers recommend checking the power steering fluid with the fluid at an operating temperature of 175°F (79.4°C).

    Answer D is right because most truck manufacturers recommend checking the power steering fluid with the fluid at an operating temperature of 175°F (79.4°C).

15. Answer A is wrong because lack of lubricant in the knuckle pin bushings would cause continual high steering effort.

    Answer B is right because air in the power steering system may cause erratic power steering effort.

    Answer C is wrong because Technician A is wrong and Technician B is right.

    Answer D is wrong because Technician A is wrong and Technician B is right.

16. Answer A is wrong because air is bled from the power steering system by turning the steering wheel fully in each direction with the engine running.

    Answer B is wrong because air is bled from the power steering system by turning the steering wheel fully in each direction with the engine running.

    Answer C is wrong because both Technician A and Technician B are wrong.

    Answer D is right because both Technician A and Technician B are wrong.

17. Answer A is wrong because the low-pressure reading could be caused by the relief valve frozen open, but Technician B is also right.

    Answer B is wrong because the low-perssure reading could be caused by the pump vanes sticking in their slots, but Technician A is also right.

    Answer C is right because both Technician A and Technician B are right.

    Answer D is wrong because both Technician A and Technician B are right.

18. Answer A is wrong because the gate valve should never be closed for more than 10 seconds.

    Answer B is right because the gate valve should never be closed for more than 10 seconds.

    Answer C is wrong because the gate valve should never be closed for more than 10 seconds.

    Answer D is wrong because the gate valve should never be closed for more than 10 seconds.

19. **Answer A is right** because a restricted return line causes higher than specified power steering pump pressure.

    Answer B is wrong because a restricted high-pressure hose causes lower than specified power steering pump pressure as indicated on the analyzer gauge.

    Answer C is wrong because worn rings on the rotary valve would not cause higher than specified power steering pump pressure.

    Answer D is wrong because a loose worm shaft bearing adjustment would not cause higher than specified power steering pump pressure.

20. Answer A is wrong because prying on the reservoir may damage or puncture the reservoir, but Technician B is also right.

    Answer B is wrong because power steering systems may have either an integral or a remote fluid reservoir, but Technician A is also right.

    Answer C is right because both Technician A and Technician B are right.

    Answer D is wrong because both Technician A and Technician B are right.

21. Answer A is wrong because the remote reservoir is connected in the return line.

Answer B is wrong because the remote reservoir may contain a filter.

Answer C is wrong because the remote reservoir must contain the specified fluid level.

Answer D is right because the remote reservoir does not contain a cooler.

22. Answer A is wrong because an overdressed pulley could cause the power steering pump pulley to become misaligned.

    Answer B is wrong because a loose fit from the pulley hug to the pump shaft could cause the power steering pump pulley to become misaligned.

    Answer C is wrong because a worn or loose pump-mounting bracket could cause the power steering pump pulley to become misaligned.

    Answer D is right because a broken engine mount will not cause the power steering pump pulley to become misaligned because all of the components would still move in unison; the whole engine assembly would move.

23. **Answer A is right** because visual inspection cannot verify proper belt tension.

    Answer B is wrong because visual inspection can identify premature belt wear due to misalignment.

    Answer C is wrong because visual inspection can identify correct orientation of a dual belt application.

    Answer D is wrong because visual inspection can identify proper belt seating in the pulley.

24. Answer A is wrong because elongated power steering pump mounting holes may cause a rattling noise with the engine running, but Technician B is also right.

    Answer B is wrong because elongated power steering pump mounting holes may cause premature belt wear, but Technician A is also right.

    Answer C is right because both Technician A and Technician B are right.

    Answer D is wrong because both Technician A and Technician B are right.

25. Answer A is wrong because it is not necessary to retime the engine when this replacement process is complete, but Technician B is also right.

    Answer B is wrong because you also inspect the gear for worn or chipped teeth, but Technician A is also right.

    Answer C is right because both Technician A and Technician B are right.

    Answer D is wrong because both Technician A and Technician B are right.

26. Answer A is wrong because you do check the pump mounting holes for wear.

    Answer B is wrong because you do remove the hoses from the pump and cap the fittings.

    Answer C is right because you never reuse the O-ring even if it is in good condition.

    Answer D is wrong because you bleed the air from the power steering system.

27. Answer A is wrong because a restricted high-pressure hose would not cause excessive belt wear.

    Answer B is wrong because a restricted filter would not cause excessive belt wear.

    Answer C is right because worn power steering pump mounting boltholes cause pulley misalignment and excessive belt wear.

    Answer D is wrong because worn rings on the rotary valve would not cause excessive belt wear.

28. Answer A is wrong because you do the rotation of the input shaft to check trueness before removal.

    Answer B is wrong because you do clean and inspect the gear case for leakage before removal.

    Answer C is wrong because checking all mounting fasteners is a process that you perform when servicing the steering gear.

    Answer D is right because you do not set the toe-in before steering gear removal.

29. Answer A is wrong because the belt tension gauge should be placed in the center of the belt span.

    Answer B is wrong because a prybar should not be placed against any part of the pump reservoir.

    Answer C is right because the pump mounting bolts should be loosened when tightening the belt.

    Answer D is wrong because the belt should be replaced if the bolt is at the end of the adjusting slot in the pump bracket.

30. Answer A is wrong because you should check the power steering cooler air passages and fins for restrictions during any PM process; that is, oil change, lubrication, and so on, but Technician B is also right.

Answer B is wrong because one should also check all power steering hoses for leaks, cracks, dents, and sharp bends, but Technician A is also right.

Answer C is right because both Technician A and Technician B are right.

Answer D is wrong because both Technician A and Technician B are right.

31. **Answer A is right** because restricted power steering cooler fins increase fluid temperature and reduce fluid viscosity, which results in premature pump wear.

    Answer B is wrong because restricted power steering cooler fins do not reduce fluid flow.

    Answer C is wrong because restricted power steering cooler fins increase fluid temperature.

    Answer D is wrong because restricted power steering cooler fins do not affect filter life.

32. Answer A is wrong because worn kingpins can cause binding during cornering but is easily correctable.

    Answer B is wrong because improper steering gear mesh can cause binding during cornering but is easily correctable.

    Answer C is wrong because a bent worm gear can cause binding during cornering but is easily correctable.

    Answer D is right because a worn tie-rod will cause the opposite conditions like drifting and hard steering correction.

33. Answer A is wrong because worn, dry knuckle pin bushings causes excessive steering effort in both directions.

    Answer B is wrong because worn, dry drag link ball studs causes excessive steering effort in both directions.

    Answer C is right because a defective or misadjusted integral valve may cause excessive steering effort in one direction.

    Answer D is wrong because low power steering pump pressure causes excessive steering effort in both directions.

34. Answer A is wrong because the sector shaft should be centered to measure the preload torque, but Technician B is also right.

    Answer B is wrong because both the input and output shafts should be disconnected to measure correct preload torque, but Technician A is also right.

    Answer C is right because both Technician A and Technician B are right.

    Answer D is wrong because both Technician A and Technician B are right.

35. Answer A is wrong because disassembly of the steering gear is not necessary when identifying true center of the steering gear pitman shaft.

    Answer B is wrong because draining lubricant from the gear housing is not necessary when identifying true center of the steering gear pitman shaft.

    Answer C is right because determining backlash areas on either side of the worm gear is a necessary step to identify the true center of the steering gear pitman shaft.

    Answer D is wrong because when identifying steering gear pitman shaft center, the worm and sector gears should not be in a full mesh position.

36. **Answer A is right** because a clogged filter could cause the steering to become harder gradually.

    Answer B is wrong because a loose steering gear mount would cause wandering and shaking in the steering.

    Answer C is wrong because Technician A is right and Technician B is wrong.

    Answer D is wrong because Technician A is right and Technician B is wrong.

37. Answer A is wrong because a worn steering shaft flex joint causes excessive steering wheel free play.

    Answer B is right because blocked airflow through the power steering cooler causes overheating of the power steering fluid.

    Answer C is wrong because a loose tie-rod end causes excessive steering wheel free play and steering wander.

    Answer D is wrong because a loose sector lash adjustment causes excessive steering wheel free play and steering wander.

38. Answer A is wrong because the misadjusted power steering gear poppet valves will not cause

steering wheel kick. Wheel kick or bum steer occurs when the steering wheel reacts to a bump that the wheels hit.

Answer B is right because misadjusted power steering gear poppet valves will cause reduced wheelcut. Wheelcut occurs when the steering wheel cannot be rotated far enough in a right or left turn. The poppet valves provide a hydraulic pressure relief on the power side of a turn, thus reducing the power assist and could prevent the wheel from rotating far enough.

Answer C is wrong because the misadjusted power steering gear poppet valves will not cause directional pull.

Answer D is wrong because the misadjusted power steering gear poppet valves will not cause nonrecovery of the steering.

39. **Answer A is right** because the pitman arm connects the steering gear to the drag link.

Answer B is wrong because the pitman arm connects the steering gear to the drag link.

Answer C is wrong because the pitman arm connects the steering gear to the drag link.

Answer D is wrong because the pitman arm connects the steering gear to the drag link.

40. Answer A is wrong because the vehicle should be road tested when the job is completed.

Answer B is right because you only replace defective tie-rod ends.

Answer C is wrong because whenever a steering component is replaced, it is recommended to do a wheel alignment.

Answer D is wrong because any replaced part should be lubricated after installation.

41. Answer A is wrong because the steering wheel is centered by adjusting the drag link length and resetting the front wheel toe.

Answer B is wrong because the steering wheel is centered by adjusting the drag link length and resetting the front wheel toe.

Answer C is wrong because the steering wheel is centered by adjusting the drag link length and resetting the front wheel toe.

Answer D is right because the steering wheel is centered by adjusting the drag link length and resetting the front wheel toe.

42. **Answer A is right** because a loose sleeve clamp on the drag link adjuster may have damaged the adjusting threads.

Answer B is wrong because a damaged steering component from a collision would not be gradual.

Answer C is wrong because Technician A is right and Technician B is wrong.

Answer D is wrong because Technician A is right and Technician B is wrong.

43. Answer A is wrong because while measuring the idler arm movement, a vertical load of 25 pounds should be applied to the idler arm.

Answer B is wrong because while measuring the idler arm movement, a vertical load of 25 pounds should be applied to the idler arm.

Answer C is right because while measuring the idler arm movement, a vertical load of 25 pounds should be applied to the idler arm.

Answer D is wrong because while measuring the idler arm movement, a vertical load of 25 pounds should be applied to the idler arm.

44. **Answer A is right** because the idler arm positions the relay rod so the tie-rods are parallel to the lower control arms. If the idler arm bushings are worn, the right tie-rod is no longer parallel to the lower control arm and this causes improper toe change and sudden steering veering in one direction when the front suspension strikes a bump.

Answer B is wrong because worn idler arm bushings would not cause a squawking noise when turning the front wheels.

Answer C is wrong because worn idler arm bushings would not cause excessive steering effort.

Answer D is wrong because worn idler arm bushings would not cause improper steering wheel return.

45. Answer A is wrong because the tie-rod clamps must NOT be positioned so the slot in the clamp is aligned with the slot in the tie-rod. The opposite is true; that is, the clamp is opposite the slot.

Answer B is right because the length of the tie-rod determines the front wheel toe.

Answer C is wrong because Technician A is wrong and Technician B is right.

Answer D is wrong because Technician A is wrong and Technician B is right.

46. **Answer A is right** because a damaged or bent steering arm can cause a change in steering wheel position.

    Answer B is wrong because a damaged or bent steering arm would not change total toe measurement. The tie-rods and cross tube or relay rod control toe-in. The steering arm is mounted to the left side knuckle and the drag link connects it to the pitman arm and steering gear.

    Answer C is wrong because a damaged or bent steering arm would not change individual toe measurement.

    Answer D is wrong because a damaged or bent steering arm would not cause a pull condition.

47. Answer A is wrong because the tie-rod in the figure is correctly threaded into the shaft, but Technician B is also right.

    Answer B is wrong because the threaded portion of the tie-rod end needs to be beyond the cross tube split, but Technician A is also right.

    Answer C is right because both Technician A and Technician B are right.

    Answer D is wrong because both Technician A and Technician B are right.

48. Answer A is wrong because this tie-rod end should be installed with the front wheels straight ahead.

    Answer B is right because this tie-rod end should be installed with the front wheels straight ahead.

    Answer C is wrong because this tie-rod end should be installed with the front wheels straight ahead.

    Answer D is wrong because this tie-rod end may be installed with the front tire supporting the vehicle weight.

49. **Answer A is right** because the drag link is connected to the tapered opening in item 9, the upper steering arm.

    Answer B is wrong because the drag link is connected to the tapered opening in item 9, the upper steering arm.

    Answer C is wrong because the drag link is connected to the tapered opening in item 9, the upper steering arm.

    Answer D is wrong because the drag link is connected to the tapered opening in item 9, the upper steering arm.

50. Answer A is wrong because wheel imbalance and wheel stops do not affect each other.

    Answer B is right because when wheel stops are missing, damaged, or out of specified adjustment range, it can cause excessive turning radius.

    Answer C is wrong because the wheel stops missing, damaged, or out of specified adjustment range will not cause the steering wheel to be off center.

    Answer D is wrong because the wheel stops missing, damaged, or out of specified adjustment range will not cause steering wheel binding on turns.

51. Answer A is wrong because turning component B out too far will not damage component B.

    Answer B is right because turning component B out too far reduces the turning radius.

    Answer C is wrong because turning component B out too far will not bend the steering arm.

    Answer D is wrong because turning component B does not affect front wheel toe.

52. Answer A is wrong because you must remove the lockbolt and the draw key before the kingpin will disassemble from the knuckle, but Technician B is also right.

    Answer B is wrong because you should grease the kingpin after reassembling the knuckle, but Technician A is also right.

    Answer C is right because both Technician A and Technician B are right.

    Answer D is wrong because both Technician A and Technician B are right.

53. Answer A is wrong because checking runout will identify a bent rim, not the front axle.

    Answer B is right because you measure the front wheel setback on both wheels and compare. A difference in measurement could be a bent axle.

    Answer C is wrong because the relay rod length will not determine axle damage.

    Answer D is wrong because toe-in would only identify a bend in the steering component.

54. Answer A is wrong because a bent axle cannot be determined by visual inspection.

    Answer B is wrong because any bend in the axle is reason for replacement.

    Answer C is wrong because both Technician A and Technician B are wrong.

    Answer D is right because both Technician A and Technician B are wrong.

55. **Answer A is right** because one must remove the knuckle cap to access the kingpin.

    Answer B is wrong because you cannot remove the spindle until you remove the kingpin.

    Answer C is wrong because the Ackerman arm can stay in place during steering knuckle removal.

    Answer D is wrong because the axle does not have to be removed.

56. **Answer A is right** because the right side and left side kingpins are the same.

    Answer B is wrong because a burred bushing will interfere with kingpin installation.

    Answer C is wrong because an improperly aligned bushing will interfere with kingpin installation.

    Answer D is wrong because omission of the reaming operation will leave burrs that will interfere with the kingpin installation.

57. Answer A is wrong because shock absorber outer tube touching the fluid reservoir will not cause premature bushing wear.

    Answer B is right because the shock absorber outer tube touching the fluid reservoir will cause a scraping noise during shock travel.

    Answer C is wrong because shock absorber outer tube touching the fluid reservoir will not cause main shaft seal failure.

    Answer D is wrong because shock absorber outer tube touching the fluid reservoir will not cause excessive fatigue in the mount assembly.

58. **Answer A is right** because leaking, worn-out front shock absorbers can cause the steering wheel to shake for a few seconds after a stretch of rough road.

    Answer B is wrong because low tire pressure would not cause wheel shake.

    Answer C is wrong because rusted rear shock absorbers would not cause wheel shake.

    Answer D is wrong because a missing jounce bumper would cause a bang on a rough road but not a shake.

59. Answer A is wrong because loose spring shackles will NOT break the center spring boltholes.

    Answer B is right because loose U-bolts can break the center spring bolthole.

    Answer C is wrong because Technician A is wrong and Technician B is right.

    Answer D is wrong because Technician A is wrong and Technician B is right.

60. **Answer A is right** because leaf spring shackles are necessary and allow free lateral movement of the spring.

    Answer B is wrong because an adjustment is necessary to compensate for off-center static loads.

    Answer C is wrong because Technician A is right and Technician B is wrong.

    Answer D is wrong because Technician A is right and Technician B is wrong.

61. Answer A is wrong because only a loose shock absorber mount would rattle over bumps.

    Answer B is wrong because underinflated tires will not cause a rattling sound.

    Answer C is right because a broken mount fastener will rattle when going over bumps.

    Answer D is wrong because a cracked air supply line will hiss, not rattle.

62. Answer A is wrong because torque rods may be substituted for shock absorbers, but Technician B is also right.

    Answer B is wrong because four to six torque rods may be used on rear spring suspension systems, but Technician A is also right.

    Answer C is right because both Technician A and Technician B are right.

    Answer D is wrong because both Technician A and Technician B are right.

63. Answer A is wrong because axle alignment adjustment on a spring suspension with torque rods can be made through an lower adjustable torque rod.

Answer B is wrong because shims can be used between the torque rod front and spring hanger bracket to align the rear axle.

Answer C is wrong because adjustment can be made with an eccentric bushing at the torque leaf.

Answer D is right because adjustment is NOT made through the upper adjustable torque rods. These rods are used to absorb braking forces and driveline angles.

64. Answer A is wrong because a bent rear axle shaft does not affect rear axle alignment.

    Answer B is wrong because a loose wheel bearing does not have that much effect on rear axle alignment.

    Answer C is right because a bent torque rod causes rear axle misalignment.

    Answer D is wrong because an improperly mounted wheel does not have that much effect on rear axle alignment.

65. Answer A is wrong because replacing a bent radius rod could correct axle alignment.

    Answer B is right because an axle shaft replacement will not correct the alignment of the axle assembly.

    Answer C is wrong because replacing a defective tracking bar could correct axle alignment.

    Answer D is wrong because turning the adjustment eccentric is a method of adjusting the rear axle alignment.

66. Answer A is wrong because rotating the lower torque rods adjusts rear axle position on some rear suspension systems, but Technician B is also right.

    Answer B is wrong because rotating eccentric bushings in the spring hangers does adjust the rear axle position on some spring designs, but Technician A is also right.

    Answer C is right because both Technician A and Technician B are right.

    Answer D is wrong because both Technician A and Technician B are right.

67. Answer A is wrong because the flat surface on the bar pin is installed at a 5-degree to 15-degree angle in relation to the torque rod, depending on the bar pin location.

    Answer B is wrong because the front and rear bar pins are installed at different angles.

    Answer C is wrong because installing another shim moves the rear axle forward since the shims are positioned between the front of the spring hanger bracket and the bar pin.

    Answer D is right because the torque rods absorb braking and acceleration torque.

68. **Answer A is right** because you can service the walking (equalizing) beam center and end bushings on the vehicle using a special tool made by Owatonna.

    Answer B is wrong because using heat on a rubber bushing is not recommended.

    Answer C is wrong because Technician A is right and Technician B is wrong.

    Answer D is wrong because Technician A is right and Technician B is wrong.

69. Answer A is wrong because there are two types of equalizing beam suspensions: leaf spring and rubber cushion, but Technician B is also right.

    Answer B is wrong because equalizing beam suspension lowers the center of gravity of the axle load, but Technician A is also right.

    Answer C is right because both Technician A and Technician B are right.

    Answer D is wrong because both Technician A and Technician B are right.

70. Answer A is wrong because worn walking beam bushings do not contribute to broken spring leaves.

    Answer B is right because worn walking beam bushing causes rear axle misalignment and steering pull or wander.

    Answer C is wrong because worn walking beam bushings do not cause broken spring center bolts.

    Answer D is wrong because worn walking beam bushings do not cause shock absorber leaks.

71. Answer A is wrong because a leaking air spring could cause the truck to sit low with the trailer attached but not high when the trailer is removed.

    Answer B is right because an improperly adjusted height control valve can cause the vehicle to sit too low or too high.

    Answer C is wrong because Technician A is wrong and Technician B is right.

    Answer D is wrong because Technician A is wrong and Technician B is right.

72. Answer A is wrong because air spring suspensions do not eliminate the need for shock absorbers.

    Answer B is right because the rear axle air springs are usually mounted on the cross channel or transverse beam.

    Answer C is wrong because Technician A is wrong and Technician B is right.

    Answer D is wrong because Technician A is wrong and Technician B is right.

73. **Answer A is right** because the air suspension system has individual height control valves on each side of the suspension, and a plugged exhaust valve may not release air pressure after unloading. This results in an increase in suspension height on one side.

    Answer B is wrong because a loose air spring seat would not cause an increase in suspension height on one side after unloading.

    Answer C is wrong because a worn support beam bushing may cause rear axle misalignment, but it would not cause an increase in suspension height on one side after unloading.

    Answer D is wrong because a defective pressure protection valve may cause low brake system air pressure, but it would not cause an increase in suspension height on one side after unloading.

74. Answer A is wrong because a rougher than normal ride cannot be caused by a slight leak at the governor. This causes higher than normal system air pressure.

    Answer B is wrong because a defective air suspension relief valve will not cause a rough ride.

    Answer C is wrong because both Technician A and Technician B are wrong.

    Answer D is right because both Technician A and Technician B are wrong.

75. Answer A is wrong because the truck must be empty.

    Answer B is wrong because it is not necessary to loosen the height control valve mounting bolts.

    Answer C is right because you must use safety stands with the specified weight rating.

    Answer D is wrong because you move the height control valve lever downward to deflate all air.

76. Answer A is wrong because the height control valve is bolted to the truck frame.

    Answer B is wrong because the height control valve linkage is usually connected to the transverse beam.

    Answer C is wrong because the height control valve does not require lubrication.

    Answer D is right because the height control valve may contain an exhaust (dump) valve.

77. Answer A is wrong because camber does not affect ride quality.

    Answer B is wrong because excessive negative caster results in reduced directional stability, but it does not adversely affect ride quality.

    Answer C is right because a fifth wheel too far ahead on the tractor causes rough ride quality on the front suspension.

    Answer D is wrong because unequal loading of the rear tandem axles may cause rough ride quality.

78. **Answer A is right** because worn equalizing beam bushings may cause wheel hop.

    Answer B is wrong because improper torque on the shock absorber mounting bolts will not cause wheel hop.

    Answer C is wrong because reduced torque on the spring U-bolts will not cause wheel hop.

    Answer D is wrong because a bent cross tube will not cause wheel hop.

79. **Answer A is right** because a dry kingpin pivot bearing will cause a squeak when turning the wheel left to right.

    Answer B is wrong because a bent front wheel will never cause a squeaking sound.

    Answer C is wrong because an overfilled fluid reservoir may cause a whining sound but not a squeak.

    Answer D is wrong because worn shock absorbers may cause a squeaking sound when bounced up and down but not when turning left to right.

80. Answer A is wrong because too high a toe-in setting cannot cause steering wheel shimmy. Toe only affects directional stability and tire wear.

    Answer B is right because improper dynamic wheel balance can cause steering wheel shimmy.

Answer C is wrong because an excessive vehicle load will not cause wheel shimmy.

Answer D is wrong because too high a toe-out setting cannot cause steering wheel shimmy. Toe only affects directional stability and tire wear.

81. **Answer A is right** because a shimmy after an alignment has been performed can be caused by too much positive caster.

    Answer B is wrong because an improper camber setting will NOT cause wheel shimmy.

    Answer C is wrong because Technician A is right and Technician B is wrong.

    Answer D is wrong because Technician A is right and Technician B is wrong.

82. Answer A is wrong because 1/4 degree positive is acceptable.

    Answer B is right because a 1/8 degree negative reading is too negative exceeding 1/4 degree positive and the 1/8 degree tolerance. The appropriate measure should be either 1/4 positive, 3/8 positive, or 1/8 positive.

    Answer C is wrong because a 3/8 degree positive is acceptable.

    Answer D is wrong because a 1/8 degree positive is acceptable.

83. Answer A is wrong because Figure 5–15 does not show a toe-out condition.

    Answer B is wrong because Figure 5–15 does not show a setback condition.

    Answer C is wrong because Figure 5–15 does not show a positive caster condition.

    Answer D is right because Figure 5–15 shows a negative camber condition.

84. Answer A is wrong because the KPI angle is not adjustable. If KPI is not within specifications, some components are bent or worn.

    Answer B is wrong because the camber angle is not adjustable. If camber is not within specifications, some components are bent or worn.

    Answer C is right because the included angle is the sum of the positive camber angle and the KPI.

    Answer D is wrong because excessive positive camber wears the outside edge of the tire.

85. **Answer A is right** because caster is typically adjusted with the use of shims.

    Answer B is wrong because camber is not adjustable on a solid axle.

    Answer C is wrong because toe-in is not adjustable with shims.

    Answer D is wrong because a shim cannot correct spring sag.

86. Answer A is wrong because you do not use more than one shim on each side of the axle.

    Answer B is right because you do not shim one side more than one degree than the other side.

    Answer C is wrong because Technician A is wrong and Technician B is right.

    Answer D is wrong because Technician A is wrong and Technician B is right.

87. Answer A is wrong because you make caster more positive by installing a shim between the axle and spring with the LOW side at the front, not the high side.

    Answer B is wrong because you make the caster more negative by installing a shim between the axle and spring with the high side at the front, not the low side.

    Answer C is wrong because both Technician A and Technician B are wrong.

    Answer D is right because both Technician A and Technician B are wrong.

88. Answer A is wrong because the distance between A and B in Figure 5–18 is a toe-out condition.

    Answer B is wrong because the distance between A and B in Figure 5–18 is a toe-out condition.

    Answer C is right because the distance between A and B in Figure 5–18 is a toe-out condition.

    Answer D is wrong because the distance between A and B in Figure 5–18 is a toe-out condition.

89. Answer A is wrong because toe setting on the average is 1/16 inch to 1/32 inch on an unloaded vehicle.

    Answer B is right because toe setting should be 1/16 inch to 1/32 inch on most M/HD trucks.

    Answer C is wrong because the adjustment should be 1/16 inch to 1/32 inch on an unloaded vehicle.

    Answer D is wrong because 3/4 to 1 inch is too large an amount.

90. Answer A is wrong because dog tracking is a condition where the rear wheels are not in proper

thrustline. This condition appears like the truck is going diagonally like a dog.

Answer B is right because proper thrustline describes when a vehicle's rear wheels track directly behind the front wheels.

Answer C is wrong because the Ackerman effect concerns the wheel angle during a turn on the front axle.

Answer D is wrong because "toe-in" is an adjustment angle of the front wheels.

91. Answer A is wrong because axle alignment adjustment on a spring suspension with torque rods can be made through an lower adjustable torque rod.

Answer B is wrong because shims can be used between the torque rod front and spring hanger bracket to align the rear axle.

Answer C is wrong because adjustment can be made with an eccentric bushing at the torque leaf.

Answer D is right because adjustment is NOT made through the upper adjustable torque rods. These rods are used to absorb braking forces and driveline angles.

92. **Answer A is right** because worn walking (equalizing) beam bushings can cause the rear axles to be out of alignment.

Answer B is wrong because a walking (equalizing) beam suspension does not use torque leaf eccentrics.

Answer C is wrong because Technician A is right and Technician B is wrong.

Answer D is wrong because Technician A is right and Technician B is wrong.

93. Answer A is wrong because a bent steering arm may cause improper toe-out-on-turns, but Technician B is also right.

Answer B is wrong because improper toe-out-on-turns causes tire scuffing, but Technician A is also right.

Answer C is right because both Technician A and Technician B are right.

Answer D is wrong because both Technician A and Technician B are right.

94. Answer A is wrong because when the outside wheel is turned 20 degrees, the inside wheel should be turned 22 degrees.

Answer B is right because when the outside wheel is turned 20 degrees, the inside wheel should be turned 22 degrees.

Answer C is wrong because when the outside wheel is turned 20 degrees, the inside wheel should be turned 22 degrees.

Answer D is wrong because when the outside wheel is turned 20 degrees, the inside wheel should be turned 22 degrees.

95. Answer A is wrong because defective rear axle bearings will not cause premature rear tire wear.

Answer B is wrong because caster is not a rear wheel measurement.

Answer C is wrong because front wheel toe-in out of specification will not affect premature rear tire wear.

Answer D is right because a bent rear axle housing by changing wheel camber or toe could cause premature rear tire wear.

96. **Answer A is right** because the tire wear in Figure 5–19 is caused by excessive toe-out.

Answer B is wrong because the tire wear in Figure 5–19 is not caused by negative camber.

Answer C is wrong because the tire wear in Figure 5–19 is not caused by excessive positive caster.

Answer D is wrong because the tire wear in Figure 5–19 is not caused by underinflation.

97. Answer A is wrong because the tire wear in Figure 5–20 is not caused by excessive toe-out.

Answer B is right because the tire wear in Figure 5–20 is not caused by excessive positive camber.

Answer C is wrong because positive caster does not cause tire wear.

Answer D is wrong because the tire wear in Figure 5–20 is not caused by overinflation.

98. Answer A is wrong because toe and camber are the two angles that affect tire wear.

Answer B is wrong because toe and camber are the two angles that affect tire wear.

Answer C is wrong because toe and camber are the two angles that affect tire wear.

Answer D is right because improper camber causes front tire tread wear.

99. Answer A is wrong because improper wheel balance can cause a vibration in the steering wheel.

    Answer B is right because worn kingpin bushings are the least likely cause of vibration in the steering wheel.

    Answer C is wrong because a bent wheel mounting surface can cause a dynamic imbalance leading to improper wheel balance and the resulting wheel vibration.

    Answer D is wrong because a shifted belt inside a tire can cause a vibration in the steering wheel.

100. **Answer A is right** because worn shock absorbers can cause cupping marks around the tire.

    Answer B is wrong because camber will cause tread wear out unevenly across the tire.

    Answer C is wrong because toe-out-on-turns is a measurement.

    Answer D is wrong because caster does not cause tire wear.

101. Answer A is wrong because too high a toe-in setting cannot cause steering wheel shimmy. Toe only affects directional stability and tire wear.

    Answer B is right because improper dynamic wheel balance can cause steering wheel shimmy.

    Answer C is wrong because an excessive vehicle load will not cause wheel shimmy.

    Answer D is wrong because too high a toe-out setting cannot cause steering wheel shimmy. Toe only affects directional stability and tire wear.

102. **Answer A is right** because a shimmy after an alignment has been performed can be caused by too much positive caster.

    Answer B is wrong because an improper camber setting will NOT cause wheel shimmy.

    Answer C is wrong because Technician A is right and Technician B is wrong.

    Answer D is wrong because Technician A is right and Technician B is wrong.

103. Answer A is wrong because a disc wheel will have right-hand threads on the right, left hand threads on the left, but Technician B is also right.

    Answer B is wrong because you always wear safety glasses and do not stand in front of a deflating tire, but Technician A is also right.

    Answer C is right because both Technician A and Technician B are right.

    Answer D is wrong because both Technician A and Technician B are right.

104. Answer A is wrong because excessive rust or corrosion buildup could cause problems when dismounting a split ring tire.

    Answer B is right because tire brand is insignificant when dismounting a split ring tire.

    Answer C is wrong because bent flanges can be a hazard when dismounting a split ring tire.

    Answer D is wrong because deep tool marks on the rings and gutter area can be a hazard when dismounting a split ring tire.

105. Answer A is wrong because dismounting and mounting split side rims is extremely dangerous and only trained professionals should perform this process according to OSHA rules and regulations, but Technician B is also right.

    Answer B is wrong because you never use a hammer on split side rim rings, but Technician A is also right.

    Answer C is right because both Technician A and Technician B are right.

    Answer D is wrong because both Technician A and Technician B are right.

106. **Answer A is right** because the equipment is installed to measure radial wheel runout.

    Answer B is wrong because the equipment is installed to measure radial wheel runout.

    Answer C is wrong because the equipment is installed to measure radial wheel runout.

    Answer D is wrong because the equipment is installed to measure radial wheel runout.

107. Answer A is wrong because you should never attempt to straighten a wheel.

    Answer B is right because the first step in trying to correct runout on a spoke wheel is to loosen the stud nut at the point of greatest clearance, and tighten the opposite stud nut. Then be sure all stud nuts are tightened to specifications.

    Answer C is wrong because the first step in trying to correct runout on a spoke wheel is to loosen the stud nut at the point of greatest clearance, and tighten the opposite stud nut. Then be sure all stud nuts are tightened to specifications.

Answer D is wrong because the first step in trying to correct runout on a spoke wheel is to loosen the stud nut at the point of greatest clearance, and tighten the opposite stud nut. Then be sure all stud nuts are tightened to specifications.

108. Answer A is wrong because overinflation will not wear the edges of the tire.

Answer B is right because overinflation will wear the center of the tire.

Answer C is wrong because tire bounce causes a cupping pattern.

Answer D is wrong because camber causes wear on the inside of the tire.

109. **Answer A is right** because you cannot mix radial tires with bias tires on a steering axle.

Answer B is wrong because mixing tires on the rear axle is dangerous because radial tires expand more than bias tires.

Answer C is wrong because Technician A is right and Technician B is wrong.

Answer D is wrong because Technician A is right and Technician B is wrong.

110. **Answer A is right** because you must inflate mounted tires in a safety cage or using a portable lock ring guard.

Answer B is wrong because you do not mount the tire on the truck before inflation because an improperly fitted rim ring could dislodge, causing injury.

Answer C is wrong because Technician A is right and Technician B is wrong.

Answer D is wrong because Technician A is right and Technician B is wrong.

111. Answer A is wrong because worn shock absorbers would not cause a popping sound or shake in the steering.

Answer B is wrong because an out-of-balance tire would cause a shake but no popping sound.

Answer C is wrong because an incorrect toe-in setting could cause tire wear but not a popping sound or shake in the steering.

Answer D is right because a separated belt in the tire could cause a popping sound and shake in the steering.

112. Answer A is wrong because the maximum wheel weight per tire should not exceed 18 ounces, but Technician B is also right.

Answer B is a wrong because if 16 ounces of weight is required in one spot, use an 8-ounce weight on each side of the rim directly across from each other, but Technician A is also right.

Answer C is right because both Technician A and Technician B are right.

Answer D is wrong because both Technician A and Technician B are right.

113. **Answer A is right** because when using a static balancing process, a wheel weight is installed 180 degrees from this heavy spot on the outside edge.

Answer B is wrong because the Technician must enter the wheel diameter, width, and offset in an electronic wheel balancer.

Answer C is wrong because Technician A is right and Technician B is wrong.

Answer D is wrong because Technician A is right and Technician B is wrong.

114. Answer A is wrong because before balancing a radial tire, you mount the tire and check for runout, but Technician B is also right.

Answer B is wrong because after the wheels are balanced you should mount the tire with the wheel weights 180 degrees away from the brake drum weights, but Technician A is also right.

Answer C is right because both Technician A and Technician B are right.

Answer D is wrong because both Technician A and Technician B are right.

115. Answer A is wrong because vulcanizing will change the dynamic balance of the tire.

Answer B is right because a vulcanized tire needs rebalance.

Answer C is wrong because Technician A is wrong and Technician B is right.

Answer D is wrong because Technician A is wrong and Technician B is right.

116. Answer A is wrong because one can perform dynamic balance with the tire on the truck.

Answer B is right because before removing a dynamically balanced tire from the truck, it should be marked and installed in the same position as removed.

Answer C is wrong because Technician A is wrong and Technician B is right.

Answer D is wrong because Technician A is wrong and Technician B is right.

117. Answer A is wrong because you cannot mix radial tires with bias tires on a steering axle.

Answer B is right because mixing tires on the rear axle is dangerous because radial tires expand more than bias tires.

Answer C is wrong because Technician A is wrong and Technician B is right.

Answer D is wrong because Technician A is wrong and Technician B is right.

118. Answer A is wrong because matching tire sizes on dual wheels prevents tire tread wear from slippage because of uneven tire surface areas, but Technician B is also right.

Answer B is wrong because dual drive wheels may be measured with a tape measure, caliper, and so on, but Technician A is also right.

Answer C is right because both Technician A and Technician B are right.

Answer D is wrong because both Technician A and Technician B are right.

119. **Answer A is right** because the measurement in Figure 5–22 is the tire diameter comparison.

Answer B is wrong because the measurement in Figure 5–22 is the tire diameter comparison.

Answer C is wrong because the measurement in Figure 5–22 is the tire diameter comparison.

Answer D is wrong because the measurement in Figure 5–22 is the tire diameter comparison.

120. **Answer A is right** because the two main designs of truck wheels used are single piece and split side rims.

Answer B is wrong because aluminum is the material used in some single piece wheels.

Answer C is wrong because disc and drum are not wheel designs.

Answer D is wrong because hub and spoke designs are used in conjunction with split side rims.

121. Answer A is wrong because when you reinstall the dual wheel assembly of a tractor axle, you do put a light coating lubricant on the spindle and place a seal protector over the threads, but Technician B is also right.

Answer B is wrong because some OEMs use oil filler plugs in the rear wheel hubs, but Technician A is also right.

Answer C is right because both Technician A and Technician B are right.

Answer D is wrong because both Technician A and Technician B are right.

122. Answer A is wrong because you do use a wheel dolly to install the dual wheel assembly.

Answer B is wrong because you use the OEM recommended lube on the spindle.

Answer C is right because you only install the specified amount of lubricant in the wheel cavity.

Answer D is wrong because you do install the adjusting nut on the spindle.

123. Answer A is wrong because rear wheel bearings do not require repacking.

Answer B is right because rear wheel bearings are lubricated with rear axle fluid.

Answer C is wrong because Technician A is wrong and Technician B is right.

Answer D is wrong because Technician A is wrong and Technician B is right.

124. Answer A is wrong because spinning a bearing with compressed air can cause the bearing to explode.

Answer B is right because spinning a bearing with compressed air can cause the bearing to explode.

Answer C is wrong because Technician A is wrong and Technician B is right.

Answer D is wrong because Technician A is wrong and Technician B is right.

125. Answer A is wrong because the lip on the rear hub seal must face toward the center of the hub.

Answer B is right because you must torque the adjusting nut to 50 ft.-lb., then back off the nut the specified amount and install the lock ring and jam nut.

Answer C is wrong because Technician A is wrong and Technician B is right.

Answer D is wrong because Technician A is wrong and Technician B is right

126. **Answer A is right** because the hub end play should be 0.001 to 0.005 in. (0.054 to 0.127 mm) after the wheel bearing is properly adjusted.

     Answer B is wrong because the hub end play should be 0.001 to 0.005 in. (0.054 to 0.127 mm) after the wheel bearing is properly adjusted.

     Answer C is wrong because the hub end play should be 0.001 to 0.005 in. (0.054 to 0.127 mm) after the wheel bearing is properly adjusted.

     Answer D is wrong because the hub end play should be 0.001 to 0.005 in. (0.054 to 0.127 mm) after the wheel bearing is properly adjusted.

127. Answer A is wrong because defective rear wheel bearings produce growling noises at low speeds.

     Answer B is right because a growling noise produced by a defective front wheel bearing is most noticeable while turning a corner.

     Answer C is wrong because Technician A is wrong and Technician B is right.

     Answer D is wrong because Technician A is wrong and Technician B is right.

128. Answer A is wrong because the lubricant level should be at the lower edge of the filler hole.

     Answer B is wrong because the lubricant level should be at the lower edge of the filler hole.

     Answer C is right because the lubricant level should be at the lower edge of the filler hole.

     Answer D is wrong because the lubricant level should be at the lower edge of the filler hole.

129. Answer A is wrong because the fifth wheel in Figure 5–24 is a sliding fifth wheel.

     Answer B is wrong because the fifth wheel in Figure 5–24 is a sliding fifth wheel.

     Answer C is wrong because the fifth wheel in Figure 5–24 is a sliding fifth wheel.

     Answer D is right because the fifth wheel in Figure 5–24 is a sliding fifth wheel.

130. **Answer A is right** because the elevating fifth wheel in Figure 5–25 is used to change an over-the-road tractor to one that is used for spotting and switching trailers.

     Answer B is wrong because the elevating fifth wheel in Figure 5–25 is used to change an over-the-road tractor to one that is used for spotting and switching trailers.

     Answer C is wrong because the elevating fifth wheel in Figure 5–25 is used to change an over-the-road tractor to one that is used for spotting and switching trailers.

     Answer D is wrong because the elevating fifth wheel in Figure 5–25 is used to change an over-the-road tractor to one that is used for spotting and switching trailers.

131. **Answer A is right** because if you locate a fifth wheel too far forward on a tractor, hard steering and tire wear can result.

     Answer B is wrong because you never weld anything to the frame flanges.

     Answer C is wrong because Technician A is right and Technician B is wrong.

     Answer D is wrong because Technician A is right and Technician B is wrong.

132. Answer A is wrong because high hitch can occur when you position the trailer kingpin (pintle) and bolster plate too high in the fifth wheel, but Technician B is also right.

     Answer B is wrong because this condition is also caused by improper bolster plate position during the coupling process, but Technician A is also right.

     Answer C is right because both Technician A and Technician B are right.

     Answer D is wrong because both Technician A and Technician B are right.

133. Answer A is wrong because when an air slide release fifth wheel is being released, you lower the landing gear to relieve pressure on the plungers.

     Answer B is right because if the locking plungers will not release, the air cylinder may be defective.

     Answer C is wrong because the tractor must never be moving when releasing the fifth wheel locking plungers.

     Answer D is wrong because the fifth wheel locking plungers may be released with the trailer loaded, and the landing gear lowered to release the pressure on the locking plungers.

134. Answer A is wrong because if the fifth wheel is too far ahead, the steering effort is increased.

     Answer B is right because if the fifth wheel is to far rearward, premature wheel lockup may occur during a hard brake application.

Answer C is wrong because Technician A is wrong and Technician B is right.

Answer D is wrong because Technician A is wrong and Technician B is right.

135. Answer A is wrong because tandem axle tractors have the maximum bending moment occur just ahead of the fifth wheel, not at the bogie centerline.

Answer B is wrong because single axle trucks with van bodies have the maximum bending moment occur just rear of the cab, not ahead of the rear axle.

Answer C is wrong because frame buckle cannot be caused by too many holes drilled in the frame web. Drilling too many holes causes frame sag.

Answer D is right because frame twist may be caused by uneven loading.

136. Answer A is wrong because drilling too many holes in the frame will not twist it.

Answer B is right because extreme operating conditions will twist a frame.

Answer C is wrong because using the wrong bolts in a repair will not twist a frame.

Answer D is wrong because a loose fifth wheel mounting will not twist a frame.

137. **Answer A is right** because towing another truck with a chain attached to one corner of the frame will cause a diamond frame condition.

Answer B is wrong because the shifting of the front or rear axles will not cause frame buckle. The shifting of axles affect alignment and tracking.

Answer C is wrong because Technician A is right and Technician B is wrong.

Answer D is wrong because Technician A is right and Technician B is wrong.

138. Answer A is wrong because holes drilled too close together in the frame may cause frame sag.

Answer B is wrong because welding the truck frame may cause frame sag.

Answer C is right because collision damage usually does not cause frame sag.

Answer D is wrong because cutting notches in the frame rails may cause frame sag.

139. Answer A is wrong because fishplate frame reinforcements may extend above or below the frame.

Answer B is wrong because channel frame reinforcements are installed on the inside or outside of the frame.

Answer C is wrong because both Technician A and Technician B are wrong.

Answer D is right because both Technician A and Technician B are wrong.

140. Answer A is wrong because you never weld brackets to the frame.

Answer B is wrong because deep section frame reinforcements are installed near the front of the frame.

Answer C is wrong because gussets may *not* be used when attaching the reinforcement plate to the frame.

Answer D is right because when you install frame reinforcement plates, the original boltholes in the frame should be used.

141. **Answer A is right** because the length of any pair of diagonal lines should not vary more than 1/4 in. (6.35 mm).

Answer B is wrong because the length of any pair of diagonal lines should not vary more than 1/4 in. (6.35 mm).

Answer C is wrong because the length of any pair of diagonal lines should not vary more than 1/4 in. (6.35 mm).

Answer D is wrong because the length of any pair of diagonal lines should not vary more than 1/4 in. (6.35 mm).

142. **Answer A is right** because you disconnect the truck batteries before welding a frame.

Answer B is wrong because you never weld across frame flanges.

Answer C is wrong because Technician A is right and Technician B is wrong.

Answer D is wrong because Technician A is right and Technician B is wrong.

143. **Answer A is right** because you never drill holes in the flanges of a truck frame, only in the webs. Drilling the flanges will weaken the frame.

Answer B is wrong because you can drill a limited number holes in the web of a truck frame.

Answer C is wrong because Technician A is right and Technician B is wrong.

Answer D is wrong because Technician A is right and Technician B is wrong.

144. Answer A is wrong because the pintle hook latch is operated by spring pressure.

Answer B is wrong because the pintle hook latch is operated by spring pressure.

Answer C is right because the pintle hook latch is operated by spring pressure.

Answer D is wrong because the pintle hook latch is operated by spring pressure.

145. Answer A is wrong because the pintle hook is bolted to the centerline of the forward trailer.

Answer B is wrong because the pintle hook is bolted to the centerline of the forward trailer.

Answer C is wrong because the pintle hook is bolted to the centerline of the forward trailer.

Answer D is right because the pintle hook is bolted to the centerline of the forward trailer.

6 Electrical/Electronic Systems

Pretest

The purpose of this pretest is to determine the amount of review you may require before taking the ASE Medium/Heavy Truck Electrical/Electronic Systems Test. If you answer all the pretest questions correctly, complete the questions and study the information in this chapter to prepare for this test.

If two or more of your answers to the pretest questions are incorrect, study the required information in *Today's Technician Medium/Heavy-Duty Truck Electrical/Electronic Systems* published by Delmar Publishers, plus a study of the questions and information in this chapter.

The pretest answers are located at the end of the pretest; these answers also are in the answer sheets supplied with this book.

1. Which of these is most likely to be done when checking a circuit for continuity?
   A. Disconnect the battery negative cable.
   B. Connect a voltmeter in series in the circuit being tested.
   C. Turn on the ignition switch and use an ammeter.
   D. Connect an ohmmeter to the disconnected component.

2. The meter in Figure 6–1 is connected to measure:
   A. current.
   B. resistance.
   C. voltage.
   D. watts.

3. A charging system check shows a low voltage output. *Technician A* says a faulty voltage regulator may cause this problem. *Technician B* says an improperly adjusted alternator belt may cause this problem. Who is right?
   A. A only
   B. B only

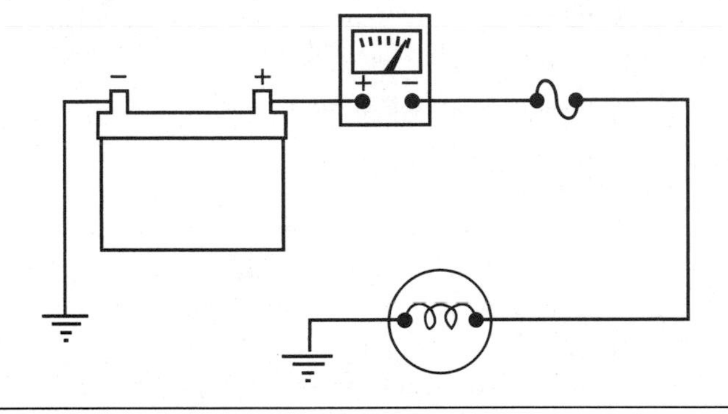

Figure 6–1 Electrical circuit diagnosis. *(Courtesy of Chrysler Corporation)*

   C. Both A and B
   D. Neither A nor B

4. The switch is closed in Figure 6–2 and the battery is fully charged. The voltmeter indicates that the voltage drop across the light bulb is 9 volts. *Technician A* says that there may be a short to ground between the switch and the light. *Technician B* says the resistance in the circuit may be too high, causing low voltage at the bulb. Who is right?
   A. A only
   B. B only

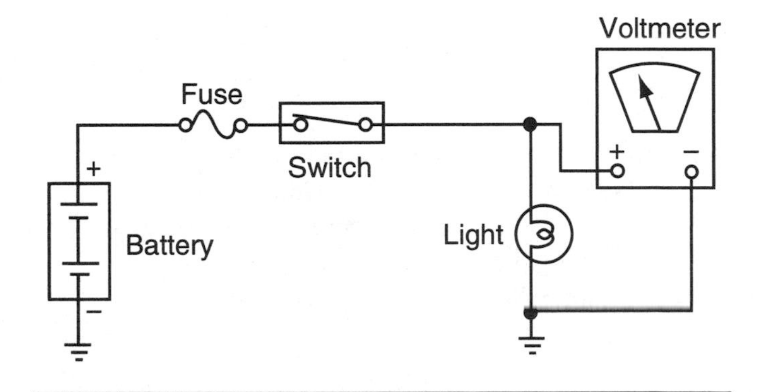

Figure 6–2 Circuit diagnosis with a voltmeter.

C. Both A and B
D. Neither A nor B

5. When checking the voltage drop on a starter motor, which of these processes is LEAST likely to be performed?
   A. Make sure the battery is fully charged.
   B. Check the voltage drop while operating the starter motor.
   C. Check the voltage drop across both battery cables.
   D. Disconnect the engine ground cable.

6. A horn operates continuously without pressing the horn button. Which of these would be the MOST likely cause?
   A. An open voltage supply wire to the relay.
   B. A shorted horn coil inside the horn assembly.
   C. The horn has a poor ground where it is mounted.
   D. A grounded wire between the horn relay and the horn ring.

7. When performing a resistance check with an ohmmeter, *Technician A* says you should use a high impedance (10 megohm input) ohmmeter when checking sensitive electronic components. *Technician B* says an ohmmeter should be connected to a circuit in which the current is flowing. Who is right?
   A. A only
   B. B only
   C. Both A and B
   D. Neither A nor B

8. The battery is disconnected on a vehicle with onboard computers. This procedure may cause:
   A. a failure of the engine to start after the battery is reconnected.
   B. a voltage surge in the electrical system.
   C. an erasure of the computer adaptive memories.
   D. damage to the onboard computers.

9. All of these statements about replacement of a halogen headlamp bulb are true EXCEPT:
   A. keep the bulb free of moisture or contaminants.
   B. change the bulb with the headlights on.
   C. handle the bulb only by its base.
   D. do not drop or scratch the bulb.

10. When using a 12V test light to test for voltage, all of the following are true EXCEPT:
    A. the test lamp may be connected to the chassis ground.
    B. the battery ground should be disconnected.
    C. the test lamp may be connected to the battery ground.
    D. the ignition can be turned on or off.

11. A truck with an erratic fuel gauge is brought in. What should be done first?
    A. Check the battery voltage.
    B. Remove the fuel gauge and send it out for repair.
    C. Check the sending unit ground connection.
    D. Verify the customer's complaint.

12. While bench testing an alternator rotor for continuity, a Technician reads 1.5 ohms across the slip rings with an ohmmeter. The specified resistance is 3.0 ohms. This reading indicates the field winding inside the rotor is:
    A. shorted.
    B. grounded.
    C. normal.
    D. open.

13. The batteries gradually become discharged in a tractor trailer combination that is driven every day. *Technician A* says the voltage regulator setting may be too low. *Technician B* says the accessory load on the electrical system may exceed the maximum alternator output. Who is right?
    A. A only
    B. B only
    C. Both A and B
    D. Neither A nor B

14. A thermal electric-type fuel gauge in the instrument panel reads full continually. The MOST likely cause of this problem is:
    A. excessive resistance in the fuel tank sending unit.
    B. a grounded wire between the dash gauge and the fuel tank sending unit.
    C. an open wire between the dash gauge and the fuel tank sending unit.
    D. an open voltage supply wire to the dash gauge.

Answers to Pretest

1. D, 2. A, 3. C, 4. B, 5. D, 6. D, 7. A, 8. C, 9. B, 10. B, 11. D, 12. A, 13. C, 14. B

General Electrical System Diagnosis

Task 1

Check continuity in electrical/electronic circuits using appropriate test equipment.

1. The bulb in Figure 6–3 is inoperative. A 12V test light is installed in place of the fuse, the test light is on, and the bulb is off. When the connector near the bulb is disconnected, the 12V test light remains on. *Technician A* says the circuit may be shorted to ground between the fuse and the disconnected connector. *Technician B* says the circuit may be open between the disconnected connector and the bulb. Who is right?
   A. A only
   B. B only
   C. Both A and B
   D. Neither A nor B

2. *Technician A* says a self-powered test light can be used to check for continuity in a circuit involving an electronic module. *Technician B* says the battery must be disconnected before testing an electronic module circuit. Who is right?
   A. A only
   B. B only
   C. Both A and B
   D. Neither A nor B

Hint

Continuity in an electric circuit may be tested with a high impedance digital voltmeter. Connect the meter ground lead to a satisfactory ground connection and connect the positive meter lead to various locations in the circuit starting at the voltage source. When the meter reads a very low voltage, the open circuit is between the meter positive lead and the previous location where a normal voltage reading was obtained. A 12V test light may be used to test continuity in a circuit, but a high impedance test light must be used when testing continuity in computer-controlled circuits. If an ohmmeter is used to test continuity, the meter leads must be connected across the terminals on the component to be tested with the component disconnected from the circuit. An ohmmeter may be damaged if it is connected to a circuit with voltage supplied to the circuit. Always disconnect the truck's battery before using a self-powered test lamp to test continuity. Never use a self-powered test lamp on an electronic module circuit, as damage to the module may result.

Task 2

Check applied voltages, circuit voltages, and voltage drops in electrical/electronic circuits using a digital (DVOM, DMM) or analog voltmeter.

3. When connecting a voltmeter to measure voltage drop in a circuit, the voltmeter leads should be connected:
   A. from the positive battery terminal to ground.
   B. in series with the circuit being tested.
   C. in parallel to the circuit being tested.
   D. across the battery terminals.

4. A high resistance is suspected in the starting motor ground circuit on a negative ground electrical system. *Technician A* says to connect the voltmeter leads from the starting motor housing to the positive battery terminal. *Technician B* says the

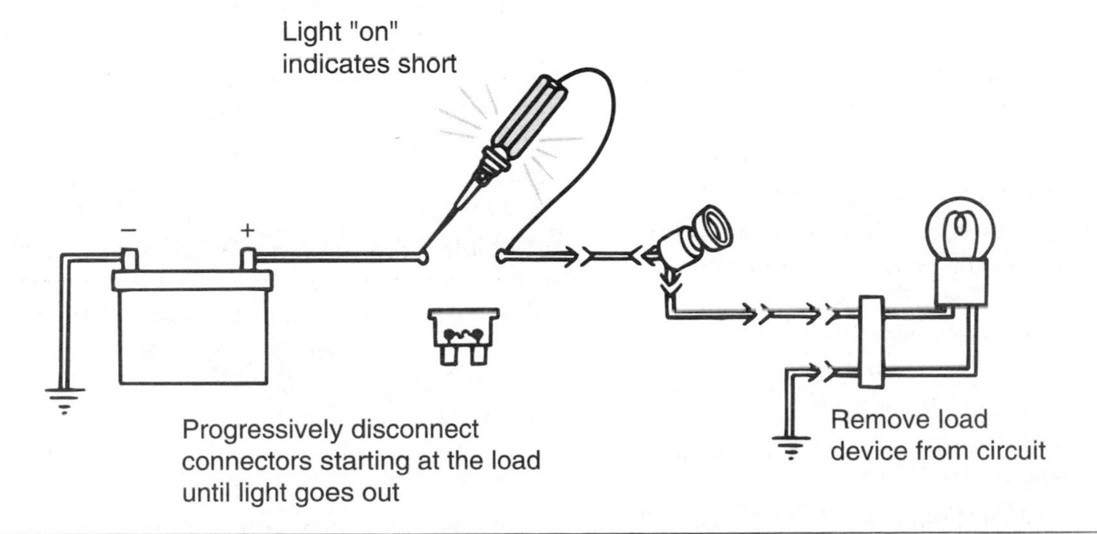

Figure 6–3 Testing for a grounded circuit. *(Courtesy of Chrysler Corporation)*

fuel injection system should be disabled and the engine cranked to read the voltage drop. Who is right?
A. A only
B. B only
C. Both A and B
D. Neither A nor B

Hint

A voltmeter is connected across a component in a circuit to measure the voltage drop across the component. Current must be flowing through the circuit during the voltage drop test. If a circuit has a normal resistance, the voltage drop is within specifications. For example, with the engine cranking, the positive battery cable may have a 0.2V drop. Excessive resistance in a cable or component causes higher than specified voltage drop and reduced current flow. The sum of the voltage drops in any series circuit will equal the source voltage or applied voltage.

Task 3

Check current flow in electrical/electronic circuits and components using an ammeter.

5. The result of a shorted field winding inside an alternator rotor is:
A. no magnetic strength around the rotor.
B. low current flow through the field winding.
C. excessive alternator voltage output.
D. damage to the voltage regulator.

6. The current flow in an electric circuit is lower than specified. *Technician A* says the voltage supplied to the circuit may be higher than specified. *Technician B* says the resistance in the circuit may be higher than specified. Who is right?
A. A only
B. B only
C. Both A and B
D. Neither A nor B

Hint

An ammeter has low resistance and the Technician must connect this meter in series in a circuit. An ammeter may be used to test for a short circuit. Be sure that the meter used is able to handle the high current draw if connected in series. Some ammeters have an inductive clamp that fits over a wire in a circuit. These ammeters measure the current flow from the strength of the magnetic field surrounding the wire. High current flow results from high voltage or low resistance. Conversely, low current flow results from high resistance or low voltage.

Task 4

Check resistance in electrical/electronic circuits and components using an ohmmeter.

7. While discussing resistance measurement with an ohmmeter, *Technician A* says an ohmmeter may be connected to a circuit in which current is flowing. *Technician B* says when testing a spark plug wire with 20,000 ohms resistance, use the X100 meter scale. Who is right?
A. A only
B. B only
C. Both A and B
D. Neither A nor B

8. *Technician A* says an ohmmeter should be used to check for a short circuit between two circuits. *Technician B* says the battery must be fully charged before checking a circuit for current draw. Who is right?
A. A only
B. B only
C. Both A and B
D. Neither A nor B

Hint

Ohmmeters use an internal battery power source. Meter damage may result if this meter is connected to a powered circuit. The Technician must select the proper scale on the meter for the component. Low resistance in a circuit causes high current flow. The cause of low resistance in a circuit could be a shorted circuit. A shorted blower motor or circuit causes a high current flow. High voltage in a circuit causes a higher than normal current flow. High resistance caused by a corroded or loose connection or worn brushes in a motor causes a lower than normal current flow.

A field coil in an alternator rotor is insulated from the metal in the rotor. A low-resistance reading between the winding slip rings and the frame indicates a short to ground condition. When checking resistance of a spark plug wire with an ohmmeter, set the scale to the X1000 scale. Spark plug wires generally have a resistance of 5,000 to 20,000 ohms. Dimming trailer lights could be caused by a loose or dirty trailer connector, which would cause high resistance in the light circuit. A frayed wire may not be able to handle the current flow necessary for proper light operation. The headlight switch affects both tractor and trailer lights and is least likely to cause the trailer lights to dim.

Task 5

Find shorts, grounds, and opens in electrical/electronic circuits.

9. In Figure 6–4, the ammeter indicates the current flow through the bulb is higher than specified. The cause of this high current could be:
   A. the fuse has an open circuit.
   B. the battery voltage is low.
   C. the light bulb filament is shorted.
   D. the bulb filament has high resistance.

10. The tractor taillight fuse blows repeatedly. When the fuse is replaced, all the taillights operate normally and have the same brilliance. The MOST likely cause of this problem is:
    A. an intermittent ground in a taillight wire.
    B. high resistance in a taillight wire.
    C. an intermittent open circuit in a taillight wire.
    D. a shorted filament in one of the taillights.

Hint

A shorted circuit may be defined as a circuit that allows current to bypass part of the normal path. A shorted rotor field winding is an example of a short circuit. The windings within a coil are insulated from each other. If this insulation breaks down and a copper-to-copper contact is made between the turns, part of the winding is bypassed thereby reducing the number of windings. A shorted circuit in a winding causes reduced effective turns, lower resistance, and increased current flow. When one turns off a switch, this is an open circuit. This open can occur in the positive or negative side of the circuit. A grounded circuit is a condition that allows current to return to ground before it has reached the intended load component. A ground is similar to a short in that the load component is bypassed. For example, if the wire leading to the taillight has an insulation breakdown that allows this wire to touch the frame, current will flow to ground and return to the source without reaching the load component.

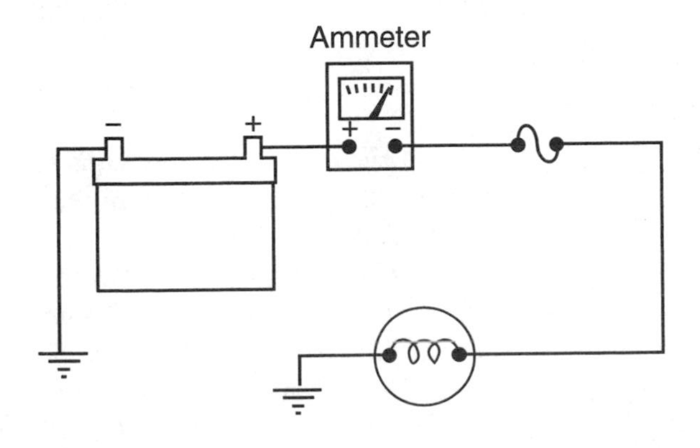

Figure 6–4 Testing for a shorted circuit. *(Courtesy of Chrysler Corporation)*

Task 6

Diagnose key-off battery drain problems.

11. While performing a parasitic battery drain test in Figure 6–5:
    A. the tester switch should be closed while starting or running the engine.
    B. a battery drain of 400 milliamperes is normal and will not discharge the battery.
    C. the actual drain is recorded immediately when the switch is opened.
    D. the driver's door should be open while measuring battery drain.

12. *Technician A* says a battery drain of 2 amperes would cause the battery to become discharged. *Technician B* says the battery drain test should be performed with the engine at normal operating temperature. Who is right?
    A. A only
    B. B only
    C. Both A and B
    D. Neither A nor B

Hint

A battery drain test should always be performed with all lights and accessories turned off. Also, the ignition

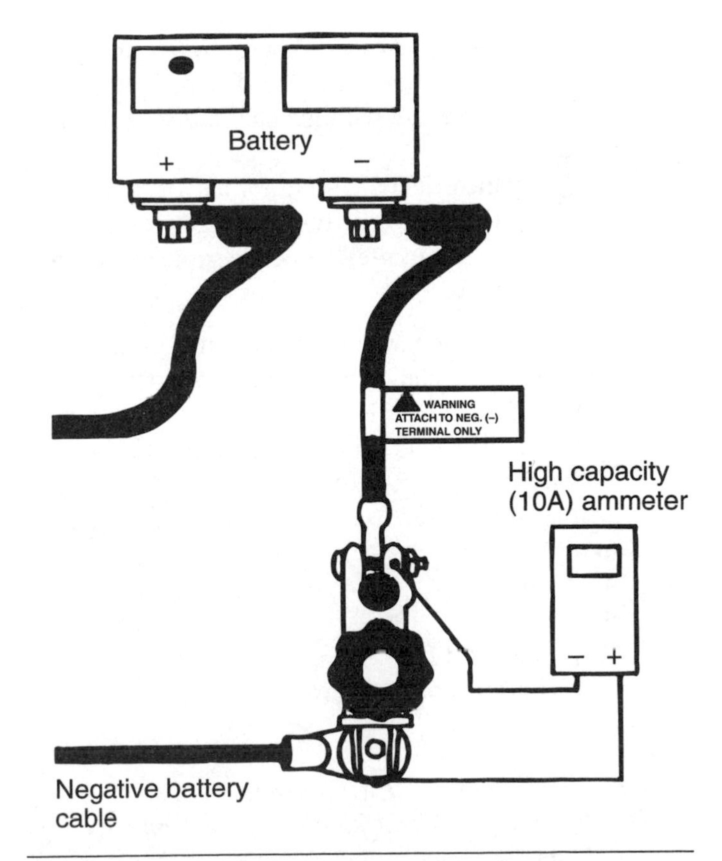

Figure 6–5 Battery drain test. *(Courtesy of Pontiac Division, General Motors Corporation)*

switch should be in the off position and all the doors should be closed.

With today's electronic control modules and accessories, some battery drain through the onboard computers is normal. A drain of 2 amps could cause a battery to go dead over a period of hours. The amount of time depends on the charge condition of the battery as well as the reserve capacity rating.

Many truck manufacturers recommend measuring battery drain with a tester connected in series at the negative battery terminal. A multimeter is connected in parallel to the tester switch and set to the milliamps scale. The test switch should be closed and the engine started. When the engine is shut off, the test switch is opened, and the milliammeter should be observed. On some computers, the drain is higher for a brief time period, and then the drain tapers off. Some computers require a brief time period to power down and enter the "sleep mode." Wait until the meter reading stabilizes before recording the battery drain. If the drain is higher than specified, disconnect individual circuits to locate the cause of the drain.

Task 7

Inspect, test, and replace fusible links, circuit breakers, and fuses.

13. While checking fuses in a tractor/trailer, the Technician finds an open fuse. The next step is to:
    A. replace the fuse with the next higher amperage fuse.
    B. check the affected circuit for a short to ground.
    C. check the affected circuit for an open.
    D. install a circuit breaker of the same amperage as the fuse.

14. After removing a 30-ampere circuit breaker from a fuse panel, a technician tests breaker continuity with an ohmmeter. *Technician A* says the current from the ohmmeter will open the circuit breaker. *Technician B* says the ohmmeter should read infinity if the circuit breaker is satisfactory. Who is right?
    A. A only
    B. B only
    C. Both A and B
    D. Neither A nor B

Hint

When replacing a fuse, always use the amperage specified by the manufacturer. Using a fuse with an amperage rating higher than specified does not properly protect the circuit. A fuse with a higher amerage rating than specified may cause a circuit overload, resulting in damage to the circuit or even a vehicle fire. When an ohmmeter is connected to a circuit breaker, fuse, or fusible link, the meter will read 0 ohm, if the component is good. An open circuit in one of these devices causes an infinity reading.

When working on a powered circuit, the battery should be disconnected first. This is to prevent an accidental short, which could damage the electrical system. When replacing a burned fuse link, always use the same size that was originally installed. This is usually four-gauge sizes smaller than the wire in the circuit being protected. Also, be sure to use fuse link wire made for this purpose. Fuse link wire has a special insulation that bubbles when overloaded. Regular wire insulation may burn and cause other damage to the electrical system.

Task 8

Inspect, test, and replace spike suppression diodes/resistors and capacitors.

15. The diode in Figure 6–6 is used to:
    A. allow the A/C compressor to run only in one direction.
    B. to protect the A/C compressor clutch from voltage spikes.
    C. to protect electronic components from voltage spikes.
    D. to limit the current flow to the A/C compressor clutch.

16. A truck with a diesel engine experiences repeated PCM failures. When the engine is running, it operates normally. *Technician A* says a voltage suppression diode may be defective in one of the electrical components. *Technician B* says one of the output solenoids or relays controlled by the PCM may have a shorted winding. Who is right?
    A. A only
    B. B only
    C. Both A and B
    D. Neither A nor B

Hint

A diode is a device that allows current to pass through it only in one direction. Current can only pass in the direction of the arrow in the diode symbol.

The compressor clutch diode, connected across the compressor clutch terminals, protects the A/C controller from voltage spikes. Voltage spikes occur whenever the clutch coils energize or de-energize. Each time the clutch's magnetic field collapses, enough voltage is created in the circuit to damage electronic components such as the powertrain control module (PCM) or A/C con-

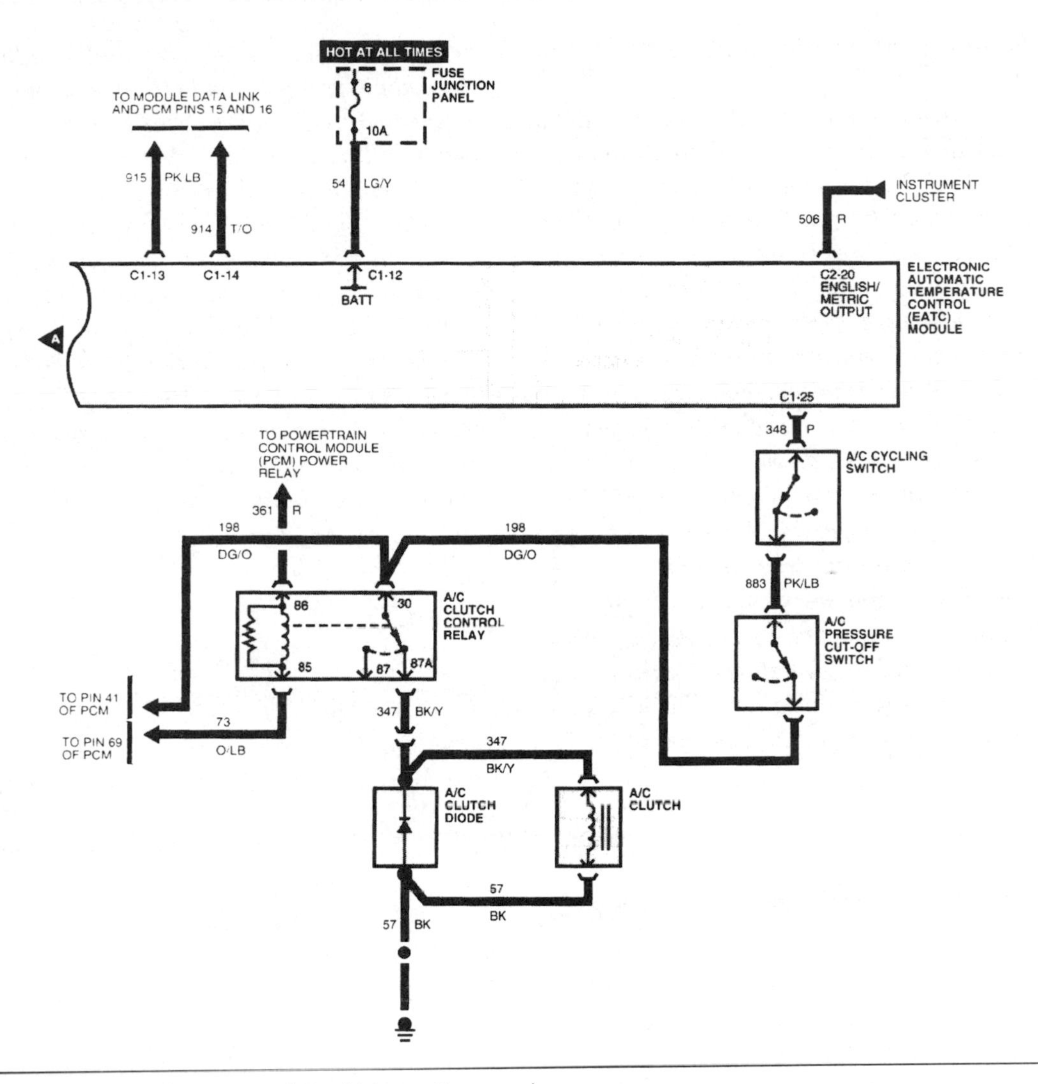

Figure 6–6 A/C clutch circuit. *(Courtesy of Ford Motor Company)*

troller. The diode completes the shortest circuit back to the other end of the coil, so the voltage spike is not applied to other components. This protects other components on the circuit from damage.

Task 9

Inspect, test, and replace relays and solenoids.

17. The horns in Figure 6–7 blow when a jumper wire is connected across the relay contact terminals, but the horns do not blow when the horn button is depressed. *Technician A* says the horn relay may be defective. *Technician B* says the circuit may be grounded between the horn relay and the horn button. Who is right?
    A. A only
    B. B only
    C. Both A and B
    D. Neither A nor B

18. Refer to Figure 6–7. The left-hand horn is inoperative, but the right-hand horn operation is normal. The cause of this problem could be:
    A. continually open horn relay contacts.
    B. an open horn relay winding.
    C. an open in fuse link B.
    D. an open between the left-hand horn and the junction C136.

Hint

A relay is an electric switch that allows a small current to control a much larger one. It consists of a control circuit and a power circuit. When the control circuit is open, no current flows to the coil, so the windings are de-energized. When current flows through the control circuit, the coil is energized, making the iron core into a magnet and drawing the armature down. This closes the circuit contacts, connecting power to the load circuit. When the control circuit is opened, the current stops

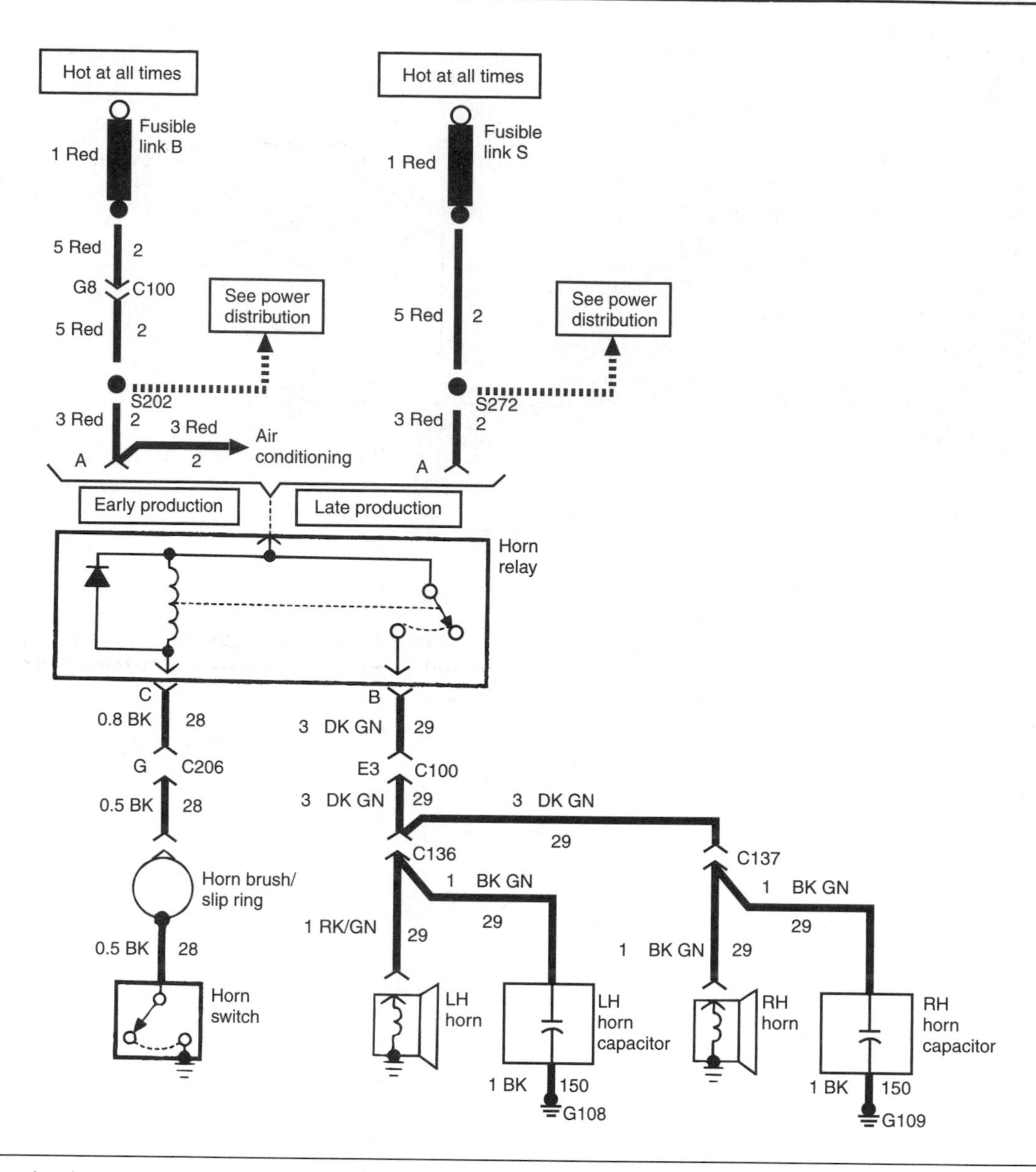

Figure 6–7 Horn circuit.

flowing in the coil, the electromagnetism disappears, and the armature is released to open the relay contacts.

Battery Diagnosis and Repair

Task 1

Perform battery hydrometer test; determine specific gravity of each cell.

19. When performing a hydrometer test on a truck battery:
    A. if the battery temperature is 50°F (10°C), the battery should be brought to room temperature before testing.
    B. if the battery temperature is 120°F (49°C), 0.020 specific gravity point should be subtracted from the hydrometer reading.
    C. the maximum variation in cell hydrometer readings is 0.100 specific gravity points.
    D. the battery is fully charged if all the cell hydrometer readings are above 1.265.

20. A maintenance-free battery is low on electrolyte, and the built-in hydrometer indicates yellow. *Technician A* says a faulty voltage regulator may cause this problem. *Technician B* says high resistance in the circuit between the alternator battery terminal and the positive battery terminal may cause this problem. Who is right?
    A. A only
    B. B only

C. Both A and B
D. Neither A nor B

Hint

When performing a battery state-of-charge test with a hydrometer, 0.004 specific gravity point should be subtracted from the hydrometer reading for every 10°F of electrolyte temperature below 80°F. During a state-of-charge test, 0.004 specific gravity points must be added to the hydrometer reading for every 10°F of electrolyte temperature above 80°F. The maximum variation in cell specific gravity readings is 0.050 specific gravity point. When all the cell readings exceed 1.265, the battery is fully charged. Some batteries have a built-in hydrometer in the top of the battery. If the hydrometer is green, the battery is sufficiently charged for testing. When the hydrometer is dark, the battery should be charged before testing. A yellow hydrometer indicates the battery is low on electrolyte, and should be replaced. The low electrolyte level may be caused by a high voltage regulator setting that causes overcharging.

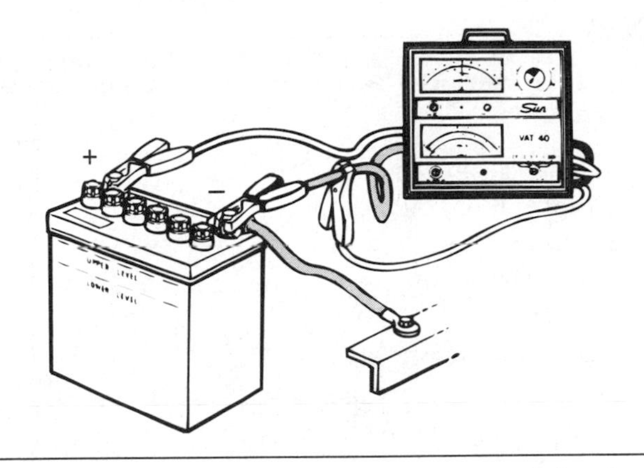

Figure 6–8 Electrical system test. *(Courtesy of Chrysler Corporation)*

Task 2

Perform battery capacity (load, high rate discharge) test and determine needed service.

21. When discussing a battery capacity test with the battery at 70°F (21°C), *Technician A* says the battery discharge rate is 2 times the battery reserve capacity rating. *Technician B* says the battery is satisfactory if the voltage remains above 9.2V. Who is right?
   A. A only
   B. B only
   C. Both A and B
   D. Neither A nor B

22. In Figure 6–8 the test being performed is:
   A. a battery drain test.
   B. a battery capacity test.
   C. a battery open circuit voltage test.
   D. a starter current draw test.

Hint

The battery discharge rate for a capacity test is usually one half of the cold cranking rating. The battery is discharged at the proper rate for 15 seconds to eliminate the surface charge. Then, after about 1 to 2 minutes, the test is repeated. The battery voltage must remain above 9.6V with the battery temperature at 70°F or above.

Task 3

Determine battery state of charge by measuring terminal post-voltage using a DVOM (DMM).

23. When performing an open circuit voltage test, the battery is fully charged if the voltage remains at or above:
   A. 12V.
   B. 12.2V.
   C. 12.4V.
   D. 12.6V.

24. *Technician A* says an open circuit voltage test may be used in place of a hydrometer test. *Technician B* says during the open circuit voltage test the headlights should be on. Who is right?
   A. A only
   B. B only
   C. Both A and B
   D. Neither A nor B

Hint

An open circuit voltage test may be used to determine a battery's state of charge. If the battery has just been charged or has recently been in for service, the surface charge must be removed before an accurate voltage measurement can be made. To remove the surface charge, crank the engine for 15 seconds. Do not allow the engine to start. If the battery is out of the vehicle, a load test of half the rated cold cranking amps for 15 seconds will remove the surface charge. Connect a voltmeter across the battery terminals and observe the reading. All the accessories and the ignition switch must be off during the open circuit voltage test. If the reading is 12.6V or more, the battery is fully charged. If the voltage reading is below 12.6V, the battery needs to be charged.

Task 4

Inspect, clean, service, or replace battery and terminal connections.

25. The battery cables are disconnected on a truck with several onboard computers. This procedure may cause all of these problems EXCEPT:
    A. the adaptive memory in the computers to be erased.
    B. radio station presets to be erased.
    C. damage to the onboard computers.
    D. erratic engine operation when the engine is restarted.

26. When replacing a battery, it is very important to:
    A. replace the battery cables.
    B. check the charging circuit.
    C. replace the starting motor.
    D. replace the alternator belt.

Hint

If battery voltage is disconnected from a computer, the adaptive memory in the computer is erased. In a PCM, disconnecting power may cause erratic engine operation or erratic transmission shifting when the engine is restarted. After the vehicle is driven for about 20 mi. (32 km), the computer relearns the system, and normal operation is restored. If the vehicle is equipped with personalized items such as memory seats or mirrors, the memory will be erased in the electronic module that controls these items. Radio station presets will also be erased. A 12V-power supply from a dry cell battery can be connected to the cigarette lighter or power point connector to maintain voltage to the electrical system when the battery is disconnected.

A battery should be cleaned with a baking soda and water solution. Always wear hand and eye protection when servicing batteries and their components. When disconnecting battery cables, always disconnect the negative battery cable first. The order is reversed when installing the battery and the cables.

Spray the cable clamps with a protective coating to prevent corrosion. A little grease or petroleum jelly also prevents corrosion. Also available are protective pads that go under the clamp and around the terminal to prevent corrosion.

When replacing a battery, it is very important to check the charging system. A faulty alternator or regulator may cause a discharged or overcharged condition, resulting in a damaged battery.

Task 5

Inspect, clean, repair, and replace battery boxes, mounts, and holddowns.

27. *Technician A* says before MIG welding on a truck with an alternator and onboard computers, disconnect the battery ground. *Technician B* says one battery holddown is sufficient to hold an 1100 series battery. Who is right?
    A. A only
    B. B only
    C. Both A and B
    D. Neither A nor B

28. When cleaning or servicing the battery, cables, holddown, and tray you should:
    A. use an air nozzle to blow off the parts.
    B. use a cleaning solvent to clean the parts.
    C. always wear safety glasses.
    D. use sulfuric acid to dissolve the residues.

Hint

The battery box, mounts, and holddowns are important to the longevity of a battery. The battery boxes protect the batteries from road debris and weather conditions. They also help to insulate the batteries in extremely cold conditions.

Task 6

Charge battery using slow or fast method as appropriate.

29. While charging the batteries in a heavy-duty truck with two batteries:
    A. the battery charge time is the same for batteries with different capacities.
    B. the battery temperature should not exceed 125°F (52°C).
    C. a high charging rate may be used to charge a battery at –20°F (-29° C).
    D. when a fast charger is used to charge two batteries in parallel, the current flow is the same through both batteries.

30. When servicing and charging batteries, observe all these precautions EXCEPT:
    A. never disconnect the negative battery cable first.
    B. charge the battery until the specific gravity is 1.265.
    C. reduce the fast charging rate after the specific gravity is above 1.225.
    D. do not charge a frozen battery.

Hint

If the battery is to be charged in the vehicle, the battery cables should be disconnected during the charging procedure. The charging time depends on the battery state of charge and capacity. If the battery temperature exceeds 125°F (52°C) while charging, the battery may be damaged. To avoid excessive gassing when fast charging a battery, reduce the charging rate when specific gravity reaches 1.225. Avoid breathing the fumes from a battery and always charge the battery in a well-ventilated area. These fumes are also very explosive because they contain hydrogen gas. While charging keep open flames and sparks away from the battery. The battery is fully charged when the specific gravity increases to 1.265. Do not attempt to fast charge a cold battery. Never charge a frozen battery as an explosion may result. Attempting to charge a sealed battery with the built-in hydrometer indicating yellow may cause excessive gassing and possible explosion.

Many fast chargers have a maximum charging rate of 50 amperes. Most slow chargers have a 2- to 10-ampere charging rate. Many shop slow chargers are capable of charging several batteries connected in series. A control knob on the slow charger allows the Technician to adjust the slow charger voltage until it is higher than the voltage of the battery or batteries.

Task 7

Jump-start a vehicle using jumper cables and a booster battery or auxiliary power supply.

31. While jump-starting a vehicle with a booster battery, *Technician A* says the accessories should be turned on in the booster vehicle. *Technician B* says the negative booster cable should be connected to an engine ground on the vehicle being boosted. Who is right?
    A. A only
    B. B only
    C. Both A and B
    D. Neither A nor B

32. While boosting a vehicle with a booster battery in a second vehicle:
    A. the bumpers on the two vehicles should be allowed to touch each other.
    B. the positive battery terminal in the boost vehicle should be connected to the negative battery terminal in the vehicle being boosted.
    C. a boost procedure may be performed on a frozen battery.
    D. the negative booster cable between the two electrical systems should be connected last.

Hint

The accessories in both vehicles must be off during the boost procedure, and the engine in the boost vehicle should be off while connecting the booster cables. Never allow the two vehicles to touch each other. The negative booster cable must be connected to an engine ground in the vehicle being boosted. Always connect the positive booster cable followed by the negative booster cable, and complete the negative cable connection last on the vehicle being boosted. After the cables are connected, start the boost vehicle, allow the engine to run for a few minutes, then attempt to start the vehicle with the discharged battery. When disconnecting the booster cables, remove the negative booster cable first on the vehicle being boosted.

Starting System Diagnosis and Repair

Task 1

Perform starter current draw test; determine needed repairs.

33. A customer has a truck towed into the shop and says the starter would not crank the engine over. What should be checked first?
    A. The ground cable connection
    B. The starter solenoid circuit
    C. The ignition switch crank circuit
    D. The battery condition

34. A starting motor fails to crank the engine and indicates a very high current draw. The MOST likely cause of this problem is:
    A. a direct ground in the starter field coils.
    B. an open circuit in the armature windings.
    C. broken brush springs and no tension on the brushes.
    D. failed magnetic switch.

Hint

High starter current draw, low cranking speed, and low cranking voltage usually indicate a faulty starter motor. Low current draw, low cranking speed, and high cranking voltage usually indicate excessive resistance in the starting circuit. The starting system circuit carries the high current flow within the system and supplies power for the actual engine cranking. Starting circuit components are the battery, battery cables, magnetic switch or solenoid, and the starter motor.

The cranking current test measures the amount of current, in amperes, that the starter circuit draws to crank the engine. This amperage reading is useful in isolating the source of certain types of starter problems.

An inoperative starter, along with a high current draw, is always an indication of a direct short to ground. Any opens, failed switches, springs, or relays would cause an operative failure and a no-current condition.

Task 2

Perform starter circuit voltage drop tests; determine needed repairs.

35. To perform a starter ground circuit test:
    A. connect a voltmeter between the battery ground terminal and the starter housing, and crank the engine.
    B. connect an ohmmeter between the starter relay housing and the starter housing, and crank the engine.
    C. connect an ohmmeter between the battery ground and the starter housing, and crank the engine.
    D. connect a voltmeter between the positive battery terminal and the starter solenoid housing and crank the engine.

36. When measuring the cranking circuit voltage drop on the insulated side of the starter circuit, connect the voltmeter leads:
    A. from the battery ground terminal to the positive battery cable connected to the starter solenoid.
    B. from the solenoid winding terminal to the positive battery terminal.
    C. from the positive battery terminal to the positive battery cable connected to the starter solenoid.
    D. from the battery negative terminal to solenoid housing.

Hint

Connect the voltmeter leads between the positive battery terminal and the positive cable on the starter solenoid; crank the engine to measure the voltage drop across the positive battery cable. A ground circuit voltage drop test is performed by connecting a voltmeter between the ground side of the battery and the starter housing. Then, the engine is cranked and the voltage read. Voltage should not exceed 100 millivolts.

Task 3

Inspect, test, and replace components and wires in the starter control circuit.

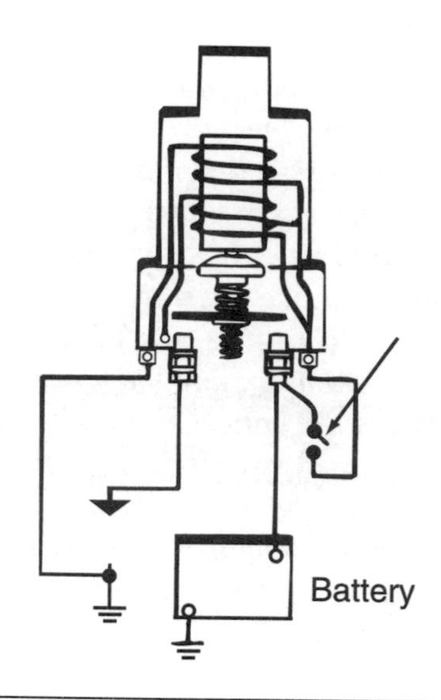

Figure 6–9 Starter solenoid.

37. In Figure 6–9 the arrow points to:
    A. the start switch.
    B. the starter relay.
    C. open pull-in winding contacts.
    D. open hold-in winding contacts.

38. Connecting a jumper wire across terminals C and D in Figure 6–10 is a method of checking:
    A. the starting motor.
    B. the solenoid contacts.
    C. the battery condition.
    D. the starter relay.

Hint

To perform a starter solenoid test, the ohmmeter leads must be connected across the solenoid terminal and the field coil terminal to test the pull-in winding. Connect the ohmmeter leads from the solenoid terminal to ground to test the hold-in winding.

Many starter motors are similar in design and operation. The starter motor consists of housing, field coils, an armature, a commutator and brushes, end frames, and a solenoid-operated shift mechanism.

A starter relay is used on many heavy-duty truck applications. This relay winding is energized when the start switch is pressed in the dash. When the relay winding is energized, the relay contacts close and supply voltage and current to the starter solenoid windings.

Task 4

Remove and replace starter; inspect flywheel ring gear or flexplate.

39. When a starting motor is removed, there are shims positioned between the starting motor mounting

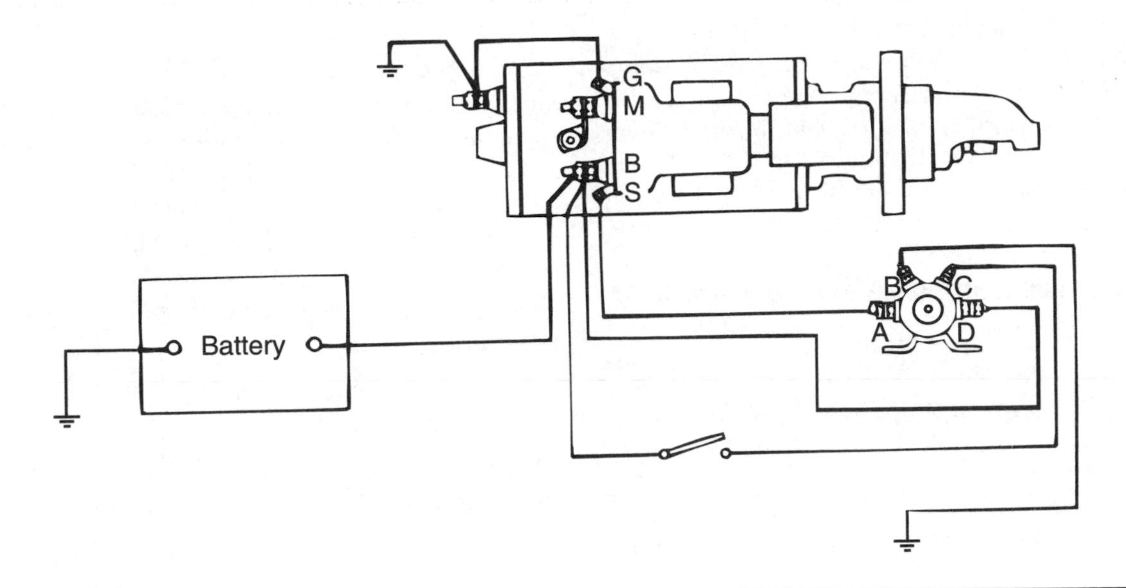

Figure 6–10 Starting motor circuit.

flange and the flywheel housing. *Technician A* says these shims provide improved electrical contact between the starting motor and the flywheel housing. *Technician B* says these shims provide easier installation of the starting motor mounting bolts. Who is right?
A. A only
B. B only
C. Both A and B
D. Neither A nor B

40. A starting motor fails to disengage for a few seconds after the engine starts. *Technician A* says the starting motor may be misaligned with the flywheel housing. *Technician B* says the solenoid return spring may be weak. Who is right?
A. A only
B. B only
C. Both A and B
D Neither A nor B

Hint

Inspect the teeth of the ring gear on the outer surface of the flywheel. If the teeth are worn or damaged, replace the ring gear or the flywheel, and inspect the starter drive teeth. If there is any damage evident on the starter drive gear, the starter drive must be replaced. When shims are positioned between the starting motor mounting flange and the flywheel housing, the same shim thickness must be reinstalled to provide proper starting motor alignment with the flywheel housing.

Task 5

Inspect, test, and replace starter relays and solenoids/switches.

41. While discussing safety starting switches, *Technician A* says the safety starting switch is connected parallel to the solenoid windings. *Technician B* says a defective safety starting switch may cause higher than specified starter current draw. Who is right?
A. A only
B. B only
C. Both A and B
D. Neither A nor B

42. The LEAST likely cause of a discharged battery is:
A. a loose alternator belt.
B. a dirty battery.
C. a defective starter solenoid.
D. a parasitic battery drain.

Hint

The starter solenoid performs two functions. When energized, the plunger movement acts on a shift lever that shifts the starter motor drive pinion into mesh with the engine flywheel teeth so that the engine can be cranked. The solenoid also closes a set of contacts that allows full battery current to flow to the starter.

The starting switch allows a small amount of current to flow through the coil of the starter relay. When the starter relay coil is energized, the relay contacts close and supply full voltage and current to flow to the solenoid windings.

The starting safety switch prevents vehicles with automatic transmissions from being started in gear. Safety switches, often called neutral safety switches, are usually connected between the starting switch and the starter relay.

Task 6

Diagnose 12/24-volt starting system problems; determine needed repairs.

43. All of these statements about the 12/24V starting system in Figure 6–11 are true EXCEPT:
    A. the batteries are connected in parallel for charging.
    B. while cranking, the negative terminal of one battery is connected to the positive terminal of the other battery.
    C. in the charge mode, the series parallel switch winding is energized.
    D. the series parallel switch is operated by electromagnetism.

44. While discussing the 12/24V starting system in Figure 6–11, *Technician A* says both batteries have the same current flow in the charge mode. *Technician B* says the ammeter indicates the current flow through both batteries. Who is right?
    A. A only
    B. B only
    C. Both A and B
    D. Neither A nor B

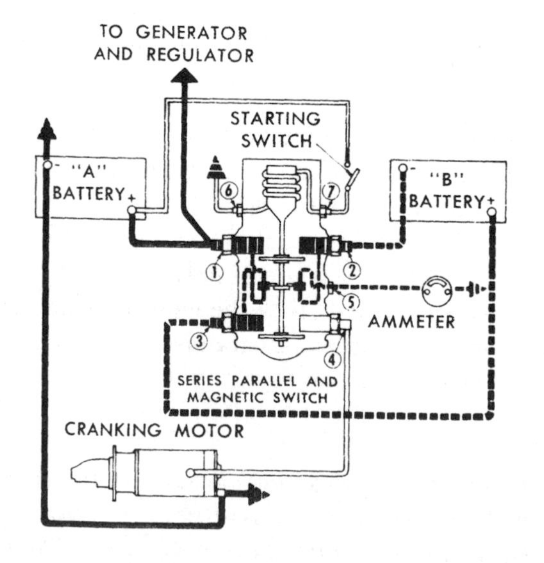

Figure 6–11 12/24V starting system. *(Courtesy of Delco-Remy, Division of GMC)*

Hint

Some heavy-duty engines that require more cranking power have a 12/24V starting system. In these systems, the batteries are connected in series to supply 24V for starting the engine. Once the engine starts, the batteries are immediately switched to a parallel connection so they have 12V potential. This action allows the use of a 12V alternator and electrical system. A series-parallel switch may be used to switch the battery connections from series to parallel.

Task 7

Inspect, clean, repair, and replace battery cables and connectors in the cranking circuit.

45. *Technician A* says the battery cables only need to be serviced when the starting or charging system is experiencing problems. *Technician B* says excessive corrosion around the battery area may be caused by a high charging voltage. Who is right?
    A. A only
    B. B only
    C. Both A and B
    D. Neither A nor B

46. All of these statements about battery and cable service are true EXCEPT:
    A. the terminals should be removed from the battery with a puller.
    B. remove the positive cable before the negative cable.
    C. install the positive cable before the negative cable.
    D. corrosion on the battery top increases the rate of self-discharge.

Hint

The battery and the cables should be serviced at regularly scheduled maintenance intervals. Because of the high amperage that the starting circuit requires, any connection problem or corrosion will add additional resistance to the circuit and may result in decreased starting performance. Starting motors draw a higher current when starting the engine in cold weather, and resistance problems in the battery and starting motor connections are even more critical under this condition.

Charging System Diagnosis and Repair

Task 1

Diagnose dash-mounted charge meters and/or indicator lights that show a no charge, low charge, or overcharge condition; determine needed repairs.

47. The charging indicator light on a truck is illuminated when the ignition switch is turned off (Figure 6–12). *Technician A* says some of the alternator diodes may be defective. *Technician B* says the wire connected to the alternator L terminal may be grounded. Who is right?
    A. A only
    B. B only
    C. Both A and B
    D. Neither A nor B

48. Refer to Figure 6–12. The charge indicator light is not on when the ignition switch is turned on, and the truck has a discharged battery. *Technician A* says the field circuit in the alternator may have an open circuit. *Technician B* says one of the stator windings may have an open circuit. Who is right?
    A. A only
    B. B only
    C. Both A and B
    D. Neither A nor B

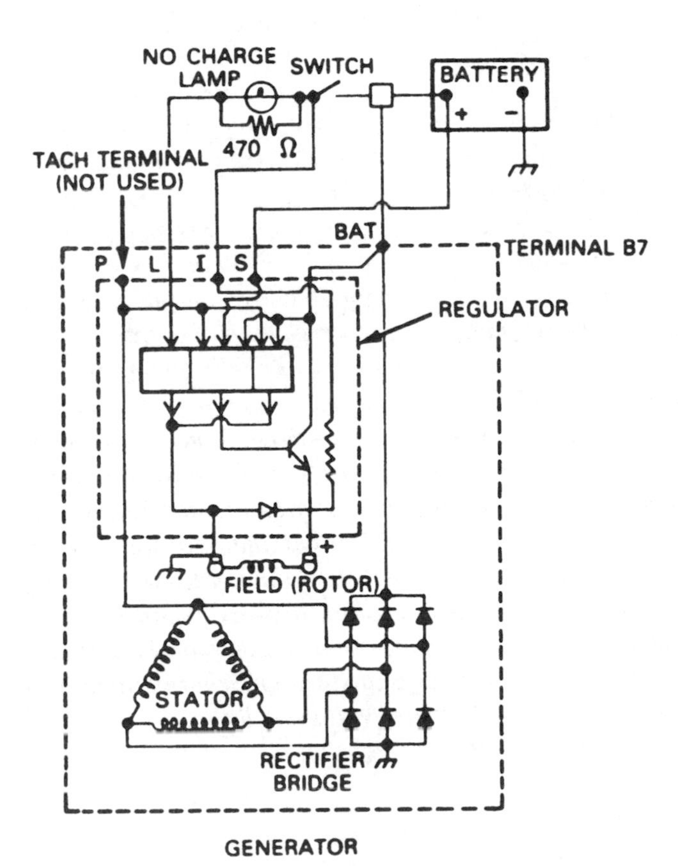

Figure 6–12 Charging circuit with integral electronic regulator. *(Courtesy of Oldsmobile Division, General Motors Corporation)*

Hint

An ammeter, a voltmeter, or charge indicator light may be mounted in the instrument panel to indicate charging circuit operation to the driver. The charge indicator light indicates the charging circuit is operational, but it does not provide any indication of the voltage or current flow in the charging circuit. This light should be on when the ignition switch is turned on, and the light should go off when the engine starts. Many ammeters indicate the amount of current flowing from the alternator through the batteries. A voltmeter indicates the charging system voltage, and this same voltage is applied to the entire electrical system when the engine is running.

Task 2

Diagnose the cause of a no-charge, low charge, or overcharge condition; determine needed repairs.

49. An undercharged battery may be caused by:
    A. a defective voltage regulator causing a high-voltage regulator setting.
    B. excessive resistance between the alternator battery terminal and the positive battery terminal.
    C. worn alternator mounting holes causing a misaligned alternator belt.
    D. a 140-ampere fuse link installed in place of a 100-ampere fuse link in the alternator battery wire.

50. A truck has a repeated discharged battery complaint. All of these defects may be the cause of the problem EXCEPT:
    A. a low-voltage regulator setting.
    B. an alternator belt bottomed in the pulley.
    C. a battery drain with the ignition switch off.
    D. a slightly higher-than-normal resistance in the charge indicator light.

Hint

Before performing a system output test, the Technician must first check for a loose alternator belt that could cause the pulley to slip, resulting in low alternator output. Corrosion and moisture on top of the battery may cause the battery voltage to drain across between the terminals. A parasitic drain, such as a glove box lamp or cell phone, could also cause a discharged battery. The voltage regulator determines the charging system voltage. A high-voltage regulator setting causes battery gassing and damages electric/electronic components. A low-voltage regulator setting causes an undercharged battery. Most electronic regulators are non-adjustable, and must be replaced if they are not providing the specified voltage. Always be sure the charging circuit wiring, including the regulator ground, does not have excessive resistance.

Task 3

Inspect, adjust, and replace alternator drive belts/gears, pulleys, fans, and mounting brackets.

51. While discussing alternator noise complaints, *Technician A* says a loose alternator belt may cause a squealing noise. *Technician B* says a shorted alternator diode may cause a whining noise. Who is right?
    A. A only
    B. B only
    C. Both A and B
    D. Neither A nor B

52. An alternator with a 90-ampere rating produces 45 amperes during an output test. The MOST likely cause of this problem is:
    A. an alternator drive belt that is bottomed in the pulley.
    B. an open circuit in the rotor field coil.
    C. a misaligned alternator pulley.
    D. a broken alternator brush lead.

Hint

An undercharged battery may be caused by a slipping alternator belt. A slipping belt may be caused by insufficient belt tension or a worn, glazed, or oil-soaked belt. The belt tension should be measured with a belt tension gauge. A slipping alternator belt may cause a squealing noise especially during acceleration. The belt must never be bottomed in the pulley, because the friction surfaces are the sides of the belt. If the alternator pulley is misaligned, rapid belt wear occurs. A misaligned alternator pulley is usually caused by worn mounting holes in the alternator housing.

Task 4

Perform charging system voltage and amperage output tests; determine needed repairs.

53. While performing an output test on a 120-ampere alternator, a jumper wire is connected to full-field the alternator. The alternator output indicated on the test ammeter is 40 amperes. The cause of this problem could be:
    A. a shorted alternator diode.
    B. an open circuit in the voltage regulator.
    C. an open circuit at one of the rotor slip rings.
    D. an open circuit in the alternator capacitor.

54. While discussing alternator output testing, *Technician A* says when the alternator is full-fielded, the voltage should be kept below 17V. *Technician B* says the output may be tested by lowering the system voltage to 11.5V. Who is right?
    A. A only
    B. B only
    C. Both A and B
    D. Neither A nor B

Hint

An undercharged battery results in slow engine cranking, or failure of the engine to start. Undercharging is not always caused by a defect in the alternator. Undercharging can be caused by a loose drive belt; loose, broken, corroded, or dirty terminals on either the battery or alternator; undersize wiring between the alternator and the battery; or by a defective battery that will not accept a charge.

When performing an alternator output test, an amp-volt tester is connected to the charging circuit. Many amp-volt testers contain an ammeter with an inductive clamp. When testing alternator output, this clamp is usually placed over the battery ground cable. The amp-volt tester also contains a carbon pile load that may be used during the output test. Some truck manufacturers recommend testing alternator output with the engine running at 2,000 rpm. All electrical accessories must be turned off during the output test. Some truck manufacturers recommend testing output by using the carbon pile load to lower the system voltage to approximately 13V. Under this condition, the alternator maximum output may be read on the ammeter. Another method of testing alternator output is to full-field the alternator by connecting a jumper wire to bypass the voltage regulator. When using this method for output testing, the carbon pile load must be used to limit the system voltage to 15V to prevent high voltage from damaging electrical/electronic components.

Task 5

Perform charging circuit voltage drop tests; determine needed repairs.

55. When performing a charging circuit voltage drop test, the voltmeter should be connected to:

A. the regulator output terminal and the alternator field terminal.
B. the battery ground cable and the alternator housing.
C. the alternator output terminal and the positive battery terminal.
D. the alternator field terminal and the vehicle frame.

56. The ground side of a truck charging circuit is to be tested for voltage drop. *Technician A* says to connect the voltmeter leads to the voltage regulator ground terminal and the battery positive terminal. *Technician B* says to connect the voltmeter leads to the alternator housing and the battery ground terminal. Who is right?
    A. A only
    B. B only
    C. Both A and B
    D. Neither A nor B

Hint

Connect the voltmeter leads from the alternator battery terminal to the positive battery terminal to measure the voltage drop in the charging system. A 10-ampere charging rate is recommended while measuring this test. Voltmeter leads should be placed on the alternator casing and the battery ground terminal to conduct a ground side voltage drop test. The voltage drop should not exceed 100 millivolts. If the voltage is above 100 millivolts, excessive resistance has built up in the form of loose mountings and/or corrosion between the alternator casing and the mounting bracket(s).

Task 6

Remove and replace the alternator.

57. Grounding the alternator battery terminal with the battery cables connected may cause all of these problems EXCEPT:
    A. burns to the Technician's hands.
    B. a burned fuse link in the alternator output wire.
    C. a burned fuse link in the voltage regulator circuit.
    D. a burned alternator output terminal.

58. *Technician A* says removing the alternator battery wire with the engine running causes extremely high voltage and possible damage to onboard electronic components. *Technician B* says removing the alternator field wire with the engine running may cause a burned fuse link in the field circuit. Who is right?
    A. A only
    B. B only
    C. Both A and B
    D. Neither A nor B

Hint

The battery ground cable must be disconnected before removing an alternator.

When the alternator is removed, inspect the mounting holes in the alternator housing and the mounting bracket for wear. If these holes are worn, replace the alternator and/or the bracket. After the alternator is installed, connect and tighten all the wiring connectors, adjust the belt tension, and then reconnect the battery ground cable.

Task 7

Inspect, repair or replace connectors and wires in the charging circuit.

59. *Technician A* says any type of wire may be used to replace a burned fusible link as long as the gauge is one size smaller than the wire in the circuit being protected. *Technician B* says that a circuit breaker should be replaced after repairing a short in a circuit. Who is right?
    A. A only
    B. B only
    C. Both A and B
    D. Neither A nor B

60. Wires for fusible links should be:
    A. the same gauge size at the circuit wire.
    B. the same color as the circuit wire.
    C. two gauge sizes larger than the circuit wire.
    D. a smaller gauge size than the circuit wire.

Hint

A thorough inspection of the connectors in the charging circuit is necessary when diagnosing any charging problem. Deterioration and corrosion of the connectors are conditions that may occur from severe driving conditions. Additional connector problems or damage may result from vehicle vibration or body flexing.

Task 8

Diagnose 12/24-volt alternator charging system problems; determine needed repairs.

61. A truck with a 12/24V system has low cranking power (Figure 6–13). *Technician A* says the winding in the series-parallel switch may be open. *Technician B* says the upper set of contacts in the series-parallel switch may be severely burned.

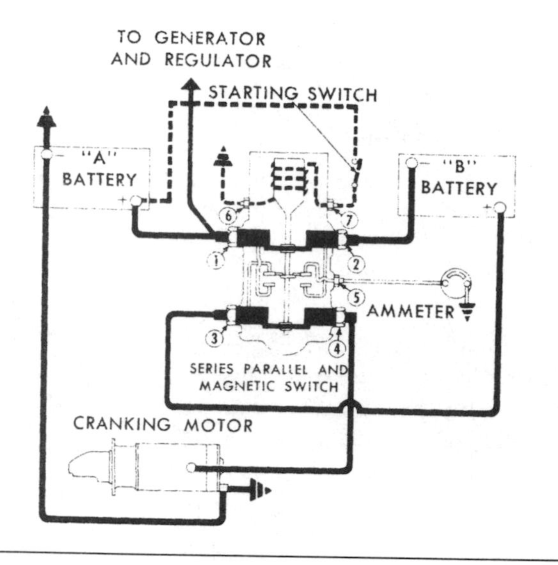

Figure 6–13 12/24V starting and charging system, start mode. *(Courtesy of Delco-Remy, Division of GMC)*

Who is right?
A. A only
B. B only
C. Both A and B
D. Neither A nor B

62. In a truck with a 12/24V system, one battery is repeatedly discharged (Figure 6–14). The cause of this problem could be:
    A. batteries with unequal internal resistance.
    B. a low-voltage regulator setting.
    C. a high-voltage regulator setting.
    D. burned lower contacts in the series-parallel switch.

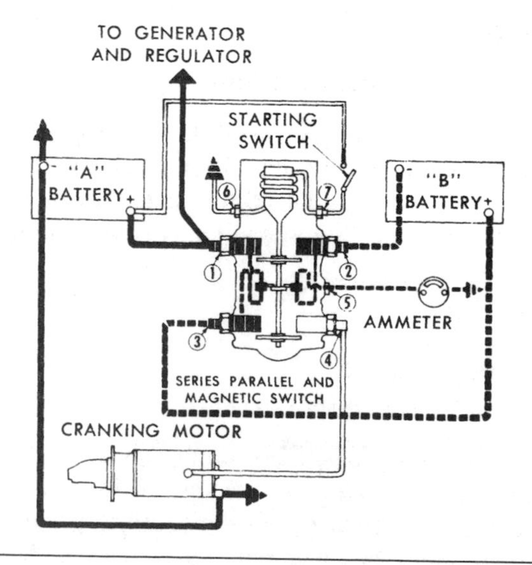

Figure 6–14 12/24V starting and charging system, charge mode. *(Courtesy of Delco-Remy, Division of GMC)*

Hint

In a 12/24V battery, starter, and charging system the batteries are connected in parallel with the engine running. The parallel battery connection provides 12V potential in the two batteries. The charging circuit and electrical accessories are 12V. To supply more electrical power for cranking, a series-parallel switch connects the batteries in series to supply 24V to the starting motor. When the engine starts, the series-parallel switch returns the batteries to a parallel connection. Because the batteries are charged in parallel, they may receive a different amount of current flow from the charging circuit depending on the internal battery resistance.

Task 9

"Flash"/Full-field alternator to restore residual magnetism.

63. The test being performed in Figure 6–15 is:
    A. full-fielding the alternator.
    B. regulator circuit testing.
    C. stator continuity testing.
    D. output diode testing.

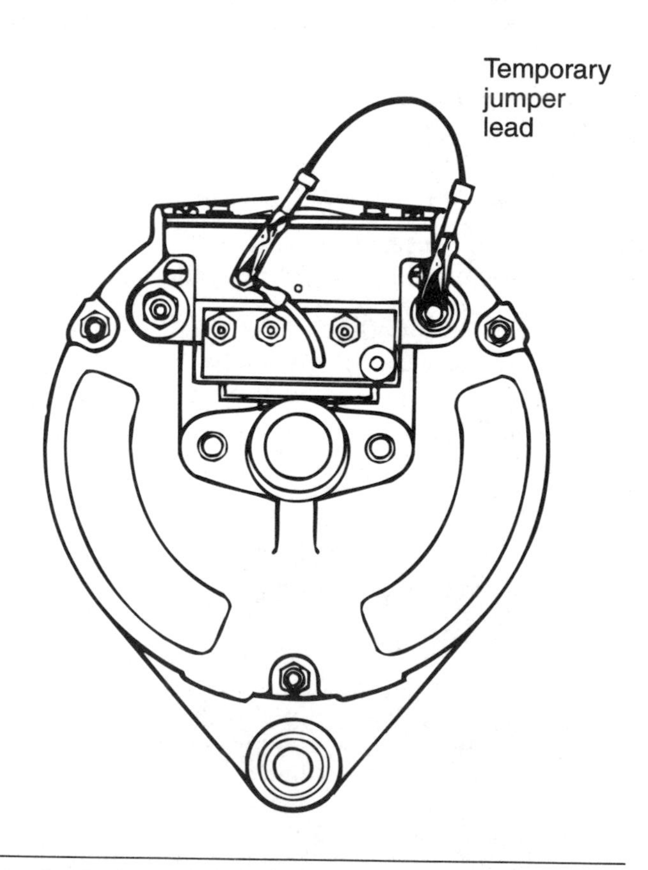

Figure 6–15 Alternator test.

64. When a satisfactory alternator is full-fielded with the engine running:
    A. the charging voltage is low.
    B. charging rate in amperes is low.
    C. the voltage regulator is limiting the alternator voltage.
    D. the voltage and current output is determined by engine speed.

Hint

By applying full battery voltage directly to the field windings in the rotor, it can be determined whether or not the regulator is the cause of the undercharging condition. Full-fielding the alternator bypasses the voltage regulator, which allows full battery voltage to flow to the rotor through the field windings. This is called full-fielding the alternator. Alternators that power the field circuit with current produced by the stator windings rely on residual magnetism in the rotor to energize the stator when the engine is being started. During handling or repair, this residual magnetism can be lost. It may be restored by completing the full-field alternator connection.

Lighting Systems Diagnosis and Repair

Headlights, Parking, Clearance, Tail, Cab, and Dash Lights

Task 1

Diagnose the cause of brighter-than-normal, intermittent, dim, or no headlight operation.

65. The left side headlight is dim only on the high beam (Figure 6–16). *Technician A* says there may be high resistance in the left-side headlight ground. *Technician B* says there may be high resistance in the dimmer switch high beam contacts. Who is right?
    A. A only
    B. B only
    C. Both A and B
    D. Neither A nor B

66. The LEAST likely cause of one dim headlight is:
    A. heavy corrosion on the headlight connector.
    B. a slightly damaged or broken headlight assembly.
    C. low alternator output.
    D. high resistance in the wires connected to the headlight.

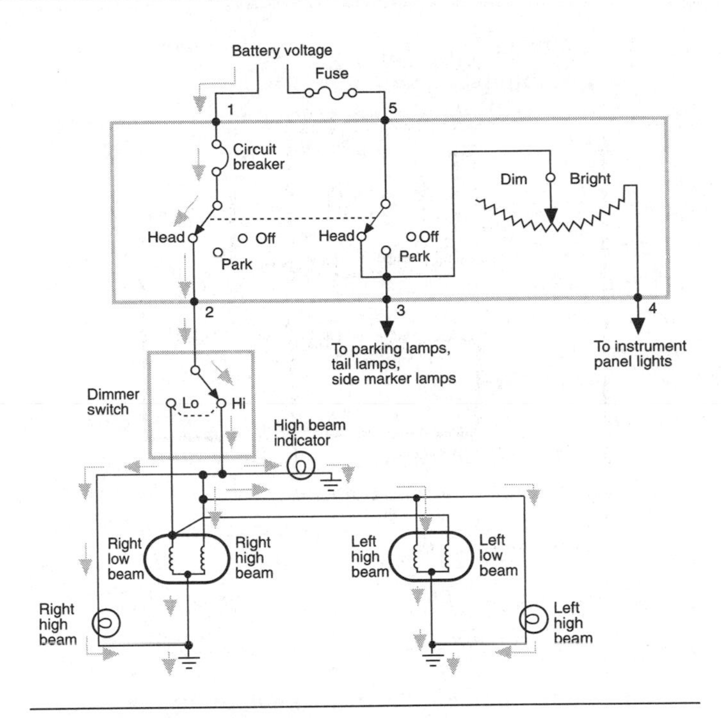

Figure 6–16 Headlight circuit.

Hint

High-resistance ground connections may cause a dim headlight on one side of the truck. Many headlight circuits have a circuit breaker in the switch that may cause inoperative headlights. Power to the headlight switch and the ground circuit of the headlight switch should be checked for resistance problems, including all fuses and circuit breakers. Chassis vibration and road conditions are known to cause connector and component problems. Also check for resistance in the complete headlight circuit and for low charging system voltage.

Task 2

Test, aim, and replace headlights.

67. *Technician A* says that headlight aim should be checked on a level floor with the vehicle unloaded. *Technician B* says the vehicle should be positioned 40 ft. (12.1 m) from the aligning screen. Who is right?
    A. A only
    B. B only
    C. Both A and B
    D. Neither A nor B

68. When aiming headlights with a headlight aligning screen:
    A. the vehicle does not need to be parallel to the aligning screen.
    B. the tire inflation pressure does not affect headlight alignment.

C. the headlight alignment may be changed by installing new sealed beams.
D. the headlight alignment is changed by turning adjusting screws on the headlight assembly.

Hint

Headlight aim should always be checked on a level floor with the vehicle unloaded. In some states, the above instructions may conflict with existing laws and regulations. If so, modify the instructions to meet the state's legal requirements.

Various types of headlight aiming equipment are available commercially. When headlight aiming equipment is not available, aiming can be checked by projecting the upper beam of each light on a screen or chart at a distance of about 25 feet ahead of the headlights. The headlights must be aimed to provide the required drop in 25 ft. (7.62 m). Some states require a 2 in. (50.8 mm) drop in 25 ft. (7.62 m).

Some manufacturers recommend coating the prongs and base of a new sealed beam with dielectric grease for corrosion protection. Sealed-beam halogen headlights are designed to give substantially more light on high beam than incandescent lights, extending the driver's range of visibility for safer night driving.

Halogen bulbs have the advantage of producing a whiter light, which helps improve visibility. Halogen bulbs also last longer, stay brighter, and use less wattage for the same amount of light produced. When replacing individual replaceable bulbs, avoid touching the glass envelope. Oil from the skin can cause the bulb to shatter when turned on.

Task 3

Test, repair, and replace headlight and dimmer switches, wires, connectors, terminals, sockets, relays, and miscellaneous components.

69. *Technician A* says some dimmer switches are floor mounted. *Technician B* says high resistance at dimmer switch contacts may reduce current flow only in the high beam circuit. Who is right?
    A. A only
    B. B only
    C. Both A and B
    D. Neither A nor B

70. The headlight switch has failed in the circuit shown in Figure 6–17. *Technician A* says the parking lights are not affected. *Technician B* says the stoplights will fail to operate. Who is right?
    A. A only
    B. B only
    C. Both A and B
    D. Neither A nor B

Hint

The headlight dimmer switch can be mounted on the floor or it can be a part of the turn signal assembly mounted on the steering column. A faulty headlight switch may cause intermittent operation or total failure in both the low and high beam circuits because both circuits are receiving voltage through the headlight switch.

A multifunction switch can serve as a turn signal, cruise control, and possible windshield wiper switch all in one assembly.

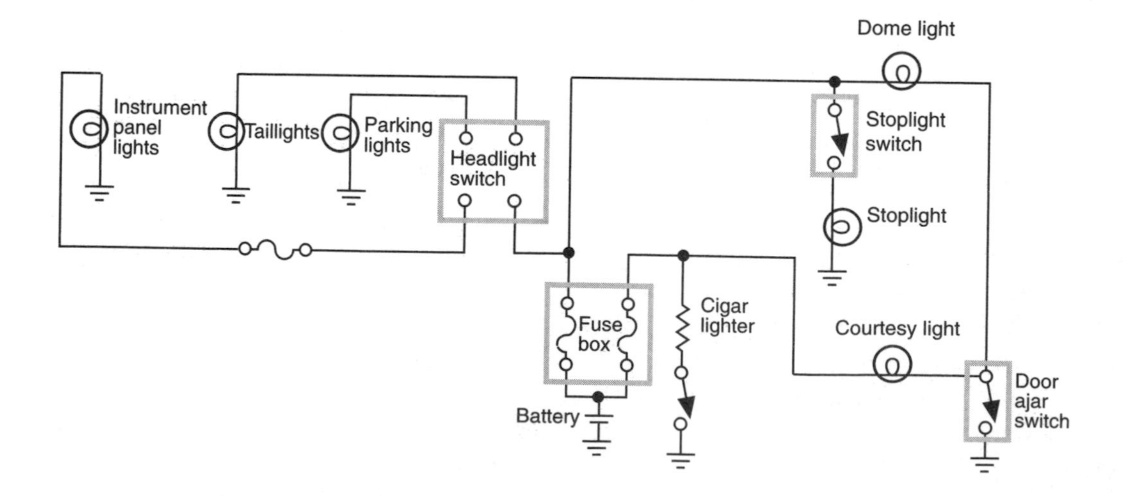

Figure 6–17 Headlight switch with park lights, taillights, stoplights, and instrument panel lights.

Task 4

Inspect, test, and repair or replace switches, bulbs, sockets, connectors, terminals, relays, and wires of parking, clearance, and taillight circuits on trucks and trailers.

71. The type of bulb in Figure 6–18C is:
    A. double contact.
    B. single-contact standard indexing.
    C. double contact with staggered indexing.
    D. single contact with staggered indexing.

72. A trailer has inoperative taillights only on one side. *Technician A* says to check the trailer connector circuit for an open. *Technician B* uses an ohmmeter to check the continuity between one taillight on the inoperative side and the trailer connector. Who is right?
    A. A only
    B. B only
    C. Both A and B
    D. Neither A nor B

Hint

Single-filament bulbs may be single-contact or double-contact types. Single-contact types are grounded through the bulb base; double-contact, single-filament types have two wires, one live and the other a ground. Both types can have either standard or staggered indexing, depending on the application.

A voltmeter or test lamp is used, beginning at the faulty taillight side and working up toward the trailer circuit connector. The open is located between the last "no-power" check and the first confirmation of power by the test equipment.

Most rear lights share a common ground. When the brakes are applied, voltage is supplied from the brake light switch to the stoplight bulbs. If any bulb is dim, there is a resistance problem in the voltage supply wire or ground wire connected to the bulb.

Task 5

Inspect, test, and repair or replace dash light circuit switches, bulbs, sockets, connectors, terminals, wires, and printed circuits.

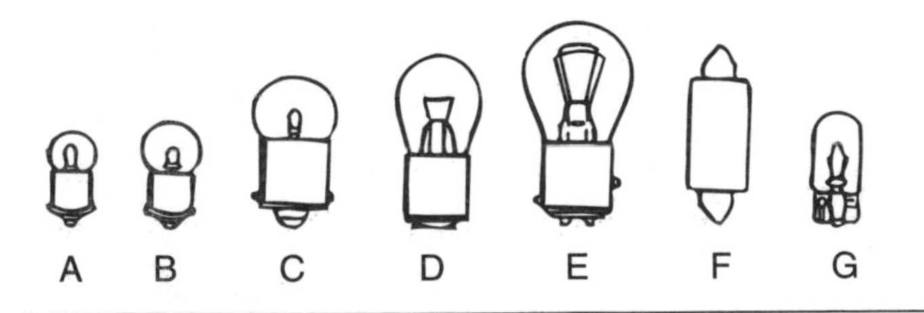

Figure 6–18 Various types of bulbs.

73. The dash lights on a truck operate intermittently. When the dash lights go out, some of the gauges in the instrument panel also provide inaccurate readings. The cause of this problem could be:
    A. a loose bulb in one of the dash lights.
    B. a loose instrument panel ground connection.
    C. an intermittent open in the dash light variable resistor.
    D. an intermittent ground on one dash light socket.

74. When diagnosing a dash light circuit:
    A. the variable resistor must be grounded to the instrument panel.
    B. the dash light bulbs are connected in series with each other.
    C. the variable resistor is connected in series with the dash light bulbs.
    D. a high resistance in one dash bulb circuit increases current flow through that bulb.

Hint

Dash-mounted bulbs, switches, and connectors are components that are affected by resistance problems. Loose terminals, corroded connectors, and shorted components are some of the problems that may cause not only lighting problems but may also damage the printed circuit boards. Proper grounding of the dash circuits should always be checked whenever a problem is suspected.

The dash lights are grounded to the instrument panel, and a variable resistor is connected in series with the dash lights. The sliding contact in this variable resistor is rotated by turning the headlight switch knob. Turning this knob clockwise increases the resistance in the dash light circuit, decreases the current flow, and dims these lights.

Task 6

Inspect, test, and repair or replace interior cab light circuit switches, bulbs, sockets, connectors, terminals, and wires.

75. The interior cab light circuit in Figure 6–19 has a short to ground. *Technician A* says the current flow through the bulb is higher than normal. *Technician B* says the light cannot be turned off. Who is right?
    A. A only
    B. B only
    C. Both A and B
    D. Neither A nor B

76. Each time a cab light fuse is replaced, it blows within 1 minute. In this cab light circuit, the door jamb switches ground the bulbs when a door is opened. The cause of this problem could be:

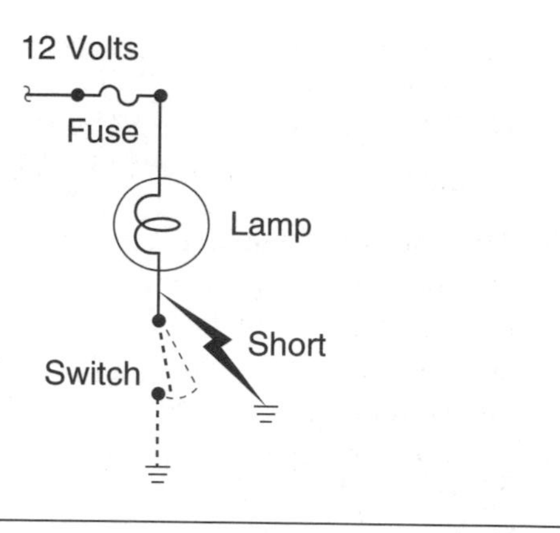

Figure 6–19 Cab light and door jamb switch.

A. a grounded circuit before one of the cab light bulbs.
B. a grounded circuit between one of the cab lights and ground.
C. shorted contacts in one of the door jamb switches.
D. an open circuit in one of the ground wires connected to a door jamb switch.

Hint

Interior cab lights and circuits are affected by resistance problems such as loose terminals, corroded connectors, and high-resistance ground connections. Grounded components may cause a fuse to blow. Never replace the blown fuse with a larger amperage fuse to correct the problem. In many cab light circuits, a door jamb switch grounds the bulb and turns on the cab lights when a door is opened. A switch in the instrument panel may also be used to turn on the cab lights.

Stoplights, Turn Signals, Hazard Lights, and Back-up Lights

Task 1

Inspect, test, adjust, and repair or replace stoplight circuit switches, bulbs, sockets, connectors, terminals, relays, and wires.

77. In the stoplight switch circuit in Figure 6–20:
    A. there is some air pressure supplied to the switch with the engine running.
    B. the switch has normally closed contacts.
    C. air pressure is supplied to the switch when the brakes are applied.
    D. air pressure is supplied to the switch from the supply air system.

78. The stoplights on a truck are illuminated continually. *Technician A* says the stoplight switch contacts may have an open circuit. *Technician B* says the wires on the stoplight switch may be shorted together. Who is right?
    A. A only
    B. B only
    C. Both A and B
    D. Neither A nor B

Hint

With the ignition switch off, voltage is available to the ignition switch and emergency light circuits. In many stoplight circuits, voltage is supplied to the brake light switch from the battery positive terminal. When the brakes are applied on a hydraulic brake system, brake pedal movement closes the stoplight switch. On air brake systems, the stoplight switch is often mounted in the tractor protection valve and senses air pressure. When the

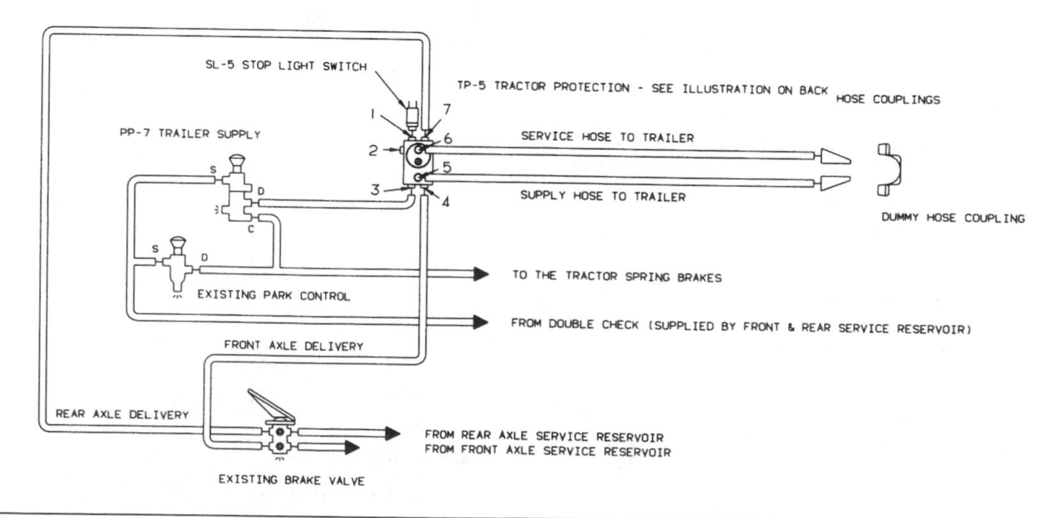

Figure 6–20 Stoplight switch in an air brake system. *(Courtesy of Allied Signal Truck Brake Systems Co.)*

brakes are applied, air pressure closes the stoplight switch contacts to illuminate the stoplights.

Task 2

Diagnose the cause of no turn signal and hazard flasher lights or lights with no flash on one or both sides.

79. In the signal light circuit in Figure 6–21 the right rear signal light is dim, and the other lights work normally. The cause of this problem could be:
    A. high resistance in the DB 180G RD wire from the turn signal switch to the rear lamp wiring.
    B. a short to ground in the DB 180 RD wire from the turn signal switch to the rear lamp wiring.
    C. high resistance in the D7 18BR RD wire from the turn signal switch to the rear lamp wiring.
    D. high resistance in the D2 18G RD wire from the turn signal switch to the rear lamp wiring.

80. The left front signal light on a truck does not flash and the left rear signal light flashes faster than normal. *Technician A* says the left front bulb may be defective. *Technician B* says there may be an open circuit between the signal light switch and the left front signal light bulb. Who is right?
    A. A only
    B. B only
    C. Both A and B
    D. Neither A nor B

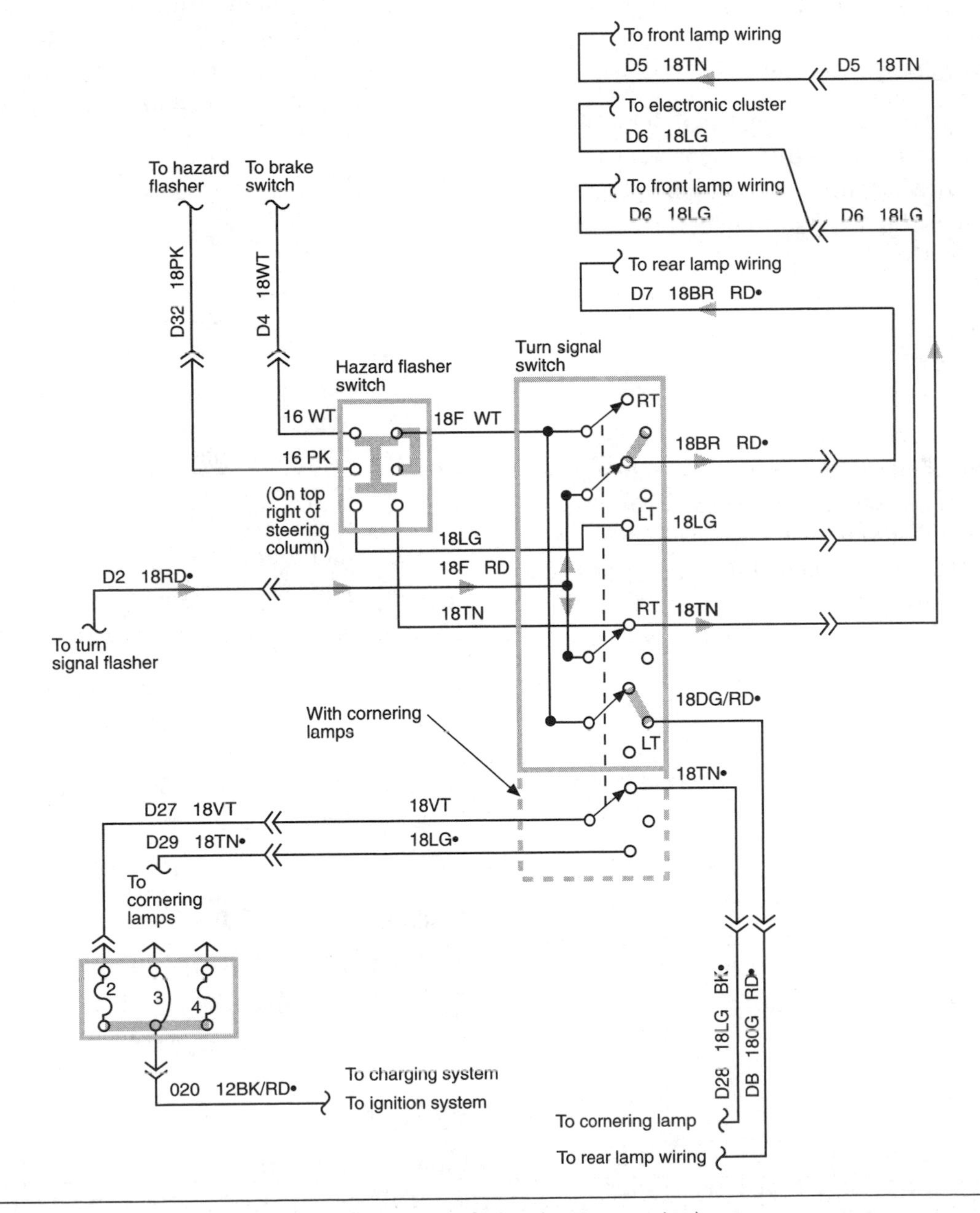

Figure 6–21 Headlight switch and related circuit. *(Courtesy of Chrysler Corporation)*

Hint

Whenever the vehicle's turn signals are not operating, the power to the circuit should be checked first. The trailer electrical connector may be disconnected to assist with the diagnosis of the problem. If only one side of the flashers works, the wiring, bulb, or socket may be causing the problem. The flasher supplies current to the turn signal switch. This switch directs this current to the right or left turn signal lamps depending on the lever position. On some trucks, current from the brake switch also passes through the turn signal circuit to the rear stop/turn signal lamps. If the brake pedal is applied during a right turn, the left brake light is on.

Task 3

Inspect, test, and repair or replace turn signal and hazard circuit flasher, switches, bulbs, sockets, connectors, terminals, relays, and wires.

81. *Technician A* says the turn signal switch may be contained in the multifunction switch. *Technician B* says the variable resistor in the dash light circuit may be contained in the multifunction switch. Who is right?
    A. A only
    B. B only
    C. Both A and B
    D. Neither A nor B

82. A tractor/trailer combination experiences repeated flasher failure, and sometimes the signal light fuse is blown when the flasher is defective. The cause of this problem could be:
    A. high resistance in the signal light ground circuit on the trailer.
    B. excessive voltage drop in the wires connected to the trailer signal lights.
    C. an intermittent ground in some of the signal light wires.
    D. corrosion on the rear tractor signal light sockets.

Hint

The flasher unit that controls the turn signals and/or the four-way flasher may be suspected when the flashers are totally inoperative. The flasher assembly may be damaged from wiring problems or excessive charging system voltage.

Task 4

Inspect, test, adjust, and repair or replace back-up light and warning device circuit switches, bulbs, sockets, horns, buzzers, connectors, terminals, and wires.

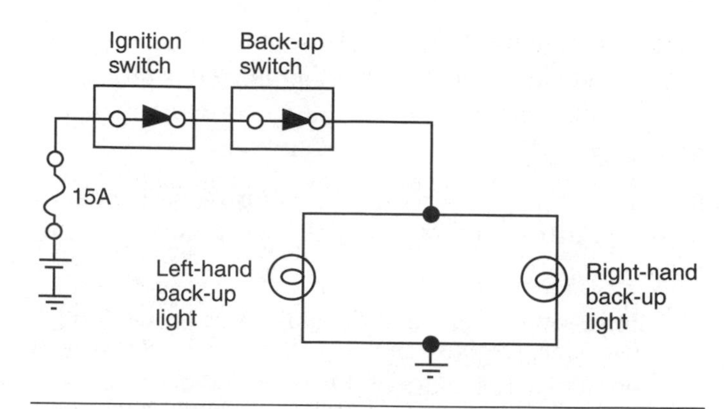

Figure 6–22 Back-up light circuit.

83. In Figure 6–22 the back-up light fuse keeps blowing. *Technician A* says to use an ohmmeter to check the circuit from the fuse to the back-up light switch. *Technician B* says to use an ohmmeter to check the circuit from the back-up light switch through the bulbs to ground. Who is right?
    A. A only
    B. B only
    C. Both A and B
    D. Neither A nor B

84. Refer to Figure 6–22. The right-hand back-up light circuit is grounded on the switch side of the bulb. *Technician A* says this condition may blow the back-up light fuse. *Technician B* says the left-hand back-up light may work normally while the right-hand back-up light is inoperative. Who is right?
    A. A only
    B. B only
    C. Both A and B
    D. Neither A nor B

Hint

Whenever basic on-off components such as the back-up light switch, are being diagnosed, the proper supply voltage is critical. If the suspected components or circuits have more than a 100-millivolt voltage drop, the ground or power circuit needs to be repaired.

Gauges and Warning Devices Diagnosis and Repair

Task 1

Diagnose the cause of intermittent, high, low, or no gauge readings; determine needed repairs. (Does not include charge indicators.)

85. In Figure 6–23, if the wire from the water temperature gauge to the sending unit is grounded:
    A. the water temperature gauge reads continually low.
    B. the water temperature gauge continually fluctuates.
    C. the water temperature gauge reads continually high.
    D. the water temperature gauge fuse blows repeatedly.

86. Refer to Figure 6–23. Ground wire number 57 on the gauge cluster has an open circuit. This defect causes all of these problems EXCEPT:
    A. an inaccurate water temperature gauge reading.
    B. an inaccurate tachometer reading.
    C. an inaccurate fuel gauge reading.
    D. an inaccurate voltmeter reading.

Hint

When gauge readings are inaccurate, sensor operation and loose connections should be checked first. On trucks with air brakes, check the dual air pressure gauge reading against the reading on an accurate test gauge. Replace the gauge assembly if either needle reading is more than 4 psi (27.5 kPa) higher or lower than the actual pressure indicated on the test gauge.

Many gauge sending units contain a thermistor that changes resistance in relation to temperature. As a thermistor is cooled, its resistance increases. Dash gauges may be thermal-electric or balancing coil-type. Some sending units such as a fuel gauge sender contain a variable resistor. A thermal-electric gauge contains a bimetallic strip connected to the gauge pointer. A heating coil is wrapped around the bimetallic strip. As the sending unit resistance decreases, the current flow increases and this action heats the bimetallic strip and pushes the gauge pointer to a higher position. In a balancing coil gauge, the sending unit controls the current flow through the two coils, and coil magnetism attracts the gauge pointer to provide the proper gauge reading.

Task 2

Test and replace gauge-circuit voltage regulators (limiters).

87. All the electric gauge readings in a truck are inaccurate. *Technician A* says the instrument voltage limiter may be defective. *Technician B* says the instrument voltage limiter ground may have high resistance. Who is right?
    A. A only
    B. B only

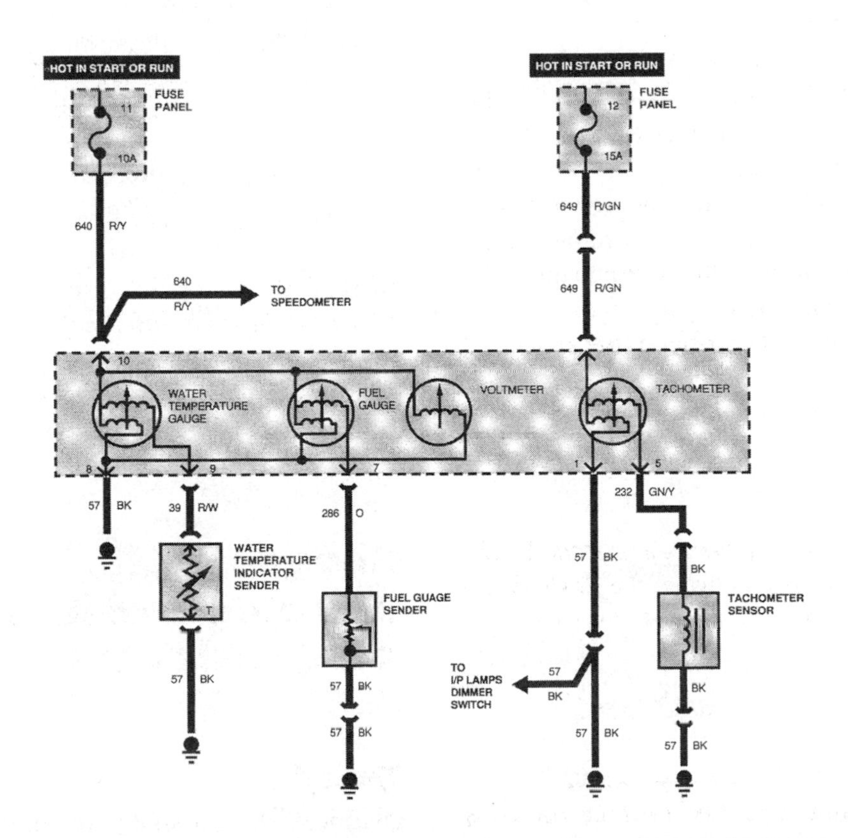

Figure 6–23 Instrument panel gauge circuit. *(Courtesy of Ford Motor Company)*

C. Both A and B
D. Neither A nor B

88. With the ignition switch on, the voltage output from the instrument voltage limiter should be:
A. 2V.
B. 4V.
C. 5V.
D. 9V.

Hint

Some instrument panel gauges require protection against heavy voltage fluctuations that could damage the gauges or cause them to give inaccurate readings. A voltage limiter provides protection against heavy voltage fluctuations by limiting voltage to the gauges to approximately 5 volts. An instrument voltage limiter contains a heating coil wound around a bimetallic strip and a set of contacts. As gauge current heats the bimetallic strip, the contacts open and close to limit the voltage to the gauges. The instrument voltage limiter must be grounded to the instrument panel. A capacitor or suppression coil is connected to most voltage limiters to reduce static on stereos and CD players. A defective voltage limiter affects all the gauge readings that are connected to the limiter.

Task 3

Inspect, test, adjust, and repair or replace gauge circuit sending units, gauges, connectors, terminals, and wires.

89. When testing a mechanical water temperature gauge for an erratic reading, *Technician A* uses an ohmmeter to test the resistance of the gauge. *Technician B* places the sensing bulb in boiling water and checks the gauge reading against a known accurate thermometer placed in the water. Who is right?
A. A only
B. B only
C. Both A and B
D. Neither A nor B

90. A thermistor-type engine temperature sensor is tested by using:
A. an ammeter connected between the two sensor terminals.
B. an ohmmeter connected across the sensor terminals at various sensor temperatures.
C. a 12V test light connected in series with one of the sensor wires.
D. a 12V test light connected from the sensor input wire to ground.

Hint

The gauge and the wiring can be tested by removing the wire from the sending unit and connecting a gauge tester from this wire to ground. The gauge tester contains a variable resistor. As the gauge tester knob is rotated with the ignition switch on, the tester resistance changes and the dash gauge reading should change accordingly. This test may be repeated with the gauge tester connected to the gauge output terminal in the dash. If the reading is satisfactory when the tester is connected to the dash gauge but unsatisfactory when the tester is connected to the sensor wire, the wire from the gauge to the sensor is defective.

If the mechanical water temperature gauge indicates an inaccurate reading, place the sensing bulb in boiling water and check the gauge reading against a known good thermometer inserted in the water.

Task 4

Inspect, test, and repair or replace warning light circuit sending units, bulbs, sockets, connectors, terminals, wires, and printed circuits/control modules.

91. The low air brake pressure warning light on a truck is illuminated, but the air pressure gauge in the dash shows 125 psi (862 kPa) air pressure in both the primary and secondary air brake systems. The cause of this problem could be:
A. a grounded wire connected to one of the low air pressure switches.
B. an open circuit in the wire to one of the low air pressure switches.
C. an air leak at one of the low air pressure switches.
D. an open circuit at the switch contacts in one of the low air pressure switches.

92. In the engine shutdown system in Figure 6–24 the:
A. the coolant temperature sensor contains a thermistor.
B. the oil pressure sensor contains a variable resistor.
C. the alarm module is connected to the engine run relay and shutdown solenoid.
D. the oil pressure sensor resistance increases as the oil pressure decreases.

Hint

Most trucks contain a number of warning lights, such as a low air brake pressure warning light. This light is illuminated by a switch in both the primary and secondary air reservoirs. If the air pressure drops below 50 to 70 psi (345 to 483 kPa), the switch contacts close and turn

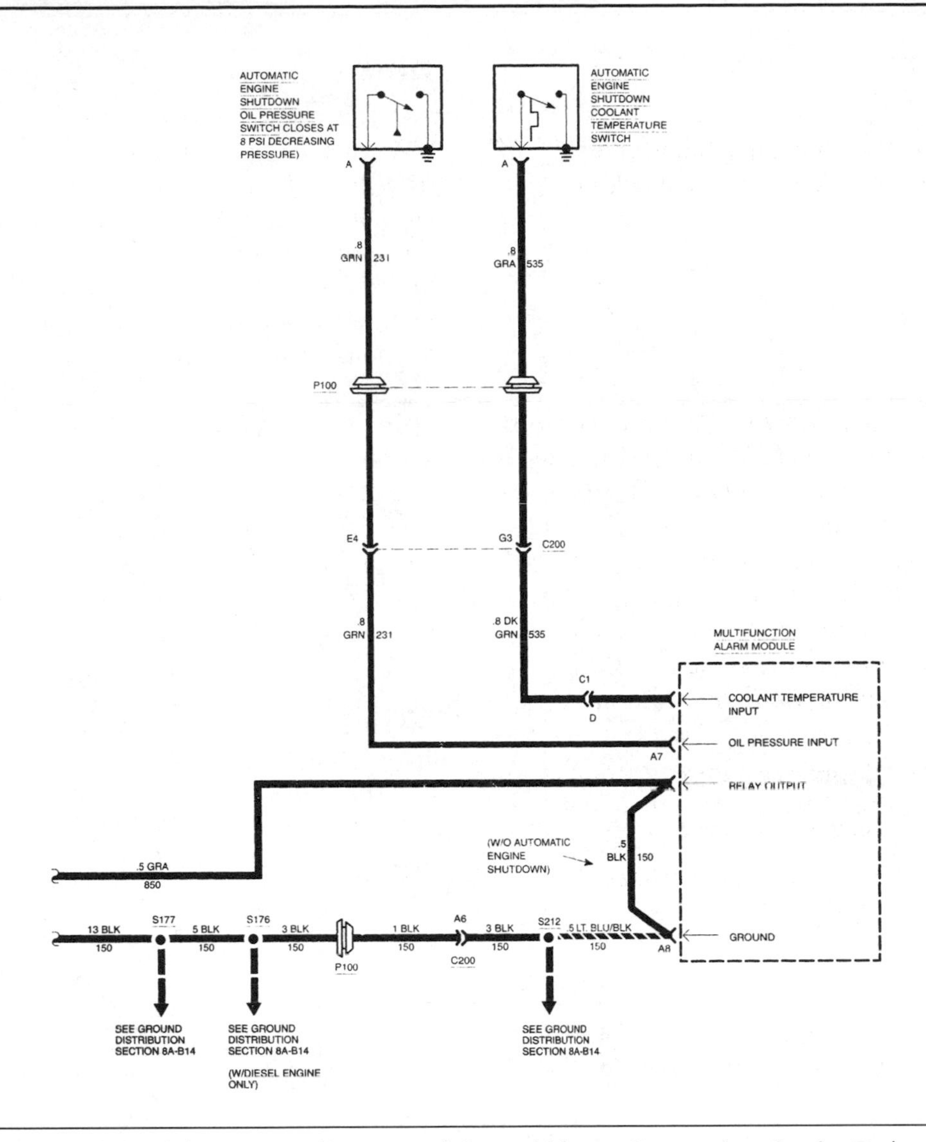

Figure 6–24 Engine alarm and shutdown system. *(Courtesy of General Motors Corporation, Service Technology Group)*

on the warning light. In some air brake systems, a buzzer is also activated when the low air pressure light is on. Some trucks have an engine shutdown system that automatically shuts off the engine if the engine has low oil pressure or high coolant temperature. In these systems an alarm module receives inputs from oil pressure and coolant temperature sensors. If the oil pressure is low or the coolant temperature is high, the alarm module shuts off the engine run relay and the fuel shutdown solenoid. The module also provides visual and/or audible alarms for the driver.

Task 5

Inspect, test, and repair or replace warning buzzer circuit sending units, buzzers, switches, relays, connectors, terminals, wires, and printed circuits/control modules.

93. While discussing the warning light and buzzer system in Figure 6–25, *Technician A* says if either low air pressure switch closes, the brake warning indicator lamp is illuminated. *Technician B* says if either low air pressure switch closes, the buzzer is activated. Who is right?
    A. A only
    B. B only
    C. Both A and B
    D. Neither A nor B

94. In the air brake warning system in Figure 6–25, the buzzer winding is open. The result of this defect is:
    A. an inoperative brake warning indicator light.
    B. blown number 14, 10-ampere fuse.
    C. excessive current flow through the low air pressure switches.
    D. an inoperative buzzer.

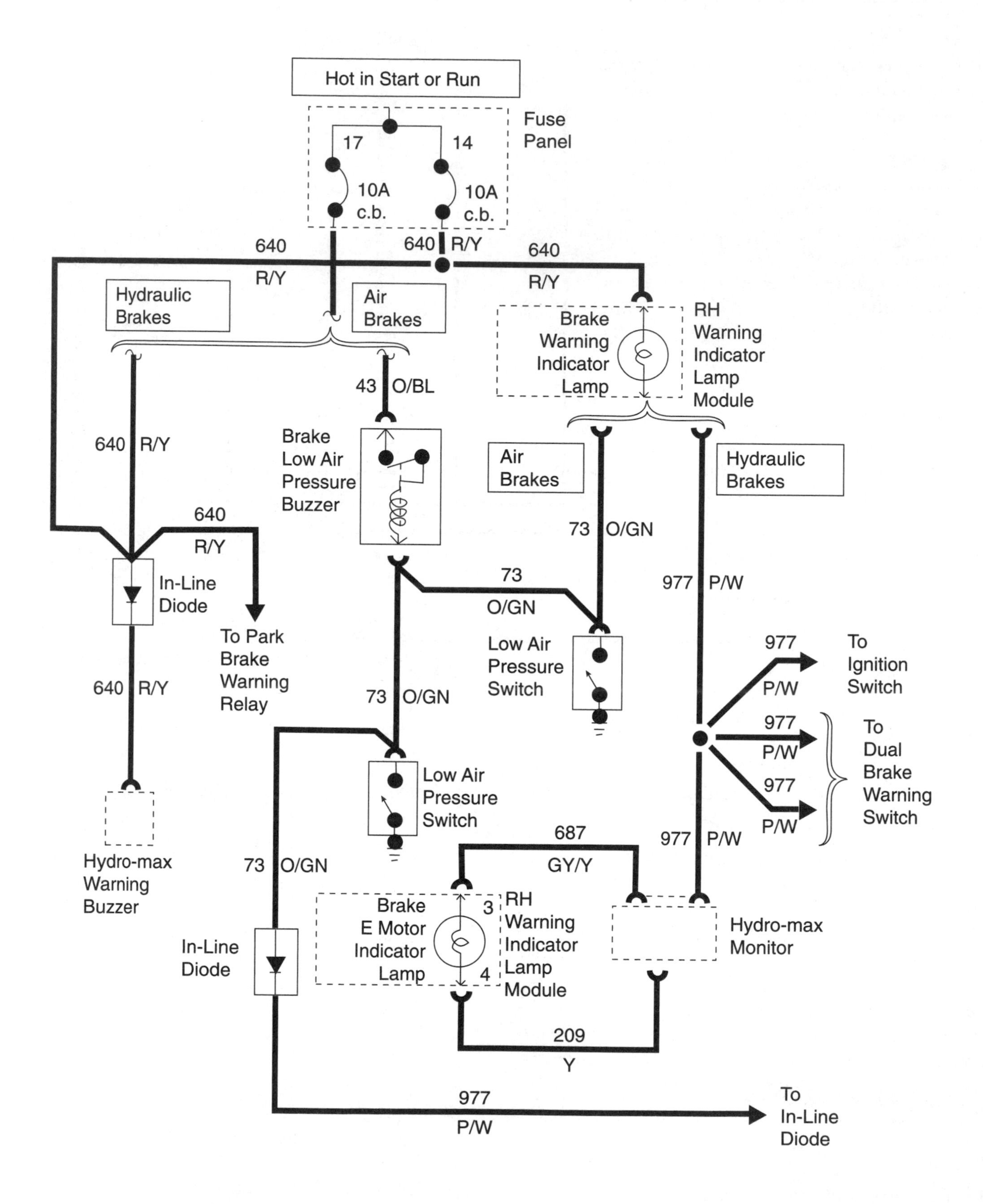

Figure 6–25 Low air pressure brake warning light and buzzer circuit. *(Courtesy of Ford Motor Company)*

Hint

If an oil pressure sending unit is suspected of being faulty, a manual check of the engine oil pressure is necessary to eliminate the possibility of worn engine components. This, along with continuity and resistance checks of the sending unit and its associated wiring, should be performed for a more complete diagnosis.

Static electricity can cause permanent damage to the instrument cluster. Before working on the cluster, be sure to remove all static electricity from your body by touching metal on the vehicle chassis. A flexible ground cable should be attached from a wrist band to a satisfactory chassis ground before working on electronic components.

Task 6

Inspect, test, replace, and calibrate electronic speedometer, odometer, and tachometer systems.

95. On some trucks with diesel engines, the tachometer signal to the instrument cluster is taken from:
    A. the negative primary ignition coil terminal.
    B. the secondary ignition coil terminal.
    C. the alternator R terminal.
    D. the vehicle speed sensor (VSS).

96. In the tachometer circuit in Figure 6–26:
    A. the voltmeter is connected parallel to the battery.
    B. both coils in the gauge are connected in series with the tachometer sensor.
    C. the tachometer sensor contains a thermistor.
    D. an open number 14, 10-ampere fuse only causes an inoperative tachometer.

Hint

The tachometer is an electromagnetic gauge, similar to an electronic speedometer with two coils. It operates from a tachometer signal off the primary ignition coil on gasoline engines. On some diesel engines, the tachometer signal is taken from the generator "R" terminal.

Miscellaneous

Task 1

Diagnose the cause of constant, intermittent, or no horn operation.

97. The electric horns on a truck will not turn off. *Technician A* says the diaphragm contacts in the horns may be welded together. *Technician B* says the horn relay may be defective. Who is right?
    A. A only
    B. B only
    C. Both A and B
    D. Neither A nor B

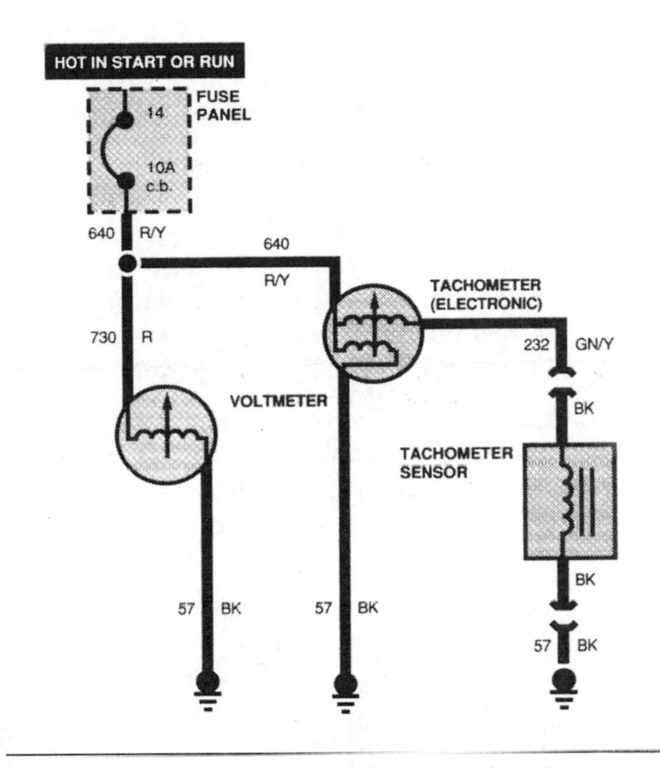

Figure 6–26 Tachometer circuit. *(Courtesy of Ford Motor Company)*

98. A tractor has an inoperative right-side air horn. *Technician A* says the air line may be kinked where it is connected to the right-side horn. *Technician B* says to check the air line where it is connected to the air manifold. Who is right?
    A. A only
    B. B only
    C. Both A and B
    D. Neither A nor B

Hint

When diagnosing any air horn system malfunction, always check the air supply lines for leakage or deterioration. Air manifold valves or switches are usually checked when the horn volume or pitch is low. In many electric horn circuits, the horn button grounds the relay winding. This action closes the relay contacts that supply voltage to the horns. In other electric horn circuits, voltage and current are supplied directly through the horn switch contacts to the horns.

Task 2

Inspect, test, and repair or replace horn circuit relays, horns, switches, connectors, terminals, and wires.

99. The horn circuit in Figure 6–27 does not provide any horn operation. All of these defects may be the cause of the problem EXCEPT:
    A. an open ground circuit in the horn relay.
    B. an open circuit in the horn relay winding.
    C. an open circuit at the horn brush/slip ring.
    D. an open fuse link in the relay power wire.

100. An electric horn on a medium-duty truck blows intermittently when turning the steering wheel. This problem may be caused by:
    A. a grounded wire between one of the horns and the horn relay.
    B. an open circuit at the horn relay contacts.
    C. the wire from the horn relay to the horn button grounding in the steering column.
    D. the power wire to the horn relay winding shorted to ground.

Hint

The horn blows continually if the wire from the horn relay winding to the horn button becomes grounded. This wire sometimes becomes grounded inside the steering column. Occasionally, this wire intermittently touches ground when turning the steering wheel, and this causes the horn to blow when rotating the wheel. Horn contacts sticking closed or shorting to ground also cause this problem.

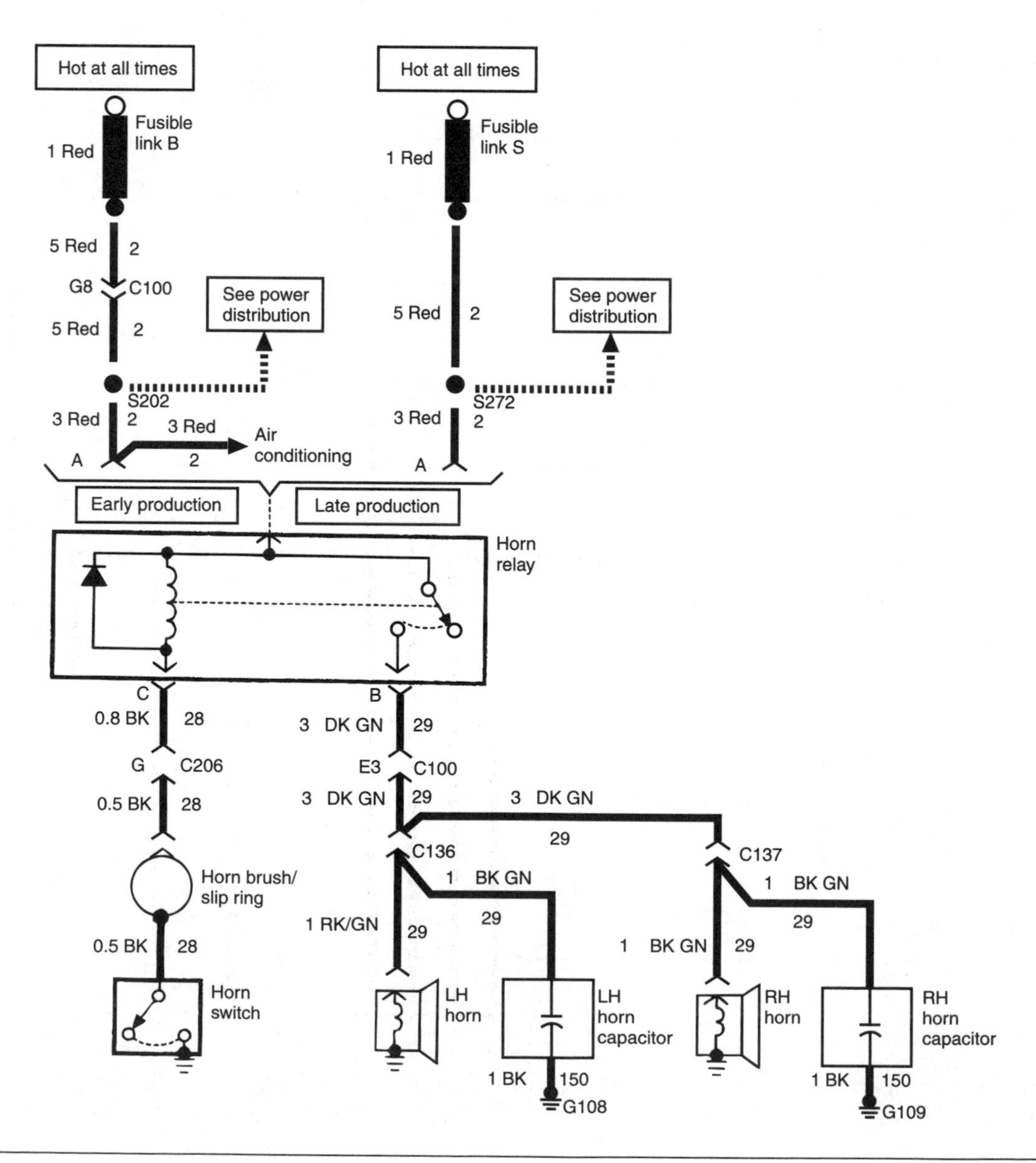

Figure 6–27 Horn circuit.

Task 3

Diagnose the cause of constant, intermittent, or no wiper operation, diagnose the cause of wiper speed control and/or park problems.

101. All of these defects may cause an inoperative electric wiper system EXCEPT:
   A. no voltage to the wiper motor.
   B. a defective wiper switch.
   C. a faulty wiper motor.
   D. the circuit breaker not operating.

102. The electric wiper system in Figure 6–28 operates only at high speed. The cause of this problem could be:
   A. an open circuit at wiper motor terminal E.
   B. an open circuit in wire 228.
   C. an open circuit at wiper motor terminal B.
   D. an open circuit at wiper motor terminal C.

Hint

Most electric wiper systems are protected by fuses, fusible links, or circuit breakers. The circuit breaker is

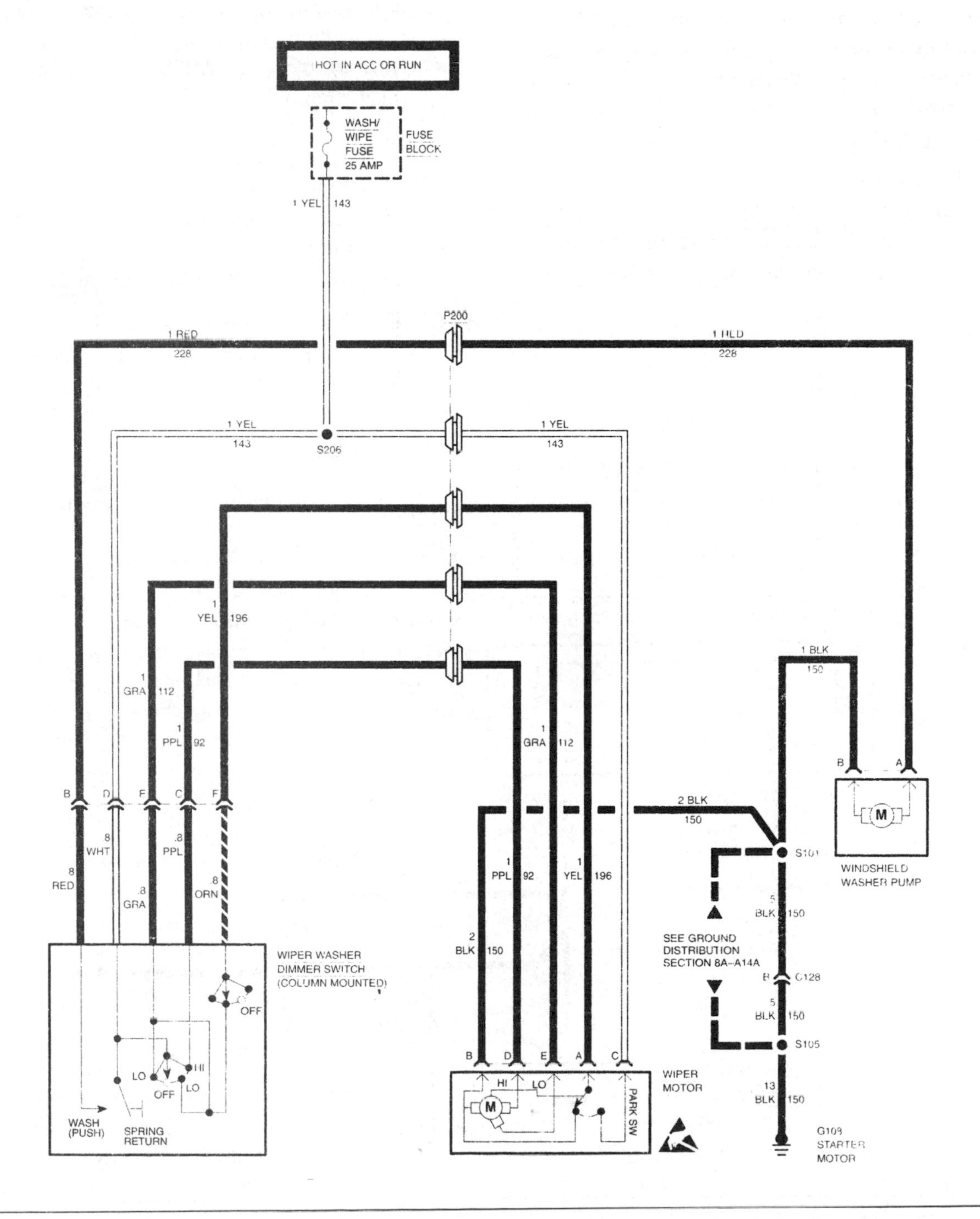

Figure 6–28 Electric wiper motor system. *(Courtesy of General Motors Corporation, Service Technology Group)*

built into some electric wiper motors. In some two-speed electric wiper motors, the speed is changed by supplying voltage to different brushes contacting the wiper motor commutator. The field pole shoes are permanent magnets in many wiper motors. Some wiper motors have field windings wound on the pole shoes. On some applications, the wiper and washer switch are designed into the multifunction switch on the steering column. Other trucks have the wiper and washer switch mounted in the instrument panel.

Task 4

Inspect, test, and repair or replace wiper motor, resistors, park switch, relays, switches, connectors, terminals, and wires.

103. The electric windshield wipers on a truck stop when the wiper switch is shut off and do not move to the park position. *Technician A* says the wiper park switch may be defective. *Technician B* says the wiper switch in the instrument panel may be defective. Who is right?
    A. A only
    B. B only
    C. Both A and B
    D. Neither A nor B

104. The air motor from an air-operated wiper system is operating erratically. *Technician A* says the motor may have a plugged exhaust port. *Technician B* says the motor may have a shorted field winding. Who is right?
    A. A only
    B. B only
    C. Both A and B
    D. Neither A nor B

Hint

Failure of the wiper motor to park after operation may be because of the wiper motor park switch. The park switch is mounted in the wiper motor. When the wiper switch is shut off, current flows through the park switch and the wiper motor until a cam on the motor drive mechanism opens the park switch. This cam is designed to open the wiper motor circuit when the wiper blades are in the park position at the bottom of the windshield.

On some electric wiper motor systems, the wiper switch is on the ground side of the circuit. On these systems, the switch controls wiper speed by switching a resistor in and out of the field circuit. This type of wiper motor has an internal relay that supplies voltage to the wiper motor circuit when the wiper switch is turned on. When the wiper switch is turned on, the relay winding is grounded through the switch contacts. In other wiper systems, the wiper switch is on the positive side of the wiper motor, and the switch controls wiper motor speed by supplying voltage to different brushes in the motor. This type of motor usually have permanent magnet-type fields.

Task 5

Inspect and replace wiper motor transmission linkage, arms, and blades.

105. In an electric wiper system with the wiper switch on the ground side of the wiper motor, the motor intermittently stops operating and restarts within 1 to 2 minutes. The MOST likely cause of this problem could be:
    A. high resistance in the power supply wire to the wiper motor.
    B. a binding wiper blade pivot.
    C. high resistance in the wires from the wiper motor to the switch.
    D. loose, corroded terminals on the wiper motor.

106. Each time the electric windshield wipers are turned on a squeaking noise is heard. The MOST likely cause of this problem is:
    A. dry bushings in the wiper motor.
    B. a dry wiper blade pivot.
    C. loose wiper arm pivots.
    D. loose wiper motor mounting bolts.

Hint

When checking an air-operated wiper motor for an erratic operating condition, the possibility of air leaks should be one of the first checks to ensure proper air operation. Binding linkage and/or wiper shafts is another common cause of sluggish or erratic operation on air-operated or electric wiper motors. Binding wiper linkages or pivots on an electric wiper system may cause excessive current draw, and this causes the circuit breaker to open the circuit. When the circuit breaker cools, it closes and the wiper motor starts again.

Task 6

Inspect, test, and repair or replace windshield washer motor or pump/relay assembly, switches, connectors, terminals, and wires.

107. In the windshield wiper and washer system in Figure 6–29 the washer pump is inoperative, but the wiper system operation is normal. The cause of this problem could be:
    A. the circuit breaker is open in the wiper/washer switch.

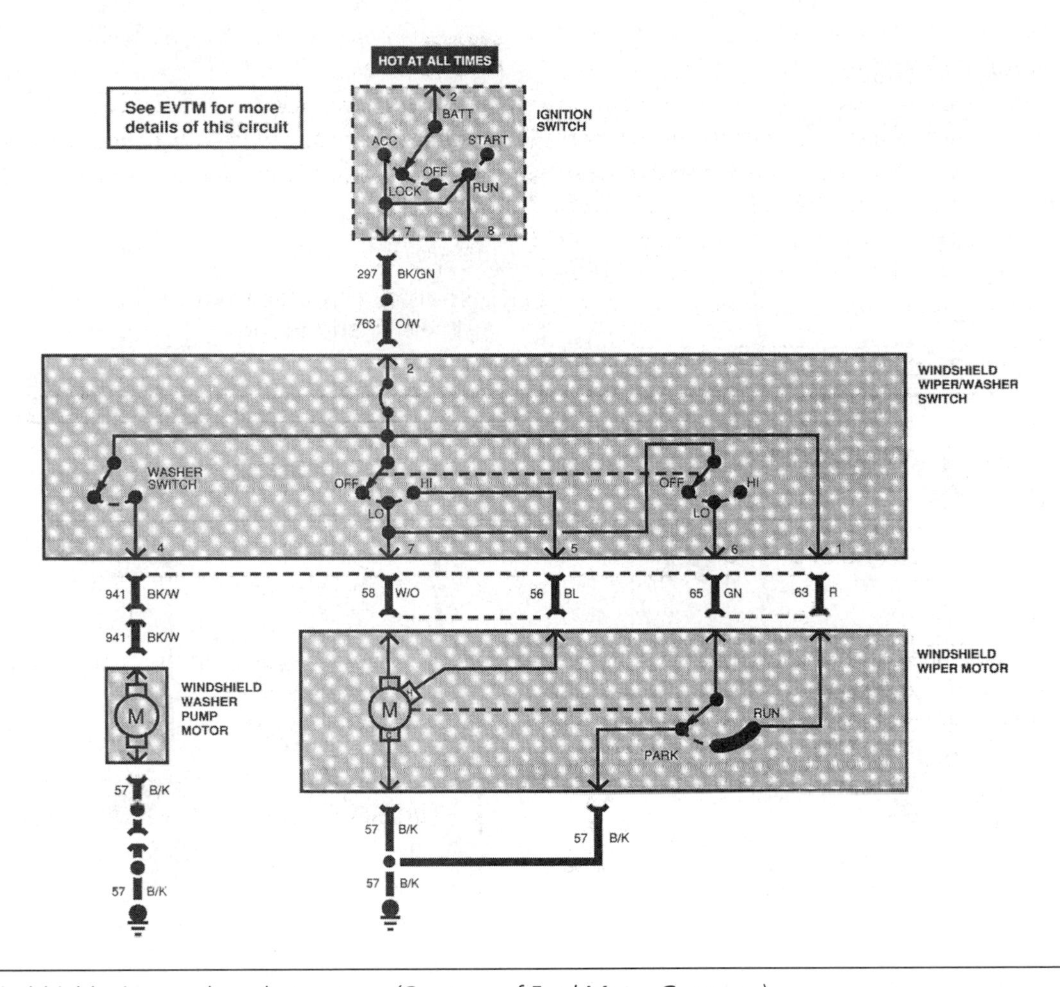

Figure 6–29 Windshield wiper and washer system. *(Courtesy of Ford Motor Company)*

B. there is an open circuit at terminal 5 on the wiper/washer switch.
C. there is an open circuit at terminal 4 on the wiper/washer switch.
D. there is an open circuit in wire number 58 connected to the wiper/washer switch.

108. Refer to Figure 6–29. The windshield washer pump runs continually when the wiper motor is switched to low speed. When the wiper switch is off, the wipers run at low speed if the washer switch is depressed. High-speed wiper operation is normal. The cause of this problem could be:
A. wires 58 and 56 are shorted together.
B. wires 941 and 58 are shorted together.
C. wire 58 is shorted to ground.
D. wire 941 is shorted to ground.

Hint

In many windshield washer pump systems, voltage is supplied to the washer pump motor when the washer switch is depressed. A ground wire is connected from the washer pump motor to the chassis. The washer pump motor is usually mounted in the bottom of the windshield washer fluid reservoir. When checking the operating condition of the washer pump always check the motor for proper voltage with the washer switch turned on.

Task 7

Inspect, test, and repair or replace side view mirror motors, heater circuit grids, relays, switches, connectors, terminals, and wires.

109. In Figure 6–30 the heaters in both side view mirrors are inoperative. All of these defects may be the cause of the problem EXCEPT:
A. an open circuit in fuse link E.
B. an open circuit in circuit breaker number 6.
C. an open circuit in wire number 960 at the switch.
D. an open circuit in the heated mirror indicator light.

110. Refer to Figure 6–30. The mirror heaters remain on continually. The cause of this problem could be:

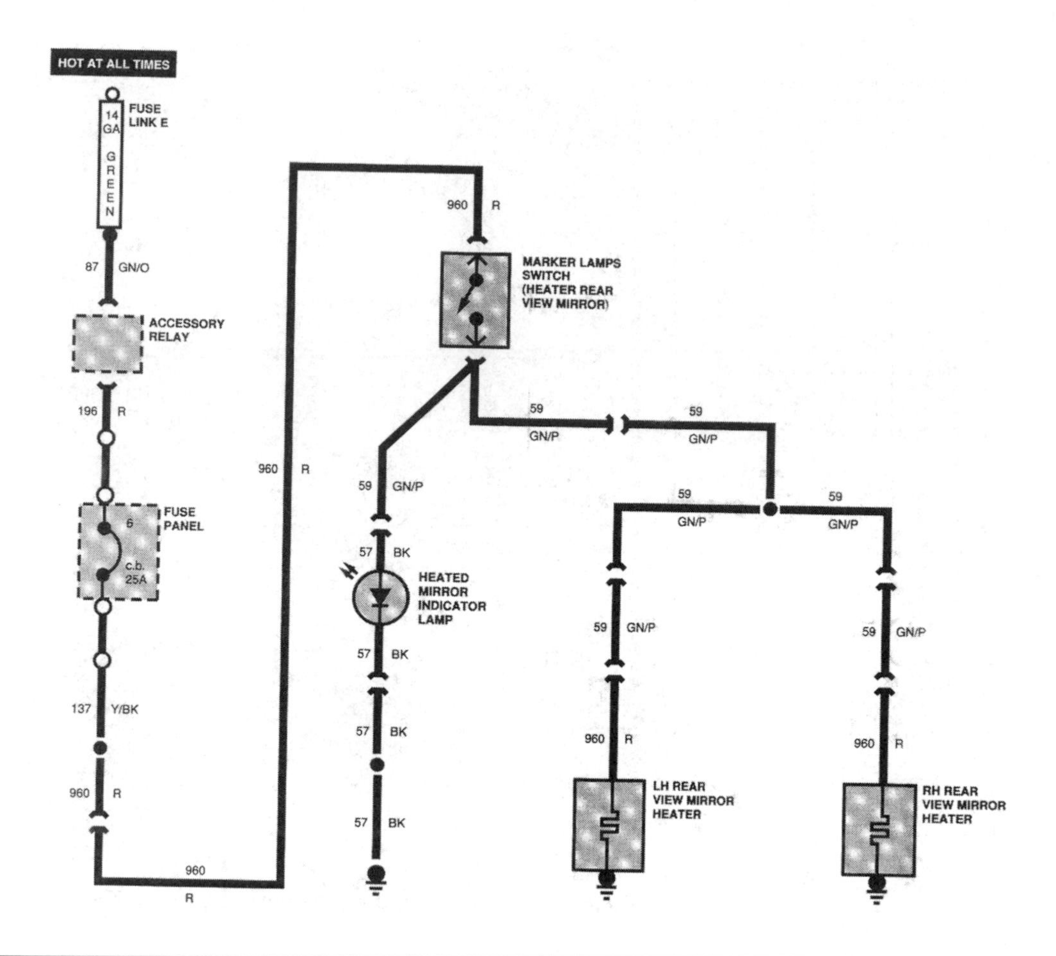

Figure 6–30 Heated mirror circuit. *(Courtesy of Ford Motor Company)*

A. wires 960 and 59 are shorted together.
B. the circuit breaker contacts are stuck closed.
C. one of the mirror heated elements is shorted.
D. wires 57 and 196 are shorted together.

Hint

In some heated mirror circuits, voltage is supplied through a fuse link, circuit breaker, and switch to the mirror heaters. When the heated mirror switch is on, it also supplies voltage to a light-emitting diode (LED) indicator light to remind the driver that the mirror heaters are on. When diagnosing mirror heaters, use a voltmeter to check for voltage at the heaters with the heater switch depressed. With the switch off the resistance of each heater may be measured with an ohmmeter.

Task 8

Inspect, test, and repair or replace heater and A/C electrical components including A/C clutches, motors, resistors, relays, switches, controls, connectors, terminals, and wires.

111. The A/C compressor clutch in Figure 6–31 does not engage at any time. All of these defects may be the cause of the problem EXCEPT:
A. an open circuit in the clutch cycle switch.
B. an open circuit in wire 349.
C. an open circuit in the A/C high-pressure switch.
D. an open circuit in wire 348.

112. The blower motor shown in Figure 6–31 does not operate at high speed, but all other blower speeds are normal. The cause of this problem could be:
A. an open circuit in the ground wire from the blower switch to ground.
B. an open circuit in the thermal limiter in the blower assembly.
C. an open circuit in the high-speed contacts in the blower switch.
D. a short to ground in the wire connected from the resistor assembly to ground.

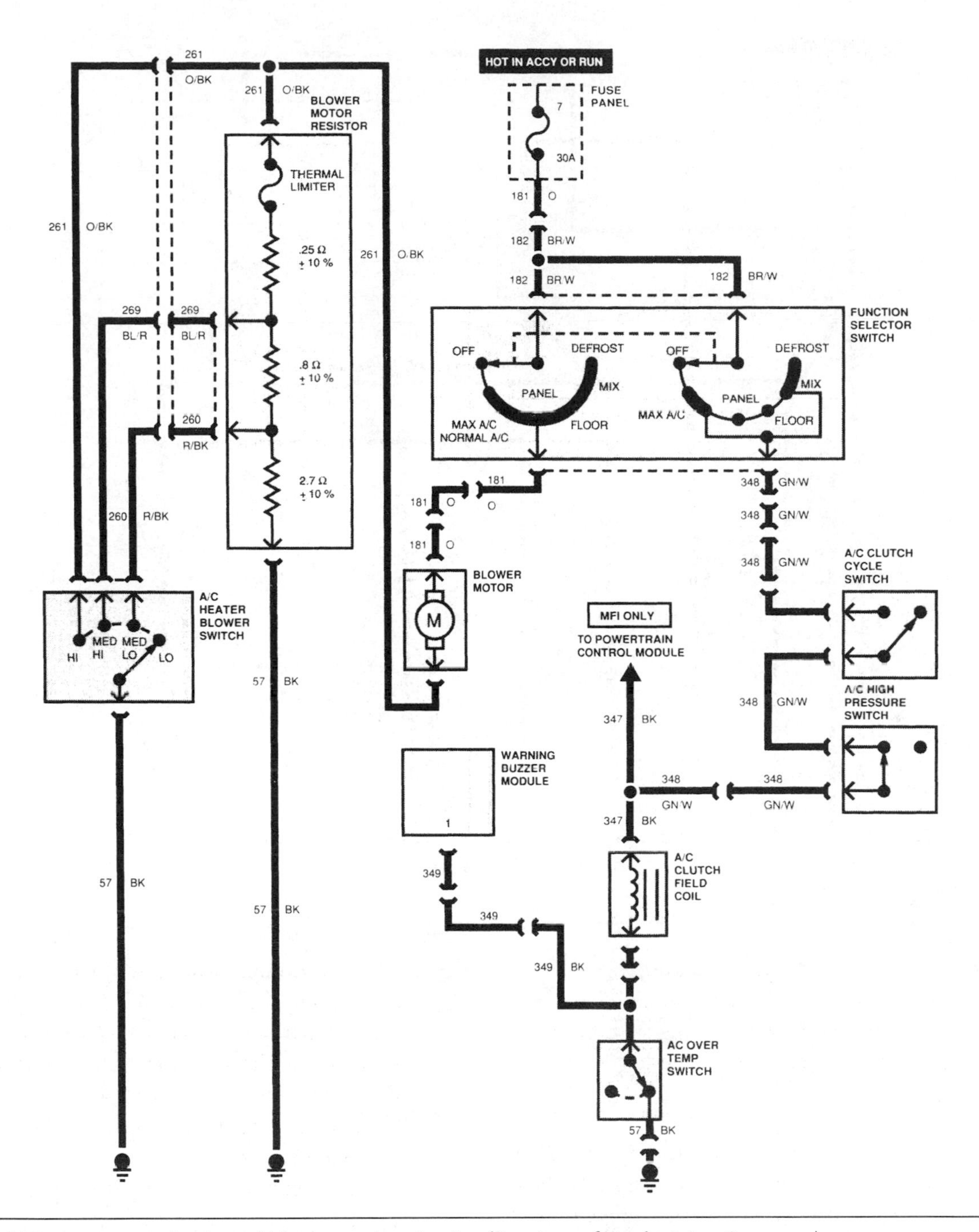

Figure 6–31 A/C compressor clutch and blower motor circuits. *(Courtesy of Ford Motor Company)*

Hint

In a blower motor circuit, blower speed is controlled by switching various resistors into the circuit. Blower motor high speed is the only part of the circuit that does not pass through the blower resistor.

In some A/C systems when the A/C button on the dash control is depressed and the ignition switch is on, the PCM grounds the compressor clutch relay. Contacts inside the clutch relay close, applying source voltage to the cycling pressure switch. The cycling pressure switch is normally closed, and supplies voltage to the clutch coil. The clutch engages so the belt can drive the compressor.

Task 9

Inspect, test, and repair or replace cigarette lighter case, integral fuse, connectors, and wires.

113. The cigarette lighter and the dome light shown in Figure 6–32 are both inoperative. *Technician A*

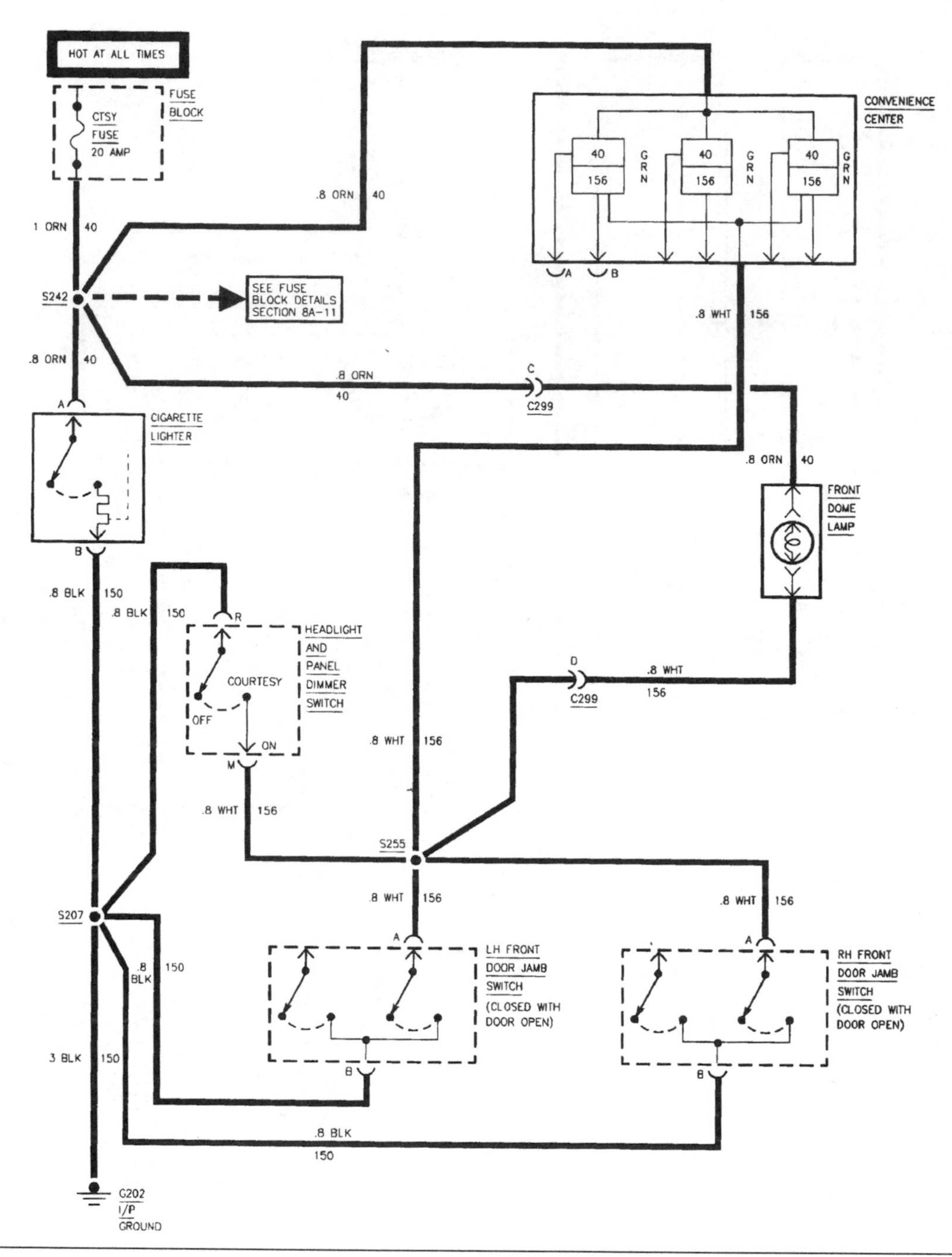

Figure 6–32 Cigarette lighter and dome light circuit. *(Courtesy of Chevrolet Motor Division, General Motors Corporation)*

says there may be an open circuit at terminal B on the lighter. *Technician B* says there may be an open circuit at ground connection G202. Who is right?

A. A only
B. B only
C. Both A and B
D. Neither A nor B

114. The dome light shown in Figure 6–32 is on continually with the ignition switch on or off. The cigarette lighter operates normally. All of these defects may be the cause of the problem EXCEPT:

A. a short to ground on wire 150 connected from the courtesy light switch to junction S207.
B. one of the door jamb switches continually closed.

C. a short to ground at junction S255.
D. the courtesy light switch continually closed.

Hint
A physical and a visual check of the cigarette lighter assembly will usually reveal the problem. The lighter assembly works by current flow through the resistance of the heating element. A short circuit in the lighter element or a foreign object in the lighter housing may cause the circuit fuse to blow. Never replace the fuse with a larger amperage fuse to solve the problem.

Task 10

Diagnose the cause of slow, intermittent, or no-power side window operation.

115. A power window shown in Figure 6–33 operates normally from the master switch, but the window does not work using the window switch. The cause of this problem could be:
    A. an open circuit between the ignition switch and the window switch.
    B. an open circuit in the window switch contacts.
    C. an open circuit in the master switch ground wire.
    D. a short to ground at the circuit breaker in the motor.

116. The power window circuit in Figure 6–33 is completely inoperative. *Technician A* says the ground connection on the master switch may have an open circuit. *Technician B* says the window off switch connected from the power circuit to the window switch may have an open circuit. Who is right?
    A. A only
    B. B only
    C. Both A and B
    D. Neither A nor B

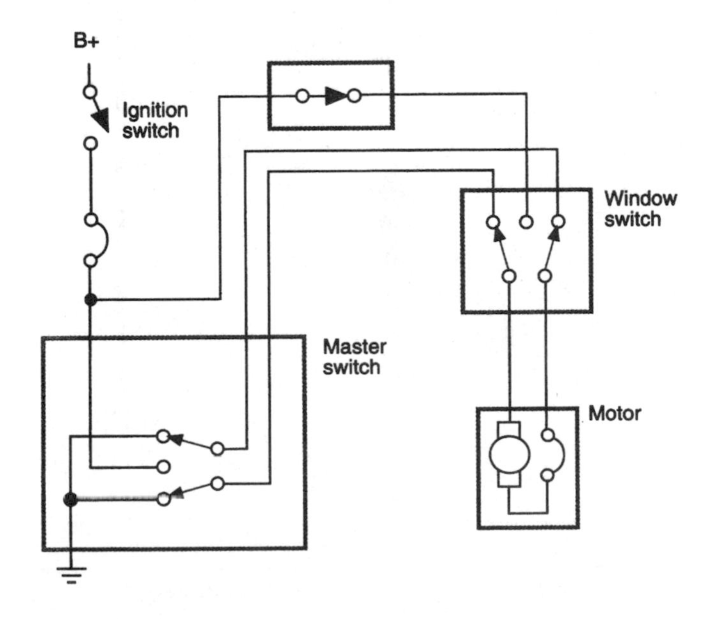

Figure 6–33 Power window circuit.

Hint
Slow operation of the side windows may be caused by binding or sticking of the window in the track. The excess strain on the motor raises the amperage required to operate the motor. This may cause the circuit fuse to blow or circuit breaker to open. Never replace the circuit protection device with a larger amperage device.

Task 11

Inspect, test, and repair or replace motors, switches, relays, connectors, terminals, and wires of power side window circuits.

117. In the power window circuit in Figure 6–34, there is an open circuit in wire 164 between the master switch and the LH power window motor. The result of this defect is:
    A. the LH window will go down, but it will not go back up.
    B. the RH window will go down, but it will not go back up.
    C. the LH window will stall part way down.
    D. the LH window is inoperative.

118. In the power window circuit in Figure 6–34 with the ignition switch on, the circuit breaker keeps opening the circuit even when the power windows are not operated. The cause of this problem could be:
    A. a short to ground at junction S501.
    B. a short to ground on wire 667 between the RH window switch and the RH power window motor.
    C. a short to ground at junction S500.
    D. a short to ground terminal A on the master switch.

Hint
When an electrical failure is suspected to be system wide, such as both windows being inoperative, then the power to the circuit must be established first before any individual component is tested. Begin testing for voltage as close to the fuse box as possible, and then progressively downstream until the open circuit is found. It must be assumed that the fuse box is getting power, since all the other systems are normal.

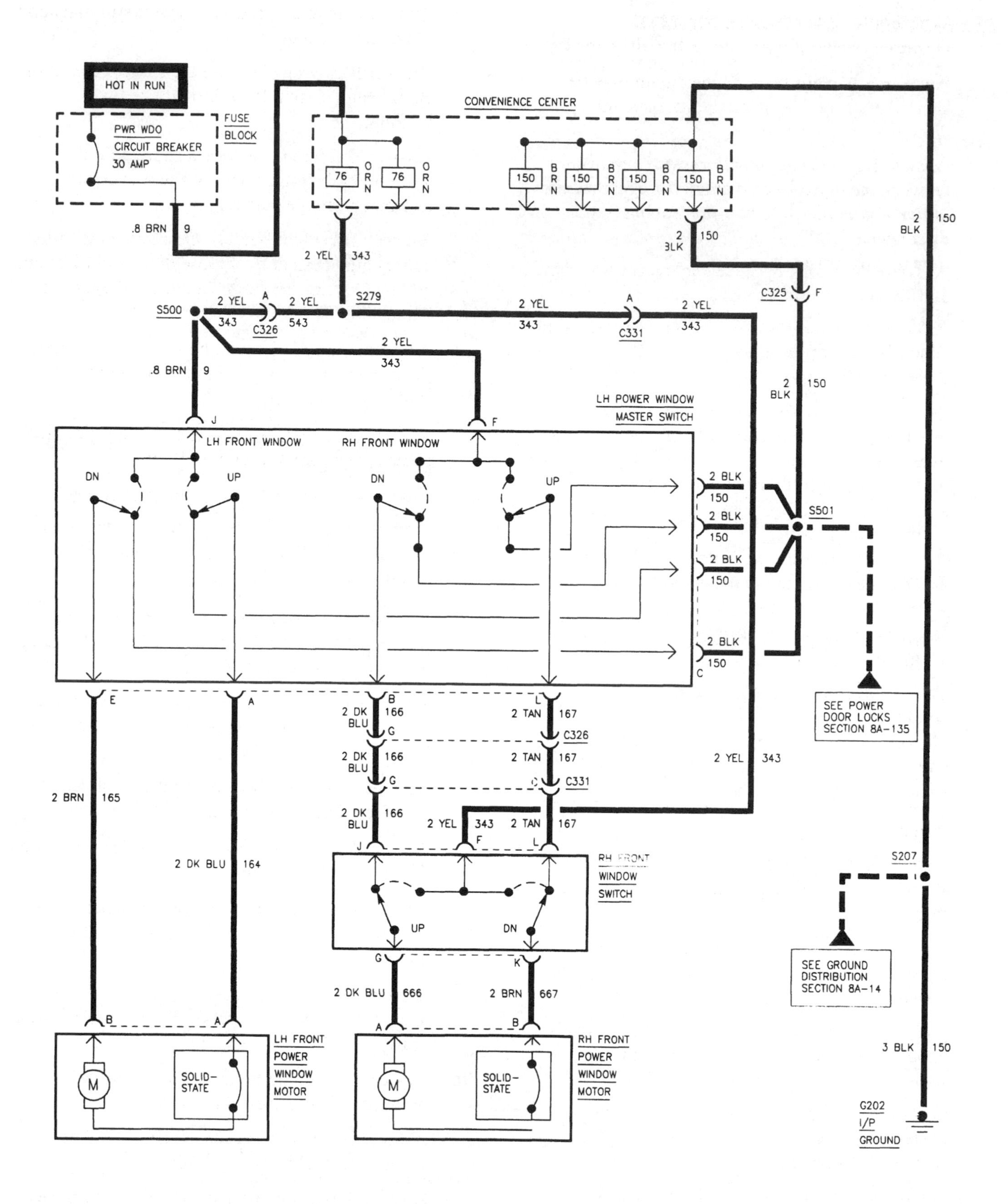

Figure 6–34 Power window circuit with convenience center. *(Courtesy of General Motors Corporation, Service Technology Group)*

Answers to Questions

1. **Answer A is right** because the circuit may be shorted to ground between the fuse and the disconnected connector.

   Answer B is wrong because an open circuit between the disconnected connector and the bulb does not affect the circuit with the connector disconnected.

   Answer C is wrong because Technician A is right and Technician B is wrong.

   Answer D is wrong because Technician A is right and Technician B is wrong.

2. Answer A is wrong because using a self-powered test light to check for continuity in a circuit involving an electronic module may damage the module.

   Answer B is wrong because the battery does not have to be disconnected before testing an electronic module circuit.

   Answer C is wrong because both Technician A and Technician B are wrong.

   Answer D is right because both Technician A and Technician B are wrong.

3. Answer A is wrong because the voltmeter leads must be connected in parallel to the circuit being tested.

   Answer B is wrong because the voltmeter leads must be connected in parallel to the circuit being tested.

   Answer C is right because the voltmeter leads must be connected in parallel to the circuit being tested.

   Answer D is wrong because the voltmeter leads must be connected in parallel to the circuit being tested.

4. Answer A is wrong because the voltmeter leads should be connected from the starting motor housing to the negative battery terminal.

   Answer B is right because the fuel injection system should be disabled and the engine cranked to read the voltage drop.

   Answer C is wrong because Technician A is wrong and Technician B is right.

   Answer D is wrong because Technician A is wrong and Technician B is right.

5. Answer A is wrong because there is still magnetic strength around the rotor.

   Answer B is wrong because the shorted condition causes high current flow through the field winding.

   Answer C is wrong because a shorted field winding in the rotor does not cause excessive alternator voltage output.

   Answer D is right because the shorted condition causes high current flow through the field winding and the voltage regulator, resulting in regulator damage.

6. Answer A is wrong because to cause low current flow, the voltage supplied to the circuit may be lower than specified.

   Answer B is right because higher than specified resistance causes low current flow.

   Answer C is wrong because Technician A is wrong and Technician B is right.

   Answer D is wrong because Technician A is wrong and Technician B is right.

7. Answer A is wrong because if an ohmmeter is connected to a circuit in which current is flowing, the meter may be damaged.

   Answer B is wrong because when testing a spark plug wire with 20,000 ohms resistance, the X1000 meter scale must be used.

   Answer C is wrong because both Technician A and Technician B are wrong.

   Answer D is right because both Technician A and Technician B are wrong.

8. Answer A is wrong because an ohmmeter is not used to check for a short circuit between two circuits.

   Answer B is right because the battery must be fully charged before checking a circuit for current draw.

   Answer C is wrong because Technician A is wrong and Technician B is right.

   Answer D is wrong because Technician A is wrong and Technician B is right.

9. Answer A is wrong because a blown fuse causes zero current flow.

   Answer B is wrong because low battery voltage reduces current flow.

Answer C is right because a shorted bulb filament increases current flow.

Answer D is wrong because high resistance in the bulb filament reduces current flow.

10. **Answer A is right** because an intermittent ground in a taillight wire may cause repeated blowing of the fuse.

    Answer B is wrong because high resistance in a taillight wire decreases current flow and this would not blow the fuse.

    Answer C is wrong because an intermittent open circuit in a taillight wire causes the taillights to go on and off.

    Answer D is wrong because a shorted filament in one of the taillights would increase the brilliance of this bulb because of increased current flow.

11. **Answer A is right** because the tester switch should be closed while starting or running the engine.

    Answer B is wrong because a battery drain of 400 milliamperes is not normal and will discharge the battery.

    Answer C is wrong because the actual drain is not recorded until the meter reading stabilizes after the switch is opened.

    Answer D is wrong because the driver's door must not be open while measuring battery drain.

12. **Answer A is right** because a battery drain of 2 amperes would cause the battery to become discharged.

    Answer B is wrong because the battery drain test may be performed at any engine temperature.

    Answer C is wrong because Technician A is right and Technician B is wrong.

    Answer D is wrong because Technician A is right and Technician B is wrong.

13. Answer A is wrong because the Technician should not replace the fuse with the next higher amperage fuse.

    Answer B is right because the Technician should check the affected circuit for a short to ground.

    Answer C is wrong because an open circuit does not cause a blown fuse.

    Answer D is wrong because the Technician should not install a circuit breaker of the same amperage as the fuse.

14. Answer A is wrong because the current from the ohmmeter will not open the circuit breaker.

    Answer B is wrong because the ohmmeter should provide a very low reading if the circuit breaker is satisfactory.

    Answer C is wrong because both Technician A and Technician B are wrong.

    Answer D is right because both Technician A and Technician B are wrong.

15. Answer A is wrong because the diode protects the electronic components.

    Answer B is wrong because voltage spikes do not damage the A/C compressor clutch.

    Answer C is right because the purpose of the diode is to protect electronic components from voltage spikes.

    Answer D is wrong because the purpose of the diode is not to limit the current flow to the A/C compressor clutch.

16. Answer A is wrong because a voltage suppression diode may be defective in one of the electrical components, but Technician B is also right.

    Answer B is wrong because one of the output solenoids or relays controlled by the PCM may have a shorted winding, but Technician A is also right.

    Answer C is right because both Technician A and Technician B are right.

    Answer D is wrong because both Technician A and Technician B are right.

17. **Answer A is right** because the horn relay may be defective.

    Answer B is wrong because if the circuit is grounded between the horn relay and the horn button, the horns would blow continually.

    Answer C is wrong because Technician A is right and Technician B is wrong.

    Answer D is wrong because Technician A is right and Technician B is wrong.

18. Answer A is wrong because continually open horn relay contacts would cause the horns to be completely inoperative.

    Answer B is wrong because an open horn relay winding would cause the horns to be completely inoperative.

Answer C is wrong because an open in fuse link B would cause the horns to be completely inoperative.

Answer D is right because an open between the left-hand horn and junction C136 would cause only the left-hand horn to be inoperative.

19. Answer A is wrong because the battery does not need to be brought to room temperature before testing.

    Answer B is wrong because the battery temperature is 120°F (49°C), 0.016 specific gravity point should be added to the hydrometer reading.

    Answer C is wrong because the maximum variation in cell hydrometer readings is 0.050 specific gravity point.

    Answer D is right because the battery is fully charged if all the cell hydrometer readings are above 1.265.

20. **Answer A is right** because a faulty voltage regulator may cause overcharging and low battery electrolyte.

    Answer B is wrong because high resistance in the circuit between the alternator battery terminal and the positive battery terminal causes undercharging of the battery.

    Answer C is wrong because Technician A is right and Technician B is wrong.

    Answer D is wrong because Technician A is right and Technician B is wrong.

21. Answer A is wrong because the battery discharge rate is two times the battery cold cranking rating.

    Answer B is wrong because the battery is satisfactory if the voltage remains above 9.6V.

    Answer C is wrong because both Technician A and Technician B are wrong.

    Answer D is right because both Technician A and Technician B are wrong.

22. Answer A is wrong because the test shown in the figure is a battery capacity test.

    Answer B is right because the test shown in the figure is a battery capacity test.

    Answer C is wrong because the test shown in the figure is a battery capacity test.

    Answer D is wrong because the test shown in the figure is a battery capacity test.

23. Answer A is wrong because if the battery is fully charged, the voltage should be at or above 12.6V.

    Answer B is wrong because if the battery is fully charged, the voltage should be at or above 12.6V.

    Answer C is wrong because if the battery is fully charged, the voltage should be at or above 12.6V.

    Answer D is right because if the battery is fully charged, the voltage should be at or above 12.6V.

24. **Answer A is right** because an open circuit voltage test may be used in place of a hydrometer test.

    Answer B is wrong because during the open circuit voltage test, the headlights should be off.

    Answer C is wrong because Technician A is right and Technician B is wrong.

    Answer D is wrong because Technician A is right and Technician B is wrong.

25. Answer A is wrong because the adaptive memory in the computers is erased.

    Answer B is wrong because the radio station resets are erased.

    Answer C is right because disconnecting the battery cables does not damage the onboard computers.

    Answer D is wrong because erratic engine operation may occur when the engine is restarted.

26. Answer A is wrong because it is not necessary to replace the battery cables when the battery is replaced.

    Answer B is right because the charging circuit should be checked to eliminate the possibility of undercharging or overcharging the battery.

    Answer C is wrong because it is not necessary to replace the starting motor when replacing the battery.

    Answer D is wrong because it is not necessary to replace the alternator belt when replacing the battery.

27. **Answer A is right** because before MIG welding on a truck with an alternator and onboard computers, disconnect the battery ground.

    Answer B is wrong because one battery holddown is not sufficient to hold an 1100 series battery.

    Answer C is wrong because Technician A is right and Technician B is wrong.

Answer D is wrong because Technician A is right and Technician B is wrong.

28. Answer A is wrong because using an air nozzle to blow off the parts blows corrosion laden with sulfuric acid around the area and causes irritation if it contacts skin or eyes.

    Answer B is wrong because a baking soda and water solution should be used.

    Answer C is right because safety glasses must be worn.

    Answer D is wrong because a baking soda and water solution should be used to dissolve the residues.

29. Answer A is wrong because the battery charge time is not the same for batteries with different capacities.

    Answer B is right because the battery temperature should not exceed 125°F (52°C).

    Answer C is wrong because a high charging rate may not be used to charge a battery at –20°F (-29°C).

    Answer D is wrong because when a fast charger is used to charge two batteries in parallel, the current flow is not likely the same through both batteries because batteries usually have different internal resistances.

30. **Answer A is right** because never disconnecting negative battery cable first is not a precaution because the negative cable should be disconnected first.

    Answer B is wrong because the battery should be charged until the specific gravity is 1.265.

    Answer C is wrong because the fast-charging rate should be reduced after the specific gravity is above 1.225.

    Answer D is wrong because a frozen battery should not be charged.

31. Answer A is wrong because the accessories should be turned off in both vehicles.

    Answer B is right because the negative booster cable should be connected to an engine ground on the vehicle being boosted.

    Answer C is wrong because Technician A is wrong and Technician B is right.

    Answer D is wrong because Technician A is wrong and Technician B is right.

32. Answer A is wrong because the bumpers on the two vehicles should not be allowed to touch each other.

    Answer B is wrong because the positive battery terminal in the boost vehicle should be connected to the positive battery terminal in the vehicle being boosted.

    Answer C is wrong because a boost procedure should not be performed on a frozen battery.

    Answer D is right because the negative battery cable between the two batteries should be connected last.

33. Answer A is wrong because the battery condition should be checked first.

    Answer B is wrong because the battery condition should be checked first.

    Answer C is wrong because the battery condition should be checked first.

    Answer D is right because the battery condition should be checked first.

34. **Answer A is right** because a direct ground in the starter field coils causes high current draw and no starter operation.

    Answer B is wrong because an open circuit in the armature windings causes no starter current draw.

    Answer C is wrong because broken brush springs and no tension on the brushes causes low starter current draw.

    Answer D is wrong because failed magnetic switch causes no starter current draw.

35. **Answer A is right** because a voltmeter should be connected between the battery ground terminal and the starter housing and the engine cranked to test the starter ground circuit.

    Answer B is wrong because an ohmmeter is not used to test the starter ground circuit.

    Answer C is wrong because an ohmmeter is not used to test the starter ground circuit.

    Answer D is wrong because a voltmeter should be connected between the battery ground terminal and the starter housing and the engine cranked to test the starter ground circuit.

36. Answer A is wrong because the voltmeter leads are connected from the battery positive terminal to the positive battery cable connected to the starter solenoid to test the voltage drop on the insulated side of the starter circuit.

Answer B is wrong because the voltmeter leads are connected from the battery positive terminal to the positive battery cable connected to the starter solenoid to test the voltage drop on the insulated side of the starter circuit.

Answer C is right because the voltmeter leads are connected from the battery positive terminal to the positive battery cable connected to the starter solenoid to test the voltage drop on the insulated side of the starter circuit.

Answer D is wrong because the voltmeter leads are connected from the battery positive terminal to the positive battery cable connected to the starter solenoid to test the voltage drop on the insulated side of the starter circuit.

37. **Answer A is right** because the arrow in the figure points to the start switch.

    Answer B is wrong because the arrow in the figure points to the start switch.

    Answer C is wrong because the arrow in the figure points to the start switch.

    Answer D is wrong because the arrow in the figure points to the start switch.

38. Answer A is wrong because connecting terminals C and D is a method of checking the starter relay.

    Answer B is wrong because connecting terminals C and D is a method of checking the starter relay.

    Answer C is wrong because connecting terminals C and D is a method of checking the starter relay.

    Answer D is right because connecting terminals C and D is a method of checking the starter relay.

39. Answer A is wrong because the shims between the starting motor flange and the flywheel housing provide proper starting motor alignment.

    Answer B is wrong because the shims between the starting motor flange and the flywheel housing provide proper starting motor alignment.

    Answer C is wrong because both Technician A and Technician B are wrong.

    Answer D is right because both Technician A and Technician B are wrong.

40. Answer A is wrong because the starting motor may be misaligned with the flywheel housing, but Technician B is also right.

    Answer B is wrong because the solenoid return spring may be weak, but Technician A is also right.

    Answer C is right because both Technician A and Technician B are right.

    Answer D is wrong because both Technician A and Technician B are right.

41. Answer A is wrong because the safety starting switch is connected in series with the solenoid windings.

    Answer B is wrong because a defective safety starting switch does not cause higher than specified starter current draw.

    Answer C is wrong because both Technician A and Technician B are wrong.

    Answer D is right because both Technician A and Technician B are wrong.

42. Answer A is wrong because the least likely cause of a discharged battery is a defective starter solenoid.

    Answer B is wrong because the least likely cause of a discharged battery is a defective starter solenoid.

    Answer C is right because the least likely cause of a discharged battery is a defective starter solenoid.

    Answer B is wrong because the least likely cause of a discharged battery is a defective starter solenoid.

43. Answer A is wrong because the batteries are connected in parallel for charging.

    Answer B is wrong because while cranking, the negative terminal of one battery is connected to the positive terminal of the other battery.

    Answer C is right because in the charge mode, the series parallel switch winding is not energized.

    Answer D is wrong because the series parallel switch is operated by electromagnetism.

44. Answer A is wrong because the two batteries may receive different current flows, because the batteries may have different internal resistance.

    Answer B is wrong because the ammeter only indicates the current flow through one battery.

    Answer C is wrong because both Technician A and Technician B are wrong.

Answer D is right because both Technician A and Technician B are wrong.

45. Answer A is wrong because the battery cables should be serviced at regularly scheduled intervals.

    Answer B is right because excessive corrosion around the battery area may be caused by a high charging voltage.

    Answer C is wrong because Technician A is wrong and Technician B is right.

    Answer D is wrong because Technician A is wrong and Technician B is right.

46. Answer A is wrong because the terminals should be removed from the battery with a puller.

    Answer B is right because the positive cable should not be removed before the negative cable.

    Answer C is wrong because the positive cable should be installed before the negative cable.

    Answer D is wrong because corrosion on the battery top increases the rate of self-discharge.

47. **Answer A is right** because some of the alternator diodes may be defective, but Technician B is also right.

    Answer B is wrong because if the wire connected to the alternator L terminal is grounded the charge indicator light is not on with the ignition switch off.

    Answer C is wrong because Technician A is right and Technician B is wrong.

    Answer D is wrong because Technician A is right and Technician B is wrong.

48. **Answer A is right** because the field circuit in the alternator may have an open circuit.

    Answer B is wrong because one open stator winding would not cause the charge indicator light to be off when the ignition switch is on.

    Answer C is wrong because Technician A is right and Technician B is wrong.

    Answer D is wrong because Technician A is right and Technician B is wrong.

49. Answer A is wrong because a defective voltage regulator causing a high voltage regulator setting results in an overcharged battery.

    Answer B is right because excessive resistance between the alternator battery terminal and the positive battery terminal may cause an undercharged battery.

    Answer C is wrong because worn alternator mounting holes causing a misaligned alternator belt may cause excessive belt wear, but this problem does not cause an undercharged battery.

    Answer D is wrong because a 140-ampere fuse link installed in place of a 100-ampere fuse link in the alternator battery wire does not cause an undercharged battery.

50. Answer A is wrong because a low-voltage regulator setting causes an undercharged battery.

    Answer B is wrong because an alternator belt bottomed in the pulley may cause an undercharged battery.

    Answer C is wrong because a battery drain with the ignition switch off may cause an undercharged battery.

    Answer D is right because a slightly higher-than-normal resistance in the charge indicator light does not cause an undercharged battery as long as there is enough current through the bulb to start the alternator charging.

51. Answer A is wrong because a loose alternator belt may cause a squealing noise, but Technician B is also right.

    Answer B is wrong because a shorted alternator diode may cause a whining noise, but Technician A is also right.

    Answer C is right because both Technician A and Technician B are right.

    Answer D is wrong because both Technician A and Technician B are right.

52. **Answer A is right** because an alternator drive belt that is bottomed in the pulley causes belt slipping and low output.

    Answer B is wrong because an open circuit in the rotor field coil causes zero output.

    Answer C is wrong because a misaligned alternator pulley does not affect output.

    Answer D is wrong because a broken alternator brush lead causes zero output.

53. **Answer A is right** because a shorted alternator diode would reduce the maximum output.

    Answer B is wrong because an open circuit in the voltage regulator does not affect alternator output, because the regulator is bypassed during the output test.

Answer C is wrong because an open circuit at one of the rotor slip rings causes zero alternator output.

Answer D is wrong because an open circuit in the alternator capacitor does not affect alternator output.

54. Answer A is wrong because when the alternator is full-fielded, the voltage should be kept below 15V.

    Answer B is wrong because when testing output by lowering the system voltage, the voltage should be lowered to 13V.

    Answer C is wrong because both Technician A and Technician B are wrong.

    Answer D is right because both Technician A and Technician B are wrong.

55. Answer A is wrong because the voltmeter leads should be connected to the alternator output terminal and the positive battery terminal.

    Answer B is wrong because the voltmeter leads should be connected to the alternator output terminal and the positive battery terminal.

    Answer C is right because the voltmeter leads should be connected to the alternator output terminal and the positive battery terminal.

    Answer D is wrong because the voltmeter leads should be connected to the alternator output terminal and the positive battery terminal.

56. Answer A is wrong because the voltmeter leads should be connected to the alternator housing and the battery ground terminal.

    Answer B is right because the voltmeter leads should be connected to the alternator housing and the battery ground terminal.

    Answer C is wrong because Technician A is wrong and Technician B is right.

    Answer D is wrong because Technician A is wrong and Technician B is right.

57. Answer A is wrong because grounding the alternator battery terminal with the battery cables connected may cause burns to the Technician's hands.

    Answer B is wrong because grounding the alternator battery terminal with the battery cables connected may cause a burned fuse link in the alternator output wire.

    Answer C is right because grounding the alternator battery terminal with the battery cables connected does not cause a burned fuse link in the voltage regulator circuit.

    Answer D is wrong because grounding the alternator battery terminal with the battery cables connected may cause a burned alternator output terminal.

58. **Answer A is right** because removing the alternator battery wire with the engine running causes extremely high voltage and possible damage to onboard electronic components.

    Answer B is wrong because removing the alternator field wire with the engine running causes zero alternator output.

    Answer C is wrong because Technician A is right and Technician B is wrong.

    Answer D is wrong because Technician A is right and Technician B is wrong.

59. Answer A is wrong because a burned fusible link must be replaced with a length of the specified fusible link wire.

    Answer B is wrong because a circuit breaker resets itself and does not have to be replaced after repairing a short in a circuit.

    Answer C is wrong because both Technician A and Technician B are wrong.

    Answer D is right because both Technician A and Technician B are wrong.

60. Answer A is wrong because wires for fusible links should be smaller than the circuit wire.

    Answer B is wrong because wires for fusible links should be smaller than the circuit wire.

    Answer C is wrong because wires for fusible links should be smaller than the circuit wire.

    Answer D is right because wires for fusible links should be smaller than the circuit wire.

61. Answer A is wrong because the winding in the series-parallel switch may be open and therefore not providing 24V for starting, but Technician B is also right.

    Answer B is wrong because the upper set of contacts in the series-parallel switch may be severely burned, and therefore not providing 24V for starting, but Technician A is also right.

Answer C is right because both Technician A and Technician B are right.

Answer D is wrong because both Technician A and Technician B are right.

62. **Answer A is right** because batteries with unequal internal resistance may cause repeated complaints of one battery being discharged.

    Answer B is wrong because a low-voltage regulator setting would result in both batteries being discharged.

    Answer C is wrong because a high-voltage regulator setting overcharges both batteries.

    Answer D is wrong because the lower contacts in the series-parallel switch are open in the charging mode.

63. **Answer A is right** because the figure shows full-fielding the alternator.

    Answer B is wrong because the figure shows full-fielding the alternator.

    Answer C is wrong because the figure shows full-fielding the alternator.

    Answer D is wrong because the figure shows full-fielding the alternator.

64. Answer A is wrong because the charging voltage is high.

    Answer B is wrong because the charging rate in amperes is high.

    Answer C is wrong because the voltage regulator is not limiting the alternator voltage.

    Answer D is right because the voltage and current output is determined by engine speed.

65. Answer A is wrong because high resistance in the left-side headlight ground would cause both the high and low beams to be dim in the left-side headlight.

    Answer B is wrong because high resistance in the dimmer switch high beam contacts would cause both the high beam headlights to be dim.

    Answer C is wrong because both Technician A and Technician B are wrong.

    Answer D is right because both Technician A and Technician B are wrong.

66. Answer A is wrong because the LEAST likely cause of one dim headlight is low alternator voltage, because this problem causes all the lights to be dim.

    Answer B is wrong because the LEAST likely cause of one dim headlight is low alternator voltage, because this problem causes all the lights to be dim.

    Answer C is right because the LEAST likely cause of one dim headlight is low alternator voltage, because this problem causes all the lights to be dim.

    Answer D is wrong because the LEAST likely cause of one dim headlight is low alternator voltage, because this problem causes all the lights to be dim.

67. **Answer A is right** because headlight aim should be checked on a level floor with the vehicle unloaded.

    Answer B is wrong because the vehicle should be positioned 25 ft. (7.62 m) from the aligning screen.

    Answer C is wrong because Technician A is right and Technician B is wrong.

    Answer D is wrong because Technician A is right and Technician B is wrong.

68. Answer A is wrong because the vehicle does need to be parallel to the aligning screen.

    Answer B is wrong because the tire inflation pressure does affect headlight alignment.

    Answer C is wrong because the headlight alignment is not changed by installing new sealed beams.

    Answer D is right because the headlight alignment is changed by turning adjusting screws on the headlight assembly.

69. Answer A is wrong because some dimmer switches are floor mounted, but Technician B is also right.

    Answer B is wrong because high resistance at dimmer switch contacts may reduce current flow only in the high beam circuit, but Technician A is also right.

    Answer C is right because both Technician A and Technician B are right.

    Answer D is wrong because both Technician A and Technician B are right.

70. Answer A is wrong because the parking lights will be affected.

    Answer B is wrong because the stoplights will fail to operate because voltage supply to these lights is independent of the headlight switch.

    Answer C is wrong because both Technician A and Technician B are wrong.

    Answer D is right because both Technician A and Technician B are wrong.

71. Answer A is wrong because the bulb is single contact with standard indexing.

    Answer B is right because the bulb is single contact with standard indexing.

    Answer C is wrong because the bulb is single contact with standard indexing.

    Answer D is wrong because the bulb is single contact with standard indexing.

72. Answer A is wrong because the wire in the trailer connector supplies voltage to both sides of the trailer.

    Answer B is wrong because a voltmeter is used to check the continuity between one taillight on the inoperative side and the trailer connector.

    Answer C is wrong because both Technician A and Technician B are wrong.

    Answer D is right because both Technician A and Technician B are wrong.

73. Answer A is wrong because a loose bulb in one of the dash lights would only affect that bulb.

    Answer B is right because a loose instrument panel ground connection may cause intermittent dash light and instrument operation.

    Answer C is wrong because an intermittent open in the dash light variable resistor would only affect the dash lights.

    Answer D is wrong because an intermittent ground on one dash light socket would blow the dash light fuse.

74. Answer A is wrong because the variable resistor must not be grounded to the instrument panel.

    Answer B is wrong because the dash light bulbs are connected parallel to each other.

    Answer C is right because the variable resistor is connected in series with the dash light bulbs.

    Answer D is wrong because a high resistance in one dash bulb circuit decreases current flow through that bulb.

75. Answer A is wrong because the current flow through the bulb does not change.

    Answer B is right because the light cannot be turned off since the short to ground continually grounds the bulb.

    Answer C is wrong because Technician A is wrong and Technician B is right.

    Answer D is wrong because Technician A is wrong and Technician B is right.

76. **Answer A is right** because a grounded circuit before one of the cab light bulbs would cause repeated blowing of the cab light fuse.

    Answer B is wrong because a grounded circuit between one of the cab lights and ground would turn on that cab light continually.

    Answer C is wrong because shorted contacts in one of the door jamb switches would turn on the cab lights continually.

    Answer D is wrong because an open circuit in one of the ground wires connected to a door jamb switch would make the cab lights inoperative when that door is opened.

77. Answer A is wrong because there is no air pressure supplied to switch with the engine running.

    Answer B is wrong because the switch has normally open contacts.

    Answer C is right because air pressure is supplied to the switch when the brakes are applied.

    Answer D is wrong because air pressure is supplied to the switch from the service air system.

78. Answer A is wrong because an open circuit in the stoplight switch causes inoperative stoplights.

    Answer B is right because the wires on the stoplight switch may be shorted together.

    Answer C is wrong because Technician A is wrong and Technician B is right.

    Answer D is wrong because Technician A is wrong and Technician B is right.

79. Answer A is wrong because the DB 180G RD wire supplies voltage to the left rear signal light.

Answer B is wrong because a short to ground in the DB 180 RD wire from the turn signal switch to the rear lamp wiring would blow the signal light fuse.

Answer C is right because the D7 18BR RD wire supplies voltage to the right rear signal light, and high resistance in this wire reduces current flow.

Answer D is wrong because high resistance in the D2 18G RD wire from the turn signal switch to the rear lamp wiring would reduce current to all the signal lights.

80. Answer A is wrong because the left front bulb may be defective, but Technician B is also right.

    Answer B is wrong because there may be an open circuit between the signal light switch and the left front signal light bulb, but Technician A is also right.

    Answer C is right because both Technician A and Technician B are right.

    Answer D is wrong because both Technician A and Technician B are right.

81. **Answer A is right** because the turn signal switch may be contained in the multifunction switch.

    Answer B is wrong because the variable resistor in the dash light circuit is not contained in the multifunction switch.

    Answer C is wrong because Technician A is right and Technician B is wrong.

    Answer D is wrong because Technician A is right and Technician B is wrong.

82. Answer A is wrong because high resistance in the signal light ground circuit on the trailer reduces current flow and will not damage the flasher.

    Answer B is wrong because excessive voltage drop in the wires connected to the trailer signal lights reduces current flow and will not damage the flasher.

    Answer C is right because an intermittent ground in some of the signal light wires causes high current flow and may damage the flasher.

    Answer D is wrong because corrosion on the rear tractor signal light sockets reduces current flow and will not damage the flasher.

83. Answer A is wrong because a voltmeter or 12V test light should be used to check the circuit from the fuse to the back-up light switch.

    Answer B is wrong because a voltmeter or 12V test light should be used to check the circuit from the back-up light switch through the bulbs to ground.

    Answer C is wrong because both Technician A and Technician B are wrong.

    Answer D is right because both Technician A and Technician B are wrong.

84. **Answer A is right** because this condition may blow the back-up light fuse.

    Answer B is wrong because this condition blows the back-up light fuse and both back-up lights are then inoperative.

    Answer C is wrong because Technician A is right and Technician B is wrong.

    Answer D is wrong because Technician A is right and Technician B is wrong.

85. Answer A is wrong because the water temperature gauge would read continually high.

    Answer B is wrong because the water temperature gauge would read continually high.

    Answer C is right because the sending unit is connected to the full coil in the gauge. A grounded wire between the gauge and the sending unit increase current flow through the full coil and attracts the pointer to the full position.

    Answer D is wrong because the high resistance of the gauge does not allow enough current flow to blow the fuse.

86. Answer A is wrong because the water temperature gauge reading would be inaccurate.

    Answer B is right because the tachometer is not connected through ground wire number 57, and the tachometer reading remains accurate.

    Answer C is wrong because the fuel gauge reading would be inaccurate.

    Answer D is wrong because the voltmeter reading would be inaccurate.

87. Answer A is wrong because the instrument voltage limiter may be defective, but Technician B is also right.

    Answer B is wrong because the instrument voltage limiter ground may have high resistance, but Technician A is also right.

Answer C is right because both Technician A and Technician B are right.

Answer D is wrong because both Technician A and Technician B are right.

88. Answer A is wrong because the voltage output from an instrument voltage limiter should be 5V.

    Answer B is wrong because the voltage output from an instrument voltage limiter should be 5V.

    Answer C is right because the voltage output from an instrument voltage limiter should be 5V.

    Answer D is wrong because the voltage output from an instrument voltage limiter should be 5V.

89. Answer A is wrong because an ohmmeter is not used to test a mechanical gauge.

    Answer B is right because the sensing bulb should be placed in boiling water and the gauge reading checked against a known accurate thermometer placed in the water.

    Answer C is wrong because Technician A is wrong and Technician B is right.

    Answer D is wrong because Technician A is wrong and Technician B is right.

90. Answer A is wrong because a thermistor-type engine temperature sensor is tested with an ohmmeter connected across the sensor terminals at various sensor temperatures.

    Answer B is right because a thermistor-type engine temperature sensor is tested with an ohmmeter connected across the sensor terminals at various sensor temperatures.

    Answer C is wrong because a thermistor-type engine temperature sensor is tested with an ohmmeter connected across the sensor terminals at various sensor temperatures.

    Answer D is wrong because a thermistor-type engine temperature sensor is tested with an ohmmeter connected across the sensor terminals at various sensor temperatures.

91. **Answer A is right** because a grounded wire connected to one of the low air pressure switches will illuminate the low air pressure light with normal air pressure.

    Answer B is wrong because an open circuit in the wire to one of the low air pressure switches does not illuminate the low air pressure warning light.

    Answer C is wrong because the air pressure on the dash gauge is normal.

    Answer D is wrong because an open circuit at the switch contacts in one of the low air pressure switches would not illuminate the low air pressure light.

92. Answer A is wrong because the coolant temperature sensor contains a heating element and an off-on switch.

    Answer B is wrong because the oil pressure sensor contains an off-on switch.

    Answer C is right because the alarm module is connected to the engine run relay and shutdown solenoid.

    Answer D is wrong because the oil pressure sensor contains an off-on switch.

93. Answer A is wrong because if either low air pressure switch closes, the brake warning indicator lamp is illuminated, but Technician B is also right.

    Answer B is wrong because if either low air pressure switch closes, the buzzer is activated, but Technician A is also right.

    Answer C is right because both Technician A and Technician B are right.

    Answer D is wrong because both Technician A and Technician B are right.

94. Answer A is wrong because an open buzzer winding only causes an inoperative buzzer.

    Answer B is wrong because an open buzzer winding only causes an inoperative buzzer.

    Answer C is wrong because an open buzzer winding only causes an inoperative buzzer.

    Answer D is right because an open buzzer winding only causes an inoperative buzzer.

95. Answer A is wrong because on some trucks with a diesel engine, the tachometer signal is taken from the alternator R terminal.

    Answer B is wrong because on some trucks with a diesel engine, the tachometer signal is taken from the alternator R terminal.

    Answer C is right because on some trucks with a diesel engine, the tachometer signal is taken from the alternator R terminal.

    Answer D is wrong because on some trucks with a diesel engine, the tachometer signal is taken from the alternator R terminal.

96. **Answer A is right** because the voltmeter is connected parallel to the battery.

    Answer B is wrong because only one coil in the gauge is connected in series with the tachometer sensor.

    Answer C is wrong because the tachometer sensor does not contain a thermistor.

    Answer D is wrong because an open number 14, 10-ampere fuse also causes an inoperative voltmeter.

97. Answer A is wrong because if the diaphragm contacts in the horns were welded together, the horns would be inoperative.

    Answer B is right because the horn relay may be defective.

    Answer C is wrong because Technician A is wrong and Technician B is right.

    Answer D is wrong because Technician A is wrong and Technician B is right.

98. **Answer A is right** because the air line may be kinked where it is connected to the right-side horn.

    Answer B is wrong because the air line where it is connected to the air manifold supplies air to both horns.

    Answer C is wrong because Technician A is right and Technician B is wrong.

    Answer D is wrong because Technician A is right and Technician B is wrong.

99. **Answer A is right** because the horn relay does not require a ground connection.

    Answer B is wrong because an open circuit in the horn relay winding causes inoperative horns.

    Answer C is wrong because an open circuit at the horn brush/slip ring causes inoperative horns.

    Answer D is wrong because an open fuse link in the relay power wire causes inoperative horns.

100. Answer A is wrong because a grounded wire between one of the horns and the horn relay causes high current flow and a blown horn fuse.

    Answer B is wrong because an open circuit at the horn relay contacts causes inoperative horns.

    Answer C is right because if the wire from the horn relay to the horn button is grounding in the steering column, the horn blows when turning the steering wheel.

    Answer D is wrong because if the power wire to the horn relay winding is shorted to ground, high current flow and a blown fuse result.

101. Answer A is wrong because no voltage to the wiper motor causes inoperative wipers.

    Answer B is wrong because a defective wiper switch causes inoperative wipers.

    Answer C is wrong because a faulty wiper motor causes inoperative wipers.

    Answer D is right because the circuit breaker not operating eliminates circuit protection, but the wipers still operate normally.

102. **Answer A is right** because an open circuit at wiper motor terminal E causes no low-speed wiper motor operation.

    Answer B is wrong because an open circuit in wire 228 only causes an inoperative washer pump.

    Answer C is wrong because an open circuit at wiper motor terminal B causes no wiper motor operation.

    Answer D is wrong because an open circuit at wiper motor terminal C causes improper parking of the wipers.

103. **Answer A is right** because the wiper park switch may be defective.

    Answer B is wrong because the wiper switch in the instrument panel is not responsible for stopping the wipers in the park position.

    Answer C is wrong because Technician A is right and Technician B is wrong.

    Answer D is wrong because Technician A is right and Technician B is wrong.

104. **Answer A is right** because the motor may have a plugged exhaust port.

    Answer B is wrong because an air-operated motor does not have a field winding.

    Answer C is wrong because Technician A is right and Technician B is wrong.

    Answer D is wrong because Technician A is right and Technician B is wrong.

105. Answer A is wrong because high resistance in the power supply wire to the wiper motor does not cause the wiper motor to intermittently stop and restart.

Answer B is right because a binding wiper blade pivot may cause excessive current draw and this opens the circuit breaker.

Answer C is wrong because high resistance in the wires from the wiper motor to the switch does not cause the wiper motor to stop intermittently and restart.

Answer D is wrong because loose, corroded terminals on the wiper motor do not cause the wiper motor to have the same consistent stop and restart time.

106. Answer A is wrong because dry bushings in the wiper motor are not the most likely cause of the squeaking noise.

Answer B is right because a dry wiper blade pivot is the most likely cause of the squeaking noise since these pivots are subjected to moisture contamination.

Answer C is wrong because loose wiper arm pivots are not the most likely cause of the squeaking noise.

Answer D is wrong because loose wiper motor mounting bolts are not the most likely cause of the squeaking noise.

107. Answer A is wrong because an open circuit breaker in the wiper/washer switch would cause inoperative wiper motor and washer pump motor.

Answer B is wrong because an open circuit at terminal 5 on the wiper/washer switch causes inoperative high speed in the wiper motor.

Answer C is right because an open circuit at terminal 4 on the wiper/washer switch causes an inoperative washer pump motor and normal wiper operation.

Answer D is wrong because an open circuit in wire number 58 connected to the wiper/washer switch causes inoperative low-speed wiper motor operation.

108. Answer A is wrong because if wires 58 and 56 are shorted together, the wiper motor operates at only one speed.

Answer B is right because the problems may be caused by wires 941 and 58 being shorted together.

Answer C is wrong because if wire 58 is shorted to ground, the circuit breaker opens the circuit when the wiper is switched to low speed.

Answer D is wrong because if wire 941 is shorted to ground, the circuit breaker opens the circuit when the washer switch is depressed.

109. Answer A is wrong because an open circuit in fuse link E causes inoperative mirror heaters.

Answer B is wrong because an open circuit in circuit breaker number 6 causes inoperative mirror heaters.

Answer C is wrong because an open circuit in wire number 960 causes inoperative mirror heaters.

Answer D is right because an open circuit in the heated mirror indicator light does not cause inoperative mirror heaters.

110. **Answer A is right** because wires 960 and 59 shorted together causes continual operation of the mirror heaters.

Answer B is wrong because the circuit breaker contacts are normally closed.

Answer C is wrong because if one of the mirror heated elements is shorted, the current flow increases and the circuit breaker may open.

Answer D is wrong because if wires 57 and 196 are shorted together, fuse link E will burn out.

111. Answer A is wrong because an open circuit in the clutch cycle switch causes an inoperative A/C compressor clutch.

Answer B is right because an open circuit in wire 349 only causes an inoperative warning buzzer.

Answer C is wrong because an open circuit in the A/C high-pressure switch causes an inoperative A/C compressor clutch.

Answer D is wrong because an open circuit in wire 348 causes an inoperative A/C compressor clutch.

112. Answer A is wrong because an open circuit in the ground wire from the blower switch to ground causes improper operation of all blower speeds.

Answer B is wrong because an open circuit in the thermal limiter in the blower assembly causes only high-speed operation.

Answer C is right because an open circuit in the high-speed contacts in the blower switch causes no high-speed operation with normal operation at the other speeds.

Answer D is wrong because a short to ground in the wire connected from the resistor assembly to ground has no effect because this wire is grounded to the chassis.

113. Answer A is wrong because an open circuit at terminal B on the lighter only causes an inoperative lighter.

Answer B is right because an open circuit at ground connection G202 causes an inoperative lighter and dome light.

Answer C is wrong because Technician A is wrong and Technician B is right.

Answer C is wrong because Technician A is wrong and Technician B is right.

114. **Answer A is right** because a short to ground on wire 150 connected from the courtesy light switch to junction S207 does not turn on the dome light unless the courtesy light switch is on.

Answer B is wrong because one of the door jamb switches continually closed turns on the dome light continually.

Answer C is wrong because a short to ground at junction S255 turns on the dome light continually.

Answer D is wrong because the courtesy light switch continually closed turns on the dome light continually.

115. **Answer A is right** because an open circuit between the ignition switch and the window switch causes the power window to operate from the master switch but not from the window switch.

Answer B is wrong because an open circuit in the window switch contacts would also prevent the window operation from the master switch.

Answer C is wrong because an open circuit in the master switch ground wire also prevents the power window operation from the master switch.

Answer D is wrong because a short to ground at the circuit breaker in the motor opens the other circuit breaker in the system.

116. **Answer A is right** because if the ground connection on the master switch has an open circuit, the power window is completely inoperative.

Answer B is wrong because if there is an open circuit in the window off switch, the power window still operates from the master switch.

Answer C is wrong because Technician A is right and Technician B is wrong.

Answer D is wrong because Technician A is right and Technician B is wrong.

117. Answer A is wrong because the LH window will be inoperative.

Answer B is wrong because the LH window will be inoperative.

Answer C is wrong because the LH window will be inoperative.

Answer D is right because the LH window will be inoperative.

118. Answer A is wrong because a short to ground at junction S501 has no effect since this junction is connected to ground.

Answer B is wrong because a short to ground on wire 667 between the RH window switch and the RH power window motor does not open the circuit breaker until the RH power window is operated.

Answer C is right because a short to ground at junction S500 causes the circuit breaker to open and close continually with the ignition switch on.

Answer D is wrong because a short to ground at terminal A on the master switch does not open the circuit breaker until the LH power window is operated.

7 Heating, Ventilation, and Air Conditioning (HVAC) Systems

Pretest

The purpose of this pretest is to determine the amount of review you may require before the ASE Medium/Heavy Truck Heating, Ventilation, and Air-Conditioning (HVAC) Test. If you answer all the pretest questions correctly, complete the questions and study the information in this chapter to prepare for this test.

If two or more of your answers to the pretest questions are incorrect, complete a study of all the chapters in *Today's Technician Medium/Heavy-Duty Truck Heating, Ventilation, and Air-Conditioning Systems* Classroom and Shop Manuals published by Delmar Publishers, plus a study of the questions and information in this chapter.

The pretest answers are located at the end of the pretest; these answers also are in the answer sheets supplied with this book.

1. A signal to the powertrain control module from which of the following sensors could cause the A/C compressor to disengage?
   A. Engine coolant temperature (ECT) sensor
   B. Intake air temperature (IAT) sensor
   C. Heated oxygen sensor (HO2S)
   D. Cooling fan control sensor

2. An electric cooling fan motor is inoperative. *Technician A* says this motor may be controlled by an independent electronic module. *Technician B* says this motor may be controlled by an air solenoid controller. Who is right?
   A. A only
   B. B only
   C. Both A and B
   D. Neither A nor B

3. The blend door actuator motor is generally mounted:
   A. using epoxy.
   B. using butylene sealer.
   C. behind the glove box.
   D. to the evaporator case.

4. The proper method of testing a vacuum actuator is to:
   A. apply shop air to the vacuum port and listen for leaks.
   B. use a hand-held vacuum pump and observe the gauge.
   C. move the control lever and verify that the actuator plunger moves.
   D. apply vacuum using an A/C evacuation pump and verify that the actuator plunger moves.

5. While discussing a vacuum check in an HVAC system, *Technician A* says the purpose of the check valve is to delay vacuum to the actuators until the coolant has reached operating temperature. *Technician B* says the purpose of the check valve is to prevent loss of vacuum to components during periods of low engine vacuum. Who is right?
   A. A only
   B. B only
   C. Both A and B
   D. Neither A nor B

6. A whistling noise coming from under the dash while the HVAC system is being operated with the blower on HIGH could indicate:
   A. a misaligned duct.
   B. a defective vacuum actuator.
   C. an improperly adjusted mode door cable.
   D. a poor electrical connection to the blend door motor.

7. A heater does not supply the cab with enough heat. The coolant level and blower are OK. *Technician A* says an improperly adjusted temperature control cable could be the cause. *Technician B* says a clogged heater core could be the cause. Who is right?
   A. A only
   B. B only
   C. Both A and B
   D. Neither A nor B

8. *Technician A* says bubbles and/or foam in the sight glass indicate the refrigerant charge is low and air has entered the system. *Technician B* says a cloudy sight glass indicates moisture in the system. Who is right?
   A. A only
   B. B only
   C. Both A and B
   D. Neither A nor B

9. There is a growling or rumbling noise at the A/C compressor. When the clutch engages, the noise stops. *Technician A* says the compressor bearing is defective. *Technician B* says the clutch bearing is defective. Who is right?
   A. A only
   B. B only
   C. Both A and B
   D. Neither A nor B

10. Before a portable container is used to transfer recycled R-12, it must be evacuated to at least:
    A. 20 in. Hg.
    B. 22 in. Hg.
    C. 27 in. Hg.
    D. 12 in. Hg.

11. Refer to Figure 7–1. When the blower switch is in the medium-2 position, how many resistors are used in the circuit to control blower motor speed?
    A. Three
    B. Two
    C. One
    D. None

12. Refer to Figure 7–1. All of the following could prevent the A/C clutch from engaging EXCEPT:
    A. a faulty 30 amp circuit breaker.
    B. ambient temperature below 40°.
    C. the ambient temperature cut-out switch stuck closed.
    D. open circuit in the black-yellow wire.

13. Particulate must be removed from an A/C system after a compressor failure. *Technician A* says nitrogen flushing is the best way to remove this particulate. *Technician B* says the best way to do this job is to use R-11 flushing. Who is right?
    A. A only
    B. B only
    C. Both A and B
    D. Neither A nor B

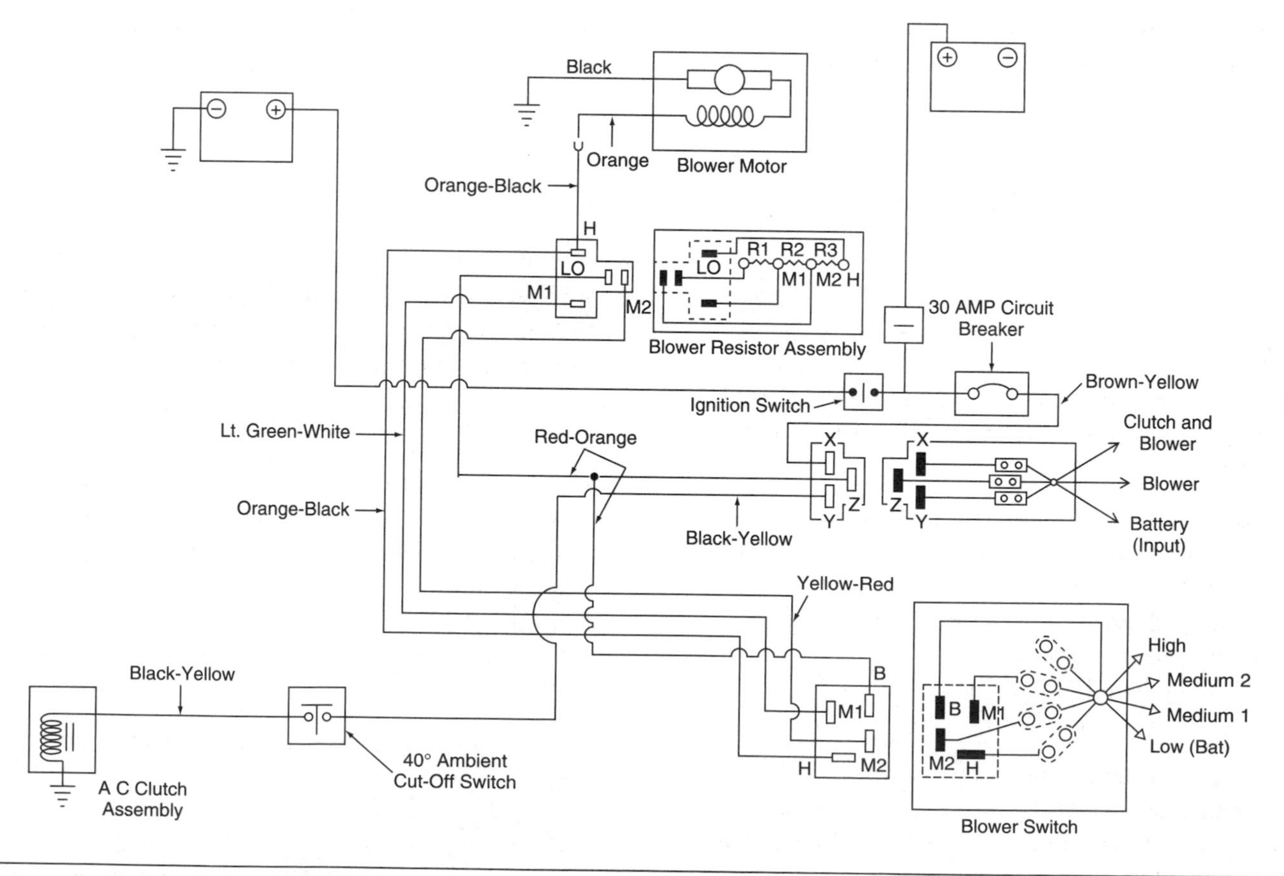

Figure 7–1 Blower motor circuit.

14. While discussing coolant boiling point, *Technician A* says when the cooling system pressure is increased, the boiling point is decreased. *Technician B* says if more antifreeze is added to the coolant mix, the boiling point is increased. Who is right?
A. A only
B. B only
C. Both A and B
D. Neither A nor B

Answers to Pretest

1. A, 2. A, 3. D, 4. B, 5. B, 6. A, 7. C, 8. C, 9. B, 10. C, 11. C, 12. C, 13. A, 14. B

HVAC Systems Diagnosis, Service, and Repair

In this chapter each task in the Heating, Ventilation, and Air-Conditioning (HVAC) Systems category is provided followed by some questions and information related to this task. If you answer any question incorrectly, study this information very carefully until you understand the correct answer. For additional information on any task, refer to *Today's Technician Medium/Heavy-Duty Truck Heating, Ventilation, and Air-Conditioning Systems* Classroom and Shop Manuals, published by Delmar Publishers.

Questions, answers, and analysis are provided at the end of this chapter and in the answer sheets provided with this book.

Task 1

Verify the need for service or repair of HVAC systems based on unusual operating noises; determine appropriate action.

1. A whistling noise coming from under the passenger side dash with the blower motor on high speed might indicate:
A. a clogged evaporator drain.
B. a cracked evaporator case.
C. a broken blend door cable.
D. a low refrigerant charge.

2. In an ATC system, a grinding noise comes from under the dash when the temperature setting is changed from warm to cold. The most likely cause of this problem is:
A. grit in the evaporator case.
B. a faulty mode door actuator.
C. bad drive gears in the blend door motor.
D. arcing in the control head.

Hint

The Service Technician must be aware of normal HVAC system operating noises in order to determine whether a system requires service. Normal noises include the sounds of A/C compressor clutch engagement, the blower motor, moving blend air and mode doors, and pressure equalization after the vehicle is shut down. Noises that could indicate the need for service include a growling sound from the water pump or A/C compressor, a whistling noise under the dash, and a grinding noise when control levers are moved.

A loose, dry, or worn A/C compressor belt will cause a squealing noise. This noise will be worse during acceleration. Worn or dry blower motor bearings may cause a squealing noise when the blower is running; this noise will occur when one first starts the engine after the truck is shut down overnight. A loose or worn clutch hub or loose compressor mounting bolt will also cause a rattling noise from the compressor.

If liquid refrigerant enters the compressor, a thumping banging noise will result. Heavy knocking compressor noises come from the following: refrigerant system blockage, incorrect pressures, or internal damage. A worn compressor pulley bearing or air clutch bearing will cause a growling noise with the compressor engaged or disengaged. If the growling noise only occurs when the system engages the clutch, internal bearings may be at fault.

Task 2

Verify the need of service or repair of HVAC systems based on unusual visual, smell, and touch conditions; determine appropriate action.

3. During an HVAC performance test, the Technician notices that the A/C compressor outlet is nearly as hot as the upper radiator hose. *Technician A* says that this is a normal condition. *Technician B* says that the system is overcharged. Who is right?
A. A only
B. B only
C. Both A and B
D. Neither A nor B

4. A faint ether-like odor coming from the panel vents in the normal A/C mode could indicate:
A. the evaporator core is leaking R-134a.
B. the evaporator core is leaking R-12.
C. the cold starting system is malfunctioning.
D. the heater core is leaking.

Hint

The Service Technician must be aware of abnormal conditions in the HVAC system in order to determine the need for system service. If the driver complains of high or low temperatures inside the cab, this is cause for a system performance test. Abnormal conditions include the smell of antifreeze inside the cab, a fogged windshield, ice buildup on A/C components, and oil or dirt buildup on A/C fittings. An antifreeze smell inside the cab may be caused by a leaking heater core. Windshield fogging is usually caused by a leaking heater core. Ice buildup on A/C components such as hoses, receiver/dryer, or TXV valve may be caused by internal restriction in these components. Evaporator freeze-up may be caused by improper control of refrigerant through the evaporator. Oil and dirt buildup on refrigerant system components usually indicates slight leaks in the system. Excessive heating of high side refrigerant system components may be caused by a refrigerant overcharge or moisture in the system. A mildew smell coming from the A/C outlets usually indicates a plugged evaporator case drain resulting in an accumulation of stagnant water in the bottom of this case. Since R-12 is odorless; a leak in the evaporator core in an R-12 system does not cause any odor. An evaporator core leak in an R-134a system causes a slight ether-like smell from the A/C outlets.

Task 3

Identify system type and conduct performance test(s) on HVAC systems; determine appropriate action.

5. Mobile A/C systems using R-12 have flared and threaded service ports, with the port size differentiating high side from low side. Service fittings on systems with R-134a use:
   A. compression fittings.
   B. SAE approved quick-connect couplings.
   C. the same fittings at R-12 systems.
   D. 10 mm threaded ports.

6. The easiest way to check the calibration of an ATC system is to:
   A. use an ohmmeter to measure the resistance of the ATC sensor.
   B. measure the voltage drop across the ATC sensor.
   C. turn the calibration screw on the ATC control unit.
   D. use a thermometer to measure the cab temperature and compare the actual temperature to the set temperature.

7. While conducting a performance test on a semiautomatic HVAC system, a Technician finds that only a small amount of air is directed to the windshield in defrost mode. The most likely cause of this problem is:
   A. a defective microprocessor.
   B. a defective blend door actuator.
   C. an open blower motor resistor.
   D. an improperly adjusted mode door cable.

Hint

First, the Technician must know if the system operates on R-12 or R-134a refrigerant. Most major R-134a components use a light blue label to indicate an R-134a design. R-12 systems use Schrader type service valves, where as R-134a systems use metric threads and quick-connect service valves.

Next, the Technician needs to identify the type of expansion device in the evaporator inlet line. Most AC systems use a thermal expansion valve (TXV) (Figure 7–2), or a fixed orifice tube (FOT) (Figure 7–3) in the evaporator inlet line to control refrigerant flow into the evaporator. There is high-pressure liquid on the inlet side of the expansion device and low-pressure liquid on the outlet side. Other systems use a suction throttling device between the evaporator and the suction side of the compressor to control refrigerant flow into the evaporator. These suction throttling devices may include a suction throttling valve (STV), pilot-operated absolute valve (POA), evaporator pressure regulator (EPR) valve, or an Evaporator Temperature Regulator (ETR) valve. Many AC systems use combination valves that usually contain two of the pressure control valves. For example, older GM class 8 tractors used an evaporator equalized valves-in-receiver (EE-VIR) assembly that contains TXV and POA valves plus a receiver/dryer (Figure 7–4 and Figure 7–5). Some Ford trucks have a combination valve containing a TXV and STV. Mack Trucks use a block-type assembly that contains an equalized TXV.

The truck industry uses three types of AC systems: mechanical, semiautomatic, or automatic. Mechanical systems use a slide-type lever or rotary switch to control the in-cab temperature manually. In a semiautomatic system, a computer electronically controls some systems, and in the automatic systems, most subsystems are computer controlled.

Manual systems rely on the driver to select the temperature, mode, and blower speed. Semiautomatic temperature control (SATC) systems regulate only the temperature of the output air and rely on the driver to select the desired mode and blower speed. A microprocessor (computer) controls a fully automatic temper-

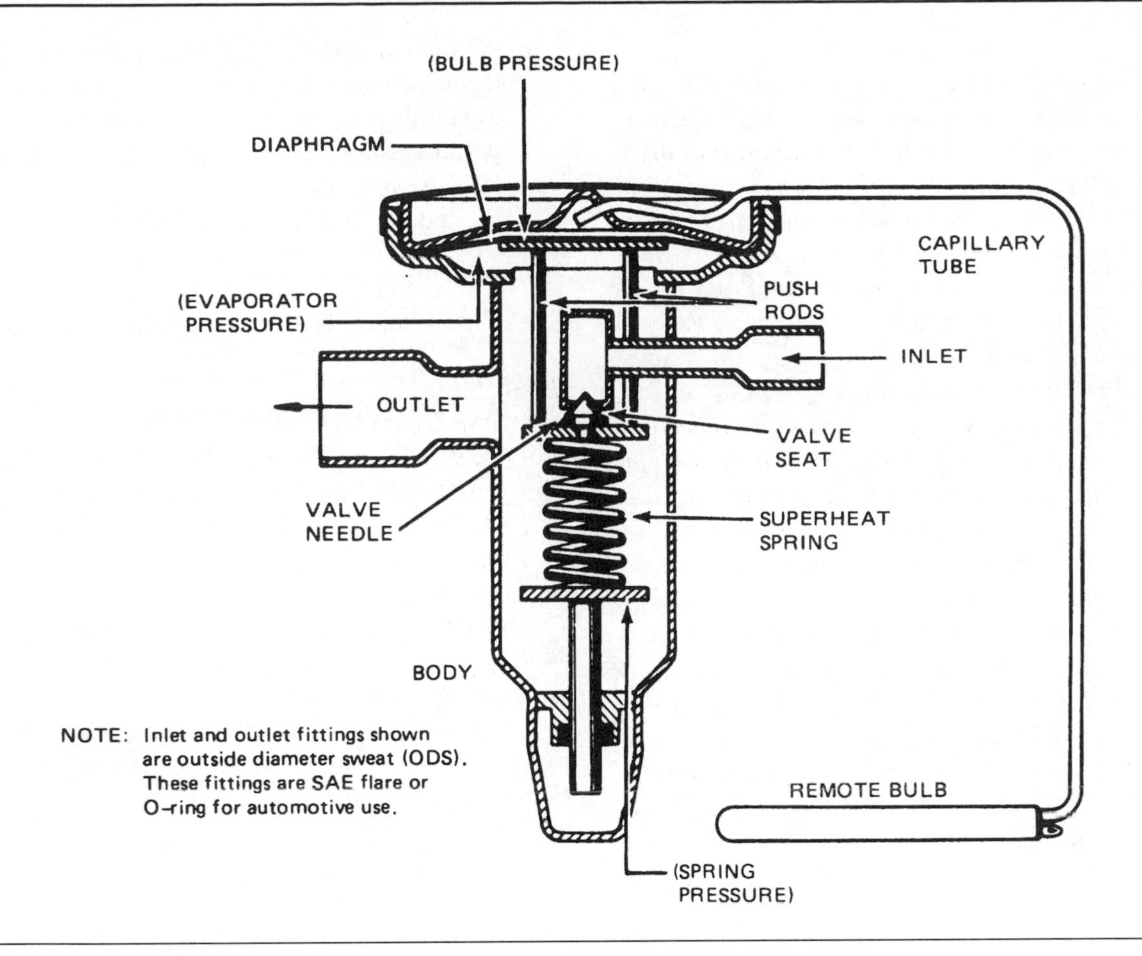

Figure 7–2 Thermal expansion valve (TXV).

ature control (ATC) system. These systems use the input from various sensors throughout the vehicle to control the blend doors automatically to adjust the interior temperature using an appropriate blower speed.

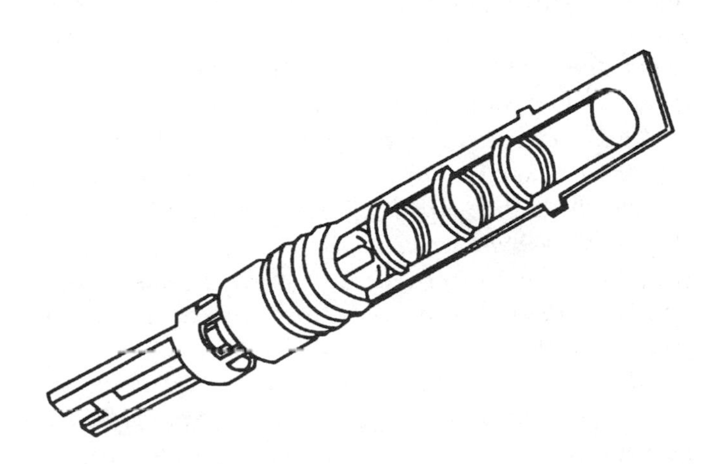

Figure 7–3 Fixed orifice tube (FOT). *(Courtesy of Ford Motor Company)*

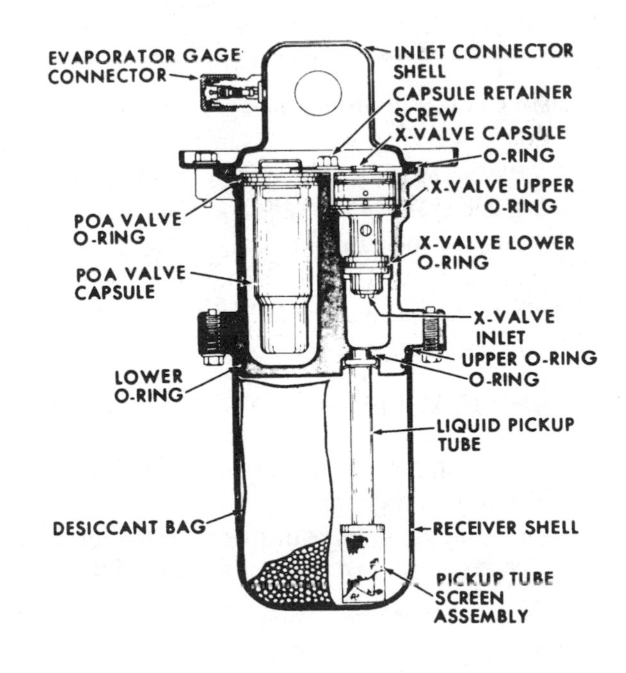

Figure 7–4 Valves in receiver assembly containing the TXV valve, pilot-operated absolute (POA) valve, and receiver/dryer. *(Courtesy of Cadillac Motor Car Division, General Motors Corporation)*

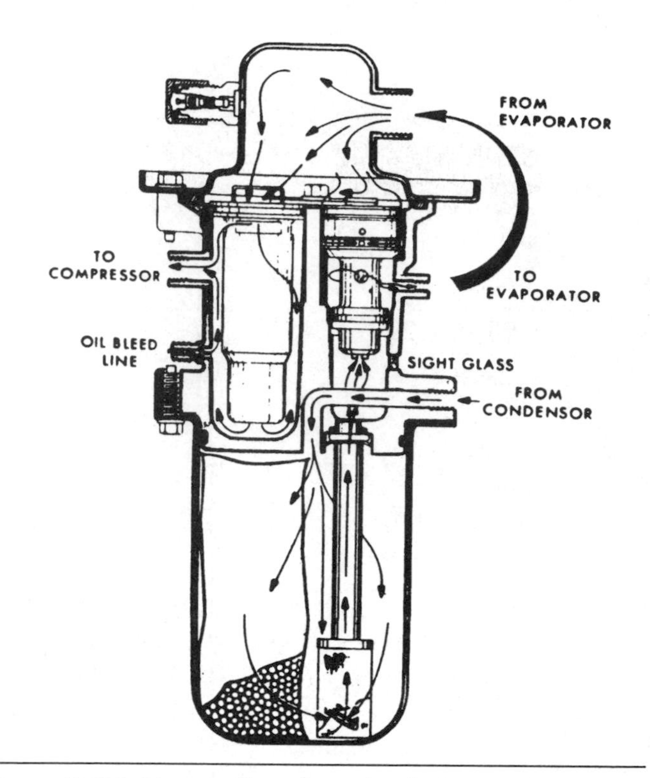

Figure 7–5 Refrigerant flow through valves in receiver (VIR) assembly. *(Courtesy of Cadillac Motor Car Division, General Motors Corporation)*

A HVAC performance test should include operation of the system in all modes, and at a variety of temperatures and blower speeds. A small pocket thermometer should be used to verify that the output air temperatures match the temperature settings.

A/C System and Component Diagnosis, Service, and Repair

A/C System General

Task 1

Diagnose the cause of temperature control problems in the A/C system; determine needed repairs.

8. Poor cooling from the A/C system that uses a STV valve can be caused by all of the following EXCEPT:
   A. the fan clutch that is always engaged.
   B. an improperly adjusted blend door cable.
   C. a low refrigerant charge.
   D. a refrigerant overcharge.

9. Inadequate airflow from one or more vents could be caused by:
   A. a misaligned air duct.
   B. a faulty blower resistor.
   C. a clogged heater core.
   D. high ambient humidity.

10. With the selector lever in the MAX A/C position, a blend air HVAC system outputs cold air for about 15 minutes, at which time the output air becomes warm. The most likely cause of this problem is:
   A. a defective thermal expansion valve.
   B. a defective coolant control valve.
   C. a defective blend air door return spring.
   D. a defective fresh air door.

Hint

The temperature of A/C system output air is generally controlled by one of two ways. In blend air systems, one mixes air cooled by the evaporator core with air warmed by the heater core. Ultimately, the output air temperature control occurs by regulating the amount of air allowed to flow through each core. In other A/C systems, you achieve temperature control by cycling the compressor clutch. Compressor clutch cycling rate is usually controlled by the engine control module or the body control module based on input from various temperature sensors.

A/C system output air temperature is affected by outside air temperature and humidity, engine coolant temperature, airflow through the condenser and evaporator, and level of refrigerant charge. Output air temperature may also be affected by mechanical or electrical failure of system components.

Task 2

Identify Refrigerant Type

11. *Technician A* says that the refrigerant containers for R-12 and R-134a are color coded. *Technician B* says that the R-134a containers use 1/2 inch, 16 ACME threads that cannot be mistakenly "hooked up" to an R-12 gauge set or recovery machine. Who is right?
   A. A only
   B. B only
   C. Both A and B
   D. Neither A nor B

12. To avoid mixing refrigerants, SAE J2197 standard specifies that R-134a service hose fittings for connection to manifold gauge sets or to recovery/recycling/charging equipment are to be:
   A. 3/16 inch, 16ACME thread.
   B. 3/16 inch, 14ACME thread.

C. 1/2 inch, 16ACME thread.
D. 1/4 inch, 16ACME thread.

Hint

The easiest way to identify the type of refrigerant that is used in a given A/C system is to observe the service fittings. SAE standard J639 defines the size and type of service fittings for R-12 and R-134a A/C systems. R-12 service fittings have external threads. R-134a systems use quick-disconnect fittings. You cannot vent either R-12 or R-134a to the atmosphere.

Task 3

Diagnose A/C system problems indicated by pressure gauge readings and sight glass/moisture indicator conditions (where applicable); determine needed service or repairs.

13. *Technician A* says bubbles and/or foam in the sight glass indicate the refrigerant charge is low and air has entered the system. *Technician B* says a cloudy sight glass indicates moisture in the system. Who is right?
A. A only
B. B only
C. Both A and B
D. Neither A nor B

14. *Technician A* says a system having high low-side pressure accompanied by a continuously running compressor indicates that the expansion valve is stuck open. *Technician B* says a system having a high low-side reading accompanied by a continuously running compressor indicates the evaporator is restricted. Who is right?
A. A only
B. B only
C. Both A and B
D. Neither A nor B

15. An A/C system with excessive high-side pressure could be the result of all of the following EXCEPT:
A. an overcharge of refrigerant.
B. an overheated engine.
C. restricted airflow through the condenser.
D. ice building up on the orifice tube screen.

16. *Technician A* says that when servicing a system with the type of service valve shown in Figure 7–6, the compressor cannot be isolated. *Technician B* says that the valve shown in Figure 7–6 permits reading of the suction and discharge pressures with the service valve back seated. Who is right?

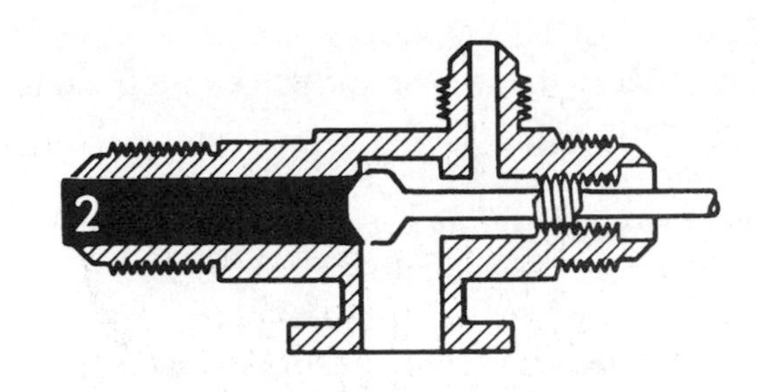

Figure 7–6 Service valve.

A. A only
B. B only
C. Both A and B
D. Neither A nor B

17. The sight glass in an R-12 system is clear, but the A/C system does not provide adequate passenger compartment cooling. *Technician A* says there may be no refrigerant charge in the system. *Technician B* says the desiccant in the receiver/dryer may be defective. Who is right?
A. A only
B. B only
C. Both A and B
D. Neither A nor B

Hint

In a normally operating A/C system, the low side pressure varies between 20 and 45 pounds per square inch, and the high-side pressure varies between 120 and 210 pounds per square inch. The following table summarizes abnormal A/C system pressures and common causes.

Low Side Pressure	*High Side Pressure*	*Possible Causes*
LOW	*LOW*	*Low refrigerant charge*
LOW	*LOW*	*Obstruction in the suction line*
LOW	*LOW*	*Clogged orifice tube*
LOW	*LOW*	*TXV valve stuck closed**
LOW	*LOW*	*Restricted line from the condenser to the evaporator**
LOW	*HIGH*	*Restricted evaporator airflow*
HIGH	*LOW*	*Internal compressor damage*
HIGH	*HIGH*	*Refrigerant overcharge*
HIGH	*HIGH*	*Restricted condenser airflow*
HIGH	*HIGH*	*High engine coolant temperature*
HIGH	*HIGH*	*TXV valve stuck open*
HIGH	*HIGH*	*Air or moisture in the refrigerant*

*Stuck-closed TXV valves or a restricted line from the condenser to the evaporator will cause frosting at the point of restriction.

In some A/C systems, a sight glass allows the Service Technician to make a quick assessment of the system condition. With the A/C compressor clutch engaged, a properly charged system will occasionally show traces of bubbles. A sight glass that appears foamy indicates that the refrigerant charge is low. When the sight glass contains bubbles and/or foam, the refrigerant charge is low and air has entered the system. Oil streaks appearing in the sight glass indicate that compressor oil is circulating through the system. A cloudy sight glass indicates that the desiccant bag in the receiver/dryer has broken down.

On R-12 systems, sight glass indications are only valid when the ambient (surrounding area) temperature is above 71°F (21°C). If the temperature is below 70°F, it is normal for bubbles to appear in the sight glass. A clear sight glass may indicate the proper refrigerant charge. A clear sight glass may also indicate an excessive refrigerant charge, or no refrigerant charge. Many R-134a systems do not have a sight glass. Most manufacturers agree that the following conditions must be present when diagnosing an R-134a system:

- *High side below 240 psi (1,570 kPa)*
- *Ambient temperature below 95°F (35°C)*
- *Temperature control set in the lowest position*
- *Humidity below 70 percent*
- *High blower speed*
- *Engine speed set at 1,500 rpm*
- *Recirculation air set*

R-134a systems have a normal charge if the sight glass shows a stream of very small bubbles that disappear when the engine speed is increased. On an R-134a system, the system is overcharged when the sight glass shows no bubbles. A constant flow of bubbles mixed with foam indicates a low charge. If the refrigerant charge is very low, fog may appear in the sight glass. R-12 mineral oil in an R-134a system causes a severely fogged sight glass.

Task 4

Diagnose A/C system problems indicated by visual, smell, and touch problems; determine needed repairs.

18. An R-12 A/C system has bubbles in the sight glass at ambient temperatures below 70°F (21°C) (Figure 7–7). At temperatures above this value, the sight glass is clear. *Technician A* says the refrigerant system is slightly low on refrigerant.

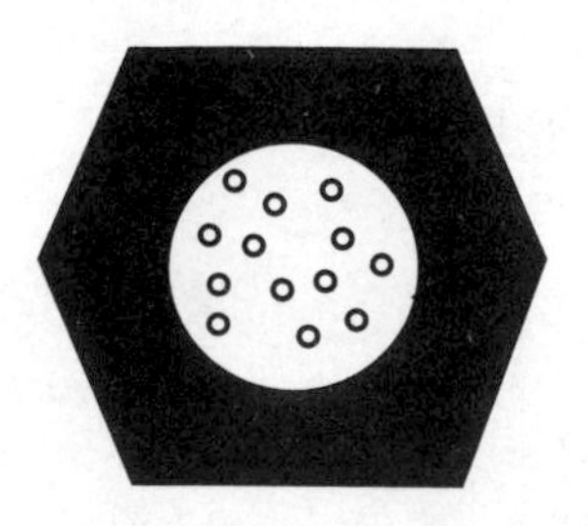

Figure 7–7 Sight glass. *(Courtesy of Chrysler Corporation)*

Technician B says there is excessive oil in the refrigerant system. Who is right?
   A. A only
   B. B only
   C. Both A and B
   D. Neither A nor B

19. On a refrigerant system with a TXV valve, the evaporator outlet is considerably warmer than the evaporator inlet. *Technician A* says the refrigerant charge may be low. *Technician B* says the compressor clutch may not be engaged. Who is right?
   A. A only
   B. B only
   C. Both A and B
   D. Neither A nor B

20. Frosting occurs on the component in Figure 7–8. *Technician A* says there may be an overcharge of refrigerant in the system. *Technician B* says the receiver dryer may be restricted. Who is right?
   A. A only
   B. B only
   C. Both A and B
   D. Neither A nor B

21. The driver complains about a rotten egg smell in the truck cab that is most noticeable when the A/C system is first turned on. The cause of this problem could be:
   A. a leaking evaporator core.
   B. a leaking condenser core.
   C. a leaking heater core.
   D. a plugged evaporator case drain.

Hint

To diagnose A/C systems efficiently, the Service Technician must use all of his senses. Restricted hoses cause a frosting or sudden temperature change at a specific point along the hose. Frost on the receiver/dryer usually indicates an internal restriction in that component. Because the receiver/dryer is located in the high-pressure liquid

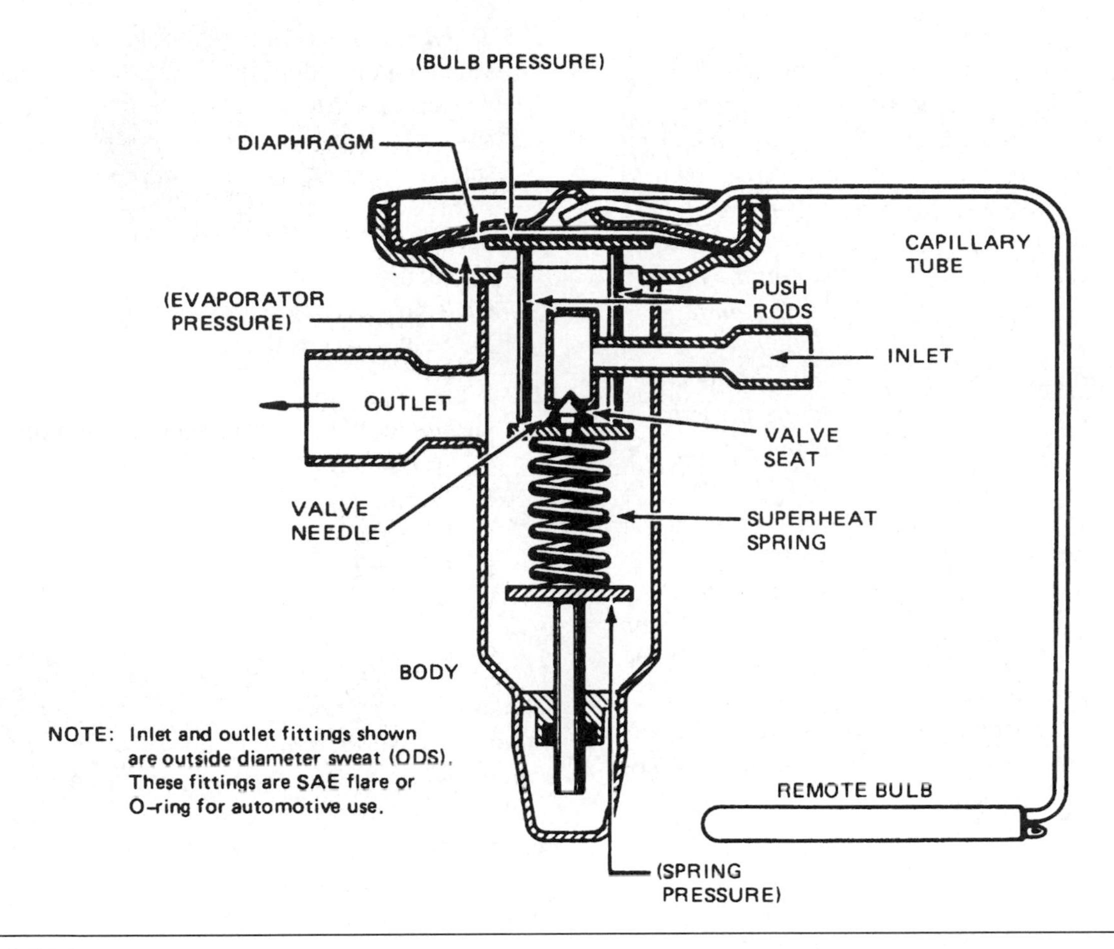

Figure 7–8 TXV valve.

line between the condenser and evaporator, it should feel warm. TXV valve frosting indicates a restricted valve or one sticking closed. Frost formation on the evaporator outlet indicates a flooded evaporator caused by an excessive refrigerant charge, or a stuck open TXV valve. Unusual noises can often guide the Technician to a faulty component. These problems may also cause frost on the compressor suction line. On a system that uses a POA valve, frosting of the suction line is normal.

It should feel cold on systems using an accumulator because of its close connection to the evaporator. Both the evaporator inlet and outlet should feel cold when operating normally. On orifice tube systems, the evaporator inlet should be slightly warmer to the touch than the outlet. If the evaporator outlet is warm, the refrigerant charge may be low. High-side refrigerant components should feel hot or warm, and low-side components should feel cool or cold. A plugged drain in the evaporator case causes a strong rotten egg smell in the cab.

Task 5

Perform A/C system leak test; determine needed repairs.

22. If R-12 comes into contact with a flame:
    A. it will explode.
    B. it will form a nontoxic gas.
    C. it will form chlorine crystals.
    D. it will form phosgene gas.

23. To detect a refrigerant leak, hold the leak detector sensor:
    A. within 3 inches of the fitting.
    B. just below the fitting.
    C. right next to the fitting.
    D. just above the fitting.

24. *Technician A* says a soapy water solution may be used to leak test A/C systems. *Technician B* says using an electronic leak detector is the most accurate method of leak testing A/C systems. Who is right?
    A. A only
    B. B only
    C. Both A and B
    D. Neither A nor B

Hint

Refrigerant systems use the following two methods to detect leaks: dye check or electronic leak detector. To check an A/C system for leaks, the Technician must first ensure that the system contains enough refrigerant to allow compressor clutch engagement. If the system is empty, install a partial refrigerant charge. On an R-12 system, you can use a special R-12 that contains a dye. After running with the dye installed for 15 minutes, the dye will appear at the leak area. Ultraviolet dye is available for installation into the refrigerant and visible under a black light detector. Electronic detectors provide an audible beeping sound when the probe is placed near the leak source. Because R-12 and R-134a are different chemically, one uses a specific electronic detector or one that does both systems. When checking for leaks, you place the leak detector probe directly below each fitting and each component, directly below the evaporator drain, and at the center panel duct. You need to check the entire system to rule out multiple leaks.

When using an electronic leak detector, you must calibrate the detector before each use. When using a flame-type leak detector, light the torch and warm the reaction plate until it will glow red, then adjust the flame until it burns pale blue. Watch the flame while moving the search hose. If a small leak is found, the flame becomes light green to yellow. A large leak causes a purple flame.

Task 6

Evacuate A/C system using appropriate equipment.

25. *Technician A* says the manifold gauge set and vacuum pump in Figure 7–9 are connected properly for A/C system evacuation. *Technician B* says stem-type service valves should be front seated during the evacuation procedure. Who is right?
    A. A only
    B. B only
    C. Both A and B
    D. Neither A nor B

26. When evacuating an A/C system, the vacuum pump should be operated a minimum of:
    A. 20 minutes.
    B. 10 minutes.
    C. 15 minutes.
    D. 30 minutes.

Hint

When evacuating an A/C system, it is important to follow the manufacturer's instructions for the specific recovery station used. Never vent refrigerant to the atmosphere; it is an illegal and environmentally irresponsible action.

When the system is completely empty, connect the manifold gauge center hose to a vacuum pump. You operate this pump for 30 minutes with the service valves open and the low side gauge valve open. After 5 minutes of operation, the low-side gauge should indicate 20 in. Hg (67.6 kPa), and the high side should read below zero, unless it is restricted by a stop pin. If the high-side gauge does not drop below zero, this indicates refrigerant

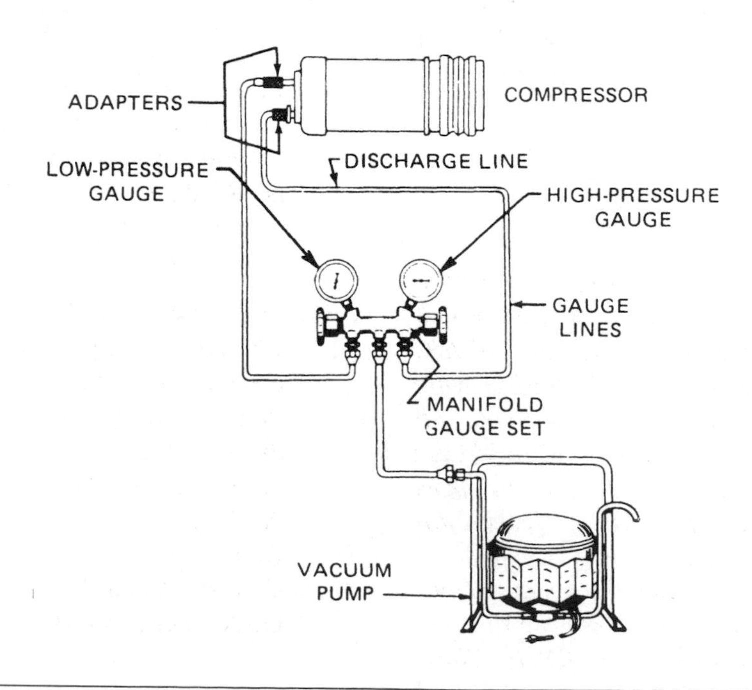

Figure 7–9 Manifold gauge set connections.

blockage. When you find blockage, you must fix this first before proceeding with the evacuation process. After 15 minutes of evacuation, the low-side should indicate 24 to 26 in. Hg (81 to 88 kPa), if there are no leaks. If you find less than this value, close the low-side gauge valve and observe the gauge. If the low-side gauge needle rises slowly, this indicates a refrigerant leak. Fix the leak and proceed. You must evacuate the system to at least 27 in. Hg. Most manufacturers recommend that the vacuum pump run for at least 30 minutes to ensure that all moisture is removed from the system.

Task 7

Internally clean A/C system components and hoses.

27. All of the following statements about nitrogen flushing the A/C system are true EXCEPT:
    A. the Technician should install a pressure regulator on the supply tank.
    B. the Technician should disconnect the A/C compressor.
    C. the Technician should remove restrictive components (i.e., STV, TXV valve) from the system.
    D. nitrogen must not be allowed to escape into the atmosphere.

28. *Technician A* says some manufacturers recommend the installation of an in-line filter between the condenser and the evaporator as an alternative to refrigerant system flushing. *Technician B* says an in-line filter containing a fixed orifice may be installed and the original orifice tube left in the system. Who is right?
    A. A only
    B. B only
    C. Both A and B
    D. Neither A nor B

Hint

If desiccant bag deterioration or catastrophic compressor failure occurs, one must clean all refrigeration system components internally. Internal cleaning of A/C system components is best accomplished by flushing with nitrogen. Before flushing, the compressor and all restricting components and filters must be removed from the system. It is important to regulate the pressure from the nitrogen supply tank to normal system pressure for each component.

Rather than system flushing, many truck manufacturers recommend using an in-line filter between the condenser and the evaporator to remove debris. These in-line filters come with or without FOT. If you use the filter that contains a FOT, you must remove the other FOT from the system.

Task 8

Charge A/C system with refrigerant.

29. *Technician A* says you can complete the high-side charging procedure with the engine running. *Technician B* says if liquid refrigerant enters the compressor, damage will result to the compressor. Who is right?
    A. A only
    B. B only
    C. Both A and B
    D. Neither A nor B

30. A heavy-duty truck is about to have the A/C system recharged. Which of these items is correct concerning that process?
    A. One completes the charging process when the system reaches the correct evaporator temperature.
    B. When the low side no longer moves from a vacuum to a pressure, the process is complete.
    C. The truck engine must be running during recharging.
    D. You can use either a high-side or a low-side charging process.

31. All of these statements about A/C recovery/recycling equipment are true EXCEPT:
    A. the equipment label must indicate UL approval.
    B. the equipment label must indicate SAE J1991 approval.
    C. any size and type of refrigerant storage container over 10 lb. may be used in this equipment.
    D. R-12 and R-134a refrigerants or refrigerant oils must not be mixed in the recovery/recycling process.

32. *Technician A* says the charging system is complete when the specified weight of refrigerant has entered the system. *Technician B* says with the system properly charged and the engine running, the average low side pressure is 25 to 45 psi (172.3 to 310 kPa). Who is right?
    A. A only
    B. B only
    C. Both A and B
    D. Neither A nor B

Hint

The Technician must complete the recovery and evacuation procedure before charging a refrigerant system. Modern recovery and charging stations do not require the A/C system to operate during system charging. It is always best to read all of the manufacturer's instructions for the specific charging station used. OEMs may recommend high-side (liquid) or low-side (vapor) charging procedures. You must close both the high-side and low-side manifold gauge valves.

Connect the center hose to the proper refrigerant container and open the container valve. With the engine not running, using the high-side (liquid) charging process, open the high-side gauge valve and observe the low-side gauge, then close the high-side gauge. If the low-side gauge does not move from a vacuum to a pressure, the refrigerant system is restricted. With no restriction present, open the high-side gauge valve to proceed with the high-side (liquid) charging procedure. Charging is complete when the correct weight of refrigerant has entered the system. Turn the compressor over by hand to make sure that no liquid refrigerant is in the compressor. Now start the engine and run an A/C performance test.

Task 9

Identify lubricant type.

33. *Technician A* says all A/C systems use a mineral-based internal lubricant. *Technician B* says PAG oil is a non-synthetic form of lubricant. Who is right?
    A. A only
    B. B only
    C. Both A and B
    D. Neither A nor B

34. The lubricant used in R-134a mobile air-conditioning systems is:
    A. a polyalkylene glycol (PAG)-based lubricant.
    B. a mineral-based petroleum lubricant.
    C. DEXRON or DEXRON II lubricant.
    D. type C-3 SAE 30-based oil lubricant.

Hint

All R-12 systems use mineral-based refrigeration oil to lubricate the compressor and prevent internal corrosion of components. Mineral-based oil is not compatible with R-134a systems. R-134a systems use polyalkylene glycol (PAG)-based refrigeration oil. The PAG lubricant is synthetic oil and is not compatible with R-12 systems.

Compressor and Clutch

Task 1

Diagnose A/C system problems that cause protection devices (pressure, thermal, and electronic) to interrupt system operation; determine needed repairs.

35. *Technician A* says component R in Figure 7–10 is a binary switch. *Technician B* says component R in Figure 7–10 sends a voltage signal to the PCM in relation to refrigerant pressure. Who is right?
    A. A only
    B. B only
    C. Both A and B
    D. Neither A nor B

36. *Technician A* says component X in Figure 7–10 must be mounted on the evaporator outlet line. *Technician B* says component X senses the pressure in the line to which it is attached. Who is right?
    A. A only
    B. B only
    C. Both A and B
    D. Neither A nor B

37. The LEAST likely cause of the high-pressure relief valve operating is:
    A. improper radiator shutter operation.
    B. a clogged condenser.
    C. inoperative cooling fan clutch.
    D. a defective A/C compressor.

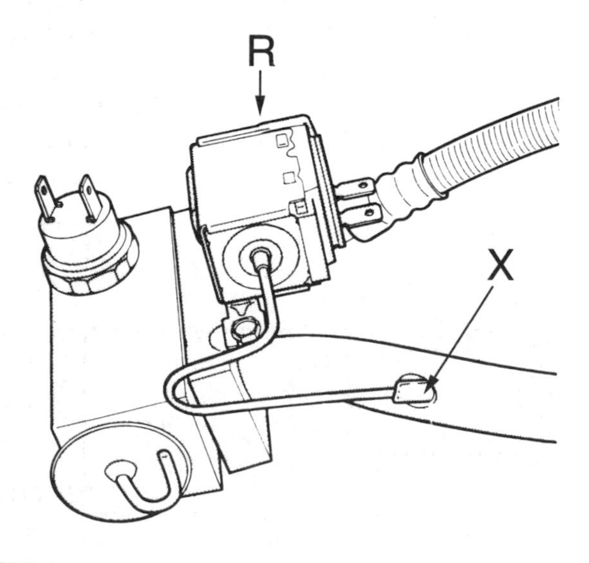

Figure 7–10 A/C system component. *(Courtesy of Chrysler Corporation)*

38. The A/C compressor clutch will not engage in any mode. The clutch engages when a Technician installs a jumper wire across the terminals of the low-pressure cut-out switch connector. *Technician A* says that the low-pressure cut-out switch must be defective. *Technician B* says that the refrigerant charge might be low. Who is right?
    A. A only
    B. B only
    C. Both A and B
    D. Neither A nor B

Hint

A variety of A/C system protection devices can be used in mobile A/C systems. The low-pressure cut-out switch will interrupt compressor operation if system pressure drops to the point that a loss of refrigerant charge occurs. The high-pressure cut-out switch interrupts compressor operation in case of extremely high system pressure. The binary switch combines the function of the low- and high-pressure cut-out switches. Some systems have a high-pressure relief valve mounted in the receiver/dryer. This valve opens and relieves system pressure if the pressure exceeds 450 to 550 psi (3,100 to 3,792 kPa). Condenser airflow restrictions cause these extremely high pressures. In gasoline and diesel engine electronic fuel management systems, the computer operates a relay that supplies voltage to the compressor clutch. All input signals go to the engine computer. In some applications, this includes a refrigerant pressure signal. If this input signals an abnormal low- or high-pressure condition, the engine computer will not engage the compressor clutch.

CCOT systems use a pressure cycling switch to cycle the compressor off and on, in relation to low-side pressure. This switch is mounted in the accumulator between the evaporator and the compressor. This switch closes and turns on the compressor when the refrigerant pressure is above 46 psi (315 kPa). The dash A/C switch supplies the power to the cycling switch. The pressure switch opens when the system pressure decreases to 25 psi (175 kPa). This cycling action maintains the evaporator temperature at 33°F (1°C).

Some refrigerant systems use a thermostatic clutch cycling switch that cycles the compressor on and off in relation to evaporator outlet temperature.

Task 2

Inspect, test, and replace A/C system pressure, thermal, and electronic protection devices.

39. *Technician A* says that the A/C compressor high-pressure relief valve must be replaced if it operates due to extreme pressure in the A/C system. *Technician B* says that the compressor must be replaced if the high-pressure relief valve operates. Who is right?
    A. A only
    B. B only
    C. Both A and B
    D. Neither A nor B

40. A compressor cycling on and off too fast is one symptom of:
    A. a defective compressor clutch.
    B. a defective control switch.
    C. an overcharged system.
    D. a low refrigerant charge.

41. With the ignition switch off, the wiring connector is removed from component 5 in Figure 7–11, and an ohmmeter is connected across the terminals on this component. *Technician A* says the ohmmeter should provide an infinity reading. *Technician B* says the contacts in component 5 should be open. Who is right?
    A. A only
    B. B only
    C. Both A and B
    D. Neither A nor B

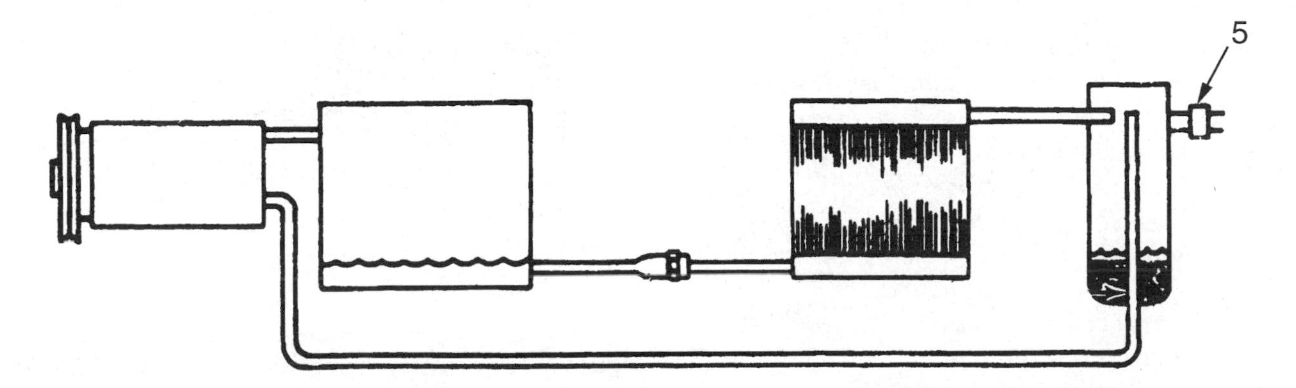

Figure 7–11 A/C refrigeration system. *(Courtesy of Chevrolet Motor Division, General Motors Corporation)*

Hint

Many A/C system protection devices are normally closed. When the specified amount of pressure or heat is applied to the device, the contacts or valve open. A/C pressure devices containing a set of contacts may be tested with an ohmmeter. Remove the wiring connector from the device and connect a pair of ohmmeter leads across the terminals. If the contacts are closed, the ohmmeter indicates a very low reading. When the contacts in the device are open, the ohmmeter provides an infinity reading.

Task 3

Inspect, adjust, and replace A/C compressor drive belts and pulleys.

42. The A/C compressor drive belt should be adjusted:
    A. using a belt tension gauge.
    B. so there is no deflection at maximum engine speed.
    C. so there is no static deflection.
    D. as tightly as possible.

43. A/C compressor drive belt edge wear could indicate any of the following conditions EXCEPT:
    A. a bent or cracked compressor mounting bracket.
    B. improperly set compressor clutch air gap.
    C. a damaged compressor pulley.
    D. a worn idler pulley bearing.

44. An A/C compressor drive belt is misaligned. *Technician A* says the compressor mounting bracket holder may be bent. *Technician B* says the compressor clutch bearing may be defective. Who is right?
    A. A only
    B. B only
    C. Both A and B
    D. Neither A nor B

Hint

When inspecting an A/C system, it is important that the Technician not overlook the A/C compressor drive belts and pulleys. Drive belt edge wear indicates a misaligned or bent pulley. If the belt is loose or bottomed out in the pulley, the belt may slip and cause inadequate cooling. One must replace cracked or frayed belts. Use a standard drive belt tension gauge to check and adjust belt tension. An improperly adjusted drive belt will wear or fail prematurely. A drive belt adjusted too loosely may slip. A slipping drive belt may cause belt squealing especially on acceleration with the A/C on and clutch engaged. A drive belt adjusted too tightly may cause internal engine wear or damage to other belt-driven components. One must replace not repair cracked or bent pulleys.

Task 4

Inspect, test, service, and replace A/C compressor clutch components or assembly.

45. The measurement in Figure 7–12 is more than specified. *Technician A* says this condition may cause an intermittent scraping noise with the engine running and the compressor clutch engaged. *Technician B* says to correct this condition, another shim should be added behind the pulley armature plate. Who is right?
    A. A only
    B. B only
    C. Both A and B
    D. Neither A nor B

46. *Technician A* says that you can test the A/C compressor clutch coil with an ohmmeter. *Technician B* says that connecting battery power to one terminal of the coil and grounding the other terminal can test the A/C compressor clutch coil. Who is right?
    A. A only
    B. B only
    C. Both A and B
    D. Neither A nor B

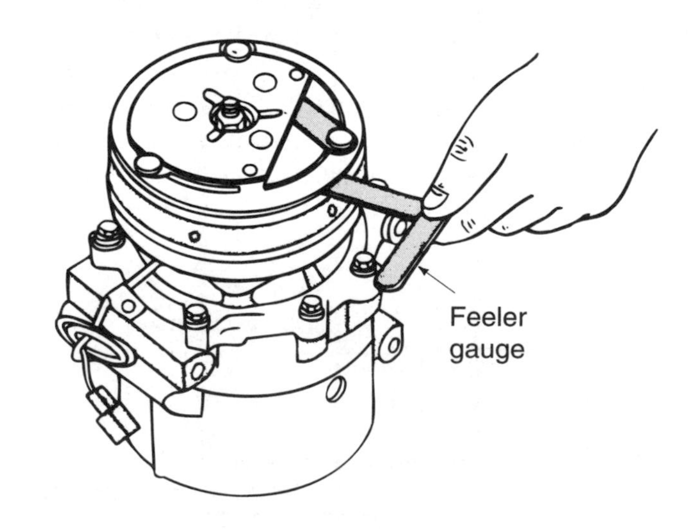

Figure 7–12 A/C compressor clutch measurement. *(Courtesy of Chrysler Corporation)*

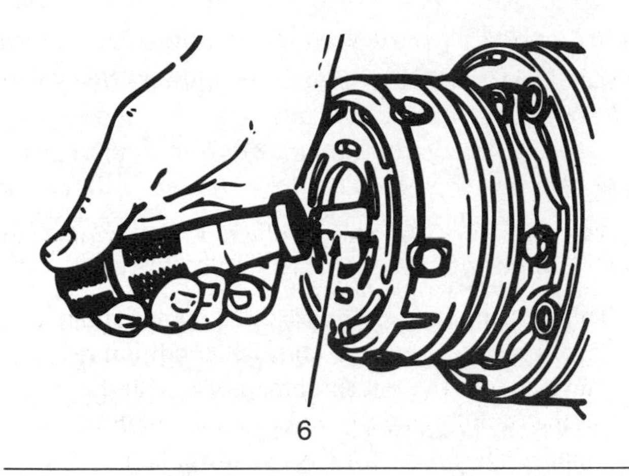

Figure 7–13 A/C system service tool. *(Courtesy of Chevrolet Motor Division, General Motors Corporation)*

47. *Technician A* says component 6 in Figure 7–13 is a seal protector that one places over the compressor shaft. *Technician B* says you must install the seal seat O-ring before the compressor shaft seal. Who is right?
    A. A only
    B. B only
    C. Both A and B
    D. Neither A nor B

48. Refer to Figure 7–14. All of the following could prevent A/C compressor clutch engagement EXCEPT:
    A. a defective blower switch.
    B. an open binary pressure switch.

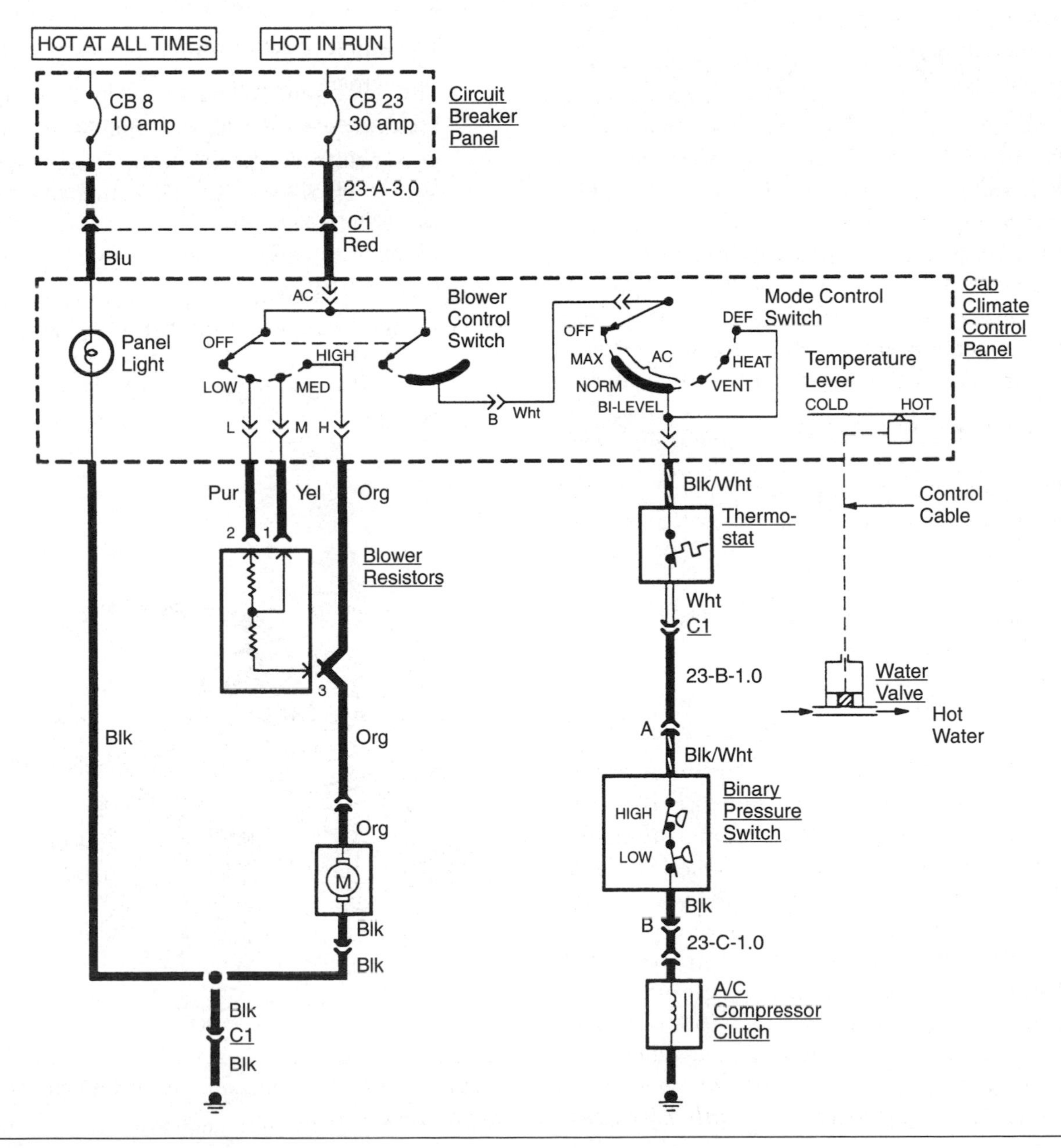

Figure 7–14 Blower and compressor clutch circuit.

C. A poor connection at the A/C thermostat.
D. a blown 10 amp fuse at CB 8.

Hint

The A/C compressor clutch assembly allows the A/C compressor to engage and disengage to modulate system pressures. The components of the compressor clutch assembly are:

- *the driven plate, which is keyed to the compressor drive shaft;*
- *the drive plate, which is integral to the drive pulley;*
- *the clutch bearing, which operates when the clutch is disengaged;*
- *the clutch coil, which creates the magnetic field that engages the clutch.*

A faulty compressor clutch bearing will make a growling noise with the engine running and the clutch disengaged. Test the compressor clutch by applying power and ground to the appropriate terminals and watching for clutch engagement. You must inspect the pulley and armature plate frictional surfaces for wear and oil contamination. Check the hub bearing for roughness, grease leakage, and looseness. A driven plate that drags on the drive plate or slips briefly on engagement indicates an improper clutch air gap. One adjusts this air gap using shims; you remove shims to decrease clearance and increase shims to increase clearance. One checks the gap on any clutch service. A Technician can check the clutch engagement dynamically with an ammeter in series with the clutch coil. A good clutch coil will read 2 to 4.15 amps.

Task 5

Inspect and correct A/C compressor lubricant level.

49. *Technician A* says that the compressor oil needs to be checked when there is evidence of a loss of system oil. *Technician B* says that when replacing refrigerant oil, it is important to use the specific type and quantity of oil recommended by the compressor manufacturer. Who is right?
    A. A only
    B. B only
    C. Both A and B
    D. Neither A nor B

50. Purging a system too fast will result in:
    A. phosgene gas formation.
    B. oil being drawn from the compressor.
    C. reed valve damage.
    D. receiver/dryer damage.

Hint

One must check the A/C compressor lubricant level and adjust it any time there is evidence of lubricant loss from the system. An excessive amount of oil in a refrigerant system reduces the system cooling efficiency. To check the compressor lubricant level, you remove the compressor from the vehicle, drain all refrigeration oil, and refill it to manufacturer's specifications.

R-12 systems require a mineral oil with a YN-9 designation, and R-134a systems with a reciprocating compressor must have a synthetic polyalkylene glycol (PAG) oil designation. Rotary compressors use a different type of PAG oil. If the oils used become intermixed, compressor damage will result.

Task 6

Inspect, test, service, or replace A/C compressor.

51. The refrigerant line leading from the evaporator to the compressor contains refrigerant as a:
    A. low-pressure gas.
    B. low-pressure liquid.
    C. high-pressure gas.
    D. high-pressure liquid.

52. The arrow shown in Figure 7–15 is pointing to the:
    A. pilot-operated absolute (POA) valve.
    B. filter.
    C. evaporator temperature regulator (ETR).
    D. muffler.

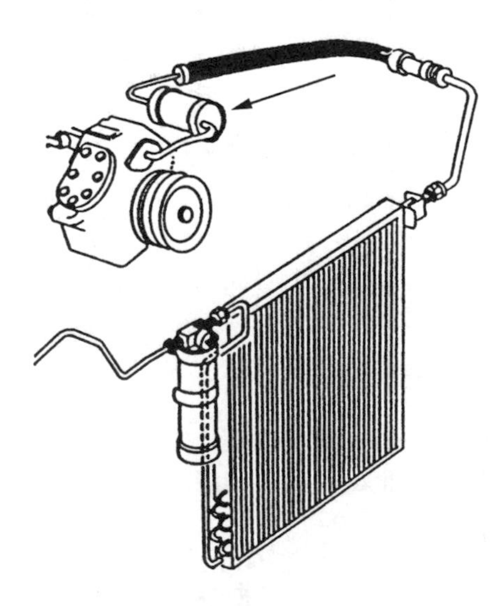

Figure 7–15 A/C system components.

53. If one of the reed valves fails in an A/C compressor, the result is low cooling because of which of these compressor reed valve functions?
    A. Stay open when the compressor is hot.
    B. Maintain a certain temperature.
    C. Maintain a certain pressure.
    D. Direct the flow of refrigerant.

54. The compressor discharge valve is designed to:
    A. open after the vaporous refrigerant is compressed, allowing the refrigerant to move to the condenser.
    B. before the vaporous refrigerant is compressed, allowing the refrigerant to move to the evaporator.
    C. regulate system variable pressure.
    D. regulate A/C system temperature.

Hint

You can diagnose A/C compressor internal damage using a standard A/C gauge set. Low high-side pressure and high low-side pressure on the manifold gauge set may indicate a defective compressor. Oil dripping from the front of the compressor indicates a faulty front seal. A rattling noise may be caused by loose compressor mounts. A growling noise that occurs when the compressor is operating is likely caused by a worn bearing in the compressor. A defective pulley will also cause a growling noise with the clutch engaged. When replacing A/C seals or O-rings, prelubricate them with the proper type of refrigeration oil.

Task 7

Inspect, repair, or replace A/C compressor mountings.

55. A cracked A/C compressor mounting plate could cause all of the following symptoms EXCEPT:
    A. drive belt wear.
    B. internal compressor damage.
    C. vibration with the A/C compressor clutch engaged.
    D. drive belt squeal or chatter.

56. An A/C compressor has a stripped mounting bolthole. *Technician A* says the threads can be restored by applying a weld to the damaged area and then drilling and retapping the hole. *Technician B* says the compressor must be replaced, not repaired. Who is right?
    A. A only
    B. B only
    C. Both A and B
    D. Neither A nor B

Hint

Damaged A/C compressor mounts or mounting plates can cause drive belt misalignment, improper drive belt tension, and compressor vibration. Welding can generally repair cracked mounts and mounting plates; however, care must be taken to align all parts properly.

Evaporator, Condenser, and Related Components

Task 1

Inspect and correct lubricant level in evaporator, condenser, receiver/dryer or accumulator/dryer, and hoses when servicing or replacing components.

57. Approximately how much refrigerant oil must be added to a newly replaced evaporator core?
    A. None
    B. 3 ounces
    C. 9 ounces
    D. 14.5 ounces

58. How much refrigeration oil should be in a typical A/C condenser?
    A. 1 ounce
    B. 5 ounces
    C. 7 ounces
    D. 11 ounces

59. *Technician A* says that the evaporator must be removed from the vehicle if the evaporator lubricant level is to be checked. *Technician B* says that first you run the A/C compressor briefly, to ensure that the refrigeration oil distributes throughout the system. Who is right?
    A. A only
    B. B only
    C. Both A and B
    D. Neither A nor B

Hint

Most A/C system components contain refrigeration oil. When replacing the evaporator, condenser, accumulator, receiver/dryer, or A/C hoses, the new component should be drained of oil, and fresh refrigeration oil should be added to manufacturer's specifications.

Task 2

Inspect, repair, or replace A/C system hoses, lines, filters, fittings, and seals.

60. What tool or piece of equipment is shown in Figure 7–16?
    A. Spring lock installation tool
    B. Bearing puller
    C. Hose end crimping tool
    D. Flare tool

61. *Technician A* says that when installing new A/C hose O-rings, a seal pick should be used to minimize skin contact with the new seal. *Technician B* says that the petroleum jelly used to lubricate the new O-rings will protect them from the oils in the skin. Who is right?
    A. A only
    B. B only
    C. Both A and B
    D. Neither A nor B

Hint

You check A/C system hoses for damage and leaks during the course of any A/C system maintenance or inspection. Hoses should be replaced if they are cracked, kinked, or abraded or if the fittings show any signs of abuse. Disassemble leaking fittings and replace the O-rings. New O-rings should be lubricated with the appropriate refrigeration oil. A/C system filters and screens are used to prevent particulate (from corrosion, compressor failure, or desiccant breakdown) from circulating through the A/C system and must be replaced if they are clogged, restricted, or damaged. Some systems have a filter in the line between the condenser and the evaporator. Some of these filters contain an orifice tube. You must install this type of filter in the proper direction.

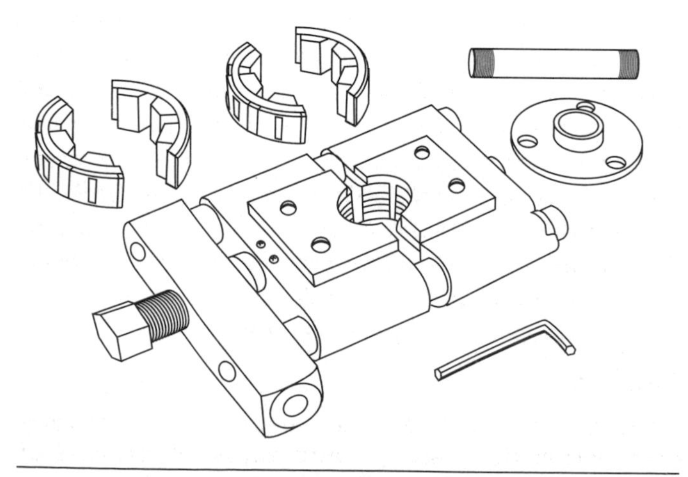

Figure 7–16 A/C system service tool.

Task 3

Inspect A/C condenser for proper airflow.

62. Which of the following is the LEAST likely cause of a shutter system that will not close completely?
    A. A blown shutter fuse
    B. An air leak in the hose to the shutter cylinder
    C. Dirt accumulation on the shutter linkage
    D. Ice buildup on the shutters

63. The A/C condenser has several bent fins and a moderate accumulation of dead insects. *Technician A* says that the condenser should be replaced or it will cause the high-pressure relief valve to operate. *Technician B* says that the condenser fins should be straightened and the dead insects removed to optimize A/C system performance. Who is right?
    A. A only
    B. B only
    C. Both A and B
    D. Neither A nor B

64. While inspecting the chassis air system, a Technician finds the radiator shutter cylinder to be leaking. *Technician A* says you must replace the cylinder. *Technician B* says you overhaul the cylinder. Who is right?
    A. A only
    B. B only
    C. Both A and B
    D. Neither A nor B

65. Any of the following can be used to clean road debris from the condenser fins EXCEPT:
    A. a mild saline solution.
    B. a soft whisk broom.
    C. compressed air.
    D. a mild soap and water solution.

Hint

The A/C condenser should be checked for proper airflow at regular intervals. During a normal A/C system inspection, any bent condenser fins should be straightened and any debris should be cleaned from the condenser. Debris in the condenser air passages causes excessive high-side and low-side pressures and reduced cooling. This problem may also cause the high-pressure relief valve to discharge refrigerant. Additionally, you should check the radiator shutter system for proper operation.

Task 4

Inspect, test, and replace A/C system condenser and mountings.

66. A routine A/C maintenance service should include all of the following EXCEPT:
    A. tightening the condenser lines.
    B. removing debris from the condenser fins.
    C. straightening the condenser fins.
    D. checking the condenser mounts.

67. *Technician A* says that reduced distance between the condenser and the radiator may affect A/C system operation. *Technician B* says that deformed or improperly aligned condenser mounting insulators could damage the condenser and refrigerant lines. Who is right?
    A. A only
    B. B only
    C. Both A and B
    D. Neither A nor B

Hint

If any refrigerant tubes are kinked cracked or leaking, the A/C condenser must be replaced. Frost on any of the condenser tubing indicates a refrigerant passage restriction. This condition results in excessive high-side and low-side pressures and inadequate cooling. If you replace the condenser, drain the new component and install fresh refrigeration oil to manufacturer specifications (typically 1 ounce). Condenser mounts and insulators should be checked for proper alignment and deformation, which could cause abrasion and fatigue damage.

Task 5

Inspect and replace receiver/dryer or accumulator/dryer.

68. *Technician A* says the refrigerant hose the Technician is touching in Figure 7–17 should feel warm or hot if the A/C system is operating properly. *Technician B* says the component to which the A/C system hoses are connected in Figure 7–17 is the accumulator. Who is right?
    A. A only
    B. B only
    C. Both A and B
    D. Neither A nor B

69. The refrigerant in an A/C system quickly changes state from liquid to vapor as it flows through the:
    A. receiver/dryer.
    B. condenser.
    C. fixed orifice.
    D. capillary tube.

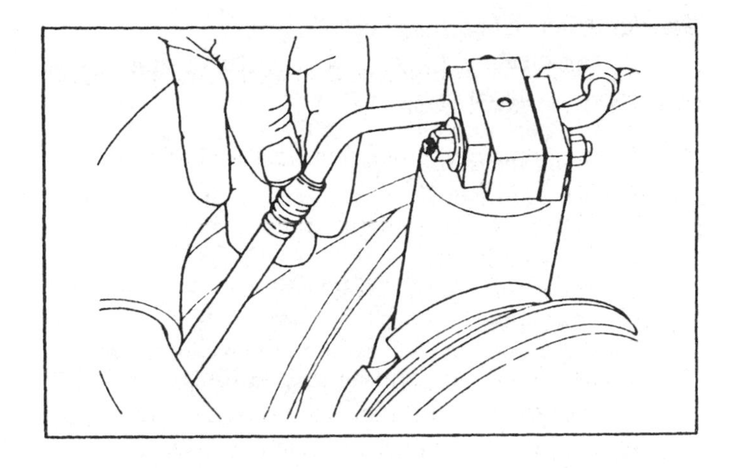

Figure 7–17 Diagnosing A/C system components by touch. *(Courtesy of Chrysler Corporation)*

70. The receiver/dryer must be replaced:
    A. if the A/C system has been open to the atmosphere for an extended period of time.
    B. when the A/C system is evacuated and charged.
    C. every 100,000 miles.
    D. if the compressor is replaced.

Hint

Most mobile A/C systems use an accumulator or a receiver/dryer to ensure an adequate supply of high-pressure liquid refrigerant to the system expansion. If the receiver/dryer inlet and outlet pipes have a significant temperature difference, the receiver/dryer is restricted. Frost forming on the receiver/dryer indicates an internal restriction. Bubbles and foam in the sight glass indicate rust and moisture contamination. Both of these devices contain a bag of desiccant designed to absorb and hold traces of moisture from the refrigerant. The accumulator is located at the outlet of the evaporator and sometimes houses the system sight glass. The receiver is located just upstream of the system expansion device. The accumulator/dryer or receiver/dryer must be replaced if the A/C system has remained open to the atmosphere for an extended period of time, if there is evidence of moisture or corrosion in the system, or if catastrophic compressor failure has occurred. When the accumulator or receiver/dryer is replaced, the new component must be drained and fresh refrigerant oil must be added to manufacturer's specifications (typically 1 ounce).

Task 6

Inspect, test, and replace cab/sleeper refrigerant solenoid, expansion valve(s); and adjust placement of thermal bulb (capillary tube).

71. The thermal bulb and capillary tube:
    A. are fastened to the condenser fins using epoxy.
    B. are located in the accumulator.
    C. are kept in contact with the evaporator inlet using insulating tape.
    D. are an integral part of the thermal expansion valve.

72. *Technician A* says the expansion valve is mounted at the evaporator outlet. *Technician B* says the frost formation on the expansion valve may indicate excessive oil in the refrigerant. Who is right?
    A. A only
    B. B only
    C. Both A and B
    D. Neither A nor B

Hint

One type of A/C system expansion device is the thermal expansion valve. The thermal expansion valve senses evaporator temperature using a capillary tube connected to a thermal bulb. As the fluid inside the thermal bulb expands, the orifice in the expansion valve opens to increase refrigerant flow through the evaporator. If the evaporator core temperature drops to near the freezing point, the fluid in the thermal bulb contracts and the expansion valve closes to restrict refrigerant flow. Different designs place this thermal bulb either embedded in the evaporator fins or affixed to the evaporator outlet with insulating tape.

In some systems, the expansion valve is housed in a combination valve or an "H" valve. In these systems, internal sensors monitor evaporator inlet and outlet temperatures and pressures and adjust the valve opening accordingly. A Technician can diagnose a faulty expansion valve using a set of A/C pressure gauges.

Task 7

Inspect and replace orifice tube.

73. The component being service in Figure 7–18 is the:
    A. high-side service valve.
    B. orifice tube.
    C. accumulator.
    D. evaporator outlet fitting.

74. A restricted orifice tube will reduce the cooling ability of an A/C system because of which of the following functions of the orifice tube?
    A. It allows water to drain from the evaporator case.
    B. It allows rapid expansion of high-pressure liquid refrigerant.
    C. It regulates refrigerant flow through the condenser.
    D. It regulates airflow through the evaporator.

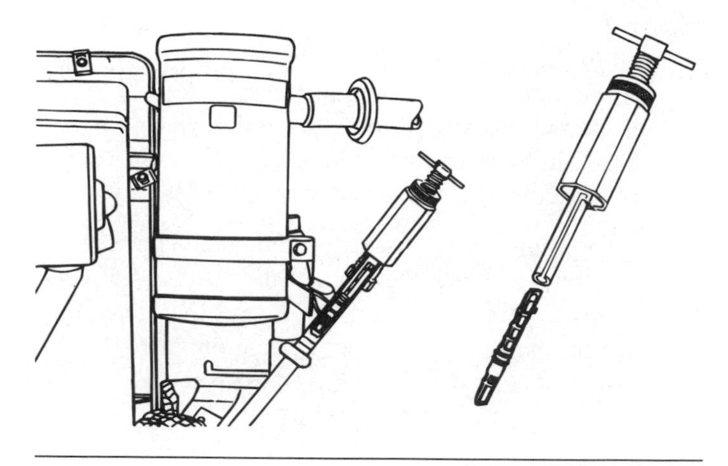

Figure 7–18 A/C system service tool. *(Courtesy of Ford Motor Company)*

75. A screen is located in the orifice tube of an A/C system. *Technician A* says that the screen is a filter used to prevent particulate from circulating through the system. *Technician B* says that the screen is used to improve atomization of the refrigerant. Who is right?
    A. A only
    B. B only
    C. Both A and B
    D. Neither A nor B

Hint

Cycling clutch-type A/C systems often use a fixed orifice tube as an expansion device. The orifice tube is located at the evaporator inlet. The orifice tube contains a fine screen to prevent the circulation of particulate through the evaporator core and back to the compressor. The orifice tube should be replaced if the screen is restricted or corroded, or in case of desiccant bag breakdown or catastrophic compressor failure.

A restricted orifice tube may cause lower than specified low-side pressure, frosting of the orifice tube, and inadequate cooling from the evaporator. If these conditions appear, place a shop towel soaked in hot water around the orifice tube. If the low-side pressure increases, there is moisture freezing in the orifice tube. To rectify this condition, you must recover, evacuate, and recharge the system. If the hot shop towel did not increase the low-side pressure, clean or replace the orifice tube.

Task 8

Inspect, test, and replace cab/sleeper evaporator.

76. Refer to Figure 7–19. *Technician A* says the air flows through the evaporator core before it flows through the heater core. *Technician B* says in the maximum A/C mode, some air flows through the heater core and some air flows through the evaporator core. Who is right?
    A. A only
    B. B only
    C. Both A and B
    D. Neither A nor B

77. In mobile A/C systems, evaporator core icing can be prevented by any of the following EXCEPT:
    A. a thermostatic switch.
    B. an EPR valve.
    C. an STV.
    D. a low-pressure cut-off switch.

78. The inside of a truck windshield has an oily film and the A/C cooling is poor. *Technician A* says a plugged HVAC heater case drain may cause this oil film. *Technician B* says this film may be caused by a leak in the evaporator core. Who is right?
    A. A only
    B. B only
    C. Both A and B
    D. Neither A nor B

Hint

The A/C evaporator core is located (along with the heater core and airflow control doors) in the evaporator case. One detects a leaking evaporator core most easily by measurement at the evaporator case drain, but it may also be detected at the panel and defroster vents. Also, you can remove the blower resistor assembly and go through that cavity. If the evaporator core has a leak, an oily film appears on the inside of the windshield, and the

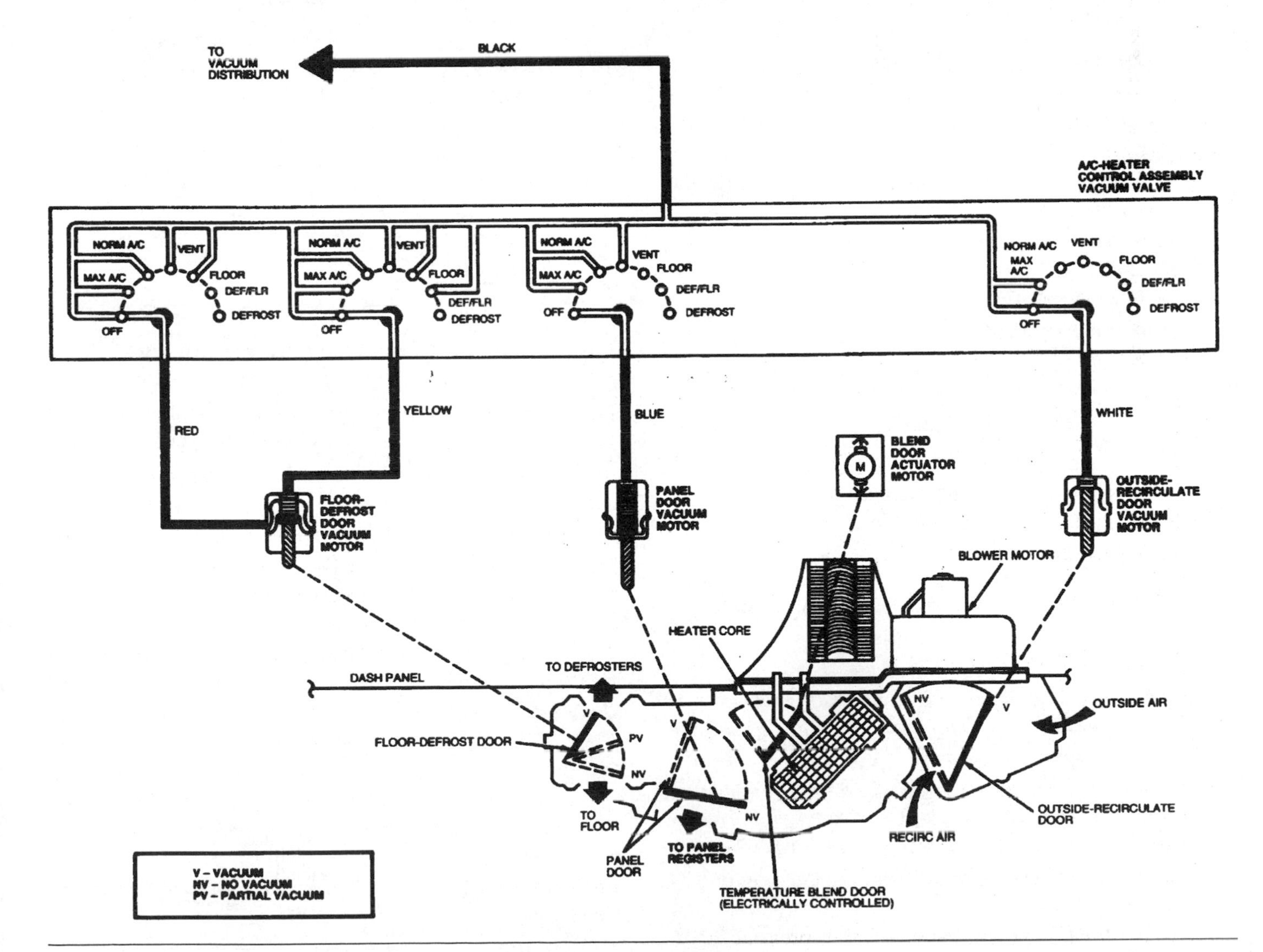

Figure 7–19 Evaporator case and A/C controls. *(Courtesy of Ford Motor Company)*

cab temperature becomes warmer than specified. If you replace an evaporator core, you drain the new component and fresh refrigeration oil must be added to manufacturer's specifications (typically 3 ounces).

To find evaporator refrigerant leaks, look for oil in the leak area. A low-side pressure that is considerably lower than specified indicates evaporator restriction.

Task 9

Inspect, clean, and replace evaporator housing and water drain; inspect and service/replace evaporator air filter.

79. Filters installed in the HVAC air delivery system:
    A. are always made of fiberglass mesh to resist corrosion.
    B. are designed to remove moisture from cab and sleeper air.
    C. are not individually replaceable.
    D. are designed to remove dust and dirt from cab and sleeper air.

80. A faint hissing noise is heard from the area of the evaporator immediately after shutting down the engine with the compressor clutch engaged. *Technician A* says that the A/C system has a leak. *Technician B* says the evaporator pressure regulator is defective. Who is right?
    A. A only
    B. B only
    C. Both A and B
    D. Neither A nor B

81. To eliminate a mildew smell from the A/C output air:
    A. remove the evaporator case, clean it with a vinegar and water solution, and dry it thoroughly before reinstallation.
    B. spray a disinfectant into the panel outlets.
    C. pour a small amount of alcohol into the air intake plenum.
    D. place an automotive deodorizer under the dash.

Hint

The evaporator case or housing contains the evaporator core, the heater core, the evaporator core drain, and the blend air and mode control doors. A clogged evaporator case drain will cause windshield fogging or a noticeable mist from the panel vents. A clogged drain can normally be opened up with a slender piece of wire or with low-pressure shop air. The evaporator drain should be checked during routine maintenance inspections. A cracked evaporator case can cause a whistling noise during high blower operation. Minor cracks can be repaired using epoxy-type adhesives. A mildew smell that is noticeable during A/C system operation can be rectified by removing the evaporator case and washing it with a vinegar and water solution or a commercially available cleaner.

Task 10

Identify, inspect, and replace A/C system service valves (gauge connections).

82. With a stem-type service valve in the position shown in Figure 7–20:
    A. you can diagnose the refrigerant system with a manifold gauge set.
    B. the refrigerant system operates normally with no pressure at the gauge ports.
    C. the refrigerant system is isolated from the compressor for compressor removal.
    D. the refrigerant system may be discharged, evacuated, and recharged.

83. *Technician A* says A/C systems with R-134a refrigerant have Schrader-type service valves. *Technician B* says that according to new environmental laws, shutoff valves must be located no more than 12 inches from test hose service ends. Who is right?
    A. A only
    B. B only
    C. Both A and B
    D. Neither A nor B

84. *Technician A* says the service valve in Figure 7–21 has a removable core. *Technician B* says the service valve in Figure 7–21 must be rear seated during A/C compressor operation. Who is right?
    A. A only
    B. B only
    C. Both A and B
    D. Neither A nor B

Hint

Three types of A/C system service valves are used in mobile A/C systems. Older systems use a three-position

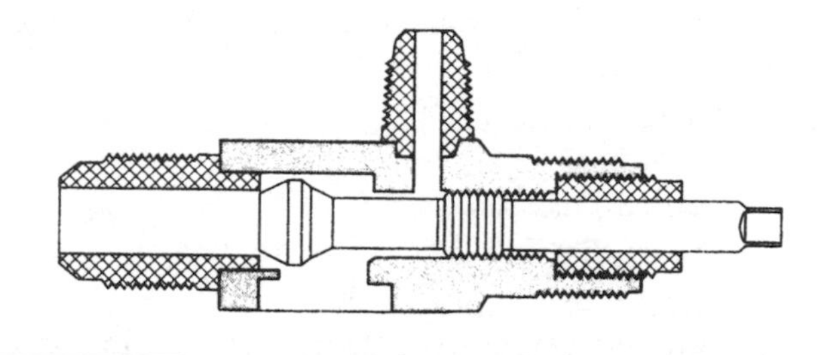

Figure 7–20 A/C system service valve.

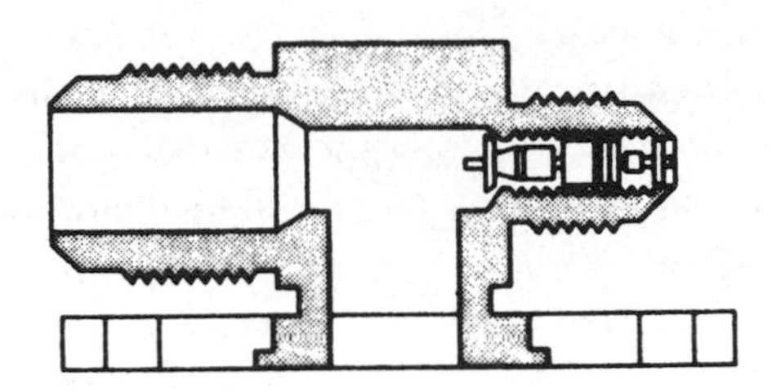

Figure 7–21 A/C system service valve.

stem-type valve. This type of service valve may be placed in the front-seated, mid-position, or back-seated position by rotating the valve stem with the proper tool. For normal operation this type of valve is placed in the back-seated position where the valve stem is rotated counterclockwise to seat the rear valve face and seal off the service gauge port. For A/C system diagnosis and service, the valve is placed in the mid-position where the service gauge port is supplied with pressure from the A/C system. The front-seated valve position isolates the compressor from the A/C system. This position allows a Technician to service the compressor without discharging the entire system. The system must never be operated with either service valve in the front-seated position.

A/C systems with R-12 refrigerant use Schrader-type service valves, and R134a systems use quick-disconnect service valves.

Task 11

Inspect and replace A/C system high-pressure relief device.

85. The A/C high-pressure relief valve shows evidence of slight oil leakage. *Technician A* says you must replace the valve and repair the leak. *Technician B* says that if the valve is just leaking a little oil and not refrigerant, replacement is not necessary. Who is right?
    A. A only
    B. B only
    C. Both A and B
    D. Neither A nor B

86. The A/C high-pressure switch is used to:
    A. boost the system high-side pressure.
    B. open the circuit to the A/C compressor clutch coil when the high-side pressure reaches its upper limit.
    C. ensure that system pressure remains at the upper limit
    D. vent refrigerant from the compressor in the event of extremely high system pressure.

87. *Technician A* says the A/C compressor high-pressure relief valve is not used in R-134a systems. *Technician B* says the compressor high-pressure relief valve resets itself when A/C system pressure returns to a safe level. Who is right?
    A. A only
    B. B only
    C. Both A and B
    D. Neither A nor B

88. A binary pressure switch provides which of the following?
    A. Only low-pressure protection for the A/C system
    B. Only high-pressure protection for the A/C system
    C. Both low- and high-pressure protection for the A/C system
    D. Neither low-pressure nor high-pressure protection for the A/C system

Hint

All mobile A/C systems are equipped with a high-pressure relief device. In most systems, this device is a self-resetting relief valve, which is threaded into the high side of the compressor. The high-pressure relief valve will vent refrigerant from the system in the event that high-side pressure exceeds safe levels.

Heating and Engine Cooling Systems Diagnosis, Service, and Repair

Task 1

Diagnose the cause of temperature control problems in the heating/ventilation system; determine needed repairs.

89. A customer complains that the air-operated heater outputs hot air when the temperature control lever is in the cold position. The LEAST likely cause of this problem is:
    A. a defective coolant control valve.
    B. a defective engine coolant temperature sensor.
    C. a defective air control solenoid.
    D. a defective blend air door control cylinder.

90. *Technician A* says that the heater control valve can be operated by engine speed. *Technician B* says that the heater control valve can be cable operated. Who is right?

A. A only
B. B only
C. Both A and B
D. Neither A nor B

91. A customer complains of being unable to control the output temperature of the HVAC system. The most likely cause of this problem is:
A. a broken blend door cable.
B. a defective compressor clutch.
C. a clogged orifice tube.
D. a defective blower switch.

92. *Technician A* says that a clogged heater core could cause insufficient heater output. *Technician B* says that an improperly adjusted coolant control valve cable could cause insufficient heater output. Who is right?
A. A only
B. B only
C. Both A and B
D. Neither A nor B

Hint

System design achieves heater temperature control by one of two methods: blend air modulation or coolant flow control. In a blend air system, coolant flows through the heater core at a constant rate. Air flowing over the heater core is mixed with outside air to achieve the desired output temperature. In a coolant flow control system, temperature control is achieved by using a coolant control valve (hot water valve) to limit the amount of coolant allowed to flow through the heater core. The coolant control valve may be vacuum or cable operated.

Task 2

Diagnose window fogging problems; determine needed repairs.

93. Which of the following is LEAST likely to cause windshield fogging in the Defrost mode?
A. A leaking heater core
B. A clogged evaporator drain
C. An exterior water leak into the plenum chamber
D. Moisture in the refrigerant

94. The inside of the windshield has a sticky film. *Technician A* says to check the engine coolant level. *Technician B* says the heater core may be leaking. Who is right?
A. A only
B. B only
C. Both A and B
D. Neither A nor B

Hint

Windshield fogging may be caused by a leaking heater core or by a clogged evaporator core drain. If windshield fogging is accompanied by the smell of antifreeze, the heater core is at fault. If you smell a pungent odor, then the evaporator drain may be clogged.

Task 3

Perform engine cooling system tests for leaks, protection level, contamination, coolant level, temperature, and conditioner concentration; determine needed repairs.

95. Coolant conditioner performs all of the following tasks EXCEPT:
A. raise the boiling point of the coolant.
B. filter rust and debris from the coolant.
C. lubricate the cooling system internally.
D. preventing cavitation corrosion of the cylinder liners.

96. To check for cooling system leaks, *Technician A* says to pressurize the system and perform a visual inspection for leaks. *Technician B* says the average medium/heavy truck cooling system may be pressurized to 20 psi (138 kPa): Who is right?
A. A only
B. B only
C. Both A and B
D. Neither A nor B

97. *Technician A* says flushing the cooling system removes acids of combustion from the cooling system. *Technician B* says flushing the cooling system removes rust from the system. Who is right?
A. A only
B. B only
C. Both A and B
D. Neither A nor B

98. When pressure testing a cooling system, there are no obvious leaks but the system cannot maintain pressure. The MOST likely cause of this problem is:
A. a leaking evaporator.
B. a leaking power steering cooler.
C. a leaking oil cooler.
D. a blown head gasket in the engine.

Hint

To test the engine cooling system for leaks, remove the radiator pressure cap following the manufacturer's instructions. Replace the cap with a standard cooling system pressure tester and pressurize the system to operating pressure. Perform a careful visual inspection for leaks. Often you can confirm cooling system contamination with a visual inspection. Rust in the system will turn the coolant an opaque reddish-brown color. If engine oil or transmission fluid has entered the system, the coolant will contain thick deposits resembling a milk shake. Coolant protection level is most easily determined using a cooling system hydrometer or optical sensor.

Task 4

Inspect and replace engine cooling and heating system hoses, lines, and clamps.

99. A spring inside the lower radiator hose is used to:
    A. preform the hose.
    B. prevent the hose from collapsing.
    C. make the hose more resilient.
    D. strengthen the hose.

100. All of these statements about cooling system service are true EXCEPT:
    A. when the cooling system pressure is increased, the boiling point increases.
    B. if one adds more antifreeze to the coolant, the boiling point decreases.
    C. a good quality ethylene glycol antifreeze contains antirust inhibitors.
    D. coolant solutions are recovered, recycled, and handled as hazardous waste.

Hint

Cooling system and heater hoses must be replaced if they are cracked or brittle, or if they show signs of bulging or abrasion. Hose clamps should be replaced if they are deformed or cracked, or if they cannot be operated smoothly.

Task 5

Inspect, test, and replace radiator, pressure cap, and coolant recovery system (surge tank).

101. Refer to Figure 7–22. *Technician A* says the tester in the figure is connected to test the radiator cap. *Technician B* says the tester in the figure is used to test the transmission cooler that is external to the radiator. Who is right?
    A. A only
    B. B only
    C. Both A and B
    D. Neither A nor B

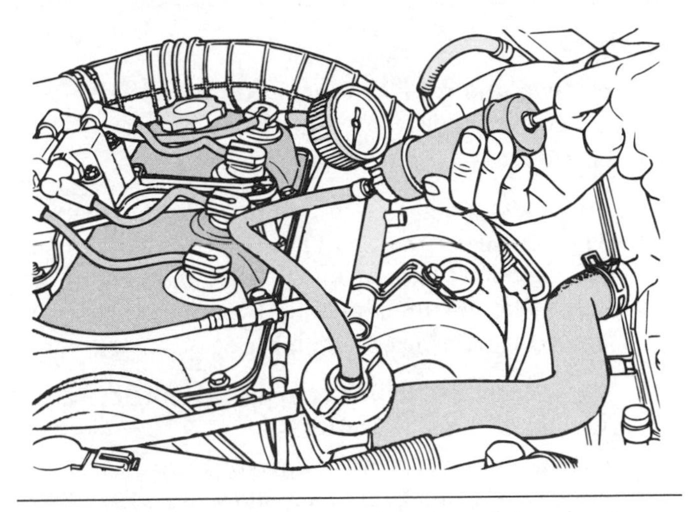

Figure 7–22 Automotive tester. *(Courtesy of Chrysler Corporation)*

102. When refilling an empty cooling system:
    A. always add the antifreeze before adding the water.
    B. always add the water before adding the antifreeze.
    C. always premix the antifreeze and water.
    D. always follow the manufacturer's instructions for bleeding the system.

103. *Technician A* says that many trucks are equipped with an engine shutdown alarm to warn the driver if coolant level and temperature are not within preset parameters. *Technician B* says that engine shutdown can be overridden for a brief period after the alarm to allow the driver to move the vehicle to a safe place. Who is right?
    A. A only
    B. B only
    C. Both A and B
    D. Neither A nor B

Hint

The Technician should inspect the radiator at every scheduled (PMI) maintenance service. You check the radiator for bent fins, kinked or cracked tubes, and leaks. With the engine at normal operating temperature, the temperature of the radiator core should be uniform. Cool spots indicate clogged tubes in the core. The pressure cap is tested using a cooling system pressure tester for pressure and vacuum. If the pressure cap seal sealing gasket or seat is damaged, the engine will overheat and coolant is lost in the recovery system. Replace the cap if it does not hold the pressure specified by the manufacturer or if there is any sign of physical damage. The coolant surge tank should be checked for leaks and for sediment buildup, and should be repaired or cleaned as necessary.

Task 6

Inspect, test, and replace water pump and drive system.

104. The LEAST likely cause of poor coolant circulation is:
 A. a defective thermostat.
 B. an eroded water pump impeller.
 C. a collapsed upper radiator hose.
 D. a collapsed lower radiator hose.

105. The water pump should be replaced any time:
 A. the fan clutch is replaced.
 B. the heater hoses are replaced.
 C. there is a small leak from the weep hole.
 D. the thermostat sticks closed.

Hint

The water pump has a small weep hole directly below its drive shaft. The water pump must be replaced if there is any sign of rust or coolant leaking from the weep hole. Any lateral free play in the water pump drive shaft or a growling noise from the front of the water pump indicates a worn bearing, and warrants replacement of the water pump. Inadequate coolant conditioner level can lead to cavitation corrosion of the water pump impeller, and cause poor coolant circulation. On air clutch applications, problems in the truck air system may be the cause. If the truck's air brakes operate normally, check at the pressure protection valve.

Task 7

Inspect, test, and replace thermostats, bypasses, housings, and seals.

106. All of the following are good methods of verifying that the thermostat opens EXCEPT:
 A. feeling the upper radiator hose.
 B. watching the temperature gauge.
 C. watching for motion in the upper radiator tank.
 D. watching the surge tank.

107. When replacing a thermostat, the side with the spring:
 A. must be installed facing the housing.
 B. must be installed facing forward.
 C. must be installed facing the heater core.
 D. must be installed facing the engine block.

Hint

The only accurate way to test a thermostat is to remove it from the vehicle, place it in a container of water with a thermometer, and heat the container until the thermostat opens. The water temperature at the point when the thermostat opens should equal the manufacturer's specification for engine operating temperature. Other indications that the thermostat has opened include visible coolant flow in the upper radiator tank, a hot upper radiator hose, and an engine temperature gauge that indicates normal operating temperature.

Task 8

Flush and refill cooling system; bleed air from system.

108. Refer to Figure 7–23. *Technician A* says the tool shown in the figure may be used to test the amount of contaminants in the coolant. *Technician B* says when more antifreeze is added to the coolant, the boiling point of the coolant solution is decreased. Who is right?
 A. A only
 B. B only
 C. Both A and B
 D. Neither A nor B

109. *Technician A* says a cooling system containing only water causes more rust formation in the system compared to a cooling system with an antifreeze and water solution. *Technician B* says the coolant is more likely to boil with only water in the system compared to a system containing a 50/50 water and antifreeze solution. Who is right?
 A. A only
 B. B only

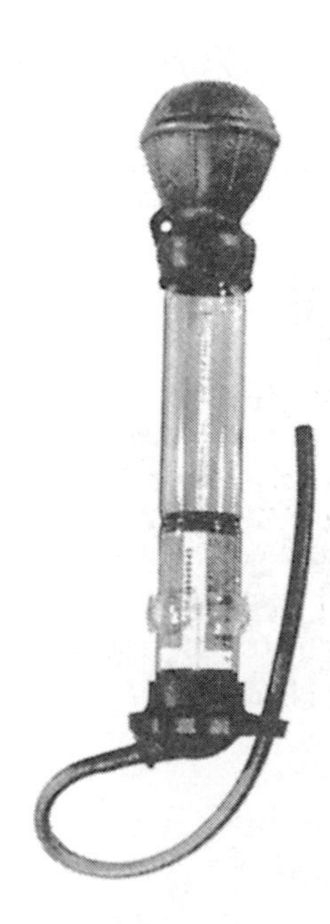

Figure 7–23 Cooling system service tool. *(Courtesy of Mac Tools)*

C. Both A and B
D. Neither A nor B

Hint
The cooling system should be flushed if there is any sign of rust or contamination in the coolant. After flushing, the entire system should be drained and fresh coolant should be added. The Technician should consult the service manual to verify cooling system capacity and bleeding procedure.

Task 9

Inspect, test, and repair or replace coolant conditioner/filter; check valves, lines, and fittings.

110. *Technician A* says the coolant conditioner cartridge is self-cleaning and does not require replacement. *Technician B* says the coolant conditioner has no effect on cavitation corrosion of cylinder liners. Who is right?
   A. A only
   B. B only
   C. Both A and B
   D. Neither A nor B

111. The component shown in Figure 7–24 is a/an:
   A. oil filter.
   B. coolant filter.
   C. transmission filter.
   D. power steering filter and cooler.

Hint
The coolant conditioner cartridge should be replaced during regularly scheduled maintenance services. The coolant conditioner internally lubricates cooling system components, maintains the neutrality of the coolant, filters particulate from the coolant, and prevents cavitation corrosion of the cylinder liners.

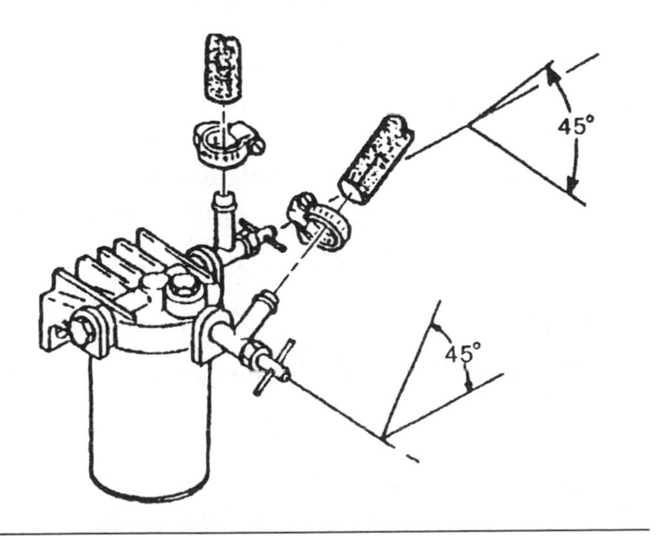

Figure 7–24 Engine accessory component. *(Courtesy of General Motors Corporation)*

Task 10

Inspect, test, and replace fan, fan hub, fan clutch, fan controls, fan thermostat, and fan shroud.

112. A cooling fan clutch can be controlled by any of the following EXCEPT:
   A. air.
   B. a thermostatic spring.
   C. changing viscosity of fan clutch fluid.
   D. a hydraulic switch.

113. When the engine cooling fan clutch is disengaged:
   A. the fan blade will remain stationary.
   B. the engine idle will drop.
   C. the radiator shutters must be closed.
   D. the fan blade may freewheel at a reduced speed.

114. An electric cooling fan motor can be controlled by any of the following EXCEPT:
   A. an electronic relay.
   B. an independent electronic module.
   C. a multiplexed electronic module.
   D. an air solenoid controller.

Hint
The Technician should inspect engine cooling fan for loose, cracked, or otherwise damaged blades. Further, the Technician should inspect the fan hub for cracks. The cooling fan should be replaced, not repaired, if any damage is found. The cooling fan clutch may be a viscous-type clutch or may be operated by a thermostatic spring. Some heavy-duty trucks use a computer-controlled clutch operated by either the chassis air system or engine oil hydraulic controls. The fan shroud should be inspected for cracks, and replaced as necessary. If fan blade and shroud damage is found, the Technician should verify that the engine mounts are in good condition before replacing the shroud and blade.

Task 11

Inspect, test, and replace heating system coolant control valve(s).

115. *Technician A* says that the best way to test a vacuum-operated coolant control valve is to disconnect the vacuum hose and observe whether coolant flows through it. *Technician B* says that the best way to test a vacuum-operated coolant

control valve is to apply and release vacuum and check if the valve arm moves freely both ways. Who is right?
A. A only
B. B only
C. Both A and B
D. Neither A nor B

116. In Figure 7–25 vacuum is supplied from a hand pump to the water valve solenoid and battery voltage is furnished to the solenoid terminals, resulting in an audible click. The system holds 16 inches of mercury. The water valve does not move. The cause of this problem could be:
A. an open winding in the solenoid.
B. a plugged vacuum hose between the solenoid and valve.
C. a vacuum leak in the solenoid.
D. a seized plunger in the water valve control solenoid.

117. Automatic temperature control systems may use a coolant control valve that is operated using any of the following EXCEPT:
A. chassis air pressure.
B. vacuum.
C. a cable.
D. a magneto.

Hint

The coolant control valve (hot water valve) controls the flow of coolant through the heater core. The coolant control valve may be vacuum operated or cable operated. One can verify proper valve operation by manually opening and closing the valve and observing the temperature change in the downstream heater hose.

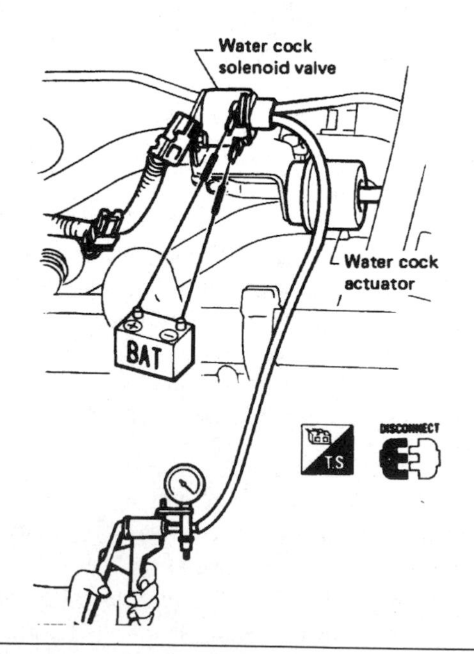

Figure 7–25 Heater coolant control valve. (*Courtesy of Nissan North America*)

Task 12

Inspect, flush, and replace heater core.

118. When replacing the heater core, foam tape is used:
A. to insulate the heater core.
B. to reduce noise from coolant surges.
C. to protect and seal around the heater core.
D. to seal the heater hose connections.

119. The driver of a truck discovers a heater core leak in the sleeper that needs replacement. *Technician A* says that it is important to add the proper amount of refrigeration oil before installation. *Technician B* says that a PAG-based lubricant is used in modern heater cores. Who is right?
A. A only
B. B only
C. Both A and B
D. Neither A nor B

120. While testing a heater core for leaks, you apply 10 psi of air pressure. The pressure bleeds to 5 psi in 3 minutes. You should conclude that:
A. the core is satisfactory.
B. the core may require repair in future.
C. the core leaks and needs repair now.
D. this is not a valid test for a leaking heater core.

Hint

In a poorly maintained cooling system, sediment may build up in the heater core causing poor heater performance. Flushing the heater core will restore heater efficiency and may reveal small heater core leaks. The heater core can be pressure tested independently from the rest of the cooling system; however, the core should never be pressurized in excess of normal operating pressure. When replacing a heater core, it is important to reinstall all foam mounting insulators to minimize the risk of vibration damage and to ensure a good seal around the heater core.

Task 13

Inspect and repair or replace radiator shutter assembly and controls.

121. *Technician A* says if the radiator shutter assembly does not close, the high-side pressure will be excessive. *Technician B* says if the radiator shutter assembly does not close, the engine takes longer to warm up. Who is right?

A. A only
B. B only
C. Both A and B
D. Neither A nor B

122. The shutters do not close completely. *Technician A* says the shutter fuse link may have an open circuit. *Technician B* says there may be an air leak in the shutter cylinder. Who is right?
A. A only
B. B only
C. Both A and B
D. Neither A nor B

Hint

Many heavy-duty trucks are equipped with a radiator shutter system to facilitate engine warmup and improve cold weather performance. Most shutter systems operate on chassis air and control occurs from the body or engine control unit. When the engine is started cold, the shutters should be fully closed. As the engine coolant temperature rises, the shutters begin to open. At normal operating temperature, the radiator shutters should be fully open.

Operating Systems and Related Controls Diagnosis and Repair

Electrical

Task 1

Diagnose the cause of failures in HVAC electrical control systems; determine needed repairs.

123. The best tool to use when troubleshooting a circuit with solid-state components is:
A. a digital multimeter (DMM).
B. a self-powered test lamp.
C. an analog volt/ohmmeter.
D. a 12V test lamp.

124. *Technician A* says that the HVAC systems on most medium-duty trucks with gasoline engines operate totally independent from the engine control system. *Technician B* says that most of the HVAC systems on these vehicles provide inputs and receive outputs from the engine control unit. Who is right?
A. A only
B. B only
C. Both A and B
D. Neither A nor B

125. A blown HVAC system fuse could indicate any of the following EXCEPT:
A. a short circuit to ground in the blower circuit.
B. a short circuit to ground in the blend door actuator.
C. a short circuit in the engine coolant temperature sensor (ECT).
D. a damaged wiring harness connector.

Hint

Diagnosing HVAC electrical control system problems is no different from diagnosing other electrical concerns. The Technician should follow a logical approach to troubleshooting, including verifying the concern and performing a thorough visual inspection. The Technician should use all available resources, including electrical schematic diagrams and service manual diagnostic routines, to locate and repair the cause of the concern.

Task 2

Inspect, test, repair, and replace A/C heater blower motors, resistors, switches, relays/modules, wiring, and protection devices.

126. Refer to Figure 7–26. The blower motor works in LO and HIGH positions but does not work on MED. *Technician A* says the problem could be an open blower resistor. *Technician B* says the problem could be the blower motor switch. Who is right?
A. A only
B. B only
C. Both A and B
D. Neither A nor B

127. Refer to Figure 7–26. The blower motor does not operate in any switch position, and 12V are supplied to the switch with the ignition switch on. The cause of this problem could be:
A. an open circuit between the blower motor and ground.
B. an open circuit in the upper resistor in the resistor assembly.
C. an open circuit between the switch and the resistor assembly.
D. an open circuit in the lower resistor in the resistor assembly.

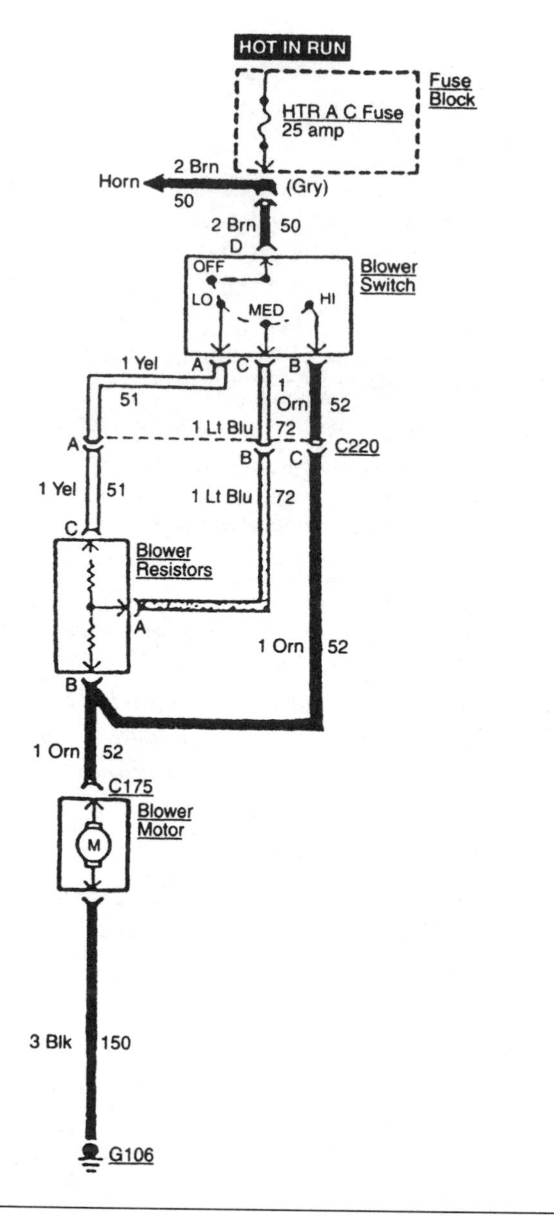

Figure 7–26 Blower motor circuit.

Hint

A 30 amp fuse or circuit breaker generally protects the HVAC blower circuit. Many blower systems are powered by one or more relays. The operator selects the desired blower motor speed by using the blower switch. The blower resistor block contains several resistors in series, and is used to step down the voltage to the blower motor, thereby providing multiple blower speeds. The resistor block usually contains a thermal fuse to prevent blower motor damage in case of a high current draw.

Task 3

Inspect, test, repair, and replace A/C compressor clutch relays/modules, wiring, sensors, switches, diodes, and protection devices.

128. During a performance test, the Technician notices that the A/C compressor clutch is slipping. *Technician A* says that the air gap was probably set improperly. *Technician B* says that the pressure plate needs to be resurfaced. Who is right?
    A. A only
    B. B only
    C. Both A and B
    D. Neither A nor B

129. Refer to Figure 7–27. The compressor clutch is inoperative in the defrost mode but operates properly in all other A/C modes. *Technician A* says the low side of the binary switch may have an open circuit. *Technician B* says the defrost switch contacts may have an open circuit. Who is right?
    A. A only
    B. B only
    C. Both A and B
    D. Neither A nor B

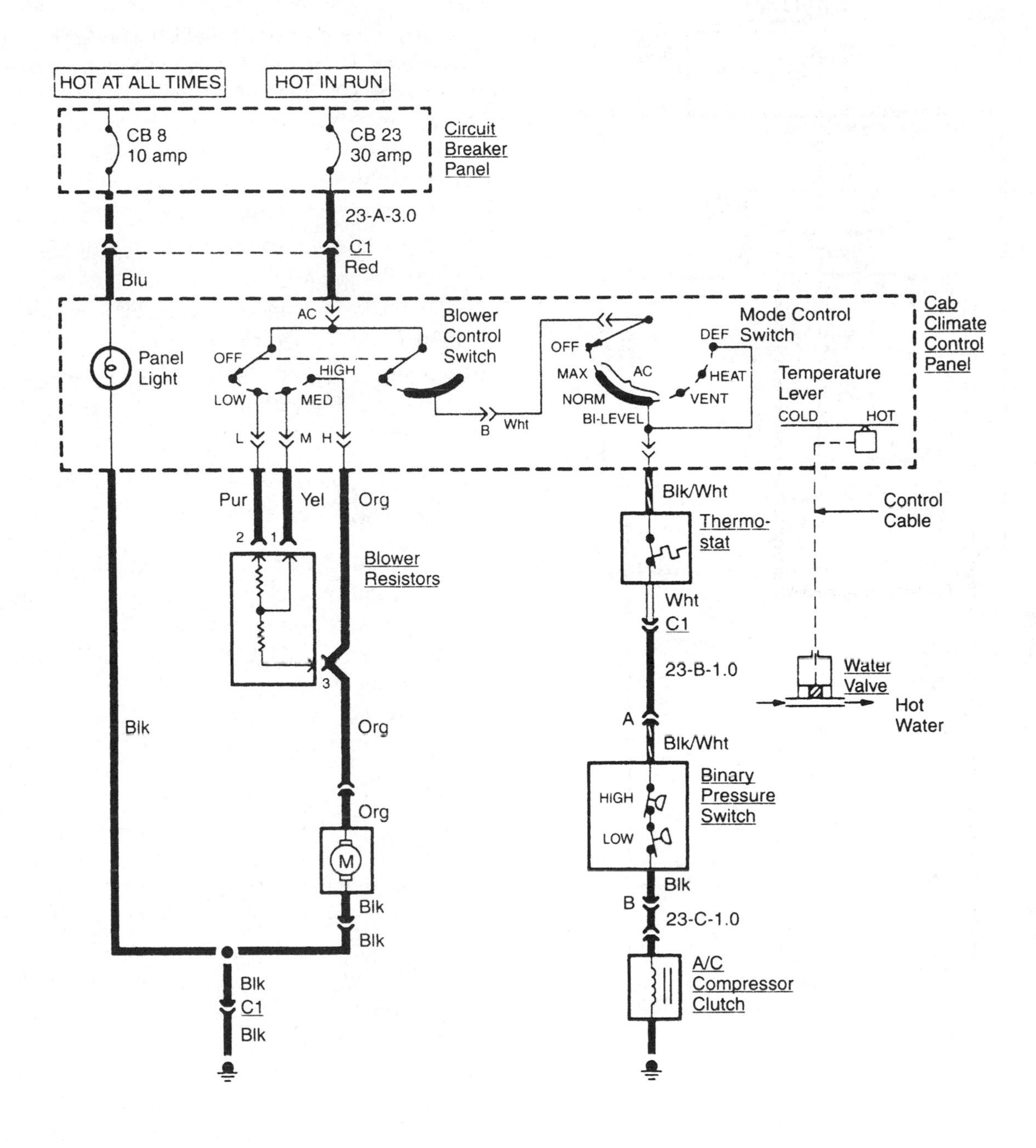

Figure 7–27 Compressor clutch circuit.

130. Refer to Figure 7–28. The ignition switch is on and the A/C switch is in the AUTO position. The ambient temperature is 75°F (24°C), and the compressor clutch is inoperative. There are 12V at terminals B, C, and S on the thermal fuse. The cause of the inoperative compressor clutch could be:
   A. an open thermal fuse.
   B. an open compressor clutch coil.
   C. a defective superheat switch.
   D. a defective thermal fuse heater.

Hint

The A/C compressor clutch coil is generally powered by an electronically controlled relay. The compressor clutch relay may be controlled by the body control unit or by the engine control unit. Many blower motor circuits con-

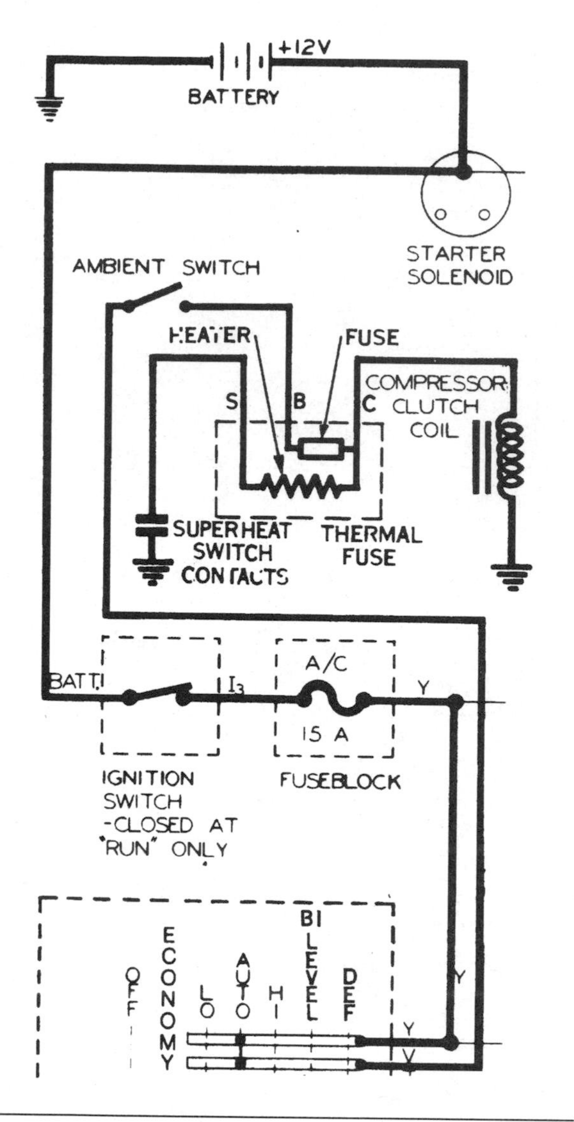

Figure 7–28 Compressor clutch circuit with thermal fuse. *(Courtesy of Cadillac Motor Division, General Motors Corporation)*

nect the motor switch and resistor strings to the ground side of the motor. When the system is on, positive current flows to one brush in the motor, then to the other brush and on to ground through the resistor assembly. When you set the blower to high, the motor brush grounds directly through the motor switch contacts at the high-speed relay. If you select one of the lower speeds, the brush grounds through that specific blower resistor.

Task 4

Inspect, test, repair, replace, and adjust A/C-related engine control systems.

131. A diesel with a distributor-type fuel injection pump and throttle fast idle solenoid experiences a stalling problem when the A/C system compressor is engaged. *Technician A* says the throttle solenoid may be out of adjustment. *Technician B* says the engine computer may not be energizing the fast idle solenoid. Who is right?
   A. A only
   B. B only
   C. Both A and B
   D. Neither A nor B

132. Refer to Figure 7–29. When measuring the voltage drop in the A/C computer ground as shown in the figure, you connect a voltmeter from computer terminal C1–24 internal ground to the external ground. With the ignition on, the maximum voltage drop should be which of these items?
   A. .1 volt
   B. .2 volt
   C. .5 volt
   D. .8 volt

133. A signal to the powertrain control module from which of the following sensors could cause the A/C compressor to disengage?
   A. Engine coolant temperature (ECT) sensor
   B. Intake air temperature (IAT) sensor
   C. Heated oxygen sensor (HO2S)
   D. Cooling fan control sensor

Hint

A/C compressor clutch operation may be dependent on signals from various engine sensors. Faulty engine-related components that affect A/C system operation can usually be diagnosed using a hand-held scan tool or a laptop PC interface. The engine control unit or body control unit will disable the A/C compressor clutch if the engine coolant temperature is too high or if the outside air temperature is too low. On vehicles with an electronically controlled automatic transmission, the compressor clutch can be disengaged briefly during shifts. The radiator shutter system is disabled (shutters are fully open) during A/C system operation.

Task 5

Inspect, test, repair, and replace engine cooling/condenser fan motors, relays/modules, switches, sensors, wiring, and protection devices.

134. An electric condenser fan may be controlled by any of the following EXCEPT:
   A. the engine control unit.
   B. the body control unit.
   C. a manual switch.
   D. an electronic relay.

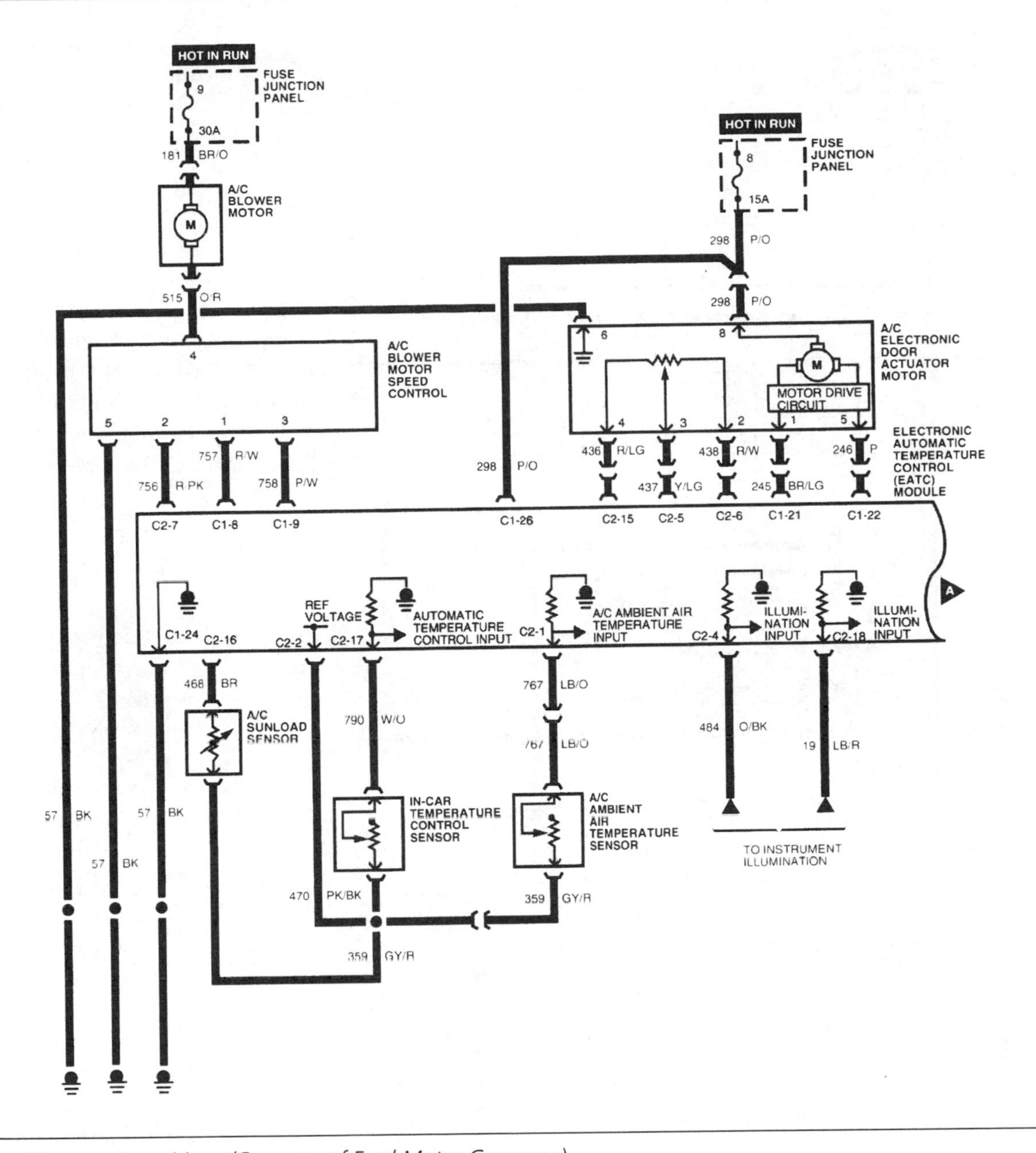

Figure 7–29 A/C computer wiring. *(Courtesy of Ford Motor Company)*

135. *Technician A* says that a cracked fan blade should be welded. *Technician B* says that a cracked fan blade can be repaired with epoxy. Who is right?
    A. A only
    B. B only
    C. Both A and B
    D. Neither A nor B

136. In the electric cooling fan circuit in Figure 7–30, the low-speed and high-speed fans do not operate unless the air conditioner is turned on. *Technician A* says the engine coolant sensor may be defective. *Technician B* says the A/C pressure fan switch may be defective. Who is right?
    A. A only
    B. B only
    C. Both A and B
    D. Neither A nor B

Hint

Some trucks are equipped with electric engine cooling fans and electric condenser fans. Electric fans are usually controlled by an electronic relay, which is in turn controlled by the engine control unit or the body control unit. The engine computer grounds the low- and high-speed fan relays in response to engine coolant temperature and compressor head temperature. When the engine coolant temperature reaches 212°F (100°C), the engine

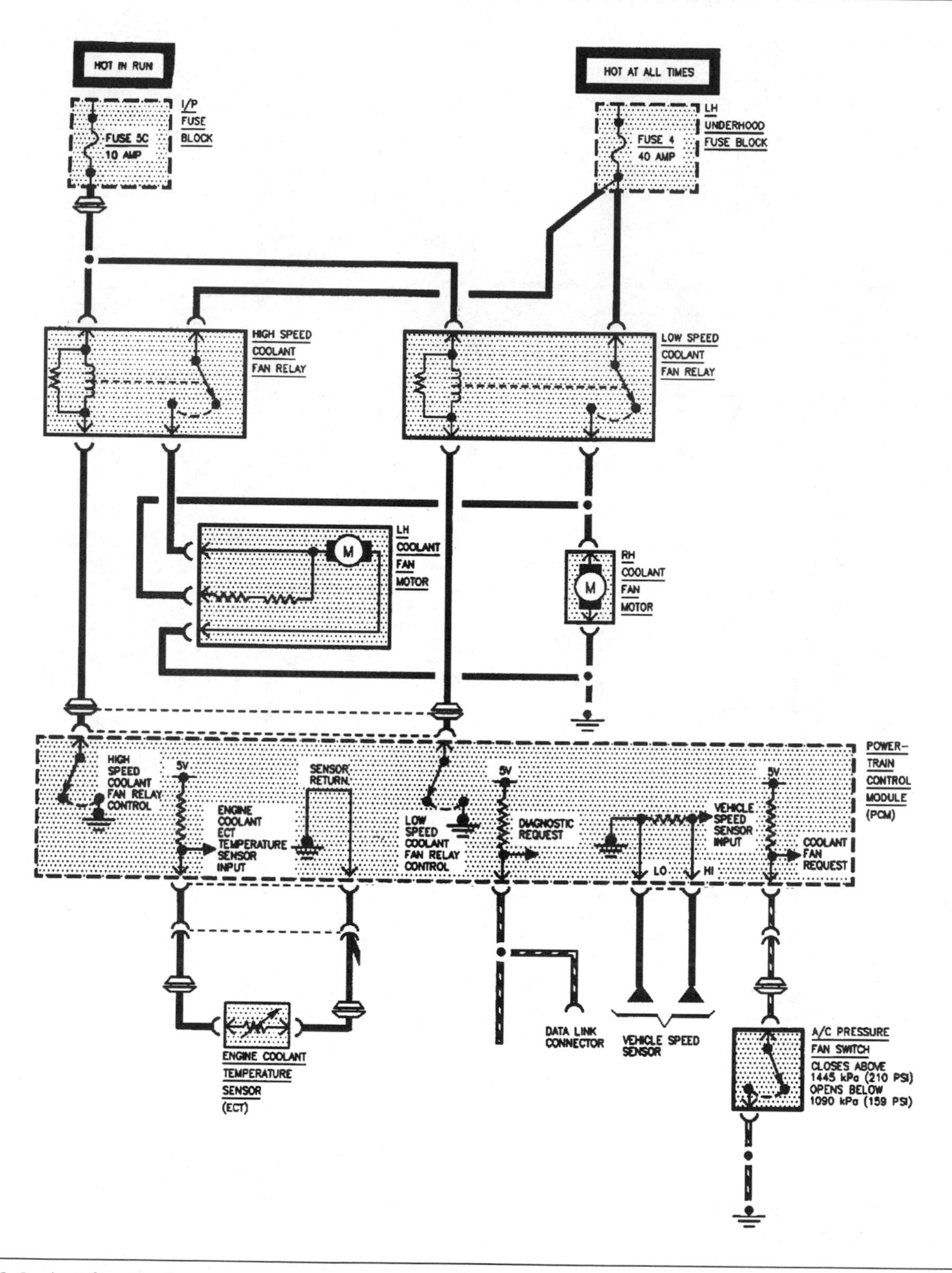

Figure 7–30 Coolant fan circuit.

computer grounds the low speed fan relay. If the coolant temperature reaches 226°F (108°C), the computer grounds the high-speed relay.

Task 6

Inspect, test, adjust, repair, and replace electric actuator motors, relays, modules, switches, sensors, wiring, and protection devices.

137. *Technician A* says that an electronic blend door motor uses a feedback device to indicate the position of the door. *Technician B* says that an electronic mode door motor uses a feedback device to indicate the position of the door. Who is right?
   A. A only
   B. B only
   C. Both A and B
   D. Neither A nor B

138. A vacuum-operated blend door actuator diaphragm:
   A. is designed to bleed vacuum at a rate of 3 in. Hg per hour.
   B. contains a small electric motor to return the actuator to the normal position.
   C. should hold vacuum indefinitely.
   D. is porous to allow moisture to evaporate.

139. An A/C system is operating in the MAX A/C mode. *Technician A* says the outside air door is closed. *Technician B* says the defroster door is open. Who is right?
   A. A only
   B. B only
   C. Both A and B
   D. Neither A nor B

Hint

Some HVAC systems control blend air and mode doors with electronic actuators. Electronic actuators contain a small motor, a geartrain, and feedback device to indicate the position of the controlled door to the controlling processor. Some electronic systems contain self-diagnostic abilities with diagnostic trouble codes (DTC). You typically diagnose these with a scan tool or laptop computer. On some systems, you can adjust the actuator doors.

Task 7

Inspect, test, service, or replace HVAC system electrical control panel assemblies.

140. All of the following statements about air-controlled HVAC systems in heavy-duty trucks are true EXCEPT:
   A. the replacement of cables with small air lines facilitates the replacement of the HVAC control panel.
   B. coolant control valves cannot be controlled using chassis air.
   C. air cylinders are used to control mode and blend air doors.
   D. air leaks may cause mode doors to move slowly or to be totally inoperative.

141. An electrical HVAC control panel must be replaced in a medium-duty truck with a supplemental restraint system. *Technician A* says to disconnect the battery and wait for the time specified by the truck manufacturer before removing the control panel. *Technician B* says the refrigerant must be discharged from the A/C system before removing the control panel. Who is right?
   A. A only
   B. B only
   C. Both A and B
   D. Neither A nor B

Hint

Electrical control panel assemblies for manual and semi-automatic HVAC systems are modular in design, allowing for replacement of individual switches and illumination bulbs without replacing the entire panel. Most electronic ATC control panels allow the Technician access only to replace illumination bulbs.

Air/Vacuum/Mechanical

Task 1

Diagnose the cause of failures in HVAC air, vacuum, and mechanical switches and controls; determine needed repairs.

142. Which of the following is the LEAST likely cause of a binding temperature control cable?
   A. A kinked cable housing
   B. Corrosion in the cable housing
   C. A deformed or overtightened cable clamp
   D. A faulty mode door

143. *Technician A* says you test control panel vacuum systems by applying vacuum with a hand pump to the upstream (output) end of the system. *Technician B* says when you test a vacuum actuator, the vacuum pump gauge should stay at a steady vacuum for 1 minute with 15 to 20 in. Hg supplied. Who is right?
   A. A only
   B. B only
   C. Both A and B
   D. Neither A nor B

Hint

The most common cause of failures in HVAC air and vacuum systems is leaking hoses and diaphragms. Air and vacuum leaks can often be located by listening for a hissing noise. To locate minor air leaks, brush a mild soap solution over fittings and connections and watch for bubbles. To check for vacuum leaks using a hand-held vacuum pump, (Mitivac) supply 20 in. Hg to one end of the vacuum hose and the other end is plugged or attached to its device. The hose should hold 15 to 20 in. Hg, without leaking.

Task 2

Inspect, test, service, or replace HVAC system air/vacuum/mechanical control panel assemblies.

144. A Technician finds that the HVAC control panel is the source of a chassis air leak. The best method of repair is:
   A. replacement of the control panel.
   B. replacement of the pintle O-rings.
   C. repacking the selector body with grease.
   D. replacing the selector levers.

145. A HVAC system with a vacuum control panel experiences no cold air out of the dash nozzles, only out of the heat outlets. Which of these items is the most likely cause?
   A. A leaking dash vacuum switch
   B. A defective A/C compressor
   C. Loss of vacuum to the control panel
   D. A defective heater control valve

Hint

HVAC vacuum and mechanical control panel assemblies require very little testing and maintenance. A leaking vacuum switch will hiss in one or more positions. Broken control cable arms or anchors will result in ineffective control levers.

Task 3

Inspect, test, adjust, or replace HVAC system air/vacuum/mechanical control cables and linkages.

146. What adjustment is being performed in Figure 7–31?
   A. The air mix door adjustment
   B. The manual coolant valve adjustment
   C. The ventilation door control rod adjustment
   D. The defroster door control rod adjustment

147. Refer to Figure 7–32. When adjusting the temperature control cable, *Technician A* says the temperature control lever must be in the maximum heat position. During this adjustment, *Technician B* says the black cable attaching flag must be removed from the flag receiver. Who is right?
   A. A only
   B. B only
   C. Both A and B
   D. Neither A nor B

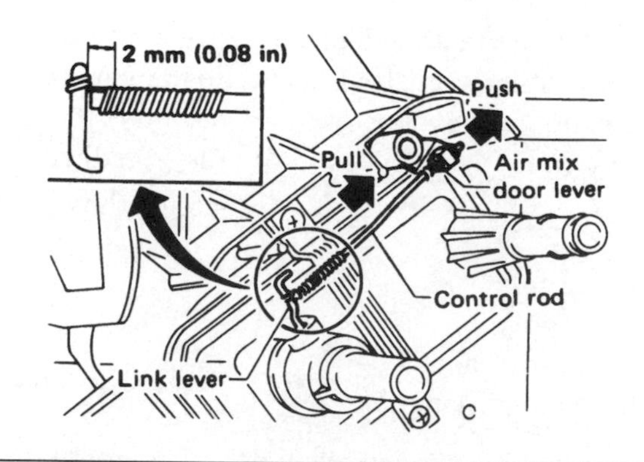

Figure 7–31 A/C system adjustment.

Hint

A Technician replaces HVAC control cables if they are kinked or seized due to internal corrosion. One must adjust control cables to allow a full range of motion for the control lever and for the output device.

Task 4

Inspect, test, and replace HVAC system vacuum actuators (diaphragms/motors) and hoses.

148. Before replacing an electric blend air door actuator, the Technician should:
   A. ensure that the batteries are removed from the vehicle.
   B. ensure that the batteries have a good ground.

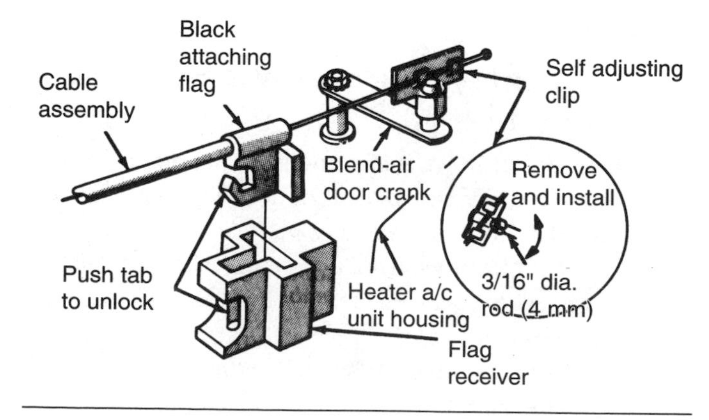

Figure 7–32 A/C system cable adjustment. *(Courtesy of Chrysler Corporation)*

C. ensure that the blend door moves freely.
D. ground himself or herself to the vehicle.

149. All of these facts about a computer-controlled A/C system are correct EXCEPT:
A. some actuator motors are calibrated automatically in the self-diagnostic mode.
B. A/C diagnostic trouble codes (DTC) represent a fault in a specific component.
C. the actuator control rods must be calibrated manually on some systems.
D. the actuator motor control rods should only require adjustment after motor replacement or adjustment.

Hint

You can evaluate the performance of vacuum actuators and hoses by using a hand-held vacuum pump. Vacuum systems are tested by applying vacuum to the upstream (input) end of the system while individual components must be tested at the component connection. Connect the vacuum pump to each vacuum actuator and supply 15 to 20 in. Hg to the actuator. Check the vacuum actuator rod to be sure it moves freely. Close the vacuum pump valve and observe the vacuum gauge. The gauge reading should remain steady for at least 1 minute. If the gauge reading drops slowly, the actuator is leaking. You replace any hoses or components that do not hold vacuum or do not operate properly.

Task 5

Identify, inspect, test, and replace HVAC system vacuum reservoir(s) check valve(s) and restrictors.

150. A check valve is installed in-line to the vacuum reservoir to:
A. delay vacuum to downstream components.
B. switch vacuum on and off to various components.
C. prevent a vacuum drop during periods of low source vacuum.
D. monitor engine vacuum.

151. The driver of a gasoline-powered truck notices that when the engine is turned off or the truck goes uphill, the air coming out of the A/C dash nozzles quickly shifts to the heater mode. *Technician A* says the A/C system may have a leaking actuator diaphragm. *Technician B* says the reservoir check valve may be defective. Who is right?
A. A only
B. B only
C. Both A and B
D. Neither A nor B

Hint

In HVAC systems that use vacuum switches and actuators, a vacuum reservoir and check valve are installed between the vacuum source and the control panel. The check valve is a one-way valve that allows the reservoir to hold constant vacuum regardless of fluctuations in the vacuum source. The vacuum reservoir supplies vacuum at a consistent level during periods of low source vacuum (during long uphill runs or engine lugging). The check valve must be replaced if it leaks or if it passes vacuum in both directions. The reservoir must be replaced if it will not hold vacuum. When vacuum is supplied with vacuum pump to the reservoir, it should hold 15 to 20 in. Hg.

Task 6

Identify, inspect, test, and replace air pressure regulator valve, lines, and hoses.

152. Before removing any chassis air hose, the Technician must:
A. start the engine.
B. drain water from the air system.
C. drain all of the air from the system.
D. remove the air compressor from the vehicle.

153. All of the following can cause the air pressure regulator to "pop-off" frequently EXCEPT:
A. a defective compressor unloader assembly.
B. a faulty automatic water drain.
C. a weak regulator control spring.
D. a defective compressor governor.

Hint

A compressor governor or an internal unloader assembly generally regulates the pressure in the chassis air system. If a governor or unloader fails, a pressure relief valve will release air from the system to prevent damage to system components. The pressure relief valve will operate at about 130 psi (896 kPa). Before replacing any chassis air system component, the system must be drained of air pressure.

Task 7

Inspect, test, adjust, repair or replace HVAC system ducts, doors, outlets.

154. The outlet air recirculation door is stuck in position A in Figure 7–33. *Technician A* says under this condition, outside air is drawn into the HVAC case. *Technician B* says under this condition, some in-vehicle air leaks past the doors. Who is right?

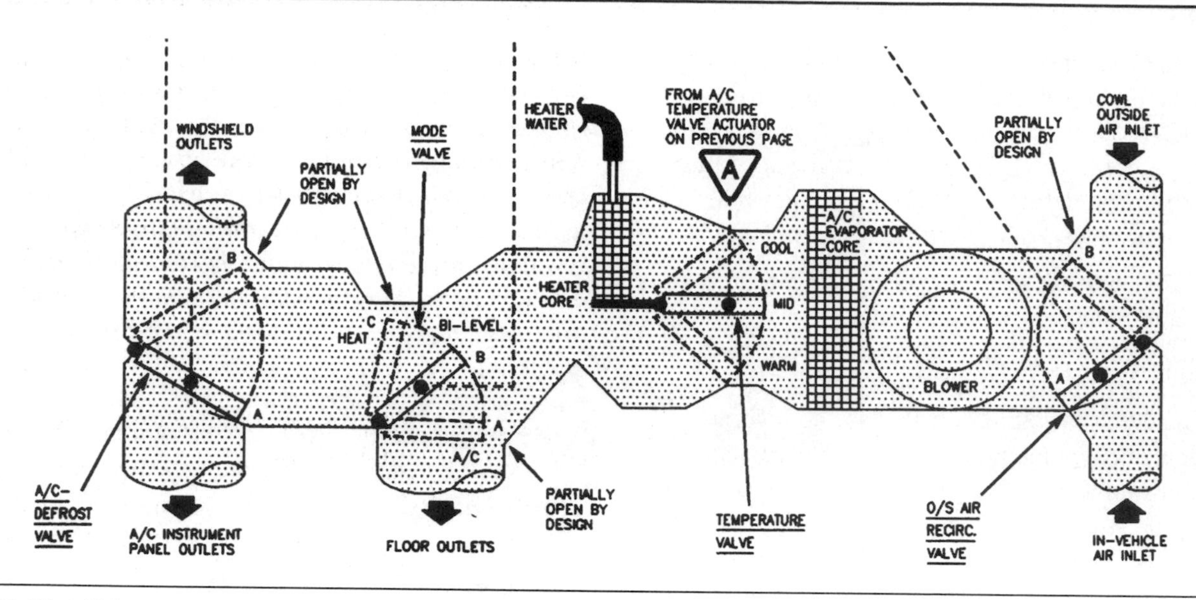

Figure 7–33 A/C-heater case.

A. A only
B. B only
C. Both A and B
D. Neither A nor B

155. Which of the following modes will cause very slow windshield defrosting in the defrost mode?
A. 50 percent of the output air is directed to the floor and 50 percent to the windshield.
B. 15 percent of the output air is directed to the floor and 85 percent to the windshield.
C. 25 percent of the output air is directed to the floor and 75 percent to the windshield.
D. 30 percent of the output air is directed to the floor and 70 percent to the windshield.

Hint

Misaligned or improperly installed HVAC ducts will cause reduced levels of system output air. Blend air and mode control doors must be adjusted to allow a full range of motion when controls are operated.

Automatic Temperature Control

Task 1

Diagnose automatic temperature control system problems; determine needed repairs.

156. A typical ATC system will delay blower motor operation until the coolant temperature reaches:
A. 90°F (32.2°C).
B. 70°F (21.1°C).
C. 100°F (37.7°C).
D. 120°F (48.8°C).

157. *Technician A* says most fully automatic temperature control systems have some form of self-diagnostic program that will display trouble codes. *Technician B* says that depending on the system design, these codes may be displayed digitally on the control assembly or on a hand-held scan tool. Who is right?
A. A only
B. B only
C. Both A and B
D. Neither A nor B

Hint

Most ATC systems provide internal diagnostic capabilities. On some ATC systems, diagnostic trouble codes may be displayed digitally on the control panel, while on other systems a hand-held scan tool or PC interface must be used to retrieve codes. The Technician should always refer to diagnostic routines in the vehicle service manual when attempting to troubleshoot ATC codes. Most importantly, the Technician must rule out the possibility of mechanical failures before searching for electronic malfunctions.

Task 2

Inspect, test, adjust or replace climate control temperature sensors.

158. In an ATC A/C system the temperature control is set at 70°F (21°C), and the in-car temperature is 80°F (27°C) after driving one hour. All refrigerant pressures are normal. *Technician A* says the in-car sensor may be defective. *Technician B* says the temperature blend door may be sticking. Who is right?
   A. A only
   B. B only
   C. Both A and B
   D. Neither A nor B

159. A diagnostic trouble code (DTC) representing the ambient sensor occurs in the circuit in Figure 7–34. The ambient sensor and connector 2 and 3 are connected and connector 7 is disconnected from the ATC control panel. An ohmmeter connected to terminals 9 and 18 in the control panel connector indicates the specified resistance. The most likely cause of this DTC is:
   A. a defective ambient sensor.
   B. a defective A/C control panel.
   C. a loose connection at connector 2 and 3.
   D. an open connection at connector 7 terminal 18.

Hint

ATC systems rely on a variety of sensors to provide feedback to the ATC control unit. The control unit uses the sensor signals to determine how much heating or cooling is required to maintain the desired cab or sleeper temperature. The ambient temperature sensor monitors outside air temperature. The engine coolant temperature sensor monitors engine coolant temperature. The ATC temperature (or interior temperature) sensor monitors the temperature of the air in the cab or the sleeper box. A sunlight sensor is used on some vehicles to monitor the intensity of the light coming through the windshield.

Task 3

Inspect, test, adjust, and replace temperature blend door/power servo system.

160. In an ATC system, the blend air door constantly moves back and forth. The most likely cause of this problem is:
   A. a binding blend air door.
   B. a faulty actuator motor.
   C. an improperly adjusted ATC sensor.
   D. a faulty actuator assembly feedback device.

161. When diagnosing a computer-controlled ATC A/C system, a diagnostic trouble code is obtained indicating a fault in the temperature blend door actuator motor. *Technician A* says the first step in the repair process is to replace the temperature blend door actuator. *Technician B* says you need to check the temperature blend door for binding. Who is right?

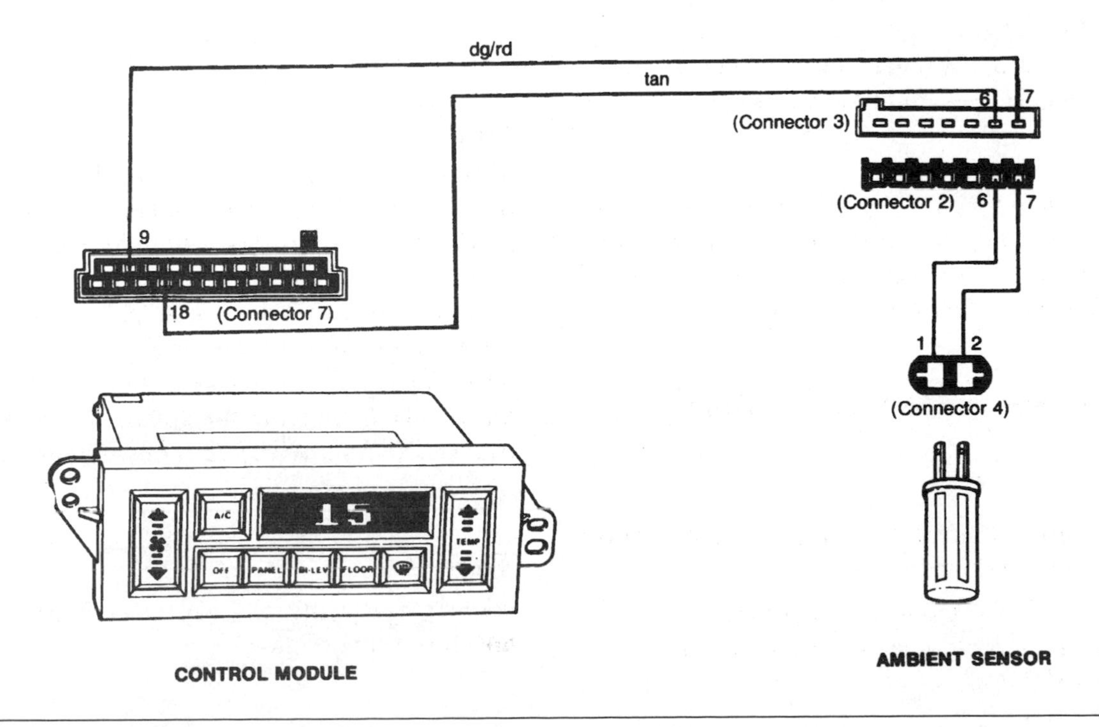

Figure 7–34 ATC control panel and ambient sensor. *(Courtesy of Chrysler Corporation)*

A. A only
B. B only
C. Both A and B
D. Neither A nor B

Hint

Most ATC systems achieve temperature modulation by blending air that has passed through the A/C evaporator core with air that has passed through the heater core. The volume of air that is allowed to pass through each core is regulated by a blend air door. The blend air door is controlled by the blend door actuator, which consists of an electric motor, a gear-train, and a feedback device. The feedback device provides precise information about the position of the blend air door to the control unit. The blend door actuator must be replaced if the drive gears are worn or damaged, if the motor develops a dead spot, or if the feedback device fails. A faulty feedback device will cause the motor to either "hunt" for the desired position to be inoperative.

Task 4

Inspect, test, and replace heater water valve and controls.

162. When replacing the coolant control valve:
A. the Technician must remove the engine thermostat.
B. the Technician does not need to drain the entire cooling system.
C. the Technician must replace the control cable.
D. there is no need to bleed the system.

163. An ATC system in a heavy-duty truck contains the type of coolant control valve in Figure 7–35. There is no heat discharged from the under-dash heater in any mode. *Technician A* says the engine thermostat may be defective. *Technician B* says there may be no vacuum supply to the coolant control valve. Who is right?
A. A only
B. B only
C. Both A and B
D. Neither A nor B

Hint

Some ATC systems use a coolant control valve to regulate the flow of coolant through the heater core. The coolant control valve on ATC systems is usually vacuum controlled and is not adjustable. A faulty coolant control valve can generally be diagnosed visually.

Task 5

Inspect, test, and replace electric, air, and vacuum motors, solenoids, and switches.

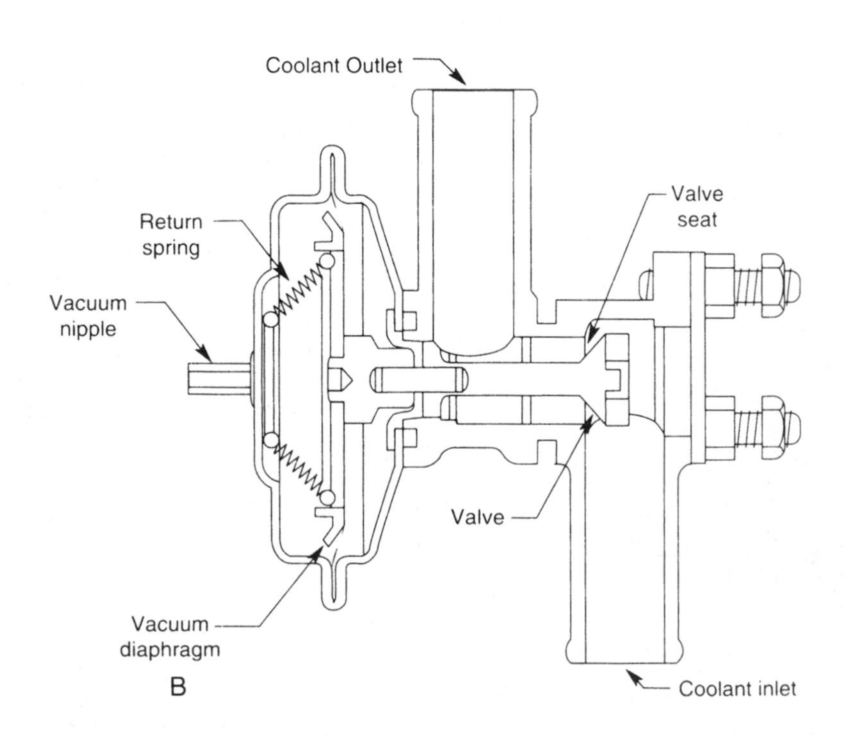

Figure 7–35 Coolant control valve.

164. In Figure 7–36 a vacuum hand pump is used to supply 20 in. Hg to the defrost door vacuum actuator, but the actuator rod does not move. When the vacuum valve on the gauge is closed, the vacuum reading remains steady for 2 minutes and then decreases slightly. *Technician A* says the defrost door may be stuck. *Technician B* says the vacuum actuator diaphragm is leaking. Who is right?
    A. A only
    B. B only
    C. Both A and B
    D. Neither A nor B

165. All of the following statements about the chassis air system are true EXCEPT:
    A. it is important to keep water drained from the system.
    B. air from the system can be used to operate the fan clutch.
    C. air from the system can be used to operate the radiator shutters.
    D. the chassis air system is integral with the air-operated fan clutch.

166. The most useful tool for diagnosing electric actuators and solenoids in an ATC system is:
    A. a 12-volt testlight.
    B. a self-powered testlight.
    C. a hand-held scan tool.
    D. an A/C system charging station.

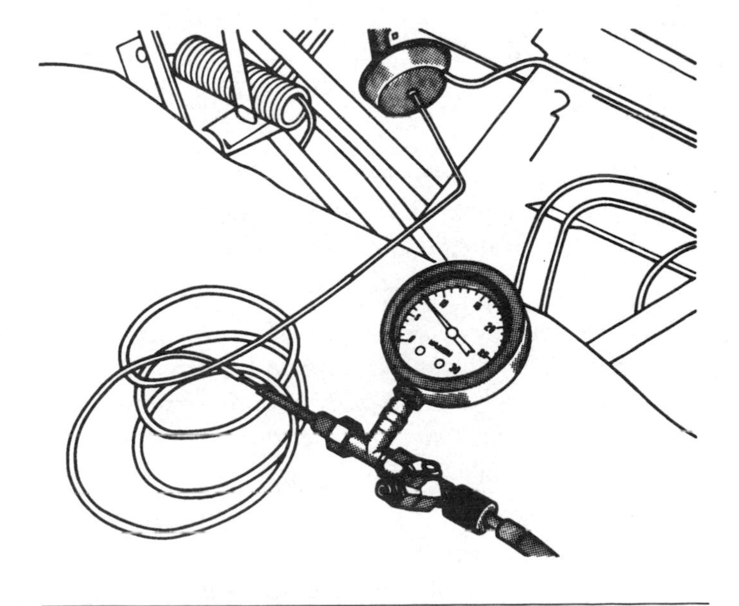

Figure 7–36 Vacuum actuator diagnosis.

Hint

In most ATC systems, the mode doors are controlled using electronic actuators that are similar to the blend door actuator. Faulty mode door actuators are usually diagnosed by following diagnostic routines in the vehicle service manual.

Task 6

Inspect, test, and replace ATC control panel.

167. *Technician A* says that many ATC systems use a microprocessor that is built into the control panel. *Technician B* says that some ATC systems use a microprocessor that can be replaced independently from the control panel. Who is right?
    A. A only
    B. B only
    C. Both A and B
    D. Neither A nor B

168. An ATC control panel is suspected of causing an A/C performance problem in a tractor. *Technician A* says you should replace the ATC control computer. *Technician B* says you should follow and perform the diagnostic steps in the truck manufacturer service manual. Who is right?
    A. A only
    B. B only
    C. Both A and B
    D. Neither A nor B

Hint

The only serviceable components in most ATC control panels are illumination bulbs. Faulty ATC control panels are generally diagnosed by following diagnostic routines in the vehicle service manual.

Task 7

Inspect, test, adjust or replace ATC microprocessor (climate control computer/programmer).

169. One segment of the digital readout on an ATC control panel is inoperative. *Technician A* says that the control panel should be replaced. *Technician B* says that the voltage supply to the control panel should be checked. Who is right?
    A. A only
    B. B only
    C. Both A and B
    D. Neither A nor B

170. When diagnosing an ATC system, all these statements are true EXCEPT:

A. when a cold engine is started, the blower motor starts immediately with the controls in the normal and floor positions.
B. before replacing an ATC computer, the computer ground should be tested.
C. some input voltages are relayed via data links from the engine computer to the ATC computer.
D. the blower motor may be controlled with a pulse width modulated (PWM) voltage signal.

Hint

The ATC control unit may be integral to the control panel or it may be a "stand-alone" component. In either event, the control unit is generally multiplexed to the engine and/or body control units, thereby providing electronic diagnostic capabilities in case of control unit failure.

Task 8

Check and adjust calibration of the ATC system.

171. *Technician A* says that the ATC control panel can be reprogrammed using a hand-held scan tool. *Technician B* says that a hand-held scan tool can be used to help diagnose ATC system failures. Who is right?
A. A only
B. B only
C. Both A and B
D. Neither A nor B

172. *Technician A* says you check the calibration of an ATC system using A/C gauges only. *Technician B* says you can recalibrate all ATC systems in the field. Who is right?
A. A only
B. B only
C. Both A and B
D. Neither A nor B

Hint

The easiest way to check the calibration of the ATC system is to use a small thermometer to monitor the interior temperature of the cab and sleeper and to compare the actual temperature to the temperature setting on the control panel. In most ATC applications, deviations from the set temperature are usually caused by a faulty sensor.

Refrigerant Recovery, Recycling, and Handling

Task 1

Maintain and verify correct operation of certified equipment.

173. *Technician A* says shops that own approved refrigerant recycling equipment may allow anyone in the shop to operate the equipment. *Technician B* says that shops that own approved refrigerant recycling equipment must maintain records of any facility to which the refrigerant is sent. Who is right?
A. A only
B. B only
C. Both A and B
D. Neither A nor B

174. Which of the following gases is most important for shielding the earth from ultraviolet radiation?
A. Methane
B. Stratospheric ozone
C. Nitrogen
D. Argon

175. Which of these characteristics does the R-134a refrigerant possess?
A. Is odorless
B. Has a faint ether-like odor
C. Has a strong rotten egg odor
D. Has a cabbage-like odor

Hint

The Clean Air Act (CAA) establishes the following rules for recordkeeping and operation of certified refrigeration service equipment:

1. *"Any person who owns approved refrigerant recycling equipment certified under the act must maintain records of the name and address of any facility to which refrigerant is sent."*
2. *"Any person who owns approved refrigerant recycling equipment must retain records demonstrating that all persons authorized to operate the equipment are certified under the act."*
3. *"Public Notification: Any person who conducts any retail sales of a Class I or Class II substance must prominently display a sign that reads: 'It is a violation of federal law to sell containers of Class I and Class II refrigerant of less than 20 pounds of such refrigerant to any one who is not properly trained and certified.'"*

4. *"Any person who sells or distributes any Class I or Class II substance that is in a container of less than 20 pounds of such refrigerant must verify that the purchaser is certified, and must retain records for a period of three years." These records must be maintained on site.*

Task 2

Identify and recover A/C system refrigerant.

176. Refrigerant recovery and storage cylinders must be Department of Transportation (DOT) approved. What level of approval is necessary for refrigerant recovery containers?
    A. DOT 39-300
    B. DOT 4BA-300
    C. DOT 4BC-400
    D. DOT 3DE-500

177. To prevent overfilling of recovery cylinders, the Service Technician must:
    A. monitor cylinder pressure as the cylinder is being filled.
    B. monitor cylinder weight as the cylinder is being filled.
    C. make sure cylinder safety relief valves are in place and operational.
    D. occasionally shake the cylinder and observe any change of pressure while filling.

Hint

According to DOT/Air Conditioning and Refrigeration Institute (ARI) guidelines, 4B4 cylinders used to store recovered refrigerant shall ultimately be painted gray with the top shoulder portion painted yellow. The refrigerant type to be stored in a given container must be clearly marked on the container's label. For recovery/recycling purposes, only cylinders that are identified for recovered refrigerant may be used. Never use a cylinder that is intended to contain new refrigerant to store recovered refrigerant.

Task 3

Recycle refrigerant.

178. *Technician A* says that during the refrigerant recycling process, the refrigerant is reprocessed to new product specifications. *Technician B* says a chemical analysis is performed on refrigerant during the recycling process. Who is right?
    A. A only
    B. B only
    C. Both A and B
    D. Neither A nor B

179. *Technician A* says that anyone who purchases R-134a must maintain records for three years indicating the name and address of the supplier. *Technician B* says recovered R-134a must be stored in a gray container with a yellow top. Who is right?
    A. A only
    B. B only
    C. Both A and B
    D. Neither A nor B

Hint

Differences among the terms recover, recycle, and reclaim must be completely understood and properly used within the industry. To recover refrigerant is to remove refrigerant in any condition from a system and store it in an external container. The refrigerant must then either be recycled on site or shipped off site for reclamation. To recycle refrigerant is to reduce contaminants in used refrigerant by oil separation with single or multiple passes through devices such as replaceable filter dryers, which reduce moisture, acidity, and particulate matter. To reclaim refrigerant is to reprocess refrigerant to new product specifications by means, which may include distillation. Chemical analysis of the refrigerant is required to ensure that appropriate product specifications are met. Reclamation usually implies the use of procedures available only at processing or manufacturing facilities.

Task 4

Handle, label, and store refrigerant.

180. Before disposing of an empty or near-empty original container that was used to ship refrigerant from the factory, you should perform which of the following?
    A. Clean it and keep it for storage of recycled refrigerant.
    B. Open the valve completely and paint an X on the cylinder.
    C. Flush it with oil and nitrogen to keep it from rusting.
    D. Recover remaining refrigerant, evacuate cylinder, and mark it empty.

181. *Technician A* says before placing refrigerant in an approved storage container, the container must be evacuated to 27 inches of mercury. *Technician B* says while placing refrigerant in an approved storage container, the safe filling level must be monitored by cylinder weight. Who is right?
    A. A only
    B. B only

C. Both A and B
D. Neither A nor B

Hint

Any portable container used for transfer of reclaimed or recycled refrigerant must conform to DOT and UL standards. Before introducing refrigerant into an approved storage cylinder, the cylinder must be evacuated to at least 27 inches of mercury. Cylinder safe filling level must be monitored by measured weight. Shutoff valves are required within 12 inches (30 cm) of service hose ends. Shutoff valves must remain closed while connecting and disconnecting hoses to vehicle air-conditioning service ports. Safety goggles should always be worn while working with or around refrigerant.

Task 5

Test recycled refrigerant for noncondensable gases.

182. The test for noncondensable gases in recovered/recycled refrigerant involves:
A. comparing the pressure of the recovered refrigerant in the container to the theoretical pressure of pure refrigerant at a given temperature.
B. comparing the atmospheric pressure to the relative humidity.
C. comparing the container pressure with the size of the container.
D. testing the refrigerant with a halogen leak detector.

183. *Technician A* says that as specified by SAE J1991, recycled refrigerant cannot contain more than 15 ppm moisture contaminants by weight. *Technician B* says that as specified by SAE J1991, recycled refrigerant cannot contain more than 330 ppm noncondensable gases (air) by weight. Who is right?
A. A only
B. B only
C. Both A and B
D. Neither A nor B

Hint

To test a refrigerant for noncondensable gases, compare the pressure of the refrigerant in a cylinder to the theoretical pressure of pure refrigerant at a given temperature. If the actual pressure in the cylinder is lower than the theoretical pressure, the refrigerant is contaminated with noncondensable gas.

Answers to Questions

1. Answer A is wrong because a clogged evaporator drain will not cause a whistling noise.

   Answer B is right because a cracked evaporator case could cause a whistling noise.

   Answer C is wrong because a broken blend door cable will not cause a whistling noise.

   Answer D is wrong because a low refrigerant charge will not cause a whistling noise.

2. Answer A is wrong because grit in the evaporator case is not likely to cause this symptom.

   Answer B is wrong because the mode door does not move when the temperature setting is changed.

   Answer C is right because bad drive gears in the blend door motor will cause this symptom.

   Answer D is wrong because arcing in the control head will not cause a grinding noise.

3. **Answer A is right** because the high-pressure refrigerant at the compressor outlet can raise the temperature of the outlet to that of the engine coolant.

   Answer B is wrong because this is a normal condition.

   Answer C is wrong because Technician A is right, and Technician B is wrong.

   Answer D is wrong because TechnicianA is right, and Techniciam B is wrong.

4. **Answer A is right** because R-134a has a faint ether-like odor.

   Answer B is wrong because R-12 is odorless.

   Answer C is wrong because HVAC system input air is not drawn from under the hood or the cab.

   Answer D is wrong because a leaking heater core will not produce an ether-like odor.

5. Answer A is wrong because compression fittings are not used in mobile A/C systems.

   Answer B is right because R-134a systems use SAE quick-connect couplings.

   Answer C is wrong because different fittings are used to prevent mixing refrigerants.

   Answer D is wrong because the size is not 10 mm.

6. Answer A is wrong because using an ohmmeter will typically not provide calibration.

   Answer B is wrong because this process only checks that sensor.

   Answer C is wrong because not all ATC units have such a dial.

   Answer D is right because the easiest way to check the calibration of an ATC system is to measure the cab temperature using a thermometer.

7. Answer A is wrong because in a semiautomatic temperature control system, mode selection is a manual operation.

   Answer B is wrong because the blend door actuator is responsible for temperature control only.

   Answer C is wrong because the blower system does not affect mode selection.

   Answer D is right because an improperly adjusted cable could affect the amount of air that is directed toward the windshield.

8. **Answer A is right** because if the fan clutch is always engaged, it will enhance the operation of the A/C system.

   Answer B is wrong because an improperly adjusted blend door cable will cause poor A/C performance.

   Answer C is wrong because a low refrigerant charge will cause poor A/C system performance.

   Answer D is wrong because a refrigerant overcharge will cause poor A/C system performance.

9. **Answer A is right** because a misaligned air duct could cause inadequate airflow from one or more vents.

   Answer B is wrong because a faulty blower resistor will affect airflow from all vents in one or more blower speed settings.

   Answer C is wrong because a clogged heater core will not affect output airflow.

   Answer D is wrong because high humidity will not affect output airflow.

10. **Answer A is right** because this symptom is typical of an A/C system with an evaporator icing problem, which could be caused by a defective thermal expansion valve.

    Answer B is wrong because a defective coolant control valve would not cause the A/C system output air to change from cold to warm.

    Answer C is wrong because the blend air door does not typically have a return spring.

    Answer D is wrong because the fresh air door does not control the output air temperature.

11. Answer A is wrong because the tanks are color coded, but Technician B is also right.

    Answer B is wrong because R-134a containers use 1/2 inch 16 ACME threads, but Technician A is also right.

    Answer C is right because both Technician A and Technician B are right.

    Answer D is wrong because both Technician A and Technician B are right.

12. Answer A is wrong because it is not the specified size.

    Answer B is wrong because it is not the specified size.

    Answer C is right because 1/2 in. 16 ACME is the specified service fitting size.

    Answer D is wrong because it is not the specified size.

13. Answer A is a good choice because bubbles and/or foam in the sight glass indicate that the refrigerant charge is low and air has entered the system, but it is wrong because Technician B is also right.

    Answer B is also a good choice because a cloudy sight glass indicates moisture in the system. However, the choice is wrong because Technician A is also right.

    Answer C is right because both Technician A and Technician B are right.

    Answer D is wrong because both Technician A and Technician B are right.

14. **Answer A is right** because if the expansion valve is stuck open, the low side pressure will be high and the compressor will run continuously.

    Answer B is wrong because a restricted evaporator causes low, low-side pressure.

    Answer C is wrong because Technician A is right and Technician B is wrong.

Answer D is wrong because Technician A is right and Technician B is wrong.

15. Answer A is wrong because a refrigerant overcharge will cause high system pressures.

    Answer B is wrong because an overheated engine will cause elevated high-side pressure due to high evaporator temperature.

    Answer C is wrong because a restricted airflow through the condenser will cause elevated high-side pressure due to high evaporator temperature.

    Answer D is right because a blockage of the orifice tube screen will cause low high-side pressure.

16. Answer A is wrong because when servicing a system with the type of service valve shown in the figure, the compressor can be isolated.

    Answer B is wrong because the valve shown in the figure only permits reading of the suction and discharge pressures when it is in the mid-position or service position.

    Answer C is wrong because both Technicians are wrong.

    Answer D is right because both Technicians are wrong for the above reasons.

17. **Answer A is right** because no refrigerant charge or a normal refrigerant charge usually provide a clear sight glass.

    Answer B is wrong because a defective desiccant in the receiver/dryer causes moisture in the system and a cloudy sight glass.

    Answer C is wrong because Technician A is right and Technician B is wrong.

    Answer D is wrong because Technician A is right and Technician B is wrong.

18. Answer A is wrong because bubbles in the sight glass in an R-12 system are normal at ambient temperatures below 70ºF (21ºC).

    Answer B is wrong because bubbles in the sight glass in an R-12 system are normal at ambient temperatures below 70ºF (21ºC).

    Answer C is wrong because Technician A and Technician B are both wrong.

    Answer D is right because Technician A and Technician B are both wrong.

19. **Answer A is right** because a low refrigerant charge may cause the evaporator outlet to be warmer than the inlet.

    Answer B is wrong because a disengaged compressor clutch results in no cooling in the evaporator and both the inlet and outlet are warm.

    Answer C is wrong because Technician A is right and Technician B is wrong.

    Answer D is wrong because Technician A is right and Technician B is wrong.

20. Answer A is wrong because frosting of the TXV valve is likely caused by a restriction in this valve.

    Answer B is wrong because frosting of the TXV valve is likely caused by a restriction in this valve.

    Answer C is wrong because Technician A and Technician B are both wrong.

    Answer D is right because Technician A and Technician B are both wrong.

21. Answer A is wrong because a rotten egg smell in the cab is caused by a plugged evaporator case drain.

    Answer B is wrong because a rotten egg smell in the cab is caused by a plugged evaporator case drain.

    Answer C is wrong because a rotten egg smell in the cab is caused by a plugged evaporator case drain.

    Answer D is right because a rotten egg smell in the cab is caused by a plugged evaporator case drain.

22. Answer A is wrong because R-12 is not flammable.

    Answer B is wrong because the gas that is formed is toxic, attacking the nervous system.

    Answer C is wrong because R-12 does not form chlorine gas in the presence of a flame.

    Answer D is right because heated R-12 will form phosgene gas.

23. Answer A is wrong because the sensor tip should be as close as possible to the fitting.

    Answer B is right because refrigerant is heavier than air, so it is most easily detected just below a leaking fitting.

Answer C is wrong because you place the sensor probe just below the fitting.

Answer D is wrong because you place the sensor probe just below the fitting.

24. Answer A is wrong because one uses this method for finding leaks in the chassis air system and soapy solutions will not show refrigerant leaks.

    Answer B is right because an electronic leak detector is the most accurate method of leak detection.

    Answer C is wrong because Technician A is wrong and Technician B is right.

    Answer D is wrong because Technician A is wrong and Technician B is right.

25. **Answer A is right** because Figure 7–9 shows the proper manifold gauge and vacuum pump connections for A/C system evacuation.

    Answer B is wrong because stem-type service valves must be mid-positioned to allow A/C system pressure or vacuum to be applied to the manifold gauge set.

    Answer C is wrong because Technician A is right and Technician B is wrong.

    Answer D is wrong because Technician A is right and Technician B is wrong.

26. Answer A is wrong because 20 minutes is not enough time to evacuate.

    Answer B is wrong because 10 minutes is not enough time to evacuate.

    Answer C is wrong because 15 minutes is not enough time to evacuate.

    Answer D is right because to ensure that the entire system is under deep enough vacuum to remove all moisture, the pump should be run for at least 30 minutes.

27. Answer A is wrong because the nitrogen pressure must be regulated to prevent damaging A/C system components.

    Answer B is wrong because the A/C compressor must be disconnected before flushing the system.

    Answer C is wrong because to avoid damaging restrictive components, they must be removed before flushing.

    Answer D is right because one can vent nitrogen to the atmosphere.

28. **Answer A is right** because some manufacturers do recommend this practice of installing an in-line filter.

    Answer B is wrong because the filter that contains an orifice replaces the original orifice tube. If you do not remove the original orifice, a restriction will result with poor performance.

    Answer C is wrong because Technician A is right and Technician B is wrong.

    Answer D is wrong because Technician A is right and Technician B is wrong.

29. Answer A is wrong because if you complete the charging process with the engine running, liquid refrigerant may enter the compressor with resulting damage.

    Answer B is right because you cannot compress liquid refrigerant and will result in damage.

    Answer C is wrong because Technician A is wrong and Technician B is right.

    Answer D is wrong because Technician A is wrong and Technician B is right.

30. Answer A is wrong because one completes the charging process when the correct weight of refrigerant has entered the system.

    Answer B is wrong because if the low side does not move from a vacuum to a pressure, there is a restriction.

    Answer C is wrong because the truck engine does not need to be running during recharging.

    Answer D is right because OEMs recommend either a high-side (liquid) or a low-side (vapor) charging process.

31. Answer A is a true statement; therefore, it is wrong because the equipment label must indicate UL approval.

    Answer B is a true statement; therefore, it is wrong because the equipment label must indicate SAE J1991 approval.

    Answer C is right as the statement is false because you must use the specified type of refrigerant storage container in this equipment.

    Answer D is a true statement; therefore, it is wrong because R-12 and R-134a refrigerants or refrigerant oils must not be mixed in the recovery/recycling process.

32. Answer A is wrong because charging is complete when the specified refrigerant weight has entered the system, but Technician B is also right.

    Answer B is wrong because the average low-side pressure is 25 to 45 psi (172 to 310 kPa), but Technician A is also right.

    Answer C is right because both Technician A and Technician B are right.

    Answer D is wrong because both Technician A and Technician B are right.

33. Answer A is wrong because only R-12 A/C systems use mineral oil.

    Answer B is wrong because PAG oil is a synthetic lubricant used in R-134a A/C systems.

    Answer C is wrong because both Technician A and Technician B are wrong.

    Answer D is right because both Technician A and Technician B are wrong.

34. **Answer A is right** because PAG oil is specifically formulated to be compatible with R-134a.

    Answer B is wrong because R-12 systems use mineral-based lubricant.

    Answer C is wrong because DEXRON is not compatible with R-134a and will damage A/C system components.

    Answer D is wrong because type C-3 SAE 30 based oil lubricant is nit used in mobile air-conditioning systems.

35. Answer A is wrong because component R is a thermal cycling switch that cycles the compressor on and off in relation to evaporator outlet temperature.

    Answer B is wrong because component A is a thermal cycling switch that cycles the compressor on and off in relation to evaporator outlet temperature.

    Answer C is wrong because both Technician A and Technician B are wrong.

    Answer D is right because both Technician A and Technician B are wrong.

36. **Answer A is right** because component X must be mounted in a well on the evaporator outlet line.

    Answer B is wrong because component X senses the temperature of the evaporator outlet line.

    Answer C is wrong because Technician A is right and Technician B is wrong.

    Answer D is wrong because Technician A is right and Technician B is wrong.

37. Answer A is wrong because improper radiator shutter operation can cause the relief valve to operate.

    Answer B is wrong because a clogged condenser will cause the relief valve to operate.

    Answer C is wrong because an inoperative fan clutch will cause the relief valve to operate.

    Answer D is right because a defective A/C compressor will not cause the relief valve to operate.

38. Answer A is wrong because the switch is not necessarily at fault.

    Answer B is right because low refrigerant charge will cause this symptom.

    Answer C is wrong because only one Technician is right.

    Answer D is wrong because one Technician is right.

39. Answer A is wrong because the high-pressure relief valve resets itself when system pressures return to normal.

    Answer B is wrong because the high-pressure relief valve is individually replaceable.

    Answer C is wrong because both Technicians are wrong.

    Answer D is right because neither Technician is right.

40. Answer A is wrong because a defective clutch will not cycle too fast.

    Answer B is wrong because a defective switch will not cause this symptom.

    Answer C is wrong because an overcharged system is more likely to cause the clutch to remain engaged.

    Answer D is right because low refrigerant charge will cause the clutch to cycle more frequently than normal.

41. Answer A is wrong because the ohmmeter should provide a very low reading.

Answer B is wrong because the contacts should be closed.

Answer C is wrong because Technician A and Technician B are both wrong.

Answer D is right because Technician A and Technician B are both wrong.

42. **Answer A is right.** Using a belt tension gauge ensures that the belt is properly adjusted.

Answer B is wrong. This method does not ensure proper belt tension.

Answer C is wrong. This method does not ensure proper belt tension.

Answer D is wrong. This method could cause engine damage due to over tightening.

43. Answer A is wrong because a cracked or bent mounting bracket can cause drive belt edge wear.

Answer B is right because a compressor clutch air gap does not affect pulley alignment.

Answer C is wrong because a damaged compressor pulley could cause drive belt edge wear.

Answer D is wrong because a worn idler pulley bearing could cause belt misalignment and result in edge wear.

44. Answer A is wrong because a bent compressor mounting bracket holder may cause drive belt misalignment, but Technician B is also right.

Answer B is wrong because a defective compressor clutch bearing may cause drive belt misalignment, but Technician A is also right.

Answer C is right because Technician A and Technician B are both right.

Answer D is wrong because Technician A and Technician B are both right.

45. **Answer A is right** because excessive clearance may cause the compressor clutch to slip due to lack of full engagement.

Answer B is wrong because adding another shim behind the armature plate increases the clearance.

Answer C is wrong because Technician A is right and Technician B is wrong.

Answer D is wrong because Technician A is right and Technician B is wrong.

46. Answer A is wrong because applying power and ground to the appropriate terminals can also test the coil, but Technician B is also right.

Answer B is wrong because the coil can also be tested with an ohmmeter, but Technician A is also right.

Answer C is right because Technician A and Technician B are both right.

Answer D is wrong because Technician A and Technician B are both right.

47. Answer A is wrong because the tool shown in the figure is used as shaft seal protector, but Technician B is also right.

Answer B is wrong because the seal seat O-ring must be installed before the shaft seal, but Technician A is also right.

Answer C is right because Technician A and Technician B are both right.

Answer D is wrong because Technician A and Technician B are both right.

48. Answer A is wrong because voltage is supplied from one terminal on the blower control switch to the mode control switch, and a defective blower control switch could prevent A/C compressor clutch engagement.

Answer B is wrong because an open binary pressure switch would prevent compressor clutch engagement.

Answer C is wrong because a poor connection at the A/C thermostat would prevent compressor clutch engagement.

Answer D is right because a blown 10 amp fuse at CB 8 prevents operation of the panel light, and this problem would not prevent compressor clutch engagement.

49. Answer A is wrong because the compressor oil needs to be checked when there is evidence of a loss of system oil, but Technician B is also right.

Answer B is a wrong because the specified type and quantity of oil recommended by the compressor manufacturer must be used, but Technician A is also right.

Answer C is right because both Technician A and Technician B are right.

Answer D is wrong because both Technician A and Technician B are right.

50. Answer A is wrong because phosgene gas forms when R-12 comes in contact with a flame.

    Answer B is right because when the system is purged too quickly, the refrigerant oil will atomize and travel from the system with the refrigerant.

    Answer C is wrong because Technician A is wrong and Technician B is right.

    Answer D is wrong because Technician A is wrong and Technician B is right.

51. **Answer A is right** because refrigerant returns to the compressor as a low-pressure gas.

    Answer B is wrong because refrigerant in an operating A/C system is never a low-pressure liquid.

    Answer C is wrong because refrigerant leaves the compressor as a high-pressure gas.

    Answer D is wrong because refrigerant in the liquid line is a high-pressure liquid.

52. Answer A is wrong because the POA valve is located near the evaporator.

    Answer B is wrong because a filter is generally used on the suction side of the compressor.

    Answer C is wrong because the ETR is located near the evaporator.

    Answer D is right because the indicated component is a muffler used to reduce compressor noise.

53. Answer A is wrong because the reed valves allow the compressor to draw refrigerant in from the system low side, compress it, and exhaust it to the system high side.

    Answer B is wrong because the valves direct the flow of refrigerant.

    Answer C is wrong because the valves direct the flow of refrigerant.

    Answer D is right because the valves direct the flow of refrigerant.

54. **Answer A is right** because the compressor discharge valve opens after the pressure compresses the vaporous refrigerant, allowing the refrigerant to move to the condenser.

    Answer B is wrong because if the discharge valve opened before the vaporous refrigerant is compressed, it would not properly raise the refrigerant pressure to allow a change from gas to a liquid. In addition, it does not go to the evaporator.

    Answer C is wrong because the reed valves just open and close; they do not regulate a variable pressure.

    Answer D is wrong because reed valves cannot sense A/C system temperature, so regulation cannot take place.

55. Answer A is wrong because a cracked mounting plate could cause drive belt wear.

    Answer B is right because a cracked mounting plate will not cause internal compressor damage.

    Answer C is wrong because a cracked mounting plate can cause a vibration with the compressor clutch engaged.

    Answer D is wrong because a cracked mounting plate can cause belt squeal or chatter.

56. **Answer A is right** because this is an acceptable method of repairing a stripped mounting bolthole.

    Answer B is wrong because the compressor can be repaired.

    Answer C is wrong because only one of the Technicians is right.

    Answer D is wrong because only one of the Technicians is right.

57. Answer A is wrong because refrigerant oil must be added to the evaporator before installation.

    Answer B is right because most manufacturers recommend that 3 ounces of oil be added to a new evaporator.

    Answer C is wrong because 9 ounces of refrigerant oil would cause an excessive system oil level.

    Answer D is wrong because the entire A/C system is likely to contain about 14 1/2 ounces of oil.

58. **Answer A is right** because the condenser generally holds 1 ounce of refrigeration oil.

    Answer B is wrong because this amount is too large.

    Answer C is wrong because this amount is too large.

    Answer D is wrong because this amount is too large.

59. Answer A is wrong because the evaporator must be removed to check the lubricant level, but Technician B is also right.

Answer B is wrong because first you must run the A/C compressor briefly to ensure that the refrigeration oil distributes throughout the system, but Technician A is also right.

Answer C is right because both Technician A and Technician B are right.

Answer D is wrong because both Technician A and Technician B are right.

60. Answer A is wrong because a spring lock tool is the tool shown.

Answer B is wrong because a bearing puller is not shown.

Answer C is right because this is the tool shown in the figure.

Answer D is wrong because a flare tool is not shown.

61. Answer A is wrong. A seal pick should never be used to install A/C hose O-rings.

Answer B is wrong. Refrigeration oil is the only acceptable lubricant for A/C system O-rings and seals.

Answer C is wrong because both Technician A and Technician B are wrong.

Answer D is right because both Technician A and Technician B are wrong.

62. **Answer A is right** because most shutter systems are air operated, not electrically operated.

Answer B is wrong because an air leak could cause the shutters to remain open.

Answer C is wrong because dirt accumulation on the shutter linkage could prevent the shutters from closing completely.

Answer D is wrong because ice buildup could prevent the shutters from closing completely.

63. Answer A is wrong because the condenser should not be replaced, rather it should be cleaned.

Answer B is right because several bent fins and a moderate accumulation of dead insects will not restrict airflow through the condenser enough to cause the high-pressure relief valve to operate.

Answer C is wrong because Technician A is wrong and Technician B is right.

Answer D is wrong because Technician A is wrong and Technician B is right.

64. **Answer A is right** because the shutter cylinder must be replaced.

Answer B is wrong because shutter cylinders cannot be overhauled.

Answer C is wrong because Technician A is right and Technician B is wrong.

Answer D is wrong because Technician A is right and Technician B is wrong.

65. **Answer A is right** because a saline solution will cause corrosion.

Answer B is wrong because a soft whisk broom can be used to remove debris from the condenser fins.

Answer C is wrong because compressed air can be used to remove debris from the condenser fins.

Answer D is wrong because a soap and water solution can be used to remove debris from the condenser fins.

66. **Answer A is right** because A/C fittings should not be disturbed if they are not leaking.

Answer B is wrong because an A/C maintenance service should include cleaning the condenser fins.

Answer C is wrong because an A/C maintenance service should include straightening bent condenser fins.

Answer D is wrong because an A/C maintenance service should include checking all component mounts and insulators.

67. Answer A is wrong because reduced space between the condenser and the radiator causes improper A/C system cooling, but Technician B is also right.

Answer B is wrong because deformed or improperly aligned condenser mounting insulators could damage the condenser and refrigerant lines, but Technician A is also right.

Answer C is right because both Technician A and Technician B are right.

Answer D is wrong both Technician A and Technician B are right.

68. **Answer A is right** because the refrigerant hose is the high-side hose connected between the condenser and the receiver/dyer, and this hose should be warm or hot if the A/C system is operating normally.

Answer B is wrong because the component to which the hoses are connected is the receiver/dryer.

Answer C is wrong because Technician A is right and Technician B is wrong.

Answer D is wrong because Technician A is right and Technician B is wrong.

69. Answer A is wrong because the refrigerant remains in liquid form as it passes through the receiver/dryer.

Answer B is wrong because the condenser carries high-pressure liquid.

Answer C is right because the sudden increase in inside diameter of the A/C line at the orifice causes a radical pressure drop allowing most of the refrigerant to vaporize instantly at that point. Any refrigerant that remains in the liquid state after it passes through the orifice is quickly vaporized in the evaporator.

Answer D is wrong because the capillary tube is a temperature sensor and does not contain refrigerant.

70. **Answer A is right** because if the A/C system is open for an extended period of time, the desiccant in the receiver/dryer will become saturated with moisture and must be replaced.

Answer B is wrong because the receiver/dryer only needs to be replaced if it is damaged or if the desiccant bag is saturated or damaged.

Answer C is wrong because there is no mileage requirement for replacement of the receiver/dryer.

Answer D is wrong because compressor replacement does not necessitate receiver/dryer replacement.

71. Answer A is wrong because you cannot sense evaporator temperature or inside car temperature from the engine compartment where the condenser is located.

Answer B is wrong because you use an accumulator with a CCOT system that does not use a thermal bulb or capillary tube.

Answer C is wrong because the capillary tube contacts the evaporator outlet.

Answer D is correct because the thermal bulb and capillary tube are part of the thermal expansion valve located near the evaporator.

72. Answer A is wrong because the expansion valve is mounted at the evaporator inlet.

Answer B is wrong because frost formation on the expansion valve indicates excessive restriction in the valve or excessive moisture in the refrigerant.

Answer C is wrong because both Technician A and Technician B are wrong.

Answer D is right because both Technician A and Technician B are wrong.

73. Answer A is wrong because the component being serviced is the orifice tube.

Answer B is right because the component being serviced is the orifice tube.

Answer C is wrong because the component being serviced is the orifice tube.

Answer D is wrong because the component being serviced is the orifice tube.

74. Answer A is wrong because the orifice tube is located in the evaporator core inlet.

Answer B is right because the orifice tube provides for the rapid expansion, and thus evaporation, of the high-pressure liquid refrigerant in the liquid line.

Answer C is wrong because the orifice tube does not affect the refrigerant flow through the condenser.

Answer D is wrong because the expansion valve does not affect airflow through the evaporator.

75. **Answer A is right** because the orifice tube screen is installed to prevent particulate from circulating through the system in the event the desiccant bag in the receiver/dryer breaks down.

Answer B is wrong because atomization is not important to the evaporation of refrigerant, only fuel.

Answer C is wrong because only Technician A is right.

Answer D is wrong because one of the Technicians is right.

76. **Answer A is right** because the air flows through the evaporator core before it flows through the heater core.

Answer B is wrong because in the maximum A/C mode, the blend air door blocks airflow through the heater core.

Answer C is wrong because Technician A is right and Technician B is wrong.

Answer D is wrong because Technician A is right and Technician B is wrong.

77. Answer A is wrong because the thermostatic switch senses temperature at the evaporator inlet and disengages the compressor clutch when evaporator icing is imminent.

Answer B is wrong because the EPR valve regulates the flow of refrigerant into the evaporator, preventing icing.

Answer C is wrong because the STV regulates the flow of refrigerant into the evaporator, preventing icing.

Answer D is right because the low-pressure cut-out switch prevents compressor damage due to a loss of refrigerant.

78. Answer A is wrong because a plugged evaporator case may cause windshield fogging, but this problem would not result in an oily film on the windshield.

Answer B is right because a refrigerant leak in the evaporator core may allow some refrigerant and oil to leak and cause a thin oily film on the windshield.

Answer C is wrong because only one Technician is right.

Answer D is wrong because one Technician is right.

79. Answer A is wrong because some evaporator filters are made of paper or a metallic mesh.

Answer B is wrong because evaporator filters are not designed to remove moisture from the air.

Answer C is wrong because evaporator filters are replaceable.

Answer D is right because they are designed to remove dust and dirt from cab and sleeper air.

80. Answer A is wrong because a faint hissing noise after shutting down the engine is caused by the equalization of refrigerant pressure in the A/C system.

Answer B is wrong because this is a normal condition and a defective EPR valve causes evaporator freezing and poor cooling.

Answer C is wrong because neither Technician is right.

Answer D is right because this is a normal condition and both Technicians are wrong.

81. **Answer A is right** because vinegar will kill the mold and mildew that are the source of the odor.

Answer B is wrong because this method will simply mask the odor and not eliminate the problem.

Answer C is wrong because this method poses a possible fire hazard.

Answer D is wrong because this method will simply mask the odor and not eliminate the problem.

82. Answer A is wrong because the service valve is front seated and this blocks the gauge port so no pressure would register on a manifold gauge set. One uses the service valve mid-position for service.

Answer B is wrong because the system could not operate normally because the passage to the compressor is blocked. Some system pressure may appear at the gauge port. You back seat the service valve for normal operation.

Answer C is right because front seating a service valve isolates the compressor from the system for service.

Answer D is wrong because pressure from the compressor is blocked. One uses the service valve mid-position for service.

83. Answer A is wrong because R-134a systems use quick-disconnect service valves.

Answer B is right because new environmental laws dictate that shut-off valves must be located no more than 12 inches from test hose service ends.

Answer C is wrong because Technician A is wrong and Technician B is right.

Answer D is wrong because Technician A is wrong and Technician B is right.

84. **Answer A is right** because the Schrader-type service valve in Figure 7–21 has a removable core.

Answer B is wrong because the Schrader-type service valve in Figure 7–21 does not provide a means to front seat or rear seat the valve.

Answer C is wrong because only Technician A is right.

Answer D is wrong because one of the Technicians is right.

85. **Answer A is right** because any leak from the system must be repaired as quickly as possible.

Answer B is wrong because if oil is leaking from the system, refrigerant is also leaking from the system.

Answer C is wrong because Technician A is right and Technician B is wrong.

Answer D is wrong because Technician A is right and Technician B is wrong.

86. Answer A is wrong because the A/C high-pressure switch does not provide a boosting function.

Answer B is right because the A/C high-pressure switch opens the electrical circuit to the compressor clutch coil when high-side pressure reaches its upper limit.

Answer C is wrong because the A/C high-pressure switch does not maintain pressure.

Answer D is wrong because the A/C high-pressure switch does perform a venting function.

87. Answer A is wrong because the high-pressure relief valve is used on R-134a systems.

Answer B is right because the high-pressure relief valve resets itself when A/C system pressure returns to a safe level.

Answer C is wrong because Technician A is wrong and Technician B is right.

Answer D is wrong because Technician A is wrong and Technician B is right.

88. Answer A is wrong because the binary switch also provides high-pressure protection.

Answer B is wrong because the binary switch also provides high-pressure protection.

Answer C is right because the binary switch disables the A/C compressor when the system pressure is too high or too low.

Answer D is wrong because it does provide protection for the compressor in a low- or high-pressure situation.

89. Answer A is wrong because a defective coolant control valve could cause this problem.

Answer B is right because a defective engine coolant temperature sensor is the least likely cause.

Answer C is wrong because a defective air control solenoid could cause this problem.

Answer D is wrong because a defective blend air door air cylinder could cause this problem.

90. Answer A is wrong because the heater control valves are not operated by engine speed.

Answer B is right because the heater control valves may be cable operated or vacuum operated.

Answer C is wrong because Technician A is wrong and Technician B is right.

Answer D is wrong because Technician A is wrong and Technician B is right.

91. **Answer A is right** because a broken blend door cable will make HVAC output temperature uncontrollable.

Answer B is wrong because a defective compressor clutch will not affect heater temperature control.

Answer C is wrong because a clogged orifice tube will not affect heater temperature control.

Answer D is wrong because an inoperative blower motor has no effect on temperature control.

92. Answer A is wrong because a clogged heater core causes insufficient heater output, but Technician B is also right.

Answer B is wrong because an improperly adjusted coolant control valve cable could cause insufficient heater output, but Technician A is also right.

Answer C is right because both Technician A and Technician B are right.

Answer D is wrong because both Technician A and Technician B are right.

93. Answer A is wrong because a leaking heater core will cause windshield fogging in the defrost mode.

Answer B is wrong because a clogged evaporator drain will cause windshield fogging in the defrost mode.

Answer C is wrong because an exterior water leak into the air intake plenum can cause windshield fogging in the defrost mode.

Answer D is right because moisture in the refrigerant will not cause windshield fogging.

94. Answer A is wrong because a sticky film on the inside of the windshield is an indication of a heater core leak, and in the case of any coolant leak the coolant level should be checked, but Technician B is also right.

Answer B is wrong because a leaking heater core could put a sticky film on the inside of the windshield, but Technician A is also right.

Answer C is right because both Technician A and Technician B are right.

Answer D is wrong because both Technician A and Technician B are right.

95. **Answer A is right** because antifreeze raises the boiling point of the coolant, not the conditioner.

    Answer B is wrong because the coolant conditioner cartridge filters particulate from the coolant.

    Answer C is wrong because the coolant conditioner internally lubricates the cooling system.

    Answer D is wrong because the coolant conditioner prevents cavitation corrosion of the cylinder liners.

96. **Answer A is right** because cooling system leaks are tested by pressurizing the system and visually watching for leaks.

    Answer B is wrong because the average medium/heavy truck cooling system is pressurized to 15 psi (103 kPa).

    Answer C is wrong because Technician A is right and Technician B is wrong.

    Answer D is wrong because Technician A is right and Technician B is wrong.

97. Answer A is wrong because the acids of combustion are in the engine oil, not in the cooling system.

    Answer B is right because flushing the cooling system removes rust from the system.

    Answer C is wrong because Technician A is wrong and Technician B is right.

    Answer D is wrong because Technician A is wrong and Technician B is right.

98. Answer A is wrong because the evaporator is not a cooling system component.

    Answer B is wrong because the power steering cooler is not a cooling system component.

    Answer C is wrong because the oil cooler is not a cooling system component.

    Answer D is right because a blown head gasket is the most likely cause of an internal engine coolant leak.

99. Answer A is wrong because the shape of a preformed hose is derived during the manufacturing process.

    Answer B is right because the spring provides internal support.

    Answer C is wrong because the resilience of the hose comes from the rubber.

    Answer D is wrong because the strength of the hose comes from the fibers inside.

100. Answer A is wrong because when the cooling system pressure is increased, the boiling point increases.

     Answer B is right because if one adds more antifreeze to the coolant, the boiling point increases.

     Answer C is wrong because a good quality ethylene glycol antifreeze contains antirust inhibitors.

     Answer D is wrong because coolant solutions must be recovered, recycled, and handled as hazardous waste.

101. Answer A is wrong because the tester is connected to pressure test the cooling system with the radiator cap removed.

     Answer B is wrong because a transmission cooler that is mounted externally from the radiator is not part of the cooling system.

     Answer C is wrong because both Technician A and Technician B are wrong.

     Answer D is right because both Technician A and Technician B are wrong.

102. Answer A is wrong because the antifreeze and water can be added in any order.

     Answer B is wrong because the antifreeze and water can be added in any order.

     Answer C is wrong because there is no need to premix the coolant.

     Answer D is right because it is important to bleed the system.

103. Answer A is wrong because most trucks are equipped with an engine shutdown alarm to warn the driver about overheating, but Technician B is also right.

Answer B is wrong because many trucks are equipped with an override switch, but Technician A is also right.

Answer C is right because both Technician A and Technician B are right.

Answer D is wrong because both Technician A and Technician B are right.

104. Answer A is wrong because a faulty thermostat can cause poor coolant circulation.

Answer B is wrong because an eroded water pump impeller will cause poor coolant circulation.

Answer C is right because the upper hose is under pressure and is unlikely to collapse.

Answer D is wrong because the lower hose carries coolant to the intake side of the water pump and could collapse.

105. Answer A is wrong because fan clutch replacement does not affect the life of the water pump.

Answer B is wrong because the water pump generally has a longer service life than the heater hoses.

Answer C is right because a small leak from the weep hole indicates that the front seal of the water pump has failed.

Answer D is wrong because thermostat operation does not indicate a need to replace the water pump.

106. Answer A is wrong because the upper radiator hose rapidly gets warm when the thermostat opens.

Answer B is wrong because the temperature gauge rises until it indicates normal operating temperature when the thermostat opens.

Answer C is wrong because there is obvious circulation motion in the upper radiator tank when the thermostat is open.

Answer D is right because thermostat opening is often not noticeable in the surge tank.

107. Answer A is wrong because the thermostat will not fit into the housing.

Answer B is wrong because it will not fit in that direction.

Answer C is wrong because the thermostat is not installed in the proximity of the core.

Answer D is right because the spring controls the valve in the thermostat, opening when it gets to the desired operating temperature. Since the engine block is the source of heat, the spring must face the engine block.

108. Answer A is wrong because the tool shown in Figure 7–23 is used to test the amount of antifreeze in the coolant.

Answer B is wrong because the boiling point of the coolant is increased if more antifreeze is added to the coolant.

Answer C is wrong because both Technician A and Technician B are wrong.

Answer D is right because both Technician A and Technician B are wrong.

109. **Answer A is right** because more rust formation occurs in the cooling system containing only water compared to a system containing an antifreeze and water solution because the antifreeze contains a rust inhibitor, but Technician B is also right.

Answer B is wrong because the coolant boiling point is higher when the cooling system contains a 50/50 water and antifreeze solution compared to a system containing only water, but Technician A is also right.

Answer C is right because both Technician A and Technician B are right.

Answer D is wrong because both Technician A and Technician B are right.

110. Answer A is wrong because the coolant conditioner cartridge must be changed at the recommended service intervals.

Answer B is wrong because the coolant conditioner does help to reduce cavitation corrosion of the cylinder liners.

Answer C is wrong because both Technician A and Technician B are wrong.

Answer D is right because both Technician A and Technician B are wrong.

111. Answer A is wrong because the component in Figure 7–24 is a coolant filter.

Answer B is right because the component in Figure 7–24 is a coolant filter.

Answer C is wrong because the component in Figure 7–24 is a coolant filter.

Answer D is wrong because the component in Figure 7–24 is a coolant filter.

112. Answer A is wrong because some fan clutches are air operated.

Answer B is wrong because some fan clutches are operated by a thermostatic spring.

Answer C is wrong because some fans use a viscous clutch.

Answer D is right because hydraulic switches are rarely used.

113. Answer A is wrong because the fan blade will freewheel at a reduced speed due to friction/viscous forces.

Answer B is wrong because the engine idle will rise slightly or be unaffected.

Answer C is wrong because the shutters may or may not be closed.

Answer D is right because the fan blade may freewheel at a reduced speed.

114. Answer A is wrong because electric cooling fans are often controlled by an electronic relay.

Answer B is wrong because a cooling fan module often controls electric cooling fans.

Answer C is wrong because a multiplexed electronic module often controls electric cooling fans.

Answer D is right because there is no air solenoid controller.

115. Answer A is wrong because some coolant control valves close when vacuum is removed.

Answer B is right because this process allows the Technician to check the specific coolant flow valve operation.

Answer C is wrong because Technician A is wrong and Technician B is right.

Answer D is wrong because Technician A is wrong and Technician B is right.

116. Answer A is wrong because the solenoid would not click when it is energized if the winding is open.

Answer B is right because a plugged vacuum hose between the solenoid and the valve diaphragm would prevent the hand pump vacuum from being supplied to the valve diaphragm.

Answer C is wrong because a vacuum leak in the solenoid would cause the hand pump vacuum to gradually decrease.

Answer D is wrong because a seized plunger would not provide a click when the solenoid is energized, and the question informs us this clicking action occurs.

117. Answer A is wrong because chassis air pressure can be used to operate a coolant control valve.

Answer B is wrong because vacuum can be used to operate a coolant control valve.

Answer C is wrong because early ATC systems used a cable to operate the coolant control valve.

Answer D is right because a magneto is used to generate the ignition spark in many small gasoline engines.

118. Answer A is wrong because the heater core does not need insulation.

Answer B is wrong because the heater core does not contribute significantly to cab noise.

Answer C is right because foam tape can be used to cushion and seal around the heater core.

Answer D is wrong because a gasket is provided to seal around heater hose connections.

119. Answer A is wrong because oil is not added to the heater core before installation.

Answer B is wrong because oil is not added to the heater core before installation.

Answer C is wrong because both Technicians are wrong.

Answer D is right because neither Technician is right.

120. Answer A is wrong because when the pressure bleeds down at all, the heater core is leaking and must be replaced.

Answer B is wrong because if the heater core leaks, it should be replaced immediately to prevent coolant loss and the risk of engine damage.

Answer C is right because when you test a heater core for leaks, you apply 10 psi of air pressure and the pressure bleeds to 5 psi in 3 minutes.

Answer D is wrong because this is an acceptable method of checking a heater core for leaks.

121. Answer A is wrong because if the shutters do not close, it has no effect on high-side pressure.

Answer B is right because when the shutters do not close, the engine is slow to warm up.

Answer C is wrong because Technician A is wrong and Technician B is right.

Answer D is wrong because Technician A is wrong and Technician B is right.

122. Answer A is wrong because shutters are not electrically operated.

Answer B is wrong because an air leak in the shutter cylinder causes the shutters to remain open.

Answer C is wrong because both Technician A and Technician B are wrong.

Answer D is right because both Technician A and Technician B are wrong.

123. **Answer A is right** because a digital multimeter (DMM) is the most precise tool for electrical/electronic components or systems and because of its 10 megohm impedance, it will not harm solid-state components.

Answer B is wrong because a self-powered test lamp could damage solid-state components.

Answer C is wrong because an analog VOM can damage solid-state components.

Answer D is wrong because a 12V test lamp can damage solid-state components.

124. Answer A is wrong because most ATC connect to the engine fuel management system.

Answer B is right because most of these systems connect closely to the engine control system.

Answer C is wrong because Technician A is wrong and Technician B is right.

Answer D is wrong because Technician A is wrong and Technician B is right.

125. Answer A is wrong because a short circuit in the blower circuit could cause a blown fuse.

Answer B is wrong because a short circuit in an actuator motor could cause a blown fuse.

Answer C is right because a shorted ECT sensor will cause a code to set, but will not blow a fuse.

Answer D is wrong because a damaged connector could cause a short circuit and blow a fuse.

126. Answer A is wrong because current to run the blower at low speed must pass through the resistor for medium-speed as well.

Answer B is right because the switch provides the path for current to the medium speed resistor.

Answer C is wrong because Technician A is wrong and Technician B is right.

Answer D is wrong because Technician A is wrong and Technician B is right.

127. **Answer A is right** because an open circuit between the blower motor and ground causes the blower motor to be inoperative in all switch positions.

Answer B is wrong because if the upper resistor is open, the blower motor still operates at high speed.

Answer C is wrong because the blower motor still operates at high speed if there is an open circuit between the switch and the resistor assembly.

Answer D is wrong because if the lower resistor is open, the blower motor still operates at high speed.

128. **Answer A is right** because if the air gap is too great, the clutch will slip.

Answer B is wrong because the drive and driven plates must be replaced.

Answer C is wrong because Technician A is right and Technician B is wrong.

Answer D is wrong because Technician A is right and Technician B is wrong.

129. Answer A is wrong because an open circuit in the low side of the binary switch would cause the compressor clutch to be inoperative in all A/C modes.

Answer B is right because an open circuit in the defrost switch contacts would cause the compressor clutch to be inoperative only in the defrost mode.

Answer C is wrong because Technician A is wrong and Technician B is right.

Answer D is wrong because Technician A is wrong and Technician B is right.

130. Answer A is wrong because the 12V at terminals B and C indicate the thermal fuse is not open, and 12V are supplied to the compressor clutch coil.

Answer B is right because an open compressor clutch coil could result in an inoperative compressor clutch.

Answer C is wrong because the superheat switch is normally open, and the 12V at terminal S indicate this switch is open as it should be.

Answer D is wrong because the 12V at terminal S indicate there is continuity through the thermal fuse.

131. Answer A is wrong because if this adjustment sets the computer-applied idle speed too low or adjustment steps were missed, the engine may stall with A/C compressor engaged, but Technician B is also right.

    Answer B is wrong because the fuel management computer typically operates this device, but Technician A is also right.

    Answer C is right because both Technician A and Technician B are right.

    Answer D is wrong because both Technician A and Technician B are right.

132. **Answer A is right** because it is the lowest value in the choices and represents the maximum voltage drop allowed across a ground at 1/10 of volt.

    Answer B is wrong because the drop is too large with resultant high resistance.

    Answer C is wrong because the drop is too large with resultant high resistance.

    Answer D is wrong because the drop is too large with resultant high resistance.

133. **Answer A is right** because the PCM disengages the compressor clutch if the coolant temperature rises to a predetermined level.

    Answer B is wrong because the IAT sensor signal is not used to control compressor operation.

    Answer C is wrong because the oxygen sensor signal is not used to control compressor operation.

    Answer D is wrong because there is no cooling fan sensor.

134. Answer A is wrong because the engine control unit can control an electric condenser fan motor.

    Answer B is wrong because the body control unit can control an electric condenser fan motor.

    Answer C is right because a manual switch does not control an electric condenser fan.

    Answer D is wrong because an electric condenser fan motor can be controlled by an electronic relay.

135. Answer A is wrong because you replace a cracked fan blade, not weld it.

    Answer B is wrong because you do not repair a cracked fan blade.

    Answer C is wrong because both Technician A and Technician B are wrong.

    Answer D is right because both Technician A and Technician B are wrong.

136. **Answer A is right** because the PCM uses the engine coolant temperature sensor signal to turn on the coolant fans. If this signal is not available, the coolant fans do not come on until the A/C system is operating and the A/C pressure fan switch signal is received by the PCM.

    Answer B is wrong because a defective A/C pressure fan switch does not cause the coolant fans to be inoperative until the A/C system is turned on.

    Answer C is wrong because Technician A is right and Technician B is wrong.

    Answer D is wrong because Technician A is right and Technician B is wrong.

137. Answer A is wrong because all electronic HVAC actuators use some type of feedback device, but Technician B is also right.

    Answer B is wrong because all electronic HVAC actuators use some sort of feedback device, but Technician A is also right.

    Answer C is right because both Technician A and Technician B are right.

    Answer D is wrong because both Technician A and Technician B are right.

138. Answer A is wrong because the blend door actuator should not bleed vacuum.

    Answer B is wrong because a vacuum-operated actuator does not contain a motor.

    Answer C is right because a vacuum actuator door should hold vacuum indefinitely.

    Answer D is wrong because the vacuum actuator diaphragm is not porous.

139. **Answer A is right** because the outside air door is closed in the MAX A/C mode.

    Answer B is wrong because the defroster door is closed in the MAX A/C mode.

    Answer C is wrong because Technician A is right and Technician B is wrong.

    Answer D is wrong because Technician A is right and Technician B is wrong.

140. Answer A is wrong because the replacement of cables with small air lines simplifies control panel replacement.

Answer B is right because coolant control valves are sometimes operated by chassis air.

Answer C is wrong because in air-controlled HVAC systems, the mode and blend air cylinders typically operate air doors.

Answer D is wrong because air leaks can cause mode and blend air doors to react sluggishly or to be inoperative.

141. **Answer A is right** because the Technician must remove the negative battery cable and wait for the specified time before removing the control panel.

Answer B is wrong because the refrigerant does not have to be discharged to remove and replace the control panel.

Answer C is wrong because Technician A is right and Technician B is wrong.

Answer D is wrong because Technician A is right and Technician B is wrong.

142. Answer A is wrong because a kinked cable housing will cause the cable to bind.

Answer B is wrong because corrosion in the cable housing will cause the cable to bind.

Answer C is wrong because a deformed or overtightened cable clamp will cause the cable to bind.

Answer D is right because the mode doors are not controlled by the temperature control cable.

143. Answer A is wrong because you test control panel vacuum systems by applying vacuum with a hand pump to the input end of the system, not output.

Answer B is right because you connect the vacuum pump to each vacuum actuator and supply 15 to 20 in. Hg to the actuator and the gauge should stay at a steady vacuum for 1 minute.

Answer C is wrong because Technician A is wrong and Technician B is right.

Answer D is wrong because Technician A is wrong and Technician B is right.

144. **Answer A is right** because replacement is the only good repair for an HVAC control panel with an air leak.

Answer B is wrong because this process will not work.

Answer C is wrong because this process will not work.

Answer D is wrong because this process will not work

145. Answer A is wrong because a leaking dash vacuum switch will hiss and cause some control problems but will not cause total loss of vacuum control to the mode doors.

Answer B is wrong because a defective A/C compressor would cause total loss of any cold air.

Answer C is right because the loss of vacuum supply to the control panel results in a fail-safe mode of all air to the lower heater outlets.

Answer D is wrong because a heater control valve failure causes either no cold air or no hot air.

146. Answer A is wrong because the figure does not show the air mix door adjustment

Answer B is right because the figure shows the manual coolant valve adjustment.

Answer C is wrong because the figure does not show the ventilation door control rod adjustment.

Answer D is wrong because the figure does not show the defroster door control rod adjustment.

147. Answer A is wrong because the temperature control lever must be rotated fully counterclockwise to the maximum cold position.

Answer B is wrong because the black cable attaching flag must be attached to the flag receiver.

Answer C is wrong because both Technician A and Technician B are wrong.

Answer D is right because both Technician A and Technician B are wrong.

148. Answer A is wrong because battery removal has no effect on this service.

Answer B is wrong because this has bearing on the removal process.

Answer C is right because if the blend air door is binding, it could damage the new actuator.

Answer D is wrong because this is not a sensitive CMOS electronic component.

149. Answer A is an accurate statement because some actuator motors are calibrated automatically in the self-diagnostic mode. Therefore, it is wrong.

Answer B is right because it is the EXCEPTION statement and A/C diagnostic trouble codes (DTC) represent a fault in a specific system, not a component.

Answer C is an accurate statement because the actuator control rods must be calibrated manually on some systems. Therefore, it is also wrong.

Answer D is an accurate statement because the actuator motor control rods should only require adjustment after motor replacement or adjustment. However, because it is a true statement, it is wrong.

150. Answer A is wrong because a vacuum check valve does not delay the passage of vacuum to components.

Answer B is wrong because a check valve does not switch vacuum on or off.

Answer C is right because it prevents a vacuum drop during periods of low source vacuum.

Answer D is wrong because a vacuum check valve is not a sensor.

151. Answer A is wrong because a leaking actuator diaphragm would cause this condition all of the time.

Answer B is right because the check valve is a one-way valve that allows the reservoir to hold constant vacuum regardless of fluctuations in the vacuum source.

Answer C is wrong because Technician A is wrong and Technician B is right.

Answer D is wrong because Technician A is right and Technician B is wrong.

152. Answer A is wrong because the engine must not be running when an air line is being removed from the vehicle.

Answer B is wrong because there is no need to drain water from the system before replacing a hose.

Answer C is right because all air must be drained from the system before attempting to replace any chassis air component.

Answer D is wrong because there is no need to remove the compressor from the vehicle when replacing an air hose.

153. Answer A is wrong because a defective unloader can cause the compressor to build up excessive system pressure.

Answer B is right because a defective moisture drain will not cause the relief valve to operate.

Answer C is wrong because a weak spring in the regulator will cause the relief valve to operate at a lower-than-normal pressure.

Answer D is wrong because a defective governor could cause this problem.

154. **Answer A is right** because when the recirculation door is in position A, outside air is drawn into the HVAC case.

Answer B is wrong because in position A in-vehicle air is blocked.

Answer C is wrong because Technician A is right and Technician B is wrong.

Answer D is wrong because Technician A is right and Technician B is wrong.

155. **Answer A is right** because a 50/50 air mix between the floor and the windshield in the defrost mode causes very slow windshield defrosting.

Answer B is wrong because a 15 solidus/85 percent air mix between the floor and the windshield in the defrost mode causes normal windhsield defrosting.

Answer C is wrong because a 25/75 percent air mix between the floor and the windshield in the defrost mode causes windshield defrosting to be slightly slower.

Answer D is wrong because a 30/70 percent air mix between the floor and the windshield in the defrost mode causes windshield defrosting to be slightly slower.

156. Answer A is wrong because this temperature is too low.

Answer B is wrong because this temperature is too low.

Answer C is wrong because this temperature is still too low.

Answer D is right because blower operation is delayed to prevent air from entering the cab at an uncomfortable temperature.

157. Answer A is wrong because most ATC systems have some type of internal diagnostic routine, but Technician B is also right.

Answer B is wrong because you can also display DTCs on the control panel or a scan tool, but Technician A is also right.

Answer C is right because both Technician A and Technician B are right.

Answer D is wrong because both Technician A and Technician B are right.

158. Answer A is wrong because a defective in-car sensor would cause this condition, but Technician B is also right.

Answer B is wrong because a sticking air blend door can also cause this condition, but Technician A is also right.

Answer C is right because both Technician A and Technician B are right.

Answer D is wrong because both Technician A and Technician B are right.

159. Answer A is wrong because the ohmmeter resistance test proved the ambient sensor and related circuit to be good.

Answer B is right because the ohmmeter resistance test proved the ambient sensor and related circuit to be good, and the control panel is the only other component.

Answer C is also wrong because an ohmmeter connected to terminals 9 and 18 in the control panel connector indicates the circuit is good.

Answer D is wrong because the ohmmeter that is connected to terminals 9 and 18 in the control panel connector indicates the circuit is good.

160. Answer A is wrong because a binding door will not cause the actuator to hunt for the proper position.

Answer B is wrong because a faulty motor will not cause the actuator to hunt for the desired position.

Answer C is wrong because ATC sensors are not adjustable.

Answer D is right because a faulty feedback device can cause the blend air door to hunt for the desired position.

161. Answer A is wrong because in a strategy-based diagnostic process of elimination, you must check for visual signs.

Answer B is right because the first process is a visual one to check the components.

Answer C is wrong because Technician A is wrong and Technician B is right.

Answer D is wrong because Technician A is wrong and Technician B is right.

162. Answer A is wrong because there is no need to remove the engine thermostat when replacing the coolant control valve.

Answer B is right because you do not have to drain the entire system.

Answer C is wrong because the cable needs to be replaced only if it is damaged.

Answer D is wrong because the system must be bled after the repair is complete.

163. Answer A is wrong because a defective engine thermostat lowers engine operating temperature and some heat is still discharged from the heater.

Answer B is wrong because vacuum closes the coolant control valve shown in Figure 7–34. With no vacuum supplied to the valve, it is open and coolant circulates through the heater core to allow hot air discharge from the heater.

Answer C is wrong because both Technician A and Technician B are wrong.

Answer D is right because both Technician A and Technician B are wrong.

164. **Answer A is right** because the vacuum actuator rod will not move if the defrost door is stuck.

Answer B is wrong because if the vacuum actuator diaphragm holds the vacuum for 1 minute or more, the diaphragm is not leaking.

Answer C is wrong because Technician A is right and Technician B is wrong.

Answer D is wrong because Technician A is right and Technician B is wrong.

165. Answer A is wrong because it is important to keep water drained from the system to reduce internal corrosion.

Answer B is wrong because some fan clutches are operated using chassis air.

Answer C is wrong because most radiator shutter systems are air operated.

Answer D is right because no accessory is integral with the chassis air; they are separated by a pressure protection valve to prevent air loss to the brakes.

166. **Answer A is right** because a 12-volt testlight is a valuable tool for diagnosing electric actuators and solenoids.

Answer B is wrong because a self-powered testlight is used to check circuit continuity.

Answer C is wrong because a hand-held scan tool is more valuable for diagnosing electronic components.

Answer D is wrong because a charging station is used to add refrigerant to the A/C system.

167. Answer A is wrong because some ATC microprocessors are built into the control panel, but Technician B is also right.

Answer B is wrong because some ATC microprocessor can be replaced independently from the control panel, but Technician A is also right.

Answer C is right because both Technician A and Technician B are right.

Answer D is wrong because both Technician A and Technician B are right.

168. Answer A is wrong because replacing the ATC control computer should only take place after following specific diagnostic steps.

Answer B is right because you should follow and perform the diagnostic steps in the truck manufacturer service manual.

Answer C is wrong because Technician A is wrong and Technician B is right.

Answer D is wrong because Technician A is wrong and Technician B is right.

169. **Answer A is right** because the control panel should be replaced when one segment of the digital readout on an ATC control panel is inoperative.

Answer B is wrong because the voltage supply must be satisfactory if some of the segments are illuminated.

Answer C is wrong because Technician A is right and Technician B is wrong.

Answer D is wrong because Technician A is right and Technician B is wrong.

170. **Answer A is right** because the blower motor does not start immediately when a cold engine is started.

Answer B is wrong because the computer ground should be tested before replacing the ATC computer.

Answer C is wrong because some input voltage signals are relayed from the engine computer via data links to the ATC computer.

Answer D is wrong because on some systems, the blower speed is controlled with a PWM voltage signal.

171. Answer A is wrong because the ATC control panel is not programmable.

Answer B is right because you can often detect ATC system failures using a scan tool.

Answer C is wrong because Technician A is wrong and Technician B is right.

Answer D is wrong because Technician A is wrong and Technician B is right.

172. Answer A is wrong because ATC system calibration is usually checked with a thermometer.

Answer B is wrong because ATC system calibration is not usually adjustable in the field.

Answer C is wrong because both Technician A and Technician B are wrong.

Answer D is right because both Technician A and Technician B are wrong.

173. Answer A is wrong because only certified A/C Technicians may operate approved recycling equipment.

Answer B is right because owners of approved refrigerant recycling equipment must keep records of the names and addresses to which the refrigerant was sent.

Answer C is wrong because Technician A is wrong and Technician B is right.

Answer D is wrong because Technician A is wrong and Technician B is right.

174. Answer A is wrong because methane is a flammable gas produced by rotting vegetation.

Answer B is right because ozone in the upper atmosphere helps shield the earth from UV radiation.

Answer C is wrong because nitrogen makes up most of the air we breathe and is important as an atmospheric fire suppressant.

Answer D is wrong because argon is an inert gas used as flux in some welding processes.

175. Answer A is wrong because it does have an odor.

Answer B is right because R-134a has a faint ether-like odor.

Answer C is wrong because R-134a has a faint ether-like odor.

Answer D is wrong because R-134a has a faint ether-like odor.

176. Answer A is wrong because the answer specifies the wrong data.

Answer B is right because DOT 4BA-300 is the level of approval necessary for refrigerant recovery containers.

Answer C is wrong because the answer specifies the wrong data.

Answer D is wrong because the answer specifies the wrong data.

177. Answer A is wrong because cylinder pressure is not an accurate measure of contents level.

Answer B is right because the total weight of the cylinder must not exceed the weight of the cylinder when it is empty plus the maximum rated net weight.

Answer C is wrong because refrigerant cylinders are not equipped with safety relief valves.

Answer D is wrong because shaking the cylinder is not an accurate measure of contents level.

178. Answer A is wrong because refrigerant is reprocessed to new product specifications during the reclaiming process, not during the recycling process.

Answer B is wrong because a chemical analysis of the refrigerant is only performed during the reclaiming process, not during the recycling process.

Answer C is wrong because both Technician A and Technician B are wrong.

Answer D is right because both Technician A and Technician B are wrong.

179. Answer A is wrong because the supplier, not the purchaser, must maintain refrigerant sales records.

Answer B is right because recovered refrigerant must be stored in gray containers with yellow tops.

Answer C is wrong because Technician A is wrong and Technician B is right.

Answer D is wrong because Technician A is wrong and Technician B is right.

180. Answer A is wrong because original refrigerant containers should not be used to store recycled refrigerant.

Answer B is wrong because if the valve is simply opened, any remaining refrigerant will be vented to the atmosphere.

Answer C is wrong because there is no reason to introduce oil into the cylinder.

Answer D is right because after any remaining refrigerant is recovered, the cylinder should be evacuated, marked, and recycled for scrap metal.

181. Answer A is wrong because cylinders must be evacuated to 27 inches of mercury before placing refrigerant in the cylinder, but Technician B is also right.

Answer B is wrong because the safe filling level must be monitored by the cylinder weight, but Technician A is also right.

Answer C is right because both Technician A and Technician B are right.

Answer D is wrong because both Technician and Technician B are right.

182. **Answer A is right** because comparing the pressure of recovered refrigerant to the theoretical pressure of pure refrigerant at a given temperature is the best method of testing for noncondensable gases in refrigerant.

Answer B is wrong because pressure cannot be compared to humidity.

Answer C is wrong because pressure cannot be compared to volume.

Answer D is wrong because a halogen leak detector cannot be used to check for noncondensable gases.

183. Answer A is wrong because recycled refrigerant must also contain less then 330 ppm of noncondensable gases, but Technician B is also right.

Answer B is wrong because recycled refrigerant must also contain less than 15 ppm of moisture, but Technician A is also right.

Answer C is right because both Technician A and Technician B are right.

Answer D is wrong because both Technician A and Technician B are right.

8 Preventive Maintenance Inspection (PMI)

Pretest

The purpose of this pretest is to determine the amount of review you may require before taking the ASE Medium/Heavy Truck Preventive Maintenance Inspection (PMI) Test. If you answer all the pretest questions correctly, complete the questions and study the information in this chapter to prepare for this test.

If two or more of your answers to the pretest questions are incorrect, study the required information in *Today's Technician Medium/Heavy-Duty Truck Series* published by Delmar Publishers, plus a study of the questions and information in this chapter.

The pretest answers are located at the end of the pretest; these answers also are in the answer sheets supplied with this book.

1. Power loss of batteries during winter driving is helped by:
   A. battery heaters.
   B. overcharging.
   C. slight overfilling with electrolyte.
   D. not maintaining a full charge.

2. One uses SAE 20 weight oil to lubricate what part of a trailer every three months?
   A. Brake camshaft support bearings
   B. Fifth wheel base and ramps
   C. Brake slack adjusters
   D. Roll-up door rollers

3. When carrying out an inspection of the trailer after the first 500 miles, *Technician A* adjusts the brakes. *Technician B* checks all areas where sealant is used and reseals as necessary when carrying out this inspection. Who is right?
   A. A only
   B. B only
   C. Both A and B
   D. Neither A nor B

4. *Technician A* says synthetic motor oil helps transmission and axle gears, bearings, and seals to last longer. *Technician B* says it can be used throughout the service life of the vehicle. Who is right?
   A. A only
   B. B only
   C. Both A and B
   D. Neither A nor B

5. All of the following statements regarding air governors are true EXCEPT:
   A. common cut-out pressure is 125 ± 5 psi.
   B. one-quarter turn of the adjusting screw will change the pressure settings approximately 4 psi.
   C. test gauges are preferred over dash gauges for checking pressure by most manufacturers.
   D. common cut-in pressure is 85 to 90 psi.

6. Which of the following would be the LEAST likely reason to tag a vehicle out of service?
   A. A faulty coolant temperature gauge
   B. Exhaust system is leaking in front of or below the driver/passenger compartment
   C. More than 20 percent of frame mounting fasteners for a fifth wheel are missing or ineffective
   D. Fuel system has a visible leak at any point

7. What is the LEAST likely operation to be performed during an inspection of a truck's electrical charging system?
   A. Checking belts for tightness
   B. Alternator current output test
   C. Voltage regulator test
   D. Voltage drop test on insulated side of charging circuit

8. All of the following would cause a constant leak from the air dryer purge valve EXCEPT:
   A. foreign particles on valve seat.
   B. purge valve seal damaged.

C. purge valve frozen.
D. restricted line.

9. What is the MOST likely mileage recommended between changing lubricant for truck axles driven on a Class A highway tractor?
A. 200,000 miles
B. 100,000 miles
C. 50,000 miles
D. 25,000 miles

10. During a PM, the inspector tests the air brake supply system and finds that the build-up time is slow. Which of these is the MOST likely cause?
A. A defective air governor
B. A leak in a brake chamber
C. An air leak in the cab
D. A bad air compressor

11. What type of fifth wheel is shown in Figure 8–1?
A. Semioscillating type
B. Full-oscillating type
C. Stabilized (four-point support)
D. No-tilt convertible

12. In Figure 8–2 what could be mounted above the roof? (depicted by arrow)
A. A/C condenser
B. Solar panels
C. Skylight
D. Air vents for sleeper cab

13. When performing a PM, the minimum tread depth on tractor drive axle tires is:
A. 1/64 in. (0.39 mm).
B. 1/32 in. (0.79 mm).
C. 2/32 in. (1.58 mm).
D. 3/32 in. (2.38 mm).

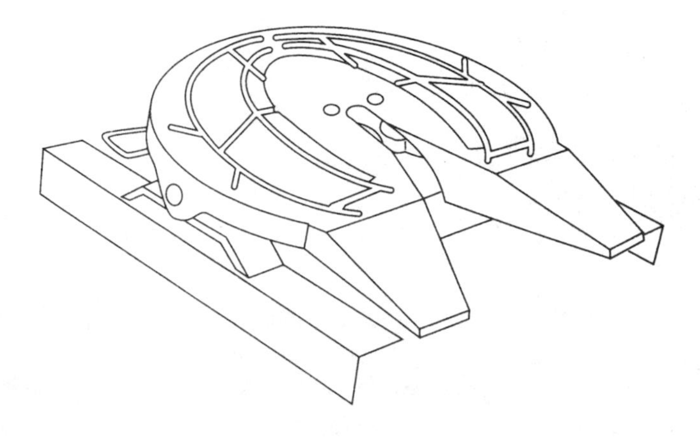

Figure 8–1 Fifth wheel.

Figure 8–2 Tractor cab.

14. A tractor has type 30 clamp-type brake chambers on the drive axles. During a PM each brake chamber has a 1.75 in. (44 mm) stroke, and an air leak is heard at the right brake chamber on the forward rear axle. *Technician A* says the tractor should be tagged out of service for excessive brake chamber stroke. *Technician B* says the tractor should be tagged out of service because of the air leak at the brake chamber. Who is right?
A. A only
B. B only
C. Both A and B
D. Neither A nor B

Answers to Pretest

1. A, 2. D, 3. A, 4. A, 5. D, 6. A, 7. D, 8. D, 9. B, 10. D, 11. A, 12. A, 13. B, 14. B

Engine Systems

Task 1

Check engine operation; record idle and governed rpm.

1. A diesel engine with an in-line injection pump and a mechanical governor has a lack of power under load (Figure 8–3). *Technician A* says wear in the governor flyweight pivots may be causing insufficient rack movement. *Technician B* says the governor flyweights may be sticking. Who is right?
   A. A only
   B. B only
   C. Both A and B
   D. Neither A nor B

2. Refer to Figure 8–4. A diesel engine with an electronic diesel control (EDC) system has:
   A. a throttle linkage connected from the accelerator pedal to the governor.
   B. a flyweight governor in the injection pump.
   C. a linear-motion solenoid controlling the rack in the injection pump.
   D. a fulcrum lever in the injection pump.

Hint

The mechanical governor used with an in-line injection pump is connected between the accelerator pedal and the injection pump rack. The governor flyweight movement controls the rack position. Because the rack rotates the injector plungers, the governor flyweight movement supplies the proper injector plunger position to provide the precise amount of fuel for all engine speed and load conditions. The governor limits the maximum engine rpm. Many engines are presently equipped with electronic governor and fuel control systems. In some of these systems, an electronic control unit (ECU) controls

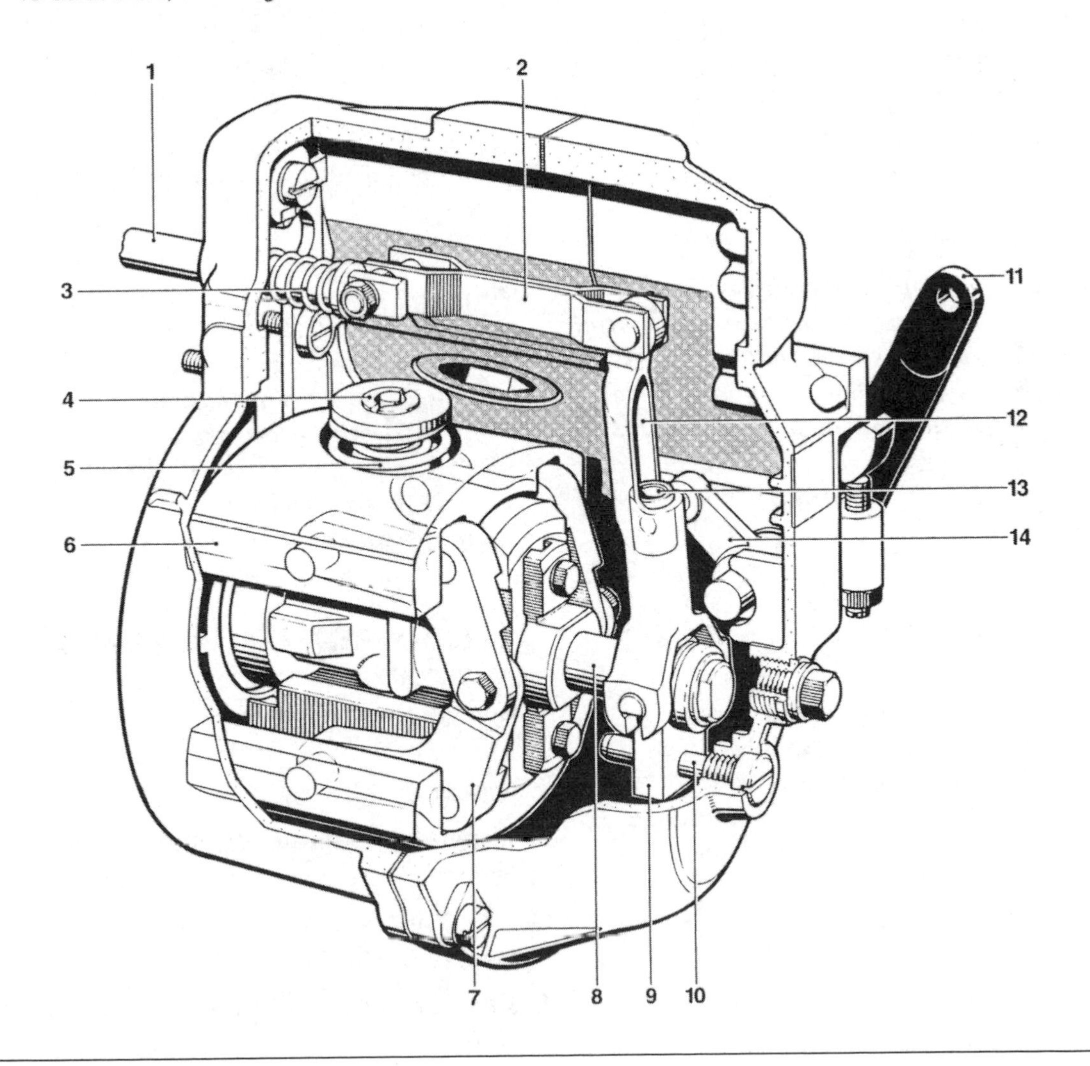

Figure 8–3 Mechanical governor. *(Reprinted with permission from Robert Bosch Corporation)*

Figure 1: Fuel-injection system with electronically controlled in-line fuel-injection pump
1 Fuel tank, 2 Supply pump, 3 Fuel filter, 4 In-line fuel-injection pump, 5 Electrical shutoff device (ELAB), 6 Fuel-temperature sensor, 7 Rack-travel sensor, 8 Actuator with linear-motion solenoid, 9 Pump-speed sensor, 10 Injector 11 Coolant-temperature sensor, 12 Accelerator-pedal sensor, 13 Switches for brakes, exhaust brake, clutch, 14 Operator panel, 15 Warning lamp and diagnosis connection, 16 Tachograph or vehicle-speed sensor, 17 ECU, 18 Air-temperature sensor, 19 Charge-air-pressure sensor, 20 Exhaust-gas turbocharger, 21 Battery, 22 Glow-plug and starter switch.

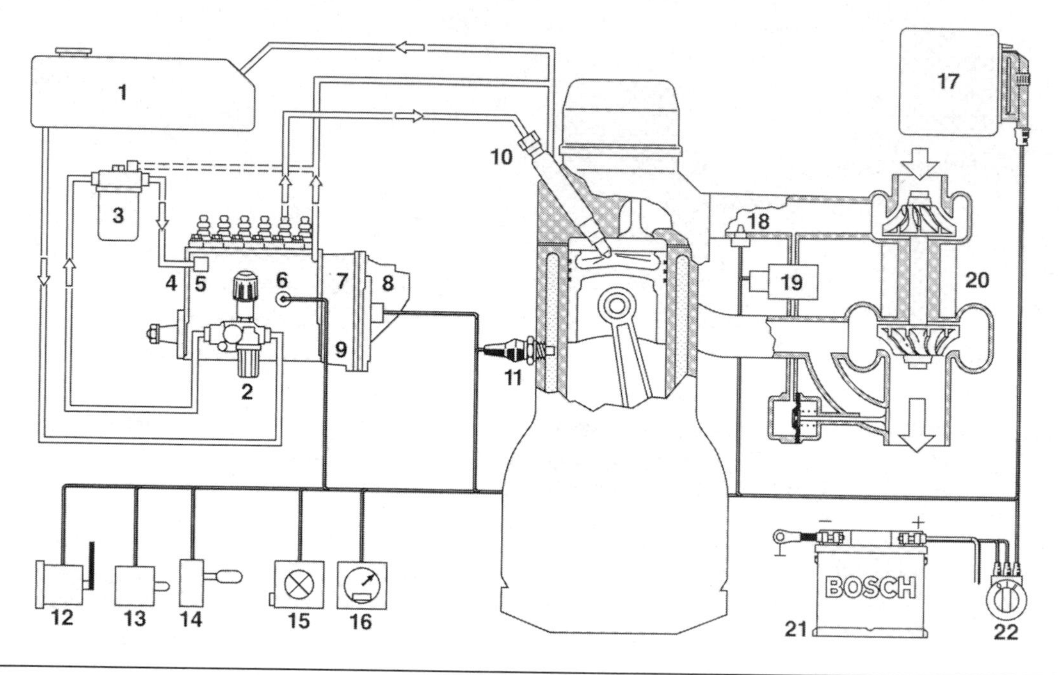

Figure 8–4 Electronic diesel control (EDC) system. *(Reprinted with permission from Robert Bosch Corporation)*

a linear-motion solenoid connected to the injection pump rack. The ECU and the linear-motion solenoid control the rack, injector plunger position, and the amount of fuel injected. In these systems the accelerator pedal is connected to an accelerator pedal sensor that sends a voltage signal to the ECU in relation to accelerator pedal movement. The accelerator pedal is not connected to the governor by a linkage.

Task 2

Inspect vibration damper.

3. *Technician A* says the vibration damper counterbalances the back-and-forth twisting motion of the crankshaft each time a cylinder fires. *Technician B* says if the seal contact area on the vibration damper hub is scored, the damper assembly must be replaced. Who is right?
   A. A only
   B. B only
   C. Both A and B
   D. Neither A nor B

Hint

As part of any PMI, the engine is inspected including the vibration damper. This vital engine part is fastened on the front of the engine, balancing the harmonic vibrations caused by the torsional loads of the rotating crankshaft. The damper is usually inspected for cracks caused by stress and/or deterioration. When an engine misses or is not properly balanced (flywheel, torque converter, vibration damper), it may be confused with a engine balance problem, wheel shimmy, or driveline vibration. A Technician should correct engine misses, unbalance concerns, and then test drive the truck to ensure the elimination of these conditions.

Task 3

Inspect belts, tensioners, and pulleys; adjust belt tension.

4. What is shown in Figure 8–5?
   A. Cooling system tester
   B. Belt tension gauge
   C. Vacuum gauge
   D. Dial indicator

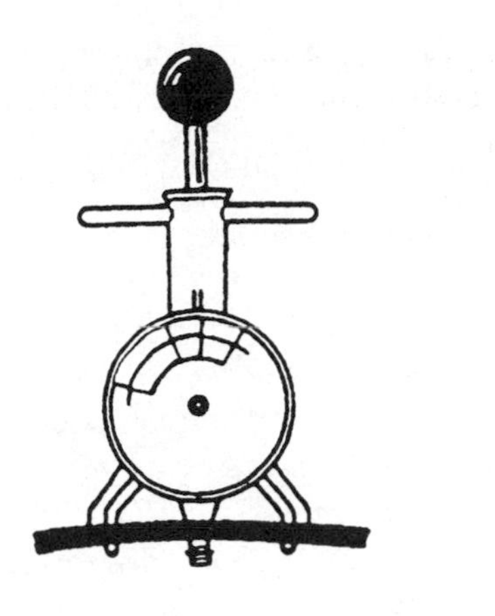

Figure 8–5 Truck service tool.

Hint

All accessory belts should be checked for obvious signs of wear, tightness, and /or correct application (if suspected). A belt gauge should be used whenever possible for the most accurate readings. If one is not available, the belt deflection should be measured at the specified location. An idler hub should be checked for loose or worn bearings, condition of the belt surfaces, and cracks from fatigue or deterioration.

Task 4

Check engine for oil, coolant, and fuel leaks (engine off).

5. A cooling system is being pressure tested. *Technician A* says that a pressure reading of 15 to 20 psi is normal. *Technician B* says that it is necessary to warm the engine to normal operating temperature before conducting a pressure test. Who is right?
   A. A only
   B. B only
   C. Both A and B
   D. Neither A nor B

Hint

A proper inspection of the cooling system should always involve a careful check of the cooling system maintenance history, including coolant change intervals and component replacement. While doing an inspection, an engine should be checked for both static and dynamic leaks. This means that the engine should be checked while it is running, and while it is off. This ensures that all gasket and mating surfaces are monitored in all possible operating and nonoperating conditions. Although a leak in the cooling system can be a serious problem, it is not looked upon as criteria for deadlining because of the capacity of the system on most heavy trucks, and because it is not a safety concern if an engine overheats.

Task 5

Inspect engine mounts for looseness and deterioration.

6. During a truck inspection, the Technician discovers the plastic radiator shroud is broken. Pieces are missing from the shroud above the fan blades. The remaining part of the shroud has impact marks near the cooling fan blades. *Technician A* says one of the engine mounts may be broken. *Technician B* says the fan blades may be out of balance. Who is right?
   A. A only
   B. B only
   C. Both A and B
   D. Neither A nor B

Hint

The most effective way to diagnose a vehicle with a suspected engine mount problem is to observe the engine assembly and its mounts while an assistant torques the engine with the brakes applied. This technique will allow a broken or deteriorated engine mount to quickly be observed and noted for replacement.

Task 6

Check engine oil level, check engine for oil, coolant, air, and fuel leaks (engine running).

7. *Technician A* says that a cooling system leak is sufficient criterion for a deadlining. *Technician B* says there must be sufficient vacuum for there to be at least one full brake application after the engine is shut off, otherwise it must be deadlined. Who is right?
   A. A only
   B. B only
   C. Both A and B
   D. Neither A nor B

8. A cooling system is pressurized with a pressure tester to locate the cause of coolant loss. After 15 minutes the tester gauge has dropped from 15 to 5 psi (103 to 34 kPa), and there are no visible leaks in the engine compartment. The cause of this problem could be:
   A. an engine thermostat stuck open.
   B. a leaking water pump seal.
   C. a leaking heater core in the cab.
   D. a leaking heater hose connection at the cylinder head.

Hint

An engine should be checked while doing an inspection for both static and dynamic leaks. This means that the engine should be checked while it is running, and while it is off. This ensures that all gasket and mating surfaces are monitored in all possible operating and nonoperating conditions. When checking for fuel leaks on an engine, it is mandatory that the engine be operating so the fuel system is pressurized, thus making a leak very apparent. A fuel leak while the engine is running is more hazardous than when an engine is off due to the higher fuel pressure with the engine running.

Task 7

Adjust intake and exhaust valves.

9. A gasoline engine in a medium-duty truck has mechanical valve lifters, and valve adjustment is provided by a screw and locknut in the end of the rocker arms. All the statements about valve adjustment in this engine are true EXCEPT:
   A. both valves in the cylinder should be closed.
   B. insufficient valve clearance may cause a clicking noise at high speed.
   C. turn the adjusting screw clockwise to decrease the valve clearance.
   D. excessive clearance may cause a clicking noise at idle speed.

Hint

When valves are properly adjusted, the specified clearance is available between the end of the rocker arm and the top of the valve stem. Valve clearance is required because heated parts expand, and if the valves have insufficient clearance they may remain open when the engine reaches normal operating temperature. Exhaust valves are subject to more heat and as a consequence, OEMs require greater valve clearance for exhaust valves. Many engine manufacturers recommend adjusting the valve with the engine at a specific operating temperature. When checking the valve adjustment on each cylinder, both valves must be closed.

Fuel System

Task 1

Check fuel tanks, mountings, lines, and caps.

10. In addition to replacing the fuel filter and priming and bleeding the system, what else should be performed when doing a PMI on a fuel system?
    A. Check fuel tank mountings
    B. Perform pour point tests on the fuel
    C. Check the heater lines for close proximity to the fuel lines
    D. Recalibrate the fuel gauge

Hint

Cloud (wax) point provides more accurate vehicle operability during cold weather than pour point. It can be defined as the temperature at which a cloud or haze of wax crystals are formed. Scientifically, this occurs at 10 degrees below the formation of the wax crystals. Flow improvement additives are available to lower cloud point.

It is important to inspect the fuel tanks for indications of the tank shifting or loose bands, and inspect the tank mounting brackets for cracks or signs of loose mounting bolts. Also, the tank fill caps should be inspected for missing or leaking gaskets. Although fuel pressure gauges are normally given in psi, fuel pressure and other gauges may be identified by universal graphic forms. A Technician/inspector must be able to recognize and read these potentially different layouts as they are exposed to them. Underground fuel storage tanks can form condensation. Checks for water need to be diligent and any water found in the fuel system must be removed.

Task 2

Inspect throttle linkages and return springs.

11. A diesel engine does not respond well to load changes. When the engine load is increased, the engine lugs down. *Technician A* says the throttle linkage may require adjusting. *Technician B* says the fuel filter may be restricted. Who is right?
    A. A only
    B. B only
    C. Both A and B
    D. Neither A nor B

Hint

Before making any critical adjustments, ensure proper fuel pressure and flow is available. Improper setting of the throttle can be detected when checking the engine response to load changes. When the engine load is increased, an improper throttle adjustment is indicated if the air-fuel ratio becomes excessively rich and there is a sudden loss of engine power and engine lugging.

Throttle linkage adjustments have become extremely critical on modern engines. The travel of the throttle controls the engine over its complete operating range. For example, on the Cummins PT fuel system with a PTG-type pump, just setting the end stops is a touchy job. On governors using air fuel control (AFC), perform throttle linkage adjustments according to manufacturer's instructions. Then make sure the throttle shaft is all the way back against the rear throttle stop screw before adjusting the lever position. Do not move the rear stop screw as this will change the throttle linkage adjustment previously made. Check the travel from idle to the full-fuel position; this must meet manufacturer's calibration specifications. If it is incorrect, adjustment of the front stop screw will correct the problem. Then, check and adjust the idle speed according to manufacturer's procedures.

Task 3

Drain water from fuel system.

12. If the fuel is contaminated with water, the water will:
    A. remain mixed with the fuel.
    B. float on top of the fuel.
    C. settle to the bottom of the fuel tank.
    D. flow through the filters into the injection system.

Hint
Any wetness or even dampness around a seam of a fuel tank should be addressed promptly and should definitely call for a failure of a safety inspection. Also, draining of the condensed water vapor in the fuel tank is mandatory before an inspection regardless of season. There should be little or no water in the fuel tank at any time.

Task 4

Replace fuel filter; prime and bleed fuel system.

13. Which of these functions does the indicated gauge in Figure 8–6 monitor?
    A. Oil pressure
    B. Amperes
    C. Water temperature
    D. Fuel pressure

Hint
An official PMI should always include a replacement of the fuel filter. The system must first be depressurized, then the work performed, and finally primed and bled with high-quality diesel fuel before being repressurized, checked for leaks, and returned to service or continue with the inspection.

Task 5

Adjust fuel injectors.

14. When performing the adjustment in Figure 8–7 on a two-cycle diesel engine, all of these statements are true EXCEPT:
    A. both exhaust valves must be fully depressed.
    B. the governor speed control lever must be in the idle position.
    C. loosen the locknut and turn the pushrod to change the adjustment.
    D. with the gauge properly installed, the gauge shoulder should not pass over the top of the injector follower.

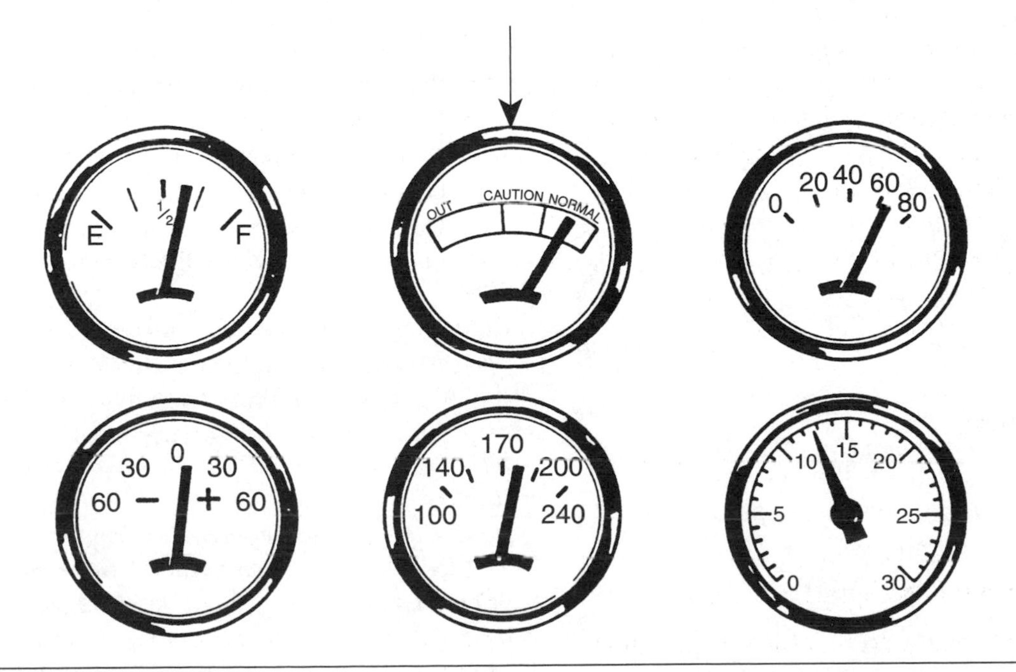

Figure 8–6 Instrument panel gauges.

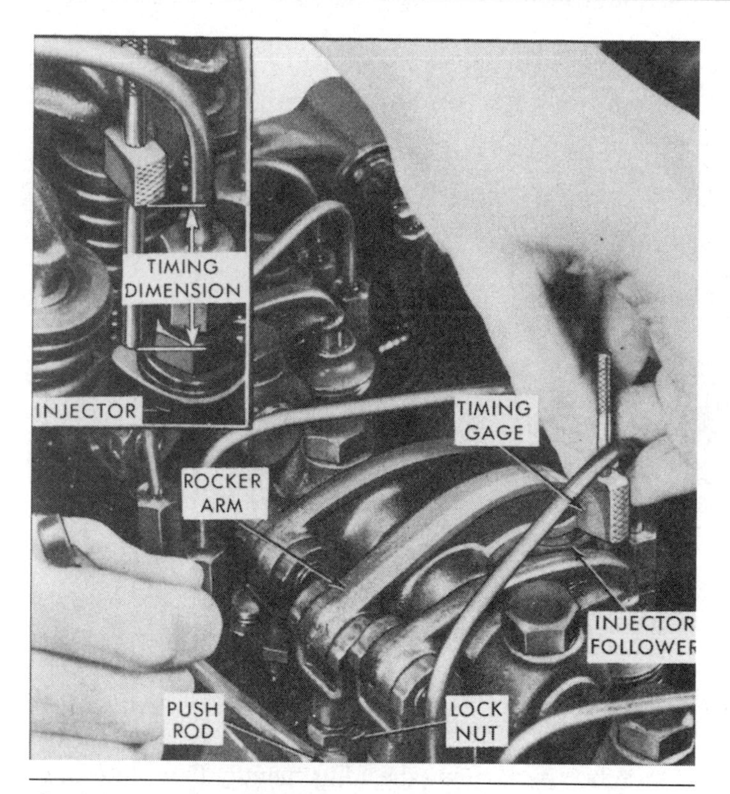

Figure 8–7 Injector timing. *(Courtesy of Detroit Diesel Allison, Division of General Motors Corporation)*

Hint

On vehicles equipped with prechambers, the fuel injector is installed in the precombustion chamber. On vehicles equipped with direct injection, the fuel injectors are installed in adapters. The nozzle and body are generally held in place by a nut and they can be removed by loosening the fuel line and removing the nut. The body and nozzle can then be lifted out. Also, when installing a fuel injector, always check the seats of both the nozzle and the precombustion chamber or adapter. Tighten the nut to the torque specifications. Excessive tightening will damage the nozzle. Excessive looseness will allow the nozzle to leak and, in some instances, cause the nozzle case to bulge and split.

Set the injector timing when the valves in the cylinder are fully depressed (open). On a Detroit Diesel Corporation (DDC) two-stroke engine, insert the dowel in the timing gauge into the injector-timing hole located in the upper flange of the mechanical unit injector (MUI). Smear a drop of engine oil on the injector follower flange and then adjust the injector pushrod so that when the timing tool is rotated it should gently wipe the oil film from the follower.

Air Induction and Exhaust System

Task 1

Check exhaust system mountings for looseness and damage.

15. A Technician is performing a PMI on a bus powered by a diesel engine. According to Commercial Vehicle Safety Alliance (CVSA) regulations, the bus must be tagged out of service if the exhaust system has a leak under the vehicle that is:
    A. 15 in. (38 cm) forward of the rearmost part of the bus.
    B. 12 in. (30 cm) forward of the rearmost part of the bus.
    C. 8 in. (20 cm) forward of the rearmost part of the bus.
    D. 6 in. (15 cm) forward of the rearmost part of the bus.

Hint

To check associated exhaust tubing and support brackets, a prybar can be typically used for the generalized pressure points. Check all exhaust components for leaks, and tap mufflers and catalytic converters with a soft hammer to check for loose internal components.

Task 2

Check engine exhaust system for leaks, restrictions, and damage.

16. A medium-duty truck with a throttle body injected gasoline engine has a loss of power and limited top speed of 45 mph (72 km/h). The fuel pump pressure and flow are within specifications. The engine runs smoothly at all speeds and never misfires. The MOST likely cause of this problem is:
    A. a restricted exhaust system.
    B. a defective ignition coil.
    C. a restricted fuel filter.
    D. a restricted fuel return line.

Hint

If any exhaust system leaking in front of or below the driver/sleeper compartment and the floor pan permits entry of exhaust fumes, or is likely to result in a fire from burning wiring, fuel supply, or other combustible parts, the vehicle must be declared out of service until it is properly repaired and tested.

Task 3

Check air induction system, piping, charge air cooler, hoses, clamps, and mountings; check for air restrictions and leaks.

17. During a PMI, *Technician A* says the air passages through the air-to-air intercooler should be checked for restrictions. *Technician B* says the air-to-air intercooler uses refrigerant to lower the intake air temperature. Who is right?
    A. A only
    B. B only
    C. Both A and B
    D. Neither A nor B

Hint

Depending on the conditions the vehicle is driven in, a truck's air induction ducts do not generally have to be cleaned out on every PMI. It is always a good idea to give a full visual inspection to the air induction system for blockage and air leaks whenever a PMI is performed.

Task 4

Inspect turbocharger for leaks; check mountings and connections.

18. A turbocharger on a tractor has less than the specified boost pressure. All of these defects may be the cause of the problem EXCEPT:
    A. turbocharger bearings that are starting to fail.
    B. improper wastegate adjustment.
    C. damaged compressor wheel.
    D. turbocharger oil seals leaking oil into the exhaust.

Hint

A turbocharger operates on the principle of pressure. If the engine is off, the boost pressure of the turbocharger would be zero. It would be impossible to check for a leak in a turbocharger if a leak is never allowed to develop.

Cooling System

Task 1

Check operation of fan clutch.

19. In the air-operated fan clutch in Figure 8–8:
    A. the fan clutch is engaged when the air solenoid shuts off air to the clutch.

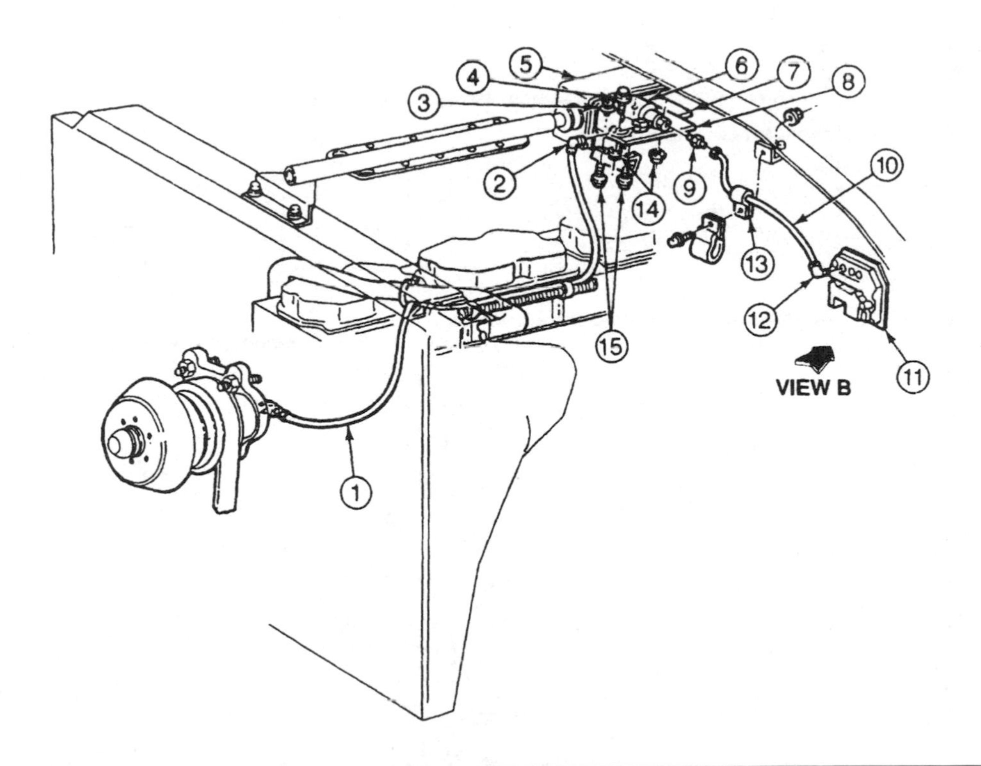

Figure 8–8 Air-operated cooling fan clutch. *(Courtesy of Ford Motor Company)*

B. the air solenoid is turned on and off by air pressure.
C. the fan clutch is engaged when the engine coolant is cold.
D. the fan clutch is engaged by air pressure.

20. Refer to Figure 8–9. *Technician A* says the water temperature switch is open at normal engine temperatures. *Technician B* says when current flows through the air solenoid, this solenoid turns on air pressure to the fan clutch to disengage the clutch. Who is right?
    A. A only
    B. B only
    C. Both A and B
    D. Neither A nor B

Hint

A fan clutch should be checked with the clutch released, and the maximum allowable play is generally recognized as 3/8 in. at the fan blade tip. Three levels of fan clutch preventative maintenance are normally performed on a heavy-duty truck: weekly air-filter draining, 10,000-mile walk-around checklist, and a 25,000-PM. The most important thing to remember is that most fan clutch failures are caused by air leaks. Whatever the PM level is being performed, an air leak check is by far the most important.

Task 2

Inspect radiator and mountings.

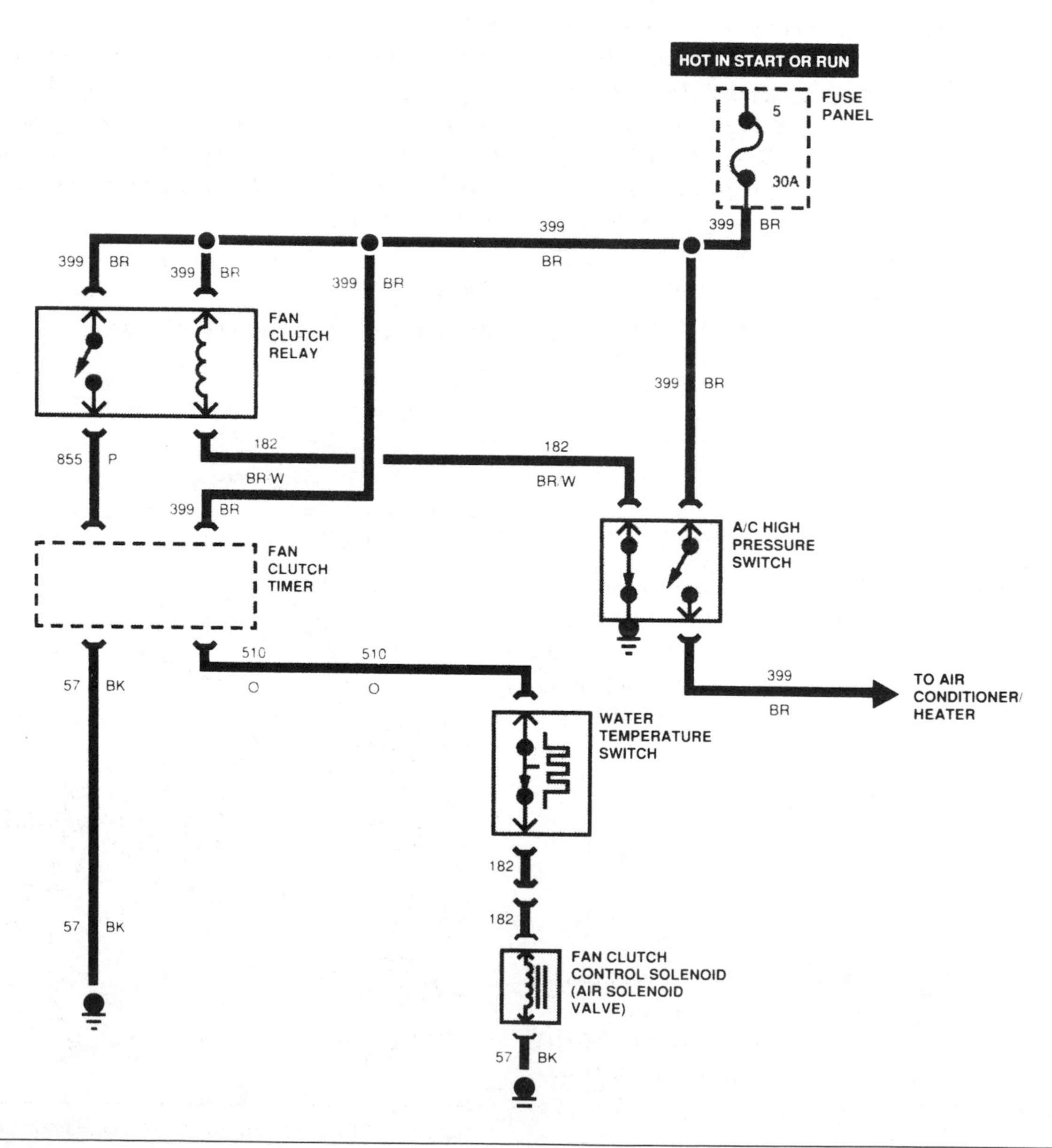

Figure 8–9 Electric circuit for air-operated fan clutch. *(Courtesy of Ford Motor Company)*

21. While discussing radiator shutters, *Technician A* says some shutters are operated by air pressure. *Technician B* says when the engine is at normal operating temperature the shutters should be closed. Who is right?
    A. A only
    B. B only
    C. Both A and B
    D. Neither A nor B

Hint

Radiator shutters are used to maintain coolant-operating temperature in cold climates, and to decrease engine warm-up time. By partially blocking off airflow through the radiator the shutters prevent excess cooling of the radiator. There are two basic ways to control shutters used on trucks. One method is a thermostatic piston that is mounted on the shutter assembly and reacts to outside temperature and radiator heat. When performing a PMI, inspect the linkage and shutters for freedom of movement, and test the thermostatic piston for proper operation.

Another method of opening and closing shutters is to use a thermostat regulated by engine temperature to supply compressed air from the engine compressor. Regulated air pressure supplied to a piston on the shutter assembly counteracts spring tension that holds the shutter open. Maintenance of this system will include inspection of all air lines for leaks and blockages and the shutter piston and thermostat. Check the shutter linkages for binding and wear.

Task 3

Inspect fan assembly and shroud.

22. While checking fan blades, *Technician A* says a cracked fan blade may cause a clicking or creaking noise with the engine idling. *Technician B* says you should not stand or position your body in line with the fan while increasing the engine speed. Who is right?
    A. A only
    B. B only
    C. Both A and B
    D. Neither A nor B

Hint

Fan assemblies must be inspected. Clean the fan and related parts with clean fuel oil and dry them with compressed air. Shielded bearings must not be washed: dirt may be washed in and the cleaning fluid cannot be entirely removed from the bearing. Examine the bearings for any indications of corrosion or pitting. Hold the inner race or cone so it does not turn and revolve the outer race or cup slowly by hand. If rough spots are found, replace the bearings. Check the fan blades for cracks; replace the fan if the blades are bent, as straightening may weaken the blades, particularly in the hub area. Look for cracks in the adjusting and support bracket castings, and check the fan shaft bearing.

Task 4

Pressure test cooling system and radiator cap.

23. The tester connected as shown in Figure 8–10 may be used to test:
    A. cooling system leaks.
    B. radiator cap vacuum valve.
    C. coolant specific gravity.
    D. external-to-the-radiator transmission cooler.

Hint

The normal system pressure for a cooling system is in the range of 15 to 20 psi (103 to 138 kPa). A pressure tester is connected to the radiator filler neck opening to test cooling system leaks. The cooling system should hold 15 psi (103 kPa) for 15 to 20 minutes if there are no leaks in the system.

Task 5

Inspect coolant hoses and clamps.

24. The upper radiator hose collapses after the engine is shut off and allowed to cool down. The cause of this problem could be:
    A. a restricted lower radiator hose.
    B. a plugged hose from the radiator filler neck to the recovery container.

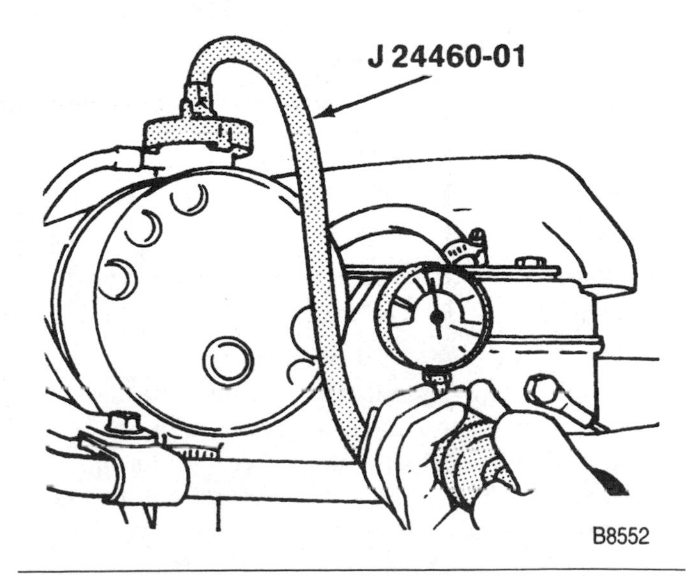

Figure 8–10 Cooling system test. *(Courtesy of General Motors Corporation, Service Technology Group)*

C. restricted radiator tubes.
D. a thermostat sticking open.

Hint

A coolant system, being one of the most important systems in a heavy truck, is always one of the first systems checked in a routine PMI inspection. A Technician will look for leaks or dried inhibitor deposits. All heater and water hoses should be sound, soft, and pliable. All clamps are checked for tightness and/or deterioration.

Task 6

Inspect coolant recovery system.

25. An excessively high coolant level in the recovery reservoir may be caused by any of these problems EXCEPT:
    A. restricted radiator tubes.
    B. a thermostat that is stuck open.
    C. a loose water pump impeller.
    D. an inoperative air-operated cooling fan clutch.

Hint

The radiator cap will maintain the pressure within the cooling system at or below approximately 15 psi (103 kPa). During heavy loads and prolonged periods of idling, this pressure may exceed the rating of the radiator cap and fluid will be vented to the coolant recovery reservoir. Once enough fluid has been vented to reduce the pressure within limits, the spring tension of the radiator cap will close the vent line. At normal temperatures, with the pressure cap removed, small fluctuations in coolant level can be expected. Inspection of this fluid level may indicate deeper problems. A constant overflowing of coolant to the recovery reservoir may indicate air pockets, uneven heating, or hot spots in the engine. Repetitious and rhythmic excessive surges in the coolant level usually indicate a leak to the combustion cylinder, such as a blown head gasket.

Task 7

Determine coolant condition and protection level (freeze point).

26. While discussing coolant solutions, *Technician A* says a cooling system may be filled with water if the truck is operating in a climate where the temperature is always above freezing. *Technician B* says a 50/50 water and antifreeze solution has a lower boiling point compared to water. Who is right?
    A. A only
    B. B only
    C. Both A and B
    D. Neither A nor B

Hint

The cooling capacity of the cooling system goes down as the engine is operated at higher altitudes. This is due to decreased atmospheric pressure lowering the boiling points of all liquids. A cooling system large enough to keep the coolant from boiling must be used.

The cooling system must be filled with a solution of antifreeze and water. This solution must contain the type and quantity of antifreeze specified by the vehicle manufacturer. A 50/50 solution of antifreeze and water provides freezing protection to -35°F (-37°C). Antifreeze contains rust inhibitor to prevent cooling system corrosion. Severe cooling system corrosion occurs if the system is filled with water.

Task 8

Service coolant filter/conditioner.

27. The component shown in Figure 8–11 is a/an:
    A. oil filter.
    B. coolant filter.
    C. transmission filter.
    D. power steering filter and cooler.

Hint

Coolant should be tested before reusing. Once the radiator or some other component is replaced, refill the system ensuring that the bleeder valves are open. When all the

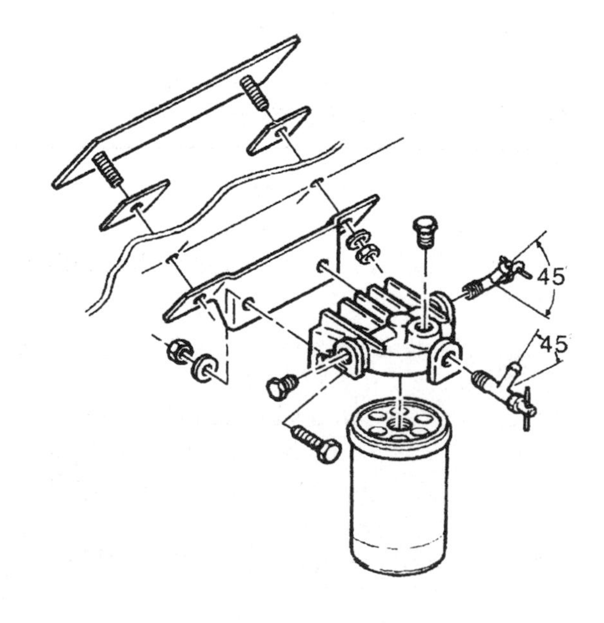

Figure 8–11 Filter. *(Courtesy of General Motors Corporation, Service Technology Group)*

air is out of the system, close the bleeder valves, start the engine, and check the coolant level. Test the antifreeze protection level at operating temperature. In modern engines, the antifreeze prevents corrosion and raises the temperature at which the coolant boils under pressure. This lengthens the life of the engine and cooling system components.

An inhibitor system is included in most types of antifreeze and no additional inhibitors are required on initial fill if a minimum antifreeze concentration of 30 percent by volume is used. Solutions of less than 30 percent concentration do not provide sufficient corrosion protection. Concentrations over 67 percent adversely affect freeze protection and heat transfer rates. Most manufacturers suggest 40 percent antifreeze to 60 percent solution. Ethylene glycol base antifreeze is recommended, such as in all Detroit Diesel engines. Commercially available inhibitor systems may be used to reinhibit antifreeze solutions.

Generally, OEMs recommend that the coolant SCA level be tested at each oil change interval. Additionally, whenever there is a substantial loss of coolant and the system has to be replenished, the SCA level should be tested. Coolant test kits permit the Technician to test for the appropriate SCA concentration, the pH level, and the total dissolved solids (TDS). The pH level determines the relative acidity or alkalinity of the coolant. Acids may form in engine coolant exposed to combustion gases or, in some cases, when cooling system metals (ferrous and copper base) degrade. The pH test is a litmus test in which a test strip is first inserted into a sample of the coolant, then removed and the color of the test strip indexed to a color chart provided with the kit.

Lubrication System

Task 1

Change engine oil and filters; visually check oil for coolant or fuel contamination.

28. All of these statements about changing the engine oil and filter are true EXCEPT:
    A. fill canister-type filters before installation.
    B. lubricate seals lightly before filter installation.
    C. an internal engine coolant leak causes a milky gray sludge in the oil.
    D. the by-pass valve opens if the oil pressure exceeds a specific limit.

Hint

After the initial lube change at 1,000 to 3,000 miles, subsequent oil changes should be made at 100,000-mile intervals for Class A or AA highway operation. Class A operation is defined as well-maintained highways of concrete or asphalt construction.

Oil filters trap particles that would otherwise cause damage to other critical components, like engine bearings. Regular replacement of filter elements and inspection of the housing and other component parts is necessary. Some filter assemblies incorporate a by-pass valve in their housing. These valve assemblies must be removed and inspected for corrosion, wear, and other signs of damage during overhaul or any time the filter assembly contains metal particles. Canister filters should be cut open and the element material examined for content. Suspended metal particles indicate a potential for bearing failure. A milky gray sludge indicates water in the oil caused by a possible head or cylinder block leak, which may already have caused bearing damage. Ensure all gasket surfaces are straight and free of nicks, which could cause an improper seal of the assembly. On element-type filter assemblies, ensure that housing grooves, center bolt threads, springs, metal gaskets, and spacers are in good condition. When reassembling filter assemblies, apply a small amount of oil to all seals to ensure a tight seal. Prefilling of canister-type filters is recommended to ensure adequate oil is present during startup to lubricate the bearings.

Task 2

Take an engine oil sample.

29. *Technician A* says an engine oil sample must be taken during every PMI. *Technician B* says to always take an engine oil sample from an engine that is at normal operating temperature. Who is right?
    A. A only
    B. B only
    C. Both A and B
    D. Neither A nor B

Hint

During a normal PMI, an engine oil sample is taken for inspection of its overall lubricating qualities, plus metal and dirt content. The oil may also be tested for corrosion-causing agents such as sulfuric and hydrochloric acids, and some esters. It is for this reason that the engine oil be at normal operating temperature to ensure that all possible contaminants and corrosives be contained in the oil in its normal operating condition.

Gasoline Engine Emission Control Systems

Task 1

Inspect positive crankcase ventilation (PCV) valve and system components.

30. When performing a PMI on a truck with a gasoline engine, the Technician notices that the air cleaner element is partially saturated with engine oil, and some oil has accumulated in the air cleaner. All of these defects may be the cause of the problem EXCEPT:
    A. a restricted PCV valve.
    B. excessive engine blowby.
    C. a leaking rocker arm cover gasket.
    D. a restricted hose from the PCV valve to the intake manifold.

Hint

The PCV valve should always be checked during any inspection of the emission control system. Improper operation of the PCV can cause erratic engine operation as well as higher-than-normal gasoline exhaust emissions, thereby possibly causing fines and/or penalties due to violations.

Task 2

Inspect emission control systems to include exhaust gas recirculation (EGR), air injection reaction (AIR), air inlet temperature control, early fuel evaporation (EFE), and fuel evaporation systems.

31. A gasoline engine experiences rough idle operation. This problem is worse when the engine is at normal operating temperature. The cylinder compression is within specifications and all the ignition components are satisfactory. The cause of this problem could be:
    A. an EGR valve stuck open.
    B. a burned exhaust valve.
    C. valve adjustment that is too tight.
    D. a bent intake valve.

Hint

One of the earliest methods used to reduce the amount of hydrocarbons in the exhaust was by forcing fresh air into the exhaust system after combustion. This additional fresh air causes further oxidation and burning of the unburned hydrocarbons. The system can be equipped with an external air pump. A pulse or aspirator air injection system may be used in place of an air pump system. The EGR valve recirculates some of the exhaust back into the intake manifold. This action reduces combustion temperature and oxides of nitrogen (NOx) emissions.

Cab and Hood

Instruments and Controls

Task 1

Inspect key condition and operation of ignition switch.

32. A diesel engine with the shutdown system in Figure 8–12 will not start. The cylinder compression is within specifications and the engine cranks normally. When the shutdown system is tested with the ignition switch on, the

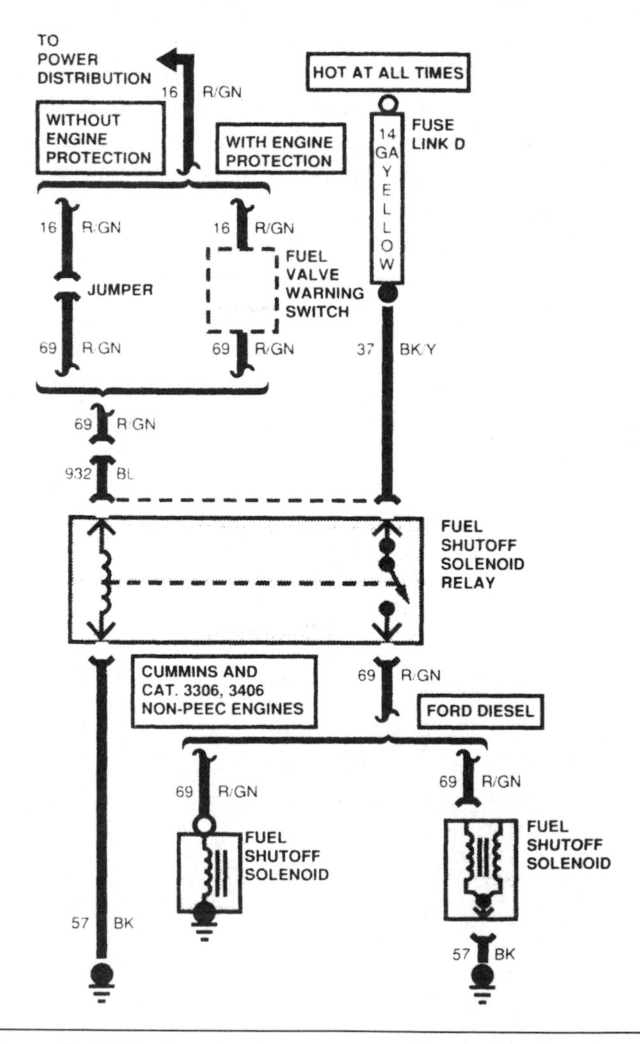

Figure 8–12 Engine shutdown system. *(Courtesy of Ford Motor Company)*

voltage supplied to the fuel shutoff relay winding is 12V, and the voltage at the ground side of the relay winding is 0.2V. The cause of this problem could be:
A. a burned-out 14 gauge yellow fuse link.
B. an open circuit in the fuel shutoff relay winding.
C. an open circuit between the fuel shutoff relay winding and ground.
D. an open circuit between the fuel shutoff relay winding and the ignition switch.

Hint

Turn on the ignition switch and perform a preliminary bulb check. Inspect the key for nicks and damage. Be sure the ignition switch furnishes power to the running systems. With the ignition on, engage the starter and determine if the engine will start.

Task 2

Check warning lights and/or alarms.

33. An engine coolant temperature gauge reading hot continuously may be caused by:
A. an open sending unit circuit.
B. a short to ground in the instrument panel power feed circuit.
C. a short circuit to or in the sending unit.
D. an open or burned out gauge.

Hint

Warning gauges and alarms may include low brake system air pressure, engine overheating, transmission temperature, rear axle temperature, and pyrometer. The pyrometer monitors exhaust gas temperature and displays this reading by varying its voltages according to exhaust temperatures.

Task 3

Check instruments; record oil pressure and voltage.

34. A gauge cluster is being inspected for proper function and illumination. *Technician A* says that a malfunctioning oil pressure gauge is not a valid reason for tagging a vehicle out of service. *Technician B* says that a proper function check of the instrument panel is required under a PMI. Who is right?
A. A only
B. B only
C. Both A and B
D. Neither A nor B

Hint

A short circuit in the sending unit wire to the gauge lowers the resistance in that circuit, thus causing the gauge needle to read high. Generally, a short to ground in a sending unit wire causes an overload of current to the gauge, which may damage the gauge. Although not considered an official criterion for deadline status, a malfunctioning oil pressure gauge may be cause for an inspector to deadline a vehicle, along with other important factors relating to that particular vehicle. Also, a proper and thorough check of a vehicle's instrument panel is always an important part of any PMI.

Task 4

Check hand throttle and manual engine shutdown operation.

35. *Technician A* says the shutdown valve is used to release a door inside the air box of the blower, completely restricting the air intake. *Technician B* says that you find shutdown valves on the air intake on two-stroke diesel bus engines. Who is right?
A. A only
B. B only
C. Both A and B
D. Neither A nor B

Hint

The shutdown valve is used to release a door inside the air box of the blower, completely restricting the air intake. The vacuum in the air box side of the rotors increases sharply, and the air pressure to the cylinders drops. Without air for combustion the engine stops.

This type of shutdown valve should only be used in case of an emergency . Constant use of this valve to perform normal shutdown will cause damage to the blowers or cylinder rings. On some turbocharged intercooled engines, air shutdown housings are mounted on the intercoolers, which are mounted on the blowers.

Task 5

Check air-conditioning (A/C), heater, and defroster controls.

36. *Technician A* says that all Technicians who perform service on the refrigerant portion of a motor vehicle air conditioner must be certified. *Technician B* always makes sure both hand valves on the manifold gauge set are open before connecting the hoses to the vehicle. Who is right?
A. A only
B. B only

C. Both A and B
D. Neither A nor B

Hint

There can be several heaters systems and their associated components in a heavy-duty truck. These can include but are not limited to: mirror heaters, cab heaters, immersion heaters, block heaters, circulation tank heaters, diesel fuel warmers, oil pan heaters, starting fluid systems, battery warmers, intake manifold heaters, coolant heaters, fuel-fired heaters, and oil dipstick heaters. Battery warmers are usually kept in the same vicinity that the batteries themselves are kept, which is at the same level, and slightly rearward of the point between the front and rear wheels on a truck tractor.

Task 6

Check operation of all accessories.

37. All these statements about VORAD collision warning systems are true EXCEPT:
    A. if another vehicle is a specific distance ahead in the path of the vehicle, the VORAD system applies the vehicle brakes.
    B. if another vehicle is a specific distance ahead in the path of the vehicle, the VORAD system illuminates warning lights to inform the driver.
    C. if another vehicle is a specific distance ahead in the path of the vehicle, the VORAD system may activate an audible warning to inform the driver.
    D. if another vehicle is in the blind spot on the right side of the tractor and trailer, the VORAD system illuminates warning lights to inform the driver.

Hint

VORAD (vehicle on-board radar) is like an electronic eye that constantly monitors other vehicles on the road. When other vehicles or objects are a specific distance ahead in the path of the vehicle, the VORAD system illuminates warning lights and in some cases provides an audible warning to the driver. This system also warns the driver when a vehicle or object is in the blind spot on the right side of the vehicle. The VORAD collision warning system gives drivers additional reaction time to respond to potential dangers. Because anticollision protection is high on the priority list of the Federal Highway Administration, collision warning systems will most likely become mandatory on heavy-duty vehicles some time in the near future.

Task 7

Using scan tool, extract engine monitoring information.

38. A scan tool is connected to the data link connector (DLC) in the cab to read engine data during a routine PMI. *Technician A* says the ignition switch must be on when disconnecting the scan tool. *Technician B* says that history DTCs can be read. Who is right?
    A. A only
    B. B only
    C. Both A and B
    D. Neither A nor B

Hint

Today's diesel engines are equipped with a variety of computerized engine control systems depending on the engine manufacturer. According To SAE Standards J1930, trouble codes are known as diagnostic trouble codes (DTCs). The DTC extraction method varies with the engine and truck manufacturer. The Technician needs to understand how to check, record, and clear DTCs. To retrieve these codes, the Technician connects a scan tool to a data link connector (DLC) on the vehicle. The ignition switch must be off when connecting or disconnecting the scan tool. DTCs may be current or history codes. Current DTCs are present at the time of diagnosis. History DTCs are caused by an intermittent fault. The defect causing a history DTC occurred sometime in the past, and this defect is not present at the time of diagnosis, but the DTC is stored in the powertrain control module (PCM) memory. DTCs may be erased with a scan tool, or by disconnecting battery voltage from the PCM. Always check the truck manufacturer's recommendations regarding DTC erasing.

Blink or flash codes may be used to obtain DTCs on some vehicles. On these systems a malfunction indicator light (MIL) in the instrument panel flashes out the DTCs. In most vehicles, the Technician has to connect specific terminals in the DLC to place the PCM in the diagnostic mode and cause the MIL light to flash the DTCs.

Safety Equipment

Task 1

Check operation of electric and air horns.

39. The air horns on a tractor are:
    A. usually mounted under the hood.
    B. supplied with air pressure through the pressure protection valve.

C. may be disassembled and serviced.
D. may be activated by a push button on the instrument panel.

Hint

Although technically electric- or air-driven components, the horns of a heavy truck are considered a safety device because they provide a warning in the vicinity of the truck, which may prevent loss of life and/or property damage.

Task 2

Check condition of safety flares, spare fuses, triangles, fire extinguisher, and all required decals.

40. *Technician A* says flares must be carried in the tractor. *Technician B* says a fire extinguisher must be mounted outside on the back of the cab. Who is right?
    A. A only
    B. B only
    C. Both A and B
    D. Neither A nor B

Hint

Check to see that the truck is equipped with all the necessary safety equipment. The truck should have spare fuses (unless equipped with circuit breakers), three red reflective triangles, and a properly charged and rated fire extinguisher within arm's reach of the driver's seat. Flares, lanterns, and flags are optional.

Commercial Vehicle Safety Alliance (CVSA) out of service criteria calls for enforcement when safety devices such as chains and hooks have been repaired improperly, this includes welding, wire, small bolts, rope, and tape. Federal regulations prohibit the usage of a vehicle that does not have at least one light illuminated on loads projecting more than 4 feet beyond the vehicle.

Rear chains and hooks that are used in double trailer systems for hookup should not be worn to the extent of a measurable reduction in link cross-sectional area or have any significant abrasions, cracks, or other faults that would affect its structural integrity. If any of the above parameters are found, then, the appropriate steps need to be taken to ensure a deadline status is invoked for the particular vehicle(s).

Task 3

Inspect seat belts and sleeper restraints.

41. The seat belts in the cab and sleeper are being inspected. *Technician A* says with the seat belt on and the truck moving, activate the parking brake to brake the truck very quickly. *Technician B* says it is not necessary to check the seat belts or restraints in the sleeper, only the cab. Who is right?
    A. A only
    B. B only
    C. Both A and B
    D. Neither A nor B

Hint

Check the operation of all seat belts and restraints. With the seat belt on and the truck moving, activate the parking brake to brake the truck very quickly and determine if the seat belt inertia reel locks.

Task 4

Inspect wiper blades and arms.

42. A roadside inspection is performed while it is raining. All of the following defects would be reason to tag the vehicle out of service EXCEPT:
    A. the wiper blade on the driver's side is cracked and badly worn.
    B. the wiper motor is inoperative.
    C. the passenger's side wiper blade is cracked and badly worn.
    D. the driver's side wiper arm is distorted so the blade does not contact the windshield properly.

Hint

CVSA safety standards out of service criteria state that any vehicle should be tagged out of service if it has an inoperative wiper or damaged parts that render it ineffective on the driver's side. This is applicable only in weather that requires the use of the windshield wipers

Hardware

Task 1

Check wiper and washer operation.

43. Component 3 in Figure 8–13 is the:
    A. windshield wiper motor.
    B. fluid level sensor.
    C. double check valve.
    D. windshield washer motor and pump.

Hint

The windshield wipers must operate smoothly at the selected speed. Pulse-type wipers must pause for various lengths of time depending on the switch setting. The

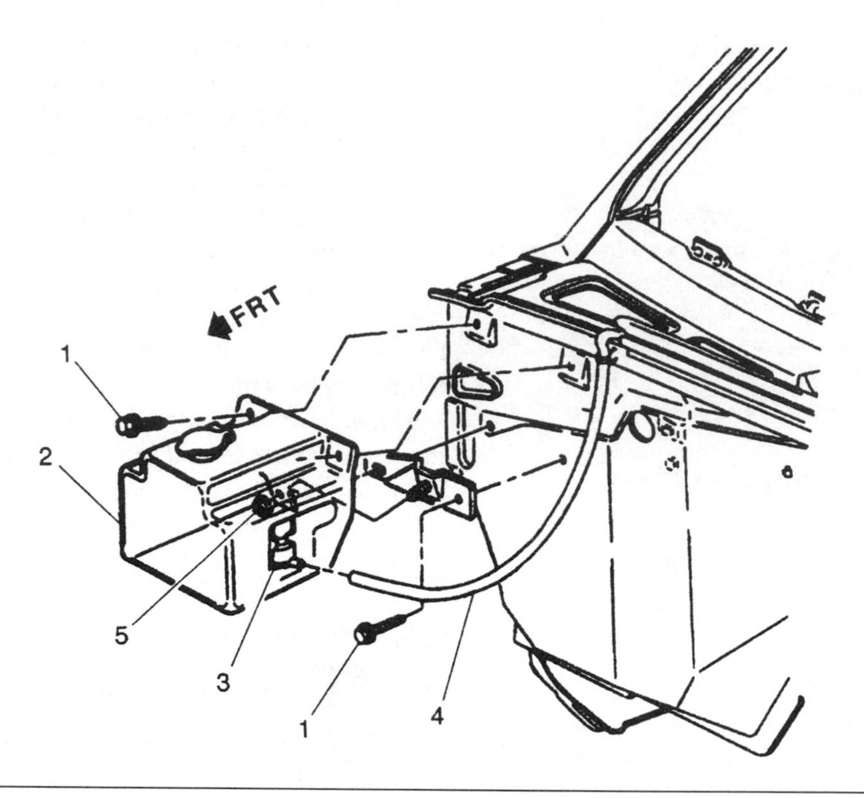

Figure 8–13 Electrical system component. *(Courtesy of General Motors Corporation, Service Technology Group)*

windshield washers must disperse water on the windshield when the washer motor is activated.

Task 2

Inspect windshield glass for cracks or discoloration; check sun visor operation.

44. A windshield is being inspected during a routine inspection. *Technician A* says that the vehicle must be deadlined, because it has cracks in the windshield greater than 1/2 inch wide on the passenger's side of the vehicle. *Technician B* says that any vision-distorting defect and/or hazardous working condition concerning the windshield glass must be within the sweep of the wiper on the driver's side to qualify for the vehicle to be deadlined. Who is right?
    A. A only
    B. B only
    C. Both A and B
    D. Neither A nor B

Hint

Check the windshield glass for cracks, dirt, illegal stickers, and any discoloration that obstructs the driver's plane of view. Clean and adjust as necessary. Check to see if the sun visor will block out sunlight. Any crack over 1/4 inch wide, intersecting cracks, discoloration not applied in manufacture, or other vision-distorting matter in the sweep of the wiper on the driver's side will take a vehicle out of service by the CVSA out-of-service standards.

Task 3

Check seat operation and mounting.

45. The pressure protection valve in an air seat system:
    A. protects the air supply in the air brake system if a leak occurs in the air seat.
    B. protects the air spring in the seat from excessive air pressure.
    C. closes when the air pressure exceeds a predetermined value.
    D. pulses on and off to regulate the air pressure.

Hint

An air-suspended seat reduces road shock transferred through the chassis, cab, and seat to the driver. Air pressure is supplied to the air-suspended seat through a pressure protection valve and a pressure reduction valve. The pressure protection valve protects the air supply in the air brake system if a leak occurs in the seat. The pressure reduction valve lowers the air pressure supplied to the seat.

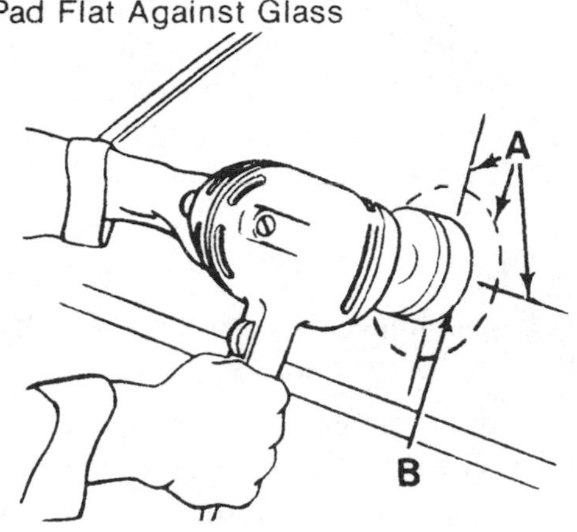

Figure 8–14 Body service. *(Courtesy of General Motors Corporation, Service Technology Group)*

Task 4

Check door glass and window operation.

46. The Technician in Figure 8–14 is:
    A. removing excess adhesive after a windshield replacement.
    B. removing a minor scratch from a side window.
    C. using a buffer and rubbing compound to remove paint abrasions.
    D. cleaning the side window molding.

Hint

Check to see if the windows will go up and down. Windows must go up and down without sticking or binding.

Task 5

Inspect steps and grab handles.

47. *Technician A* says a loose grab handle may cause personal injury. *Technician B* says the steps must be level and free from debris. Who is right?
    A. A only
    B. B only
    C. Both A and B
    D. Neither A nor B

Hint

Inspect that all steps are in good condition and secure. The grab handles must be securely attached to the cab.

Task 6

Inspect mirror mountings, brackets, and glass.

48. When inspecting a tractor for physical damage, a Technician notes that the mirror mounts for the passenger side are loose and in danger of coming off the door. What should the Technician do?
    A. Record the observed problems on a PMI
    B. Inform the service department that the bracket needs servicing
    C. Record the malfunctioning bracket on a PMI, then service it immediately
    D. Service it after a thorough inspection of the entire vehicle is performed

Hint

A typical fault such as a loose mirror mounting is always recorded on a PMI. This is in accordance with the Department of Transportation's Federal Motor Carrier Safety Regulations, and the Commercial Vehicle Safety Alliance standard inspection procedures.

Task 7

Record all observed physical damage.

49. While discussing van and over-top trailer bodies, *Technician A* says if one floor cross member is broken and sagging below the lower rail, the vehicle should be tagged out of service. *Technician B* says if a drop frame is visibly twisted, the vehicle should be placed out of service. Who is right?
    A. A only
    B. B only
    C. Both A and B
    D. Neither A nor B

Hint

Check the vehicle area for hazards on the vehicle and any physical damage such as dents or cracks. On van and over-top trailer bodies, if three or more floor cross members are broken and sag below the lower rail, the vehicle should be tagged out of service.

Task 8

Lubricate all grease fittings.

50. *Technician A* says that the Department of Transportation (DOT) does not require PM records on lubrication. *Technician B* says that maintenance records are an invaluable source of information in the life of a heavy truck. Who is right?
    A. A only
    B. B only
    C. Both A and B
    D. Neither A nor B

Hint

While supplying grease through all the grease fittings, these fittings should be checked to be sure they are not missing, damaged, broken, or plugged.

Task 9

Inspect and lubricate door and hood hinges, latches, strikers, linkages, and cables.

51. Item G is Figure 8–15 is:
    A. a hood hinge.
    B. a door hinge.
    C. a door check strap.
    D. a door protector.

Hint

Lubrication of door latches, locks, and hinges are always performed as part of a PMI.

Task 10

Inspect tilt cab mountings, hinges, latches, linkages and cables, and ride height; service as needed.

52. A scheduled inspection is being carried out on a cab over engine (COE) vehicle. *Technician A* says that if a component is not on the checklist, then it is not mandatory to inspect it. *Technician B* says you always check the cab ride height of a cab-over vehicle. Who is right?
    A. A only
    B. B only
    C. Both A and B
    D. Neither A nor B

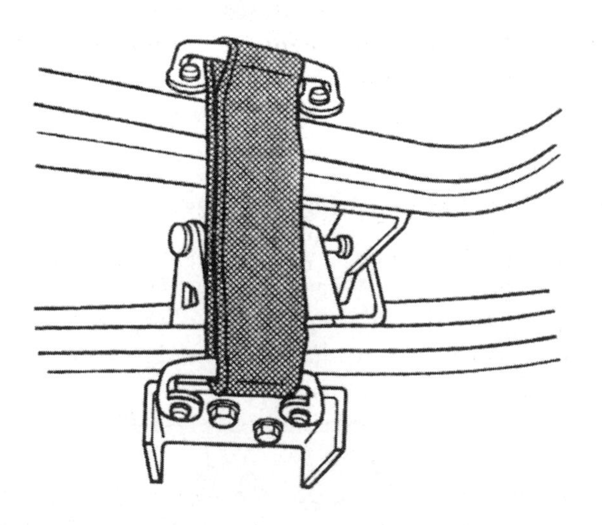

Figure 8–15 Chassis component. *(Courtesy of General Motors Corporation, Service Technology Group)*

Hint

With most hydraulic cab lift systems, there are two circuits: the push circuit that raises the cab from the lowered position to the desired tilt position, and the pull circuit that brings the cab from a fully tilted position up and over the center.

Task 11

Inspect tilt cab hydraulic cylinders for leakage; service as needed.

53. The "push" circuit in a cab over engine design:
    A. pushes the cab back over the engine from a fully extended position.
    B. pushes the cab to the 45-degree tilt position.
    C. allows the cab to descend to the 60-degree overhead position.
    D. raises the cab from the lowered position to the desired tilt position.

Hint

Inspect the hydraulic cylinders for excessive leakage and determine if the cab can be raised and lowered.

Air-conditioning and Heating (HVAC)

Task 1

Inspect A/C condenser and lines for condition and visible leaks; check mountings.

54. Refrigerant leaks at condenser line fittings are indicated by:
    A. an oily film around the fitting.
    B. a clear fluid dripping from the fitting.
    C. a hissing sound coming from the fitting.
    D. corrosion on the fitting.

Hint

Depending on the rate of leakage, the Clean Air Act establishes regulations concerning the repair of significant leaks. The Technician/inspector must report and log all refrigerant leaks into a PM schedule, so these repairs are addressed in a timely manner. All Technicians must be certified to perform service on an A/C system. When connecting a manifold gauge set to an A/C system, it is necessary to close the hand valves on the gauges to prevent refrigerant from escaping.

There should be no confusion between changes in vehicle air-conditioning and changes in refrigeration units (reefers). While both have undergone major changes, for the most part the refrigerants used in each and the regulations governing them are different.

The condenser is normally mounted just in front of the truck radiator, or if rooftop mounted, in the center of the roof above the driver/passenger area. In either position, it receives the full flow of ram air from the movement of the truck.

Task 2

Inspect A/C compressor and lines for condition and visible leaks; check mountings.

55. An A/C compressor has a growling noise only when the compressor clutch is engaged. The cause of this noise could be:
    A. loose compressor mountings.
    B. a worn compressor drive belt.
    C. a defective bearing in the compressor.
    D. liquid refrigerant entering the compressor.

Hint

Depending on the rate of leakage, the Clean Air Act establishes regulations concerning the repair of significant leaks. The Technician/inspector must report and log all refrigerant leaks into a PM schedule, so these repairs are addressed in a timely manner.

Electrical/Electronics

Battery and Starting Systems

Task 1

Inspect battery box(s), cover(s), and mountings.

56. *Technician A* says moisture and corrosion on battery tops cause faster battery self-discharge. *Technician B* says excessive corrosion on the battery box and cover indicates undercharging. Who is right?
    A. A only
    B. B only
    C. Both A and B
    D. Neither A nor B

Hint

An inspection of a battery should always begin with a thorough visual inspection of the case and the immediate surrounding area for any signs of leakage. The terminals should be in good condition, and the electrolyte level should be satisfactory. Then, after the surface charge has been removed, an open circuit voltage test and a load test may be performed, indicating the battery's current state of charge and capacity. On battery and starting motor ground circuits, a voltage drop reading above 0.2V indicates excessive resistance. These sources of high resistance may include ground cables, mounting bolts, and/or corrosion.

Task 2

Inspect battery holddowns, cables, and connections.

57. *Technician A* says to connect the battery positive cable first when hooking up a battery. *Technician B* says one always disconnects the ground cable first when disconnecting a battery. Who is right?
    A. A only
    B. B only
    C. Both A and B
    D. Neither A nor B

Hint

Battery terminals are either two tapered posts or threaded studs on top of the case or two internally threaded connectors on the side. Some newer batteries have both types of terminals so that one battery fits either application. When disconnecting a battery, the ground cable should always be disconnected first, then the positive cable; therefore, when reconnecting a battery, the positive cable should always be connected first and the ground cable last.

A battery can be cleaned with a baking soda and water solution. Always wear hand and eye protection when servicing batteries and their components. If the built-in hydrometer indicates light yellow or clear, the electrolyte level is low, and the battery should be replaced. A high-voltage regulator setting that causes overcharging may cause a low electrolyte level.

Task 3

Check/record battery state of charge.

58. A fully charged battery has a specific gravity of:
    A. 1.200.
    B. 1.225.
    C. 1.235.
    D. 1.265.

Hint

The battery state of charge may be measured with a hydrometer if the battery has removable cell filler caps. A fully charged battery has a specific gravity of 1.265. Open circuit voltage measured across the battery terminals with no load on the battery is also an indication of the battery state of charge. A fully charged 12V battery should have at least 12.6V. Batteries with sealed tops usually have built-in hydrometers. If this hydrometer is green, the battery is sufficiently charged for testing.

Task 4

Perform battery (load, high rate discharge) test.

59. A load test is performed on a battery with a 1.265 specific gravity on all cells and the battery temperature at 70°F (21°C) At the end of the load test the battery voltage is 9V. This indicates the battery:
 A. is sulfated.
 B. should be charged and retested.
 C. is defective.
 D. should be slowly discharged and recharged.

Hint

The battery discharge rate for a capacity test is usually one half of the cold cranking rating. The battery is discharged at the proper rate for 15 seconds to eliminate the surface charge. Then, after about 1 to 2 minutes, the test is repeated. The battery voltage must remain above 9.6V with the battery temperature at 70°F or above. Batteries may be kept from freezing by maintaining a full charge.

Task 5

Inspect starter, mounting, and connections.

60. *Technician A* says replacing a starting motor cable with a smaller diameter than the original cable has no effect on starting motor operation. *Technician B* says a replacement starting motor cable that is 3 ft. (91 cm) longer than the original cable results in excessive voltage drop at the starting motor. Who is right?
 A. A only
 B. B only
 C. Both A and B
 D. Neither A nor B

Hint

The starter circuit carries the high current flow and supplies power for the actual engine cranking. Components of the starting circuit are the battery cables, magnetic switch or solenoid, and the starter motor. Many starter relays are an integral part of the starter and they are replaced as a unit. Other starter relays may be mounted as a separate unit on the fender well. Any time maintenance is to be performed on the starter, remove the ground (negative lead) from the battery to prevent accidental shorting.

Task 6

Engage starter; check for unusual noises, starter drag, and starting difficulty.

61. A dragging starting motor may be caused by all of these problems EXCEPT:
 A. worn starter bushings.
 B. an overcharged battery.
 C. a defective starter armature.
 D. worn brushes and commutator in the starter.

Hint

A dragging starting motor may be caused by defects in the starting motor, a discharged or defective battery, or excessive resistance in battery and starting motor cables. If the starting motor provides a whirring noise but fails to engage, the starter drive is defective or not engaging with the flywheel ring gear.

Task 7

Check starter current draw.

62. The jumper wire in Figure 8–16 is connected to test the operation of the:
 A. magnetic switch.
 B. starting motor.
 C. starting switch.
 D. starter solenoid.

Hint

Although both are technically accurate, an inductive clamp ammeter is often used instead of a series ammeter. It is clamped onto the ground battery cable instead of connected in series with it. Therefore, the cable is not disconnected.

High starter current draw, low cranking speed, and low cranking voltage usually indicate a faulty starter motor. Low current draw, low cranking speed, and high cranking voltage usually indicate excessive resistance in the starting circuit.

Charging System

Task 1

Inspect alternator, mountings, and wiring.

63. A check of the general condition of the alternator is being performed. *Technician A* says to check the "ears" for bad mounting holes (out of round). *Technician B* always checks the alternator belt tension before making any performance tests. Who is right?

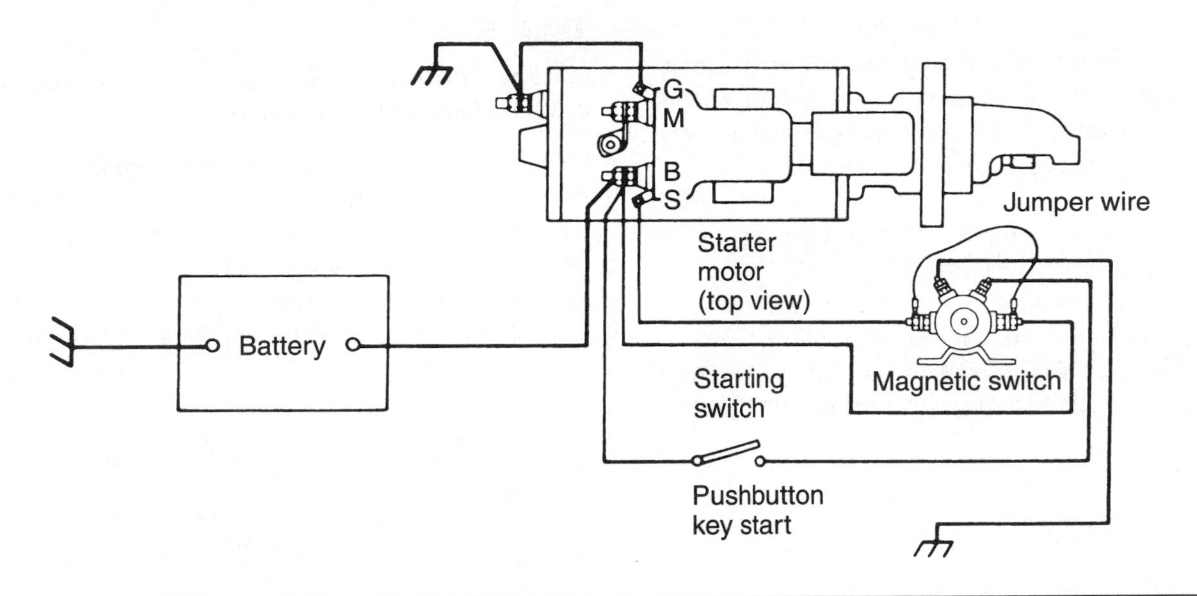

Figure 8–16 Starting circuit test. *(Courtesy of Freightliner Corporation)*

A. A only
B. B only
C. Both A and B
D. Neither A nor B

Hint

If the alternator mounting ear holes are badly distorted due to overtorquing and/or age, then the alternator should be replaced. Severely out-of-round mounting holes on an alternator causes severe misalignment of the drive belt, and may cause premature bearing failure due to uneven forces. A complaint of an unusual noise requires a thorough visual inspection. A visual inspection will usually determine whether a loose component is causing a vibration or any unusual sounds.

Task 2

Perform alternator current output test.

64. When testing the maximum output of an alternator, the engine should be run at:
    A. 2,000 rpm or above.
    B. 1,000 rpm or above.
    C. 3,000 rpm or above.
    D. 1,500 rpm or above.

Hint

The alternator belt tension should be checked before performing an output test. An amp-volt tester must be connected to the charging circuit. Most testers have an inductive ammeter clamp that is placed over the battery ground cable during the test. Some truck manufacturers may recommend installing this clamp over the positive battery cable or over the alternator battery wire. Some older amp-volt testers have an ammeter that is connected in series between one of the battery terminals and the battery cable. Some truck manufacturers recommend performing the alternator output test by full-fielding the alternator. This is done by connecting a jumper wire to the field circuit so the voltage regulator is bypassed, and full-field current flows through the alternator field circuit, which creates a very strong magnetic field around the alternator rotor. Many truck manufacturers recommend checking alternator output by using the carbon pile load in the amp-volt tester to lower the voltage to about 13V. When performing the output test, the engine rpm should be 2,000 or more, and the accessories should be turned off. If the alternator does not produce the rated output in amperes, the alternator usually has internal defects.

Task 3

Perform voltage regulator test.

65. What is the LEAST likely operation to be performed during an inspection of a truck's electrical charging system?
    A. Checking the belts for tightness
    B. Alternator current output test
    C. Voltage regulator test
    D. Voltage drop test on the insulated side of the charging circuit

Hint

A lower-than-specified voltage regulator setting causes an undercharged battery. If the voltage regulator setting is higher than specified, electrical components on the vehicle may be damaged from excessive current flow. A higher-than-specified voltage regulator setting also causes excessive battery gassing. Most voltage regulators are electronic and nonadjustable.

Lighting System

Task 1

Check operation of interior lights.

66. A heavy-duty truck is having its electrical system inspected. The interior lights glow very dimly even at a higher engine speed. *Technician A* says the ballast resistor should be inspected. *Technician B* says to inspect some of the cab ground circuits for corrosion and high resistance. Who is right?
    A. A only
    B. B only
    C. Both A and B
    D. Neither A nor B

Hint

The lighting systems of the heavy-duty truck can be divided into two categories: exterior and interior. Exterior lights are generally mounted on the sides, roof, and front of the vehicle. On the sides (depending on cab design) are turn signal lights, side marker indicators, and sometimes an intermediate turn signal. On the roof are clearance lights (usually on either end), identification lights (usually in the middle), and sometimes additional utility lights. In the front are single or dual headlights, fog lights, turn signal lights, and sometimes additional utility lights.

Task 2

Check all exterior lights, lenses, and reflectors; check headlight alignment.

67. When checking headlight aiming, *Technician A* says that the vehicle should be on level ground at a distance of 35 feet from the headlight alignment surface. *Technician B* says to always use headlight aiming equipment when adjusting headlights. Who is right?
    A. A only
    B. B only
    C. Both A and B
    D. Neither A nor B

Hint

Various types of headlight-aiming equipment are available commercially. When using aiming equipment, follow the instructions provided by the equipment manufacturer. Where headlight-aiming equipment is not available, headlight aiming can be checked by projecting the upper beam of each light on a screen or chart at a distance of 25 feet ahead of the headlights. The vehicle should be exactly perpendicular to the chart. An operational check of the brake light switch is performed during a check of the electrical system.

Faulty and/or improper ground connections are the most common sources of dim lights. This is especially true when conditions are confined to limited areas of any particular circuit.

Task 3

Inspect and test seven-pin connector, cable, and holder.

68. The trailer lights of a double combination vehicle are being inspected. *Technician A* says that the wiring connector for all trailer lights is always located under the front frame rail of the first trailer. *Technician B* says that it is mandatory to check the last trailer first in a multiple combination vehicle. Who is right?
    A. A only
    B. B only
    C. Both A and B
    D. Neither A nor B

Hint

The wiring connector for all trailer lights is located inside the cab, directly behind the driver's seat. Access is gained by usually removing a plastic cover held in place by four screws. In sleeper models, the connector is usually in the luggage compartment. Whenever a lighting inspection of a vehicle and/or combination vehicle(s) is to be performed, the inspector usually starts at the front and works back from there. This is because the source of power is usually up front and possible simple diagnosing would be made easier there.

In most modern box trailers, there are no splices for corrosion to attack, and the junction box has been eliminated for simplicity and overall economy. Generally, color-coded wires are molded in abrasion-resistant jackets that form tough, waterproof cables. Light plugs are molded on the cable for a positive seal, are filled with grease, and have spherical sealing rings to ensure positive connections.

Frame and Chassis

Air Brakes

Task 1

Check parking brake operation.

69. *Technician A* says that the control on the left in Figure 8–17 is the trailer control brake valve. *Technician B* says that the control on the right in the same figure is the tractor parking brake valve. Who is right?
    A. A only
    B. B only
    C. Both A and B
    D. Neither A nor B

Hint

Trucks with air brakes have spring brake chambers at all, or some of, the wheels. Mechanical parking brakes have a drum and shoe assembly mounted between the transmission and the propeller shaft. Some OEMs use an electric-hydraulic parking brake, which is a hybrid of the two previous designs. It has the same type of brake assembly as a transmission-mounted mechanical parking brake. It also has only one spring brake chamber or actuator that applies the brake. This parking brake actuator is mounted separately from the brake assembly in a component box under the driver's door. The actuator has a large coil spring that applies the brake. The coil spring is not directly linked to the brake assembly. Instead, a cable runs between the brake and the spring.

The tractor spring brake control is usually mounted on the dash of the tractor, is a push-pull type control (push to release, pull to apply), and normally has a yellow-colored diamond-shaped plastic knob. This brake valve applies and releases the tractor's spring brakes.

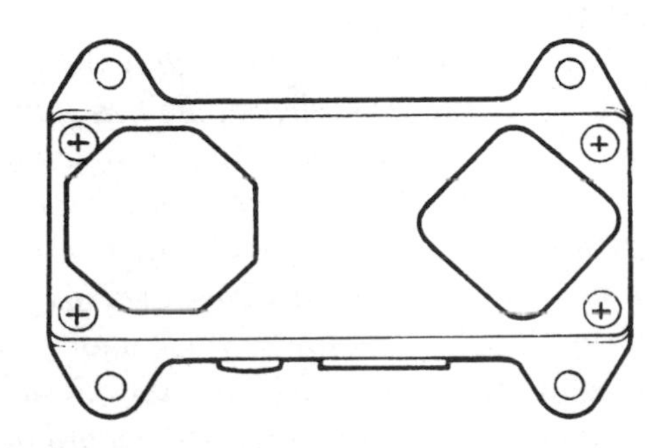

Figure 8–17 Air brake controls.

Task 2

Record air governor cut-out setting (psi).

70. An acceptable governor cut-out pressure is:
    A. 110 psi (758 kPa).
    B. 125 psi (862 kPa).
    C. 140 psi (965 kPa).
    D. 150 psi (1,034 kPa).

Hint

The governor simply controls the compressor loaded and unloaded cycles and determines the system pressure. System pressure in most trucks is set at values between 110 and 130 psi (758 and 896 kPa) with 120 psi (827 kPa) being typical. System pressure is known as cut-out pressure, the pressure at which the governor outputs the unloader signal to the compressor. The unloader signal is maintained until pressure in the supply tank drops to the cut-in value. Cut-in pressure is required by FMVSS 121 to be no more than 25 psi (172 kPa) less than the cut-out value. The difference on most systems ranges between 20 and 25 psi (138 and 172 kPa).

Governor operation can be easily checked. One method is to drop the air pressure in the supply tank to below 60 psi (414 kPa) and with the vehicle's engine running, build up the pressure. A master gauge should be used to record the cut-out pressure value. This should be exactly at the specified value. If not, remove the dust boot at the top of the governor, release the locknut, and turn the adjusting screw either CW to lower or CCW to raise the cut-out pressure.

If the unloader signal is not delivered to the compressor unloader assembly, high system pressures will result. If the safety valve on the supply tank opens, this is usually an indication of governor or compressor unloader malfunction. The only adjustment on an air governor is the cut-out pressure value. If the difference between governor cut-out and cut-in is out of specification, the governor must be replaced.

Task 3

Check air dryer drain valve operation.

71. While inspecting an air dryer, *Technician A* says the purge valve should open each time the compressor enters the unloaded cycle. *Technician B* says oil being discharged from the air dryer purge valve is a normal condition. Who is right?

A. A only
B. B only
C. Both A and B
D. Neither A nor B

Hint

The purge valve in an air dryer opens each time the compressor enters the unloaded cycle. When the purge valve opens, water and other contaminants are discharged from the dryer.

Task 4

Check air system for leaks (brakes released).

72. When leak testing the tractor service brake relay valve intake valve:
    A. a soapy water solution should be applied to the inlet port on the relay valve.
    B. a soapy water solution should be applied to the area around the inlet and exhaust valve assembly.
    C. the brakes should be applied.
    D. a 2 in. (50.8 mm) bubble in 3 seconds indicates normal intake valve leakage.

Hint

Air leaks in the air brake system may be tested with a soapy water solution applied to the area to be tested. On many components, such as air brake system valve exhaust ports, safety standards specify that any leak should not exceed a certain size bubble in a given time. Other components such as lines and fittings should not indicate any leaks. When testing specific components, the regulations require that the brakes be applied with specific application pressure.

System air build-up time is defined by federal legislation, specifically FMVSS 121. This legislation defines the required build-up times and values. A common check performed by enforcement agencies requires that the supply circuit on a vehicle be capable of raising air system pressure from 85 to 100 psi in 25 seconds or less. Failure to achieve this build-up time indicates a worn compressor, defective compressor unloader assembly, supply circuit leakage, or a defective governor.

Task 5

Check air system for leaks (brakes applied).

73. With the brakes applied at 80 psi (551 kPa), leakage at the brake application valve exhaust port should not exceed:
    A. a 2 in. (50.8 mm) bubble in 1 second.
    B. a 2 in. (50.8 mm) bubble in 2 seconds.
    C. a 3 in. (76.2 mm) bubble in 3 seconds.
    D. a 1 in. (25.4 mm) bubble in 3 seconds.

Hint

Air leaks in the air brake system may be tested with a soapy water solution applied to the area to be tested. On many components, such as air brake system valve exhaust ports, safety standards specify that any leak should not exceed a certain size bubble in a given time. Other components such as lines and fittings should not indicate any leaks. When testing specific components, the regulations require that the brakes be applied with specific application pressure.

According to the Commercial Vehicle Safety Alliance, a vehicle with more than 20 percent of its air brakes defective shall be declared out of service.

Task 6

Test one-way and double check valves.

74. While discussing one-way and two-way check valves, *Technician A* says a one-way check valve may be tested by disconnecting the supply line and checking for back leakage. *Technician B* says a two-way check valve supplies air pressure from the lowest of two pressure sources connected to the valve. Who is right?
    A. A only
    B. B only
    C. Both A and B
    D. Neither A nor B

Hint

The supply tank feeds the primary and secondary reservoirs of the brake system. Each is pressure protected by means of a one-way check valve. One-way check valve operation can be verified by removing the supply and checking back leakage. One-way check valves are used variously throughout an air brake system to pressure protect and isolate portions of the circuit.

Two-way (double) check valves play an important role as a safeguard in dual circuit air brake systems. The typical two-way check valve is a T with two inlets and a single outlet. It outputs the higher of the two source pressures to the outlet port and blocks the lower pressure source. The valve will shuttle in case of a change in the source pressure. In other words it always prioritizes the higher source pressure. Two-way check valves provide a means of providing the primary circuit with secondary circuit air and vice versa in case of a circuit failure. In the event that both source pressures are equal, as would be the case in a properly functioning dual air brake circuit, the valve will prioritize the first source to act on it.

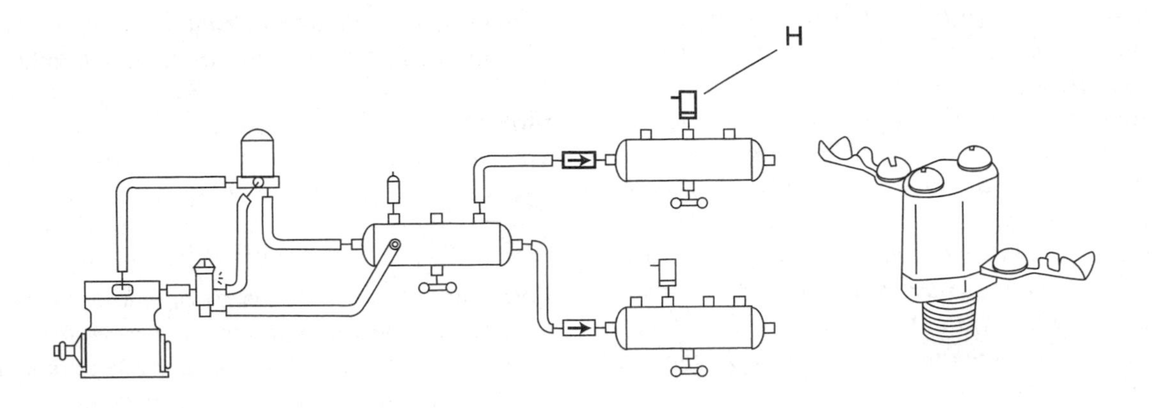

Figure 8–18 Air brake air supply system. *(Courtesy of Allied Signal Truck Brake Systems Co.)*

Task 7

Check low air pressure warning devices.

75. Component H in Figure 8–18 is a:
    A. low-pressure warning device.
    B. safety valve.
    C. two-way check valve.
    D. moisture ejector valve.

Hint

FMVSS 121 requires that a driver receive a visible alert when the system pressure drops below 60 psi (414 kPa). In most cases this is accompanied by an audible alert, usually a buzzer. A low air pressure warning device is fitted to both the primary and secondary circuits. This is a simple electrical switch that can be plumbed anywhere into a system requiring monitoring. The switch is electrically closed whenever the air pressure being monitored is below 60 psi (414 kPa). When the air pressure value exceeds 60 psi (414 kPa), the switch opens.

Verifying the operation of a low air pressure warning switch can be done by pumping the service application valve until the system pressure drops to the trigger value. A dash-located gauge must monitor both the primary and secondary circuit air pressures. The required visible warning is usually a dash warning light possibly coupled with a buzzer.

Task 8

Check air governor cut-in pressure.

76. The governor cut-out pressure is 120 psi (827 kPa), and the cut-in pressure is 80 psi (552 kPa). The unloader line and unloader mechanism are in satisfactory condition. To repair this problem it would be necessary to:
    A. adjust the governor.
    B. replace the compressor cylinder head.
    C. replace the governor.
    D. clean and service the governor.

Hint

The governor cut-in pressure should be 25 psi (172 kPa) less than the cut-out pressure. The governor cut-in pressure is not adjustable. If the cut-out pressure is within specifications and the governor has an improper cut-in pressure, be sure the unloader line connected to the compressor and the unloader mechanism in the compressor cylinder head are operating normally. If these components are satisfactory, replace the governor.

Task 9

Check spring brake inversion system.

77. An inversion valve is being tested on a combination vehicle. *Technician A* says to drain both the wet tank and the primary axle service reservoir before beginning the test. *Technician B* always verifies that secondary system pressure is at least 100 psi before conducting the test. Who is right?
    A. A only
    B. B only
    C. Both A and B
    D. Neither A nor B

Hint

To check for proper functioning of the inversion (spring brake control) valve, the air system should be fully charged (minimum 100 psi), and the spring parking brakes must be released. Next, locate and drain the wet tank and then the primary axle service reservoir com-

pletely by means of the draincocks. The primary gauge needle will drop to 0 psi indicating a pressure loss for that system, which is normal. The secondary gauge needle must remain at system pressure (minimum 100 psi) to properly start the check of the inversion valve. Depress the foot valve in a normal manner. The spring brakes on the rear axle(s) will apply as the air is exhausted from the brake air chambers. Release the foot valve; the spring brakes will release as air is supplied to them from the secondary reservoir. If the spring brakes do not partially apply and release, the inversion has failed the test and should be replaced.

Task 10

Check tractor protection valve.

78. The tractor protection valve closes to protect the tractor air supply if the air pressure in the supply line drops below:
    A. 20 psi (138 kPa).
    B. 55 psi (379 kPa).
    C. 75 psi (517 kPa).
    D. 80 psi (552 kPa).

Hint

The tractor protection valve protects the tractor air supply under a trailer breakaway condition or severe air leakage. Two air lines connect the tractor and trailer air brake systems. The trailer supply line supplies the trailer with air for braking and any other pneumatic systems such as its suspension. The trailer service line is the service brake signal line. Both these lines are plumbed to the tractor protection valve. When the trailer air supply dash valve is pulled outward to the off position, the tractor protection valve is also off in this condition. Under this condition, no air will exit the tractor protection valve service signal line to the trailer when the driver makes a service brake application. When the trailer air supply valve is pushed inward to the open position, air will open the tractor protection valve and supply the trailer pneumatic systems with air. In this condition, when the driver makes a service brake application, the service brake signal air will be transmitted through the tractor protection valve to apply its service brakes.

It should be noted that the trailer air supply controls the trailer park brakes. Whenever air supply is interrupted, such as when the trailer is being parked, or unintentionally, such as in a breakaway, the spring brakes will apply regardless of how much pressure is in the trailer air reservoirs.

The tractor protection valve is designed to isolate the tractor air system from that of the trailer at a predetermined value that ranges from 20 to 45 psi (138 to 310 kPa), depending on the system. It should be noted that at these pressures, the spring brakes on both the tractor and trailer would be partially applied.

Task 11

Test air pressure build-up time.

79. During a PM, the inspector tests the air brake supply system and finds that the build-up time is slow. Which of these is the MOST likely cause?
    A. A defective air governor
    B. A leak in a brake chamber
    C. An air leak in the cab
    D. A defective air compressor

Hint

System air build-up times are defined by federal legislation, specifically FMVSS 121. This legislation defines the required build-up times and values. A common check performed by enforcement agencies requires that the supply circuit on a vehicle be capable of raising air system pressure from 85 to 100 psi (586 to 689 kPa) in 25 seconds or less. Failure to achieve this build-up time indicates a worn compressor, defective compressor unloader assembly, supply circuit leakage, or a defective governor.

Task 12

Inspect coupling air lines, holders, and gladhands.

80. A pressure drop test is being performed on a straight single truck's air brake system. *Technician A* says that the pressure drop should be no more than 6 psi in 2 minutes. *Technician B* says you hold down the brake pedal for the first minute then release it. Who is right?
    A. A only
    B. B only
    C. Both A and B
    D. Neither A nor B

Hint

Before performing a brake system pressure drop test on a heavy truck, a Technician should ensure the engine is stopped, and the system is at full pressure. Then, make and maintain a brake application. A block of wood may be used to hold the brake pedal down during these tests. Allow pressure to stabilize for 1 minute, then begin timing for 2 minutes while watching the dash gauge. Pressure drop for single vehicles is 4 psi within 2 minutes, and for combination vehicles 6 psi within 2 minutes.

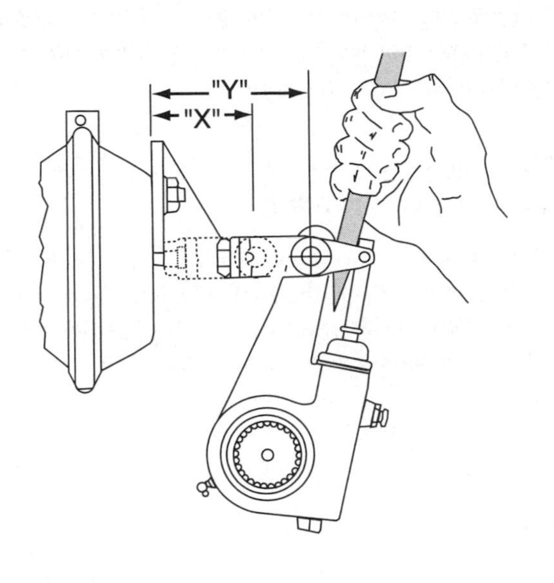

Figure 8–19 Brake adjustment. *(Courtesy of Rockwell Automotive)*

Task 13

Check brake chambers and air lines for secure mountings and damage.

81. The adjustment shown in Figure 8–19 is:
    A. the applied stroke.
    B. free stroke.
    C. slack adjuster stroke.
    D. spring brake stroke.

Hint

Service brake and spring brake chambers should be checked for leaks, physical damage, loose mounting, and damaged air lines. The brake chamber free stroke is the brake chamber pushrod stroke when the brakes are applied with a prybar installed in the pushrod clevis. The applied stroke is the brake chamber pushrod stroke when the brakes are applied with 80 to 90 psi (552 to 620 kPa). The free stroke or applied stroke must not exceed specifications. The specified stroke varies depending on the type and diameter of the brake chamber.

Air pressure gauge operation can be verified by using a master gauge, a good quality, liquid-filled gauge that uses a Bourdon principle of operation. When troubleshooting vehicle air pressure management problems, the vehicle gauges should not be relied on.

Task 14

Service desiccant pack in air dryer.

82. When discussing air dryer service, *Technician A* says the air brake system pressure must be decreased to zero before changing a spin-on dryer cartridge. *Technician B* says in most air dryers with an integral cartridge, the cartridge can be changed with the dryer installed on the tractor. Who is right?
    A. A only
    B. B only
    C. Both A and B
    D. Neither A nor B

Hint

If the air dyer has a spin-on cartridge, this cartridge may be changed without removing the air dryer assembly from the truck. The Technician must loosen all the reservoir draincocks and decrease the system pressure to zero before loosening the air dryer spin-on cartridge. When the dryer has an integral desiccant cartridge, the dryer usually has to be removed from the truck to replace the cartridge.

Task 15

Inspect and record front and rear brake lining/pad condition and thickness.

83. The percentage of defective brakes required to declare a truck out of service is:
    A. 75 percent or more
    B. 50 percent or more
    C. 40 percent or more
    D. 20 percent or more

Hint

Truck brake shoes are, in most cases, fixed anchor assemblies mounted to the axle spider and actuated by an S-camshaft. Brake shoes for a heavy-duty truck can have their friction facings or linings mounted to the shoe by bonding, riveting, or by fasteners. Current application shoes are remanufactured and replaced as an assembly, that is, with the new friction facing already installed. When reusing shoes, one must check for arc deformities usually caused by prolonged operation with out-of-adjustment brakes. The lining blocks are tapered and seldom require machine arcing.

The friction rating of linings is coded by letter codes. Combination lining sets of shoes have different friction ratings on the primary and secondary shoes. When combination friction lining sets are used, care should be taken to install the lining blocks in the correct locations on the brake shoes.

It is good practice to replace the brake linings on all four wheels of a tandem drive axle truck on a PM schedule. When the linings on a single wheel are damaged, such as in case of wet seal axle lube failure, the linings of both wheel assemblies on the axle should be replaced to maintain brake balance.

The friction pads in air disc brake assemblies should also be changed in paired sets, that is, both wheels on an axle. Vented disc brakes may have thicker inner pads than outer pads, while solid disc brakes usually have equal thickness inner and outer pads. Heat transfers more uniformly in the solid disc assemblies than vented ones, requiring thicker inboard pads.

Task 16

Inspect condition of front and rear brake drums/rotors.

84. The drum shown Figure 8–20 indicates:
    A. a concave drum.
    B. a convex drum.
    C. an offset drum.
    D. a shifted drum.

Hint

Worn brake drums may be scored, bell-mouthed, concave, or convex. A convex condition has a drum diameter greater at the friction surface edges as compared to the center of the friction surface. A concave condition has a drum diameter greater at the center of the friction surfaces as compared to the friction surface edges. A bell-mouthed condition occurs when the drum diameter is greater at the edge of the drum next to the backing plate compared to the edge next to the wheels. The out-of-round condition is a variation in the drum diameter at measurements taken 180 degrees apart. Recent practice and the low cost of drums usually result in drum replacement with a brake job.

Brake rotors must be visually inspected for heat checking, scoring, and cracks. One must measure rotors for thickness with a micrometer. Use a dial indicator to check rotor runout and parallelism. In highway applications, rotors tend to outlast drums and are often reused after a brake job. They must be turned within legal service specifications using a heavy-duty rotor lathe.

Figure 8–20 Brake drum.

Task 17

Check operation of front and rear brake manual slack adjusters; adjust as needed.

85. *Technician A* says one always lubricates the slack adjusters whenever an inspection of the brakes is scheduled. *Technician B* says to always check the rear slack adjuster pushrods for any stress cracks. Who is right?
    A. A only
    B. B only
    C. Both A and B
    D. Neither A nor B

Hint

Manual slack adjusters are adjusted by rotating the adjusting screw until the shoes are forced into the drums, then backing off the adjusting screw to set the specified free play. When a wheel-up adjustment is performed, free play is set to minimum drag of the shoe/drum relationship.

Task 18

Check operation and adjustment of front and rear brake automatic slack adjusters.

86. In the automatic slack adjuster shown in Figure 8–21:
    A. the rack turns the one-way clutch and worm shaft to adjust the brake shoes.
    B. the external teeth of the gear in the center of the slack adjuster are meshed with the teeth on the rack.
    C. as the brake linings wear, the brake chamber pushrod stroke and the slack adjuster arm movement decrease.
    D. the brake shoes are adjusted each time the brakes are applied and released.

Hint

The slack adjuster is critical in maintaining the required free play and adjustment angle. Current slack adjusters are required to be automatically adjusting, but their operation must be verified routinely.

Slack adjusters connect the brake chambers with the foundation assemblies on each wheel. They are connected to the brake chamber pushrod by means of a clevis yoke and pin. Slack adjusters are spline mounted to the S-cams and positioned by shims and an external snap ring.

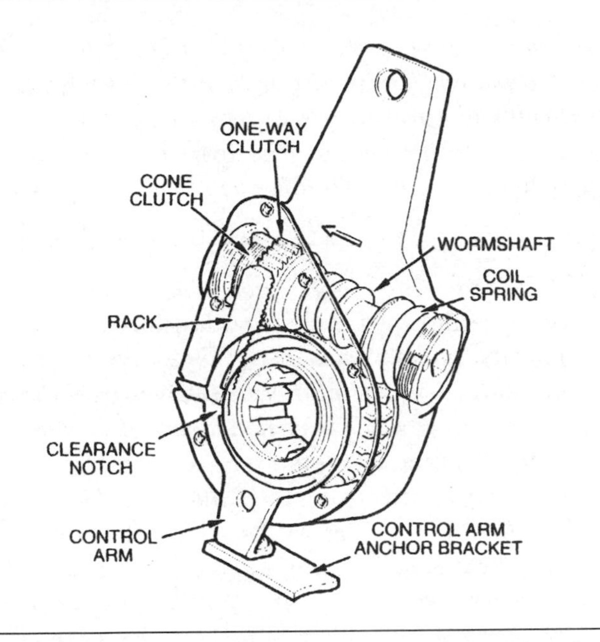

Figure 8–21 Automatic slack adjuster. *(Courtesy of Rockwell Automotive)*

The slack adjuster converts the linear force of the brake chamber rod into rotary force or torque and multiplies it. The distance between the slack adjuster S-cam axis and the clevis pin axis defines the leverage factor. The greater this distance, the greater the leverage.

The objective of brake adjustment whether manual or automatic is to maintain a specified drum-to-lining clearance and the specified brake chamber stroke. Grease or automatic lubing systems lubricate slack adjusters. The seals in slack adjusters are always installed with the lip angle facing outward. When grease is pumped into the slack adjuster, grease will easily exit the seal lip when the internal lubrication circuit has been charged.

Automatic slack adjusters may require periodic adjustment.

Task 19

Lubricate all grease fittings.

87. While discussing chassis lubrication, *Technician A* says standard nonvented grease fittings may be replaced with pressure relief-type grease fittings. *Technician B* says when the air chamber bracket is lubricated, grease should flow out of the end of the bracket tube next to the cam head. Who is right?
    A. A only
    B. B only
    C. Both A and B
    D. Neither A nor B

Hint

If a slack adjuster passes inspection, the following procedure for lubrication should be followed. Apply a thin film of chassis grease to the slack adjuster splines. After reassembly, pressure lubricate the slack adjuster according to the manufacturer's instructions. Pressure lubricate the air chamber bracket until grease flows out of the slack adjuster end of the tube. Grease should not flow out of the end of the tube toward the cam head. If it does, the seal is defective, and it must be replaced. Also, do not replace the existing grease fitting(s) with the pressure relief type. Only standard nonvented fittings are to be used with spring-loaded lip seals.

Hydraulic Brakes

Task 1

Check master cylinder fluid level and condition.

88. When handling brake fluids and servicing hydraulic brake systems:
    A. DOT 4 brake fluid may be mixed with DOT 5 brake fluid.
    B. DOT 3 brake fluid is nonhydroscopic.
    C. brake fluids should be stored in a hot, humid location.
    D. brake system components are lubricated by the brake fluid.

Hint

When checking hydraulic brake fluid in a vehicle, not only should the level be checked, but any leaks from the master cylinder, reservoir, or brake line fittings as well. The appearance of the fluid is an important part of the inspection as well. A cloudy or opaque look to the fluid will indicate some sort of contamination (dirt, rust, etc.).

Task 2

Inspect brake lines, fittings, flexible hoses, and valves for leaks and damage.

89. *Technician A* says if a flexible hose ruptures completely on one of the front wheels, the brakes will completely fail. *Technician B* says a leak at a brake line may cause pedal fade with the brakes applied. Who is right?
    A. A only
    B. B only
    C. Both A and B
    D. Neither A nor B

Hint

A proper inspection of a heavy truck's hydraulic braking system should always include a visual check around the wheel cylinders for contamination, which generally indicates leakage. Brake lines should be in good condition at all times with no corrosion or damage that may impede the system's performance.

Task 3

Check parking brake operation, inspect parking brake application and holding devices.

90. What is indicated in Figure 8–22?
    A. Hydraulic liftgate cylinder
    B. By-pass solenoid for reefer unit
    C. Remote automatic kingpin locking relay
    D. Hydraulic spring parking brake actuator

Hint

The parking brake must be able to hold the vehicle and load on any incline. Hydraulic brake systems may have a mechanical parking brake on the propeller shaft, or spring brakes on the rear wheels.

Task 4

Check operation of hydraulic system; pedal travel, pedal effort, residual (holding) pressure.

91. Excessive brake pedal effort may be caused by all of these defects EXCEPT:
    A. a defective brake booster.
    B. seized wheel caliper or wheel cylinder pistons.
    C. improper brake pedal free play adjustment.
    D. glazed brake linings.

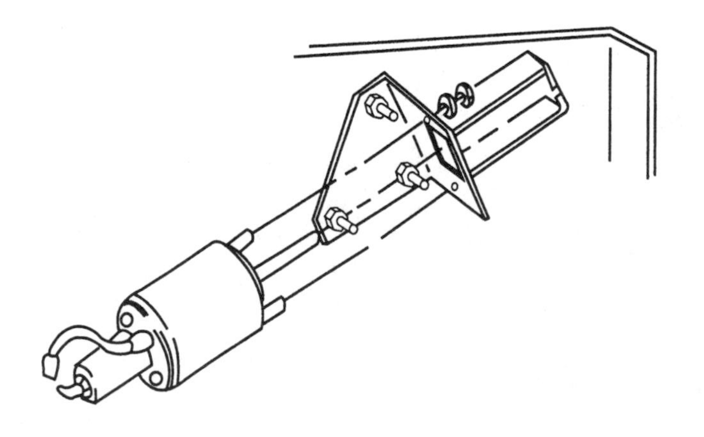

Figure 8–22 Brake system component.

Hint

Pressure values within the hydraulic circuit may be tested with pressure gauges. Starting the vehicle engine and applying the brakes can pressurize the hydraulic system circuit by foot pressure. External leaks may be verified by cleaning the externally visible portions of the circuit and applying the brakes. Internal leaks are more difficult to locate. Internal leakage within a master cylinder can be verified by using gauges plumbed to each portion of the hydraulic circuit.

If the failure occurs in the primary circuit, the primary piston will be forced through its travel without generating any fluid pressure until it contacts the secondary piston and mechanically actuates the secondary circuit. If the failure has occurred in the secondary circuit, the primary circuit functions normally, but the actuation of the secondary piston results in no pressure delivered to the secondary circuit. When a failure occurs in either circuit, the pressure differential light in the dash will illuminate the first time the brakes are applied following the failure. In either case, the vehicle should be brought to an immediate standstill and not operated until a repair has been undertaken.

Deteriorated fluid, deteriorated seals, or mixture of incompatible fluids may cause sludge and particulate in the brake fluid that may plug fill and compensating ports, resulting in slow application times, slow release times, and brake failure.

Task 5

Inspect wheel cylinders/calipers for leakage and damage.

92. The steering pulls to the right when the brakes are applied on a truck with hydraulic drum brakes on all wheels. The cause of this problem could be:
    A. seized wheel cylinder pistons on the left front wheel.
    B. seized wheel cylinder pistons on the right front wheel.
    C. brake fluid contaminated with moisture.
    D. a restricted right front brake hose.

Hint

A proper inspection of a heavy truck's hydraulic braking system should always include a visual check around the wheel cylinders for contamination, which generally indicates leakage. Brake lines should be in good condition at all times with never any corrosion or damage that may impede the system's performance; therefore, a thorough inspection of such lines is always considered an important part in inspecting the hydraulic brake system.

Task 6

Inspect power brake booster(s), hoses, and control valves.

93. A truck with a vacuum brake booster has excessive pedal effort. *Technician A* says the vacuum hose connected to the brake booster may be restricted. *Technician B* says the brake pedal pushrod may require adjusting. Who is right?
    A. A only
    B. B only
    C. Both A and B
    D. Neither A nor B

Hint

One criterion for the Commercial Vehicle Safety Alliance (CVSA) to put a vehicle out of service is if the vacuum reserve on the vehicle is below certain standards. These general standards may vary from vehicle to vehicle, but are generally accepted as insufficient if the vehicle cannot make one full brake application after it has been shut off.

In a hydro-max power brake booster, if flow from the hydraulic pump is interrupted, an electric pump backs up the system. When replacing hydraulic hoses in the boost system, the lines must conform to the SAE J189 standard. Cranking the engine over with the ignition system disabled bleeds hydro-boost systems with reservoir filled with the recommended fluid. The refilling procedure may have to be repeated.

Task 7

Inspect and record front and rear brake lining/pad condition and thickness.

94. A truck with hydraulic brakes experiences brake drag and excessive brake lining wear on all wheels. The cause of this problem could be:
    A. a defective metering valve.
    B. complete lack of brake pedal free play.
    C. seized wheel cylinder pistons on the right rear forward axle.
    D. a defective proportioning valve.

Hint

Most disc brake assemblies today have wear indicators, which produce a noise when the wear limit is exceeded. Servicing disc brake pads usually involves removing the caliper assembly, ensuring that float pins are not seized, backing off the automatic adjusting mechanism, and installing a new pair of brake pads. When a self-adjusting mechanism is used, care should be taken to ensure it is properly activated on reassembly. Whenever the brake pads are replaced, the brake rotors should be both measured and visually inspected.

When a brake job is performed, the brake shoes, return springs, and fastening hardware are replaced. The friction face codes should be observed when replacing brake shoes. Most brake shoes today use bonded friction blocks, but riveted and bolted types are still in existence. Ensure that primary and secondary shoes are installed in their correct locations.

Task 8

Inspect condition of front and rear brake drums/rotors.

95. While discussing brake drums, *Technician A* says the number stamped on the drum is the maximum wear diameter. *Technician B* says the maximum machining diameter may be 0.030 in. (0.762 mm) more than the maximum wear diameter before the drum has to be discarded. Who is right?
    A. A only
    B. B only
    C. Both A and B
    D. Neither A nor B

Hint

The critical brake drum specifications are the maximum wear limit, machine limit, and maximum permissible diameter. Drums are measured with a drum for out-of-round, bell-mouthed, convex, concave, and taper conditions. Before machining drums, they should be inspected for heat checks and cracks. Disc brake rotors may be reused if they are within the manufacturer's specifications. They should be measured for thickness with a micrometer and checked for parallelism and runout with a dial indicator. If within machine limits, the rotor may be turned on a rotor lathe.

Task 9

Adjust drum brakes.

96. The service or adjustment being performed in Figure 8–23 is:
    A. brake lining-to-drum clearance.
    B. brake shoe adjustment.
    C. parking brake adjustment.
    D. brake shoe-to-backing plate clearance.

Hint

You adjust drum brakes as follows:

1. *Remove the backing plugs for measuring the shoe-to-drum clearance. Be sure the parking brake chambers are caged, if used.*

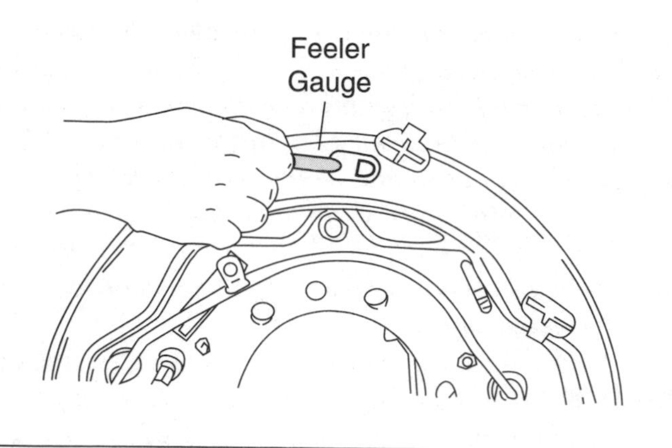

Figure 8–23 Brake service. *(Courtesy of Ford Motor Company)*

2. *Apply and release the brake pedal several times to center the brake shoes.*
3. *Insert a feeler gauge through the backing plate opening and measure the clearance between the brake shoe and the drum. On some trucks this clearance should be .040 to .100 inch (1.01 to 2.54 mm).*
4. *If the shoe clearance is not correct, adjust the manual override wheel on the adjuster cylinder to obtain the correct clearance.*

Drivetrain

Task 1

Check operation of clutch brake.

97. *Technician A* says that a clutch brake needs to be inspected for wear and fatigue in the same manner as the pressure plate. *Technician B* says that as long as the component is in place, it will function properly. Who is right?
    A. A only
    B. B only
    C. Both A and B
    D. Neither A nor B

Hint

To check the operation of the clutch brake, depress the clutch pedal in the cab and note the point during the pedal stroke when the clutch brake engages. You observe this by viewing clutch brake movement through the inspection cover.

Task 2

Check clutch linkage for looseness or binding.

98. Excessive clutch pedal free travel could cause:
    A. a growling noise with the clutch pedal depressed and the engine running.
    B. hard transmission shifting.
    C. a growling noise with the clutch pedal released and the engine running.
    D. excessive wear on the clutch facings.

Hint

Whenever you service the release bearing and other lubrication points on the clutch, all pivot points on the clutch linkage should be checked for free movement. On a push-type clutch, you simply adjust the clutch pedal free travel. On pull-type internal clutches with or without a clutch brake, a two-step adjustment is necessary to obtain proper free travel. Note that pull-type clutch internal adjustments must be completed before any external linkage adjustment.

Task 3

Check clutch adjustment.

99. When a self-adjusting clutch is found to be out of adjustment, check all of the following EXCEPT:
    A. correct placement of the actuator arm.
    B. bent adjuster arm.
    C. seized adjusting ring.
    D. clutch pedal linkage.

Hint

If the clutch pedal free travel is between 1 and 2 inches, the travel meets most specifications. Measuring the distance the pedal can be pushed down before a thrust load is exerted on the clutch release bearing checks clutch pedal free travel.

"Riding" the clutch pedal is another name for operating the vehicle with the clutch partially engaged. This is very destructive to the clutch, as it permits slippage and generates excessive heat. Riding the clutch also puts constant thrust load on the release bearing, which can thin out the lubricant and cause excessive wear on the pads. Release bearing failures are often the result of this type of driving practice. The best way to determine if clutch disc failure is due to driver error or mechanical failure is to speak with the driver of the vehicle.

Manually adjusted clutches have an adjusting ring that permits the clutch to be manually adjusted to compensate for wear on the friction linings. The ring is positioned behind the pressure plate and is threaded into the clutch cover. A lock strap or lock plate secures the ring so that it cannot move. The levers are seated in the ring. When the lock strap is removed, the adjusting ring is rotated in the cover so that it moves toward the engine.

Task 4

Check transmission(s) case, seals, filter, and cooler for cracks and leaks.

100. While inspecting a transmission for leaks, the Technician notices that the gaskets appear to be blown out of their mating sealing surfaces. What should the Technician check first?
   A. The transmission breather
   B. The shifter cover
   C. The release bearing
   D. The rear seal

Hint

The converter-cooler-lubrication circuit originates at the main pressure regulator valve. Converter-in oil flows to the torque converter. Oil must flow through the converter continuously to keep it filled and to carry off the heat generated by the converter. The converter pressure regulator valve controls converter-in pressure by bypassing excessive oil to the sump. Converter-out oil leaving the torque converter flows to an external cooler (supplied by the vehicle or engine manufacturer). A flow of air or water over or through the cooler removes the heat from the transmission oil. A thorough rebuild of a transmission includes a flushing of the transmission cooling system.

Task 5

Inspect transmission(s) breather(s).

101. The best way to clean a transmission case breather is to:
   A. replace the breather.
   B. soak the breather in gasoline.
   C. use solvent, then forced air.
   D. use a rag to wipe the orifice clean.

Hint

Two of the more simple items on a transmission that often get overlooked during servicing are the transmission case and breather(s). The transmission case must be checked for any signs of fatigue. Cracking is a symptom that is usually accompanied by fluid leakage. Plugged breathers are also associated with fluid leakage but not at the location of the breather. When a breather becomes plugged, fluid is often forced past seals in the transmission. If a Technician jumps to a conclusion upon noticing fluid leakage at a seal, he or she may mistakenly replace the seal and think the problem is solved. A thorough job requires the Technician to check the transmission breathers during any transmission diagnosis.

Task 6

Inspect transmission(s) mounts.

102. To thoroughly inspect a transmission mount, a Technician should:
   A. remove the mount and put opposing tension on the two mounting plates while inspecting for cracking or other signs of damage.
   B. remove the mount and place the mount into a vise while inspecting for cracking or other signs of damage.
   C. visually inspect the mount while still in the vehicle.
   D. replace the mount if it is suspect.

Hint

Transmission mounts and insulators play an important role in keeping drivetrain vibration from transferring to the chassis of the vehicle. If the vibration were allowed to transmit to the chassis of the vehicle, the life of the vehicle would be greatly reduced. Driving comfort is another reason why insulators are used in transmissions. The most important reason is the ability of the insulators to absorb shock and torque. If the transmission was mounted directly to a stiff and rigid frame, the entire torque associated with hauling heavy loads would be absorbed by the transmission and its internal components, causing increased damage and a much shorter service life. Broken transmission mounts are not readily identifiable by any specific symptoms. They should be visually inspected any time a Technician is working near them.

Task 7

Check transmission(s) oil level.

103. When checking transmission fluid level on a manual transmission, what is the proper procedure?
   A. Follow the guidelines stamped on the transmission dipstick
   B. Check for proper oil level by using your finger to feel for oil through the filler plug hole
   C. Make sure the oil level is even with the bottom of the filler plug hole
   D. Remove the specified rear housing retaining bolt, and check for proper fluid level at the bottom of this bolthole

Hint

Cold temperature operations may cause a transmission or drive axle lubrication problem known as channeling. High-viscosity oils typically used in transmission and axles thicken and may not flow at all when the vehicle is first started. Rotating gears can push lubricant aside, leaving voids or channels where no lubricant is actually touching the gears. It is not until heat generated from the underlubricated gears melts the stiffened lube that it starts to flow and do its job.

Task 8

Inspect U-joints, yokes, drivelines, and center bearings for looseness, damage, and proper phasing.

104. The procedure being performed in Figure 8–24 is:
  A. lubricating the rear U-joint.
  B. lubricating the slip joint.
  C. lubricating the center bearing.
  D. lubricating the rear bearing.

Hint

Defective drive shaft conditions such as loose end yokes, excessive radial looseness, slip spline radial looseness, bent shaft tubing, or missing plugs in the slip yoke cause vibrations, U-joint, and shaft support (center) bearing problems. A PMI will normally include the inspection of all of the above, plus checking the center bearing for damage and/or leakage, and the trunnions of the U-joints for failure(s) and/or excessive wear.

Center support bearings are used when the distance between the transmission and the rear axle is too great to span with a single drive shaft. The center support bearing is fastened to the frame and aligns the two connecting drive shafts. It consists of a stamped steel bracket that is used to align and fasten the bearing to the frame. A rubber mount inside the bracket surrounds the bearing, which supports the drive shaft.

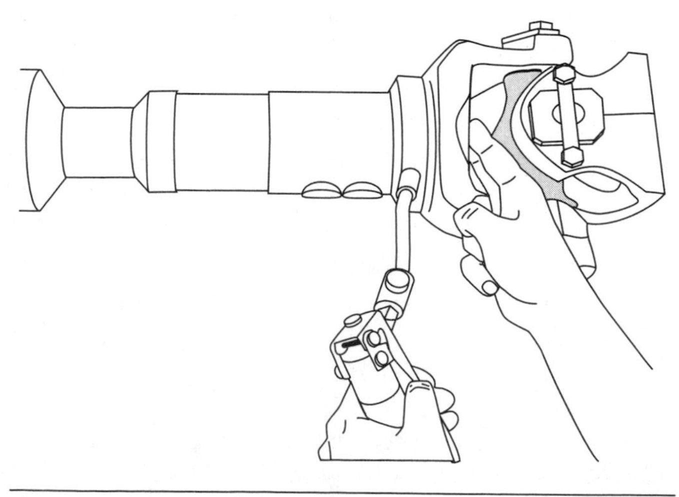

Figure 8–24 Chassis lubrication.

Task 9

Inspect axle housing(s) for cracks and lube leaks.

105. On a tractor with a tandem rear axle, the right-hand inner hub seal in the forward axle has been replaced three times because it was leaking grease. There are no other problems with the wheels or brakes on this wheel. There are no other leaks in the forward differential, and the wheel bearings are adjusted to specifications each time the seal is replaced. *Technician A* says the brake drum may be out of balance. *Technician B* says the seal seating area on the spindle may be scored. Who is right?
  A. A only
  B. B only
  C. Both A and B
  D. Neither A nor B

Hint

It is important to recognize normal wear to eliminate unnecessary downtime and parts replacement. A Technician/inspector should be able to spot potential problems using test drives and/or direct inspection. Incorrect lubrication will greatly affect the life of bearings, gears, and thrust washers. Conditions or symptoms such as contamination or improper lubricant may be factors in a drive axle failure.

Task 10

Inspect axle breather(s).

106. While discussing rear axle breathers, *Technician A* says the breather is located on the side of the rear axle housing. *Technician B* says on some off-road vehicles, a hose attached to the rear axle breather prevents water from entering the rear axle. Who is right?
  A. A only
  B. B only
  C. Both A and B
  D. Neither A nor B

Hint

Carefully investigate any source of leaks found on the drive axle. Replacing the seal if the drive axle breather is plugged does not cure leaking seals. A plugged breather results in high pressure, which can result in leakage past seals.

Task 11

Lubricate all grease fittings.

107. All these statements about chassis lubrication are true EXCEPT:
    A. some tractors have an air-operated automatic chassis lubrication system.
    B. overlubrication may damage seals in components with grease fittings.
    C. underlubrication may cause premature wear on certain components.
    D. an electric pump motor activates the air-operated automatic chassis lubrication system.

Hint

In air-driven automatic chassis lubrication systems, progressive feeders, piston distributors, or metering valves at the end of the dispensing lines are strategically located at the chassis grease points such as major suspension points, kingpins, fifth wheel, etc. A solenoid valve usually activates the air-driven system pump.

Overgreasing can be as damaging as undergreasing. Some maintenance personnel judge adequate lubrication by the amount of lubricant oozing from around the bearing. However, grease guns exert tremendous pressure, and oozing may mean that a bearing seal has blown out. When this happens, the lubricant will be contaminated and the part will be destroyed. The slip joint should always be greased in a proper manner, which is until grease is forced out of the relief hole. Then the hole should be covered while continually applying the grease, until the grease begins to ooze out around the seal.

Task 12

Check drive axle(s) oil level.

108. While discussing tandem rear axle lubrication, *Technician A* says the proper fluid level is when you can feel the lubricant with your finger in the differential filler plug hole. *Technician B* says in some rear axles, a temperature sensor is located in the differential cover below the filler plug. Who is right?
    A. A only
    B. B only
    C. Both A and B
    D. Neither A nor B

Hint

Some tractors are equipped with rear dead axles. These axles provide additional load-carrying capacity. If mounted ahead of a drive axle, they are called pusher axles; if mounted behind a live axle, they are referred to as tag axles.

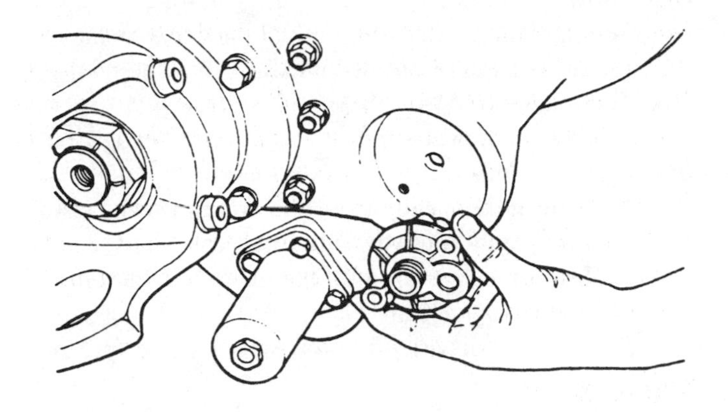

Figure 8–25 Rear axle component. *(Courtesy of Rockwell Automotive)*

Task 13

Change drive axle(s) lube and filter; check magnetic plugs.

109. The rear axle component in the Technician's hand in Figure 8–25 is:
    A. an oil filter adapter.
    B. an oil pump cover.
    C. an oil pump assembly.
    D. a solenoid adapter.

Hint

An experienced Technician can usually determine the operating condition of the drive axle differential by the fluid. Pay special attention to the condition of the fluid during the scheduled fluid changes. Most drive axles are equipped with a magnetic plug that is designed to attract any metal particles suspended in the gear oil. A nominal amount of "glitter" is normal because of the high torque environment of the drive axle. However, too much "glitter" indicates a problem that requires further investigation.

Drain and flush the factory-fill axle lubricant of a new or reconditioned axle after the first 1,000 miles (621 km) and never later than 3,000 miles (1,864 km). This is necessary to remove fine particles of wear material generated during break-in that would cause accelerated wear on gears and bearings if not removed. Draining the lubricant while the unit is still warm ensures that any contaminants are still suspended in the lubricant. Flush the axle with clean axle lubricant of the same viscosity as used in service. Do not flush axles with solvents such as kerosene. Avoid mixing lubricants of a different viscosity or oils made by different manufacturers.

Task 14

Check two-speed axle unit; check magnetic plugs.

110. During a routine drive axle oil change, a Technician notices a few metal particles on the magnetic plug of the drive axle. What should the Technician do?
    A. Inform the customer that further investigation is needed.
    B. Inform the customer of the condition and tell them to monitor the amount of particles.
    C. Do not follow up with the customer, as some metal particles are normal.
    D. Begin to disassemble the drive axle to find the cause.

Hint

Although some vehicles are equipped with electrical shift units, most axles are equipped with pneumatic shift systems. There are two air-activated shift system designs predominantly used to select the range of a dual range tandem axle or to engage a differential lockout. Usually the air shift unit is not serviceable. If it is found defective it should be replaced.

Task 15

Change transmission oil; check magnetic plugs.

111. What is the most common cause of bearing failure in a transmission?
    A. Extended high torque operating conditions
    B. Dirt and contaminants in the lubricant
    C. Improper rear axle ratio for load and driving conditions
    D. Poor quality of lubricant

Hint

Most manufacturers suggest a specific grade and type of transmission oil, heavy-duty engine oil, or straight mineral oil, depending on the ambient air temperature during operation. Do not use mild EP gear oil or multipurpose gear oil when operating temperatures are above 230°F (110°C). Many of these gear oils break down above 230°F (110°C) and coat seals, bearings, and gear with deposits that may cause premature failures. If these deposits are observed especially on seal areas where they can cause oil leakage, change to heavy-duty engine oil or mineral gear oil to ensure maximum component life.

Task 16

Take transmission(s) oil sample.

112. An automatic transmission in a truck must be overhauled because the fluid had a burned smell and some of the clutch plates were burned. *Technician A* says the transmission cooler must be flushed. *Technician B* says if the torque converter end play is satisfactory, no further converter service is required. Who is right?
    A. A only
    B. B only
    C. Both A and B
    D. Neither A nor B

Hint

When changing the transmission lubricant, always follow the manufacturer's exact hydraulic fluid specifications. For example, several transmission manufacturers recommend DEXRON, DEXRON II, and type C-3 (ATD approved SAE 10W or SAE 30) oils for their automatic transmissions. Type C-3 fluids are the only fluids usually approved for use in off-highway applications. Type C-3 SAE 30 is specified for all applications where the ambient temperature is consistently above 86°F (30°C). Some, but not all, DEXRON II fluids also qualify as type C-3 fluids. If type C-3 fluids must be used, be sure all materials used in tubes, hoses, external filters, seals, and so forth, are C-3 compatible.

Task 17

Take drive axle(s) oil sample.

113. A Technician notices a whitish milky substance when changing the fluid in an axle. This evidence of water is likely caused by:
    A. common condensation.
    B. infrequent driving and short trips.
    C. the axle being submerged in water.
    D. the vehicle being frequently driven during rainy or wet conditions.

Hint

During a normal PMI, a transmission and drive axle oil sample is taken for inspection of its overall lubricating qualities, and metal and dirt content. The oil sample may be tested for corrosion-causing agents such as sulfuric and hydrochloric acids, and some esters. It is for this reason that the transmission/drive axle oil should be at normal operating temperature to ensure that all possible contaminants and corrosives be contained in the oil in its normal operating condition.

Suspension and Steering Systems

Task 1

Check steering wheel operation for free play or binding.

114. Excessive steering wheel free play may be caused by all of these defects EXCEPT:
    A. a worn tie-rod end.
    B. a worn drag link end.
    C. a dry kingpin bushing.
    D. loose steering gear mounting bolts.

Hint

In addition to checking the vehicle's service records for previous suspension work, the Technician should first inspect the vehicle for worn suspension and steering components. Diagnosis is the first step in the service process, and if done correctly, will ensure a correct repair on the vehicle the first time. The best method for checking looseness in steering joints is with the wheels on the floor and two people working as a team. One should rock the steering wheel back and forth continuously, while the inspector visually checks for looseness at each of the pivoted ends of the entire steering system.

Task 2

Check power steering pump and hoses for leaks and mounting; check fluid level.

115. All of the following could cause the power steering pump pulley to become misaligned EXCEPT:
    A. an overpressed pulley.
    B. a loose fit from pulley hub to pump shaft.
    C. a worn or loose pump mounting bracket.
    D. a broken engine mount.

Hint

Reservoir O-rings, drive shaft seals, high-pressure fittings, and dipstick cap are all possible leak sources. If leaks occur at any of the seal locations, replace the seal. When a leak is present at the high-pressure fitting, first tighten it to the specified torque. If this fails, replace the O-ring at this fitting.

Task 3

Change power steering fluid and filter.

116. While doing routine maintenance, a Technician notices that the power steering system has a filter. *Technician A* says the filter should be replaced on a PM according to the OEM. *Technician B* says the filter should be replaced after a component failure. Who is right?
    A. A only
    B. B only
    C. Both A and B
    D. Neither A nor B

Hint

Most OEMs recommend the use of power steering fluid or automatic transmission fluid in power steering systems. Low fluid level causes increased steering effort and erratic steering. This condition may also cause a growling or cavitation noise in the pump. Foaming in the remote reservoir may indicate air in the power steering system. Most OEM truck manufacturers recommend checking the power steering fluid level at operating or working temperature of 140°F to 160°F (60°C to 71°C).

With the engine at 1,000 rpm or less, turn the steering wheel slowly and completely in each direction several times to raise the fluid temperature. Check the reservoir for foaming as a sign of aerated fluid. The fluid level in the reservoir should be at the hot full mark on the dipstick.

Task 4

Inspect steering gear for leaks and mounting.

117. A sector shaft seal has been replaced twice because of fluid leaks at this seal. The cause of this repeated seal failure could be:
    A. worn sector shaft bearings.
    B. excessive worm shaft end play.
    C. loose steering column U-joints.
    D. loose pitman arm on the sector shaft.

Hint

A loose steering gear mount may cause steering column binding and increased steering effort. Elongated mounting holes in the power steering pump bracket may cause a noise while in operation. Worn holes in the power steering pump-mounting bracket could cause premature belt wear.

Task 5

Inspect steering shaft U-joints, pinch bolts, splines, Pitman arm-to-steering sector shaft, tie-rod ends, and linkage assist power steering cylinders.

118. *Technician A* says to test a steering linkage joint for excessive wear, you should compress the joint with a C-clamp or large pliers. *Technician B* says simply pushing against the joint with force that you can create with your hands should be enough to identify excessive wear. Who is right?
    A. A only
    B. B only
    C. Both A and B
    D. Neither A nor B

Hint

Loose steering shaft U-joints and worn tie-rod ends or drag link ends cause excessive play and poor steering response.

To check a suspected loose tie-rod end or drag link end, simply push against the joint with the force you can create with your hands. This should be enough to identify excessive wear.

A loose pitman arm on the steering gear output shaft may cause a steering shimmy below 30 mph. The pitman arm can be directly responsible for directional stability. A damaged pitman arm can be directly responsible for the steering wheel being off center. A loose sleeve clamp on the drag link adjuster may allow enough play in the sleeve to damage the adjusting threads on the steering arm. Under this condition, the drag link end may shift inside the drag link and change the steering wheel position so the steering wheel is no longer centered when driving straight ahead. Tie-rod clamp interference with the I-beam on a full turn can cause a noise when turning over bumps, and a damaged tie-rod assembly. A misadjusted drag link could cause the steering gear to operate off center and cause a steering wheel shimmy.

Task 6

Check kingpin wear.

119. The procedure being performed in Figure 8–26 is:
    A. knuckle vertical play.
    B. knuckle pin fit upper bushing.
    C. lower bushing deflection.
    D. kingpin axial play.

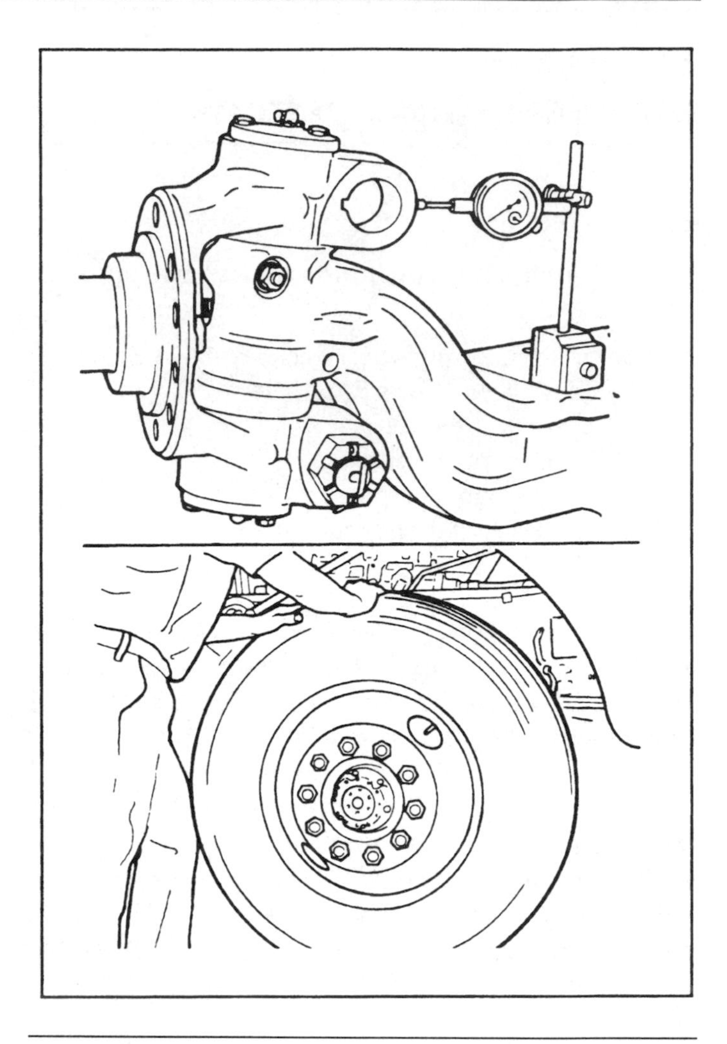

Figure 8–26 Knuckle pin measurement. *(Courtesy of Rockwell Automotive Inc.)*

Hint

Upper knuckle pin bushing wear is measured by moving the top of the tire and wheel inward and outward with a dial indicator placed against the inner side of the top end of the steering knuckle. Lower knuckle pin bushing wear is measured by moving the bottom of the tire and wheel inward and outward with a dial indicator placed against the inner side of the lower end of the steering knuckle. If knuckle pin bushing wear exceeds specifications, replace the knuckle pin bushings.

Task 7

Check front and rear wheel bearings for looseness and noise.

120. The operation being performed in Figure 8–27 is:
    A. checking wheel hub runout.
    B. aligning wheel hub with axle hub.
    C. installing front wheel bearings.
    D. installing front wheel bearing seals.

Hint

When adjusting wheel bearings torque the wheel bearing adjusting nut 200 ft.-lb. (271 Nm) while rotating the wheel to seat the bearings. Back off the adjusting nut 1 turn, and tighten this nut to 50 ft.-lb. (68 Nm), and then back off the adjusting nut the specified amount usually

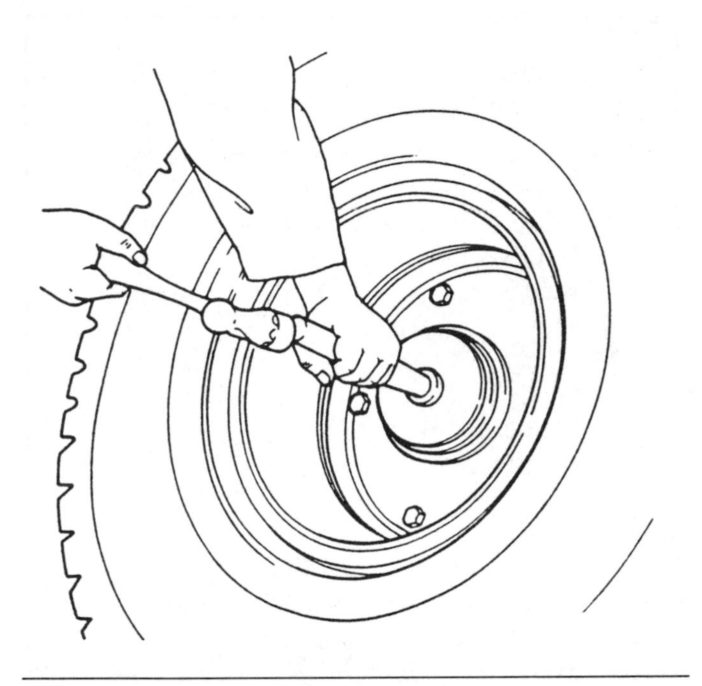

Figure 8–27 Wheel hub service.

one-quarter turn. Install the locking device, and measure the hub end play, which must be 0.001 to 0.005 in. (0.254 to 0.127 mm). Defective rear wheel bearings produce growling noises at LOW speeds. A growling noise produced by a defective front wheel bearing is most noticeable while turning a corner. Some rear wheel bearings do not require repacking because they are lubricated with rear axle fluid. Never spin dry a wheel bearing with compressed air because this action may cause the bearing to disintegrate, resulting in personal injury.

Task 8

Check oil level in all non-drive hubs; check for leaks.

121. The bearings in the wheel hub shown in Figure 8–28:
    A. must be lubricated with wheel bearing grease before assembly.
    B. are permanently lubricated with the specified oil.
    C. require a specified hub end play.
    D. require inner and outer hub seals.

Hint

Hubs should be prelubed when installing the wheel assembly. On many drive axles, the bearing and hub assembly must be filled to a prescribed level in the calibrated inspection cover. In many drive axles, the bearing and hub assembly is supplied with lubricant from the differential carrier.

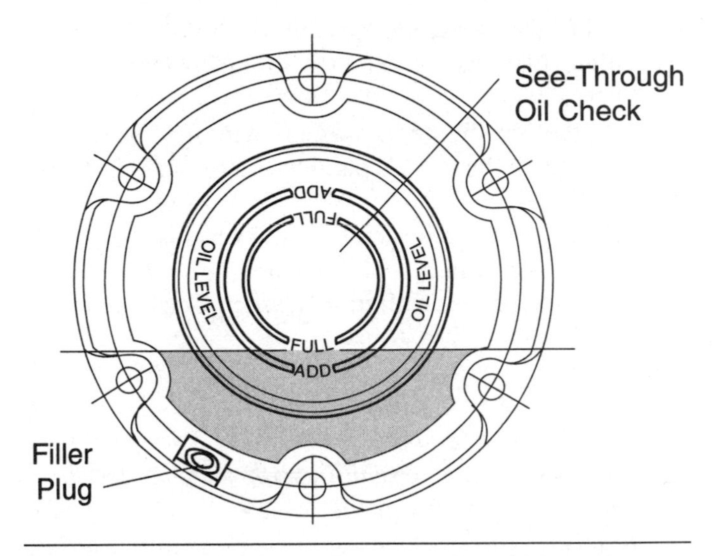

Figure 8–28 Hub lubricant. *(Courtesy of Ford Motor Company)*

Task 9

Remove and inspect front and rear wheel bearings; reassemble and adjust.

122. When adjusting wheel bearings, *Technician A* says to torque the adjusting nut to 200 ft.-lb. (271 Nm), back this nut off 1 turn, torque the nut to 50 ft.-lb. (68 Nm), then back off the nut the specified amount and install the locking device. *Technician B* says after a wheel bearing adjustment, the hub end play should be 0.005 to 0.010 in. (0.127 to 0.254 mm). Who is right?
    A. A only
    B. B only
    C. Both A and B
    D. Neither A nor B

Hint

Wheel bearings should be inspected at each brake job and at each PMI service that requires wheel removal. Most current trucks use wet bearings lubricated with the specified oil, which is usually gear lube.

When reinstalling wet bearings, they should be prelubed with the same oil to be used in the axle hub. When grease-packed bearings are used, the bearing cone must be packed with grease. This procedure may be performed by hand or using a grease gun and cone packer. Grease-lubricated wheel bearings require the use of a high-temperature axle grease.

Task 10

Inspect front and rear springs, hangers, shackles, spring U-bolts, and insulators.

123. All of these procedures should be done when replacing a radius rod on a spring suspension EXCEPT:
   A. replace alignment washers with new ones.
   B. secure the truck's parking brakes and chock the tires.
   C. ensure the replacement radius rod is of the correct length.
   D. check the axle alignment when finished.

Hint

The leaf spring type suspension uses semielliptic leaf springs to cushion load and road shocks. The springs are mounted on saddle assemblies above the equalizer beams and are pivoted at the front end on spring pins and brackets. The rear ends of the springs have no rigid attachment to the spring brackets but are free to move forward and backward to compensate for spring deflection.

Task 11

Inspect shock absorbers for leaks and mounting.

124. On a tandem axle with a leaf spring suspension, the right wheel on the forward drive axle has a wheel hop condition. This problem may be caused by all of these defects EXCEPT:
   A. worn spring bushings and shackles.
   B. a worn-out shock absorber.
   C. improper shim adjustment between the radius rod and the spring bracket.
   D. a loose equalizer spring bracket.

Hint

When a vehicle hits a bump, the wheel and suspension move upward in relation to the chassis. This causes wheel jounce and rebound. Shock absorbers dampen or control spring action from jounce and rebound, reduce body sway, and improve directional stability and driver comfort. Worn-out shock absorbers allow the front end to bounce, causing the steering wheel to shake for a few seconds. The shock absorber piston shield often gets bent or damaged and scrapes against the shock absorber when the suspension moves up and down. Shock absorbers stop spring oscillation and are still needed on air spring suspensions. It is advisable to replace shock absorbers in pairs. With one end disconnected inspect shock absorbers for resistance, bent pistons, bushing failure, and bent piston sleeves.

Task 12

Inspect air suspension springs, mounts, hoses, valves, linkage, and fittings for leaks and damage.

125. A truck with air spring suspension sits low when you hook up the trailer and sits high without the trailer. *Technician A* says a leaking air spring could be the problem. *Technician B* says an improperly adjusted height control valve could be the problem. Who is right?
   A. A only
   B. B only
   C. Both A and B
   D. Neither A nor B

Hint

Air spring suspensions provide a smooth shock and vibration-free ride and automatically adjust riding height. A height control valve controls the air supply to the air springs and maintains a constant ride height. These systems may have single- or dual-height control valves. The control and relay valves provide a means of deflating the air springs when uncoupling a trailer from a tractor. If the air springs are blown, the suspension height is less than specified. Rubber cushions inside the air springs bottom out if the air spring is blown.

Misadjusted height control valves cause improper ride height. In a rear axle air suspension system, the air springs are mounted between the frame and under the outer end of the main support beams, and these beams are attached to the rear axle. The front of the main support beam is mounted in a frame bracket.

Trailer air suspension systems have a rigid beam on each side of the suspension. The front of this beam is retained in a bracket with a bushing. Some systems have an eccentric pivot bolt in this bracket for axle alignment. These brackets are welded to the trailer axle and retained in the rigid beams by two bushings and retaining bolts. The air springs are mounted between the rear end of the rigid beam and the frame.

Task 13

Check and record suspension ride height.

126. When adjusting the height control valve in an air suspension system:
   A. the height control valve arm may be moved upward to lower the suspension height.
   B. steel locating pins are installed in the height control valve arms.

C. the air brake system pressure must be 70 psi (483 kPa) or more.
D. the truck must have an average load on the chassis.

Hint

An air spring suspension has a height control valve that automatically controls the amount of air in the air spring and maintains a constant ride height. An improperly adjusted height control valve can cause the vehicle to sit too low or too high.

Adjustment of the height (leveling) control valve requires the following:

1. *The truck must be empty and air brake system pressure must be 70 psi (483 kPa) or more.*
2. *Disconnect the height control valve linkages.*
3. *Move the height control valve upward to raise the vehicle frame until you can place the safety stands set at the specified height under each side of the frame.*
4. *Move the height control valve lever downward to deflate all air. Hold them in the 45-degree downward position for 10 seconds.*
5. *Return the control valve slowly to the center position. Insert the wood locating pins into the adjusting block.*
6. *Loosen the 1/4 inch locknut on each height control valve. Connect the linkage and tighten the lock to specifications. Remove the wood locating pins.*

Task 14

Lubricate all grease fittings.

127. The component in Figure 8–29 is part of:
A. an air dryer system.
B. an on-board lubrication system.
C. a fuel filter system.
D. an air suspension system.

Hint

One of the greatest boons to PM chassis lubricating is the on-board system. These systems are either manually or automatically operated. On automatic chassis lube systems (ACLS), a uniform shot of grease is injected to critical points as often as every few minutes. To distribute the grease, a reservoir is typically fitted to the vehicle's frame. Although becoming much more prevalent on today's modern trucks, an ACLS is not a mandatory requirement by any federal act.

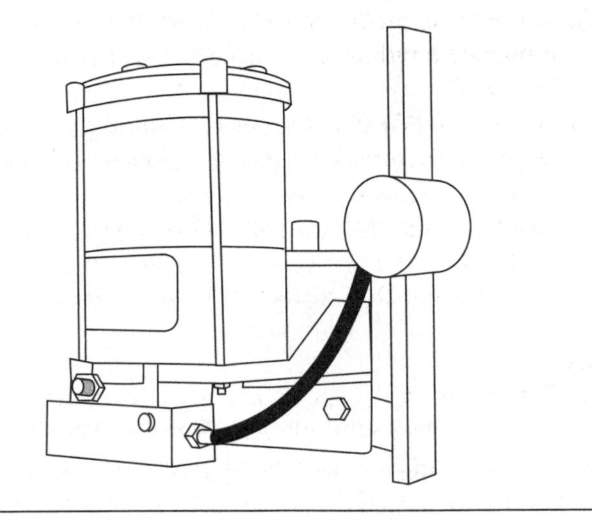

Figure 8–29 Heavy-duty truck system component.

To distribute grease, on-board systems use the same basic components. A ***reservoir*** *holds the grease supply until ready to be distributed. The system uses a* ***pump****, which delivers the lubricant through a network of* ***grease lines****.* ***Metering*** *valves dispense the grease, automatic timer mechanisms direct the flow of grease, and electrical motors and/or other power sources supply the compressed air.*

Task 15

Check toe-in.

128. Figure 8–30 indicates which of these front suspension conditions?
A. Positive camber
B. Negative caster
C. Toe-out
D. Toe-in

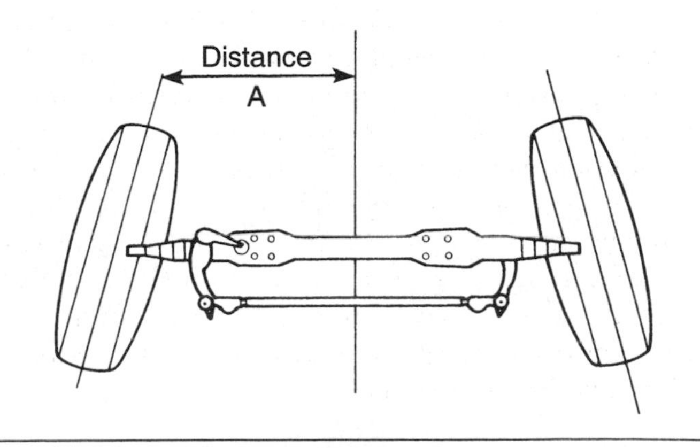

Figure 8–30 Front suspension measurement.

Hint

Toe-in is when the distance between the rear inside edges of the front tires is greater than the distance between the front inside edges of the front tires. Toe-out occurs when the distance between the inside front tires edges exceeds the distance between the inside rear tire edges. A toe-in error of 1/8 inch is the equivalent of dragging the tires crosswise for 11 feet for each mile driven. These cross-wise movement causes severe feathered tire tread wear.

Task 16

Check tandem axle alignment and spacing.

129. All of these statements about rear axle alignment are true EXCEPT:
    A. the thrust line should be at or very close to the geometric centerline.
    B. worn leaf spring shackles may cause improper rear axle alignment.
    C. worn equalizing beam bushings may cause improper rear axle alignment.
    D. improper rear axle alignment does not affect steering quality.

Hint

The vehicle thrustline is a line projected forward at a 90-degree angle from the center of rear axle. When a vehicle's rear wheels track directly behind the front wheels, the thrustline is positioned at the geometric centerline of the vehicle. Worn equalizing beam bushings may cause the rear axles to be out of alignment. A bent radius rod pulls one side of the rear axle ahead in relation to the opposite side of the rear axle. This action moves the rear axle thrust line so it is no longer at the geometric centerline of the vehicle. Improper rear axle alignment causes steering pull and excessive tire wear.

Turning the adjustment eccentric in the spring hanger bracket is one method of adjusting the rear axle alignment. On other rear suspensions, shims between the radius rods and the spring hanger brackets provide a means of rear axle alignment. It is not necessary to replace the axle alignment shims when replacing a radius rod on a truck. If the shims are not worn beyond use, then they should be reused to ensure a correct realignment of the axle. Also, the number of shims and their respective positions should be noted and returned exactly as removed.

Tires and Wheels

Task 1

Inspect tires for irregular wear patterns and tread direction.

130. The excessive wear on the inside edge of the front tire tread in Figure 8–31 was caused by:
    A. excessive toe-in.
    B. excessive positive camber.
    C. excessive negative camber.
    D. excessive toe-out.

Hint

The two front suspension measurements that may cause excessive tire wear are camber and toe. Camber is an inward or outward tilt of the top of the tires when viewed from the front of the vehicle. If the centerline of the tire is straight up and down or perfectly vertical, the tire has a camber setting of zero degrees. If the tire centerline is tilted toward the vehicle, the camber is negative. If the tire centerline is tilted away from the vehicle, the camber is positive. Excessive positive camber causes wear on the outside edge of the tire, and excessive negative camber causes wear on the inside edge of the tire. Improper toe adjustment causes feathered tire tread wear.

Task 2

Inspect tires for cuts and sidewall damage.

131. While discussing tire inspection, *Technician A* says if a tire on a steer axle has a sidewall cut that exposes the cords, tire replacement is necessary.

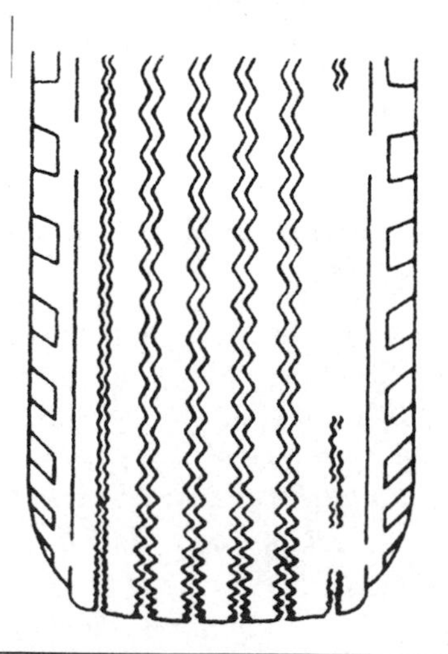

Figure 8–31 Tire tread wear on one edge. *(Courtesy of General Motors Corporation, Service Technology Group)*

Technician B says if a drive axle tire has a sidewall bulge 5/8 in. (15.8 mm) in height above the sidewall surface, the tire must be replaced. Who is right?

A. A only
B. B only
C. Both A and B
D. Neither A nor B

Hint

Inspecting all the tires is a very important part of every PMI and every pretrip inspection. The CVSA provides uniform standards for complete tire inspection. When certain tire defects are found, the tire must be replaced before the vehicle is placed in service.

Task 3

Inspect valve caps.

132. During an inspection, a tractor must be placed out of service if any of these tire conditions are present EXCEPT:
   A. if a tire is mounted or inflated so it contacts any part of the vehicle.
   B. if a drive axle tire has a tread depth of less than 1/32 in. (0.79 mm) at two adjacent tread grooves at three separate locations.
   C. if a steer axle tire has a tread depth of 3/32 in. (2.38 mm) at two adjacent tread grooves at any location on the tire.
   D. if the tire is labeled "Not for Highway Use."

Hint

If valve caps on any of the tires are missing during a pretrip inspection, they must be replaced. If air can be heard or felt escaping from a tire, the tire must be repaired or replaced before the vehicle is placed in service.

Task 4

Measure (and record) tread depth; probe for imbedded stones and glass.

133. *Technician A* says the vehicle is tagged out of service if the weight carried exceeds the tire load limit. *Technician B* says the vehicle is tagged out of service if a tire has 50 percent of the tread width loose or missing in 12 in. (30.4 cm) of the tire circumference. Who is right?
   A. A only
   B. B only
   C. Both A and B
   D. Neither A nor B

Hint

A vehicle must be deadlined with less than 2/32 in. (1.58 mm) tread depth on a steer axle when measured anywhere on the tire in any two adjacent tread grooves. Minimum tread depth on drive axle tires is 1/32 in. (0.79 mm) measured in two adjacent tread grooves at three separate locations. Deadlining may also occur when any part of the breaker strip or casing ply is showing on the tread, and when the sidewall is cut, worn, or damaged so that the ply cord is exposed.

Task 5

Check (and record) air pressure.

134. The tire tread wear in Figure 8–32 was caused by:
   A. underinflation.
   B. overinflation.
   C. improper toe adjustment.
   D. excessive positive camber setting.

Hint

The proper restraining device or cage should be used while inflating a tire. This allows safe inflating without the risk of personal injury or death. Feather-edged wear toward the outside of the tire is from excessive toe-out. Excessive positive camber causes wear on the outside edge of the tire tread. Underinflation will wear the edges of the tire. Overinflation will wear the center of the tire. Never mount the tire on the truck before inflation because an improperly fitted rim ring could dislodge, causing injury. Dismounting and mounting split side rims is extremely dangerous and only trained professionals should perform this process according to OSHA rules and regulations. Never use a hammer on split side rim rings.

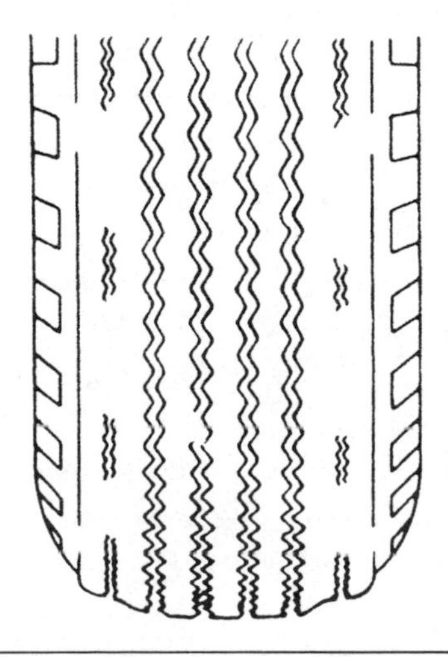

Figure 8–32 Tire tread wear on both edges *(Courtesy of General Motors Corporation, Service Technology Group)*

Task 6

Check for loose lugs and/or slipped wheels.

135. The tires are being inspected on a truck. *Technician A* says on a cast spoke wheel, the assembly includes a one-piece casting that includes the hub, spokes, and rim. *Technician B* says that disc wheels generally experience more alignment and balance problems than do the more stable spoke designs. Who is right?
    A. A only
    B. B only
    C. Both A and B
    D. Neither A nor B

Hint

A cast spoke wheel consists of a one-piece casting that includes the hub and the spokes. Spokes are made of ductile iron, cast steel, or aluminum. Tires are mounted on a separate rim that is clamped onto the spokes. The proper installation and torquing sequence is critical to true running. Generally, spoke wheels experience greater alignment and balance problems than disc designs, but with proper installation and torquing, it is possible that spoke wheels can run virtually wobble-free.

When overloading does occur, it normally causes cracks or small fractures from lug hole to lug hole. Cracked wheels must be replaced.

Task 7

Retorque lugs in accordance with manufacturers' specifications.

136. *Technician A* says that retorquing of lug nuts is not required when performing a PMI. *Technician B* says that an increased load-carrying capacity is an advantage of switching to low-profile tires. Who is right?
    A. A only
    B. B only
    C. Both A and B
    D. Neither A nor B

Hint

A PMI requires retorquing of the lug nuts because this inspection requires removal of the entire wheel and tire assembly to ensure that proper safety and manufacturing guidelines are adhered to. When lower profile tires are used, a decrease in load-carrying capacity is observed and recorded.

Task 8

Inspect wheels and spacers for cracks or damage.

137. Cracks between wheel lug holes indicate:
    A. overinflation of tires.
    B. overloading.
    C. using larger than recommended tire sizes.
    D. incorrect wheel size for application.

Hint

Wide base wheels are referred to as the following: high flotation, super single, wide body, duplex, or jumbo wheels. One wide base wheel and tire replaces traditional dual wheels and tires. Instead of 18 wheels, a tractor/trailer needs only 10. All tires (new and retread) sold in the United States must have a DOT number cured into the lower sidewall on one side of the tire.

Always check the tire size on the new and old tires, but the brand really does not matter. With the stud nuts slightly loose on demountable rims, strike the wheel clamps with a hammer to loosen these clamps. Different makes of rim components must not be interchanged.

Task 9

Check tire matching (diameter and tread) on dual tire installations.

138. Which of these criteria is the LEAST likely to be considered tire mismatching?
    A. Having radials and bias ply on the same axle
    B. Having dual wheels of the same basic tread design
    C. Installation of a new tire next to an old tire on the same axle
    D. Installation of tires with different circumferences and diameters on the same axle

Hint

Although not necessarily a desirable condition, tires of a different tread design that are mounted on the same axle can be tolerable, assuming these tires have the same circumference and diameter, a similar load rating, and are relatively in the same general condition. Do not mix radial tires with bias tires on a steering axle. Mixing tires on the rear axle is dangerous because radial tires expand more than bias tires. Matching tire sizes on dual wheels prevents tire tread wear from slippage while driving on uneven surface areas. Dual drive wheels may be measured with a tape measure, caliper, etc.

Frame and Fifth Wheel

Task 1

Inspect fifth wheel mounting bolts, air lines, and locks.

139. While discussing fifth wheels, *Technician A* says the position of a sliding fifth wheel may be changed with the vehicle in motion. *Technician B* says a sliding fifth wheel may be welded to the tractor frame. Who is right?
   A. A only
   B. B only
   C. Both A and B
   D. Neither A nor B

Hint

The fifth wheel couples the tractor to the trailer kingpin. During the tractor-to-trailer coupling process, the fifth wheel should be adjusted so the trailer bolster plate makes initial contact with the fifth wheel at a point 8 inches (20.32 cm) to the rear of the center on the fifth wheel mounting bracket. After the tractor is coupled to the trailer, the driver should get under the tractor and trailer and use a flashlight to visually inspect the position of the kingpin in the fifth wheel jaws.

Task 2

Test operation of fifth wheel locking device.

140. When inspecting a fifth wheel, the tractor is tagged out of service if:
   A. 10 percent of the fifth wheel-to-frame fasteners are missing or ineffective.
   B. there is more than 1/8 in. (3.17 mm) horizontal movement between the pivot bracket pin and the bracket.
   C. 15 percent of the latching fasteners per side are ineffective.
   D. there is more than 1/2 in. (12.7 mm) horizontal movement between the upper and lower fifth wheel halves with the trailer coupled.

Hint

When checking the fifth wheel adjustment, it should not exceed more than 1/8 inch horizontal play at any time. The jaws should also be checked for ease of operation, and the sliders should work smoothly. A sliding fifth is designed to move forward or rearward on its mounting plate. This type of fifth wheel is mounted on tracks and locked in position. The locking mechanism may be released mechanically with a lever, or by air pressure supplied to an air cylinder. Test the operation of either type.

Task 3

Check mud flaps and brackets.

141. *Technician A* says that a circle inspection includes items such as torn mud flaps, leaking hoses, and outdated permits, to name a few. *Technician B* says to perform a circle inspection first before any other inspection as part of a PMI. Who is right?
   A. A only
   B. B only
   C. Both A and B
   D. Neither A nor B

Hint

Mud flaps and brackets must be inspected to be sure they are in satisfactory condition. If a mud flap is missing, stones may fly off a tire and hit a vehicle behind or beside the tractor.

Task 4

Check pintle hook.

142. The vehicle to which a pintle hook or drawbar is attached should be tagged out of service for any of these defects EXCEPT:
   A. wear in the pintle hook horn that exceeds 10 percent of the horn section.
   B. loose or missing mounting bolts.
   C. any welded repairs to the drawbar eye.
   D. cracks in the pintle hook mounting surface.

Hint

Most trailers are built with a 47-inch kingpin height. This should be confirmed, however, for your specific application. The maximum allowable fifth wheel height is determined by subtracting the trailer height and tractor frame height from the maximum height of 13 feet 6 inches. As a final check, the tire clearance should be considered, keeping in mind that most spring suspension has a 2-inch deflection under full load.

Task 5

Lubricate all grease fittings.

143. When discussing chassis lubrication, *Technician A* says when lubricating brake camshaft and bushings, grease should not ooze out of the slack adjuster end of the brake chamber bracket.

Technician B says excessive lubrication may damage seals in some components. Who is right?
A. A only
B. B only
C. Both A and B
D. Neither A nor B

Hint

To distribute grease, most on-board systems use the same basic components. These components are as follows: a reservoir, which holds the grease supply until ready to be distributed; a pump, which delivers the lubricant through a network of grease lines; metering valves, which dispense the grease; automatic timer mechanisms, which direct the flow of grease; and electrical motors and/or other power sources (such as compressed air).

Task 6

Inspect frame and frame members for cracks and damage.

144. When inspecting a frame rail as shown in Figure 8–33, *Technician A* says the frame rail marked A is an incorrect repair. *Technician B* says the frame reinforcement on the frame rail marked is called a channel reinforcement or fishplating. Who is right?
A. A only
B. B only
C. Both A and B
D. Neither A nor B

Hint

Many states have set their heavy truck length limits at 65 feet overall, which includes the combination of a tractor and one or two trailers. Heat-treated frame rails can be identified in one of two ways: the rails are thicker (10 mm or more), or there is an identification label warning of significance regarding the special material. Angle cutting will distribute the cut load and bolt pattern over a greater overall area than a cut made at right angles to the frame. Reinforcement plates must be long enough to extend beyond the critical area so that the ends can be cut on an angle instead of square across the frame section. This will make the cross section of the rail inherently stronger across the section. Frame defects include sidesway, sag, buckle, diamond shaped, and twist.

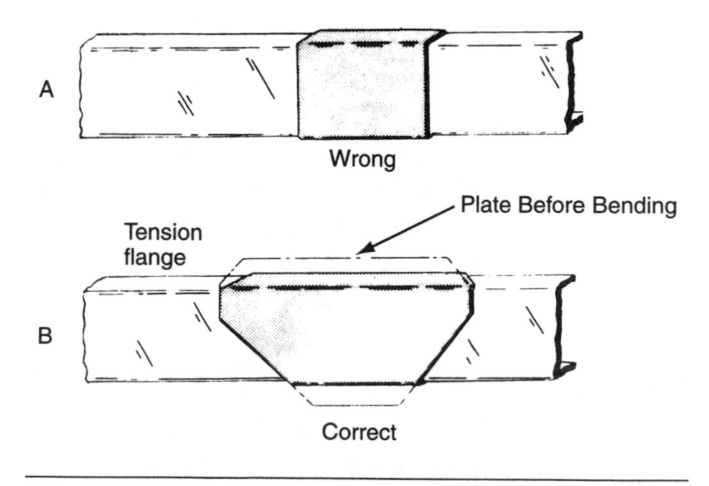

Figure 8–33 Frame repairs. *(Courtesy of Mack Trucks, Inc.)*

Drive Test

Task 1

Check operation of clutch and gearshift.

145. A truck's clutch is being measured for its free travel. *Technician A* says the measurement should be between 1 and 2 inches. *Technician B* says that on a hydraulic clutch, the measurement should be 2 to 3 inches. Who is right?
A. A only
B. B only
C. Both A and B
D. Neither A nor B

Hint

An effective test drive should include a check of the operation of the following systems: clutch and gearshift; all instruments, gauges, and lights; steering wheel; electronic controls; governor; cruise control; back-up warning; exhaust brake; and service brake.

Task 2

Check operation of all instruments, gauges, and lights.

146. During a PMI, the antilock brake system (ABS) warning light is illuminated with the engine running. *Technician A* says there is a mechanical defect in the ABS. *Technician B* says the ABS will still function normally. Who is right?
A. A only
B. B only
C. Both A and B
D. Neither A nor B

Hint

Check that all instruments, gauges, and indicator lights in the instrument panel for normal, consistent readings.

Task 3

Check steering wheel for play or binding.

147. A binding steering wheel may be caused by:
    A. loose tie-rod ends.
    B. loose worn steering shaft U-joints.
    C. loose drag link ends.
    D. worn kingpin bushings.

Hint

Steering systems use U-joints to allow the shafts to mount on an angle. The turn signal switch mounts to the left side of the steering column and a boot seals the steering column where it passes through the body. Loose steering wheel-to-steering shaft fasteners cause excessive play and poor steering response.

Task 4

Check operation of electronic controls.

148. While diagnosing an engine computer system with a scan tool:
    A. a diagnostic trouble code (DTC) indicates a defect in a specific component.
    B. always begin the diagnosis with the engine cold.
    C. connect and disconnect the scan tool with the ignition switch on.
    D. an engine coolant temperature sensor contains a thermistor.

Hint

Connect the appropriate engine/transmission scan tool, OEM special tool, or laptop computer to the truck's data link connector (DLC). During the road test, activate the scan tool's parameter specification readout and observe all the engine and transmission data. Record any improper readings during the tests.

Task 5

Check road speed governor.

149. In Figure 8–34 the governor has positioned the injector plunger in the:
    A. zero fuel delivery position.
    B. partial fuel delivery position.
    C. maximum fuel delivery position.
    D. deceleration fuel delivery position.

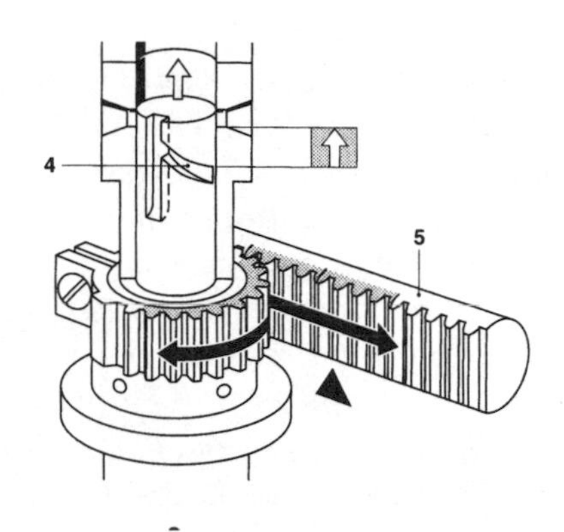

Figure 8–34 Injection pump rack and plunger. *(Reprinted with permission from Robert Bosch Corporation)*

Hint

Connect the appropriate engine scan tool, OEM special tool, or laptop computer to the truck's diagnostic connector. During the road test, activate the tool's parameter specification readout and record the road speed governor rpm. Always follow the tool's operating instructions when checking rpm and other values.

Task 6

Check cruise control.

150. An electronic cruise control maintains the vehicle speed at 45 mph (72 km/h) when the selected speed is 60 mph (96 km/h). The MOST likely cause of this problem is:
    A. a defective cruise control module.
    B. a defective vehicle speed sensor (VSS).
    C. improper size of tires on the drive axles.
    D. a defective cruise control switch.

Hint

Connect the appropriate engine scan tool, OEM special tool, or laptop computer to the truck's DLC. During the road test, activate the tool's parameter specification readout and record the cruise set, coast, and disengagement. Always follow the tool's operating instructions when checking rpm and other values.

Task 7

Observe exhaust for excessive smoke.

151. Excessive black smoke in diesel exhaust may be caused by:
    A. a restricted air intake.
    B. misfiring injectors that do not inject fuel.
    C. coolant entering the combustion chambers.
    D. oil entering the combustion chambers.

Hint

The color and consistency of the exhaust can reveal problems associated with the combustion in the cylinders. Oil entering the combustion chambers causes blue smoke in the exhaust. Unburned fuel or an improper grade of fuel produces a gray or black smoke and soot formation if this condition becomes excessive. This condition may occur if the intake system has provided insufficient air for the amount of fuel delivered. Water or coolant entering the combustion chambers produces a white steam in the exhaust. White smoke may also indicate that injectors are misfiring. This happens when combustion does not occur and fuel vapors are present in the exhaust. White smoke occurs either when the engine starts the first time or when the ambient temperature is too low for proper combustion.

Task 8

Test service brakes.

152. When performing a test drive on a vehicle, *Technician A* says you test the service brakes by applying and releasing them several times in quick succession. *Technician B* says that although a test drive is not a mandatory procedure, it is always a good idea to perform one after a vehicle has been serviced. Who is right?
 A. A only
 B. B only
 C. Both A and B
 D. Neither A nor B

Hint

Brakes top the list of maintenance concerns that fleets cite when asked to name the top five trailer maintenance problems. A proper inspection of a heavy truck's hydraulic braking system should always include a visual check around the wheel cylinders for contamination, which generally indicates leakage. Brake lines should be in good condition at all times with no corrosion or damage that may impede the system's performance.

When testing the service brakes on an air brake system, one procedure is to apply and release the brakes several times to analyze their effective holding and releasing action. This common test is always one of the first tests performed during a drive test procedure.

The complete air brake system should be inspected for damaged brake chambers, lines, fittings, valves, and hoses. Brake adjustment must be checked at each wheel, and the spring brake operation must be satisfactory.

Task 9

Verify engine/exhaust brake operation.

153. An exhaust brake is being inspected on a truck. *Technician A* says that the exhaust brake linkage should be manually actuated while the brake operation is visually inspected. *Technician B* says that it is not necessary to check the entire exhaust system for leaks as long as the vehicle sounds quiet. Who is right?
 A. A only
 B. B only
 C. Both A and B
 D. Neither A nor B

Hint

The periodic servicing and/or inspection procedures for exhaust brakes are simple to perform. Every week the brake operation should be visually checked while the exhaust brake linkage is manually actuated. The wiring and plumbing connections should be checked to make sure they are secure, and the exhaust system should be checked for leakage.

When the compression brake is energized, it causes the engine to perform like a power-absorbing air compressor. The compression brake system uses engine oil pressure to open the exhaust valves before the compression stroke is complete and combustion does not occur in the cylinder. The compressed air is released into the engine exhaust system.

Task 10

Check operation of back-up warning.

154. *Technician A* says the back-up warning system provides a warning to anyone standing behind the truck. *Technician B* says the back-up warning system provides a visual warning. Who is right?
 A. A only
 B. B only
 C. Both A and B
 D. Neither A nor B

Hint

The back-up warning provides a noise that is quite audible outside the vehicle when the transmission is placed in reverse. The purpose of this signal is to warn anyone standing or parked behind the truck that the truck is going to back up.

Answers to Questions

1. Answer A is wrong because wear in the governor flyweight pivots may be causing insufficient rack movement, but Technician B is also right.

   Answer B is wrong because the governor flyweights may be sticking, but Technician A is also right.

   Answer C is right because both Technician A and Technician B are right.

   Answer D is wrong because both Technician A and Technician B are right.

2. Answer A is wrong because an EDC system does not have a throttle linkage connected from the accelerator pedal to the governor.

   Answer B is wrong because an EDC system does not have a flyweight governor in the injection pump.

   Answer C is right because an EDC system does have a linear-motion solenoid controlling the rack in the injection pump. This linear-motion solenoid is controlled by the electronic control unit (ECU) to deliver the proper amount of fuel under all engine speed and load conditions.

   Answer D is wrong because an EDC system does not have a fulcrum lever in the injection pump.

3. **Answer A is right** because the vibration damper counterbalances the back-and-forth twisting motion of the crankshaft each time a cylinder fires.

   Answer B is wrong because if the seal contact area on the vibration damper hub is scored, the damper assembly must be replaced, the hub may be machined, and a sleeve pressed on the hub.

   Answer C is wrong because Technician A is right and Technician B is wrong.

   Answer D is wrong because Technician A is right and Technician B is wrong.

4. Answer A is wrong because a cooling system tester is hooked up to the inlet spout of the radiator.

   Answer B is right because a belt tension gauge is shown in the figure.

   Answer C is wrong because a vacuum gauge uses a small vacuum hose for hookup.

   Answer D is wrong because a dial indicator does not have a handle, but does have a small pointer.

5. **Answer A is right** because a cooling system pressure reading of 15 to 20 psi is normal.

   Answer B is wrong because a cooling system pressure tester does not generally require a normal engine operating temperature for operation. A normal cooling system pressure tester has a manually operated pump to raise system pressure without requiring to raise its temperature.

   Answer C is wrong because Technician A is right and Technician B is wrong.

   Answer D is wrong because Technician A is right and Technician B is wrong.

6. **Answer A is right** because a broken engine mount may cause the fan blades to strike the radiator shroud.

   Answer B is wrong because out-of-balance fan blades do not cause these blades to hit the radiator shroud.

   Answer C is wrong because Technician A is right and Technician B is wrong.

   Answer D is wrong because Technician A is right and Technician B is wrong.

7. Answer A is wrong because a cooling system leak is not considered a safety priority.

   Answer B is right because there must be sufficient vacuum for there to be at least one full brake application after the engine is shut off; otherwise it must be deadlined.

   Answer C is wrong because Technician A is wrong and Technician B is right.

   Answer D is wrong because Technician A is wrong and Technician B is right.

8. Answer A is wrong because an engine thermostat stuck open does not cause a coolant leak.

   Answer B is wrong because a leaking water pump seal causes a visible leak under the pump.

   Answer C is right because a leaking heater core in the cab could cause a loss of coolant with no visible leaks in the engine compartment.

   Answer D is wrong because a leaking heater hose connection at the cylinder head causes a visible leak in the engine compartment.

9. Answer A is wrong because both valves in the cylinder should be closed.

Answer B is right because this statement is not true. Insufficient valve clearance may cause cylinder misfiring, but it does not cause clicking at high speed.

Answer C is wrong because the adjusting screw is turned clockwise to decrease the valve clearance.

Answer D is wrong because excessive clearance may cause a clicking noise at idle speed.

10. **Answer A is right** because you also check fuel tank mountings.

    Answer B is wrong because pour point tests are never performed as part of a PMI.

    Answer C is wrong because the heater lines have nothing to do with a fuel system PMI.

    Answer D is wrong because fuel gauges are not recalibrated as part of any PMI.

11. Answer A is wrong because improper engine response to load changes may indicate the throttle linkage requires adjusting, but Technician B is also right.

    Answer B is wrong because improper engine response to load changes may indicate the fuel filter is restricted, but Technician A is also right.

    Answer C is right because both Technician A and Technician B are right.

    Answer D is wrong because both Technician A and Technician B are right.

12. Answer A is wrong because the water does not remain mixed with the fuel.

    Answer B is wrong because water does not float on top of the fuel.

    Answer C is right because the water settles to the bottom of the fuel tank.

    Answer D is wrong because the water usually does not flow through the filters.

13. Answer A is wrong because the indicated gauge in the figure measures fuel pressure.

    Answer B is wrong because the indicated gauge in the figure measures fuel pressure.

    Answer C is wrong because the indicated gauge in the figure measures fuel pressure.

    Answer D is right because the indicated gauge in the figure measures fuel pressure.

14. Answer A is wrong because both exhaust valves must be fully depressed.

    Answer B is wrong because the governor speed control lever must be in the idle position.

    Answer C is wrong because the locknut is loosened and the pushrod turned to change the adjustment.

    Answer D is right because with the gauge properly installed, the gauge shoulder should just pass over the top of the injector follower.

15. **Answer A is right** because if there is an exhaust leak 15 in. (cm) forward of the rearmost part of the bus, the bus must be tagged out of service.

    Answer B is wrong because if there is an exhaust leak 15 in. (38 cm) forward of the rearmost part of the bus, the bus must be tagged out of service.

    Answer C is wrong because if there is an exhaust leak 15 in. (38 cm) forward of the rearmost part of the bus, the bus must be tagged out of service.

    Answer D is wrong because if there is an exhaust leak 15 in. (38 cm) forward of the rearmost part of the bus, the bus must be tagged out of service.

16. **Answer A is right** because a restricted exhaust system may cause a loss of power and limited top speed.

    Answer B is wrong because a defective ignition coil causes engine misfiring, especially under load.

    Answer C is wrong because a restricted fuel filter causes engine cutout and surging at high speeds.

    Answer D is wrong because a restricted fuel return line causes high fuel pressure, a rich air-fuel ratio, and excessive fuel consumption.

17. **Answer A is right** because the air passages through the air-to-air intercooler should be checked for restrictions.

    Answer B is wrong because the air-to-air intercooler does not use refrigerant to lower the intake air temperature.

    Answer C is wrong because Technician A is right and Technician B is wrong.

    Answer D is wrong because Technician A is right and Technician B is wrong.

18. Answer A is wrong because turbocharger bearings that are starting to fail may cause reduced boost pressure.

    Answer B is wrong because improper wastegate adjustment may cause reduced boost pressure.

Answer C is wrong because a damaged compressor wheel may cause reduced boost pressure.

Answer D is right because turbocharger oil seals leaking oil into the exhaust cause blue smoke in the exhaust, but this problem does not cause reduced boost pressure.

19. **Answer A is right** because the fan clutch is engaged when the air solenoid shuts off air to the clutch.

    Answer B is wrong because the air solenoid is turned on and off by electric current flow.

    Answer C is wrong because the fan clutch is not engaged when the engine coolant is cold.

    Answer D is wrong because the fan clutch is engaged when the air pressure to the clutch is shut off by the air solenoid. When the solenoid supplies air pressure to the fan clutch, the clutch is disengaged.

20. Answer A is wrong because the water temperature switch is closed at normal engine temperatures.

    Answer B is right because when current flows through the air solenoid, this solenoid turns on air pressure to the fan clutch to disengage the clutch.

    Answer C is wrong because Technician A is wrong and Technician B is right.

    Answer D is wrong because Technician A is wrong and Technician B is right.

21. **Answer A is right** because some shutters are operated by air pressure.

    Answer B is wrong because at normal operating engine temperature, the shutters should be open.

    Answer C is wrong because Technician A is right and Technician B is wrong.

    Answer D is wrong because Technician A is right and Technician B is wrong.

22. Answer A is wrong because a cracked fan blade may cause a clicking or creaking noise with the engine idling, but Technician B is also right.

    Answer B is wrong because you should not stand or position your body in line with the fan while increasing the engine speed, but Technician A is also right.

    Answer C is right because both Technician A and Technician B are right.

    Answer D is wrong because both Technician A and Technician B are right.

23. **Answer A is right** because the pressure tester in the figure does check for cooling system leaks.

    Answer B is wrong because the pressure tester in the figure does not check the radiator cap vacuum valve.

    Answer C is wrong because the pressure tester in the figure does not check coolant specific gravity.

    Answer D is wrong because the pressure tester in the figure does not check external-to-the-radiator transmission coolers.

24. Answer A is wrong because a restricted lower radiator hose does not cause a collapsed upper radiator hose.

    Answer B is right because a plugged hose from the radiator filler neck to the recovery container may cause a collapsed upper radiator hose.

    Answer C is wrong because restricted radiator tubes do not cause a collapsed upper radiator hose.

    Answer D is wrong because a thermostat sticking open does not cause a collapsed upper radiator hose.

25. Answer A is wrong because restricted radiator tubes may cause excessive coolant level in the recovery reservoir.

    Answer B is right because a thermostat that is stuck open does not cause excessive coolant level in the recovery reservoir.

    Answer C is wrong because a loose water pump impeller may cause excessive coolant level in the recovery reservoir.

    Answer D is wrong because an inoperative air-operated cooling fan clutch may cause excessive coolant level in the recovery reservoir.

26. Answer A is wrong because a cooling system filled with water causes severe corrosion of the system.

    Answer B is wrong because a 50/50 water and antifreeze solution has a higher boiling point compared to water.

    Answer C is wrong because both Technician A and Technician B are wrong.

    Answer D is right because both Technician A and Technician B are wrong.

27. Answer A is wrong because the component in Figure 7–24 is a coolant filter.

    Answer B is right because the component in Figure 7–24 is a coolant filter.

    Answer C is wrong because the component in Figure 7–24 is a coolant filter.

    Answer D is wrong because the component in Figure 7–24 is a coolant filter.

28. Answer A is wrong because canister-type filters should be filled before installation.

    Answer B is wrong because seals should be lubricated lightly before filter installation.

    Answer C is wrong because an internal engine coolant leak causes a milky gray sludge in the oil.

    Answer D is right because the by-pass valve is not opened by high engine oil pressure. This valve opens if the oil filter is partially restricted and the oil pressure drop across the filter exceeds a predetermined value.

29. Answer A is wrong because an engine oil sample must be taken during every PM, but Technician B is also right.

    Answer B is wrong because you always take an engine oil sample from an engine that is at normal operating temperature, but Technician A is also right.

    Answer C is right because both Technician A and Technician B are right.

    Answer D is wrong because both Technician A and Technician B are right.

30. Answer A is wrong because a restricted PCV valve may cause oil to enter the air cleaner.

    Answer B is wrong because excessive engine blowby may cause oil to enter the air cleaner.

    Answer C is right because a leaking rocker arm cover gasket does not cause oil to enter the air cleaner.

    Answer D is wrong because a restricted hose from the PCV valve to the intake manifold may cause oil to enter the air cleaner.

31. **Answer A is right** because an EGR valve stuck open causes rough idle operation.

    Answer B is wrong because a burned exhaust valve causes low compression, and the question says the compression is within specifications.

    Answer C is wrong because a valve adjustment that is too tight causes low cylinder compression, and the question says the compression is within specifications.

    Answer D is wrong because a bent intake valve causes low cylinder compression, and the question says the compression is within specifications.

32. **Answer A is right** because a burned-out 14 gauge yellow fuse link may be opening the circuit to the fuel shutoff relay contacts and causing the fuel shutoff solenoid not to be energized.

    Answer B is wrong because an open circuit in the fuel shutoff relay winding results in 0V at the ground side of the relay winding, and the question says the voltage at this point is 0.2V, which is normal.

    Answer C is wrong because an open circuit between the fuel shutoff relay winding and ground would cause a 12V reading at the ground side of the relay winding, and the question says there is 0.2V at this point.

    Answer D is wrong because an open circuit between the fuel shutoff relay winding and the ignition switch causes a 0V reading at the switch side of the relay winding, and the question says the voltage at this point is 12V.

33. Answer A is wrong because having an open circuit in the sending circuit would cause a low temperature gauge reading.

    Answer B is wrong because a short to the power feed circuit would open (blow) the fuse protecting it.

    Answer C is right because a short circuit to ground or in the sending unit will cause a continuous high temperature reading.

    Answer D is wrong because an open or burned out gauge will cause no operation of that gauge.

34. Answer A is wrong because a malfunctioning oil pressure gauge is not a valid reason for tagging a vehicle out of service, but Technician B is also right.

    Answer B is wrong because a proper function check of the instrument panel is required under a PMI, but Technician A is also right.

    Answer C is right because both Technician A and Technician B are right.

Answer D is wrong because both Technician A and Technician B are right.

35. Answer A is wrong because the shutdown valve is used to release a door inside the air box of the blower, completely restricting the air intake, but Technician B is also right.

    Answer B is wrong because you find shutdown valves on the air intake on two-stoke diesel bus engines, but Technician A is also right.

    Answer C is right because both Technician A and Technician B are right.

    Answer D is wrong because both Technician A and Technician B are right.

36. **Answer A is right** because all Technicians who perform service on the refrigerant portion of a motor vehicle air conditioner must be certified.

    Answer B is wrong because both hand valves on the manifold gauge set must be closed before connecting the hoses to the A/C system.

    Answer C is wrong because Technician A is right and Technician B is wrong.

    Answer D is wrong because Technician A is right and Technician B is wrong.

37. **Answer A is right** because when another vehicle is a specific distance ahead in the path of the vehicle, the VORAD system does not apply the vehicle brakes.

    Answer B is wrong because when another vehicle is a specific distance ahead in the path of the vehicle, the VORAD system illuminates warning lights to inform the driver.

    Answer C is wrong because when another vehicle is a specific distance ahead in the path of the vehicle, the VORAD system may activate an audible warning to inform the driver.

    Answer D is wrong because when another vehicle is in the blind spot on the right side of the tractor and trailer, the VORAD system illuminates warning lights to inform the driver.

38. Answer A is wrong because the ignition switch must be off when disconnecting the scan tool.

    Answer B is right because history DTCs can be read.

    Answer C is wrong because Technician A is wrong and Technician B is right.

    Answer D is wrong because Technician A is wrong and Technician B is right.

39. Answer A is wrong because air horns are usually mounted on the cab roof.

    Answer B is right because air pressure is supplied through the pressure protection valve.

    Answer C is wrong because air horns are not considered serviceable.

    Answer D is wrong because air horns are activated by a pull cord.

40. Answer A is wrong because carrying flares in the tractor is optional.

    Answer B is wrong because a fire extinguisher must be mounted within reach of the driver.

    Answer C is wrong because both Technician A and Technician B are wrong.

    Answer D is right because both Technician A and Technician B are wrong.

41. **Answer A is right** because with the seat belt on and the truck moving, activate the parking brake to brake the truck very quickly.

    Answer B is wrong because all seat belts and restraints of any type must be checked on a PMI.

    Answer C is wrong because Technician A right and Technician B is wrong.

    Answer D is wrong because Technician A right and Technician B is wrong.

42. Answer A is wrong because when the wiper blade on the driver's side is cracked and badly worn, the vehicle should be tagged out of service.

    Answer B is wrong because when the wiper motor is inoperative, the vehicle should be tagged out of service.

    Answer C is right because when the passenger's side wiper blade is cracked and badly worn, the vehicle should not be tagged out of service.

    Answer D is wrong because if the driver's side wiper arm is distorted so the blade does not contact the windshield properly, the vehicle should be tagged out of service.

43. Answer A is wrong because component 3 is the windshield washer motor and pump.

    Answer B is wrong because component 3 is the windshield washer motor and pump.

Answer C is wrong because component 3 is the windshield washer motor and pump.

Answer D is right because component 3 is the windshield washer motor and pump.

44. Answer A is wrong because the vehicle cannot be tagged out of service for cracks in the passenger's side of the windshield.

    Answer B is right because any vision-distorting defect and/or hazardous working condition concerning the windshield glass must be within the sweep of the wiper on the driver's side to qualify for the vehicle to be deadlined.

    Answer C is wrong because Technician A is wrong and Technician B is right.

    Answer D is wrong because Technician A is wrong and Technician B is right.

45. **Answer A is right** because the pressure protection valve protects the air supply in the air brake system if a leak occurs in the air seat.

    Answer B is wrong because the pressure protection valve does not protect the air spring in the seat from excessive air pressure.

    Answer C is wrong because the pressure protection valve does not close when the air pressure exceeds a predetermined value.

    Answer D is wrong because the pressure protection valve does not pulse on and off to regulate the air pressure.

46. Answer A is wrong because the Technician is not removing excess adhesive after a windshield replacement.

    Answer B is right because the Technician is removing a minor scratch from a side window.

    Answer C is wrong because the Technician is not using a buffer and rubbing compound to remove paint abrasions.

    Answer D is wrong because the Technician is not cleaning the side window molding.

47. Answer A is wrong because a loose grab handle may cause personal injury, but Technician B is also right.

    Answer B is wrong because the steps must be level and free from debris, but Technician A is also right.

    Answer C is right because both Technician A and Technician B are right.

    Answer D is wrong because both Technician A and Technician B are right.

48. **Answer A is right** because you do record the observed problems on a PMI.

    Answer B is wrong because informing the service department will not accomplish any service on the vehicle.

    Answer C is wrong because servicing the bracket personally does not fall under the federal regulations.

    Answer D is wrong because servicing the bracket personally does not fall under the federal regulations.

49. Answer A is wrong because three floor cross members have to be broken and sagging below the lower rail for the vehicle to be tagged out of service.

    Answer B is right because if a drop frame is visibly twisted, the vehicle should be placed out of service.

    Answer C is wrong because Technician A is wrong and Technician B is right.

    Answer D is wrong because Technician A is wrong and Technician B is right.

50. Answer A is wrong because most PM records on heavy trucks are required by the DOT.

    Answer B is right because maintenance records are an invaluable source of information in the life of a heavy truck.

    Answer C is wrong because Technician A is wrong and Technician B is right.

    Answer D is wrong because Technician A is wrong and Technician B is right.

51. Answer A is wrong because component G is a door check strap.

    Answer B is wrong because component G is a door check strap.

    Answer C is right because component G is a door check strap.

    Answer D is wrong because component G is a door check strap.

52. Answer A is wrong because generally, if any components or systems are suspected of requiring servicing, then those non-checklisted areas are more thoroughly inspected and/or serviced immediately.

Answer B is right because you always check the cab ride height of a cab-over vehicle on a PMI.

Answer C is wrong because Technician A is wrong and Technician B is right.

Answer D is wrong because Technician A is wrong and Technician B is right.

53. Answer A is wrong because the pull circuit allows the cab to come back over the engine.

Answer B is wrong because the pull circuit allows the cab to the 45-degree tilt position.

Answer C is wrong because the pull circuit allows the cab to descend to the 60-degree overhead position.

Answer D is right because the push circuit raises the cab from the lowered position to the desired tilt position.

54. **Answer A is right** because refrigerant leaks at condenser line fittings are indicated by an oily film on the fittings.

Answer B is wrong because refrigerant leaks at condenser line fittings are indicated by an oily film on the fittings.

Answer C is wrong because refrigerant leaks at condenser line fittings are indicated by an oily film on the fittings.

Answer D is wrong because refrigerant leaks at condenser line fittings are indicated by an oily film on the fittings.

55. Answer A is wrong because loose compressor mountings may cause a rattling noise.

Answer B is wrong because a worn compressor drive belt may cause a squealing noise.

Answer C is right because a defective bearing in the compressor may cause a growling noise only when the compressor clutch is engaged.

Answer D is wrong because liquid refrigerant entering the compressor causes a thumping or knocking noise with the compressor clutch engaged.

56. **Answer A is right** because moisture and corrosion on battery tops causes faster battery self-discharge.

Answer B is wrong because excessive corrosion on the battery box and cover indicates overcharging.

Answer C is wrong because Technician A is right and Technician B is wrong.

Answer D is wrong because Technician A is right and Technician B is wrong.

57. Answer A is wrong because the battery positive cable must be disconnected first when hooking up a battery, but Technician B is also right.

Answer B is wrong because one always disconnects the ground cable first when disconnecting a battery, but Technician A is also right.

Answer C is right because both Technician A and Technician B are right.

Answer D is wrong because both Technician A and Technician B are right.

58. Answer A is wrong because a fully charged battery has a specific gravity of 1.265.

Answer B is wrong because a fully charged battery has a specific gravity of 1.265.

Answer C is wrong because a fully charged battery has a specific gravity of 1.265.

Answer D is right because a fully charged battery has a specific gravity of 1.265.

59. Answer A is wrong because the test results do not indicate a sulfated battery.

Answer B is wrong because the question says the battery is fully charged at 1.265 specific gravity.

Answer C is right because the battery voltage should remain above 9.6V if the battery is satisfactory.

Answer D is wrong because the question says the battery is fully charged at 1.265 specific gravity.

60. Answer A is wrong because replacing a starting motor cable with a smaller diameter than the original cable causes excessive voltage drop at the starting motor and reduces cranking speed.

Answer B is right because a replacement starting motor cable that is 3 ft. (91 cm) longer than the original cable results in excessive voltage drop at the starting motor.

Answer C is wrong because Technician A is wrong and Technician B is right.

Answer D is wrong because Technician A is wrong and Technician B is right.

61. Answer A is wrong because worn starter bushings cause a dragging starter.

    Answer B is right because an overcharged battery does not cause a dragging starter.

    Answer C is wrong because a defective starter armature causes a dragging starter.

    Answer D is wrong because worn brushes and commutator in the starter cause a dragging starter.

62. **Answer A is right** because the jumper wire is connected to test the magnetic switch.

    Answer B is wrong because the jumper wire is connected to test the magnetic switch.

    Answer C is wrong because the jumper wire is connected to test the magnetic switch.

    Answer D is wrong because the jumper wire is connected to test the magnetic switch.

63. Answer A is wrong because you do check the "ears" for worn mounting holes, but Technician B is also right.

    Answer B is wrong because the belt tension should be checked before the output test, but Technician A is also right.

    Answer C is right because both Technician A and Technician B are right.

    Answer D is wrong because both Technician A and Technician B are right.

64. **Answer A is right** because 2,000 rpm or above are sufficient to generate enough rotational speed to produce the maximum current output from the alternator.

    Answer B is wrong because 1,000 rpm are not sufficient rpm to produce maximum output.

    Answer C is wrong because 3,000 rpm are not necessary to produce maximum current output.

    Answer D is wrong because 1,500 rpm are not sufficient to produce maximum current output.

65. Answer A is wrong because a belt tightness check is one of the first operations performed in an inspection.

    Answer B is wrong because a current output test is always performed for an inspection.

    Answer C is wrong because a voltage regulator is routinely performed during an inspection.

    Answer D is right because a voltage drop test on insulated side of charging circuit is LEAST likely to be performed.

66. Answer A is wrong because ballast resistors are used for ignition systems.

    Answer B is right because you do inspect some of the cab ground circuits for corrosion and high resistance.

    Answer C is wrong because Technician A is wrong and Technician B is right.

    Answer D is wrong because Technician A is wrong and Technician B is right.

67. Answer A is wrong because the vehicle should be on level ground at a distance of 25 feet from the alignment screen or chart.

    Answer B is wrong because you do not always use headlight-aiming equipment when adjusting headlights. One can use a screen or chart.

    Answer C is wrong because both Technician A and Technician B are wrong.

    Answer D is right because both Technician A and Technician B are wrong.

68. Answer A is wrong because the wiring connector for all trailer lights is always located inside the cab directly behind the driver's seat, not under the front frame rail of the first trailer.

    Answer B is wrong because it is customary to begin at the front and work from this point to the last vehicle.

    Answer C is wrong because both Technician A and Technician B are wrong.

    Answer D is right because both Technician A and Technician B are wrong.

69. Answer A is wrong because the valve on the left is the trailer supply valve. It is octagon shaped and red in color.

    Answer B is right because the control on the right in the same figure is the system park or parking brake valve and applies the tractor and trailer spring brakes. It is diamond shaped and always yellow in color.

    Answer C is wrong because Technician A is wrong and Technician B is right.

    Answer D is wrong because Technician A is wrong and Technician B is right.

70. Answer A is wrong because an acceptable governor cut-out pressure is 125 psi (862 kPa).

    Answer B is right because an acceptable governor cut-out pressure is 125 psi (862 kPa).

    Answer C is wrong because an acceptable governor cut-out pressure is 125 psi (862 kPa).

    Answer D is wrong because an acceptable governor cut-out pressure is 125 psi (862 kPa).

71. **Answer A is right** because the purge valve should open each time the compressor enters the unloaded cycle.

    Answer B is wrong because oil being discharged from the air dryer purge valve likely indicates worn piston rings in the compressor.

    Answer C is wrong because Technician A is right and Technician B is wrong.

    Answer D is wrong because Technician A is right and Technician B is wrong.

72. Answer A is wrong because a soapy water solution should be applied to the area around the inlet and exhaust valve assembly.

    Answer B is right because a soapy water solution should be applied to the area around the inlet and exhaust valve assembly.

    Answer C is wrong because the brakes should not be applied.

    Answer D is wrong because a 2 in. (50.8 mm) bubble in 3 seconds indicates excessive intake valve leakage. The maximum leakage is a 1 in. (25.4 mm) bubble in 3 seconds.

73. Answer A is wrong because leakage at the brake application valve exhaust port with the brakes applied should not exceed a 1 in. (2.54 mm) bubble in 3 seconds.

    Answer B is wrong because leakage at the brake application valve exhaust port with the brakes applied should not exceed a 1 in. (2.54 mm) bubble in 3 seconds.

    Answer C is wrong because leakage at the brake application valve exhaust port with the brakes applied should not exceed a 1 in. (2.54 mm) bubble in 3 seconds.

    Answer D is right because leakage at the brake application valve exhaust port with the brakes applied should not exceed a 1 in. (2.54 mm) bubble in 3 seconds.

74. **Answer A is right** because a one-way check valve may be tested by disconnecting the supply line and checking for back leakage.

    Answer B is wrong because a two-way check valve supplies air pressure from the highest of two pressure sources connected to the valve.

    Answer C is wrong because Technician A is right and Technician B is wrong.

    Answer D is wrong because Technician A is right and Technician B is wrong.

75. **Answer A is right** because component H is a low-pressure warning device.

    Answer B is wrong because component H is a low-pressure warning device.

    Answer C is wrong because component H is a low-pressure warning device.

    Answer D is wrong because component H is a low-pressure warning device.

76. Answer A is wrong because the governor cut-in pressure is not adjustable.

    Answer B is wrong because the question says the unloader mechanism in the cylinder head is operating normally.

    Answer C is right because the governor should be replaced.

    Answer D is wrong because the governor is replaced as a unit.

77. Answer A is wrong because you do drain both the wet tank and the primary axle service reservoir before beginning the test, but Technician B is also right.

    Answer B is wrong because you always verify that secondary system pressure is at least 100 psi (689 kPa) before conducting the test, but Technician A is also right.

    Answer C is right because both Technician A and Technician B are right.

    Answer D is wrong because both Technician A and Technician B are right.

78. **Answer A is right** because the tractor protection valve closes to protect the tractor air supply if the supply line pressure drops below 20 psi (138 kPa).

    Answer B is wrong because the tractor protection valve closes to protect the tractor air supply if the supply line pressure drops below 20 psi (138 kPa).

Answer C is wrong because the tractor protection valve closes to protect the tractor air supply if the supply line pressure drops below 20 psi (138 kPa).

Answer D is wrong because the tractor protection valve closes to protect the tractor air supply if the supply line pressure drops below 20 psi (138 kPa).

79. Answer A is wrong because a defective air governor will not cause a slow buildup, only cut-in and off problems.

    Answer B is wrong because a leak in a brake chamber will not cause slow buildup because it is on the application side, not the supply.

    Answer C is wrong because an air leak in the cab will not cause slow buildup because the air supply is protected by pressure protection valves.

    Answer D is right because a defective air compressor is the MOST likely cause of slow buildup.

80. Answer A is wrong because the pressure drop on a straight truck is 4 psi in 2 minutes.

    Answer B is wrong because the brake pedal must be held down for the 2-minute test.

    Answer C is wrong because both Technician A and Technician B are wrong.

    Answer D is right because both Technician A and Technician B are wrong.

81. Answer A is wrong because the adjustment shown in the figure is the free stroke.

    Answer B is right because the adjustment shown in the figure is the free stroke.

    Answer C is wrong because the adjustment shown in the figure is the free stroke.

    Answer D is wrong because the adjustment shown in the figure is the free stroke.

82. **Answer A is right** because the air brake system pressure must be decreased to zero before changing a spin-on dryer cartridge.

    Answer B is wrong because most air dryers with an integral cartridge must be removed from the tractor to change the cartridge.

    Answer C is wrong because Technician A is right and Technician B is wrong.

    Answer D is wrong because Technician A is right and Technician B is wrong.

83. Answer A is wrong because a vehicle with 75 percent defective brakes should not be driven or put into service.

    Answer B is wrong because 50 percent defective brakes is not the standard the CVSA has set for deadlining.

    Answer C is wrong because a vehicle with 40 percent defective brakes is not the standard by which the CVSA bases its deadlining.

    Answer D is right because a vehicle should be tagged out of service if 20 percent of the brakes are defective.

84. Answer A is wrong because a concave drum would be bowed out, not in. The concave condition indicates drum diameter greater at the friction surface center as compared to the edges of the friction area.

    Answer B is right because a convex drum condition is shown. The convex condition indicates a drum diameter greater at the friction surface edges as compared to the center of the friction surface.

    Answer C is wrong because an offset is not a defective drum condition.

    Answer D is wrong because a shifted drum is not a defective drum condition.

85. **Answer A is right** because you always lubricate the slack adjusters whenever an inspection of the brakes is scheduled.

    Answer B is wrong because all slack adjuster pushrods must be checked during a brake inspection.

    Answer C is wrong because Technician A is right and Technician B is wrong.

    Answer D is wrong because Technician A is right and Technician B is wrong.

86. **Answer A is right** because the rack turns the one-way clutch and worm shaft to adjust the brake shoes.

    Answer B is wrong because the external teeth of the gear in the center of the slack adjuster are meshed with the teeth on the worm shaft.

    Answer C is wrong because as the brake linings wear, the brake chamber pushrod stroke and the slack adjuster arm movement increase.

Answer D is wrong because the brake shoes are only adjusted when the linings wear enough to provide a specific increase in pushrod stroke.

87. Answer A is wrong because standard nonvented grease fittings should not be used to replace pressure relief-type grease fittings.

Answer B is wrong because when the air chamber bracket is lubricated, grease should not flow out of the end of the bracket tube next to the cam head because the seal should prevent this action.

Answer C is wrong because both Technician A and Technician B are wrong.

Answer D is right because both Technician A and Technician B are wrong.

88. Answer A is wrong because DOT 4 brake fluid should not be mixed with DOT 5 brake fluid.

Answer B is wrong because DOT 3 brake fluid is hydroscopic.

Answer C is wrong because brake fluids should be stored in a cool, dry location.

Answer D is right because brake system components are lubricated by the brake fluid.

89. Answer A is wrong because if a flexible hose ruptures completely on one of the front wheels, the half of the master cylinder that supplies the rear brakes is still operational.

Answer B is right because a leak at a brake line may cause pedal fade with the brakes applied.

Answer C is wrong because Technician A is wrong and Technician B is right.

Answer D is wrong because Technician A is wrong and Technician B is right.

90. Answer A is wrong because a hydraulic liftgate cylinder would be much larger.

Answer B is wrong because the indicated object is not a solenoid.

Answer C is wrong because the kingpin does not have an automatic locking relay.

Answer D is right because a hydraulic spring parking brake actuator is shown in the figure.

91. Answer A is wrong because a defective brake booster causes excessive pedal effort.

Answer B is wrong because seized wheel caliper or wheel cylinder pistons cause excessive pedal effort.

Answer C is right because improper brake pedal free play adjustment does not cause excessive pedal effort.

Answer D is wrong because glazed brake linings cause excessive pedal effort.

92. **Answer A is right** because seized wheel cylinder pistons on the left front wheel result in very little braking force on the left front wheel and normal braking force on the right front wheel, resulting in steering pull to the right.

Answer B is wrong because seized wheel cylinder pistons on the right front wheel cause steering pull to the left.

Answer C is wrong because brake fluid contaminated with moisture does not cause steering pull.

Answer D is wrong because a restricted right front brake hose causes steering pull to the left.

93. **Answer A is right** because a restricted vacuum hose connected to the brake booster causes excessive pedal effort.

Answer B is wrong because improper brake pedal pushrod adjustment does not cause excessive brake pedal effort.

Answer C is wrong because Technician A is right and Technician B is wrong.

Answer D is wrong because Technician A is right and Technician B is wrong.

94. Answer A is wrong because a defective metering valve does not cause brake drag and excessive lining wear.

Answer B is right because a complete lack of brake pedal free play may cause the master cylinder cup seal to cover the compensating ports, resulting in pressure buildup, brake drag, and excessive lining wear.

Answer C is wrong because seized wheel cylinder pistons on the right rear forward axle could only cause brake drag and excessive lining wear on this wheel.

Answer D is wrong because a defective proportioning valve does not cause brake drag and excessive lining wear.

95. **Answer A is right** because the number stamped on the drum is the maximum wear diameter.

Answer B is wrong because the maximum machining diameter must be at least 0.030 in. (0.762 mm) less than the maximum wear diameter, or the drum has to be discarded.

Answer C is wrong because Technician A is right and Technician B is wrong.

Answer D is wrong because Technician A is right and Technician B is wrong.

96. **Answer A is right** because the measurement in the figure is brake lining-to-drum clearance.

Answer B is wrong because the measurement in the figure is brake lining-to-drum clearance.

Answer C is wrong because the measurement in the figure is brake lining-to-drum clearance.

Answer D is wrong because the measurement in the figure is brake lining-to-drum clearance.

97. **Answer A is right** because a clutch brake needs to be inspected for wear and fatigue in the same manner as the pressure plate.

Answer B is wrong because the friction face needs to be inspected.

Answer C is wrong because Technician A is right and Technician B is wrong.

Answer D is wrong because Technician A is right and Technician B is wrong.

98. Answer A is wrong because excessive clutch pedal free travel does not cause a growing noise with the clutch pedal depressed and the engine running.

Answer B is right because excessive clutch pedal free play causes improper clutch release and hard transmission shifting.

Answer C is wrong because excessive clutch pedal free travel does not cause a growling noise with the clutch pedal released and the engine running.

Answer D is wrong because excessive clutch pedal free travel does not cause excessive wear on the clutch facings.

99. Answer A is wrong because checking the placement of the actuator arm is not a valid inspection.

Answer B is wrong because checking a frozen adjusting ring is not a valid inspection.

Answer C is wrong because checking the linkage is not a valid inspection for a self-adjusting clutch.

Answer D is right because clutch pedal linkage is not checked on a self-adjusting clutch.

100. **Answer A is right** because when the breather is plugged, internal pressure can build inside the transmission and push the oil out.

Answer B is wrong because the breather could be plugged.

Answer C is wrong because the breather could be plugged.

Answer D is wrong because the breather could be plugged

101. Answer A is wrong because replacing the breather is not cleaning the breather.

Answer B is wrong because soaking the breather in gasoline is not a recommended procedure.

Answer C is right because the best way to clean a transmission breather is to use solvent and then forced air.

Answer D is wrong because wiping the orifice will not clean any dirt trapped inside the breather.

102. Answer A is wrong because removal of the mount is not necessary to thoroughly inspect it.

Answer B is wrong because removal of the mount is not necessary to thoroughly inspect it.

Answer C is right because you visually inspect the mount while still in the vehicle.

Answer D is wrong because replacing the mount if it is only suspected is not an accepted practice.

103. Answer A is wrong because the proper procedure is to be sure the lubricant is level with the bottom of the filler plug hole.

Answer B is wrong because the proper procedure is to be sure the lubricant is level with the bottom of the filler plug hole.

Answer C is right because the proper procedure is to be sure the lubricant is level with the bottom of the filler plug hole.

Answer D is wrong because the proper procedure is to be sure the lubricant is level with the bottom of the filler plug hole.

104. Answer A is wrong because the rear U-joint is not lubricated from that location.

Answer B is right because lubricating the slip joint is what is shown in the figure.

Answer C is wrong because the center bearing is not lubricated from that location.

Answer D is wrong because the rear bearing (if applicable) is not lubricated from that location.

105. Answer A is wrong because a brake drum out of balance may cause a wheel vibration, but the question says there are no other problems with this wheel.

Answer B is right because the seal seating area on the spindle may be scored, resulting in repeated seal failure.

Answer C is wrong because Technician A is wrong and Technician B is right.

Answer D is wrong because Technician A is wrong and Technician B is right.

106. Answer A is wrong because the breather is located on the top of the rear axle housing.

Answer B is right because on some off-road vehicles, a hose attached to the rear axle breather prevents water from entering the rear axle.

Answer C is wrong because Technician A is wrong and Technician B is right.

Answer D is wrong because Technician A is wrong and Technician B is right.

107. Answer A is wrong because some tractors have an air-operated automatic chassis lubrication system.

Answer B is wrong because overlubrication may damage seals in components with grease fittings.

Answer C is wrong because underlubrication may cause premature wear on certain components.

Answer D is right because the air-operated automatic chassis lubrication system is activated by an air solenoid and not by an electric pump motor.

108. Answer A is wrong because the proper fluid level is when the lubricant is level with the bottom of the differential filler plug hole.

Answer B is right because in some rear axles, a temperature sensor is located in the differential cover below the filler plug.

Answer C is wrong because Technician A is wrong and Technician B is right.

Answer D is wrong because Technician A is wrong and Technician B is right.

109. **Answer A is right** because the component is a differential oil filter adapter.

Answer B is wrong because the component is a differential oil filter adapter.

Answer C is wrong because the component is a differential oil filter adapter.

Answer D is wrong because the component is a differential oil filter adapter.

110. Answer A is wrong because a few particles indicate normal wear, not a problem needing immediate resolution.

Answer B is right because you do inform the customer of the condition and tell them to monitor the amount of particles.

Answer C is wrong because it is always best to inform the customer of any condition on their vehicle that may require extra attention.

Answer D is wrong because a few particles indicate normal wear, not a problem needing immediate resolution.

111. Answer A is wrong because an extended high torque operating condition may be a cause of bearing failure, but not the leading cause.

Answer B is right because dirt and contaminants in the lubricant are the most common cause of transmission bearing failure.

Answer C is wrong because an improper rear axle ratio for the load and driving conditions may cause excessive wear on rear axle and transmission components, but this is not the leading cause of transmission bearing failure.

Answer D is wrong because poor quality of lubricant may be a cause of bearing failure, but not the leading cause.

112. **Answer A is right** because the transmission cooler must be flushed during a transmission overhaul.

Answer B is wrong because if the torque converter end play is satisfactory, the torque converter must be flushed to remove clutch plate material and contaminated fluid.

Answer C is wrong because Technician A is right and Technician B is wrong.

Answer D is wrong because Technician A is right and Technician B is wrong.

113. Answer A is wrong because it is normal for drive axles to acquire slight condensation, which usually "boils off" during normal driving conditions.

Answer B is right because infrequent driving and short trips do not bring the axle to temperature, and moisture does not burn off.

Answer C is wrong because although submerging the axles in water can introduce water into the axle, it is not the most likely cause.

Answer D is wrong because any water that happens to make its way into the axle would normally be "boiled off" during normal driving conditions.

114. Answer A is wrong because a worn tie-rod end causes excessive steering wheel free play.

Answer B is wrong because a worn drag link end causes excessive steering wheel free play.

Answer C is right because a dry kingpin bushing does not cause excessive steering wheel free play.

Answer D is wrong because loose steering gear mounting bolts cause excessive steering wheel free play.

115. Answer A is wrong because an overpressed pulley could cause the power steering pump pulley to become misaligned.

Answer B is wrong because a loose fit from the pulley hub to pump shaft could cause the power steering pump pulley to become misaligned.

Answer C is wrong because a worn or loose pump mounting bracket could cause the power steering pump pulley to become misaligned.

Answer D is right because a broken engine mount will not cause the power steering pump pulley to become misaligned.

116. Answer A is a wrong because the filter should be replaced on a PM according to the OEM, but Technician B is also right.

Answer B is wrong because the filter should be replaced after a component failure, but Technician A is also right.

Answer C is right because both Technician A and Technician B are right.

Answer D is wrong because both Technician A and Technician B are right.

117. **Answer A is right** because worn sector shaft bearings may cause repeated sector shaft seal failure.

Answer B is wrong because excessive worm shaft end play does not cause sector shaft seal failure.

Answer C is wrong because loose steering column U-joints do not cause sector shaft seal failure.

Answer D is wrong because a loose pitman arm on the sector shaft does not cause sector shaft seal failure.

118. Answer A is wrong because you never compress the joint with a C-clamp or large pliers.

Answer B is right because a loose tie-rod or drag link end is checked by simply pushing against the joint with hand force.

Answer C is wrong because Technician A is wrong and Technician B is right.

Answer D is wrong because Technician A is wrong and Technician B is right.

119. Answer A is wrong because vertical play is checked with the indicator on top of the knuckle.

Answer B is right because upper knuckle bushing wear measurement is shown in the figure.

Answer C is wrong because the indicator should be mounted on the lower bushing area.

Answer D is wrong because there is no kingpin axial play test.

120. Answer A is wrong because wheel hub runout is not checked.

Answer B is wrong because there is no aligning a wheel hub to an axle hub.

Answer C is wrong because wheel bearings are usually installed by hand.

Answer D is right because the figure shows the installation of a front wheel bearing seal.

121. Answer A is wrong because the bearings in the figure are lubricated with the specified oil.

Answer B is wrong because the bearings are not permanently lubricated with the specified oil.

Answer C is right because the bearings require a specified hub end play.

Answer D is wrong because the bearings only have an inner hub seal.

122. **Answer A is right** because the proper wheel bearing adjustment procedure is to torque the adjusting nut to 2,000 ft.-lb. (271 Nm), back this nut off 1 turn, torque the nut to 50 ft.-lb. (68 Nm), then back off the nut the specified amount and install the locking device.

Answer B is wrong because after a wheel bearing adjustment, the hub end play should be 0.001 to 0.005 in. (0.0254 to 0.127 mm).

Answer C is wrong because Technician A is right and Technician B is wrong.

Answer D is wrong because Technician A is right and Technician B is wrong.

123. **Answer A is right** because you do not replace alignment washers with new ones, therefore the exception.

Answer B is wrong because securing the vehicle is one of the first operations performed.

Answer C is wrong because replacement rods come in many different sizes.

Answer D is wrong because an axle alignment check is always in order when replacing a radius rod.

124. Answer A is wrong because worn spring bushings and shackles may cause wheel hop.

Answer B is wrong because a worn-out shock absorber may cause wheel hop.

Answer C is right because an improper shim adjustment between the radius rod and the spring bracket does not cause wheel hop.

Answer D is wrong because a loose equalizer spring bracket may cause wheel hop.

125. Answer A is wrong because a leaking air spring will cause the trailer to sit low, not high.

Answer B is right because an improperly adjusted height control valve could be the problem.

Answer C is wrong because Technician A is wrong and Technician B is right.

Answer D is wrong because Technician A is wrong and Technician B is right.

126. Answer A is wrong because the height control valve arm may be moved upward to raise the suspension height.

Answer B is wrong because wood locating pins are installed in the height control valve arm.

Answer C is right because the air brake system pressure must be 70 psi (483 kPa) or more.

Answer D is wrong because the truck must be unloaded or empty.

127. Answer B is wrong because the component in the figure is part of an on-board chassis lubrication system.

Answer B is right because the component in the figure is part of an on-board chassis lubrication system.

Answer C is wrong because the component in the figure is part of an on-board chassis lubrication system.

Answer D is wrong because the component in the figure is part of an on-board chassis lubrication system.

128. Answer A is wrong because toe-in is shown in the figure.

Answer B is wrong because toe-in is shown in the figure.

Answer C is wrong because toe-out is indicated with the wheels pointing out, not in.

Answer D is right because toe-in is shown in the figure.

129. Answer A is wrong because the thrust line should be at or very close to the geometric centerline.

Answer B is wrong because worn leaf spring shackles may cause improper rear axle alignment.

Answer C is wrong because worn equalizing beam bushings may cause improper rear axle alignment.

Answer D is right because improper rear axle alignment causes steering pull.

130. Answer A is wrong because the excessive tire tread wear in the figure was caused by excessive negative camber.

Answer B is wrong because the excessive tire tread wear in the figure was caused by excessive negative camber.

Answer C is right because the excessive tire tread wear in the figure was caused by excessive negative camber.

Answer D is wrong because the excessive tire tread wear in the figure was caused by excessive negative camber.

131. Answer A is wrong because if a tire on a steer axle has a sidewall cut that exposes the cords, tire replacement is necessary, but Technician B is also right.

Answer B is wrong because if a drive axle tire has a sidewall bulge 5/8 in. (15.8 mm) in height above the sidewall surface, the tire must be replaced, but Technician A is also right.

Answer C is right because both Technician A and Technician B are right.

Answer D is wrong because both Technician A and Technician B are right.

132. Answer A is wrong because if a tire is mounted or inflated so it contacts any part of the vehicle, this problem must be corrected before placing the vehicle in service.

Answer B is wrong because if a drive axle tire has a tread depth of less than 1/32 in. (0.79 mm) tread depth at two adjacent tread grooves at three separate locations, this problem must be corrected before placing the vehicle in service.

Answer C is right if a steer axle tire has a tread depth of 3/32 in. (2.38 mm) tread depth at two adjacent tread grooves at any location on the tire, the tire is satisfactory. Minimum tread depth is 2/32 in. (1.58 mm).

Answer D is wrong because if the tire is labeled "Not for Highway Use," this problem must be corrected before placing the vehicle in service.

133. **Answer A is right** because the vehicle is tagged out of service if the weight carried exceeds the tire load limit.

Answer B is wrong because the vehicle is tagged out of service if a tire has 70 percent of the tread width loose or missing in 12 in. (30.4 cm) of the tire circumference.

Answer C is wrong because Technician A is right and Technician B is wrong.

Answer D is wrong because Technician A is right and Technician B is wrong.

134. **Answer A is right** because the tire tread wear in the figure was caused by underinflation.

Answer B is wrong because the tire tread wear in the figure was caused by underinflation.

Answer C is wrong because the tire tread wear in the figure was caused by underinflation.

Answer D is wrong because the tire tread wear in the figure was caused by underinflation.

135. Answer A is wrong because on a cast spoke wheel, the assembly includes a one-piece casting that includes the hub and the spokes, not a rim.

Answer B is wrong because spoke wheels, not disc, generally experience more alignment and balance problems than do the more stable disc wheel design.

Answer C is wrong because both Technician A and Technician B are wrong.

Answer D is right because both Technician A and Technician B are wrong.

136. Answer A is wrong because retorquing of lug nuts is required when performing a PMI.

Answer B is wrong because an increased load-carrying capacity is not an advantage of switching to low-profile tires. Low-profile tires decrease the load-carrying capacity.

Answer C is wrong because both Technician A and Technician B are wrong.

Answer D is right because both Technician A and Technician B are wrong.

137. Answer A is wrong because overinflation causes cracks in the rim base.

Answer B is right because overloading causes lug hole cracking.

Answer C is wrong because this causes cracks in the rim base.

Answer D is wrong because this generally does not cause cracks.

138. Answer A is wrong because mounting tires of a different construction on the same axle is mismatching.

Answer B is right because you may have dual wheels with similar tread designs.

Answer C is wrong because installation of a new tire next to an old tire is considered mismatching.

Answer D is wrong because mounting dual wheels with tires of different circumferences and diameters is considered mismatching.

139. Answer A is wrong because the position of a sliding fifth wheel must not be changed with the vehicle in motion.

Answer B is wrong because a sliding fifth wheel must be bolted to the tractor frame.

Answer C is wrong because both Technician A and Technician B are wrong.

Answer D is right because both Technician A and Technician B are wrong.

140. Answer A is wrong because the tractor is not tagged out of service unless 20 percent of the fifth wheel-to-frame fasteners are missing or ineffective.

Answer B is wrong because the tractor is not tagged out of service unless there is more than 3/8 in. (9.52 mm) horizontal movement between the pivot bracket pin and the bracket.

Answer C is wrong because the tractor is not tagged out of service unless 25 percent of the latching fasteners per side are ineffective.

Answer D is right because the tractor is tagged out of service if there is more than 1/2 in. (12.7 mm) horizontal movement between the upper and lower fifth wheel halves with the trailer coupled.

141. Answer A is wrong because a circle inspection includes items such as torn mud flaps, leaking hoses, and outdated permits, but Technician B is also right.

Answer B is wrong because performing a circle inspection first before any other inspection is part of a PMI, but Technician A is also right.

Answer C is right because both Technician A and Technician B are right.

Answer D is wrong because both Technician A and Technician B are right.

142. **Answer A is right** because this answer is not true. The vehicle is not tagged out of service unless wear in the pintle hook horn exceeds 20 percent of the horn section.

Answer B is wrong because the vehicle is tagged out of service if there are loose or missing pintle hook mounting bolts.

Answer C is wrong because the vehicle is tagged out of service if there are any welded repairs to the drawbar eye.

Answer D is wrong because the vehicle is tagged out of service if there are cracks in the pintle hook mounting surface.

143. Answer A is wrong because when lubricating the brake camshaft and bushings, grease should ooze out of the slack adjuster end of the brake chamber bracket because of the direction of the seal lip.

Answer B is right because excessive lubrication may damage seals in some components.

Answer C is wrong because Technician A is wrong and Technician B is right.

Answer D is wrong because Technician A is wrong and Technician B is right.

144. Answer A is wrong because the frame rail marked A is an incorrect repair, because the ends should be cut at a 45-degree angle and extend past the stress point, but Technician B is also right.

Answer B is wrong because the frame reinforcement on the frame rail marked B is called a channel reinforcement or fishplating, but Technician A is also right.

Answer C is right because both Technician A and Technician B are right.

Answer D is wrong because both Technician A and Technician B are right.

145. **Answer A is right** because the measurement should be between 1 and 2 inches.

Answer B is wrong because the manufacturer's recommendations should always be checked on hydraulic clutches.

Answer C is wrong because Technician A is right and Technician B is wrong.

Answer D is wrong because Technician A is right and Technician B is wrong.

146. Answer A is wrong because there is an electrical defect in the ABS.

Answer B is wrong because the ABS will be inoperative and normal non-ABS braking is maintained.

Answer C is wrong because both Technician A and Technician B are wrong.

Answer D is right because both Technician A and Technician B are wrong.

147. Answer A is wrong because loose tie-rod ends do not cause a binding steering wheel.

Answer B is right because loose worn steering shaft U-joints may cause a binding steering wheel.

Answer C is wrong because loose drag link ends do not cause a binding steering wheel.

Answer D is wrong because worn kingpin bushings do not cause a binding steering wheel.

148. Answer A is wrong because a diagnostic trouble code (DTC) indicates a defect in a specific circuit, not in a specific component.

Answer B is wrong because the diagnosis should begin with the engine at normal operating temperature.

Answer C is wrong because the scan tool should be connected and disconnected with the ignition switch off.

Answer D is right because an engine coolant temperature sensor contains a thermistor.

149. Answer A is wrong because the governor has positioned the injector plunger in the maximum fuel delivery position.

Answer B is wrong because the governor has positioned the injector plunger in the maximum fuel delivery position.

Answer C is right because the governor has positioned the injector plunger in the maximum fuel delivery position.

Answer D is wrong because the governor has positioned the injector plunger in the maximum fuel delivery position.

150. Answer A is wrong because a defective cruise control module is not the most likely cause of improper cruise control set speed.

Answer B is right because a defective vehicle speed sensor (VSS) is the most likely cause of improper cruise control set speed.

Answer C is wrong because improper size of tires on the drive axles is not the most likely cause of improper cruise control set speed.

Answer D is wrong because a defective cruise control switch is not the most likely cause of improper cruise control set speed.

151. **Answer A is right** because a restricted air intake may cause black smoke in the exhaust.

Answer B is wrong because misfiring injectors that do not inject fuel do not cause black smoke in the exhaust.

Answer C is wrong because coolant entering the combustion chambers causes white or gray smoke in the exhaust.

Answer D is wrong because oil entering the combustion chambers causes blue smoke in the exhaust.

152. **Answer A is right** because you test the service brakes by applying and releasing them several times.

Answer B is wrong because a test drive is a mandatory procedure of a PMI.

Answer C is wrong because Technician A is right and Technician B is wrong.

Answer D is wrong because Technician A is right and Technician B is wrong.

153. **Answer A is right** because the exhaust brake linkage should be manually actuated while the brake operation is visually inspected.

Answer B is wrong because the entire exhaust system should always be checked for leaks whenever an inspection of an exhaust brake is performed.

Answer C is wrong because Technician A is right and Technician B is wrong.

Answer D is wrong because Technician A is right and Technician B is wrong.

154. **Answer A is right** because the back-up warning system provides a warning to anyone standing behind the truck.

Answer B is wrong because the back-up warning system provides an audible warning.

Answer C is wrong because Technician A is right and Technician B is wrong.

Answer D is wrong because Technician A is right and Technician B is wrong.

Glossary

ABS An abbreviation for Anti-lock Brake System.

Absolute Pressure The aero point from which pressure is measured.

Ackerman Principle The geometric principle used to provide toe-out on turns. The ends of the steering arms are angled so that the inside wheel turns more than the outside wheel when a vehicle is making a turn.

Actuator A device that delivers motion in response to an electrical signal.

Adaptor The welds under a spring seat to increase the mounting height or fit a seal to the axle.

Adapter Ring A part that is bolted between the clutch cover and the flywheel on some two-plate clutches when the clutch is installed on a flat flywheel.

A/D Converter An abbreviation for Analog-to-Digital Converter.

Additive An additive intended to improve a certain characteristic of the material.

Adjustable Torque Arm A member used to retain axle alignment and, in some cases, control axle torque. Normally one adjustable and one rigid torque arm are used per axle so the axle can be aligned. This rod has means by which it can be extended or retracted for adjustment purposes.

Adjusting Ring A device that is held in the shift signal valve bore by a press fit pin through the valve body housing. When the ring is pushed in by the adjusting tool, the slots on the ring that engage the pin are released.

After-Cooler A device that removes water and oil from the air by a cooling process. The air leaving an after-cooler is saturated with water vapor, which condenses when a drop in temperature occurs.

Air Bag An air-filled device that functions as the spring on axles that utilize air pressure in the suspension system.

Air Brakes A braking system that uses air pressure to actuate the brakes by means of diaphragms, wedges, or cams.

Air Brake System A system that uses compressed air to activate the brakes.

Air Compressor (1) An engine-driven mechanism for supplying high pressure air to the truck brake system. There are basically two types of compressors: those designed to work on in-line engines and those that work on V-type engines. The in-line type is mounted forward and is gear driven, while the V-type is mounted toward the firewall and is camshaft driven. With both types the coolant and lubricant are supplied by the truck engine. (2) A pump-like device in the air conditioning system that compresses refrigerant vapor to achieve a change in state for the refrigeration process.

Air Conditioning The control of air movement, humidity, and temperature by mechanical means.

Air Dryer A unit that removes moisture.

Air Filter/Regulator Assembly A device that minimizes the possibility of moisture-laden air or impurities entering a system.

Air Hose An air line, such as one between the tractor and trailer, that supplies air for the trailer brakes.

Air-Over-Hydraulic Brakes A brake system utilizing a hydraulic system assisted by an air pressure system.

Air-Over-Hydraulic Intensifier A device that changes the pneumatic air pressure from the treadle brake valve into hydraulic pressure which controls the wheel cylinders.

Air Shifting The process that uses air pressure to engage different range combinations in the transmission's auxiliary section without a mechanical linkage to the driver.

Air Slide Release A release mechanism for a sliding fifth wheel, which is operated from the cab of a tractor by actuating an air control valve. When actuated, the valve energizes an air cylinder, which releases the slide lock and permits positioning of the fifth wheel.

Air Spring An airfilled device that functions as the spring on axles that utilize air pressure in the suspension system.

Air Spring Suspension A single or multi-axle suspension relying on air bags for springs and weight distribution of axles.

Air Timing The time required for the air to be transmitted to or released from each brake, starting the instant the driver moves the brake pedal.

Altitude Compensation System An altitude barometric switch and solenoid used to provide better driveability at more than 4,000 feet (1220 meters) above sea level.

Ambient Temperature Temperature of the surrounding or prevailing air. Normally, it is considered to be the temperature in the service area where testing is taking place.

Amboid Gear A gear that is similar to the hypoid type with one exception: the axis of the drive pinion gear is located above the centerline axis of the ring gear.

Amp An abbreviation for ampere.

Ampere The unit for measuring electrical current.

Analog Signal A voltage signal that varies within a given range (from high to low, including all points in between).

Analog-to-Digital Converter (A/D converter) A device that converts analog voltage signals to a digital format; this is located in a section of the processor called the input signal conditioner.

Analog Volt/Ohmmeter (AVOM) A test meter used for checking voltage and resistance. Analog meters should not be used on solid state circuits.

Annulus The largest part of a simple gear set.

Anticorrosion Agent A chemical used to protect metal surfaces from corrosion.

Antifreeze A compound, such as alcohol or glycerin, that is added to water to lower its freezing point.

Anti-lock Brake System (ABS) A computer controlled brake system having a series of sensing devices at each wheel that control braking action and prevent wheel lockup.

Anti-lock Relay Valve (ARV) In an anti-lock brake system, the device that usually replaces the standard relay valve used to control the rear axle service brakes and performs the standard relay function during tractor/trailer operation.

Antirattle Springs Springs that reduce wear between the intermediate plate and the drive pin, and helps to improve clutch release.

Antitrust Agent An additive used with lubricating oils to prevent rusting of metal parts when the engine is not in use.

Application Valve A foot-operated brake valve that controls air pressure to the service chambers.

Applied Moment A term meaning a given load has been placed on a frame at a particular point.

Area The total cross section of a frame rail including all applicable elements; usually given in square inches.

Armature The rotating component of a (1) starter or other motor, (2) generator, (3) compressor clutch.

Articulating Upper Coupler A bolster plate kingpin arrangement that is not rigidly attached to the trailer, but provides articulation and/or oscillation (such as a frameless dump) about an axis parallel to the rear axle of the trailer.

Articulation Vertical movement of the front driving or rear axle relative to the frame of the vehicle to which they are attached.

ASE An abbreviation for Automotive Service Excellence, a trademark of National Institute for Automotive Service Excellence.

Aspect Ratio A tire term calculated by dividing the tire's section height by its section width.

ATEC System A system that includes an electronic control system, torque converter, lockup clutch, and planetary gear train.

Atmospheric Pressure The weight of the air at sea level; 14.696 pounds per square inch (psi) or 101.33 kilopascals (kPa).

Automatic Slack Adjuster The device that automatically adjusts the clearance between the brake linings and the brake drum or rotor. The slack adjuster controls the clearance by sensing the length of the stroke of the push rod for the air brake chamber.

Autoshift Finger The device that engages the shift blocks on the yoke bars that correspond to the tab on the end of the gearshift lever in manual systems.

Auxiliary Filter A device installed in the oil return line between the oil cooler and the transmission to prevent debris from being flushed into the transmission causing a failure. An auxiliary filter must be installed before the vehicle is placed back in service.

Auxiliary Section The section of a transmission where range shifting occurs; housing the auxiliary drive gear, auxiliary main shaft assembly, auxiliary countershaft, and the synchronizer assembly.

Axis of Rotation The center line around which a gear or part revolves.

Axle (1) A rod or bar on which wheels turn. (2) A shaft that transmits driving torque to the wheels.

Axle Range Interlock A feature designed to prevent axle shifting when the interaxle differential is locked out, or when lockout is engaged. The basic shift system operates the same as the standard shift system to shift the axle and engage or disengage the lockout.

Axle Seat A suspension component used to support and locate the spring on an axle.

Axle Shims Thin wedges that may be installed under the leaf springs of single axle vehicles to tilt the axle and correct the U-joint operating angles. Wedges are available in a range of sizes to change pinion angles.

Backing Plate A metal plate that serves as the foundation for the brake shoes and other drum brake hardware.

Battery Terminal A tapered post or threaded studs on top of the battery case, or internally threaded provisions on the side of the battery for connecting the cables.

Beam Solid Mount Suspension A tandem suspension relying on a pivotal mounted beam, with axles attached at the ends for load equalization. The beam is mounted to a solid center pedestal.

Beam Suspension A tandem suspension relying on a pivotally mounted beam, with axles attached at ends for lead equalization. Beam is mounted to center spring.

Bellows A movable cover or seal that is pleated or folded like an accordion to allow for expansion and contraction.

Bending Moment A term implying that when a load is applied to the frame, it will be distributed across a given section of the frame material.

Bias A tire term where belts and plies are laid diagonally or crisscrossing each other.

Bimetallic Two dissimilar metals joined together that have different bending characteristics when subjected to different changes of temperature.

Blade Fuse A type of fuse having two flat male lugs sticking out for insertion in the female box connectors.

Bleed Air Tanks The process of draining condensation from air tanks to increase air capacity and brake efficiency.

Block Diagnosis Chart A troubleshooting chart that lists symptoms, possible causes, and probable remedies in columns.

Blower Fan A fan that pushes or blows air through a ventilation, heater, or air conditioning system.

Bobtail Proportioning Valve A valve that senses when the tractor is bobtailing and automatically reduces the amount of air pressure that can be applied to the tractor's drive axle(s). This reduces braking force on the drive axles, lessening the chance of a spin-out on slippery pavement.

Bobtailing A tractor running without a trailer.

Bogie The axle spring suspension arrangement on the rear of a tandem axle tractor.

Bolster Plate The flat, load-bearing surface under the front of a semitrailer, including the kingpin, which rests firmly on the fifth wheel when coupled.

Bolster Plate Height The height from the ground to the bolster plate when the trailer is level and empty.

Boss A heavy cast section that is used for support, such as the outer race of a bearing.

Bottoming A condition that occurs when: (1) the teeth of one gear touch the lowest point between teeth of a mating gear, (2) the bed or frame of the vehicle strikes the axle, such as that in the case of overloading.

Bottom U-Bolt Plate A plate that is located on the bottom side of the spring or axle and is held in place when the U-bolts are tightened to the clamp spring and axle together.

Bracket An attachment used to secure parts to the body or frame.

Brake Control Valve A dual brake valve that releases air from the service reservoirs to the service lines and brake chambers. The valve includes a piston which pushes on diaphragms to open ports; these vent air to service lines in the primary and secondary systems.

Brake Disc A steel disc used in a braking system with a caliper and pads. When the brakes are applied, the pad on each side of the spinning disc is forced against the disc, thus imparting a braking force. This type of brake is very resistant to brake fade.

Brake Drum A case metal bell-like cylinder attached to the wheel that is used to house the brake shoes and provide a friction surface for stopping a vehicle.

Brake Fade A condition that occurs when friction surfaces become hot enough to cause the coefficient of friction to drop to a point where the application of severe pedal pressure results in little actual braking.

Brake Lining A special friction material used to line brake shoes or brake pads. It withstands high temperatures and pressure. The molded material is either riveted or bonded to the brake shoe, with a suitable coefficient of friction for stopping a vehicle.

Brake Pad The friction lining and plate assembly that is forced against the rotor to cause braking action in a disc brake system.

Brake Shoe The curved metal part, faced with brake lining, which is forced against the brake drum to produce braking action.

Brake Shoe Rollers A hardware part that attaches to the web of the brake shoes by means of roller retainers. The rollers, in turn, ride on the end of an S-cam.

Brake System The vehicle system that slows or stops a vehicle. A combination of brakes and a control system.

Breakaway Valve A device that automatically seals off the tractor air supply from the trailer air supply when the tractor system pressure drops to 30 or 40 psi (207 to 276 kPa).

British Thermal Unit (Btu) A measure of heat quantity equal to the amount of heat required to raise 1 pound of water 1°F.

Broken Back Drive Shaft A term often used for nonparallel drive shaft.

Btu An abbreviation for British Thermal Unit.

Bump Steer Erratic steering caused from rolling over bumps, cornering, or heavy braking. Same as orbital steer and roll steer.

CAA An abbreviation for Clean Air Act.

Caliper A disc brake component that changes hydraulic pressure into mechanical force and uses that force to press the brake pads against the rotor and stop the vehicle. Calipers come in three basic types: fixed, floating, and sliding, and can have one or more pistons.

Camber The attitude of a wheel and tire assembly when viewed from the front of a car. If it leans outward, away from the car at the top, the wheel is said to have positive camber. If it leans inward, it is said to have negative camber.

Cam Brakes Brakes that are similar in operation and design to the wedge brake, with the exception that an S-type camshaft is used instead of a wedge and rubber assembly.

Cartridge Fuse A type of fuse having a strip of low melting point metal enclosed in a glass tube. If an excessive current flows through the circuit, the fuse element melts at the narrow portion, opening the circuit and preventing damage.

Caster The angle formed between the kingpin axis and a vertical axis as viewed from the side of the vehicle. Caster is considered positive when the top of the kingpin axis is behind the vertical axis.

Cavitation A condition that causes bubble formation.

Center of Gravity The point around which the weight of a truck is evenly distributed; the point of balance.

Ceramic Fuse A fuse found in some import vehicles that has a ceramic insulator with a conductive metal strip along one side.

CFC An abbreviation for chlorofluorocarbon.

Charging System A system consisting of the battery, alternator, voltage regulator, associated wiring, and the electrical loads of a vehicle. The purpose of the system is to recharge the battery whenever necessary and to provide the current required to power the electrical components.

Charge the Trailer To supply the trailer air tanks with air by means of a dash control valve, tractor protection valve, and a trailer relay emergency valve.

Charging Circuit The alternator (or generator) and associated circuit used to keep the battery charged and to furnish power to the vehicle's electrical systems when the engine is running.

Check Valve A valve that allows air to flow in one direction only. It is a federal requirement to have a check valve between the wet and dry air tanks.

Chlorofluorocarbon (CFC) A compound used in the production of refrigerant that is believed to cause damage to the ozone layer.

Circuit The complete path of an electrical current, including the generating device. When the path is unbroken, the circuit is closed and current flows. When the circuit continuity is broken, the circuit is open and current flow stops.

Clean Air Act (CAA) Federal regulations, passed in 1992, that have resulted in major changes in air-conditioning systems.

Climbing A gear problem caused by excessive wear in gears, bearings, and shafts whereby the gears move sufficiently apart to cause the apex (or point) of the teeth on one gear to climb over the apex of the teeth on another gear with which it is meshed.

Clutch A device for connecting and disconnecting the engine from the transmission or for a similar purpose in other units.

Clutch Brake A circular disc with a friction surface that is mounted on the transmission input spline between the release bearing and the transmission. Its purpose is to slow or stop the transmission input shaft from rotating in order to allow gears to be engaged without clashing or grinding.

Clutch Housing A component that surrounds and protects the clutch and connects the transmission case to the vehicle's engine.

Clutch Pack An assembly of normal clutch plates, friction discs, and one very thick plate known as the pressure plate. The pressure plate has tabs around the outside diameter to mate with the channel in the clutch drum.

COE An abbreviation for cab-over-engine.

Coefficient of Friction A measurement of the amount of friction developed between two objects in physical contact when one of the objects is drawn across the other.

Coil Springs Spring steel spirals that are mounted on control arms or axles to absorb road shock.

Combination A truck coupled to one or more trailers.

Compression Applying pressure to a spring or any springy substance, thus causing it to reduce its length in the direction of the compressing force.

Compressor (1) A mechanical device that increases pressure within a container by pumping air into it. (2) That component of an air-conditioning system that compresses low temperature/pressure refrigerant vapor.

Condensation The process by which gas (or vapor) changes to a liquid.

Condenser A component in an air-conditioning system used to cool a refrigerant below its boiling point causing it to change from a vapor to a liquid.

Conductor Any material that permits the electrical current to flow.

Constant Rate Springs Leaf-type spring assemblies that have a constant rate of deflection.

Control Arm The main link between the vehicle's frame and the wheels that acts as a hinge to allow wheel action up and down independent of the chassis.

Controlled Traction A type of differential that uses a friction plate assembly to transfer drive torque from the vehicle's slipping wheel to the one wheel that has good traction or surface bite.

Converter Dolly An axle, frame, drawbar, and fifth wheel arrangement that converts a semitrailer into a full trailer.

Coolant Liquid that circulates in an engine cooling system.

Coolant Heater A component used to aid engine starting and reduce the wear caused by cold starting.

Coolant Hydrometer A tester designed to measure coolant specific gravity and determine the amount of antifreeze in the coolant.

Cooling System Complete system for circulating coolant.

Coupling Point The point at which the turbine is turning at the same speed as the impeller.

Crankcase The housing within which the crankshaft and many other parts of the engine operate.

Cranking Circuit The starter and its associated circuit, including battery, relay (solenoid), ignition switch, neutral start switch (on vehicles with automatic transmission), and cables and wires.

Cross Groove Joint disc-shaped type of inner CV joint that uses balls and V-shaped grooves on the inner and outer races to accommodate the plunging motion of the half-shaft. The joint usually bolts to a transaxle stub flange; same as disc-type joint.

Cross-Tube A system that transfers the steering motion to the opposite, passenger side steering knuckle. It links the two steering knuckles together and forces them to act in unison.

C-Train A combination of two or more trailers in which the dolly is connected to the trailer by means of two pintle-hook or coupler-drawbar connections. The resulting connection has one pivot point.

Cycling (1) Repeated on-off action of the air conditioner compressor. (2) Heavy and repeated electrical cycling that can cause the positive plate material to break away from its grids and fall into the sediment chambers at the base of the battery case.

Dampen To slow or reduce oscillations or movement.

Dampened Discs Discs that have dampening springs incorporated into the disc hub. When engine torque is first transmitted to the disc, the plate rotates on the hub, compressing the springs. This action absorbs the shocks and torsional vibration caused by today's low rpm, high torque, engines.

Dash Control Valves A variety of handoperated valves located on the dash. They include parking brake valves, tractor protection valves, and differential lock.

Data Links Circuits through which computers communicate with other electronic devices such as control panels, modules, some sensors, or other computers in the form of digital signals.

Dead Axle Non-live or dead axles are often mounted in lifting suspensions. They hold the axle off the road when the vehicle is traveling empty, and put it on the road when a load is being carried. They are also used as air suspension third axles on heavy straight trucks and are used extensively in eastern states with high axle weight laws. An axle that does not rotate but merely forms a base on which to attach the wheels.

Deadline To take a vehicle out of service.

Deburring To remove sharp edges from a cut.

Dedicated Contract Carriage Trucking operations set up and run according to a specific shipper's needs. In addition to transportation, they often provide other services such as warehousing and logistics planning.

Deflection Bending or moving to a new position as the result of an external force.

Department of Transportation (DOT) A government agency that establishes vehicle standards.

Detergent Additive An additive that helps keep metal surfaces clean and prevents deposits. These additives suspend particles of carbon and oxidized oil in the oil.

DER An abbreviation for Department of Environmental Resources.

Diagnostic Flow Chart A chart that provides a systematic approach to the electrical system and component troubleshooting and repair. They are found in service manuals and are vehicle make and model specific.

Dial Caliper A measuring instrument capable of taking inside, outside, depth, and step measurements.

Differential A gear assembly that transmits power from the drive shaft to the wheels and allows two opposite wheels to turn at different speeds for cornering and traction.

Differential Carrier Assembly An assembly that controls the drive axle operation.

Differential Lock A toggle or push-pull type air switch that locks together the rear axles of a tractor so they pull as one for off-the-road operation.

Digital Binary Signal A signal that has only two values: on and off.

Digital Volt/Ohmmeter (DVOM) A type of test meter recommended by most manufacturers for use on solid state circuits.

Diode The simplest semiconductor device formed by joining P-type semiconductor material with N-type semiconductor material. A diode allows current to flow in one direction, but not in the opposite direction.

Direct Drive The gearing of a transmission so that in its highest gear, one revolution of the engine produces one revolution of the transmission's output shaft. The top gear or final drive ratio of a direct drive transmission would be 1:1.

Disc Brake A steel disc used in a braking system with a caliper and pads. When the brakes are applied, the pad on each side of the spinning disc is forced against the disc, thus imparting a braking force. This type of brake is very resistant to brake fade. A type of brake that generates stopping power by the application of pads against a rotating disc (rotor).

Dispatch Sheet A form used to keep track of dates when the work is to be completed. Some dispatch sheets follow the job through each step of the servicing process.

Dog Tracking Off-center tracking of the rear wheels as related to the front wheels.

DOT An abbreviation for Department of Transportation.

Downshift Control The selection of a lower range to match driving conditions encountered or expected to be encountered. Learning to take advantage of a downshift gives better control on slick or icy roads and on steep downgrades. Downshifting to lower ranges increases engine braking.

Double Reduction Axle An axle that uses two gear sets for greater overall gear reduction and peak torque development. This design is favored for severe service applications, such as dump trucks, cement mixers, and other heavy haulers.

Drag Link A connecting rod or link between the steering gear, Pitman arm, and the steering linkage.

Drawbar Capacity The maximum, horizontal pulling force that can be safely applied to a coupling device.

Driven Gear A gear that is driven or forced to turn by a drive gear, shaft, or some other device.

Drive or Driving Gear A gear that drives another gear or causes another gear to turn.

Drive Line The propeller or drive shaft, universal joints, and so forth, that links the transmission output to the axle pinion gear shaft.

Drive Line Angle The alignment of the transmission output shaft, driveshaft, and rear axle pinion centerline.

Drive Shaft An assembly of one or two universal joints connected to a shaft or tube; used to transmit power from the transmission to the differential.

Drive Train An assembly that includes all power transmitting components from the rear of the engine to the wheels, including clutch/torque converter, transmission, drive line, and front and rear driving axles.

Driver Controlled Main Differential Lock A type of axle assembly has greater flexibility over the standard type of single reduction axle because it provides equal amounts of drive line torque to each driving wheel, regardless of changing road conditions. This design also provides the necessary differential action to the road wheels when the truck is turning a corner.

Driver's Manual A publication that contains information needed by the driver to understand, operate, and care for the vehicle and its components.

Drum Brake A type of brake system in which stopping friction is created by the shoes pressing against the inside of the rotating drum.

Dual Hydraulic Braking System A brake system consisting of a tandem, or double action master cylinder which is basically two master cylinders usually formed by aligning two separate pistons and fluid reservoirs into a single cylinder.

ECU An abbreviation for electronic control unit.

Eddy Current A small circular current produced inside a metal core in the armature of a starter motor. Eddy currents produce heat and are reduced by using a laminated core.

Electricity The movement of electrons from one place to another.

Electric Retarder Electromagnets mounted in a steel frame. Energizing the retarder causes the electromagnets to exert a dragging force on the rotors in the frame and this drag force is transmitted directly to the drive shaft.

Electromotive Force (EMF) The force that moves electrons between atoms. This force is the pressure that exists between the positive and negative points (the electrical imbalance). This force is measured in units called volts.

Electronically Programmable Memory (EPROM) Computer memory that permits adaptation of the ECU to various standard mechanically controlled functions.

Electronic Control Unit (ECU) The brain of the vehicle.

Electronics The technology of controlling electricity.

Electrons Negatively charged particles orbiting around every nucleus.

Elliot Axle A solid bar front axle on which the ends span the steering knuckle.

EMF An abbreviation for electromotive force.

End Yoke The component connected to the output shaft of the transmission to transfer engine torque to the drive shaft.

Engine Brake A hydraulically operated device that converts the vehicle's engine into a power absorbing retarding mechanism.

Engine Stall Point The point, in rpms, under load is compared to the engine manufacturer's specified rpm for the stall test.

Environmental Protection Agency (EPA) An agency of the United States government charged with the responsibilities of protecting the environment and enforcing the Clean Air Act (CAA) of 1990.

EPA An abbreviation for the Environmental Protection Agency.

EPROM An abbreviation for Electronically Programmable Memory.

Equalizer A suspension device used to transfer and maintain equal load distribution between two or more axles of a suspension. Formerly called a rocker beam.

Equalizer Bracket A bracket for mounting the equalizer beam of a multiple axle spring suspension to a truck or trailer frame while allowing for the beam's pivotal movement. Normally there are three basic types: flange-mount, straddle-mount, and under- or side-mount.

Evaporator A component in an air conditioning system used to remove heat from the air passing through it.

Exhaust Brake A valve in the exhaust pipe between the manifold and the muffler. A slide mechanism which restricts the exhaust flow, causing exhaust back pressure to build up in the engine's cylinders. The exhaust brake actually transforms the engine into a low pressure air compressor driven by the wheels.

External Housing Damper A counterweight attached to an arm on the rear of the transmission extension housing and designed to dampen unwanted driveline or power-train vibrations.

Extra Capacity A term that generally refers to: (1) a coupling device that has strength capability greater than standard, (2) an oversized tank or reservoir for a fluid or vapor.

False Brinelling The polishing of a surface that is not damaged.

Fanning the Brakes Applying and releasing the brakes in rapid succession on a long downgrade.

Fatigue Failures The progressive destruction of a shaft or gear teeth material usually caused by overloading.

Fault Code A code that is recorded into the computer's memory. A fault code can be read by plugging a special break-out box tester into the computer.

Federal Motor Vehicle Safety Standard (FMVSS) A federal standard that specifies that all vehicles in the United States be assigned a Vehicle Identification Number (VIN).

Federal Motor Vehicle Safety Standard No. 121 (FMVSS 121) A federal standard that made significant changes in the guidelines that cover air brake systems. Generally speaking, the requirements of FMVSS 121 are such that larger capacity brakes and heavier steerable axles are needed to meet them.

FHWA An abbreviation for Federal Highway Administration.

Fiber Composite Springs Springs that are made of fiberglass, laminated, and bonded together by tough polyester resins.

Fifth Wheel A coupling device mounted on a truck and used to connect a semitrailer. It acts as a hinge point to allow changes in direction of travel between the tractor and the semitrailer.

Fifth Wheel Height The distance from the ground to the top of the fifth wheel when it is level and parallel with the ground. It can also refer to the height from the tractor frame to the top of the fifth wheel. The latter definition applies to data given in fifth wheel literature.

Fifth Wheel Top Plate The portion of the fifth wheel assembly that contacts the trailer bolster plate and houses the locking mechanism that connects to the kingpin.

Final Drive The last reduction gear set of a truck.

Fixed Valve Resistor An electrical device that is designed to have only one resistance rating for controlling voltage, which should not change.

Flammable Any material that will easily catch fire or explode.

Flare To spread gradually outward in a bell shape.

Flex Disc A term often used for flex plate.

Flex Plate A component used to mount the torque converter to the crankshaft. The flex plate is positioned between the engine crankshaft and the T/C. The purpose of the flex plate is to transfer crankshaft rotation to the shell of the torque converter assembly.

Float A cruising drive mode in which the throttle setting matches engine speed to road speed, neither accelerating nor decelerating.

Floating Main Shaft The main shaft consisting of a heavy-duty central shaft and several gears that turn freely when not engaged. The main shaft can move to allow for equalization of the loading on the countershafts. This is key to making a twin countershaft transmission workable. When engaged, the floating main shaft transfers torque evenly through its gears to the rest of the transmission and ultimately to the rear axle.

FMVSS An abbreviation for Federal Motor Vehicle Safety Standard.

FMVSS 121 An abbreviation for Federal Motor Vehicle Safety Standard No. 121.

Foot Valve A foot-operated brake valve that controls air pressure to the service chambers.

Foot-Pound An English unit of measurement for torque. One foot-pound is the torque obtained by a force of 1 pound applied to a foot-long wrench handle.

Forged Journal Cross Part of a universal joint.

Frame Width The measurement across the outside of the frame rails of a tractor, truck, or trailer.

Franchised Dealership A dealer that has signed a contract with a particular manufacturer to sell and service a particular line of vehicles.

Fretting A result of vibration that the bearing outer race can pick up the matching pattern.

Friction Plate Assembly An assembly consisting of a multiple disc clutch that is designed to slip when a predetermined torque value is reached.

Front Axle Limiting Valve A valve that reduces pressure to the front service chambers, thus eliminating front wheel lockup on wet or icy pavements.

Front Hanger A bracket for mounting the front of the truck or trailer suspensions to the frame. Made to accommodate the end of the spring on spring suspensions. There are four basic types: flange-mount, straddle-mount, under-mount, and side-mount.

Full Trailer A trailer that does not transfer load to the towing vehicle. It employs a tow bar coupled to a swiveling or steerable running gear assembly at the front of the trailer.

Fully Floating Axles An axle configuration whereby the axle half shafts transmit only driving torque to the wheels and not bending and torsional loads that are characteristic of the semi-floating axle.

Fully Oscillating Fifth Wheel A fifth wheel type with fore/aft and side-to-side articulation.

Fusible Link A term often used for fuse link.

Fuse Link A short length of smaller gauge wire installed in a conductor, usually close to the power source.

GCW An abbreviation for gross combination weight.

Gear A disk-like wheel with external or internal teeth that serves to transmit or change motion.

Gear Pitch The number of teeth per given unit of pitch diameter, an important factor in gear design and operation.

General Over-the-Road Use A fifth wheel designed for multiple standard duty highway applications.

Gladhand The connectors between tractor and trailer air lines.

Gross Combination Weight (GCW) The total weight of a fully equipped vehicle including payload, fuel, and driver.

Gross Trailer Weight (GTW) The sum of the weight of an empty trailer and its payload.

Gross Vehicle Weight (GVW) The total weight of a fully equipped vehicle and its payload.

Ground The negatively charged side of a circuit. A ground can be a wire, the negative side of the battery, or the vehicle chassis.

Grounded Circuit A shorted circuit that causes current to return to the battery before it has reached its intended destination.

GTW An abbreviation for gross trailer weight.

GVW An abbreviation for gross vehicle weight.

Halogen Light A lamp having a small quartz/glass bulb that contains a fuel filament surrounded by halogen gas. It is contained within a larger metal reflector and lens element.

Hand Valve (1) A valve mounted on the steering column or dash, used by the driver to apply the trailer brakes independently of the tractor brakes. (2) A hand operated valve used to control the flow of fluid or vapor.

Harness and Harness Connectors The organization of the vehicle's electrical system providing an orderly and convenient starting point for tracking and testing circuits.

Hazardous Materials Any substance that is flammable, explosive, or is known to produce adverse health effects in people or the environment that are exposed to the material during its use.

Heads Up Display (HUD) A technology used in some vehicles that superimposes data on the driver's normal field of vision. The operator can view the information, which appears to "float" just above the hood at a range near the front of a conventional tractor or truck. This allows the driver to monitor conditions such as limited road speed without interrupting his normal view of traffic.

Heater Control Valve A valve that controls the flow of coolant into the heater core from the engine.

Heat Exchanger A device used to transfer heat, such as a radiator or condenser.

Heavy-Duty Truck A truck that has a GVW of 26,001 pounds or more.

Helper Spring An additional spring device that permits greater load on an axle.

High CG Load Any application in which the load center of gravity (CG) of the trailer exceeds 40 inches (102 centimeters) above the top of the fifth wheel.

High-Resistant Circuits Circuits that have an increase in circuit resistance, with a corresponding decrease in current.

High-Strength Steel A low-alloy steel that is much stronger than hot-rolled or cold-rolled sheet steels that normally are used in the manufacture of car body frames.

Hinged Pawl Switch The simplest type of switch; one that makes or breaks the current of a single conductor.

HUD An abbreviation for heads up display.

Hydraulic Brakes Brakes that are actuated by a hydraulic system.

Hydraulic Brake System A system utilizing the properties of fluids under pressure to activate the brakes.

Hydrometer A tester designed to measure the specific gravity of a liquid.

Hypoid Gears Gears that intersect at right angles when meshed. Hypoid gearing uses a modified spiral bevel gear structure that allows several gear teeth to absorb the driving power and allows the gears to run quietly. A hypoid gear is typically found at the drive pinion gear and ring gear interface.

I-Beam Axle An axle designed to give great strength at reasonable weight. The cross section of the axle resembles the letter "I."

ICC Check Valve A valve that allows air to flow in one direction only. It is a federal requirement to have a check valve between the wet and dry air tanks.

Inboard Toward the centerline of the vehicle.

In-Line Fuse A fuse that is in series with the circuit in a small plastic fuse holder, not in the fuse box or panel. It is used, when necessary, as a protection device for a portion of the circuit even though the entire circuit may be protected by a fuse in the fuse box or panel.

In-Phase The in-line relationship between the forward coupling shaft yoke and the driveshaft slip yoke of a two-piece drive line.

Input Retarder A device located between the torque converter housing and the main housing designed primarily for over-the-road operations. The device employs a "paddle wheel" type design with a vaned rotor mounted between stator vanes in the retarder housing.

Installation Templates Drawings supplied by some vehicle manufacturers to allow the technician to correctly install the accessory. The templates available can be used to check clearances or to ease installation.

Insulator A material, such as rubber or glass, that offers high resistance to the flow of electrons.

Integrated Circuit A component containing diodes, transistors, resistors, capacitors, and other electronic components mounted on a single piece of material and capable to perform numerous functions.

Jacobs Engine Brake A term sometimes used for Jake brake.

Jake Brake The Jacobs engine brake, named for its inventor. A hydraulically operated device that converts a power producing diesel engine into a power-absorbing retarder mechanism by altering the engine's exhaust valve opening time used to slow the vehicle.

Jumper Wire A wire used to temporarily bypass a circuit or components for electrical testing. A jumper wire consists of a length of wire with an alligator clip at each end.

Jumpout A condition that occurs when a fully engaged gear and sliding clutch are forced out of engagement.

Jump Start The procedure used when it becomes necessary to use a booster battery to start a vehicle having a discharged battery.

Kinetic Energy Energy in motion.

Kingpin (1) The pin mounted through the center of the trailer upper coupler (bolster plate) that mates with the fifth wheel locks, securing the trailer to the fifth wheel. The configuration is controlled by industry standards. (2) A pin or shaft on which the steering spindle rotates.

Landing Gear The retractable supports for a semitrailer to keep the trailer level when the tractor is detached from it.

Lateral Runout The wobble or side-to-side movement of a rotating wheel or of a rotating wheel and tire assembly.

Lazer Beam Alignment System A two- or four-wheel alignment system using wheel-mounted instruments to project a lazer beam to measure toe, caster, and camber.

Lead The tendency of a car to deviate from a straight path on a level road when there is no pressure on the steering wheel in either direction.

Leaf Springs Strips of steel connected to the chassis and axle to isolate the vehicle from road shock.

Less Than Truckload (LTL) Partial loads from the networks of consolidation centers and satellite terminals.

Lite Beam Alignment System An alignment system using wheel-mounted instruments to project light beams onto charts and scales to measure toe, caster, and camber, and note the results of alignment adjustments.

Limited-Slip Differential A differential that utilizes a clutch device to deliver power to either rear wheel when the opposite wheel is spinning.

Linkage A system of rods and levers used to transmit motion or force.

Live Axle An axle on which the wheels are firmly affixed. The axle drives the wheels.

Live Beam Axle A non-independent suspension in which the axle moves with the wheels.

Load Proportioning Valve (LPV) A valve used to redistribute hydraulic pressure to front and rear brakes based on vehicle loads. This is a load- or height-sensing valve that senses the vehicle load and proportions the braking between front and rear brakes in proportion to the load variations and degree of rear-to-front weight transfer during braking.

Lockstrap A manual adjustment mechanism that allows for the adjustment of free travel.

Lock-up Torque Converter A torque converter that eliminates the 10 percent slip that takes place between the impeller and turbine at the coupling stage of operation. It is considered a four-element (impeller, turbine, stator, lockup clutch), three-stage (stall, coupling, and locking stage) unit.

Longitudinal Leaf Spring A leaf spring that is mounted so it is parallel to the length of the vehicle.

Low-Maintenance Battery A conventionally vented, lead/acid battery, requiring normal periodic maintenance.

LTL An abbreviation for less than truckload.

Magnetorque An electromagnetic clutch.

Maintenance-Free Battery A battery that does not require the addition of water during normal service life.

Maintenance Manual A publication containing routine maintenance procedures and intervals for vehicle components and systems.

Main Transmission A transmission consisting of an input shaft, floating main shaft assembly and main drive gears, two counter shaft assemblies, and reverse idler gears.

Manual Slide Release The release mechanism for a sliding fifth wheel, which is operated by hand.

Metering Valve A valve used on vehicles equipped with front disc and rear drum brakes. It improves braking balance during light brake applications by preventing application of the front disc brakes until pressure is built up in the hydraulic system.

Moisture Ejector A valve mounted to the bottom or side of the supply and service reservoirs that collects water and expels it every time the air pressure fluctuates.

Mounting Bracket That portion of the fifth wheel assembly that connects the fifth wheel top plate to the tractor frame or fifth wheel mounting system.

Multiaxle Suspension A suspension consisting of more than three axles.

Multiple Disc Clutch A clutch having a large drum-shaped housing that can be either a separate casting or part of the existing transmission housing.

NATEF An abbreviation for National Automotive Education Foundation.

National Automotive Education Foundation (NATEF) A foundation having a program of certifying secondary and post secondary automotive and heavy-duty truck training programs.

National Institute for Automotive Service Excellence (ASE) A nonprofit organization that has an established certification program for automotive, heavy-duty truck, auto body repair, engine machine shop technicians, and parts specialists.

Needlenose Pliers This tool has long tapered jaws for grasping small parts or for reaching into tight spots. Many needlenose pliers also have cutting edges and a wire stripper.

NIASE An abbreviation for National Institute for Automotive Service Excellence, now abbreviated ASE.

NIOSH An abbreviation for National Institute for Occupation Safety and Health.

NLGI An abbreviation for National Lubricating Grease Institute.

NHTSA An abbreviation for National Highway Traffic Safety Administration.

Nonlive Axle Non-live or dead axles are often mounted in lifting suspensions. They hold the axle off the road when the vehicle is traveling empty, and put it on the road when a load is being carried. They are also used as air suspension third axles on heavy straight trucks and are used extensively in eastern states with high axle weight laws.

Nonparallel Driveshaft A type of drive shaft installation whereby the working angles of the joints of a given shaft are equal; however the companion flanges and/or yokes are not parallel.

Nonpolarized Gladhand A gladhand that can be connected to either service or emergency gladhand.

Nose The front of a semitrailer.

No-tilt Convertible Fifth A fifth wheel with fore/aft articulation that can be locked out to produce a rigid top plate for applications that have either rigid and/or articulating upper couplers.

OEM An abbreviation for original equipment manufacturer.

Off-road With reference to unpaved, rough, or ungraded terrain on which a vehicle will operate. Any terrain not considered part of the highway system falls into this category.

Ohm A unit of measured electrical resistance.

Ohm's Law The basic law of electricity stating that in any electrical circuit, current, resistance, and pressure work together in a mathematical relationship.

On-road With reference to paved or smooth-graded surface terrain on which a vehicle will operate, generally considered to be part of the public highway system.

Open Circuit An electrical circuit whose path has been interrupted or broken either accidentally (a broken wire) or intentionally (a switch turned off).

Operational Control Valve A valve used to control the flow of compressed air through the brake system.

Oscillation The rotational movement in either fore/aft or side-to-side direction about a pivot point. Generally refers to fifth wheel designs in which fore/aft and side-to-side articulation are provided.

OSHA An abbreviation for Occupational Safety and Health Administration.

Out-of-Phase A condition of the universal joint which acts somewhat like one person snapping a rope held by a person at the opposite end. The result is a violent reaction at the opposite end. If both were to snap the rope at the same time, the resulting waves cancel each other and neither would feel the reaction.

Out-of-Round A wheel or tire defect in which the wheel or tire is not round.

Output Driver An electronic on/off switch that the computer uses to control the ground circuit of a specific actuator. Output drivers are located in the processor along with the input conditioners, microprocessor, and memory.

Output Yoke The component that serves as a connecting link, transferring torque from the transmission's output shaft through the vehicle's drive line to the rear axle.

Oval A condition that occurs when a tube is not round, but is somewhat egg-shaped.

Overall Ratio The ratio of the lowest to the highest forward gear in the transmission.

Overdrive The gearing of a transmission so that in its highest gear one revolution of the engine produces more than one revolution of the transmission's output shaft.

Overrunning Clutch A clutch mechanism that transmits power in one direction only.

Overspeed Governor A governor that shuts off the fuel or stops the engine when excessive speed is reached.

Oxidation Inhibitor (1) An additive used with lubricating oils to keep oil from oxidizing even at very high temperatures. (2) An additive for gasoline to reduce the chemicals in gasoline that react with oxygen.

Pad A disc brake lining and metal back riveted, molded, or bonded together.

Parallel Circuit An electrical circuit that provides two or more paths for the current to flow. Each path has separate resistors and operates independently from the other parallel paths. In a parallel circuit, amperage can flow through more than one resistor at a time.

Parallel Joint Type A type of drive shaft installation whereby all companion flanges and/or yokes in the complete drive line are parallel to each other with the working angles of the joints of a given shaft being equal and opposite.

Parking Brake A mechanically applied brake used to prevent a parked vehicle's movement.

Parts Requisition A form that is used to order new parts, on which the technician writes the names of what part(s) are needed along with the vehicle's VIN or company's identification folder.

Payload The weight of the cargo carried by a truck, not including the weight of the body.

Pipe or Angle Brace Extrusions between opposite hangers on a spring or air-type suspension.

Pitman Arm A steering linkage component that connects the steering gear to the linkage at the left end of the center link.

Pitting Surface irregularities resulting from corrosion.

Planetary Drive A planetary gear reduction set where the sun gear is the drive and the planetary carrier is the output.

Planetary Gear Set A system of gearing that is somewhat like the solar system. A pinion is surrounded by an internal ring gear and planet gears are in mesh between the ring gear and pinion around which all revolve.

Planetary Pinion Gears Small gears fitted into a framework called the planetary carrier.

Plies The layers of rubber-impregnated fabric that make up the body of a tire.

Pogo Stick The air and electrical line support rod mounted behind the cab to keep the lines from dragging between the tractor and trailer.

Polarity The particular state, either positive or negative, with reference to the two poles or to electrification.

Pole The number of input circuits made by an electrical switch.

Pounds per Square Inch (psi) A unit of English measure for pressure.

Power A measure of work being done.

Power Flow The flow of power from the input shaft through one or more sets of gears, or through an automatic transmission to the output shaft.

Power Steering A steering system utilizing hydraulic pressure to reduce the turning effort required of the operator.

Power Synchronizer A device to speed up the rotation of the main section gearing for smoother automatic downshifts and to slow down the rotation of the main section gearing for smoother automatic upshifts.

Power Train An assembly consisting of a drive shaft, coupling, clutch, and transmission differential.

Pressure The amount of force applied to a definite area measured in pounds per square inch (psi; English) or kilopascals (kPa; metric).

Pressure Differential The difference in pressure between any two points of a system or a component.

Pressure Relief Valve (1) A valve located on the wet tank, usually preset at 150 psi (1,034 kPa). Limits system pressure if the compressor or governor unloader valve malfunctions. (2) A valve located on the rear head of an air-conditioning compressor or pressure vessel that opens if an excessive system pressure is exceeded.

Printed Circuit Board An electronic circuit board made of thin nonconductive plastic-like material onto which conductive metal, such as copper, has been deposited. Parts of the metal are then etched away by an acid, leaving metal lines that form the conductors for the various circuits on the board. A printed circuit board can hold many complex circuits in a very small area.

Programmable Read Only Memory (PROM) An electronic component that contains program information specific to different vehicle model calibrations.

PROM An abbreviation for Programmable Read Only Memory.

Priority Valve A valve that ensures that the control system upstream from the valve will have sufficient pressure during shifts to perform its automatic functions.

Proportioning Valve A valve used on vehicles equipped with front disc and rear drum brakes. It is installed in the lines to the rear drum brakes, and in a split system, below the pressure differential valve. By reducing pressure to the rear drum brakes, the valve helps to prevent premature lockup during severe brake application and provides better braking balance.

Psi An abbreviation for pounds per square inch.

Pull Circuit A circuit that brings the cab from a fully tilted position up and over the center.

Pull-Type Clutch A type of clutch that does not push the release bearing toward the engine; instead, it pulls the release bearing toward the transmission.

Pump/Impeller Assembly The input (drive) member that receives power from the engine.

Push Circuit A circuit that raises the cab from the lowered position to the desired tilt position.

Push-Type Clutch A type of clutch in which the release bearing is not attached to the clutch cover.

P-type Semiconductors Positively charged materials that enables them to carry current. They are produced by adding an impurity with three electrons in the outer ring (trivalent atoms).

Quick Release Valve A device used to exhaust air as close as possible to the service chambers or spring brakes.

Radial A tire design having cord materials running in a direction from the center point of the tire, usually from bead to bead.

Radial Load A load that is applied to an axis of rotation at 90°.

RAM An abbreviation for random access memory.

Ram Air Air that is forced into the engine or passenger compartment by the forward motion of the vehicle.

Random Access Memory (RAM) The memory used during computer operation to store temporary information. The microcomputer can write, read, and erase information from RAM in any order, which is why it is called random.

Range Shift Cylinder A component located in the auxiliary section of the transmission. This component, when directed by air pressure via low and high ports, shifts between high and low range of gears.

Range Shift Lever A lever located on the shift knob allows the driver to select low or high gear range.

Rated Capacity The maximum, recommended safe load that can be sustained by a component or an assembly without permanent damage.

Ratio Valve A valve used on the front or steering axle of a heavy-duty truck to limit the brake application pressure to the actuators during normal service braking.

RCRA An abbreviation for Resource Conservation and Recovery Act.

Reactivity The characteristic of a material that enables it to react violently with air, heat, water, or other materials.

Read Only Memory (ROM) A type of memory used in microcomputers to store information permanently.

Rear Hanger A bracket for mounting the rear of a truck or trailer suspension to the frame. Made to accommodate the end of the spring on spring suspensions. There are usually four types: flange-mount, straddle-mount, under-mount, and side-mount.

Recall Bulletin A bulletin that pertains to special situations that involve service work or replacement of parts in connection with a recall notice.

Reference Voltage The voltage supplied to a sensor by the computer, which acts as a base line voltage; modified by the sensor to act as an input signal.

Relay An electric switch that allows a small current to control a much larger one. It consists of a control circuit and a power circuit.

Relay/Quick Release Valve A valve used on trucks with a wheel base 254 inches (6.45 meters) or longer. It is attached to an air tank to main supply line to speed the application and release of air to the service chambers. It is similar to a remote control foot valve.

Refrigerant A liquid capable of vaporizing at a low temperature.

Refrigerant Management Center Equipment designed to recover, recycle, and recharge an air-conditioning system.

Release Bearing A unit within the clutch consisting of bearings that mount on the transmission input shaft but does not rotate with it.

Reserve Capacity Rating The ability of a battery to sustain a minimum vehicle electrical load in the event of a charging system failure.

Resistance The opposition to current flow in an electrical circuit.

Resisting Bending Moment A measurement of frame rail strength derived by multiplying the section modulus of the rail by the yield strength of the material. This term is universally used in evaluating frame rail strength.

Resource Conservation and Recovery Act (RCRA) A law that states that after using a hazardous material, it must be properly stored until an approved hazardous waste hauler arrives to take them to the disposal site.

Reverse Elliot Axle A solid-beam front axle on which the steering knuckles span the axle ends.

Revolutions per Minute (rpm) The number of complete turns a member makes in one minute.

Right to Know Law A law passed by the federal government and administered by the Occupational Safety and Health Administration (OSHA) that requires any company that uses or produces hazardous chemicals or substances to inform its employees, customers, and vendors of any potential hazards that may exist in the workplace as a result of using the products.

Rigid Disc A steel plate to which friction linings, or facings, are bonded or riveted.

Rigid Fifth Wheel A platform that is fixed rigidly to a frame. This fifth wheel has no articulation or oscillation. It is generally used in applications where the articulation is provided by other means, such as an articulating upper coupler of a frame-less dump.

Rigid Torque Arm A member used to retain axle alignment and, in some cases, to control axle torque. Normally, one adjustable and one rigid arm are used per axle so the axle can be aligned.

Ring Gear (1) The gear around the edge of a flywheel. (2) A large circular gear such as that found in a final drive assembly.

Rocker Beam A suspension device used to transfer and maintain equal load distribution between two or more axles of a suspension.

Roll Axis The theoretical line that joins the roll center of the front and rear axles.

Roller Clutch A clutch designed with a movable inner race, rollers, accordion (apply) springs, and outer race. Around the inside diameter of the outer race are several cam-shaped pockets. The clutch assembly rollers and accordion springs are located in these pockets.

Rollers A hardware part that attaches to the web of the brake shoes by means of roller retainers. The rollers, in turn, ride on the end of an S-cam.

ROM An abbreviation for read only memory.

Rotary Oil Flow A condition caused by the centrifugal force applied to the fluid as the converter rotates around its axis.

Rotation A term used to describe the fact that a gear, shaft, or other device is turning.

rpm An abbreviation for revolutions per minute.

Rotor (1) A part of the alternator that provides the magnetic fields necessary to create a current flow. (2) The rotating member of an assembly.

Runout A deviation of the specified normal travel of an object. The amount of deviation or wobble a shaft or wheel has as it rotates. Runout is measured with a dial indicator.

Safety Factor (SF) (1) The amount of load which can safely be absorbed by and through the vehicle chassis frame members. (2) The difference between the stated and rated limits of a product, such as a grinding disk.

Screw Pitch Gauge A gauge used to provide a quick and accurate method of checking the threads per inch of a nut or bolt.

Secondary Lock The component or components of a fifth wheel locking mechanism that can be included as a backup system for the primary locks. The secondary lock is not required for the fifth wheel to function and can be either manually or automatically applied. On some designs, the engagement of the secondary lock can only be accomplished if the primary lock is properly engaged.

Section Height The tread center to bead plane on a tire.

Section Width The measurement on a tire from sidewall to sidewall.

Self-Adjusting Clutch A clutch that automatically takes up the slack between the pressure plate and clutch disc as wear occurs.

Semiconductor A solid state device that can function as either a conductor or an insulator, depending on how its structure is arranged.

Semifloating Axle An axle type whereby drive power from the differential is taken by each axle half-shaft and transferred directly to the wheels. A single bearing assembly, located at the outer end of the axle, is used to support the axle half-shaft.

Semioscillating A term that generally describes a fifth wheel type that oscillates or articulates about an axis perpendicular to the vehicle centerline.

Semitrailer A load-carrying vehicle equipped with one or more axles and constructed so that its front end is supported on the fifth wheel of the truck tractor that pulls it.

Sensing Voltage The voltage that allows the regulator to sense and monitor the battery voltage level.

Sensor An electronic device used to monitor relative conditions for computer control requirements.

Series Circuit A circuit that consists of two or more resistors connected to a voltage source with only one path for the electrons to follow.

Series/Parallel Circuit A circuit designed so that both series and parallel combinations exist within the same circuit.

Service Bulletin A publication that provides the latest service tips, field repairs, product improvements, and related information of benefit to service personnel.

Service Manual A manual, published by the manufacturer, that contains service and repair information for all vehicle systems and components.

Shift Bar Housing Available in standard- and forward-position configurations, a component that houses the shift rails, shift yokes, detent balls and springs, inter-lock balls, and pin and neutral shaft.

Shift Fork The Y-shaped component located between the gears on the main shaft that, when actuated, cause the gears to engage or disengage via the sliding clutches. Shift forks are located between low and reverse, first and second, and third and fourth gears.

Shift Rail Shift rails guide the shift forks using a series of grooves, tension balls, and springs to hold the shift forks in gear. The grooves in the forks allow them to interlock the rails, and the transmission cannot be accidentally shifted into two gears at the same time.

Shift Tower The main interface between the driver and the transmission, consisting of a gearshift lever, pivot pin, spring, boot and housing.

Shift Yoke A Y-shaped component located between the gears on the main shaft that, when actuated, cause the gears to engage or disengage via the sliding clutches. Shift yokes are located between low and reverse, first and second, and third and fourth gears.

Shock Absorber A hydraulic device used to dampen vehicle spring oscillations for controlling body sway and wheel bounce, and/or prevent spring breakage.

Short Circuit An undesirable connection between two worn or damaged wires. The short occurs when the insulation is worn between two adjacent wires and the metal in each wire contacts the other, or when the wires are damaged or pinched.

Single-Axle Suspension A suspension with one axle.

Single Reduction Axle Any axle assembly that employs only one gear reduction through its differential carrier assembly.

Slave Valve A valve to help protect gears and components in the transmission's auxiliary section by permitting range shifts to occur only when the transmission's main gearbox is in neutral. Air pressure from a regulator signals the slave valve into operation.

Slide Travel The distance that a sliding fifth wheel is designed to move.

Sliding Fifth Wheel A specialized fifth wheel design that incorporates provisions to readily relocate the kingpin center forward and rearward, which affects the weight distribution on the tractor axles and/or overall length of the tractor and trailer.

Slipout A condition that generally occurs when pulling with full power or decelerating with the load pushing. Tapered or worn clutching teeth will try to "walk" apart as the gears rotate, causing the sliding clutch and gear to slip out of engagement.

Slip Rings and Brushes Components of an alternator that conducts current to the rotor. Most alternators have two slip rings mounted directly on the rotor shaft; they are insulated from the shaft and from each other. A spring loaded carbon brush is located on each slip ring to carry the current to and from the rotor windings.

Solenoid An electromagnet that is used to perform work, made with one or two coil windings wound around an iron tube.

Solid-State Device A device that requires very little power to operate, is very reliable, and generates very little heat.

Solid Wires A single-strand conductor.

Solvent A substance which dissolves other substances.

Spade Fuse A term used for blade fuse.

Spalling Surface fatigue occurs when chips, scales, or flakes of metal break off due to fatigue rather than wear. Spalling is usually found on splines and U-joint bearings.

Specialty Service Shop A shop that specializes in areas such as engine rebuilding, transmission/axle overhauling, brake, air conditioning/heating repairs, and electrical/electronic work.

Specific Gravity The scientific measurement of a liquid based on the ratio of the liquid's mass to an equal volume of distilled water.

Spiral Bevel Gear A gear arrangement that has a drive pinion gear that meshes with the ring gear at the centerline axis of the ring gear. This gearing provides strength and allows for quiet operation.

Splined Yoke A yoke that allows the drive shaft to increase in length to accommodate movements of the drive axles.

Spontaneous Combustion A process by which a combustible material ignites by itself and starts a fire.

Spread Tandem Suspension A two-axle assembly in which the axles are spaced to allow maximum axle loads under existing regulations. The distance is usually more than 55 inches.

Spring A device used to reduce road shocks and transfer loads through suspension components to the frame of the trailer. There are usually four basic types: multileaf, monoleaf, taper, and air springs.

Spring Chair A suspension component used to support and locate the spring on an axle.

Spring Deflection The depression of a trailer suspension when the springs are placed under load.

Spring Rate The load required to deflect the spring a given distance, usually one inch.

Spring Spacer A riser block often used on top of the spring seat to obtain increased mounting height.

Stability A relative measure of the handling characteristics which provide the desired and safe operation of the vehicle during various maneuvers.

Stabilizer A device used to stabilize a vehicle during turns; sometimes referred to as a sway bar.

Stabilizer Bar A bar that connects the two sides of a suspension so that cornering forces on one wheel are shaped by the other. This helps equalize wheel side loading and reduces the tendency of the vehicle body to roll outward in a turn.

Staff Test A test performed when there is an obvious malfunction in the vehicle's power package (engine and transmission), to determine which of the components is at fault.

Stand Pipe A type of check valve which prevents reverse flow of the hot liquid lubricant generated during operation. When the universal joint is at rest, one or more of the cross ends will be up. Without the stand pipe, lubricant would flow out of the upper passage ways and trunnions, leading to partially dry startup.

Starter Circuit The circuit that carries the high current flow within the system and supplies power for the actual engine cranking.

Starter Motor The device that converts the electrical energy from the battery into mechanical energy for cranking the engine.

Starting Safety Switch A switch that prevents vehicles with automatic transmissions from being started in gear.

Static Balance Balance at rest, or still balance. It is the equal distribution of the weight of the wheel and tire around the axis of rotation so that the wheel assembly has no tendency to rotate by itself regardless of its position.

Stationary Fifth Wheel A fifth wheel whose location on the tractor frame is fixed once it is installed.

Stator A component located between the pump/impeller and turbine to redirect the oil flow from the turbine back into the impeller in the direction of impeller rotation with minimal loss of speed or force.

Stator Assembly The reaction member or torque multiplier supported on a free wheel roller race that is splinted to the valve and front support assembly.

Steering Gear A gear set mounted in a housing that is fastened to the lower end of the steering column used to multiply driver turning force and change rotary motion into longitudinal motion.

Steering Stabilizer A shock absorber attached to the steering components to cushion road shock in the steering system, improving driver control in rough terrain and protecting the system.

Stepped Resistor A resistor designed to have two or more fixed values, available by connecting wires to either of the several taps.

Still Balance Balance at rest; the equal distribution of the weight of the wheel and tire around the axis of rotation so that the wheel assembly has no tendency to rotate by itself regardless of its position.

Stoplight Switch A pneumatic switch that actuates the brake lights. There are two types: (1) a service stoplight switch that is located in the service circuit, actuated when the service brakes are applie;. (2) an emergency stoplight switch located in the emergency circuit and actuated when a pressure loss occurs.

Storage Battery A battery to provide a source of direct current electricity for both the electrical and electronic systems.

Stranded Wire Wire that is made up of a number of small solid wires, generally twisted together, to form a single conductor.

Structural Member A primary load-bearing portion of the body structure that affects its over- the-road performance or crash-worthiness.

Sulfation A condition that occurs when sulfate is allowed to remain in the battery plates for a long time, causing two problems: (1) it lowers the specific gravity levels, increasing the danger of freezing at low temperatures; (2) in cold weather a sulfated battery may not have the reserve power needed to crank the engine.

Suspension A system whereby the axle or axles of a unit are attached to the vehicle frame, designed in such a manner that road shocks from the axles are dampened through springs reducing the forces entering the frame.

Suspension Height The distance from a specified point on a vehicle to the road surface when not at curb weight.

Swage To reduce or taper.

Sway Bar A component that connects the two sides of a suspension so that cornering forces on one wheel are shared by the other. This helps equalize wheel side loading and reduces the tendency of the vehicle body to roll outward in a turn.

Switch A device used to control on/off and direct the flow of current in a circuit. A switch can be under the control of the driver or can be self-operating through a condition of the circuit, the vehicle, or the environment.

Synchromesh A mechanism that equalizes the speed of the gears that are clutched together.

Synchro-transmission A transmission with mechanisms for synchronizing the gear speeds so that the gears can be shifted without clashing, thus eliminating the need for double-clutching.

System Protection Valve A valve to protect the brake system against an accidental loss of air pressure, buildup of excess pressure, or back-flow and reverse air flow.

Tachometer An instrument that indicates rotating speeds, sometimes used to indicate crankshaft rpm.

Tag Axle The rearmost axle of a tandem axle tractor used to increase the load- carrying capacity of the vehicle.

Tapped Resistor A resistor designed to have two or more fixed values, available by connecting wires to either of the several taps.

Tandem One directly in front of the other and working together.

Tandem Axle Suspension A suspension system consisting of two axles with a means for equalizing weight between them.

Tandem Drive A two-axle drive combination.

Tandem Drive Axle A type of axle that combines two single axle assemblies through the use of an interaxle differential or power divider and a short shaft that connects the two axles together.

Three-Speed Differential A type of axle in a tandem two-speed axle arrangement with the capability of operating the two drive axles in different speed ranges at the same time. The third speed is actually an intermediate speed between the high and low range.

Throw (1) The offset of a crankshaft. (2) The number of output circuits of a switch.

Tie-Rod Assembly A system that transfers the steering motion to the opposite, passenger side steering knuckle. It links the two steering knuckles together and forces them to act in unison.

Time Guide Prepared reference material used for computing compensation payable by the truck manufacturer for repairs or service work to vehicles under warranty, or for other special conditions authorized by the company.

Timing (1) A procedure of marking the appropriate teeth of a gear set prior to installation and placing them in proper mesh while in the transmission. (2) The combustion spark delivery in relation to the piston position.

Toe A suspension dimension that reflects the difference in the distance between the extreme front and rear of the tire.

Toe In A suspension dimension whereby the front of the tire points inward toward the vehicle.

Toe Out A suspension dimension whereby the front of the tire points outward from the vehicle.

Top U-Bolt Plate A plate located on the top of the spring and is held in place when the U-bolts are tightened to clamp the spring and axle together.

Torque To tighten a fastener to a specific degree of tightness, generally in a given order or pattern if multiple fasteners are involved on a single component.

Torque and Twist A term that generally refers to the forces developed in the trailer and/or tractor frame that are transmitted through the fifth wheel when a rigid trailer, such as a tanker, is required to negotiate bumps, like street curbs.

Torque Converter A component device, similar to a fluid coupling, that transfers engine torque to the transmission input shaft and can multiply engine torque by having one or more stators between the members.

Torque Limiting Clutch Brake A system designed to slip when loads of 20 to 25 pound-feet (27 to 34N) are reached protecting the brake from overloading and the resulting high heat damage.

Torque Rod Shim A thin wedge-like insert that rotates the axle pinion to change the U-joint operating angle.

Torsional Rigidity A component's ability to remain rigid when subjected to twisting forces.

Torsion Bar Suspension A type of suspension system that utilizes torsion bars in lieu of steel leaf springs or coil springs. The typical torsion bar suspension consists of a torsion bar, front crank, and rear crank with associated brackets, a shackle pin, and assorted bushings and seals.

Total Pedal Travel The complete distance the clutch or brake pedal must move.

Toxicity A statement of how poisonous a substance is.

Tracking The travel of the rear wheels in a parallel path with the front wheels.

Tractor A motor vehicle, without a body, that has a fifth wheel and is used for pulling a semitrailer.

Tractor Protection Valve A device that automatically seals off the tractor air supply from the trailer air supply when the tractor system pressure drops to 30 or 40 psi (207 to 276 kPa).

Tractor/Trailer Lift Suspension A single axle air ride suspension with lift capabilities commonly used with steerable axles for pusher and tag applications.

Trailer A platform or container on wheels pulled by a car, truck, or tractor.

Trailer Hand Control Valve A device located on the dash or steering column and used to apply only the trailer brakes; primarily used in jackknife situations.

Trailer Slider A movable trailer suspension frame that is capable of changing trailer wheelbase by sliding and locking into different positions.

Transfer Case An additional gearbox located between the main transmission and the rear axle to transfer power from the transmission to the front and rear driving axles.

Transistor An electronic device produced by joining three sections of semiconductor materials. Like the diode, it is very useful as a switching device, functioning as either a conductor or an insulator.

Transmission A device used to transmit torque at various ratios and that can usually also change the direction of the force of rotation.

Transverse Vibrations A condition caused by an unbalanced driveline or bending movements in the drive shaft.

Treadle A dual brake valve that releases air from the service reservoirs to the service lines and brake chambers. The valve includes a piston which pushes on diaphragms to open ports; these vent air to service lines in the primary and secondary systems.

Treadle Valve A foot-operated brake valve that controls air pressure to the service chambers.

Tree Diagnosis Chart A chart used to provide a logical sequence for what should be inspected or tested when troubleshooting a repair problem.

Triaxle Suspension A suspension consisting of three axles with a means of equalizing weight between axles.

Trunnion The end of the universal cross; they are case hardened ground surfaces on which the needle bearings ride.

TTMA An abbreviation for Truck and Trailer Manufacturers Association.

Turbine The output (driven) member that is splined to the forward clutch of the transmission and to the turbine shaft assembly.

TVW An abbreviation for: (1) total vehicle weight; (2) towed vehicle weight.

Two-Speed Axle Assembly An axle assembly having two different output ratios from the differential. The driver selects the ratios from the controls located in the cab of the truck.

U-Bolt A fastener used to clamp the top U-bolt plate, spring, axle, and bottom U-bolt plate together. Inverted (nuts down) U-bolts cross springs when in place; conventional (nuts up) U-bolts wrap around the axle.

UNEP An abbreviation for United Nations Environment Program. Mandates the complete phaseout of CFC-based refrigerants by 1995.

Underslung Suspension A suspension in which the spring is positioned under the axle.

United Nations Environmental Program (UNEP) A protocol that mandated the complete phase-out of CFC-based refrigerants by the year 1995.

Universal Gladhand A term often used for non-polarized gladhand.

Universal Joint (U-joint) A component that allows torque to be transmitted to components that are operating at different angles.

Upper Coupler The flat load-bearing surface under the front of a semitrailer, including the kingpin, which rests firmly on the fifth wheel when coupled.

Vacuum Air below atmospheric pressure. There are three types of vacuums important to engine and component function: manifold vacuum, ported vacuum, and venturi vacuum. The strength of either of these vacuums depend on throttle opening, engine speed, and load.

Validity List A list supplied by the manufacturer of valid bulletins.

Valve Body and Governor Test Stand A specialized piece of test equipment. The valve body of the transmission is removed from the vehicle and mounted into the test stand. The test stand duplicates all vehicle running conditions, so the valve body can be thoroughly tested and calibrated.

Variable Pitch Stator A stator design often used in torque converters in off-highway applications such as dirt and stone aggregate dump or haul trucks, or other specialized equipment used to transport unusually heavy loads in rough terrain.

Vehicle Body Clearance (VBC) The distance from the inside of the inner tire to the spring or other body structures.

Vehicle On-board Rada r (VORAD) A system similar to an electronic eye that constantly monitors other vehicles on the road to give the driver additional reaction time to respond to potential dangers.

Vehicle Retarder An optional type of braking device that has been developed and successfully used over the years to supplement or assist the service brakes on heavy-duty trucks.

Vertical Load Capacity The maximum, recommended vertical downward force that can be safely applied to a coupling device.

VIN An abbreviation for Vehicle Identification Number.

Viscosity The ability of an oil to maintain proper lubricating quality under various conditions of operating speed, temperature, and pressure. Viscosity describes oil thickness or resistance to flow.

Volt The unit of electromotive force.

Voltage Generating Sensors These are devices which produce their own input voltage signal.

Voltage Limiter A device that provides protection by limiting voltage to the instrument panel gauges to approximately 5 volts.

Voltage Regulator A device that controls the amount of current produced by the alternator or generator and thus the voltage level in the charging circuit.

VORAD An acronym for Vehicle On-board Radar.

Vortex Oil Flow The circular flow that occurs as the oil is forced from the impeller to the turbine and then back to the impeller.

Watt The measure of electrical power.

Watt's Law A basic law of electricity used to find the power of an electrical circuit expressed in watts. It states that power equals the voltage multiplied by the current, in amperes.

Wear Compensator A device mounted in the clutch cover having an actuator arm that fits into a hole in the release sleeve retainer.

Wedge-Actuated Brakes A brake system using air pressure and air brake chambers to push a wedge and roller assembly into an actuator that is located between adjusting and anchor pistons.

Wet Tank A supply reservoir.

Wheel Alignment The mechanics of keeping all the parts of the steering system in the specified relation to each other.

Wheel and Axle Speed Sensors Electromagnetic devices used to monitor vehicle speed information for an antilock controller.

Wheel Balance The equal distribution of weight in a wheel with the tire mounted. It is an important factor which affects tire wear and vehicle control.

Wingdings (1) The three separate bundles in which wires are grouped in the stator. (2) The coil of wire found in a relay or other similar devices. (3) That part of an electrical clutch that provides a magnetic field.

Work (1) Forcing a current through a resistance. (2) The product of a force.

Yield Strength The highest stress a material can stand without permanent deformation or damage, expressed in pounds per square inch (psi).

Yoke Sleeve Kit This can be installed instead of completely replacing the yoke. The sleeve is of heavy walled construction with a hardened steel surface having an outside diameter that is the same as the original yoke diameter.

Zener Diode A variation of the diode, this device functions like a standard diode until a certain voltage is reached. When the voltage level reaches this point, the zener diode will allow current to flow in the reverse direction. Zener diodes are often used in electronic voltage regulators.